PHYSICS
FOR
TECHNICIANS

A SYSTEMS APPROACH

PHYSICS

FOR

TECHNICIANS

CENTER FOR OCCUPATIONAL
RESEARCH AND DEVELOPMENT

JOHN WILEY & SONS
New York   Chichester   Brisbane   Toronto   Singapore

Interior and Cover designed by Carolyn Joseph
Cover art by Etienne Weill/Image Bank

Library of Congress Cataloging in Publication Data:

Physics for technicians.

  Revision of: Unified technical concepts: physics
for technicians.
  Bibliography: p. 969
  Includes index.
  1. Physics—Study and teaching (Higher)—United
States. 2. Technology—Study and teaching (Higher)—
United States. I. Center for Occupational Research
and Development. II. Unified technical concepts.

QC44.A67 1988   530'.07'11   87-16043
ISBN 0-471-85428-X

Printed in the United States of America

10 9 8 7 6 5 4 3 2 1

Preface

In today's technologically oriented society, technicians play a role of ever-increasing importance. The services that they perform vary from the repair and maintenance of modern automobiles to the installation and adjustment of optical-fiber telecommunication networks. It is obvious that the education of technicians must include many hours of training in the tasks peculiar to the specific systems that they will service. In addition to this specific training, there are several general areas that make up the necessary background for the technician. For many, physics is a key element in that background.

In the traditional approach to technical physics, the usual subject matter is organized into separate sections on mechanics, electricity and magnetism, optics, thermodynamics, and so forth. In the presentation of individual topics, there is an emphasis on **derivations** in the lecture class and **proof of principles** in the laboratory class. This traditional approach is quite proper for scientists and engineers, but fails to meet the needs of most technicians because several years of study are required to recognize the similarities between the basic quantities of physics in the various energy systems—mechanical, fluid, electromagnetic, and thermal—and to apply them effectively to the maintenance, modification, and repair of complex technical systems. A one-year course in traditional technical physics may not succeed in providing technicians with this important insight. Additionally, it may fail because the *strong emphasis on derivations causes technicians to lose sight of the important physical principles and the* **application** *of these principles to real-world systems.*

The Unified Technical Concepts (UTC) approach attempts to meet the needs of technicians by unifying the basic physical concepts in four major systems—mechanical, fluid, electromagnetic, and thermal. In Unified Technical Concepts—Physics for Technicians, *developed for technician training in postsecondary schools, each of the concept modules, or chapters, begins with a development of the basic physics relevant to the physical concept in question. Then, the basic physical concept is identified as a unifying principle that is extended to a variety of problems in mechanical, fluid, electromagnetic, and thermal systems. For example, the concept module on force first covers the basic idea of a force, defining it as a push or pull, and applies it to linear and rotational mechanical systems. But, in addition, the idea of force as a unifying concept is introduced and extended to fluid, electromagnetic, and thermal systems. The concept module shows that a temperature difference, a pressure difference, and a voltage difference all behave like a force, even though these quantities are not forces and do not have the same dimensions as force (mass times length divided by time squared). In this manner, the concept of force is extended beyond the simple idea of a push or pull in mechanical systems and becomes a useful, unifying concept that aids the student in understanding motion in fluid, electromagnetic, and thermal systems. This general pattern of defining and explaining each concept and then demonstrating its application as a unifying concept in the four systems is continued throughout the text.*

The Unified Technical Concepts—Physics for Technicians *text—the forerunner to this text (*Physics for Technicians—A Systems Approach*)—includes thirteen concepts or chapters. These are:* **Force, Work, Rate, Momentum, Resistance, Power, Energy, Force Transformers, Energy Convertors, Transducers, Vibrations and Waves, Time Constants,** *and* **Radiation.** *Nine of these concepts are easily recognized as fundamental to physics, but four of the topics —***Force Transformers, Energy Convertors, Transducers, and Time Constants***—are generally not given extensive coverage in traditional technical physics texts even though they are important concepts for technician.*

One of the major strengths of the Unified Technical Concepts *text has been its emphasis on the laboratory experience. The laboratory exercises presented in the application modules enable the student to observe the physical principles that are presented in the concept modules. These application modules emphasize the hands-on learning experience that is more useful to the technician than the "blackboard approach" of most traditional physics courses. The concept and application modules of* Unified Technical Concepts *were developed as integrated components of a technical physics course and were not intended to stand alone, either as a text for a lecture course or as a laboratory manual.*

Unified Technical Concepts *has been used in postsecondary technical training programs since 1978 and has been well received. In addition, the UTC approach has been adopted as the framework for the development of a two-year high school course called* Principles of Technology. *Currently, some 43 states and provinces in North America are teaching* Principles of Technology *to high school vocational students who are interested in careers as technicians. Both of these training programs—*Unified Technical Concepts Physics *and* Principles of Technology*—have been successful in helping students learn technical physics.*

The development of Physics for Technicians—A Systems Approach—*has evolved from recommendations for changes in* Unified Technical Concepts. *The necessity for these changes has been driven by the increased level of preparation of students entering postsecondary schools and the increasing number of students who are completing* Principles of Technology *in the secondary schools. Based on recommendations from teachers who have been using* Unified Technical Concepts, *the* Physics for Technicians *text incorporates the following changes:*

- **The math content has been upgraded to a precalculus level.** *The development of the physical relationships employs notation and concepts that are common to calculus—such as the summation over a given element—without going into the calculus itself.*

- **The physics has been upgraded to include more equations and more derivations.** *Most of this material was previously found in the preparatory section of the application modules of* Unified Technical

Concepts. *Movement of this material into the text provides a better foundation for the lecture portion of the course and allows the labs that accompany* Physics for Technicians *to be "streamlined." This also makes* Physics for Technicians *more useful as a reference book for the working technician.*

- **Magnetism has been added to electricity and included in the discussion of the electromagnetic energy system throughout the text.**

- **A chapter on optics and lasers has been added to the thirteen chapters that formed the core of Unified Technical Concepts.**

- **Exercises for the students have been significantly increased in number and level at the end of each section.**

The revisions that are included in Physics for Technicians *do not reflect a change in philosophy. Rather, they build on the strengths of* Unified Technical Concepts. *Although* Physics for Technicians *can be used without the application modules, the laboratory exercises remain a very important part of the UTC program. It is strongly recommended that a minimum of sixty laboratory exercises be completed during the one-year course.*

It is hoped that the revisions made in Unified Technical Concepts—*those revisions that have led to* Physics for Technicians—*will meet the need for a practical course in technical physics. It is further hoped that* Physics for Technicians *will serve as a useful text and reference book for students and practicing technicians.*

Leno S. Pedrotti
Project Director
Unified Technical Concepts
Vice President
Center for Occupational
Research and Development

Acknowledgments

Extensive revisions and rewriting of Unified Technical Concepts—Physics for Technicians *led to this edition of* Physics for Technicians—A Systems Approach. *Significant recognition and credit are due Paul W. Schreiber, research physicist and teacher, for his arduous labors as principal author of the text. Recognition is also due Norman L. Baker, staff member, for his conscientious work as chief technical editor.*

L.S.P

Contents

APPENDIXES

PHYSICS
FOR
TECHNICIANS

Force

INTRODUCTION

The concept of force is based on the common experiences of every person. A force is a push or pull that may cause an object to move. Force—a physical quantity that causes movement—is the "prime mover" in the mechanical energy system.

*This chapter discusses the concept of force and forcelike quantities as they apply to four energy systems: **mechanical, fluid, electromagnetic,** and **thermal.** The forcelike quantities discussed are force, torque, pressure difference, voltage difference, magnetomotive force, and temperature difference.*

- Force *causes motion of an object in a translational mechanical system. This force serves as the model for all forcelike quantities.*
- Torque *causes rotation of an object in a rotational mechanical system. Torque is a forcelike quantity.*
- Pressure difference *causes a fluid to move in a fluid system. Pressure difference is a forcelike quantity.*
- Voltage difference *causes charge to move in an electrical system. Voltage difference is a forcelike quantity.*

- Magnetomotive force *causes the existence of a magnetic field.*
  Magnetomotive force is a forcelike quantity.

- Temperature difference *causes heat energy to move in a thermal system.*
  Temperature difference is the forcelike quantity.

The discussion that follows includes the mathematical formulation and units
used to describe each forcelike quantity in each energy system and stresses
the characteristics of forcelike quantities that apply to all energy systems.

1.1 OBJECTIVES

FORCE AND TORQUE IN MECHANICAL SYSTEMS

Upon completion of this section, the student should be able to

- Define the physical quantities
  Force
  Torque
  and, where applicable, state their units in both SI units (international systems of units) and English Units.

- State Newton's second law of motion.

- Define
  Concurrent forces
  Coplanar forces

- Determine the resultant of concurrent forces.

- Use the conditions for the equilibrium of concurrent forces to find two unknowns.

- Given two of the following quantities in a mechanical rotational system, determine the third:
  Force
  Lever arm
  Torque

- Determine the resultant of a system of parallel forces.

- Use the conditions for the equilibrium of parallel forces to find two unknowns.

- Define
  Center of gravity
  Two-force member
  Multiply-force member

- Apply the conditions for equilibrium to coplanar, nonconcurrent forces and find three unknowns.

FIGURE 1.1 Force applied to an object.

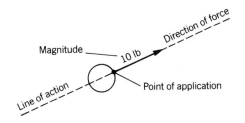

DISCUSSION

Force is the model for forcelike quantities. In this discussion, force is examined in the translational system. The results are then used in the rotational system, where an applied force produces torque.

Force

Definition of Force   A linear mechanical force is a push or pull applied along a straight line called its line of action. The force is represented graphically by a line drawn along the line of action. The length of the line is proportional to the magnitude or strength of the force. An arrowhead on the line denotes which way the force is directed.

A quantity that has both magnitude and direction is called a **vector.** If a quantity has only magnitude, such as time or volume, it is called a **scalar.** Since force has magnitude and direction, it is a vector quantity.

Figure 1.1 illustrates the result when a single force is applied to an object at rest. The force causes the object to move with increasing speed in the direction of the force. If the object is already moving when a force is applied, the force may cause it to slow down, speed up, or change direction. Which one of these happens depends on the direction of the applied force with respect to the original direction of motion of the object.

The equation that relates the change in the body's motion to the force acting on it is

$$\mathbf{F} = m\frac{\Delta \mathbf{v}}{\Delta t} \tag{1.1}$$

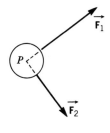

where

  $\mathbf{F}$ = force (a vector)
  m = body's mass or quantity of matter present in the object (a scalar)
  $\Delta \mathbf{v}$ = change in the body's velocity (a vector)
  Δt = change in time (a scalar)

The bold letters for the force ($\mathbf{F}$) and the change in velocity ($\Delta \mathbf{v}$) indicate that they are vector quantities. In the illustrations, an arrow placed above a letter indicates a vector quantity. The direction of the change in the velocity must have the same direction as the force. This physical law was discovered by Sir Isaac Newton. It is known as Newton's law of motion.

In the physical world, complex systems of forces must be dealt with precisely.

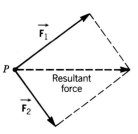

FIGURE 1.2 Adding forces.

Figure 1.2 shows two forces applied to an object at point P and a vector diagram of

TABLE 1.1   SI and English Units

	Length	Time	Mass	Force
SI system of units	meter m	second sec	kilogram kg	newton[a] N = kg·m/sec²
English system of units	foot ft	second sec	slug (no symbol)[a] lb·sec²/ft	pound lb

[a]Derived units.

the forces. The resultant force is the vector sum of the two individual forces. Techniques for adding forces in different directions will be given in this chapter.

The presence of forces acting on an object does not mean necessarily that there will be a change in the state of motion of that object. A change in motion results only when the resultant force (also called the "net force") is not zero. If the net force is zero, the object remains in its original state of motion; that is, objects at rest remain at rest and objects in motion remain in motion in a straight line at a constant speed. This condition is called **equilibrium.**

Systems of Units   The English system of units is based on length in feet, time in seconds, and **force in pounds.** The weight of an object in pounds (lb) is the gravitational force exerted on that object by the earth. The English unit of mass is the slug, a derived unit. The slug is defined as the amount of mass that will change its speed by one foot per second when a one-pound force is applied to it for one second. The slug is not a common unit; it is defined here only to aid in the comparison of English and SI units. When English units are used, an object's **weight** (at the earth's surface) usually is specified in pounds.

The Système Internationale, or international system of units (SI), is based on length in meters, time in seconds, and **mass in kilograms.** The SI unit of force is a derived unit called a newton (N). A newton is defined as the amount of force that will change the speed of a one-kilogram mass by one meter per second if applied for one second. Since weight is a force, weight is measured in the SI system in newtons. When SI units are used, an object's **mass** usually is specified in kilograms.

Table 1.1 shows the relationships of the units of length, time, mass, and force in the SI and English systems.

The Law of Universal Gravitation

Newton's Law   Newton's law of universal gravitation states that "Every object in the universe attracts every other object with a force directly proportional to each of their masses and inversely proportional to the square of the distance separating them."

If a ball is thrown upward, it is pulled back to earth by the gravitational force exerted by the earth on the ball. Also, the earth itself moves around the sun in an elliptical orbit because of the gravitational force that the sun exerts on it; otherwise, the earth would move away in a straight line.

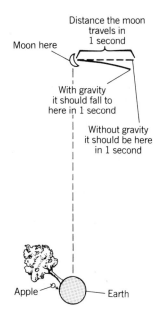

FIGURE 1.3 Gravitational forces.

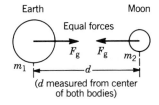

FIGURE 1.4 Gravitational attraction.

Newton was the first person to conclude that the same force that causes the moon to move about the earth also causes an apple to fall from a tree to the earth. The moon also is falling toward the earth; otherwise, it would move away in a straight line. It falls just enough each second to remain in its orbit around the earth, as illustrated in Figure 1.3. From this original insight, Newton was able to explain the motion of the planets about the sun and of the moon about the earth.

Weight and Mass   An important distinction exists between weight and mass. Weight is "the force on a body due to gravity" and depends on its location relative to an astronomical body. An object would weigh more on the earth than on the moon, for example. Mass, however, is the same for an object everywhere in the universe. It can be defined as "the quantity of matter in a body, or the property of a body's resistance to change of motion." Mass is thus a measure of the inertia, or sluggishness, of an object in stopping or changing its motion. An object weighs less on the moon than on the earth, yet its mass is the same.

Force of Gravity   Newton's law of gravitation, or gravitational force, is expressed in terms of the masses of the objects involved and their separation distance, indicated by

$$F_g = \frac{Gm_1 m_2}{d^2} \tag{1.2}$$

where

F_g = force of gravity, in N
G = universal gravitational proportional constant, 6.67×10^{-11} m³/kg·sec²
m_1 = mass of object 1, in kg
m_2 = mass of object 2, in kg
d = distance between objects, in m

Equation 1.2 can be used to explain the motion of all celestial bodies in our solar system. Figure 1.4 illustrates the gravitational force between two (approximately) spherical bodies, the earth and the moon. The distance between them is measured from their centers. The force of gravitational attraction on one sphere is equal and opposite to that on the other. Equation 1.2 is used in solving a problem in Example 1.1.

EXAMPLE 1.1   Gravitational force
. . . .

Given: A ball with a mass of 50 kg is located at the surface of the earth. The earth has a radius of $d = 6.4 \times 10^6$ m and a mass of $m_e = 6.02 \times 10^{24}$ kg.

Find: The force the earth exerts on the ball and the force the ball exerts on the earth.

Solution: $F_g = \dfrac{Gm_1 m_2}{d^2}$

$$F_g = \frac{(6.67 \times 10^{-11} \text{ m}^3/\text{kg·sec}^2)\,(6.02 \times 10^{24} \text{ kg})\,(50 \text{ kg})}{(6.4 \times 10^6 \text{ m})^2}$$

$$F_g = 490 \text{ N}$$

The force that the earth exerts on the ball is equal to the force the ball exerts on the earth; they are oppositely directed.

Figures 1.5 and 1.6 illustrate how gravitational force changes with separation distance. Within a planet of uniform composition and density, gravitational force is directly proportional to the radial distance from the center and is maximum at its surface. Because of the distance squared factor in the denominator, however, the force from the surface outward decreases rapidly as the distance increases (Figure 1.5). This relationship is called the "inverse square law" and means that the earth's gravitational pull on an outside object varies inversely with the square of its distance from the center of the earth. For example, an object that weighs one pound at the surface of the earth will weigh 1/4 pound when it is twice as far from the center of the earth (Figure 1.6).

Gravitational force is a relatively weak force. A mass the size of the earth exerts a force of only 150 lb on a person weighing 150 lb. Two people cannot detect or measure their mutual gravitational attraction, no matter how close they might get. Other forces in nature, such as the electrical force, are much stronger than the gravitational force.

The gravitational force equation (Equation 1.2) for objects on the surface of the earth can be rewritten as

$$F_g = mg \qquad (1.3)$$

where

F_g = force of gravity
m = mass of object on surface of the earth
$g = \dfrac{Gm_e}{d^2}$, where m_e is the mass of the earth

The force F_g is the body's weight and will be designated by the symbol w. Thus, the weight of an object is the product of its mass and the constant g:

FIGURE 1.5 Variation of gravitational force with distance. (left)

FIGURE 1.6 Inverse square variation of force. (right)

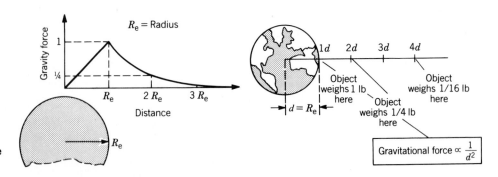

$$w = mg \qquad (1.4)$$

If this equation is compared with Newton's second law of motion (Equation 1.1), the g can be associated with $\Delta v/\Delta t$. The change in velocity divided by the corresponding change in time ($\Delta v/\Delta t$) is called the **acceleration** of the body. The acceleration of a freely falling body (only the force of gravity acting on it) is equal to g and has units of m/sec^2 or ft/sec^2.

The acceleration due to gravity (g) is almost a constant for positions close to the surface of the earth. The following calculated values bear this out:

$g = 9.800$ m/sec^2 for $d = 4000$ mi (at the surface of the earth)

$g = 9.795$ m/sec^2 for $d = 4001$ mi (one mile above the surface)

$g = 9.751$ m/sec^2 for $d = 4010$ mi (ten miles above the surface)

The accepted value of g at the earth's surface is 9.8 m/sec^2 in the international system of units and 32 ft/sec^2 in the English system of units. An object would have to be high above the earth's surface for the g value to be noticeably different.

An acceleration of 9.8 m/sec^2 means that the velocity of the object will increase by 9.8 m/sec during each second it falls. Thus, if a body starts from rest, its velocity after one second of fall is 9.8 m/sec, after two seconds of fall its velocity is 19.6 m/sec, and after three seconds the velocity is 29.4 m/sec. In general the velocity is equal to the acceleration (g) times the time.

Example 1.2 illustrates the use of Equation 1.4 to determine the weight (w) of an object.

EXAMPLE 1.2   Calculation of weight
. . . .

Given: A brick with a mass of 2 kg.

Find: Weight of the brick.

Solution: $w = mg$

$w = (2 \text{ kg}) (9.8 \text{ m/sec}^2)$

$w = 19.6 \text{ kg·m/sec}^2$

$w = 19.6 \text{ N} \qquad (1 \text{ N} = 1 \text{ kg·m/sec}^2)$

Table 1.2 lists some conversion factors that are useful when working with weight and mass.

Colinear Forces

Resultant Force   Forces that have the same line of action are called **colinear**

TABLE 1.2 Weight and mass
conversion factors

1 lb (force) = 4.45 N (force)
1 slug (mass) = 14.6 kg (mass)
1 kg weighs 9.8 N or 2.2 lb (on earth)
1 slug weighs 32.2 lb or 143 N (on earth)

forces. An example of colinear forces is shown in Figure 1.7. A sky diver falls toward the earth at increasing speed because of the force of gravity exerted on his body by the earth. He eventually reaches a steady maximum speed called the "terminal speed." The terminal speed occurs when the upward resistance force of the air (drag) is equal to the downward pull of gravity (weight). The sky diver then continues to fall at a constant speed in a state of equilibrium. When the parachute opens, the much larger area of the parachute causes a greater upward force due to increased wind resistance. This force slows the sky diver until a new equilibrium is reached at a lower speed and the sky diver lands safely.

The resultant force when several colinear forces act on a body is obtained by adding the forces algebraically. Forces in one direction are arbitrarily called positive and forces in the other direction are negative. Thus, the equation to find the resultant force is

$$\mathbf{F}_t = \mathbf{F}_1 + \mathbf{F}_2 + \mathbf{F}_3 + \cdots \tag{1.5}$$

or

$$\mathbf{F}_t = \Sigma \, \mathbf{F}_n \tag{1.6}$$

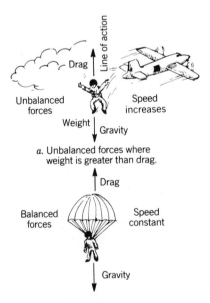

a. Unbalanced forces where weight is greater than drag.

b. Balanced forces where weight is equal to drag.

FIGURE 1.7 Balanced and unbalanced forces.

where

$\mathbf{F}_t$ = total force (resultant)

$\mathbf{F}_n$ = $\mathbf{F}_1, \mathbf{F}_2, \mathbf{F}_3 \ldots$ are the individual forces acting on the body

Σ = a symbol that means "sum all the forces"

Example 1.3 shows how to use Equation 1.5 or 1.6 in a problem.

EXAMPLE 1.3 Tug-of-war problem
. . . .

Given: Three people are playing tug-of-war. Two people are on one side and each is pulling with a force of fifty pounds. The person on the other side is pulling with a force of sixty-five pounds.

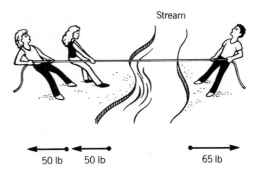

Find: The net force and the initial direction of motion.

Solution: Establish right as positive and left as negative. The forces acting on the rope are

-50 lb
-50 lb
$+65$ lb

From Equation 1.5

$\mathbf{F}_t = \mathbf{F}_1 + \mathbf{F}_2 + \mathbf{F}_3$

$\mathbf{F}_t = (-50 \text{ lb}) + (-50 \text{ lb}) + (65 \text{ lb})$

$\mathbf{F}_t = -100 \text{ lb} + 65 \text{ lb}$

$\mathbf{F}_t = -35 \text{ lb}$

The net force is 35 lb to the left, and the motion will be in that direction. The person on the right gets wet!

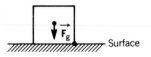

a. Space diagram

b. Free-body diagram

FIGURE 1.8 Forces on a body in equilibrium

Equilibrium   If a body is at rest and remains at rest (equilibrium) when acted on by a system of colinear forces, then the resultant (F_t) must be zero. If the resultant were not zero, the body would move according to Newton's second law. Thus, the condition for equilibrium for colinear forces is

$$\mathbf{F_t} = \Sigma \mathbf{F_n} = 0 \qquad (1.7)$$

An example of a body in equilibrium is shown in Figure 1.8. To analyze the forces, a free-body diagram is constructed as shown in Figure 1.8*b*. The body is isolated and all forces acting on it are applied to the body. A positive direction—say upward—is assigned. For this case, the force of gravity acts downward ($-F_g$) and the surface pushes upward on the body ($+F_N$). Since the body is in equilibrium

$$\mathbf{F_t} = (+F_N) + (-F_g) = 0$$

or

$$F_N = F_g$$

Thus, the normal force (F_N) is equal in magnitude to the weight, but it acts in the upward direction as shown in the diagram. If F_N is applied in the **wrong** direction, the equation for equilibrium becomes

$$\mathbf{F_t} = (-F_N) + (-F_g) = 0$$

or

$$F_N = -F_g$$

The negative sign indicates that the force was applied in the wrong direction in the free-body diagrams. This is important because there are cases in more complex problems in which it is not known which way the force is directed before the problem is solved. Negative signs always show that the calculated forces act opposite to the assumed direction.

Coplanar, Concurrent Forces

Definition of Concurrent Forces   When the lines of action of a number of forces acting on a body meet at a common point, the forces are called **concurrent** forces.

FIGURE 1.9 Concurrent forces.

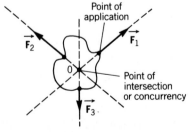

a. Forces acting on a body

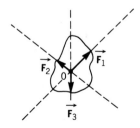

b. Forces translated along lines of action

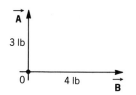

FIGURE 1.10 Two forces at right angles.

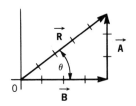

FIGURE 1.11 Resultant of two forces at right angles.

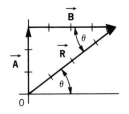

FIGURE 1.12 Another way to add forces at right angles.

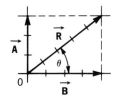

FIGURE 1.13 Parallelogram method of adding forces.

If all the forces lie in the same plane, they are called **coplanar forces.** The resultant of a system of coplanar, concurrent forces can be found by adding the forces as vector quantities. This resultant force will always produce the same effect on the motion of the body as do the individual forces.

If the forces are not applied at the same point on the body, the forces may be extended along their lines of action to the point of intersection as illustrated in Figure 1.9.

Resultant of Two Forces   Two forces **A** and **B** at right angles are shown in Figure 1.10. Bold letters are used for the vector quantities **A** and **B.** There are two methods of determining the resultant force graphically.

1.   *Triangle method.*   Force **A** is laid off with its tail at the head of force **B,** as in Figure 1.11. All vectors are drawn according to a convenient scale.

   The closing side of the triangle, **R,** represents the resultant. The magnitude and direction of the resultant, **R,** can be measured with a scale and protractor, respectively. The same resultant would be obtained if force **B** were added to force **A,** as in Figure 1.12. This indicates that the order of adding forces is not important, since the same resultant will always be obtained.

2.   *Parallelogram method.*   To find the resultant force using this method, the given forces **A** and **B** are first laid off to scale. Then horizontal and vertical construction lines are drawn parallel to **A** and **B** as shown in Figure 1.13. The resultant, **R,** is the diagonal of the parallelogram, as shown. As in the triangle method, the magnitude of **R** and the angle θ can be measured with a scale and a protractor, respectively. In these examples, forces **A** and **B** were chosen to be 3 lb and 4 lb, respectively. In all the graphical solutions, the result **R** is 5 lb, and the angle θ is 37° above the horizontal axis.

   Because the forces act at right angles to each other and can be added as shown in Figure 1.12, the Pythagorean theorem can be used to compute the magnitude of the resultant. This gives

$$R^2 = A^2 + B^2$$

or

$$R = \sqrt{A^2 + B^2} \tag{1.8}$$

where

   R = magnitude of the resultant
   A = magnitude of the vertical force
   B = magnitude of the horizontal force

The direction of the resultant is indicated by the angle θ. The angle θ can be determined from its tangent:

$$\tan \theta = \frac{A}{B} \tag{1.9}$$

or

$$\theta = \tan^{-1} \frac{A}{B}$$

Example 1.4 shows the use of Equations 1.8 and 1.9 in solving a problem.

EXAMPLE 1.4 Two forces at right angles—mathematical solution
. . . .

Given: A vertical force *A* of 3 lb and a horizontal force *B* of 4 lb, as shown in Figure 1.10.

Find: **a.** The resultant *R*.
            b. The angle θ.

Solution: **a.** The resultant *R* can be found by Equation 1.8:

$$R = \sqrt{A^2 + B^2}$$

Because *A* = 3 lb and *B* = 4 lb, Equation 1.8 becomes

$$R = \sqrt{(3\ lb)^2 + (4\ lb)^2} = \sqrt{9\ lb^2 + 16\ lb^2}$$

$$R = \sqrt{25\ lb^2}$$

$$R = 5\ lb$$

b. The angle θ can be computed from either its sine, cosine, or tangent:

$$\sin \theta = \frac{A}{R} = \frac{3\ lb}{5\ lb} = 0.6$$

$$\cos \theta = \frac{B}{R} = \frac{4\ lb}{5\ lb} = 0.8$$

$$\tan \theta = \frac{A}{B} = \frac{3\ lb}{4\ lb} = 0.75$$

The angle θ for each of these values, as obtained with a calculator or looked up in trig tables, is

$$\theta = 37°$$

Notice that these results are in agreement with those obtained graphically. In summary, a force of 5 lb acting at an angle of 37° with the horizontal will produce the same effect as the two forces of 3 lb vertically and 4 lb horizontally. The resultant is not the arithmetical sum of 3 lb and 4 lb. In other words, the two forces are not equivalent to a single force of 7 lb.

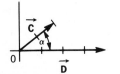

FIGURE 1.14 Forces not at right angles.

The two arrows, **C** and **D,** in Figure 1.14 represent two forces at an angle α whose resultant is to be determined. The forces are not at right angles to each other.

The two graphical methods used in the previous case will be applied here, since the techniques are very similar.

FIGURE 1.15 Triangle method for adding forces not at right angles.

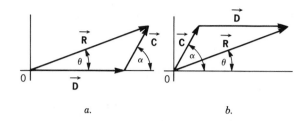

a.   *b.*

1. *Triangle method.* Choose a convenient scale and lay off force **C** with its tail at the head of force **D** as in Figure 1.15*a*. The force **D** can also be drawn with its tail at the head of **C** as in Figure 1.15*b* with the same results. The magnitude of the resultant, **R**, and its direction, θ, can then be measured with a scale and protractor, respectively.

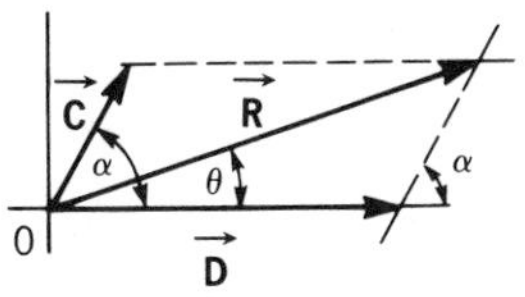

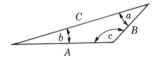

FIGURE 1.16 Parallelogram method for adding forces not at right angles.

2. *Parallelogram method.* Draw dashed construction lines parallel to **C** and **D**, as in Figure 1.16. The resultant is the diagonal of the parallelogram as shown in this figure. The magnitude of the resultant, **R**, and the angle θ can be determined by using a scale and a protractor, respectively. In these examples, force **C** and **D** are 2 lb and 4 lb, respectively, and are applied at 45° to each other. In all the graphical solutions, the resultant, **R**, is 5.60 lb with an angle θ of 14.6°.

For two forces not at right angles, there are two mathematical approaches that can be used.

FIGURE 1.17 Drawing for cosine law.

1. *Cosine law and sine law.* The cosine law and sine law are given by Equations 1.10 and 1.11 for the oblique triangle shown in Figure 1.17. Capital letters designate sides and small letters designate angles.

$$C^2 = A^2 + B^2 - 2AB \cos c \qquad \text{(cosine law)} \qquad (1.10)$$

$$\frac{A}{\sin a} = \frac{B}{\sin b} = \frac{C}{\sin c} \qquad \text{(sine law)} \qquad (1.11)$$

Applying the two laws to the triangle shown in Figure 1.15*a* yields

$$R^2 = C^2 + D^2 - 2CD \cos (180 - \alpha)$$

or

$$R = \sqrt{C^2 + D^2 - 2CD \cos (180 - \alpha)} \qquad (1.12)$$

After R is calculated, the angle θ can be determined from the sine law. This gives

$$\frac{R}{\sin (180 - \alpha)} = \frac{C}{\sin \theta}$$

or

$$\sin \theta = \frac{C \sin (180 - \alpha)}{R} \qquad (1.13)$$

The two laws are used to solve a problem in Example 1.5.

EXAMPLE 1.5   Two forces not at right angles—mathematical solution

. . . .

Given: The following values for use with Figure 1.15a: $C = 2$ lb, $D = 4$ lb, and $\alpha = 45°$.

Find: **a.**   The magnitude of the resultant, **R.**

        b.   The angle θ.

Solution: **a.**   The resultant **R** can be found by using Equation 1.12 as follows:

$$R^2 = C^2 + D^2 - 2CD \cos(180° - \alpha)$$

$$R^2 = (2\text{ lb})^2 + (4\text{ lb})^2 - 2(2\text{ lb})(4\text{ lb})\cos(180° - 45°)$$

$$R^2 = 4\text{ lb}^2 + 16\text{ lb}^2 - 16\text{ lb}^2 \cos 135°$$

$$R^2 = 20\text{ lb}^2 - 16\text{ lb}^2(-0.707)$$

$$R^2 = 20\text{ lb}^2 + 11.3\text{ lb}^2 = 31.3\text{ lb}^2$$

$$R = \sqrt{31.3\text{ lb}^2} = 5.59\text{ lb}$$

        b.   The angle θ in Figure 1.15a can be computed from the law of sines as follows:

$$\frac{C}{\sin \theta} = \frac{R}{\sin(180° - \alpha)}$$

$$\sin \theta = \frac{C}{R}\sin(180° - \alpha)$$

$$\sin \theta = \frac{2\text{ lb}}{5.59\text{ lb}}\sin(180° - 45°)$$

$$\sin \theta = 0.2530$$

$$\theta \approx 14.6°$$

---

2. *Rectangular resolution method.*   Using the cosine law and sine law to find the resultant is often awkward and, when applied to more than two forces, becomes very difficult to manage. The rectangular resolution method explained here is easier to handle and can be used with any number of forces. Indeed, this method is probably the best for computational purposes to determine the resultant force in any situation; it can be outlined in three steps:

   a.   **Resolve** all forces into their **rectangular components** along the x and y axes. (The method gets its name from the bold print words.)

   b.   Find the algebraic sum of all x and y components. The problem now has been reduced to one net force along the x axis and one net force along the y axis.

c. Combine these two net forces to find the resultant, just as for the right triangle explained in Figure 1.11 or 1.12.

As an example, the resultant of the two forces C and D used previously are computed in Example 1.6 by the rectangular resolution method.

EXAMPLE 1.6   Two forces not at right angles—the rectangular resolution method
· ·

Given: The following values (same as in Example 1.5) for use with the first two sketches below: $C = 2$ lb, $D = 4$ lb, and $\alpha = 45°$.

Find: **a.** The resultant R.
             b. The angle θ.

Solution: **a.** The resultant **R** can be found by the rectangular resolution method.

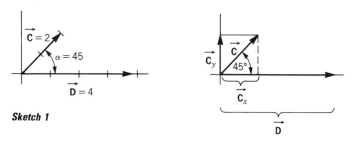

Sketch 1

Sketch 2

Since force **D** already lies on the x axis (Sketch 1), resolve force **C** into its x and y components (Sketch 2). x components directed to the right are usually considered positive and those to the left negative. Similarly, y components directed upward are positive and those downward are negative. The x component of the 2-lb force **C** is 2 cos 45°, which is 2(0.707) = 1.414 lb. This component added to the 4-lb force **D** already in the positive x direction yields a net force of 1.414 lb + 4 lb = 5.414 lb to the right. The y component of the 2-lb force **C** is 2 sin 45°, which is 2(0.707) = 1.414 lb upward. Because the 4-lb force **D** has no vertical component, the algebraic sum of all the y components is simply 1.414 lb. These resultant x and y components are shown in Sketch 3. The symbols Σx and Σy stand for the sum of the x and y components, respectively. The resultant now can be calculated as before, as shown in Sketch 4.

$\vec{C_y} = \Sigma Y$

1.414 lb

5.414 lb

$(\vec{C_x} + \vec{D}) = \Sigma X$

Sketch 3

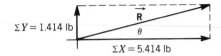

Sketch 4

$$R = \sqrt{(\Sigma x)^2 + (\Sigma y)^2}$$

$$R = \sqrt{(5.414 \text{ lb})^2 + (1.414 \text{ lb})^2} = \sqrt{29.31 \text{ lb}^2 + 2.000 \text{ lb}^2}$$

$$R = \sqrt{31.1 \text{ lb}^2}$$

$$R = 5.60 \text{ lb}$$

This value agrees with the result calculated previously.

b. The angle θ can be computed from any of the trigonometric functions, but only one will be used here.

$$\tan \theta = \frac{1.414 \text{ lb}}{5.414 \text{ lb}}$$

$$\tan \theta = 0.2611$$

$$\theta \approx 14.6°$$

θ also agrees with the result calculated previously.

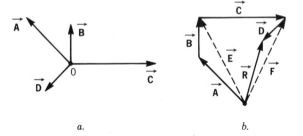

FIGURE 1.18 Adding more than two forces.

a.   *b.*

Resultant of Many Forces

When the resultant of a number of concurrent forces is to be found, the triangle method becomes the polygon method. This process, illustrated in Figure 1.18*a*, shows four coplanar forces **A, B, C,** and **D** acting concurrently at a point 0. In Figure 1.18*b*, forces **A** and **B** are combined graphically to give the resultant **E.** Resultant **E** is combined with force **C** to give resultant **F.** Finally, force **F** is combined graphically with force **D** giving the final resultant **R.**

This process could be simplified, because forces **E** and **F** need not have been drawn. Instead, the given forces are drawn in succession, with the tail of each at the head of the force preceding it. The resultant is then the force that completes this polygon and is drawn from the tail of the first force to the head of the last. To illustrate that the order of adding the forces is unimportant, they are added in a

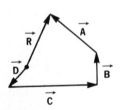

FIGURE 1.19 Adding forces.

FIGURE 1.20 Method of rectangular resolution.

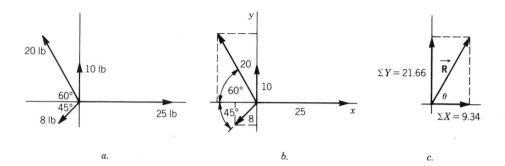

a.   b.   c.

different sequence in Figure 1.19 (**D + C + B + A**). The same resultant, **R**, would occur for any other sequence. For an interesting exercise, this claim should be verified. When the resultant has been obtained in this manner, its magnitude, *R*, and its direction, θ, can be determined with the aid of a scale and a protractor.

Because the "law of sine and cosine method" requires computations involving oblique triangles and is tedious, only the method of rectangular resolution will be demonstrated. The resultant of the four forces shown in Figure 1.18*a* is computed in Example 1.7. The magnitude and angles of the forces are indicated in Figure 1.20*a*.

EXAMPLE 1.7   Rectangular resolution of more than two forces
. . . .

Given: The *x* and *y* components of the forces shown in Figure 1.20*b*.

Force vector	x component	y component
10 lb	0	+ 10 lb
25 lb	+ 25 lb	0
20 lb	− 20 lb cos 60 = − 10 lb	+ 20 lb sin 60 = 17.32 lb
8 lb	− 8 lb cos 45 = − 5.66 lb	− 8 lb sin 45 = − 5.66 lb

Find: **a.** The algebraic sum of *x* and *y* components.

              b. The resultant *R*.

Solution: **a.** The algebraic sum of the *x*-component forces is 9.34 lb directed toward the right, and the algebraic sum of the *y*-component forces is 21.66 lb directed upward.

              b. Magnitude and direction of the resultant can be determined as before for a right triangle as shown in Figure 1.20*c*.

$$R = \sqrt{(\Sigma x)^2 + (\Sigma y)^2} = \sqrt{(9.34 \text{ lb})^2 + (21.66 \text{ lb})^2} = \sqrt{556.39 \text{ lb}^2}$$

$$R = 23.6 \text{ lb}$$

$$\tan \theta = \frac{\Sigma y}{\Sigma x}$$

$$\tan \theta = \frac{21.7 \text{ lb}}{9.34 \text{ lb}}$$

$$\tan \theta = 2.32$$

$$\theta = 66.7°$$

These results are accurate. The results obtained graphically will agree well with these results if the drawings and measurements with scale and protractor are done carefully.

Equilibrium   If a body is in equilibrium, then the resultant of the forces acting on it must be zero. As shown in Examples 1.6 and 1.7, the resultant can be expressed as

$$R = \sqrt{(\Sigma F_x)^2 + (\Sigma F_y)^2}$$

where

F_x = force in the x direction
F_y = force in the y direction

Since the square of a number is always positive, the conditions necessary for equilibrium are

$$\Sigma F_x = 0$$

$$\Sigma F_y = 0$$

These two conditions are also sufficient for the equilibrium of concurrent forces in a plane. Because there are only two independent equations, there can only be two unknowns in a given problem. An unknown may be a magnitude or a direction. The solution of a problem with two unknowns is given in Example 1.8.

EXAMPLE 1.8   Equilibrium of concurrent forces in a plane
. . . .

Given: Two wires, as shown in the sketch, support a 100-lb weight in equilibrium.

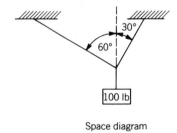

Sketch 1   *Sketch 2*

Find: The tension in the two wires.

Solution: First draw a free-body diagram at the point of intersection. The condition $\Sigma F_x = 0$ gives

$$-F_1 \sin 60 + F_2 \sin 30 = 0$$

$$F_1 = F_2 \frac{\sin 30}{\sin 60} = 0.577\, F_2$$

Therefore, $F_1 = 0.577\, F_2$. The condition $\Sigma F_y = 0$ gives

$$F_1 \cos 60 + F_2 \cos 30 - 100\ \text{lb} = 0$$

$$0.5\, F_1 + 0.866\, F_2 = 100\ \text{lb}$$

Substitution for F_1 from the equation $F_1 = 0.577\, F_2$ gives

$$(0.5)(0.577)\, F_2 + 0.866\, F_2 = 100\ \text{lb}$$

Solving for F_2

$$F_2 = 86.6\ \text{lb}$$

Solving for F_1

$$F_1 = (0.577)(86.6\ \text{lb}) = 50\ \text{lb}$$

An example of a statically **indeterminate** problem (more than two unknowns) is shown in Figure 1.21. There are three unknown tensions in the wires and only two equations for equilibrium. For this case, the tensions will also depend on the physical properties of the materials used for each wire and the elongation of each wire. With only the two equations $\Sigma F_x = 0$ and $\Sigma F_y = 0$, the problem as stated cannot be solved.

Forcelike Quantities

Force is the prime mover in a translational mechanical system. Any change in the state of motion of an object occurs because of forces acting on that object. A resultant force on a mass causes the velocity of the mass to change.

Forcelike quantities are those physical quantities in other energy systems (mechanical rotational, fluid, electromagnetic, thermal) that act as the prime movers in those systems. In an energy system, an unbalanced forcelike quantity

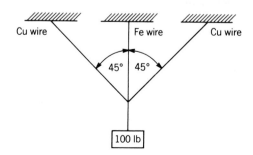

FIGURE 1.21 Statically indeterminate problem.

causes a change of some other quantity in that system. Although these forcelike quantities do not act in their respective systems in exactly the same way that force acts in a translational mechanical system, all have certain characteristics in common with force. Force was chosen as the first prime mover to be discussed because it is the most basic and most closely related to everyday experience. Other forcelike quantities are actually the result of forces acting within a system. The remainder of this chapter discusses forcelike quantities in other energy systems and brings a unification to physics by pointing out the analogies that relate these quantities to the more familiar and basic concept of force.

Torque

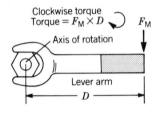

FIGURE 1.22 Torque applied to a wrench.

Definition of Torque Another mechanical system of interest is the mechanical rotational system. The prime mover, or forcelike quantity, in this system is torque. Net torque causes a change in the state of rotational motion about an axis, just as net force causes a change in the state of linear or translational motion. Torque results when a force is applied at some distance from an axis of rotation, as shown in Figure 1.22. The **perpendicular distance** from the line of action of the force to the axis is called the **lever arm,** or the **moment arm,** of the force. The torque is often called the ''moment of the force.'' Torque or moment of force is defined by

$$\tau = Fl \tag{1.14}$$

where

τ = torque (ft·lb or N·m)
F = force (lb or N)
l = lever arm or moment arm (ft or m)

The use of this equation to determine torque is illustrated in Example 1.9.

EXAMPLE 1.9 Calculation of torque
....

Given: A force of 5 lb (22.25 N) acts with a lever arm of 2 ft (0.61 m). The lever arm is perpendicular to the line of action of the force.

Find: Torque.

Solution: *English units*

$\tau = Fl$

$\tau = -(5\ \text{lb})(2\ \text{ft})$

$\tau = -10\ \text{ft·lb}$

SI units

$\tau = Fl$

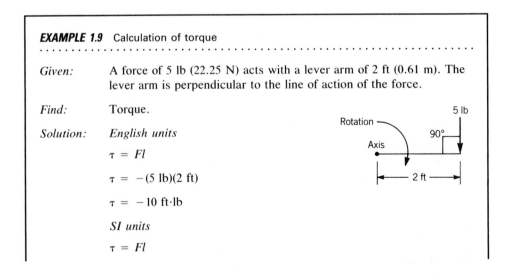

$$\tau = -(22.25 \text{ N})(0.61 \text{ m})$$

$$\tau = -13.6 \text{ N·m}$$

(The negative sign indicates a clockwise (cw) rotation.)

Just as the direction of a force must be specified, the direction of the rotation resulting from a torque must be specified. The usual convention is to specify torques that produce counterclockwise rotation as positive and those that produce clockwise rotation as negative. When solving problems involving total torque in a system, the torques are added algebraically as in

$$\tau_t = \tau_1 + \tau_2 + \tau_3 + \cdots$$

or
$$\tau_t = \Sigma \, \tau_n \tag{1.15}$$

where

τ_t = total torque

$\tau_n = \tau_1, \tau_2, \tau_3, \ldots$ are individual torques acting on the system

Examples 1.10 and 1.11 show how to use this equation and Equation 1.3 in solving torque problems.

EXAMPLE 1.10 Beam problem
....

Given: Two people are sitting on a beam as shown.

Find: Total torque and direction of rotation.

Solution: Choose counterclockwise as positive. Torque of person on right is

$$\tau_R = F_R l_R$$

$$\tau_R = (85 \text{ lb})(6 \text{ ft}) \text{ cw}$$

$$\tau_R = 510 \text{ lb·ft cw (clockwise)}$$

Torque of person on left is

$$\tau_L = F_L l_L$$

$$\tau_L = (120 \text{ lb})(4 \text{ ft}) \text{ ccw}$$

$$\tau_L = 480 \text{ lb·ft ccw (counterclockwise)}$$

Total torque is

$$\tau_t = \tau_R + \tau_L$$

$$\tau_t = 510 \text{ lb·ft (cw)} + 480 \text{ lb·ft (ccw)}$$

$$\tau_t = -510 \text{ lb·ft} + 480 \text{ lb·ft}$$

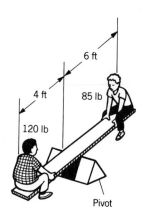

$$\tau_t = -30 \text{ lb·ft}$$

$$\tau_t = 30 \text{ lb·ft (cw)}$$

The beam will rotate in a clockwise direction, with a torque of 30 ft·lb. The 120-lb person will be lifted and the 85-lb person lowered.

EXAMPLE 1.11   Pulley system problem
. . . .

Given: A pulley system has weights attached to provide forces as shown.

Find: Total torque and direction of rotation.

Solution: Choose counterclockwise as positive.

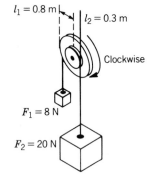

$$\tau_1 = Fl$$

$$\tau_1 \text{ (ccw)} = +F_1 l_1$$

$$\tau_1 = +(8 \text{ N})(0.8 \text{ m})$$

$$\tau_1 = +6.4 \text{ N·m}$$

$$\tau_2 \text{ (cw)} = F_2 l_2$$

$$\tau_2 = -(20 \text{ N})(0.3 \text{ m})$$

$$\tau_2 = -6.0 \text{ N·m}$$

$$\tau_t = \tau_1 + \tau_2$$

$$\tau_t = +6.4 \text{ N·m} - 6.0 \text{ N·m}$$

$$\tau_t = +0.4 \text{ N·m}$$

The pulley system will turn in a counterclockwise direction with a torque of 0.4 N·m.

Coplanar, Parallel Forces

Vertical forces are commonly encountered in engineering practice because of gravitational forces. The solution of problems involving parallel forces is often simplified by using the resultant force. The resultant must have the proper magnitude and direction, and its line of action must be located in the plane so that it produces the same torque as the individual forces about any point in the plane. Since the forces are parallel, the magnitude of the resultant is the algebraic sum of the individual forces. For example, upward-acting forces are positive and downward-acting forces are negative. Mathematically this may be expressed as

$$\mathbf{F}_t = \mathbf{F}_1 + \mathbf{F}_2 + \mathbf{F}_3 + \dots$$

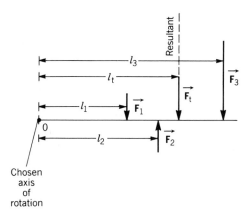

FIGURE 1.23 Resultant of three parallel forces.

or

$$\mathbf{F}_t = \Sigma \mathbf{F}_n$$

The line of action of **F** is parallel to the individual forces. To locate its position in the plane, consider Figure 1.23. The magnitude and direction of the resultant is

$$\mathbf{F}_t = (-F_1) + (+F_2) + (-F_3)$$

If $\mathbf{F}_t$ is negative, the direction of the resultant is downward as shown in Figure 1.23.

Let 0 be any point in the plane. If the resultant is to produce the same torque about point 0 as the individual forces, then

$$-F_t l_t = -F_1 l_1 + F_2 l_2 - F_3 l_3$$

or

$$l_t = \frac{-F_1 l_1 + F_2 l_2 - F_3 l_3}{-F_t}$$

If F_t is zero, the l_t is not defined. For this case, the system of forces cannot be replaced by a single resultant force. Two forces (F_c) equal in magnitude, having opposite directions, and separated by a distance, l, as shown in Figure 1.24 are required to produce the same torque as the individual forces. Since the forces are equal in magnitude and opposite in direction, the condition

$$\Sigma \, \mathbf{F}_n = 0$$

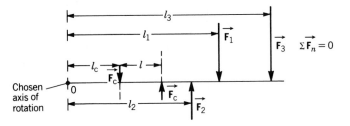

FIGURE 1.24 Resultant of parallel forces involving a couple.

remains satisfied. To produce the same torque, the condition

$$-F_c l_c + F_c(l + l_c) = -F_1 l_1 + F_2 l_2 - F_3 l_3$$

or

$$F_c l = -F_1 l_1 + F_2 l_2 + F_3 l_3$$

must be satisfied. The quantity $F_c l$ is called a couple, and its torque is independent of its position denoted by l_c. Thus, the couple may be placed at any location in the plane and it will produce the same torque about any point as the original system of forces. Any value of F_c or l may be used as long as their product equals the torque produced by the system of forces. Two sample calculations are given in Examples 1.12 and 1.13.

EXAMPLE 1.12   Resultant of a system of parallel forces
. . . .

Given: The mechanical system with four parallel forces shown in the sketch.

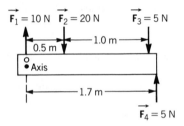

Find: The magnitude and point of application of the resultant force that effectively replaces the four forces.

Solution: To effectively replace the four forces, the resultant must equal the net value of the four forces and be applied at a point to produce the same net torque.

a.   Magnitude of the resultant F_t.

$$\mathbf{F_t} = \Sigma\mathbf{F_n} = \mathbf{F_1} + \mathbf{F_2} + \mathbf{F_3} + \mathbf{F_4}$$

$$\mathbf{F_t} = +\,10\text{ N} - 20\text{ N} - 5\text{ N} + 5\text{ N}$$

$$\mathbf{F_t} = -10\text{ N}$$

The resultant acts downward with a magnitude of 10 N.

b.   To find the point of application of $\mathbf{F_t}$, calculate torques about point 0. Equate the torques as follows:

$$-10\text{ N }l_t = -\,(20\text{ N})\,(0.5\text{ m}) - 5\text{ N }(1.5\text{ m}) + 5\text{ N }(1.7\text{ N})$$

$$-10\text{ N }l_t = -\,9\text{ N·m}$$

$$l_t = \frac{-9 \text{ N·m}}{-10 \text{ N}} = +0.9 \text{ m}$$

The resultant is shown in the following sketch.

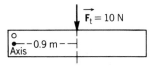

EXAMPLE 1.13   Resultant couple of system of parallel forces whose resultant is zero

. . . .

Given: The mechanical system shown in the sketch.

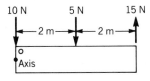

Find: The resultant couple.

Solution: **a.** $\mathbf{F}_t = 0$ (The resultant force is zero.)

               b. $\mathbf{F}_c l = (10 \text{ N}) (0 \text{ m}) - (5 \text{ N}) (2\text{m}) + (15 \text{ N}) (4 \text{ m})$

                   $\mathbf{F}_c l = 50 \text{ N·m}$

                   The force system is equivalent to a coupling having a torque of 50 N·m. The couple can be located anywhere along the body and produce the same result.

Center of Gravity   An important result occurs when a single force is used to represent the weight of an object. The line of action of this force passes through a point called the center of gravity. If the object is supported at a single point, the line of action must pass through this point because the weight and supporting forces become a system of two colinear forces. Equilibrium occurs only if the two forces are equal, opposite in direction, and have the same line of action.

The center of gravity (C of G) of an object is defined as that point at which the entire weight of an object can be considered to be concentrated. An object will balance if supported or hung from its center of gravity. Example 1.14 shows how to find the center of gravity of a nonuniform density body of irregular shape.

EXAMPLE 1.14   Determination of the center of gravity of an irregular body
....

Given: The irregular object shown in the sketch.

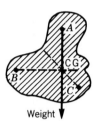

Find: Center of gravity.

Solution: **a.** Suspend the object near the periphery point *A;* drop a weighted line from *A* and mark its location.

                   b. Suspend the object near the periphery, say point *B;* drop a weighted line from *B* and mark where it crosses line *A*.

                   c. Suspend the object near the periphery, say point *C;* drop a weighted line from *C*. It should cross the point of intersection of lines *A* and *B*. That point is the center of gravity of the irregular body.

Equilibrium   When a body is acted upon by a system of coplanar parallel forces, the conditions for equilibrium become

$$\Sigma \mathbf{F}_n = 0 \tag{1.16}$$

$$\Sigma \tau_n = 0 \tag{1.17}$$

These conditions ensure that the body is in translational and rotational equilibrium. When a body is in equilibrium, it is customary to call torque the "moment of the force." Thus, the sum of the moments must be zero for equilibrium.

Figure 1.25 shows a system of forces that meets both conditions for equilibrium. Equilibrium in this situation can be verified by the use of Equations 1.16 and 1.17 in Example 1.15.

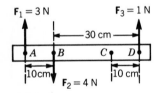

FIGURE 1.25   A body in translational and rotational equilibrium.

EXAMPLE 1.15   Equilibrium system calculation
....

Given: The mechanical system in Figure 1.25.

Find: Whether the system is in equilibrium.

Solution: **a.** $\Sigma F_n = F_1 + F_2 + F_3$ (Define "up" to be positive.)

$\Sigma F_n = 3\text{ N} + (-4\text{ N}) + 1\text{ N}$

$\Sigma F_n = 0$ (First condition for equilibrium is met.)

b. $\Sigma \tau_n = F_1 l_1 + F_1 l_2 + F_3 l_3$ (Define counterclockwise to be positive.)

1. Choose point A as the pivot point—the axis.

$\Sigma \tau_n = (3\text{ N})(0\text{ m}) - (4\text{ N})(0.1\text{ m}) + (1\text{ N})(0.4\text{ m})$

$\Sigma \tau_n = -0.4\text{ N·m} + 0.4\text{ N·m}$

$\Sigma \tau_n = 0$ (Second condition for equilibrium is met.)

Notice that F_1 causes no rotation about point A, F_2 gives a negative torque, and F_3 gives an equal and opposite torque.

2. Choose point D as the pivot point.

$\Sigma \tau_n = -(3\text{ N})(0.4\text{ m}) + (4\text{ N})(0.3\text{ m}) + (1\text{ N})(0\text{ m})$

$\Sigma \tau_n = -1.2\text{ N·m} + 1.2\text{ N·m}$

$\Sigma \tau_n = 0$

3. Choose point C as the pivot point.

$\Sigma \tau_n = -(3\text{ N})(0.3\text{ m}) + (4\text{ N})(0.2\text{ m}) + (1\text{ N})(0.1\text{ m})$

$\Sigma \tau_n = -0.9\text{ N·m} + 0.8\text{ N·m} + 0.1\text{ N·m}$

$\Sigma \tau_n = 0$

Rotational equilibrium is shown to be independent of the choice of pivot point. Remember, then, that you can always select the pivot point to simplify calculations.

Example 1.15 shows that if a body is in rotational equilibrium about some axis, it is in equilibrium about any axis; however, it is only necessary to calculate torques about one point. In systems in which a fixed axis of rotation is defined, that axis is usually chosen. (Any convenient point may be used in testing for equilibrium of torques.) In systems involving the action of gravitational forces, the center of gravity may be chosen.

Since there are only two independent equations for equilibrium, there can be only two unknowns. An unknown can be a magnitude or a location of a force. When this condition is satisfied in a problem, it is said to be statically determinant. For this case, the solution does not depend on the physical properties of the materials. Examples 1.16 and 1.17 are typical problems for parallel, coplanar forces.

EXAMPLE 1.16   Equilibrium beam calculation

. . . .

Given: The system shown in the sketch.

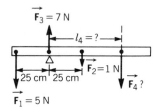

Find: F_4 and l_4 that will cause the system to be in equilibrium about the pivot or fulcrum Δ.

Solution: **a.** $\Sigma \mathbf{F}_n = 0 = \mathbf{F}_1 + \mathbf{F}_2 + \mathbf{F}_3 + \mathbf{F}_4$ (Define down to be negative.)

$$0 = -(5 \text{ N}) - (1 \text{ N}) + 7 \text{ N} + \mathbf{F}_4$$

$$\mathbf{F}_4 = +5 \text{ N} + 1 \text{ N} - 7 \text{ N}$$

$$\mathbf{F}_4 = -1 \text{ N} \quad (\mathbf{F}_4 \text{ is 1 N downward.})$$

     b. $\Sigma \tau_n = 0 = F_1 l_1 + F_2 l_2 + F_3 l_3 + F_4 l_4$ [Choose ccw (counterclockwise) as positive and the fulcrum as reference point.]

$$0 = (5 \text{ N})(25 \text{ cm}) - (1 \text{N})(25 \text{ cm}) + (7 \text{ N})(0) + (1 \text{ N})(l_4)$$

$$0 = 125 \text{ N·cm} - 25 \text{ N·cm} - 1 \text{ N} \times l_4$$

$$0 = 100 \text{ N·cm} - 1 \text{ N} \times l_4$$

$$1 \text{ N} \times l_4 = 100 \text{ N·cm}$$

$$l_4 = \frac{100 \text{ N·cm}}{1 \text{ N}}$$

$$l_4 = 100 \text{ cm}$$

$$l_4 = 1 \text{ m}$$

Since l_4 equals one meter, $\mathbf{F}_4$ is one meter to the right of the pivot.

EXAMPLE 1.17   Equilibrium beam calculation
· ·

Given: The system shown in the sketch.

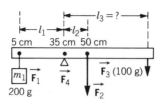

m_2 = Mass of meter stick = 50 g

CG = Center of gravity

$\mathbf{F}_4$ = Upward force at fulcrum or pivot

Find: Where a 100-g weight should be placed to achieve equilibrium.

Solution: m_1 = 200 g, m_2 = 50 g, m_3 = 100 g

l_1 = 30 cm, l_2 = 15 cm, l_3 = ?   (Choose fulcrum as reference point.)

$\Sigma\tau = 0 = F_1 l_1 + F_2 l_2 + F_3 l_3 + F_4 l_4$

$0 = +m_1 g l_1 - m_2 g l_2 - m_3 l_3 g + 0$ ($\mathbf{F}_4$ is the upward force at the fulcrum. Since l_4 is zero, $\mathbf{F}_4$ need not be calculated.)

$\Sigma\tau = g\,(+m_1 l_1 - m_2 l_2 - m_3 l_3)$ (Since g is in every term, it can be divided out of the equation.)

$\Sigma\tau = +\,(200\text{ g})(30\text{ cm}) - (50\text{ g})(15\text{ cm}) - (100\text{ g})(l_3)$

$\Sigma\tau = +\,6000\text{ g·cm} - 750\text{ g·cm} - 100\text{ g } l_3$

$0 = +\,5250\text{ g·cm} - 100\text{ g } l_3$

$100\text{ g } l_3 = 5250\text{ g·cm}$

$$l_3 = \frac{5250\text{ g·cm}}{100\text{ g}}$$

$l_3 = 52.5\text{ cm}$

The 100-g weight should be placed 52.5 cm to the right of the reference point (fulcrum) or at the 87.5-cm mark on the meterstick.

FIGURE 1.26 Two-force member.

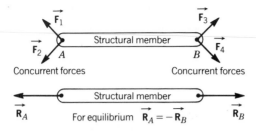

Coplanar, Nonconcurrent Forces

Two-Force Member In general, forces in a plane may not be concurrent or parallel. An example is shown in Figure 1.26. The two-force member is an important case because forces are applied at only two points. This occurs in an approximate way in many structures. A resultant can easily be found at point *A* and at point *B*. The problem then reduces to two forces. If the body is in equilibrium, the two forces must be equal in magnitude but opposite in direction. Thus, the structural member is in either tension or compression and there is no tendency for the member to bend. For two-force members, the forces are always **along** the member.

Resultant Figure 1.27 shows a different case. Here, the structural member will have a bending moment and shear stress. In both cases, a single resultant can be found that will produce the same translational and rotational changes in the body's motion. To do this mathematically, it is usually advantageous to break up each force into its *x* components and *y* components. The magnitude of the resultant then becomes

$$R = \sqrt{(\Sigma F_x)^2 + (\Sigma F_y)^2}$$

The quantities $\Sigma \mathbf{F}_x$ and $\Sigma \mathbf{F}_y$ also determine the direction of the resultant. The angle that the line of action takes with the *x* -axis is

$$\tan \theta = \frac{\Sigma F_y}{\Sigma F_x}$$

In addition to having the correct magnitude and direction, the resultant must also produce the same torque about any point as did the individual forces. This condition may be expressed as

$$Rl = F_1 l_1 + F_2 l_2 + F_3 l_3 + \ldots = \Sigma F_n l_n$$

where *l* and l_n are the moment arms (perpendicular distances to the lines of action

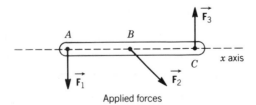

FIGURE 1.27 Three-force member.

of the corresponding forces). It is advantageous to use the component forces to calculate torques about a point through which the greatest number of lines of action pass, such as points *A, B,* or *C* in Figure 1.27. If *R* = 0, then forces can be replaced by an equivalent couple.

Equilibrium If a body is in equilibrium, there are three equations that must be satisfied. They are

$$\Sigma \mathbf{F}_x = 0 \tag{1.18}$$

$$\Sigma \mathbf{F}_y = 0 \tag{1.19}$$

$$\Sigma \tau_n = 0 \tag{1.20}$$

Thus, there may be three unknowns if the problem is determined by statics alone. Example 1.18 illustrates how unknown forces can be determined.

EXAMPLE 1.18 Force on a boom
. . . .

Given: The structure shown in the sketch.

Find: The reaction forces at points *A* and *D*.

Solution: Forces are applied at only two points on the member *DC*. Thus, the net force is in the same direction as the members. Forces are applied at three points on the member *ACB*. The net force will not be the same direction as the member, except in special cases. Member *ACB* will have a tendency to bend. The member *ACB* may be taken out as a free body as shown in the sketch.
Sum of forces in the *x* direction yields

$$F_{1x} - F_{2x} = 0$$

$$F_{1x} = F_{2x}$$

Sum of forces in the *y* direction yields

$$\mathbf{F}_{1y} - 1000 \text{ lb} = 0$$

$$\mathbf{F}_{1y} = 1000 \text{ lb}$$

Sum of torques about any point yields (pick point *A*)

$$20 \text{ ft } (F_{2x}) - (10 \text{ ft})(-1000 \text{ lb}) = 0; \qquad 20 F_{2x} = 10,000$$

$$F_{2x} = -500 \text{ lb} \qquad (\mathbf{F}_{2x} \text{ is force at point } D \text{ where } \theta \text{ is zero.})$$

$$\mathbf{F}_{1x} = 500 \text{ lb}$$

The resultant at *A* is

$$F_1 = \sqrt{F_{1x}^2 + F_{1y}^2} = \sqrt{(500 \text{ lb})^2 + (1000 \text{ lb})^2} = 1118 \text{ lb}$$

$$\theta = \tan^{-1} \frac{F_{1y}}{F_{1x}} = \tan^{-1} \frac{1000 \text{ lb}}{500 \text{ lb}} = 63.4°$$

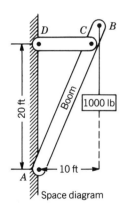

Space diagram

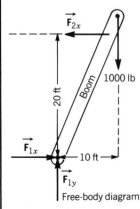

Free-body diagram

Example 1.19 shows how to calculate the forces in the members of a simple truss (all members have forces applied at only two points).

EXAMPLE 1.19   Forces in a simple truss
· ·

Given: The free-body diagram of a simple truss as shown in the sketch.

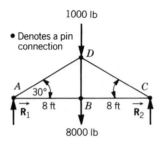

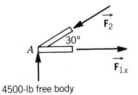

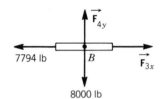

Find: The forces in *AB*, *BC*, *AD*, *DB*, and *DC*.

Solution: If the loads are given, solve for the reaction forces.

Parallel forces

$$\Sigma \mathbf{F}_n = 0$$

$$\Sigma \tau_n = 0$$

Sum of torques about point *B* yields, $(\Sigma \tau_n)_B = 0$,

$$-7 R_1 + 7 R_2 = 0$$

$$R_1 = R_2$$

Sum of vertical forces yields, $\Sigma \mathbf{F}_y = 0$,

$$2 R_1 - 9000 \text{ lb} = 0$$

$$R_1 = R_2 = 4500 \text{ lb}$$

Take a free body out about point A.

$\Sigma \mathbf{F}_x = 0$

$\Sigma \mathbf{F}_y = 0$

Sum of vertical forces yields

$4500 \text{ lb} - F_2 \sin 30° = 0$

$F_2 = \dfrac{4500 \text{ lb}}{\sin 30°} = 9000 \text{ lb}$ (compression)

Sum of horizontal forces yields

$F_{1x} - 9000 \text{ lb} \cos 30° = 0$

$F_{1x} = 9000 \text{ lb} \cos 30° = 7794 \text{ lb}$ (tension)

Take a free body out about point B.

Concurrent forces

$\Sigma \mathbf{F}_x = 0$

$\Sigma \mathbf{F}_y = 0$

Sum of vertical forces yields

$F_{4y} = 8000 \text{ lb}$ (tension)

Sum of horizontal forces yields

$F_{3x} = 7794 \text{ lb}$ (tension)

From symmetry

Force in AD = Force in DC = 9000 lb   (compression)

Force in AB = Force in BC = 7794 lb   (tension)

Force in BD = 8000 lb   (tension)

SECTION 1.1 EXERCISES

1. Define the physical quantities force and torque. State appropriate units for each in both the SI and English systems of units.

2. With the help of Newton's law of universal gravitation, explain how each object in the universe affects all other objects.

3. How do weight and mass differ? Is the mass of an object the same on earth as on any other plant? Can the same be said of weight?

4. Calculate the gravitational force exerted by the earth on the moon. (The moon's mass is 7.34×10^{22} kg, and it is 3.86×10^8 m from earth. The earth's mass is 6.02×10^{24} kg.)

5. Describe how force varies with distance in the inverse square law.

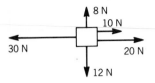

6. An object is acted upon by three simultaneous forces: 40 lb to the right, 180 lb to the left, and 200 lb to the right. Determine the resultant force and direction of motion.

7. The diagram on the left shows five forces acting on an object initially at rest. Determine the net force and the direction of motion.

8. A toy rocket experiences an upward thrust of 100 N. What is the net acceleration of the rocket if the mass of the rocket is 2 kg? Do not neglect the force due to gravity.

9. Determine the resultant of the coplanar, concurrent forces graphically and analytically for each of the following cases.

a.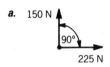

b.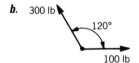

c.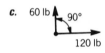

d.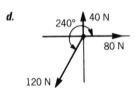

e.   25 lb at 30°
    32 lb at 120°
    15 lb at 300°

f.   4.2 N at 37°
    6.8 N at 58°
    1.3 N at 198°

10. Solve for the tension in each wire for the system of concurrent, coplanar forces at the left.

11. A force of 10 lb is used to produce a torque of 32.8 ft·lb. The force acts in a direction perpendicular to the lever arm. Determine the length of the lever arm.

12. Convert the values in Problem 11 to SI units and solve for lever arm in meters. Compare the answer to that from Problem 11.

13. Three masses are hung from a massless meterstick, as shown in the sketch to the left. Find the net torque and the direction of rotation.

14. Three masses are hung from a massless meterstick as shown in the following sketch. What force must be applied to the right end of this stick to keep the meterstick from rotating? Be sure to specify the direction in which the force must be applied.

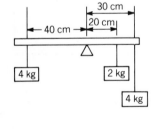

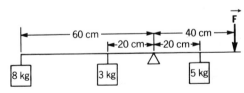

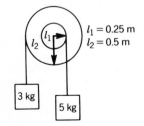

15. A frictionless, massless pulley has 3-kg and 5-kg masses attached as shown in the sketch to the left. Calculate the resultant torque and determine the direction of rotation.

16. As illustrated in the following sketch, a lever is to be used to lift an object weighing 6000 N. If the distance from the pivot to the object is 1 m and the lever is 6 m long, can the object be lifted if the maximum force that can be applied to the other side of the lever is 1000 N?

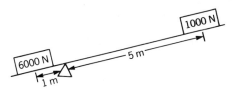

17. Is it possible to satisfy both conditions of equilibrium, $\Sigma F_y = 0$ and $\Sigma \tau = 0$, by applying one additional force to the system shown in the following sketch? Justify your answer.

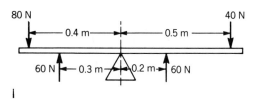

18. Is it possible to bring the system shown in the following sketch to equilibrium by the addition of a single force? Justify your answer.

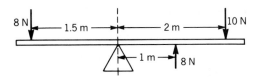

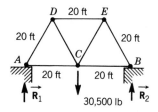

19. Calculate the reaction forces at *A* and *B* in the structure shown in the sketch to the left. Calculate the tension or compression in each of the massless, two-force members, *AD*, *AC*, *DC*, *DE*, *CE*, *CB*, and *BE*. (Be sure to use symmetry to help you reduce the number of calculations.)

20. *For the crane shown in the following sketch, solve for the vertical and horizontal components of the reaction forces at *A* and *B* and the reaction at *D* (no vertical component at *D*). Solve for the stress in *CE* (this is a two-force member). Consider all force members to be massless.

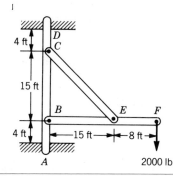

* Challenge question.

1.2 .... OBJECTIVES..

PRESSURE— FORCELIKE QUANTITY IN FLUID SYSTEMS

Upon completion of this section, the student should be able to

- Define the forcelike quantity in fluid systems, **pressure,** and state appropriate units in the SI and English systems of units.

- Given two of the following quantities in a fluid system, determine the third:
  Force
  Area
  Pressure

- Distinguish between mass density and weight density.

- Distinguish between gage pressure and absolute pressure.

- State Archimedes' principle.

- Explain what causes a buoyant force on a floating object.

- Given two of the following quantities in a fluid system, determine the third:
  Pressure
  Height of fluid
  Weight density

- Given the height of a manometer fluid and its density, calculate the differential pressure in the manometer.

DISCUSSION

We have studied the action of **force** in translational systems and **torque** in rotational systems. Both are important in mechanical energy systems. Now let us study the role of the forcelike quantity **pressure** in fluid systems.

Pressure

Definition of Pressure   The prime mover, or forcelike quantity, in fluid systems (hydraulic or pneumatic) is called "pressure." Pressure is defined as the force per unit area exerted by a fluid and is given by

$$p = \frac{F}{A} \tag{1.21}$$

where

p = pressure (lb/inch^2 or N/m^2)
F = force (lb or N)
A = area (inch^2 or m^2)

Example 1.20 illustrates the use of Equation 1.21 in solving a problem that involves fluid pressure in a hydraulic jack.

EXAMPLE 1.20   Fluid pressure in a hydraulic jack

· ·

Given: A hydraulic jack is rated at 4000-lb lifting capacity and has a large piston with a diameter of 2 inch.

Find: The fluid pressure in the jack at maximum load.

Solution: *Area*

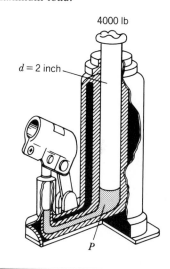

$A = \pi r^2$

$A = (3.14)\,(1 \text{ inch})^2$

$A = 3.14 \text{ inch}^2$

$p = \dfrac{F}{A}$

$p = (4000 \text{ lb}) \div (3.14 \text{ inch}^2)$

$p = 1274 \text{ lb/inch}^2$

Density

Mass Density and Weight Density   The property of matter described as **density** is important in fluid systems. Density is defined as body mass or weight per unit volume. When the mass per unit volume is given (g/cm^3 or kg/m^3, for example), it is called **mass density.** The sympol ρ, without subscript, will be used for mass density. For weight density, the sympol ρ_w, with subscript w, will be used.

The mass density (ρ) is defined by

$$\rho = \frac{m}{V} \tag{1.22}$$

where

    ρ = mass density of substance
    m = mass of substance
    V = volume of substance

Table 1.3 lists the mass density of some common substances.

Weight density (ρ_w) is the weight (*w* or gravitational force) per unit volume (*V*). The mathematical expression is

$$\rho_w = \frac{w}{V} = \left(\frac{m}{V}\right) g = \rho\, g \tag{1.23}$$

TABLE 1.3   Mass density of common substances

Solids		Liquids		Gases at 0°C and 1.013×10^5 N/m²	
Substance	Density (kg/m³)	Substance	Density (kg/m³)	Substance	Density (kg/m³)
Aluminum	2.70×10^3	Ethyl		Air	1.239
Copper	8.89×10^3	alcohol	7.89×10^2	Carbon dioxide	1.977
Ice	9.22×10^2	Ether	7.40×10^2	Helium	1.79×10^{-1}
Steel	7.80×10^3	Glycerin	1.26×10^3	Hydrogen	9.0×10^{-2}
Tungsten	1.90×10^4	Mercury	1.36×10^4	Nitrogen	1.25
Zinc	7.14×10^3	Water	1.00×10^3	Oxygen	1.429

where

ρ = mass density

g = acceleration due to gravity

ρ_w = weight density expressed in N/m³ or lb/ft³

Since the weight density depends on g, its value is not independent of location.

The specific gravity (SG) of a substance is the ratio of the density of a substance to the density of water at 0°C. This relationship is

$$SG = \frac{\rho_{sub}}{\rho_{water}} = \frac{(\rho_{sub})g}{(\rho_{water})g} = \frac{(\rho_w)_{sub}}{(\rho_w)_{water}} \tag{1.24}$$

where

SG = specific gravity of a substance

ρ_{sub} = mass density of the substance (kg/m³ or g/cm³)

ρ_{water} = mass density of water (1000 kg/m³ or 1.0 g/cm³)

ρ_w = weight density

If the SI unit, 1.0 g/cm³, is used for the density of water, the specific gravity will be equal numerically to the density of the substance. Specific gravity has no units because it represents a ratio of two numbers with like units. Since specific gravity has no units, it must have the same value regardless of the system of units in which one is working.

Pressure in Liquids

Liquid Pressure   Pressure in a liquid at rest increases with depth. At a given location, it is the same in all directions. A diver practicing in a deep swimming pool experiences an increase in pressure in her eardrums as she dives to the bottom of the pool. The pressure depends not only on the depth but also on the weight density of the liquid. Pressures act over a surface, whereas the weight is a body force. For a liquid at rest, the principles of statics apply, that is, the sum of

FIGURE 1.28 Liquid pressure in an open container.

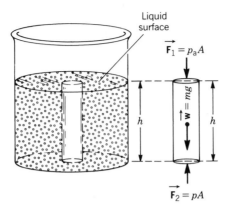

Liquid surface

$\vec{F}_1 = p_a A$

$w = mg$

h

h

$\vec{F}_2 = pA$

a. Space diagram

b. Free-body diagram (only vertical forces)

the surface forces and body forces must equal zero in any direction. In addition, there can be no net torque on any free body that is purely liquid.

Consider an open container as shown in Figure 1.28. A free body can be isolated, and the surface and body forces applied. For this discussion, only the vertical forces are shown acting on the free body. Setting the sum of the vertical forces equal to zero (equilibrium) gives

$$F_2 - F_1 - mg = 0$$

or

$$+ pA - p_a A - mg = 0 \qquad (1.25)$$

where

p = pressure at a depth h
A = top and bottom surface area of the free body
p_a = atmospheric pressure

The mass of the free body can be expressed in terms of the density and volume (Equation 1.22):

$$m = \rho V = \rho A h$$

Substituting this expression for the mass (m) in Equation 1.25, one obtains

$$pA = p_a A + \rho g h A$$

or

$$p = p_a + \rho g h = p_a + \rho_w h \qquad (1.26)$$

This value of p is called the **absolute pressure.** Many instruments, however, measure only the pressure above atmospheric pressure. The reading obtained is called the **gage pressure.** It may be expressed as

$$p_G = p_{abs} - p_a \qquad (1.27)$$

TABLE 1.4 Unit conversions for pressure

1 Pa = 1 N/m²
1 bar = 10⁵ Pa
1 inch of water = 0.0361 psi
1 inch of Hg = 0.491 psi
1 mm of Hg = 0.133 kPa ≈ 1/8 kPa
1 inch of water = 0.249 kPa ≈ 1/4 kPa
1 inch of Hg = 3.3858 kPa ≈ 3.4 kPa
1 psi = 6.895 kPa ≈ 7 kPa
1 atmosphere = 14.7 psi

The gage pressure for a liquid at rest in an open tank is

$$p_G = \rho g h = \rho_w h \qquad (1.28)$$

In engineering practice, the pressure is expressed in many ways depending on the instrument used. The metric (SI) units of pressure are the pascal and bar. A pascal (Pa) is a newton per square meter, which is a relatively small unit of pressure. Therefore, kPa (1000 N/m²) is frequently used. It is also common to express the pressure in pounds per square inch (psi), dynes per square centimeter, or the height of a liquid column such as 60 cm of mercury or 20 inch of water. A set of unit relationships for different pressures is given in Table 1.4.

It is often important to calculate a pressure difference (Δp) in a liquid at rest in a gravitational field. A free body is shown in Figure 1.29 for this case. The difference in pressure between the bottom and top surface is

$$\Delta p = p_2 - p_1 = (p_a + \rho_w h_2) - (p_a + \rho_w h_1)$$

or

$$\Delta p = \rho_w (h_2 - h_1) = \rho_w \Delta h \qquad (1.28a)$$

where

Δp = pressure difference ($p_2 - p_1$)
Δh = difference in height
ρ_w = weight density
p_a = atmospheric pressure above the liquid surface

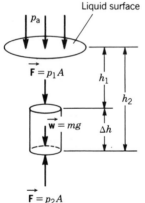

Liquid surface

p_a

$\vec{F} = p_1 A$ h_1

h_2

$\vec{w} = mg$ Δh

$\vec{F} = p_2 A$

FIGURE 1.29 Free-body diagram for pressure calculation.

Note the similarity between Equations 1.28 and 1.28a. Example 1.21 illustrates the use of Equation 1.28 to calculate the gage pressure for an open tank. Example 1.22 shows how to calculate a pressure difference for a liquid at rest using Equation 1.28a.

EXAMPLE 1.21 Water pressure problem
. . . .

Given: The height of the water in a water system is 100 ft (30.5 m) above a faucet. The mass density of water (SI units) is 1000 kg/m³. The weight density of water (English units) is 62.4 lb/ft³.

Find: The gage pressure at the faucet is SI and English units.

Solution: **SI units**

Use Equation 1.28.

$$p_G = \rho g h$$

$$p_G = (1000 \text{ kg/m}^3)(9.8 \text{ m/sec}^2)(30.5 \text{ m})$$

$$p_G = 2.99 \times 10^5 \frac{\text{kg} \cdot \text{m}^2/\text{sec}^2}{\text{m}^3}$$

$$p_G = 2.99 \times 10^5 \left(\frac{\text{kg} \cdot \text{m}}{\text{sec}^2}\right)\left(\frac{1}{\text{m}^2}\right)$$

$$p_G = 2.99 \times 10^5 \text{ N/m}^2$$

The units of pressure are force per area.

English units

Use Equation 1.28.

$$p_G = \rho_w h$$

$$p_G = (62.4 \text{ lb/ft}^3)(100 \text{ ft})$$

$$p_G = 6240 \text{ lb/ft}^2$$

$$p_G = (6240 \text{ lb/ft}^2)\left(\frac{1 \text{ ft}^2}{144 \text{ inch}^2}\right)$$

$$p_G = 43.3 \text{ psi}$$

The units have been converted to the more familiar pounds per square inch. This is also force per unit area.

EXAMPLE 1.22 Calculation of pressure difference
· ·

Given: Two points (1 and 2) in a column of mercury are separated by 2.036 inch. [Weight density (ρ_w) of mercury is 0.491 lb/inch³.]

Find: The difference in pressure Δp between points 1 and 2.

Solution: Use Equation 1.28a, $\Delta p = p_2 - p_1 = \rho_w \Delta h$.

$$\Delta p = (0.491 \text{ lb/inch}^3)(2.036 \text{ inch})$$

$$\Delta p = p_2 - p_1 = 1 \text{ lb/inch}^2$$

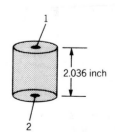

FIGURE 1.30 Unequal pressures across valve.

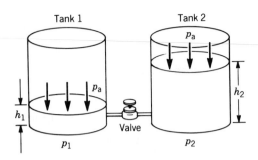

FIGURE 1.31 Pressures across ends of valve.

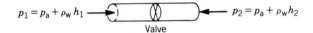

FIGURE 1.32 Equal pressures across valve.

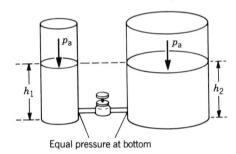

If the sum of all the forces on a volume of liquid is not equal to zero, motion will occur according to Newton's second law. In the vertical direction, the value of Δp must be greater or less than the weight of the water for vertical motion to occur. In the horizontal direction, motion occurs if Δp is not zero. For this reason, pressure acts like a forcelike quantity in a fluid system. Figure 1.30 shows two open tanks with different liquid levels connected by a pipe with a valve. The pressures p_1 and p_2 at the ends of the pipe are shown in Figure 1.31. Using Equation 1.26, the value of Δp across the valve becomes

$$\Delta p = p_2 - p_1 = \rho_w (h_2 - h_1)$$

Thus, when the valve is opened, liquid will flow until the water levels are equal. When $h_2 = h_1$, Δp is zero and equilibrium is established. Equilibrium depends only on the heights being equal and not on the volume of liquid in the tanks. The system shown in Figure 1.32 is in equilibrium, although the right tank contains more liquid.

Buoyant Force If a body is floating on a liquid, it will be buoyed up by a force that compensates for its weight. Figure 1.33 shows a rectangular solid floating in a liquid and the corresponding free-body diagram with the vertical forces applied.

Setting the sum of the vertical forces equal to zero, in accordance with $\Sigma F_y = 0$, yields

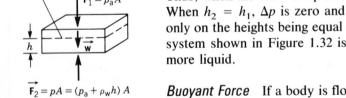

a. Space diagram

Water surface

b. Free-body diagram

FIGURE 1.33 Buoyant force on a floating object.

$$F_2 - w - F_1 = 0$$

$$(p_a + \rho_w h)\,A - w - p_a A = 0$$

$$p_a A + \rho_w h A - w - p_a A = 0$$

$$w = \rho_w h A \tag{1.29}$$

where

p_a = atmospheric pressure at top of liquid surface
h = measure of body submerged in liquid
w = weight of body
A = area of the upper and lower surfaces
ρ_w = weight density of the liquid

The value of hA is the volume of the liquid displaced by the body, and from the definition of weight density (Equation 1.23) $\rho_w h A$ is the weight of the liquid displaced. Thus, the weight of the object is equal to the weight of the liquid displaced. It can be shown that the buoyant force is equal to the weight of the liquid displaced for a body of any shape. If a body does not float, its apparent weight will be reduced by the buoyant force. The principle was first discovered by Archimedes and is known as **Archimedes' principle.**

Equation 1.29 can be used to measure the density of an unknown liquid. Suppose a block having a known weight density $(\rho_w)_B$ floats on a liquid. By measuring the height of the block (a) and the distance it is submerged (h), it is determined that h/a of the block's volume is below the surface. Thus, the volume of the liquid displaced (V_L) is h/a times the volume of the block (V_B). Letting the weight of the block (w_B) equal the weight of the liquid displaced (w_L) gives

$$w_L = w_B$$

or

$$(\rho_w)_L V_L = (\rho_w)_B V_B$$

where

$(\rho_w)_L$ = weight density of the liquid
$(\rho_w)_B$ = weight density of the block

Thus, solving for $(\rho_w)_L$ yields

$$(\rho_w)_L = \frac{V_B}{V_L}(\rho_w)_B = \frac{a\,A}{h\,A}(\rho_w)_B = \frac{a}{h}(\rho_w)_B \tag{1.30}$$

If the liquid is water, Archimedes' principle can be used to measure the specific gravity of solids that sink in water. One measures the weight of an object (w_B) and its apparent weight when submerged in water $(w_B)_w$. The apparent weight $(w_B)_w$ is equal to

$$(w_B)_w = w_B - w_w$$

or

$$w_w = w_B - (w_B)_w$$

where

w_W = weight of the water displaced by the total submerged block

Since the volume of the body (V_B) is equal to the volume of the water displaced (V_W), one may write from the definition of specific gravity (Equation 1.24) that

$$SG = \frac{(\rho_w)_B}{(\rho_w)_W} = \frac{(\rho_w)_B V_B}{(\rho_w)_W V_B} = \frac{w_B}{w_W} = \frac{w_B}{w_B - (w_B)_W} \quad (1.31)$$

where

$(\rho_w)_W$ = weight density of water
$(\rho_w)_B$ = weight density of the solid
V_B = volume of the block
w_B = weight of the block
w_W = weight of water equal to volume of block
$(w_B)_W$ = apparent weight of block in water

Sample problems applying Equations 1.30 and 1.31 are given in Examples 1.23 and 1.24.

EXAMPLE 1.23   Density of a liquid
· ·

Given: A rectangular solid whose height (*a*) is 1 cm sinks a distance (*h*) equal to 0.8 cm, in a certain liquid. The density of the solid is 1.5 g/cm³.

Find: The density of the fluid.

Solution: Equation 1.30 states

$$(\rho_w)_L = \frac{a}{h} (\rho_w)_B$$

$$(\rho_w)_L = \left(\frac{1.0 \text{ cm}}{0.8 \text{ cm}}\right)(1.5 \text{ g/cm}^3)$$

$$(\rho_w)_L = 1.87 \text{ g/cm}^3$$

The density of the fluid must be greater than the density of the solid if the solid is to float. That's why a steel ball ($\rho_m = 7.80$ g/cm³) floats in the liquid mercury ($\rho_m = 13.6$ g/cm³).

EXAMPLE 1.24   Specific gravity of a solid that sinks in water
· ·

Given: A body's weight in air is 10.0 lb, and its weight when submerged in water is 5.0 lb.

Find: The specific gravity of the body.

Solution: Equation 1.31 gives for the specific gravity

$$SG = \frac{w_B}{w_B - (w_B)_W}$$

$$SG = \frac{10.0}{10.0 - 5.0} = \frac{10.0}{5.0}$$

$$SG = 2$$

Pressure in Gases

The pressure of gas on a surface depends on the number and speed of molecules colliding against the surface. Therefore, one way to increase the pressure of a gas in a container is either to pump more molecules into the container or to force the same molecules into a small container. To decrease the pressure, some of the molecules can be released from the container or placed into a larger container. Another way to increase pressure is to heat the gas inside a closed container. This procedure causes the molecules to increase their speed, creating an increased pressure. Cooling of the container has the reverse effect on the molecules and, consequently, decreases the pressure.

The force per unit area that is exerted on any body immersed in air at one atmosphere is approximately 14.7 psi. One standard atmosphere is defined as the air pressure at sea level at 0°C (32°F). Typical values for 1 standard atmosphere are

1 standard atmosphere = 14.7 lb/inch² = 14.7 psi

$\qquad\qquad\qquad$ = 2116 lb/ft²

$\qquad\qquad\qquad$ = 1.013×10^5 N/m²

$\qquad\qquad\qquad$ = 1.013×10^6 dynes/cm²

$\qquad\qquad\qquad$ = pressure at the bottom of a column of mercury 760 mm high (29.92 inch)

$\qquad\qquad\qquad$ = pressure at the bottom of a column of water 33.91 ft high (10.33 m)

Barometric pressure is atmospheric pressure at a particular location, often above sea level and at temperatures other than 0°C (32°F).

A partial vacuum is a space in which there are very few atoms or molecules. A perfect vacuum is a theoretical space in which no atoms are present. The term vacuum is generally taken to mean a space containing air or gas at well below atmospheric pressure, although, literally speaking, this space should be referred to as a partial vacuum. A vacuum is sometimes called a "negative pressure," which is a gage pressure.

Pressure Measurements

There are many different instruments to measure pressure. The simplest device

FIGURE 1.34 Mercury and water U-tube manometers.

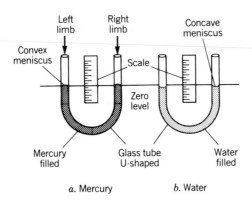

a. Mercury   b. Water

FIGURE 1.35 Measuring a pressure above atmospheric pressure. (left)

FIGURE 1.36 Measuring a pressure below atmospheric pressure. (right)

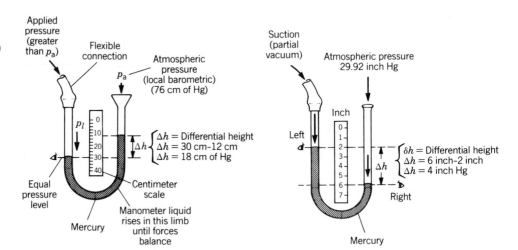

for measuring moderate pressure is the U-tube manometer. A simple U-tube manometer consists of a tube of glass or other transparent material bent in the shape of a "U." Both ends may be left open or one end may be closed. For most applications, it is more convenient to use a manometer with both ends open. For this case, the U-tube is usually filled about halfway with a liquid such as water or mercury as shown in Figure 1.34.

When one side of the manometer is connected to an unknown pressure, p, the liquid levels will change. In Figure 1.35 the unknown pressure applied to the left limb is greater than atmospheric pressure, and in Figure 1.36 the applied pressure is less than atmospheric pressure. For either case, equilibrium occurs when the pressure difference on a small volume of liquid, at the bottom of the U-tube, is zero, as shown in Figure 1.37.

Since F_1 must equal F_2 for equilibrium, one obtains

$$F_1 = F_2$$

FIGURE 1.37 Equal pressures at the bottom of the U-tube arms.

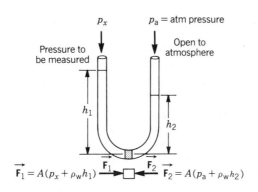

$$A(p_x + \rho_w h_1) = A(p_a + \rho_w h_2)$$

or

$$p_x = p_a + \rho_w(h_2 - h_1) = p_a + \rho_w \Delta h \qquad (1.32)$$

If h_2 is greater than h_1, p_x is greater than atmospheric pressure; and if h_1 is greater than h_2, p_x is less than atmospheric pressure. Example 1.25 illustrates the use of Equation 1.32.

EXAMPLE 1.25   Pressure calculation from manometer reading
. . . .

Given: The information shown in Figure 1.37 and the weight density of mercury, 0.491 lb/inch³.

Find: **a.**   Calculate the partial vacuum pressure (p) in psi.
b.   Calculate the gage pressure (p_G) in psi.

Solution: Given Equation 1.32

   a. $p_x = p_a + \rho_w \Delta h$

   Since h_1 (left) is greater than h_2 (right), pressure is less than atmospheric pressure. Therefore, Δh is negative.

   $p_x = 14.7 \text{ psi} - (0.491 \text{ lb/inch}^3)(4 \text{ inch})$

   $p_x = 12.7 \text{ psi}$

   b.   Gage pressure is (Equation 1.27)

   $p_G = p_x - p_a$

   $p_G = 12.7 \text{ psi} - 14.7 \text{ psi}$

   $p_G = -2 \text{ psi}$ (A negative pressure relative to atmospheric pressure.)

SECTION **1.2** **EXERCISES**

1. Define the forcelike quantity, pressure, in fluid systems. State appropriate units in both the SI and English systems of units.

2. What is the **average** pressure on the face of a dam if the water exerts a total force of 12×10^8 N and the surface area of the dam is 600 m²?

3. A 130-lb woman wearing spike-heel shoes, with heels measuring $1/2 \times 1/2$ inch, is standing on a linoleum floor. Calculate how much pressure will be applied to the linoleum if she puts all her weight on one heel.

4. Calculate the water pressure in pounds per square inch at the bottom of a lake 250 ft deep. (Weight density of water is 62.4 lb/ft³.)

5. Calculate the depth of water in a tank if the gage pressure at the bottom is 10^5 N/m². (Density of water is 1000 kg/m³.)

6. What is the lifting capacity of a hydraulic jack if the fluid pressure at maximum load is 1600 psi and the piston that lifts the load has a radius of 2 inch?

7. Describe Archimedes' principle in your own words.

8. If the weight of an object and the weight of an equal volume of water were known, could the specific gravity be determined? If not, what other information would be needed? If so, describe the solution in an equation format.

9. The specific gravity of aluminum is 2.7. The weight of water is 62.4 lb/ft³. Calculate the weight in pounds of a solid piece of aluminum with a volume of 2 ft³.

10. What is the volume of a floating object if it displaces 2 kg of water when it is placed in water and its mass density is 850 kg/m³?

11. How much would a man have to weigh, when standing on one foot, to exert the same pressure on the floor as did the woman in Problem 3? Use 18 inch² as the area of the bottom of the man's foot. Considering the necessary weight of the man, is it reasonable to assume that a man could exert the same pressure as did the woman?

12. What is the density of a liquid if, at the bottom of a tank filled with this liquid, the pressure is 10^5 N/m²? The depth of the tank is 6 m.

13. The porthole of a small submarine breaks when an **average** force of 10^6 N is distributed over its area. How deep below the surface of the ocean can this submarine go before the glass in the porthole breaks? The area of the porthole is 2 m².

14. The differential height of a mercury column in a barometer is 76 cm. Calculate the differential height in meters if oil (density 0.90 g/cm³ is used instead of mercury (density 13.6 g/cm³).

15. If a mercury manometer is used to measure a differential pressure of 2.0 psi, calculate the differential height between the two levels of mercury columns in inches.

16. In Problem 15, what would the difference in column levels be if water were used instead of mercury?

17. A pressure was applied to one arm of an open-tube manometer (mercury filled), and the level difference was found to be 12.0 inch. Compute the applied pressure in pounds per square inch.

18. *A cube of wood floats as shown in the diagram to the left. The bottom side of the wood cube is 2 cm below the water surface. The density of the oil is 0.60 g/cm³ and the density of water is 1.0 g/cm³. If the cube is 10 cm on each side, what is the density of the wood?

* Challenge question.

1.3 ....

POTENTIAL DIFFERENCE AND MAGNETOMOTIVE FORCE—FORCELIKE QUANTITIES IN ELECTROMAGNETIC SYSTEMS

OBJECTIVES...

Upon completion of this section, the student should be able to

- Define the forcelike quantities **voltage** and **magnetomotive force** and give appropriate units for each.

- State Coulomb's law.

- Define the electric field.

- Draw electric field lines around point charges and charged conducting plates.

- Describe how field lines indicate direction and changing strength of the electric force on charged particles located in the field.

- Draw a parallel-plate capacitor with electric field lines.

- Describe the relationships between plate charge, potential difference, and capacitance.

- Describe the relationship between potential difference, plate separation, and electric field strength.

- Describe the nature of the source of permanent magnetism.

- Define magnetic induction.

- Calculate the force on a moving charge.

- Define magnetic field strength.

- Explain the operation of electric generators.

- Explain the operation of electric motors.

DISCUSSION

We have studied the actions of forcelike quantities in mechanical systems (force and torque) and fluid systems (pressure). There are forcelike quantities in electromagnetic systems also. Let's examine these. They are voltage and magnetomotive force.

Electric Systems

Voltage   The prime mover, or forcelike quantity, in electrical systems is the voltage, sometimes called "potential difference" or "electromotive force." Voltage occurs in an electrical system because of the separation of electrical charges.

Charge   The origin of charge lies in the atomic structure of matter. All matter is composed of atoms. The structure of an atom is represented in Figure 1.38. The nucleus contains most of the atomic mass in the form of neutrons and protons. The protons have a positive charge and the neutrons are neutral (no charge). The nucleus is surrounded by a cloud composed of the much less massive electrons.

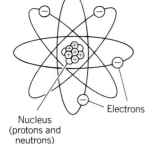

Nucleus
(protons and
neutrons)
Electrons

FIGURE 1.38   The atom.

TABLE 1.5   Methods used to produce a potential difference

Friction	Rubbing a cat's fur, for example, can create static electricity. Friction strips away electrons from the cat's fur, leaving a positive charge on the fur and a negative charge on the hand. A potential difference exists between the fur and the hand.
Chemical	Chemical energy in dry cells (flashlight battery type) or in wet cells (auto battery type) separates the charge to form a positive pole and a negative pole. A potential difference exists between the two poles.
Magnetic mechanical	A coil of wire moving in a magnetic field produces a force on electric charges that separates the positive and negative charges. A potential difference exists between the ends of the coil.
Light	Light energy falling on certain materials causes the separation of electrons from their atoms. A potential difference exists across certain regions in the material. Light meters make use of this process and indirectly record light intensities.

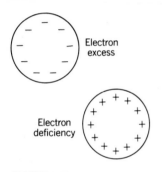

FIGURE 1.39   Separated charges.

Each electron has a negative charge of the same magnitude as the positive charge of a proton. A normal atom is neutral (has no net charge) because it has the same number of electrons and protons. Electrons are bound to the atom by electrical forces of attraction that obey the rule: unlike charges attract and like charges repel. These forces result in a voltage whenever the positive and negative charges are separated. Four common ways of creating a potential difference are listed in Table 1.5.

Neutral objects are those that contain the same quantity of positive and negative charge.

In Figure 1.39, some of the electrons have been moved from one object to another. The object with excess electrons is said to have a negative charge and the one from which the electrons have been removed, therefore, has a positive charge. If a path is provided for the electrons from one object to the other, an electric current will flow; that is, charge will be transferred from one object to the other until both have the same electrical potential. Potential difference moves charge in an electrical system in the same way that pressure moves fluid in a fluid system. Thus, potential difference plays the role of a forcelike quantity in electrical systems just as pressure does in a fluid system.

Electric current is the rate at which charge moves through a system. Current exists in a circuit whenever two points of different electrical potential are connected by an electrical conductor. If there is no potential difference, the system is in equilibrium and no charge flows.

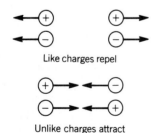

FIGURE 1.40   Forces between electric charge.

Coulomb's Law   Different masses, such as the sun and the earth, can only attract each other, they can never repel one another. Thus, gravitational forces—the forces between different masses—are attractive. Electrical charges, on the other hand, both attract and repel one another. Figure 1.40 shows that like charges repel and unlike charges attract. A positive charge repels another positive charge but attracts a negative charge. Similarly, a negative charge repels another negative charge but attracts a positive charge. The magnitude of the force between charged

TABLE 1.6 Some elementary charged particles

Particle	Charge (coulombs)	Mass (kilograms)
Electron	-1.602×10^{-19} C	9.108×10^{-31} kg
Proton	$+1.602 \times 10^{-19}$ C	1.672×10^{-27} kg
Alpha	$+3.204 \times 10^{-19}$ C	6.648×10^{-27} kg
Neutron	No charge	1.675×10^{-27} kg

particles was determined experimentally by Coulomb. Coulomb's law defines the unit of charge in terms of previously defined units of force and distance. When SI units are used Coulomb's law is

$$F = \frac{kQ_1Q_2}{d^2} \tag{1.33}$$

where

F = an attractive or repulsive force in newtons

k = a constant equal to 9×10^9 N·m²/C²

Q_1, Q_2 = charges of the particles in coulombs (C)

d = separation in meters

Elementary Particles Experimental methods have been developed to measure the charge and mass of elementary particles such as the proton and electron. Table 1.6 lists the experimentally determined values for several commonly known elementary particles. (The letter C is the abbreviation for coulomb.)

The electron and the proton have the same charge, although the proton is about 1836 times heavier. The alpha particle has twice the charge of the proton and is about four times heavier. The neutron has no charge; it is an electrically neutral particle. Table 1.6 also shows that all the charged particles listed are very light (almost massless) and carry very little charge. For example, about 6.24×10^{18} electrons (6,240,000,000,000,000,000 electrons) are required to equal one coulomb of electrical charge. The charge on an electron or proton, the smallest charge known, is a very small part of one coulomb of charge.

The use of Equation 1.33 in solving problems that involve electrical forces is illustrated in Examples 1.26 and 1.27.

EXAMPLE 1.26 Attractive forces between unlike charges
· ·

Given: An electron is separated from a proton by a distance of 1 cm.

Find: The force with which the electron pulls on the proton, and the force with which the proton pulls on the electron.

Solution: The electron pulls on the proton with the same force that the proton pulls on the electron. This force is given by Equation 1.33. Thus,

$$F_1 = F_2 = F = \frac{kQ_1Q_2}{d_2}$$

where

$$k = 9 \times 10^9 \text{ N·m}^2/\text{C}^2$$
$$Q_1 = Q_{proton} = 1.602 \times 10^{-19} \text{ C} \quad \text{(See Table 1.6.)}$$
$$Q_2 = Q_{electron} = -1.602 \times 10^{-19} \text{ C} \quad \text{(See Table 1.6.)}$$
$$d = 1 \text{ cm} = 10^{-2} \text{ m}$$

$$F = 9 \times 10^9 \text{ N·m}^2/\text{C}$$

$$\times \frac{(1.602 \times 10^{-19} \text{ C}) \times (-1.602 \times 10^{-19} \text{ C})}{(10^{-2})^2 \text{m}^2}$$

$$F = -2.3 \times 10^{-24} \text{ N} \quad \text{(The negative sign indicates an \textbf{attractive} force.)}$$

Since about 4.5 N are required to lift a 1-lb weight, the electrical force here is very small. Electrical forces between charged particles are often very small.

EXAMPLE 1.27   Repulsive forces between like charges
. . . .

Given: Two electrons in adjoining copper atoms are separated by 10^{-8} cm.

Find: The force with which the electrons repel each other.

Solution: $$F_1 = F_2 = F = \frac{kQ_1Q_2}{d^2}$$

where

$$k = 9 \times 10^9 \text{ N·m}^2/\text{C}^2$$
$$Q_1 = Q_2 = 1.602 \times 10^{-19} \text{ C} \quad \text{(See Table 1.6.)}$$
$$d = 10^{-8} \text{ cm} = 10^{-10} \text{ m}$$

$$F = \frac{(9 \times 10^9 \text{ N·m}^2/\text{C}^2)(1.602 \times 10^{-19})^2 \text{ C}^2}{(10^{-10})^2 \text{m}^2}$$

$$F = 2.3 \times 10^{-8} \text{ N}$$

This force is ten million billion times larger than the force in Example 1.26, but is still small compared to one newton.

Notice that although the forces are often very small, as shown in both Examples 1.26 and 1.27, they often act on very small particles (see masses of electron and proton in Table 1.6, for example) and, therefore, are quite effective in pushing or pulling these particles around.

Definition of Electric Field   An electric charge in space will affect the space around it; the charge sets up an **electric field.** The electric field intensity or strength *E* at a position *P* is defined as the electric force per unit positive charge at position *P,* as shown by

$$\mathbf{E} = \frac{\mathbf{F}}{Q} \qquad (1.34)$$

where

    E = electric field intensity in N/C at position *P*
    F = force in newtons on a charge *Q* at position *P*
    Q = charge at position *P*

The electric field intensity **E** and force **F** are written with arrows to show that each is a vector. Since a vector has both a magnitude and a direction, the electric field intensity **E** will have both a magnitude and a direction, just as does the force **F.** The magnitude of **E** (in newtons/coulomb) will be given always by the ratio *F/Q* in Equation 1.34. Its direction will be given by the direction of the electric force **F** on a small positive test charge *Q* placed at the position in question. The use of Equation 1.34 is illustrated in Example 1.28.

EXAMPLE 1.28   Electric field near a charged sphere
. . . .

Given: A hollow metal sphere is charged to a high voltage, thereby producing an electric field in the space around it. The sphere has a charge of + 10 C and a radius of 1 cm.

Note 1.   Any charged conducting sphere can be replaced by a single-point charge of the same magnitude at the center of the sphere, so long as force and field calculations are made external to the sphere; thus, point P is effectively 10 cm away from a 10-C point charge. Now arbitrarily choose 0.01 C as the "small" test charge at P. The choice of 0.01 C for the test charge is not important, since its value does not affect the final value of E calculated above. (See Note 2.)

Find: The electric field intensity **E** at a position 9 cm due east of the sphere.

Solution: From Equation 1.33

a. $F = \dfrac{k\, Q_{sph} Q_{test}}{d^2}$

where

$$k = 9 \times 10^9 \text{ N·m}^2/\text{C}^2$$
$$Q_{sph} = 10 \text{ C}$$
$$Q_{test} = 0.01 \text{ C}$$
$$d = 1 \text{ cm} + 9 \text{ cm} = 10 \text{ cm} = 0.1 \text{ m}$$
$$d_2 = 0.01 \text{ m}^2$$

$$F = \frac{(9 \times 10^9 \text{ N·m}^2/\text{C}^2)\,(10 \text{ C})\,(0.01 \text{ C})}{0.01 \text{ m}^2} = 9 \times 10^{10} \text{ N}$$

and from Equation 1.34

b. $\mathbf{E} = \dfrac{\mathbf{F}}{Q_{test}} = \dfrac{9 \times 10^{10} \text{ N}}{1/100 \text{ C}} = 9 \times 10^{12} \text{ N/C}$

The complete answer for the required electric field intensity **E** at point P would then be 9×10^{12} N/C in an eastward direction. The direction of the field at a point is always taken as the direction of the force on a small positive test charge placed at that point. Thus

Note 2. Observe that the choice of 0.01 C for the magnitude of the charge does not affect the final answer. The effect of Q_{test} on the answer in part (a) is canceled out in part (b); therefore, a value of $Q_{test} = 1$ C could be taken, making the math simpler and still giving the same answer.

As illustrated in Example 1.28, the direction of the electric field at any point is the same as the direction of the force on a positive charge at that point. The advantage of using the E field (if one knows it) is that one can measure the strength of the electric force without knowing anything about the charges that cause the force. Since forces that arise from many charges can be mathematically quite complicated, working with the electric field is usually simpler than working with the detailed charge configurations and the resulting forces between them.

FIGURE 1.41 Density of lines represents relative strength of electric field.

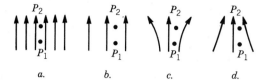

Field Lines In portraying an electric field around a charge distribution, whether the distribution is a single-point charge or an assembly of charges on a conductor, imaginary lines are used to represent the electric field, in both direction and strength. These lines always begin on positive charges and end on negative charges. The direction of a line at any position in space around the charges is the direction of the force on a positive charge at that position. The strength of the field is represented by the density of lines; that is, the closer the lines the stronger is the field, as shown by the four separate drawings in Figure 1.41. Different representations of an electric field near two positions P_1 and P_2 are shown in Figure 1.41.

Figures 1.41a and 1.41b represent uniform or constant-strength fields, since the field lines remain the same distance apart. Figure 1.41a represents a field twice as strong as that in Figure 1.41b since the lines are spaced twice as close together. In each case the field is constant, and the force on a positive charge at points P_1 and P_2 would be identical and directed upward. In Figure 1.41c, the field weakens in going from P_1 to P_2 since the lines spread. In Figure 1.41d, the field intensifies in going from P_1 to P_2 since the lines crowd closer together. In Figure 1.41c, the direction of the field at P_2 is mostly north but somewhat east. In Figure 1.41d, the field is mostly north but slightly west. These principles are used to solve a problem in Example 1.29.

EXAMPLE 1.29 Electric field problem

. . . .

Given: The electric field configuration as shown in the following sketch. A point charge of $+2 \times 10^{-7}$ C at position P_2 experiences a force of 6 N.

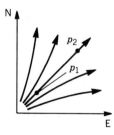

Find:
a. The electric field at P_2.
b. Whether the force on the same charge is larger or smaller at position P_1.

Solution:
a. $\mathbf{E} = \dfrac{\mathbf{F}}{Q} = \dfrac{6 \text{ N}}{2 \times 10^{-7} \text{ C}} = 3 \times 10^7$ N/C, direction northeast.

b. The force on the same charge is greater at P_1, since field E is larger (lines are closer together.)

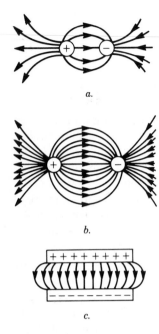

a.

b.

c.

FIGURE 1.42 Electric field lines around different charge configurations.

Electric field lines around various charge configurations are shown in Figure 1.42. Electric field lines at any position always point in the direction of force on a small positive charge placed at that point. Figure 1.42*a* shows the field lines around an equal positive and negative charge pair. In Figure 1.42*b* field lines are shown around another equal positive and negative charge pair as in Figure 1.42*a*, but here the charges are larger, leading to a stronger electric field (thus lines are drawn closer together). Figure 1.42*c* shows the cross section of two oppositely charged parallel plates. There are fringing, or bulging, effects of the field at the plate edges. The field lines are evenly spaced at the center and throughout the interior region between the plates. These features near the center indicate a constant strength field, as in Figures 1.41*a* and 1.41*b*.

The number of lines of force coming from an isolated charge in air can be easily calculated. Consider an isolated positive charge at the center of an imaginary sphere of radius R as shown in Figure 1.43. The magnitude of the electric field (N/C or lines/m²) is determined by Coulomb's law. If a unit positive charge is placed on the sphere, the electric field, force per unit positive charge, becomes

$$E = \frac{kQ \cdot 1}{R^2}$$

At any point on the sphere, the magnitude of E is constant and the direction of E is normal to the surface. If E, the number of lines per square meter, is multiplied by the total area in square meters, then one obtains the total number of lines (N) coming from the positive charge Q. Mathematically this can be expressed as

$$N = EA = \left[\frac{kQ}{R^2}\right](4\pi R^2) = 4\pi kQ \tag{1.35}$$

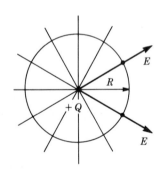

FIGURE 1.43 Field lines around an isolated point charge.

Voltage

Definition of Voltage   The voltage difference between two points separated by a very small distance (Δl) is defined as

$$\Delta V = E (\cos \theta) \Delta l \tag{1.36}$$

where

ΔV = a small voltage difference in volts (V)
E = electric field strength in N/C
θ = angle between the vector **E** and the direction of Δl

Figure 1.44 illustrates these quantities. Equation 1.36 defines the SI unit of voltage. It is called the volt. One volt is equal to one newton-meter per coulomb.

If the electric field is constant in magnitude and direction, and Δl is taken in the direction of the field, $\cos \theta = 1$. Then Equation 1.36 reduces to

$$\Delta V = E\Delta l$$

where Δl may be a large quantity. Notice that this equation is very similar to the case of fluid pressure, Δp, in a constant gravitational field (Equation 1.28):

$$\Delta p = \rho_w \Delta h$$

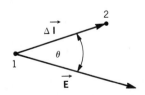

FIGURE 1.44 Parameters that relate electric field and voltage.

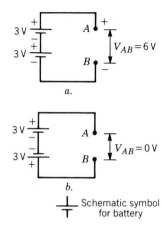

FIGURE 1.45 Series connections for batteries.

The force (weight) per unit volume (ρ_w) and the force per unit charge (E) play similar roles in determining the forcelike quantities Δp and ΔV.

The potential difference (voltage difference) between any two points in an electrical system is measured in volts. If a circuit contains more than one voltage source in series, the total voltage is the sum of the individual sources. In Figure 1.45a, two batteries are connected in series, with a positive terminal of one connected to the negative terminal of the other. In this circuit, both voltage sources will cause charge flow in the same direction if the circuit is completed. The total voltage produced is the sum of the individual voltages. In this case, a net voltage difference acts as a forcelike quantity to cause charge flow.

In Figure 1.45b, the lower battery has been reversed. In this circuit the two voltages oppose each other, and the total voltage is their difference—in this case zero. This zero value means that no charge will flow if points A and B are connected. The absence of a net voltage difference means that no forcelike quantity is present to cause charge flow.

Capacitors

Definition of Capacitance A capacitor is defined as two conducting surfaces separated by a dielectric (insulating) medium. The simplest version of a capacitor, and one that illustrates the important relationships, is the so-called parallel-plate air capacitor. Figure 1.46a shows the charging circuit, and Figure 1.46b shows a three-dimensional view of the capacitor alone.

In Figure 1.46a, a battery with voltage ΔV is used to charge a capacitor to a potential difference of ΔV volts with a positive charge of $+Q$ coulombs on one plate and a negative charge of $-Q$ coulombs on the other plate. The total number of field lines coming from positive charge in the top plate is

$$N = 4\pi kQ \tag{1.37}$$

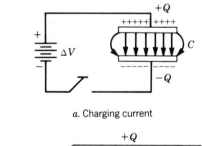

a. Charging current

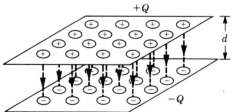

FIGURE 1.46 Charging a parallel-plate capacitor.

b. Parallel-plate capacitor with field lines

If the fringing of field lines at the edge of the plate is neglected, the electric field between the plates becomes

$$E = \frac{N}{A} = \frac{4\pi k Q}{A}$$

where

k = a constant equal to 9×10^9 N·m²/C²
A = area of the plate in m²
Q = charge on either plate

From the definition of potential difference (ΔV), the voltage difference between the plates is

$$\Delta V = E \,(\cos \theta)\, \Delta l = E \, \Delta l \qquad (1.38)$$

or

$$E = \frac{\Delta V}{\Delta l}$$

where

E = electric field magnitude
ΔV = voltage difference
Δl = distance between the plates in m

Equating the two expressions for E yields

$$\frac{\Delta V}{\Delta l} = \left(\frac{4\pi k}{A}\right) Q$$

or

$$\Delta V = \left(\frac{4\pi k \Delta l}{A}\right) Q$$

The quantity $4\pi k \cdot \Delta l / A$ is a constant that depends only on the dimensions of the capacitor and the physical constant $4\pi k$. The reciprocal of this constant is called the capacitance (C) of the capacitor. The unit of capacitance is the farad (F). The relationship between Q, ΔV, and C is

$$C = A/4\pi k \Delta l$$

$$\Delta V = Q/C \qquad (1.39)$$

or

$$Q = (\Delta V)C \qquad (1.40)$$

where

ΔV = voltage difference in volts
Q = charge in coulombs
C = capacitance in farads

Example 1.30 illustrates the application of the equations in a typical capacitor problem.

EXAMPLE 1.30   Capacitor problem
. . . .

Given: A parallel-plate capacitor whose plates are separated by 1 cm is charged to a voltage of 50 V.

Find: **a.** The electric field strength between the plates.

        b. The force F_e exerted on an electron by the electric field if the electron is passing through the air space between the capacitor.

        c. The capacitance if the plates each carry a charge of 50×10^{-9} C.

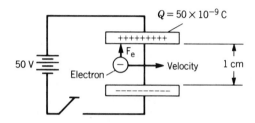

Solution: **a.** From Equation 1.36

$$E = \frac{\Delta V}{\Delta l}$$

$$E = \frac{50 \text{ V}}{10^{-2} \text{ m}} = 5000 \text{ V/m} = 5000 \text{ N/C}$$

        b. From Equation 1.33

$$E = \frac{F_e}{Q}$$

$$Q = 1.6 \times 10^{-19} \text{ C} \qquad \text{(See Table 1.6.)}$$

$$E = 5000 \text{ N/C}$$

$$F_e = EQ = (5000 \text{ N/C})(1.6 \times 10^{-19} \text{ C})$$

$$F_e = 8 \times 10^{-16} \text{ N}, \qquad \text{directed upward toward the } + \text{ plate}$$

        c. From Equation 1.38

$$Q = C \, \Delta V$$

$$C = \frac{Q}{\Delta V} = \frac{50 \times 10^{-9} \text{ C}}{50 \text{ V}} = 10^{-9} \frac{\text{C}}{\text{V}} = 10^{-9} \text{ F}$$

$$C = 1 \text{ nF} \qquad (\text{nF} = \text{nanofarad} = 10^{-9} \text{ farad})$$

FIGURE 1.47 Capacitor with insulating material between parallel plates.

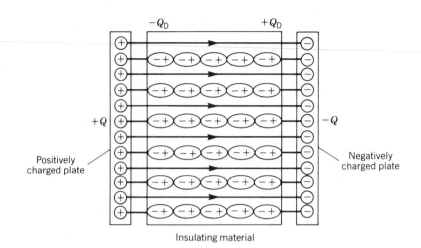

Insulating material

Dielectrics   If an insulator is placed between the plates of a capacitor, the positive and negative charges (nuclei and electrons) in the insulator will be displaced relative to each other because of the coulomb forces. These charges will move only a short distance (order of the molecular diameter). As a result of these movements, there will be a net positive charge on the surface of one side of the insulator and a net negative charge on the opposite side of the insulator. This is shown diagrammatically in Figure 1.47.

As can be seen from the diagram, part of the field lines starting at the positive charges will terminate on the induced negative charges on the surface of the insulator. For this reason, the electric field strength inside the insulator will be less (fewer lines per square meter). The total number of lines in the insulator will be given by Equation 1.37:

$$N = 4\pi k \, (Q - Q_D)$$

and the electric field inside the insulator becomes

$$E = \frac{N}{A} = \frac{4\pi k}{A} \, (Q - Q_D) \tag{1.41}$$

where

   Q_D = charge on the surface of the insulator
   A = area of each surface

Thus, one can see that the effect of the insulator (called a dielectric) is to **reduce** the field strength inside it. If no field exists inside the dielectric, there would be no surface charge. Thus, it is reasonable to assume that the surface charge increases linearly with the electric field strength (E) as well as the area (A) of the surface. Mathematically this can be written as

$$Q_D = k_D \, AE \tag{1.42}$$

where k_D is a constant. Substituting this expression for Q_D in Equation 1.41 yields

$$E = \frac{4\pi k}{A} \, (Q - Q_D) = \frac{4\pi k Q}{A} - 4\pi k \, k_D E$$

Grouping terms in E and factoring gives

$$E (1 + 4\pi k k_{\mathrm{D}}) = \frac{4\pi k Q}{A} \qquad (1.43)$$

The constant quantity

$$k' = (1 + 4\pi k k_{\mathrm{D}})$$

is called the dielectric constant. Thus, E may be written as

$$E = \frac{4\pi k Q}{A \, k'} \qquad (1.44)$$

The voltage difference, $\Delta V = E \Delta l$, becomes

$$\Delta V = \frac{4\pi k Q \Delta l}{k' A}$$

or

$$\Delta V = Q/C$$

where

$$C = \frac{k' A}{4\pi k \Delta l} \qquad (1.45)$$

Thus, the capacity of an air capacitor is increased by the factor k' when a dielectric of constant k' is placed between the plates. Table 1.7 lists the value of the dielectric constant (k') for several common substances. An application of Equation 1.45 is given in Example 1.31.

EXAMPLE 1.31   Parallel-plate capacitor
. . . .

Given: The dimensions of the plates of a parallel-plate capacitor are 0.10 by 0.10 m. The separation between the plates is 0.00020 m.

Find: The capacitance in farads for air and for quartz.

Solution: The capacitance in air is (Equation 1.45)

$$C = \frac{k' A}{4\pi k \Delta l} \qquad (k = 9 \times 10^9 \ \mathrm{N \cdot m^2/C^2} \text{ and } k' = 1, \text{ see Table 1.7.})$$

a. $C = \dfrac{(1)(0.10 \ \mathrm{m})(0.10 \ \mathrm{m})}{4\pi \ (9 \times 10^9 \ \mathrm{N \cdot m^2/C^2})(2.0 \times 10^{-4} \ \mathrm{m})}$

  $C = 4.42 \times 10^{-10} \ \mathrm{F}$ (air)

  $C = 0.442 \ \mathrm{nF}$ (air)

b. $C_{\mathrm{quartz}} = (c_{\mathrm{air}})(k'_{\mathrm{quartz}})$

  $C = (4.42 \times 10^{-10})(4)$ (See Table 1.7 for k'_{quartz}.)

  $C = 1.77 \times 10^{-9} \ \mathrm{F}$ (quartz)

  $C = 1.77 \ \mathrm{nF}$

TABLE 1.7   Common dielectric constants

Substance	Dielectric Constant
Air	1
Glass	5–10
Mica	3–6
Porcelain	7
Quartz	4
Sulfur	3.6
Vacuum	1
Water	81

Since the capacitance is usually a very small number, it is expressed in micro-farads (10^{-6} farad), nanofarads (10^{-9} farad), or picofarads (10^{-12} farad).

Applications

Example 1.31 indicates that a charged particle, such as an electron, experiences a force because of the existence of an electric field in the region between the parallel plates of the capacitor. Such a force can be used to deflect moving electrons and thereby change their paths. This basic idea is used to control the path of electrons that move from the electron gun to the screen in a television tube. Figure 1.48 illustrates deflection of an electron beam with two sets of charged capacitor plates in a cathode-ray tube.

The electron beam leaves the accelerating electrodes along the central axis of the tube. If the vertical deflecting plates are charged so that the top plate is positive and the bottom plate is negative, at the instant the electron passes through, it will be deflected upward. Then, if the horizontal deflecting plates are charged with front plate positive and rear plate negative, as the electron passes between these plates it will be deflected toward the front positive plate. The combined deflections will result in the electron beam striking the screen in the upper left quadrant, as shown. By changing voltage on the horizontal and vertical deflecting plates in a prearranged manner, the electron beam can be directed to various positions on the screen. In each instance, as in Example 1.30, the exis-

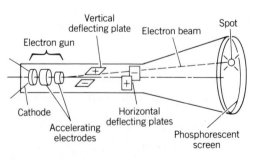

FIGURE 1.48 Forces on moving electrons in a cathode-ray tube.

tence of an electric field between the capacitor plates gives rise to an electric force on the charged particles (electron beam). This force, in turn, deflects the motion of the particles.

Magnetic Fields

Magnetomotive Force   The forcelike quantity in a magnetic circuit is the magnetomotive force (mmf). The magnetomotive force is the physical quantity that causes a magnetic field to be established. In this respect, it is similar to the electromotive force in an electric circuit. The magnetomotive force is established when an electric current flows. This usually occurs in the form of current loops such as a current flowing in a coil of wire. The magnetomotive force is defined as

$$mmf = NI \tag{1.46}$$

where

    mmf = magnetomotive force (a forcelike quantity)
      N = number of turns in coil
      I = current in coil

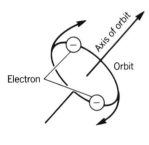

a.

b.

FIGURE 1.49 Whirling electrons create magnetic fields.

Permanent Magnetism   Our first experience with magnetism and magnetic forces usually comes from contact with bar magnets and compasses. Bar magnets and compasses have north and south poles that strongly attract one another. Two north poles or two south poles strongly repel one another. Also, some magnets, such as lodestone, are permanent in behavior, that is, they exhibit magnetic properties seemingly forever. On the other hand, some magnets are temporary in behavior. They exhibit magnetic properties only while a current is flowing near a magnetic substance such as soft iron. Electromagnets—wires wrapped around soft iron cores—are good examples of temporary magnets.

The origin of natural, permanent magnetism, as found in the substance called lodestone, or of temporary magnets, as produced in electromagnets, can be traced in either case to the existence of microscopic atomic currents deep within matter itself. A tiny current, such as that formed by an electron whirling around inside an atom, gives rise to a magnetic force field in a region surrounding the electron's orbit.

In Figure 1.49*a*, an electron is whirling around in an orbit with axis tilted as shown. In Figure 1.49*b*, the magnetic field created by the movement of the electron in its orbit is shown as if it were produced by a tiny magnet aligned along the same axis with appropriate N and S poles.

In effect, as illustrated in Figure 1.49, the tiny electron current can be thought of as a tiny bar magnet with appropriate N and S poles giving rise to a magnetic field around the bar magnet. In natural or permanent magnets such as lodestone, therefore, these tiny magnets must be permanently aligned in such a manner that they add up to one large, effective magnet. For permanent magnets, then, the key idea is in the natural alignment of clusters of atoms—called magnetic domains—which in turn assures that the overall addition of billions and billions of the tiny equivalent magnets will result in one large, effective magnet. (See Figure 1.50*a*.) On the other hand, initially unmagnetized materials somehow must lack the

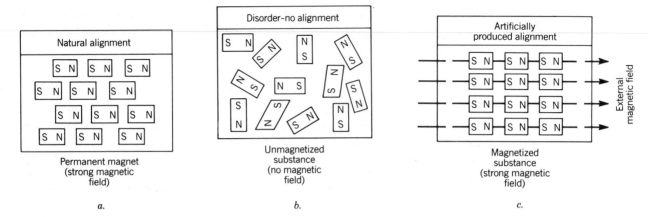

FIGURE 1.50   Permanent and artificial magnets.

precise alignment present in the magnetic domains of permanent magnets; that is, the alignment must be random in nature. (See Figure 1.50*b*.) The addition of billions and billions of randomly ordered, unaligned, tiny equivalent magnets results in an overall effect of no magnetism at all, as with ordinary soft iron. But in unmagnetized substances—such as soft iron—which can be magnetized, these billions and billions of initially disordered tiny magnets can be rearranged and aligned (as in permanent magnets) by an external magnetic force threading through the substance. In this manner, an unmagnetized substance often can be magnetized as shown in Figure 1.50*c*.

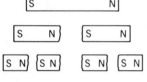

FIGURE 1.51   Subdividing a magnet.

As indicated in the introductory discussion of the origin of magnetism, all magnetic substances can be represented in terms of two poles, a north (N) and south (S) pole. Any attempt to subdivide a bar magnet—as shown in Figure 1.51— with the intention of separating and isolating one pole from the other is doomed to failure. Subdivision of a magnet simply results in a smaller magnet of less strength but always with two poles, N and S, just as in the original magnet. Although scientists have been trying for years, no one has yet succeeded in isolating a free magnetic pole, either a south or north pole. Thus, one is forced to work with both N and S poles when studying magnetic forces. When electric forces were studied, no trouble was encountered in separating positive and negative charges.

If two simple bar magnets are arranged with their north poles (or south poles) next to one another, the magnets strongly repel each other. This observation is summarized by the statement "Like magnetic poles repel." On the other hand, if the same two bar magnets are arranged with their N and S poles next to each other, they will attract each other strongly. This is summarized by the statement "Unlike magnetic poles attract."

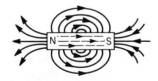

FIGURE 1.52   Magnetic field lines around a magnet.

Field Lines   Figure 1.52 shows a single bar magnet with clearly labeled N and S poles. Drawn in the space surrounding the magnet is a series of lines to represent the direction of force on a tiny N pole of a very long magnet. (The force on the distant south pole is unimportant here.) The series of lines, just as in the discussion of electric charges and electric force, constitutes a field, in this case a

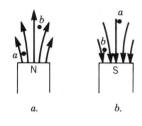

a.   *b.*

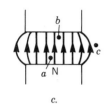

c.

FIGURE 1.53 Spacing of magnetic field lines indicates relative field strength.

magnetic field. Figure 1.52 shows the magnetic field lines that surround a simple bar magnet. Solid lines represent the field outside the magnet and dashed lines represent the field inside the magnet. The lines always point in the direction of the force on a small north pole located at the given point in space. At the left end (N) of the bar magnet in Figure 1.52, field lines are directed away from the N pole (since two N poles repel). At the right end (S), the field lines are directed toward the S pole (since N and S poles attract). The strength of the magnetic field is indicated by the density of lines drawn. Closely spaced lines indicate a strong field. Widely spaced lines indicate a weak field. Uniformly spaced lines indicate a magnetic field of constant strength. Diverging or converging field lines indicate a decreasing or increasing magnetic field strength, respectively.

In Figure 1.53*a*, the magnetic field weakens in going from point *a* to *b* since the field lines spread at *b*. In Figure 1.53*b*, the field intensifies in going from point *a* to point *b* since the lines are more densely packed at *b*. In Figure 1.53*c*, the field remains constant in going from *a* to *b*, since the lines there remain equally spaced. In going from *a* to *c* in Figure 1.53*c*, the field would, of course, weaken.

Patterns of magnetic field lines for various geometries are shown in Figure 1.54. Figures 1.54*a* and 1.54*b* illustrate the magnetic field lines near adjoining N–S and N–N poles. Figure 1.54*c* maps the field lines near a horseshoe magnet; that is, a simple bar magnet bent in a U-shape. Figure 1.54*d* shows the field lines along a long, straight wire carrying electron current. Figures 1.54*e* and 1.54*f* show the field lines around an air-core and iron-core solenoid (coil).

The patterns of the field lines in Figures 1.54*a* and 1.54*b* are vastly different, whereas the patterns in Figures 1.54*a* and 1.54*c* are quite similar. Notice that the magnetic field created by a long wire carrying a current, *I*, is a series of concentric rings encircling the wire, as shown in Figure 1.54*d*. The direction of the magnetic field lines can be determined easily as follows: If the thumb of the **left hand** is placed along the wire in the direction of **electron** current flow, the fingers of the left hand wrap themselves around the wire in the direction of the circular magnetic field lines. In Figure 1.54*d*, therefore, the thumb would point downward in the direction of electron current flow, and the fingers would encircle the wire in a counterclockwise direction (looking down from the top), as shown.

For the air-core and iron-core solenoids (Figures 1.54*e* and 1.54*f*), the pattern of magnetic field lines is similar to that shown in Figure 1.52 for a simple bar magnet. For electron current flowing around the core of the solenoid, as shown, each solenoid behaves like a simple bar magnet, with an N pole at the left end and an S pole at the right end. Which end of the solenoid behaves like an N pole can be determined from the following simple rule. If the fingers of the **left hand** are wrapped around the solenoid in the direction of electron current flow, the extended thumb will point to the end of the solenoid that behaves like a N pole. If this rule is applied to Figures 1.54*e* and 1.54*f*, the left ends of the solenoids are the indicated N poles.

A comparison of the magnetic field sketches for the air-core and iron-core solenoids shows that the field is intensified in the center of the iron-core solenoid by the addition of the iron core. A stronger magnetic field can be obtained with a solenoid of given current and windings simply by inserting an iron core through the solenoid.

FIGURE 1.54 Mappings of magnetic field patterns.

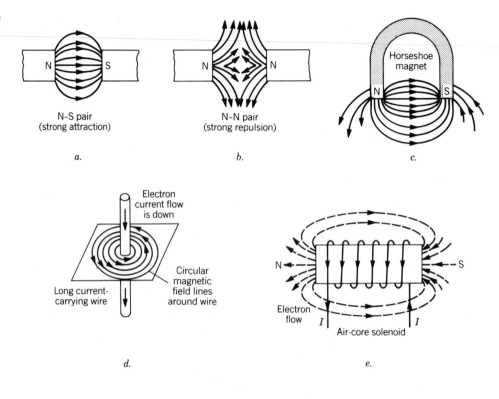

N-S pair
(strong attraction)

a.

N–N pair
(strong repulsion)

b.

Horseshoe magnet

c.

Electron current flow is down

Long current-carrying wire

Circular magnetic field lines around wire

d.

N — S

Electron flow I I

Air-core solenoid

e.

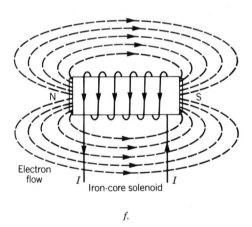

N — S

Electron flow I I

Iron-core solenoid

f.

Magnetic Induction If a wire carrying an electric current is placed in a magnetic field, one observes a force acting on the wire that changes as the orientation of the wire is changed. In one direction, no force is observed. This unique direction is the direction of the magnetic field (B). (See Figure 1.55.) In other orientations with respect to the direction of the field, the observed force is normal to the plane formed by the wire and the field direction as shown in Figure 1.55.

The direction of the force can be deduced by the so-called right-hand or left-

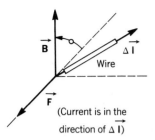

FIGURE 1.55 Force on a wire carrying a current.

hand rule. The right-hand rule is used when conventional current (positive current) is considered. The left-hand rule is used when electron current (negative current) is considered. In Figure 1.55, assume that a conventional current, I, flows in the direction of the wire $\Delta \mathbf{l}$.

The direction of the force can be obtained by curving the fingers of the **right hand** from the wire direction, $\Delta \mathbf{l}$, to the field direction with the thumb extended. The thumb then points in the direction of the observed force as shown. If that same direction ($\Delta \mathbf{l}$) were the direction of **electron flow,** then the fingers of the **left hand,** curving from the wire direction $\Delta \mathbf{l}$ to the field direction **B,** with thumb extended, would indicate the observed force to be in a direction **opposite** to that in Figure 1.55. The magnitude of the measured force is maximum where **B** and $\Delta \mathbf{l}$ form a right angle. The force also increases linearly with the current. From these measurements, the magnitude of the magnetic induction or magnetic flux density (**B**) is defined as

$$B = \frac{F}{I\Delta l} \tag{1.47}$$

where

B = magnetic flux density in Webers (Wb)/m² (1 Wb = 1 N·m/A)
I = current through the wire in amperes (A)
F = force in N
Δl = length of the wire in m

For other angles, the magnitude of the force is found to be

$$F = IB\Delta l \sin \theta \tag{1.48}$$

where

θ = angle between **B** and $\Delta \mathbf{l}$

Magnetic Force The current in a wire is equal to the total charge passing through a cross section of the wire per second. If there are n charges in the wire per unit length having a drift velocity v and charge q, then the current is

$$I = qnv \tag{1.49}$$

This can be seen from Figure 1.56. All the charges in the volume between cross sections A and B will pass through area B in one second. This total charge is qnv.

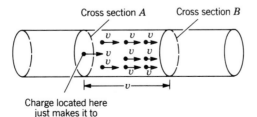

FIGURE 1.56 Charge flow and current in a wire.

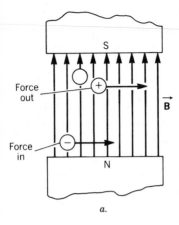

Force out

Force in

a.

b.

c.

FIGURE 1.57   Direction of force on a moving charge depends on sign of charge.

The force then becomes (substituting for I in Equation 1.48)

$$F = qnvB(\Delta l) \sin \theta$$

Since the total number of charges is $n\Delta l$, the force on each charge becomes

$$F_q = \frac{F}{n\Delta l} = qvB \sin \theta \tag{1.50}$$

The direction of the force depends on the sign of q as illustrated in Figure 1.57. In the upper half of the figure, a **positively charged** particle of magnitude $+Q$ is shown moving through a uniform magnetic field of strength B with a velocity v. The magnetic field is directed upward. In this configuration, the force, F, exerted on the positive charge is directed out of the page toward the reader (Figure 1.57b). If the particle is negatively charged, everything else being the same, the force on the **negative charge** would be directed into the page, away from the reader (Figure 1.57c). The magnetic force, F, for a charged particle, Q, moving with velocity, v, in a magnetic field, B, is always directed at right angles to both the direction of the field, B, and the velocity, v.

Magnetic Field Strength   In a region in which there is no current flowing, one can define a magnetic scalar potential difference between two points. The definition, which is similar to the definition of potential difference, is defined as

$$\Delta\Omega = \frac{B}{\mu} \Delta l \cos \theta \tag{1.51}$$

where

$\Delta\Omega$ = magnetic potential difference in A·turn
B = magnetic induction in Wb/m^2
Δl = a small displacement in m
μ = a constant that depends on the material. It is called the permeability and is measured in Wb/m·A·turn.
($\mu = 1.257 \times 10^{-6}$ Wb/m·A·turn for air)

Note: Wb is the symbol for weber and is defined as

$$\text{Wb} = \frac{\text{J}}{\text{A·turn}} = \frac{\text{N·m}}{\text{A·turn}} = \frac{\text{N·m·sec}}{\text{C·turn}} = \frac{\text{kg·m}^2}{\text{C·turn·sec}}$$

The quantity B/μ is called the magnetic field strength. This may be expressed as

$$H = \frac{B}{\mu} \tag{1.52}$$

where

H = magnetic field strength in A·turn/m

Equation 1.51 may now be written as

$$\Delta\Omega = H\Delta l \cos \theta \tag{1.53}$$

This equation has the same form as the expression for the voltage difference

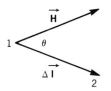

FIGURE 1.58 Parameters that relate magnetic field and magnetic potential.

between two neighboring points given earlier by Equation 1.36. The corresponding relationship between **H** and $\Delta\mathbf{l}$ is shown in Figure 1.58.

It is found experimentally that if all the $\Delta\Omega$'s are summed around a closed path, the sum's value is equal to the magnetomotive force (mmf). This can be expressed as

$$\Sigma(\Delta\Omega) = H\Sigma(\Delta l)\cos\theta = NI \tag{1.54}$$

since NI is the mmf. The simplest case occurs when H is constant in magnitude and Δl is taken in the same direction as H so that $\cos\theta = 1$. With these conditions, Equation 1.54 may be written as

$$\Sigma(\Delta\Omega) = H\Sigma(\Delta l) = NI \tag{1.55}$$

Since $\Sigma(\Delta l)$ is the total path length, Equation 1.55 reduces to

$$Hl = NI$$

or

$$H = \frac{NI}{l} \tag{1.56}$$

Example 1.32 shows the application of Equation 1.56 to a long, straight wire carrying a constant current.

EXAMPLE 1.32   Magnetic induction for a long wire
. . . .

Given: A long, straight wire carrying a current of 10 A in air.

Find: The magnitude of the magnetic field strength (H) and the magnetic flux density (B) at a distance of 1 m from the wire.

Solution: The magnetic field lines form concentric circles around the wire (see sketch). For a given radius the magnitudes of H and B are constant. For this case

$$H = \frac{NI}{l} \quad \text{(Equation 1.56)}$$

$$l = 2\pi R = 2\pi(1) = 2\pi \text{ m}$$

$$N = 1$$

$$H = \frac{(1)(10 \text{ A})}{(2\pi \text{ m})}$$

$$H = 1.59 \text{ A·turn/m}$$

$$B = \mu H \quad \text{(Equation 1.52)}$$

$$\mu = 1.257 \times 10^{-6} \text{ Wb/m·A·turn}$$

$$B = (1.257 \times 10^{-6} \text{ Wb/m·A·turn})(1.59 \text{ A·turn/m})$$

$$B = 2 \times 10^{-6} \text{ Wb/m}^2$$

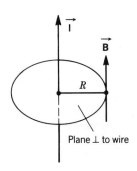

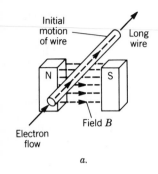

Initial
motion
of wire

Long
wire

N   S

Field *B*

Electron
flow

a.

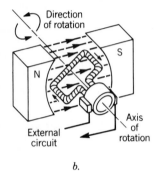

Direction
of rotation

S

N

External
circuit

Axis
of
rotation

b.

FIGURE 1.59   Elements of
an electric generator.

Applications

Generators   The force on a moving charge in a magnetic field lies at the heart of the principle of operation of the electric generator and electric motor. In the generator, external mechanical forces are used to push a conducting wire through a strong magnetic field, thereby inducing magnetic forces on the charge parallel to the wire and generating a current in the wire. In the motor, electric current (moving charge) is forced through a conducting wire located in a strong magnetic field. The magnetic forces in this case act on the charges in the wire, but in a direction perpendicular to the wire, thereby pushing the wire. In generators, external mechanical power is converted into electric power. In motors, external electric power is converted to mechanical power.

The basic idea behind the generator is illustrated in Figure 1.59. In Figure 1.59*a*, a long, straight conducting wire filled with positively charged nuclei and negatively charged electrons, initially carrying no current, is pushed downward into a magnetic field *B* directed from left to right. When the wire is forced downward, each charge in the wire also moves downward with the wire. Thus, each charge will experience a force, given by Equation 1.50, that is equal to

$$F = qvB$$

Notice that the angle between the field and the velocity of the charge is 90°, giving sin 90° = 1. In accordance with Figure 1.57, this force will be in the direction of the wire out of the paper for positive charges and into the paper for negative charges. The positive charges (the nuclei) are massive compared to the electrons, besides they are locked in place and can't move. The electrons are much lighter. Some electrons (valence electrons) are free to move. Thus, a current of electrons will flow in a direction along the wire into the paper. Motion of the wire in the magnetic field, therefore, generates an electron current in the wire.

If the wire shown in Figure 1.59*a* is part of a loop shown in Figure 1.59*b* and if the loop rotates as shown, electron current can be generated in the left and right portions of the loop. Current is generated in opposite directions, as shown, to provide continuous current flow to an external circuit. This design, then, is the simplest example of a working generator.

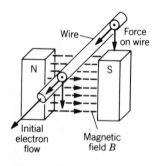

Wire

Force
on wire

N   S

Initial
electron
flow

Magnetic
field *B*

FIGURE 1.60   Elements of
an electric motor.

Motors   The direct-current electric motor can be thought of as a generator working in reverse. In Figure 1.60, the wire is situated in the magnetic field as shown. An electron is forced through the wire out of the page. This process again constitutes an electric charge moving in a magnetic field; in accordance with Figure 1.57, a magnetic force acts on the moving charge in a downward direction. Since this moving charge is perpendicular to the wire, the wire experiences a force and moves downward. Electrical power used to force current through the conductor is transformed into the mechanical power in the moving wire. If the wire in Figure 1.60 is made part of a loop and fastened to a rotating shaft, somewhat as shown in Figure 1.59*b*, the mechanical power generated in the rotating loop can be used to do useful external work.

SECTION 1.3 EXERCISES

1. Define the forcelike quantities voltage and magnetomotive force. State appropriate units for each.

2. What is the electric field intensity due to a charged sphere at a point 30 cm away from the center of the sphere if the charge on the sphere is 2.5×10^{-4} C?

3. How far apart are the plates of a parallel-plate capacitor if the charge on the plates is plus and minus 1.5×10^{-6} C, the capacitor is valued at 2 μF, and the electric field $E = 2500$ N/C?

4. A test charge is located 40 cm away from a sphere that has a net charge of $(+)5 \times 10^{-6}$ C. If the test charge experiences a force of 10 N radially away from the center of the sphere, what is the value of the charge on the test charge?

5. Determine the voltage V_{AB} in each of the following circuits.

        a.   *b.*

6. State the rules of attraction and repulsion between north and south magnetic poles.

7. Sketch the magnetic field lines in each of the following sketches to the left.

8. Indicate the locations of north and south poles for the solenoid in the figure to the left.

9. An electron moves with constant speed from point A toward point C in the two following magnetic fields. In each case, does the magnetic force on the electron increase or decrease along its path? In which direction is the electron deflected?

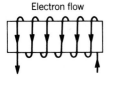

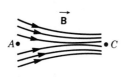 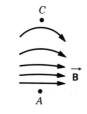

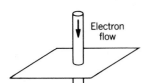

10. A long wire carrying an **electron** current, as shown in the sketch, is located between an N–S pole pair. What is the direction of the magnetic force on the wire?

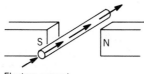

Electron current

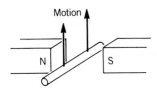

11. A long, straight conducting wire is forced upward in the magnetic field between the N–S pair shown in the sketch to the left. What is the direction of **electron** current flow in the conductor caused by the magnetic force?

12. *A capacitor consists of two parallel plates. Each plate has an area of 0.01 m². If the separation between the plates is 0.001 m and the dielectric constant is 5, what is the capacitance of the capacitor? If the capacitor is connected to 100 V, what is the charge on either plate?

13. *The current in a long, straight wire is 100 A. What is the magnetic induction, B, at a point 0.50 m from the wire? What is the magnetic field strength, H?

14. *A parallel-plate capacitor of 10-μF capacitance is charged to a voltage of 100 V. The plates are separated by one 1 mm of air. What is the electric field in the region between the plates in volts per meter? In newtons per coulomb? How much charge in coulombs was deposited on each plate in the charging process? If an alpha particle (see Table 1.6) is moving along a midplane between the capacitor plates, what is the force on the alpha particle? Is the force on the alpha particle the same at the ends of the capacitor as it is near the center?

---

* Challenge question.

1.4 .... OBJECTIVES..

TEMPERATURE DIFFERENCE— FORCELIKE QUANTITY IN THERMAL SYSTEMS

Upon completion of this section, the student should be able to

- Define the forcelike quantity **temperature difference** and give appropriate units in the SI and English systems of units.

- Distinguish between temperature and heat energy.

- Describe how a liquid-in-glass thermometer works.

- Given a temperature in either degrees Celsius or degrees Fahrenheit, determine the equivalent temperature on the other scale.

- Distinguish between the meaning of 1 C° and 1°C.

- Describe a thermocouple and explain how it measures temperature.

- Describe a pyrometer and explain how it measures temperature.

- Describe how pressure in fluid systems, voltage in electrical systems, magnetomotive force in magnetic systems, and temperature in thermal systems are similar to force and torque in mechanical systems.

DISCUSSION

So far we have studied forcelike quantities in mechanical systems (force and torque), in fluid systems (pressure), and in electrical systems (voltage and magnetomotive force). Now it is time to study thermal systems and the forcelike quantity temperature difference.

Temperature Difference

The prime mover, or forcelike quantity, in thermal systems is "temperature difference," often written as ΔT. Heat energy always flows naturally from warmer regions to colder regions. The rate of heat flow in any system is dependent on

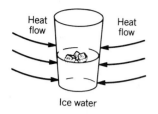

FIGURE 1.61 Heat flow.

temperature difference just as fluid flow is dependent on pressure difference and electric current is dependent on voltage difference.

A glass of water sitting in a warm room (Figure 1.61) will gain heat energy from its surroundings until all the ice melts and the water reaches the same temperature as the room. No further heat energy will flow in the final equilibrium condition because no temperature difference exists.

Molecular Motion All matter is composed of particles (molecules, atoms, ions, and electrons) that are in constant random motion. The energy associated with the motion is called kinetic energy. In addition, there is potential energy due to the relative positions of the particles. The total energy (sum of kinetic and potential energy) is called the internal energy of the substance. Heat energy is the energy flowing into, through, or out of the body due to a temperature difference. Thus, heat energy may increase or decrease the internal energy of the body.

Temperature is a measure of the average random energy of a molecule. If the temperature of a substance increases, the random energy of the molecules increases and the corresponding internal energy is greater. Thus, energy flows from a region where the average random molecular energy is greater to a region where the average random molecular energy is less. The energy flow (heat) will continue until the average random molecular energy is the same in both regions. The bodies are then said to be in thermal equilibrium.

Thermometers A thermometer is an instrument to measure the temperature of a body. Usually a portion of the thermometer is brought into thermal equilibrium with the body whose temperature is being measured. The indicated temperature is the equilibrium temperature of the sample and the thermometer. The energy flow to the thermometer should be small so that the temperature of the body is not significantly changed by the presence of the thermometer. The operation of all thermometers is based on a physical property that changes with temperature.

Volume is one physical property of a liquid that changes with temperature. The volume of mercury, for example, increases if the temperature is increased and decreases if the temperature is decreased. The increased volume of warmed mercury within the confined glass tube of the thermometer causes the level to rise. By selecting certain reference points (such as the freezing and boiling points of water) and uniformly subdividing the distance between them, a temperature scale is established (Figure 1.62).

Temperature Scales

Reference Points Until recently the Fahrenheit temperature scale was used almost exclusively in the United States. There is now an intensive effort to adopt the centigrade, or Celsius, scale used throughout the rest of the world. In defining either scale, it is necessary to determine two reference-point temperatures. Once these reference points are determined, the position of the zero point and the size of the temperature unit are obtained. One of these reference points is called the freezing point. This is the temperature at which water changes to ice. The other point is called the boiling point at a standard pressure of one atmosphere. This is

FIGURE 1.62 Liquid-in-glass thermometer.

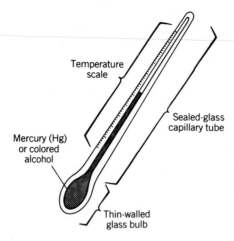

the temperature at which water boils and changes to steam. On the Celsius scale, the freezing point is given a value of 0° and the boiling point 100°. On the Fahrenheit scale these points are assigned the values 32° and 212°, respectively.

Temperature Scales   When constructing a liquid-in-glass thermometer with a Celsius scale, first insert the thermometer in an ice-water mixture and carefully mark the tube at the height of the liquid column. Then immerse the thermometer in boiling water at standard atmospheric pressure and again mark the height of the liquid column on the tube. The distance between these marks is then divided into 100 equal parts and numbered from 0° to 100°, as shown in Figure 1.63. If required, the scale can be extended above 100° and below 0°. The Fahrenheit scale, also illustrated in Figure 1.63, is determined in a similar way by dividing the length of the column between the freezing and boiling points into 180 equal divisions and extending the scale in either direction. The temperature 0°C is equivalent to 32°F and the temperature 100°C is equivalent to 212°F.

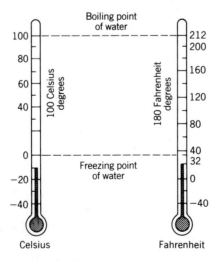

FIGURE 1.63 Comparison of temperature scales.

There is a simple method of recalling the relationship between Fahrenheit and Celsius temperature scales. If the temperature in Fahrenheit is known and is to be converted to Celsius, simply draw two simple thermometers and establish ratios as shown in Example 1.33.

EXAMPLE 1.33   Scale ratio between celsius and fahrenheit
. . . .

Given: T_C to T_F.

Find: Their equivalent in the other system.

Solution: Equate the ratios from both sides:

$$\frac{T_C - 0}{100 - 0} = \frac{T_F - 32}{212 - 32}$$

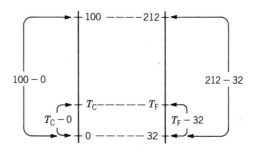

Solving this ratio of T_C or T_F yields

$$T_C = \tfrac{5}{9}(T_F - 32)$$

$$T_F = \tfrac{9}{5}T_C + 32$$

When only the temperature difference (ΔT) is important and not the actual temperature reading, conversion is made using the relationship 9 F° = 5 C°, or 1.8 F° = 1 C°. Temperatures are specified in degrees Celsius (°C) or degrees Fahrenheit (°F). Temperature differences are expressed in Celsius degrees (C°) or Fahrenheit degrees (F°), as shown in Example 1.34.

EXAMPLE 1.34   Temperature conversion
. . . .

Given: A block of ice at 32°F is left in a room at 78°F.

Find: a. Temperature difference in Fahrenheit degrees.
                   b. Temperature difference in Celsius degrees.

c.   Room temperature in degrees Celsius.

Solution: **a.** $\Delta T = T_2 - T_1$

$\qquad = 78°\text{F} - 32°\text{F}$

$\Delta T = 46\ \text{F}°$

The temperature difference is 46 Fahrenheit degrees, written as 46 F°, *not* 46°F.

b. $\Delta T\ (\text{C}°) = \tfrac{5}{9}\ \Delta T\ \text{F}°$

$\qquad = \tfrac{5}{9}\ (46\ \text{F}°)$

$\Delta T = 25.6\ \text{C}°$

The temperature difference is 25.6 Celsius degrees, written as 25.6 C°, *not* 25.6°C.

c. $T\ (°\text{C}) = \tfrac{5}{9}\ (78°\text{F} - 32°\text{F})$

$\qquad = \tfrac{5}{9}\ (46°\text{F})$

$T = 25.6°\text{C}$

The temperature of the room is 25.6 degrees Celsius.

There is an experimental and theoretical lowest temperature known as absolute zero corresponding to the lowest possible internal energy of a body. Absolute zero is $-273.2°\text{C}$ or, rounded off, $-273°\text{C}$. Since it is sometimes inconvenient to work with negative temperatures, temperature scales have been defined whose zero point is absolute zero. These scales are called absolute temperature scales. Scales using the Fahrenheit degree and the Celsius degree as their smallest division are both in common use and are illustrated in Figure 1.64. The scale using the Celsius degree is called the Kelvin scale, and the scale using the Fahrenheit degree is called the Rankine scale.

Absolute temperatures can be obtained with the aid of

$$T\ (°\text{K}) = T\ (°\text{C}) + 273 \qquad\qquad (1.57)$$

$$T\ (°\text{R}) = T\ (°\text{F}) + 460 \qquad\qquad (1.58)$$

FIGURE 1.64 Kelvin and Rankine temperature scales.

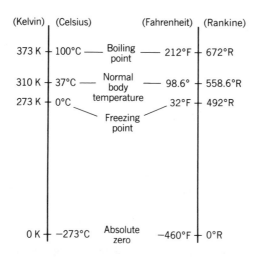

FIGURE 1.65 Thermocouples measure temperature.

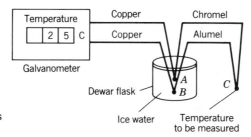

Other Thermometers

Thermocouple The thermocouple shown in Figure 1.65 works on the principle that when wires of any two unlike metals are formed into a complete circuit (as shown in Figure 1.65), an emf or voltage difference is generated whenever the junctions *A* and *B* and the junction *C* are at different temperatures.

In the circuit shown, the junctions *A* and *B* are kept at 0°C by ice and water in a Dewar flask (essentially a thermos bottle). The junction *C* is placed in contact with the body whose temperature is to be measured. Since these junctions will be at different temperatures, a voltage is generated, read by the galvanometer, and converted into a temperature readout.

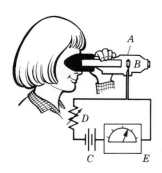

FIGURE 1.66 Using an optical pyrometer to measure furnace temperature.

Pyrometer The optical pyrometer shown in Figure 1.66 is used primarily to measure hot temperatures from a distance, for example, by looking at a hot furnace through a special telescope containing a lamp filament and a red filter *A*. By adjusting the rheostat *D*, the current flow through the lamp filament *B* is increased and the filament starts to glow. The current through the filament is increased until the intensity of the filament just matches the background intensity of the oven. The brightness temperature of the oven can then be read on meter *E*. The brightness temperature may be different from the temperature of the body. This error will be discussed in Chapter 13.

SECTION   *1.4   EXERCISES*

1. Convert the following temperatures in °F to temperatures in °C.
    a. 413°F
    b. 26°F
    c. −13°F

2. Convert the following temperatures in °C to temperatures in °F.
    a. 25.5°C
    b. 5°C
    c. −40°C

3. Explain why a temperature **difference** of 25 F° is not the same as a temperature reading of 25°F.

4. At what value on the Fahrenheit temperature scale and the Celsius temperature scale are the temperatures the same?

5. Describe in a short paragraph the similarities of the forcelike quantities force, torque, pressure, voltage, magnetomotive force, and temperature difference. Which one(s) is a true force? What one(s) behaves like a force and is, therefore, "forcelike"?

S   U   M   M   A   R   Y

Forcelike quantities are the prime movers in all physical systems. Although one would be incorrect to say that net force, net torque, pressure differential, voltage, magnetomotive force, and temperature difference affect their respective energy systems in exactly the same way, they play strikingly analogous (similar) roles. The basic and very important similarities of these physical quantities can be of value in applying knowledge gained in one energy system to the solution of analogous problems in other energy systems. Table 1.8 summarizes the forcelike quantities in the four energy systems.

Table 1.8 Forcelike quantities

| Energy System | Forcelike Quantity | Units | | Quantity Affected |
		English	SI	
Mechanical				
Translational	Force (F)	lb	N	Mass moved through a distance
Rotational	Torque (τ)	ft·lb	N·m	Mass rotated through an angle
Fluid	Pressure (p)	psi	N/m^2	Fluid volume Fluid mass
Electrical	Potential difference (ΔV)	.V	V	Charge
Magnetic	Magnetomotive force (mmf)	A·turn	A·turn	Magnetic field established
Thermal	Temperature difference (ΔT)	F°	C°	Heat energy

Work

INTRODUCTION

*Years ago, work was accomplished mainly through the physical efforts of men and animals. Today, less than three percent of the energy required by industry is provided by muscular exertion. Modern industries are highly mechanized. Many machines are controlled by mechanical devices rather than by men. **Work** has become a technical term, and scientists and technicians must define it in a technical way.*

*In Chapter 1, forcelike quantities were defined as those physical quantities that produce **changes** in mechanical, fluid, electromagnetic, and thermal systems. If no net force, net torque, pressure difference, voltage difference, change in magnetomotive force, or temperature difference is present, the system is in **equilibrium** and there will be **no change** in the system.*

In equilibrium, the forcelike quantities in a system are balanced in such a way that they produce no change. In mechanical systems, equilibrium occurs when the sum of all forces and the sum of all torques are zero. This means that objects at rest remain at rest and objects in motion remain in motion with no change. In fluid systems, equilibrium occurs when the pressure difference in the fluid is zero or when pressure differences balance out the body forces,

indicating that there is no fluid flow. In electrical systems, equilibrium means that no charge moves between two points if the sum of the voltages between them is zero. Similarly, in magnetic systems, the magnetic field is constant unless there is a change in the magnetomotive force. Thermal equilibrium exists in a closed system only when there is no temperature difference in that system. This condition means that heat energy is not moving in the system. A system in equilibrium experiences no change, although separate forcelike quantities (which add to zero collectively) are present in the system.

If equilibrium does not exist in a system, that system changes in response to the unbalanced forcelike quantities present. **Work** is done any time that an unbalanced **forcelike quantity** produces **change** in a system. The amount of work done depends on both the magnitude of the forcelike quantity that produces the change and the magnitude of the change produced.

To describe the change in a system due to the work done on or by the system, the term **energy** is used. The change in energy is set equal to the work done on or by the system. Thus, the units of energy are the same as the units of work. A system gains energy when work is done on it, and it loses energy when it does work on another system. If a body or object possesses energy, it can do work on a second body or object under the proper circumstances. Thus, energy is often defined as the capacity to do work.

The general formula for determining the amount of work done in any system is given by

Work = Forcelike Quantity × Displacementlike Quantity   (2.1)

In this chapter, the general formula for work is interpreted and applied in mechanical, fluid, and electromagnetic systems. The discussion of work in thermal systems will show that this general equation has limitations when describing the movement of heat energy. In each energy system where the general equation is applicable, the meaning of displacementlike quantity will be clarified.

2.1 ....

WORK IN MECHANICAL SYSTEMS

OBJECTIVES ....

Upon completion of this section, the student should be able to

- Define work and energy in general terms that apply to any energy system.

- Distinguish between work and energy in the following systems:
  Mechanical translational
  Mechanical rotational

- Define the following units of work and energy and identify the system of units in which they are used:
  Foot-pound
  Newton-meter
  Joule

- Define the term **radian** and explain how it is used to determine work done in mechanical rotational systems.

- Calculate the work done in mechanical translational systems for constant force.

- Show that the work done can be interpreted as an area under a curve.

- Calculate the work done in compressing a spring.

- Calculate the work done in moving a weight up an inclined plane.

- Calculate the work done in mechanical rotational systems for constant torques.

- Calculate the work done in twisting a rod.

DISCUSSION

Work in Mechanical Translational Systems

Definition of Work   In a mechanical translational system, the general equation for work takes the form of

$$\Delta W = F\Delta l \cos \theta \qquad (2.2)$$

where

ΔW = a small amount of work done by a force (N·m or ft·lb)
  F = magnitude of the force (N or lb)
  Δl = small distance through which the force acts (m or ft)
  θ = angle between the direction of $\Delta \mathbf{l}$ and the force as shown in Figure 2.1.

In the SI system of units, one newton-meter (N·m) is called one joule (J).

The quantity $F \cos \theta$ in Equation 2.2 is the component of $\mathbf{F}$ in the direction of $\Delta \mathbf{l}$. If F is perpendicular to $\Delta \mathbf{l}$ then **no work** is done. This is the simplest calculation for work done. For example, no work occurs when a planet travels around the sun in a circular path. The gravitational force acting on the planet is always perpen-

FIGURE 2.1   Relationship of force and displacement in work.

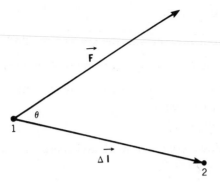

dicular to the direction of motion. This is approximately true for the earth, since its orbit is nearly a circle. Note that the physical quantity work does not have direction associated with it. Work is a scalar quantity like time or volume.

The next simplest case for work done occurs when the component of **F** in the direction of $\Delta \mathbf{l}$ is constant. Equation 2.2 then becomes

$$\Delta W = F_c \, \Delta l \qquad (2.3)$$

where

F_c = magnitude of the component of **F** in the direction of $\Delta \mathbf{l}$

Since the total work is the sum of all the small ΔW's, the total work becomes

$$W = \Sigma \Delta W = \Sigma F_c \Delta l = F_c \Sigma \Delta l = F_c l$$

or

$$W = F_c l \qquad (2.4)$$

where

$\Sigma \Delta l = l$ = the total path the body moves

If F_c is not constant, then the calculation of the total work is more difficult. However, one can interpret the total work as an area under a curve. To see this, consider Figure 2.2, where F_{cn} is plotted as a function of l. One sees here that the force F_c changes from point 1 to point 2 of the displacement l. The total distance l may be divided into many small Δl's. For each small Δl, $\Delta W = F_{cn}\Delta l$, where F_{cn} is a constant force over the small displacement Δl. However, the area of the small rectangle is also equal to $F_{cn}\Delta l$. One can see, if all the ΔW's are added, that the total sum of the rectangular areas is approximately equal to the total area under the curve. As Δl is made smaller and smaller, the two areas approach each other. Thus, the total work is equal to the area under the curve:

$$\Delta W = \Sigma F_{cn}\Delta l = \text{Area under curve} \qquad (2.5)$$

In this equation Δl is made very small (approaches 0). Equation 2.5 defines the work done, but it does not indicate the form of the change in energy produced by the work.

FIGURE 2.2 Interpreting work as area under a curve.

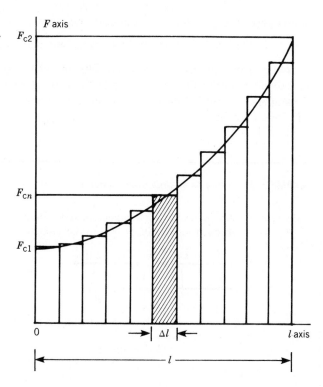

To calculate the total work, one may plot F_{cn} as a function of l and then estimate the area under the curve. A simple, but important, case occurs when the plot is a straight line as shown in Figure 2.3. In this case, force varies linearly with displacement. The area under the curve is the area of the rectangle plus the area of the triangle. Thus,

$$W = F_{c1}l + \tfrac{1}{2} l(F_{c2} - F_{c1})$$

$$W = (l) \left(\frac{F_{c1} + F_{c2}}{2} \right) \tag{2.6}$$

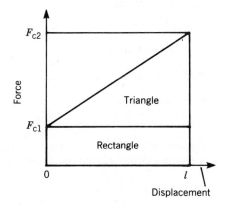

FIGURE 2.3 Area under the curve when force varies linearly with the displacement.

For this case, the general work equation becomes

Work = Displacement Quantity × Average Forcelike Quantity

It should be clear that the displacementlike quantity here is nothing more than the actual displacement l that occurs while the force F acts.

Examples 2.1 and 2.2 illustrate the use of the basic concepts in calculating the work done on a body.

EXAMPLE 2.1　Work as an area under a curve

. . . .

Given:　　An object having a single constant force of 4 N applied to it moves through a distance of 5 m in the direction of the force.

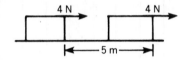

Sketch 1

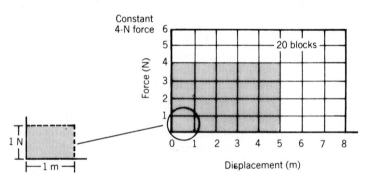

Sketch 2

Find:　　**a.**　The mechanical work done.

　　　　b.　How the mechanical work equals the area underneath a force–distance curve.

Solution:　**a.**　$W = Fl$

　　　　　　$W = (4 \text{ N}) (5 \text{ m})$

　　　　　　$W = 20 \text{ N·m}$

　　　　　　$W = 20 \text{ J}$

　　　　b.　Draw a force–displacement graph.

　　　　　　The area of each block of graph corresponds to 1 J of work (1 N·1 m); therefore, the mechanical work done by the force is

　　　　　　$(20 \text{ blocks}) \left(\dfrac{1 \text{ J}}{\text{block}} \right) = 20 \text{ J}$

EXAMPLE 2.2 Work done in extending a spring
. . . .

Given: The spring shown in the following sketch. The initial force in the unstretched spring is zero. While extending the spring by the amount 0.20 m, the force increases linearly with distance as given by

$$F = kl$$

where k is a constant equal to 1000 N/m.

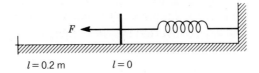

$l = 0.2$ m $l = 0$

Sketch 1

Find: The work done to extend the spring.

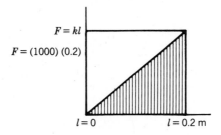

Sketch 2

Solution: Draw a sketch of force versus displacement.
Work is the area under the curve or

Work = Displacement × Average Force

$$W = (l) \times \left(\frac{kl}{2}\right)$$

$$W = (0.20 \text{ m}) \left(\frac{(1000 \text{N/m} \ (0.20 \text{ m}))}{2}\right) = 20 \text{ N·m}$$

$$W = 20 \text{ J}$$

For any length l the work done is

$$W = \tfrac{1}{2} kl^2$$

where k is the spring constant and l is the distance the unstretched spring is compressed or stretched.

Work Involving Frictional Forces   Frictional forces are common in all energy systems. Frictional forces always act in a direction opposite to the velocity or impending motion. In many cases the frictional force has a constant magnitude. The work done against a frictional force then becomes

$$W = F_f l \qquad (2.7)$$

where
  F_f = magnitude of the frictional force
  l = total path length moved through while the frictional force acts

The work depends on the total path length. Thus, no unique value of work can be assigned between two given points because there are many possible path lengths between two points. The work done on, or by, a body against a frictional force is always converted into heat energy. Example 2.3 illustrates the dependence of the path on the work done between two points against a frictional force.

EXAMPLE 2.3   Work done against friction
. . . .

Given: A body is moved from point 1 to point 2 as shown in the sketch. The magnitude of the frictional force is 10 N.

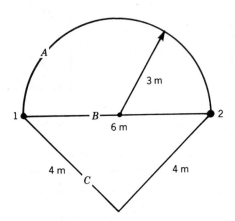

Find: **a.**   The work done if the body is moved along path *A*.
              b.   The work done if the body is moved along path *B*.
              c.   The work done if the body is moved along path *C*.

Solution: From Equation 2.7, the work done is

$$W = F_f l$$

              a.   The path length l for the semicircle is $l = \pi R$.

$$l = \pi R = (3.14)\ (3\ \text{m}) = 9.42\ \text{m}$$

$$W = (10\ \text{N})\ (9.42\ \text{m}) = 94.2\ \text{J}$$

b. The path length is 6 m.

$$W = (10\ \text{N})\ (6\ \text{m}) = 60\ \text{J}$$

c. The path length is 8 m.

$$W = (10\ \text{N})\ (8\ \text{m}) = 80\ \text{J}$$

Clearly then, work done against a frictional force depends on the path length followed from starting point to finishing point.

Work Involving Gravitational Forces   Near the surface of the earth the gravitational force acting on a body may be considered constant in magnitude and directed toward the earth's surface. Work is done against the force of gravity when the height of a body is increased.

To illustrate the work done against gravity, consider Figure 2.4. A man has done mechanical work by lifting a weight from the floor. The force exerted by his muscles has moved the weight through a distance. If he had not been able to lift the weight, no work would have been done. The man can tug and strain at the weight but, if he is unable to lift it, he does no work on the weight. When he holds the weight above his head, he is doing no work on the weight because it is not being moved by a force. How tired he becomes is not involved in the definition of mechanical work. The energy expended by his muscles in just holding the weight is converted to heat and to chemical energy, not to mechanical work. To summarize, then, he does work *only* while lifting (moving) the weight.

Example 2.4 illustrates the use of Equation 2.4 in determining work.

FIGURE 2.4 Doing work against gravity.

EXAMPLE 2.4 Work done while lifting a weight
· ·

Given: A man lifts a 200-lb weight a distance of 6 ft directly upward.

Find: The mechanical work done.

Solution: $W = F_c l$

                $W = (200 \text{ lb}) (6 \text{ ft})$

                $W = 1200 \text{ ft·lb}$

The unit of mechanical work in the English system of units is the foot-pound—the amount of work done when an unbalanced force of one pound is applied through a distance of one foot. Notice that the unit of work (ft·lb) is the product of the units of force and distance.

Since the force of gravity is essentially constant in magnitude and direction, Equation 2.2 can be simplified. From Figure 2.5 one may write

$$\Delta h = \Delta l \cos \theta$$

where
    Δh = change in the height of the body
    θ = angle between upward force and displacement

Thus, Equation 2.2 becomes for any path

$$\Delta W = F\Delta h = mg\Delta h = w\Delta h$$

where
    m = mass of the body
    g = acceleration due to gravity
    w = weight of the body

Since mg is a constant near the earth's surface, the total work done becomes

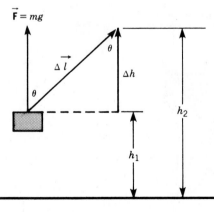

FIGURE 2.5 Gravitational work done in lifting a weight.

$$W = \Sigma mg\Delta h = mg\Sigma\Delta h$$

The sum of all the Δh's is the total height through which the body is moved. The work can then be expressed as

$$W = mgh = wh \tag{2.8}$$

where

h = total change in height
$mg = w$ = body weight

Thus, the work done against gravity from a surface parallel to the earth's surface depends only on the body's weight and the height above the surface. The results obtained using Equation 2.8 are compared with Equation 2.2 in Example 2.5.

EXAMPLE 2.5   Work done on an inclined plane
. . . .

Given: The frictionless inclined plane shown in the sketch.
              The weight of the body is 3000 N.
              The displacement (l) along the incline is 30 m.
              The vertical height (Δh) is 10 m.

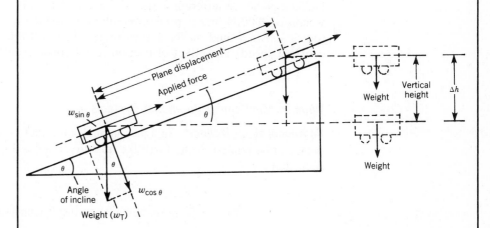

Find: **a.**   The force F_c required to move the cart up the inclined plane.
              b.   The work done using Equation 2.4, $W = F_c l$, with no friction.
              c.   The work done using Equation 2.8, $W = mgh$.

Solution: **a.**   To keep a body moving up the inclined plane requires an applied force equal to the component of the weight down the inclined plane. From the diagram, the component force is

$$F_c = W \sin \theta$$

$$\sin \theta = \Delta h/l$$

$$\sin \theta = \frac{10 \text{ m}}{30 \text{ m}}$$

$$F_c = (3000 \text{ N}) \left(\frac{10 \text{ m}}{30 \text{ m}}\right) = 1000 \text{ N}$$

b. The work done in moving the body along the inclined plane is (Equation 2.4)

$$W = F_c l$$

$$W = (1000 \text{ N}) (30 \text{ m}) = 30,000 \text{ J}$$

c. The work done according to Equation 2.8 is

$$W = mgh = w_T h, \qquad \text{where } w_T \text{ is the total body weight}$$

$$W = (3000 \text{ N}) (10 \text{ m}) = 30,000 \text{ J}$$

The inclined plane is a simple machine that allows less force than the weight of an object to be used on the object as it is moved along the plane of the incline. Figure 2.6 shows an object moved through a vertical distance by moving it along inclined planes at different angles of incline. The object raised (Figure 2.6) has a weight of 2000 N and is pulled on rollers to reduce friction to a minimum. The forces shown would be the forces required to pull the object up the plane at a constant speed. The effect of friction is not considered.

Work in Mechanical Rotational Systems

Rotational Work Defined   In a mechanical rotational system, the amount of work done can be related to the forcelike quantity τ and a displacementlike quantity θ such that

$$\Delta W = \tau \times \theta \tag{2.9}$$

Work = Forcelike Quantity × Displacementlike Quantity

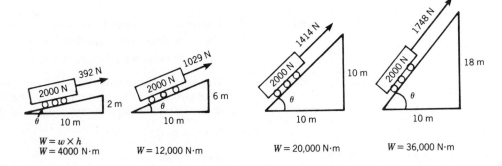

FIGURE 2.6 Work done depends on change in height of body.

$W = w \times h$
$W = 4000$ N·m

$W = 12,000$ N·m

$W = 20,000$ N·m

$W = 36,000$ N·m

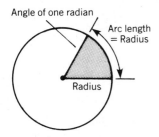

FIGURE 2.7 Definition of radian measure.

where

ΔW = a small amount of work in N·m or ft·lb

τ = torque in N·m or ft·lb

$\Delta\theta$ = a small angular rotation through which the torque acts in radians

Equation 2.9 will give the amount of work done only if $\Delta\theta$ is measured in radians. As shown in Figure 2.7, an angle of one radian is the angle that, if placed at the center of a circle, would intercept an arc length equal to the radius of the circle. The size of an angle of one radian is shown in Figure 2.7. This angle of one radian is approximately 57.3°. However, an identity derived from the formula for the circumference of a circle is easier to remember. The circumference of a circle encloses an angle of 360°. Since circumference is found by multiplying the radius by 2π, and an arc length of one radius corresponds to an angle of one radian, 2π radians = 360°. This relationship and another that results when both sides are divided by two are shown in Table 2.1.

The radian is not actually a unit in the same sense as a meter or a pound. It is the ratio of the **length** of an arc to the **length** of the radius. Lengths are expressed in the same units, which cancel to leave a dimensionless quantity. Thus, when angles expressed in radians are used in calculations, the term "radian" is often dropped. An angle in radians may be expressed mathematically as

$$\Delta\theta = \frac{\Delta l}{R} \tag{2.10}$$

where

Δl = arc length subtended by the angle θ

R = corresponding radius of the circle

As we have seen, the basic equation for work is

$$\Delta W = F_c\Delta l \tag{2.11}$$

If Δl is a small arc of a circular path, Equation 2.10 yields

$$\Delta l = R\Delta\theta$$

Substituting this value for Δl in Equation 2.11 gives

$$\Delta W = F_c R\Delta\theta \tag{2.12}$$

However, F_c is in the direction of Δl. Thus, from the definition of torque ($\tau = F_c R$) one obtains

$$\Delta W = \tau\Delta\theta \tag{21.3}$$

TABLE 2.1 Conversion factors

2π radians = 360°

π radians = 180°

Notice that Equation 2.13 has the same form as the basic equation for work. The application of Equation 2.13 is illustrated in Examples 2.6 and 2.7.

EXAMPLE 2.6   Work done to turn a pulley
. . . .

Given: A force of 70 lb is required to turn a pulley with a radius of 1.5 ft at a constant speed. The pulley turns through 5 revolutions.

Find: The work done on the pulley.

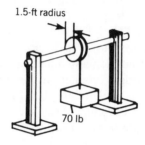

Solution: Recall from Chapter 1 that

$$\tau = Fl \quad \text{(Torque} = \text{Force} \times \text{Lever Arm)}$$

$$\tau = (70 \text{ lb}) (1.5 \text{ ft})$$

$$\tau = 105 \text{ ft·lb}$$

$$\Delta\theta = (5 \text{ rev}) \left(\frac{2\pi \text{ rad}}{1 \text{ rev}}\right) = 10\pi \text{ rad}$$

From Equation 2.13

$$\Delta W = \tau \Delta\theta$$

$$\Delta W = (105 \text{ ft·lb}) (10\pi) \quad \text{(''rad'' is dropped)}$$

$$\Delta W = (105 \text{ ft·lb}) (10) (3.14)$$

$$\Delta W = 3297 \text{ ft·lb}$$

3297 ft·lb of work is done by the applied torque.

EXAMPLE 2.7   Work done in twisting a rod
. . . .

Given: A rod supported at one end with a disk at the other end as shown in the sketch. Forces F_1 and F_2 are applied to the disk as shown to

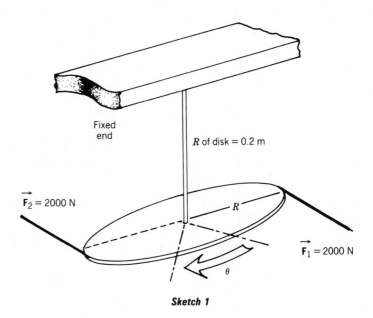

Fixed end

R of disk = 0.2 m

$\vec{F}_2$ = 2000 N

R

$\vec{F}_1$ = 2000 N

θ

Sketch 1

form a couple. From experiments, it is found that:

$$\tau = k\theta$$

where

$k = 10,000$ N·m
θ = angle of twist at the disk

The final value of the applied forces is 2000 N.

Find: **a.** The angle of twist at the disk.

b. The work done in twisting the rod.

Solution: **a.** The forces form a couple. The torque of a couple (see Chapter 1) is

$$\tau = Fl$$

where

$l = 2R$
$\tau = (2000 \text{ N}) (0.4 \text{ m}) = 800$ N·m

The angle θ is

$$\theta = \tau/k$$

$$\theta = \frac{800 \text{ N·m}}{10,000 \text{ N·m}} = 0.08 \text{ rad} \qquad (0.08 \text{ rad} = 4.58°)$$

b. To find the amount of work done, plot a graph of τ versus θ. The work is equal to the area under the curve.

Work = Displacementlike Quantity × Average Forcelike Quantity

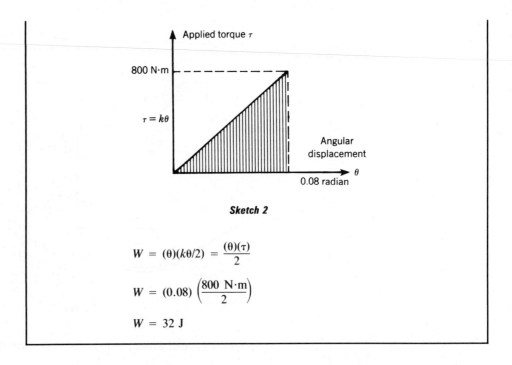

Sketch 2

$$W = (\theta)(k\theta/2) = \frac{(\theta)(\tau)}{2}$$

$$W = (0.08)\left(\frac{800 \text{ N·m}}{2}\right)$$

$$W = 32 \text{ J}$$

Many times, a stopping force or stopping torque is applied to bring motion to a stop. When that happens, the work done by the stopping force (or torque) is converted to heat energy. Example 2.8 provides an illustration of work done on a rotating flywheel to bring it to rest.

EXAMPLE 2.8   Work done in stopping a rotating flywheel
. . . .

Given: A heavy flywheel, 1.2 m in diameter, rotates 484° before stopping, all while a constant braking force of 80 N is applied at its outer edge.

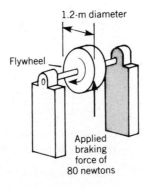

Find: Amount of work done by the braking force to stop the rotation.

Solution: Since the diameter is 1.2 m, the radius is 0.6 m. The work done is

$$W = \tau\theta = Fl\theta$$

$$W = (80 \text{ N})(0.6 \text{ m})(484°)\left(\frac{\pi \text{ rad}}{180°}\right)$$ (Remember! Angles in degrees must be converted to radians.)

$$W = (48 \text{ N·m})(8.45 \text{ rad})$$

$$W = 405.5 \text{ N·m}$$

The work done to stop the flywheel is 405.5 N·m or 405.5 J. In the process, heat energy equal to 405.5 J is produced—and lost.

Notice that the units of torque and work are the same—N·m in SI units and ft·lb in English units. This does not mean that torque and work are equivalent. Care must be taken to identify a quantity expressed in these units as either torque or work. The multiplication of torque (N·m or ft·lb) by an angle of rotation in radians (no units) yields work (N·m or ft·lb). In the SI system, both work and energy usually are expressed in joules to eliminate the possible misunderstanding.

SECTION 2.1 EXERCISES

1. Define work and energy, and state the similarities and differences of the two concepts.
2. Define the following units of work and energy, and state the energy systems in which each is used.
   a. Foot-pound
   b. Newton-meter
   c. Joule
3. Explain the two meanings of the English unit "ft·lb."
4. Define radian and complete the following relationships:
   a. 360° = _____ radians
   b. π rad = _____ °
   c. 90° = _____ rad
   d. 1 rad = _____ °
5. A force of 70 N is applied to slide a box a distance of 500 m. How much work is done?
6. Forty-five foot-pounds of work are required to lift an object a distance of 15 ft. How much does the object weigh?
7. A torque of 280 ft·lb does 12 ft·lb of work. What is the angle of rotation in radians and degrees?
8. How much work is done in removing a nut from a bolt if an average torque of 70 N·m must be supplied for 4.5 revolutions? How many ft·lb of work are required?
9. A man applies a constant horizontal force to a 5-kg mass to accelerate the mass

horizontally across a frictionless table. If the man does 20 J of work to move the mass 3 m, what is the constant horizontal force applied to the block?

10. What braking force applied at the outer edge of a heavy flywheel is required to bring the flywheel to rest in one revolution if the flywheel has a radius of 1 m and 600 J of work is needed to stop the flywheel?

11. The spring constant of a spring is 950 N/m. How much work is done while compressing the spring 0.05 m?

12. The torque constant of a rod is 10 N·m/rad. How much work is done if the total twist angle is 0.2 radian?

13. *A box is moved up an inclined plane with a 200-lb force. The total distance traveled is 20 ft. The weight of the box is 500 lb and the box is raised 5.0 ft. What is the efficiency of the inclined plane? (The efficiency is equal to the effective work accomplished in raising the box divided by the actual work done in moving the box up the inclined plane.)

14. The general equation for work is: Work equals Forcelike Quantity times Displacementlike Quantity.

 a. What is the forcelike quantity in the mechanical translational system? In the mechanical rotational system?

 b. What is the displacementlike quantity in the mechanical translational system? In the mechanical rotational system?

15. *Under each of the drawings shown in Figure 2.6, the result of a calculation for work is shown.

 a. Verify each of the results given, showing the proper equation (in symbols) used and the correct numerical values substituted for the symbols.

 b. What is the efficiency of each of the inclined planes in Figure 2.6?

 c. Why is this answer different from that obtained for Problem 13?

---

* Challenge question.

2.2
WORK IN FLUID SYSTEMS

OBJECTIVES

Upon completion of this section, the student should be able to

- Identify the forcelike quantity and displacementlike quantity involved in the general work equation for fluid systems.

- Calculate the work done in fluid systems when a constant pressure is applied to move or change a volume of fluid.

- Calculate the work done to fill a water tank.

- Calculate the work done by a piston that moves a certain volume of water in a cylinder or pipe.

DISCUSSION

In a fluid system, work is accomplished when a pressure is applied to move or change a volume of the fluid. A fluid may be a liquid or a gas. Liquids, for many

FIGURE 2.8 Work done
while compressing gas in a
cylinder.

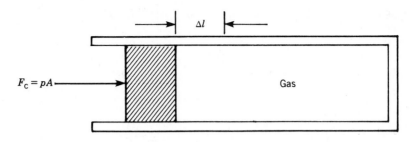

FIGURE 2.8 Work done
while compressing gas in a
cylinder.

practical purposes, may be considered incompressible; but gases are easily compressed. For both cases, however, the general formula—Work = Forcelike Quantity × Displacementlike Quantity—takes the form of Equation 2.14. Here the forcelike quantity is the pressure p and the displacementlike quantity is the volume ΔV moved or changed:

$$\Delta W = p\Delta V \qquad (2.14)$$

Work = Forcelike Quantity × Displacementlike Quantity

The validity of Equation 2.14 for the work done in compressing a gas in a cylinder can be easily shown. This situation is illustrated in Figure 2.8. From the definition of work, one has

$$\Delta W = F_c\Delta l$$

where

F_c = constant force applied
Δl = a small displacement of the piston

Since the force is the pressure times the area, the equation for a small amount of work reduces to

$$\Delta W = pA\Delta l = p\Delta V$$

where

A = area of the piston
p = pressure applied
$\Delta V = = A\Delta l$ = change of gas volume due to compression

As the gas is compressed, the pressure changes with the change in volume and temperature. Thus, additional information is required before the work for **large** displacements can be calculated.

Equation 2.14 also applies to liquids. For example, consider the work done on the system shown in Figure 2.9. The work done by F_1 in moving the piston a distance Δl_1 is

FIGURE 2.9 Work done by
piston in moving fluid
through pipe.

FIGURE 2.10　Work done on
a fluid while moving it and
raising its elevation.

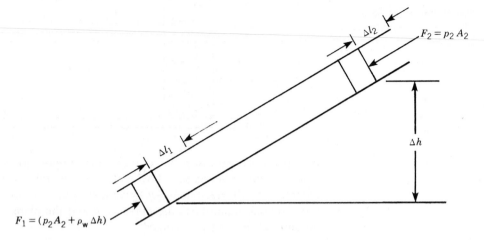

$$\Delta W_1 = F_1 \Delta l_1 = p A_1 \Delta l_1 = p_l \Delta V_1$$

Also, the work done by F_2 is

$$\Delta W_2 = F_2 \Delta l_2 = -p A_2 \Delta l_2 = -p_2 \Delta V_2$$

The work ΔW_2 is negative because Δl_2 and F_2 are in opposite directions ($\cos 180° = -1$). In this example, $\Delta V_1 = \Delta V_2$ since the liquids are incompressible. Thus, the net work may be expressed as

$$\Delta W = \Delta W_1 + \Delta W_2 = (p_1 - p_2)\,\Delta V = (\Delta p)\,(\Delta V) \qquad (2.15)$$

If the level of the liquid changes and the speed of the fluid is constant (constant-diameter pipe), then the work done in moving the liquid between two levels can be easily calculated. To do this, consider Figure 2.10.

The net work is

$$\Delta W = (p_1 - p_2)\,\Delta V = (p_2 + \rho_w \Delta h - p_2)\,\Delta V$$

$$\Delta W = (\rho_w \Delta h)\,\Delta V \qquad (2.16)$$

Equation 2.16 relates the net work done on the fluid to the work done against the gravitational force. The work done on the fluid may also change the fluid velocity. This effect will be discussed in Chapter 6.

In a system involving liquids, which are incompressible, the volume of a given mass of fluid remains the same; that is, the volume of fluid entering a pipe is equal to the volume leaving the pipe during the same time interval. Only fluid pressure changes in the system. Example 2.9 illustrates the use of Equation 2.16 in solving a problem. Example 2.10 uses a form of Equation 2.16 to solve a work problem where pressure is a linear function of height.

EXAMPLE 2.9　Work done to fill a water tank
. . . .

Given:　　A water tank system of 500–ft³ capacity. The density of water is 62.4 lb/ft³.

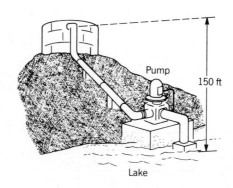

Find: The amount of work required to fill it with water from a lake 150 ft below.

Solution: From Chapter 1

$$\Delta p = \rho_w \Delta h$$

$$\Delta p = (62.4 \text{ lb/ft}^3)\ (150 \text{ ft})$$

$$\Delta p = 9360 \text{ lb/ft}^2$$

The pressure difference Δp between the top of the tank and the surface of the lake is 9360 lb/ft².

$$\Delta W = (\Delta p)\ (\Delta V)$$

$$\Delta W = (9360 \text{ lb/ft}^2)\ (500 \text{ ft}^3)$$

$$\Delta W = 4{,}680{,}000 \text{ ft·lb}$$

The work required to fill the tank is 4.68×10^6 ft·lb.

EXAMPLE 2.10   Work done in filling a tank
. . . .

Given: Water is pumped from a reservoir into an open tank as shown. Initially the tank is empty. The tank is filled to a height of 10 ft. The area of a cross section is 25 ft².

Find: The work done.

Solution: $$W = \begin{bmatrix} \text{Displacementlike} \\ \text{Quantity} \end{bmatrix} \times \begin{bmatrix} \text{Average Forcelike} \\ \text{Quantity} \end{bmatrix}$$

$$\Delta p = \rho_w h \qquad \text{(linear function of } h\text{)}$$

$$W = (\Delta V) \times \left(\frac{\Delta p_1 + \Delta p_2}{2} \right)$$

$$\Delta p_1 = 0$$

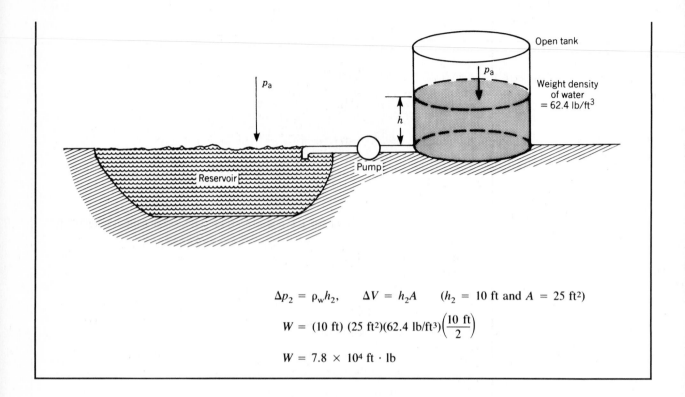

$$\Delta p_2 = \rho_w h_2, \qquad \Delta V = h_2 A \qquad (h_2 = 10 \text{ ft and } A = 25 \text{ ft}^2)$$

$$W = (10 \text{ ft})\,(25 \text{ ft}^2)(62.4 \text{ lb/ft}^3)\left(\frac{10 \text{ ft}}{2}\right)$$

$$W = 7.8 \times 10^4 \text{ ft} \cdot \text{lb}$$

As previously stated, there must be some movement in the system, or work has not been done. For example, air pressure of forty pounds per square inch in a tire is not work in itself, but work is done in putting air into the tire. The change in air volume involved in the work is the difference between the volume of air under pressure in the tire and the volume that the same quantity of air would occupy if released to the atmosphere. The pressure involved is the pressure of the air in the tire at any time, which increases gradually from one atmosphere (14.7 psi) to the final pressure in the tire.

Work done in filling a tire with air is partly stored in increased internal energy of the air molecules within the tire. This shows up as a rise in temperature, and partly in the stretched tire itself. A mechanic must exercise caution in filling an automobile tire (or tube) with air after repairing it, because this stored energy can be released quickly. If the tire and wheel are not put together correctly, the tire can explode and cause serious injury. Businesses involved in changing and repairing large tires provide a heavy safety frame that holds the tire while it is being filled with air.

SECTION **2.2**   *EXERCISES*

1. The general equation for work is: Work equals Forcelike Quantity times Displacementlike Quantity. Identify the forcelike quantity and displacementlike quantity for work in fluid systems. Then write down the basic equation for work in fluid systems.

2. It requires 5,000,000 J to fill a tank through the top with water from a reservoir located some distance below the tank. What must the effective pressure difference between the top of the tank and the surface of the reservoir be if the tank has a volume of 300 m³?

3. What is the approximate distance from the reservoir to the top of the tank in Problem 2?

4. A pump consumes 13,000 J of energy when moving 8 m³ of water under constant pressure. What pressure (N/m² and lb/ft²) did the pump supply to the water?

5. An air conditioner is capable of providing an air pressure of 0.25 psi. How much work must be done to fill a room 20 × 30 × 8 ft with cool air?

2.3

WORK IN ELECTROMAGNETIC SYSTEMS

OBJECTIVES

Upon completion of this section, the student should be able to

- Identify the appropriate forcelike quantity and displacementlike quantity involved in the general work equation for
  Electrical systems
  Magnetic systems

- Write down the equations to be used to calculate the work done in
  Electrical systems to move electric charge
  Magnetic systems to establish a magnetic field

- Calculate the work done in electrical systems to move charge at constant voltage differences.

- Calculate the work done to charge a capacitor.

- Calculate the work done to establish a magnetic field.

DISCUSSION

Electrical Systems

Work in Electrical Systems   In an electrical system, work and energy are measured in joules (J). The general formula for work takes the form

$$\Delta W = (\Delta V)(\Delta Q) \tag{2.17}$$

Work = Forcelike Quantity × Displacementlike Quantity

where

ΔV = potential difference through which charge is moved—the forcelike quantity

ΔQ = quantity of charge moved—the displacementlike quantity

Equation 2.17 can be derived from the basic equations for work (Equation 2.2):

$$\Delta W = F \cos \theta \, \Delta l$$

The force on a unit charge was defined as the electric field E. Thus, the force on a charge ΔQ becomes

$$F = E \, \Delta Q$$

Substituting this force in the equation for work yields

$$\Delta W = (E \cos \theta \, \Delta l) \, \Delta Q$$

However, in Chapter 1, $E \cos \theta \, \Delta l$ was defined as the potential difference ΔV. Thus, by substitution, the equation for ΔW becomes

$$\Delta W = (\Delta V)(\Delta Q)$$

Potential difference, measured in volts (V), is the forcelike quantity. Charge, measured in coulombs (C), is that substance in electrical circuits that is moved by a potential difference. Thus, charge is the displacementlike quantity.

A net charge of one coulomb represents 6,250,000,000,000,000,000 more electrons at the negative pole than at the positive pole! Since this number is so large, it usually is written with the aid of scientific notation as 6.25×10^{18}. One joule of work or energy in the electrical system is the amount of energy required to move one coulomb of charge through a potential difference of one volt. Table 2.2 gives useful conversion factors for electrical energy.

Example 2.11 shows the use of Equation 2.17 in solving a problem that deals with charging a battery.

EXAMPLE 2.11   Energy required to charge a battery
. . . .

Given: A 6-V automobile battery stores 8000 C.

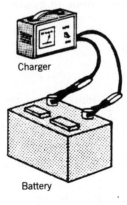

Charger

Battery

Find: The energy required to charge it.

Solution: $\Delta W = (\Delta V)(\Delta Q)$

               $\Delta W = (6 \text{ V})(8000 \text{ C})$

               $\Delta W = 48,000 \text{ J}$

               48,000 J or 48 kJ are required to charge the battery.

TABLE 2.2 Conversion factors for electrical systems

1 J = 1 V × 1 C
1 C = charge of 6.25 × 10^{18} electrons

In most electrical systems, the measurement of electrical charge is unnecessary. Instead, electrical current is measured. Current is the rate at which charge flows and is defined by

$$I = \frac{\Delta Q}{\Delta t} \tag{2.18}$$

where

I = current in amperes (A)
ΔQ = charge transferred in coulombs (C)
Δt = time in seconds (sec) required for transfer

This equation will be examined in more detail in Chapter 3. It is presented here only because of its importance in determining work done in an electrical system. Example 2.12 shows the use of Equations 2.17 and 2.18 in solving for the electrical work done by a motor.

EXAMPLE 2.12 Work done by an electric motor
. . . .

Given: A dc electric motor, operating on a voltage of 12 V and drawing a current of 4 A, runs for 2 min.

Find: Work done or electrical energy consumed.

Solution: From Equation 2.18

$$\Delta Q = I\Delta t$$

$$\Delta Q = (4 \text{ A})(120 \text{ sec})$$

$$\Delta Q = 480 \text{ C}$$

The charge transferred is 480 C.
From Equation 2.17

$$\Delta W = (\Delta V)(\Delta Q)$$

$$\Delta W = (12 \text{ V})(480 \text{ C})$$

$$\Delta W = 5760 \text{ J}$$

Equation 2.17 can also be used to calculate the total work required to charge a capacitor. This is illustrated in Example 2.13.

EXAMPLE 2.13   Work done in charging a capacitor

. . . .

Given: A 10-μF capacitor is charged to a potential difference of 1000 V.

Find: The work done in charging the capacitor.

Solution: From Chapter 1

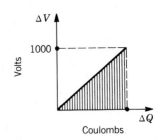

$$\Delta V = \frac{\Delta Q}{C}$$

Plot ΔV versus ΔQ.

$$\Delta Q = C(\Delta V)$$

$$\Delta Q = (10^{-7} \text{ F})(10^3 \text{ V})$$

$$\Delta Q = 10^{-4} \text{ C}$$

Work is the area under the curve.

Work = Displacementlike Quantity × Average Forcelike Quantity

$$\Delta W = (\Delta Q)\left(\frac{\Delta V}{2}\right)$$

$$\Delta W = (10^{-4} \text{ C})\left(\frac{10^3 \text{ V}}{2}\right) \qquad (1 \text{ C·V} = 1 \text{ J})$$

$$\Delta W = 0.05 \text{ J}$$

Magnetic Systems

Work in Magnetic Systems   In a magnetic system the work and energy are measured in joules. The equation for work takes the form

Work = Forcelike Quantity × Displacementlike Quantity

$$\Delta W = (NI)\Delta\phi \qquad\qquad (2.19)$$

where

   NI = magnetomotive force in ampere turns—the forcelike quantity
   $\Delta\phi$ = increase in the magnetic flux in webers—the displacementlike quantity

To show that Equation 2.19 is equivalent to mechanical work requires the use of rate equations, so this equivalence will be established in Chapter 3.

   The magnetic flux depends on the current since B, the flux density, depends on the current as shown in Chapter 1. The use of Equation 2.19 to calculate the energy stored in the magnetic field of a solenoidal coil is illustrated in Example 2.14.

EXAMPLE 2.14   Work required to establish a magnetic field
· ·

Given: A solenoidal coil of 200 turns has an air core with a permeability of
             $\mu = 1.26 \times 10^{-6}$ N/A². The solenoid has an inner radius of 0.30 m
             and an outer radius of 0.31 m as shown in the sketch. The height of
             the rectangular cross section of the core is 0.05 m. A current of 100
             A is flowing through the coil.

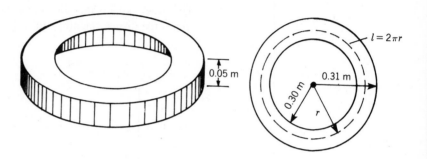

Find: The work required to establish the magnetic field in and around the
             coil.

Solution: The equation for a small amount of work is

             $$\Delta W = (NI)\Delta\phi = (\text{mmf})\Delta\phi$$

             where *NI* (ampere-turns) is equal to the magnetomotive force mmf.
             To calculate the total work, mmf must be expressed as a function of
             ϕ and the area under the curve calculated. To calculate ϕ, start with
             the equation for *B* (Chapter 1)

             $$B = \frac{\mu NI}{l} = \frac{\mu(\text{mmf})}{l}$$

             The average radius, *r*, as seen from the diagram, is 0.305 m. Thus, *B*
             becomes

             $$B = \frac{(1.26 \times 10^{-6} \text{ N/A}^2)(\text{mmf})}{(2\pi)(0.305 \text{ m})} = [6.57 \times 10^{-7} \, NI] \frac{\text{N}}{\text{A}^2\cdot\text{m}}$$

             $$1 \, \frac{\text{N}}{\text{A}\cdot\text{m}} = 1 \text{ Wb/m}^2$$

             (Note that the N in the term N/A² and N/(A·m) refers to newtons,
             not number of turns.)
             The total flux is the flux density (lines/m²) times the cross-sectional
             area of the core.

             $$\phi = BA$$

$$\phi = \left(6.57 \times 10^{-7} \frac{N}{A^2 \cdot m}\right)(0.05 \text{ m})(0.01 \text{ m})(\text{mmf})$$

$$\phi = \left(3.29 \times 10^{-10} \frac{N \cdot m}{A^2}\right)(\text{mmf})$$

Solving for the magnetomotive force mmf $= NI$ gives

$$\text{mmf} = NI = \left(3.04 \times 10^9 \frac{A^2}{N \cdot m}\right)\phi$$

The work done is the area under the curve of the forcelike quantity times the displacementlike quantity.

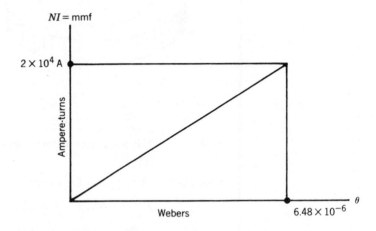

Work is the area under the curve.

Work = Displacementlike Quantity × Average Forcelike Quantity

$$W = [\Delta\phi] \times \left(\frac{\text{mmf}}{2}\right)$$

$$W = \left[\left(3.29 \times 10^{-10} \frac{N \cdot m}{A^2}\right)(200 \text{ turn})(100 \text{ A})\right]\left[\frac{(200 \text{ turn})(100 \text{ A})}{2}\right]$$

$$W = 6.58 \times 10^{-2} \text{ J} = 0.066 \text{ J} \qquad \text{(This is work done to establish the magnetic field.)}$$

SECTION 2.3   EXERCISES

1. Identify the forcelike quantity and displacementlike quantity used to calculate work in electrical systems. Write out the equation for work, using appropriate symbols.

2. Identify the appropriate forcelike quantity and displacementlike quantity used to calcu-

late work in magnetic systems. Write out the equation for work using appropriate symbols.

3. If a 12-V (dc) electric refrigerator consumes 1800 J of energy over a certain period of time to keep food cold, how much charge (coulombs) has been transferred?

4. It requires 10,000 J to charge a 12-V battery. If the current used to charge the battery is 1 A, how long will it take to charge the battery?

5. What is the operating voltage of a dc electric motor if the motor consumes 8000 J of energy in 3 min and draws an average current of 3.7 A?

6. A 10-μF capacitor is connected across a 1000-V source. How much energy is stored in the capacitor?

7. The relationship between the magnetic flux ϕ and the magnetomotive force mmf is

$$\phi = \left(2.0 \times 10^{-10} \, \frac{\text{N·m}}{\text{A}^2}\right)(\text{mmf})$$

where

$$\text{mmf} = NI$$
$$N = 100 \text{ turns}$$
$$I = \text{current in A}$$

    a. If the current is increased from 0 to 200 A, how much energy is stored in the magnetic field?

    b. How much work is done by the magnetomotive force to establish the field?

2.4

WORK IN THERMAL SYSTEMS

OBJECTIVES

Upon completion of this section, the student should be able to

- Understand the relationship between work, heat, and internal energy in a thermal system.

- Define the following terms and explain their usefulness in determining heat transferred:
  Specific heat
  Heat capacity
  Latent heat
  Sensible heat

- Outline several experimental methods that are used to measure specific heat and latent heat.

- Calculate the efficiency of systems by comparing work output with work input.

DISCUSSION

Heat Energy

In a thermal system, application of the general work equation—Work equals Forcelike Quantity times Displacementlike Quantity—does not follow the anal-

FIGURE 2.11   Heat energy (ΔQ) added to the confined gas increases the internal energy of the gas (ΔU) and does work (ΔW) by moving the piston.

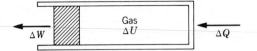

ogy described for the other systems. In thermal systems, the forcelike quantity is identified as a temperature change and the displacementlike quantity as heat energy. To this point, these analogies have provided a systematic unification with the other energy systems. In the application of the work equation, the analogy fails. Like all analogies, there may be a situation in which the point of view of the analogy is incapable of clearly describing the situation.

Work does have an important role in thermal systems. For example, consider a gas in a cylinder with a piston as shown in Figure 2.11. If heat energy flows into the gas, the gas can expand, and work will be done. In addition, the temperature of the gas may increase with a corresponding increase in internal energy. The equation relating these quantities is

$$\Delta Q = \Delta W + \Delta U = p\Delta V + \Delta U \qquad (2.20)$$

where

ΔQ = heat energy added to the gas in joules
ΔW = work done by the gas in joules
ΔU = change in the internal energy of the gas in joules

Equation 2.20 is a statement of the conservation of energy.

Specific Heat   If the volume of the cylinder remains constant, no work is done and

$$\Delta Q = \Delta U = mc_V\Delta T \qquad (2.21)$$

where

c_V = a constant known as the specific heat at constant volume
m = mass of the gas
ΔT = change in temperature of the gas

If the process described above takes place at constant pressure, the gas will expand when heat is added. This will result in work done by the gas according to the equation

$$\Delta W = p\Delta V$$

For this case, the heat added can be written as

$$\Delta Q = mc_p\Delta T \qquad (2.22)$$

where

c_p = specific heat at constant pressure
m = mass of gas
ΔT = change in temperature of the gas

The specific heat at constant pressure is **greater** than the specific heat at constant volume because of the work done by the gas. For other processes, the specific heat takes on different values, but c_V and c_p are of special practical interest.

In the case of liquids and solids, the expansion is so small that it is usually unnecessary to distinguish between c_p and c_V. The equation then becomes

$$\Delta Q = mc\Delta T \tag{2.23}$$

where c is the specific heat of a solid or liquid.

The Equation 2.23, the quantity mc is the product of the mass and the specific heat of the body that undergoes the temperature change. The product mc is called the "heat capacity" of the object and represents the amount of heat energy required to raise the temperature of that body by one degree (Fahrenheit in English units and Celsius in SI units).

Specific heat indicates how many units of heat energy are required to raise a unit mass of some substance through a unit of temperature difference. For example, water has a specific heat of 1 cal/g·C°, or 1 Btu/lb·F°. Since water has a large specific heat, that is, a capacity to soak up large amounts of heat energy at the expense of only a small increase in temperature, it makes an ideal coolant. That is why water is used as the coolant in most systems in which considerable thermal energy is generated.

Units for Heat Energy   In SI units, quantity of heat is measured in calories (cal) or kilocalories (kcal). One **calorie** is the amount of heat required to raise the temperature of one gram of water one Celsius degree. Since the calorie measures small amounts of heat, kilocalories often are used to measure larger quantities. Measures of the heat values of foods, incorrectly given as calories in diet books, actually are kilocalories. Bread claimed to have 60 calories per slice actually has 60 kilocalories per slice.

In English units, quantity of heat is measured in British thermal units (Btu). One **British thermal unit** is the amount of heat required to raise the temperature of one pound of water one Fahrenheit degree. This unit is used in the United States to measure the cooling capacity of air conditioners or the heating capacity of stoves. Table 2.3 lists some useful conversions.

Values of Specific Heats   Specific heats of numerous substances have been determined experimentally and can be found in sources such as Table 2.4. All values listed are specific heats at constant volume (c_V) with the exception of air, which is for constant pressure (c_p).

TABLE 2.3   Conversion factors

1 Btu = 252 cal
1 kcal = 1000 cal
1 cal = 3.09 ft·lb
1 Btu = 778 ft·lb
1 cal = 4.18 J

TABLE 2.4   Specific heats of
common substances

Substance	Specific Heat
Air	0.24 (c_p)
Aluminum	0.22
Brass	0.091
Copper	0.093
Glass	0.21
Iron (steel)	0.115
Stone (average)	0.192
Tin	0.055
Water	1.0
Wood (average)	0.42

Specific heats are stated in Table 2.4 as the ratio of the specific heat of the given substance to that of water. The value of the specific heat of any substance is the same in either SI or English units. Units for specific heat in Table 2.4 are Btu/lb·F°, kcal/kg·C°, or cal/g·C°. Specific heat of water was adopted as a standard because of its large heat capacity and availability.

The use of Equation 2.23 in thermal systems is illustrated in Examples 2.15 and 2.16.

EXAMPLE 2.15   Energy required to heat water
. . . .

Given: A mass of 417 lb (approximately 50 gal) of water is heated from 70°F to 130°F (1 ft³ = 7.48 gal).

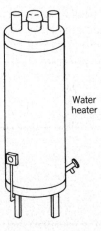

Water
heater

Find: The amount of heat energy required. Assume no heat loss in heating the water tank or atmosphere.

Solution: From Equation 2.23

$$\Delta Q = mc\Delta T$$

$$\Delta Q = (417 \text{ lb})(1 \text{ Btu/lb·F°})(130°F - 70°F)$$

$$\Delta Q = (417 \text{ lb})(1 \text{ Btu/lb·F°})(60 \text{ F°})$$

$$\Delta Q = 25,020 \text{ Btu}$$

EXAMPLE 2.16   Heat energy removed to cool an object

. . . .

Given: A 25-kg brass ball at 100°C is cooled to 30°C in an insulated container of water. The water temperature increases from 10°C to 30°C.

Find:
   a. The amount of heat energy removed from the ball and added to the water.
   b. The mass of water in the container.

   Assume no heat loss to container or atmosphere.

Solution:
   a. $\Delta Q = mc\Delta T$

      $\Delta Q = (25 \text{ kg})(0.091 \text{ kcal/kg·C°})(100°C - 30°C)$

      $\Delta Q = 159.3 \text{ kcal, or } 159,300 \text{ cal}$

   b. $m = \dfrac{Q}{c\Delta T}$

      $m = \dfrac{(159,300 \text{ cal})}{(1 \text{ cal/g·C°})(20 \text{ C°})}$

      $m = 7965 \text{ g}$

Change of State

If heat is supplied to matter, two possible changes can occur. The first possible change, described in Example 2.16, is that the temperature of the material will be raised while its state remains fixed. The second possible change is that the temperature will remain fixed, but the state of the substance will change; that is, the substance will change from a solid to a liquid, or from a liquid to a gas. Heat can change the temperature or the state, but not both at the same time.

Sensible and Latent Heat   If the temperature of a material is changing, heat supplied to the material is called "sensible heat," since the effect can be sensed by means of a thermometer. If the state of the substance is changing, the heat is

FIGURE 2.12 Sensible and latent heat are both involved in changing ice to steam.

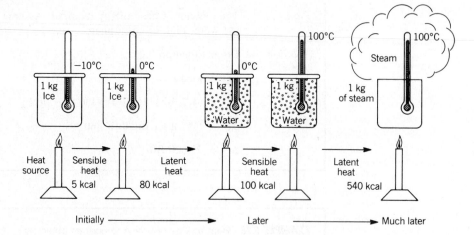

called "latent heat." The word "latent" means "hidden." Figure 2.12 shows a container of ice to which heat is applied. The resulting temperatures at various times in the heating process are shown. Latent heat is energy used in changing matter from one state to another. Change of state is a very important principle used in air conditioners, in refrigerators, and in many other cooling and heating systems.

The normal melting point of a crystalline solid is the temperature at which the solid melts when subjected to one standard atmosphere of pressure. The melting occurs when heat is supplied to the sample. Freezing occurs at the same temperature when heat is removed from the substance. The energy required to melt a unit mass of the material is called the **latent heat of fusion.** The energy required to freeze a unit mass is called the **latent heat of solidification.** The latent heat of fusion is equal to the latent heat of solidification if the pressure is the same. At one standard atmosphere of pressure, the heat of fusion or solidification for water is 80 cal/g at 0°C.

The temperature at which a liquid boils at one standard atmosphere is called the normal boiling point. In this case, heat is supplied to the substance. The temperature at which the substance condenses is equal to the boiling temperature if the pressure is the same. Condensation occurs when heat is removed. The energy required to vaporize a unit mass of the substance is called the **latent heat of vaporization.** The heat required to condense one unit mass is the **latent heat of condensation.** The two latent heats are the same if the pressure is the same. The heat of vaporization or condensation for water at one standard atmosphere is 539 cal/g at 100°C.

The latent heat is usually given the symbol L. The energy supplied during melting or vaporization is associated with the difference in the internal energy of the two phases and the work done by the substance in pushing back the atmosphere. The total heat required in a change of state may be expressed as

$$Q = mL \qquad (2.24)$$

TABLE 2.5 Latent heats at standard pressure

Substance	Normal Melting Point (°C)	Heat of Fusion (cal/g)	Normal Boiling Point (°C)	Heat of Vaporization (cal/g)
Nitrogen	−210	6.1	−196	48
Mercury	−39	2.8	357	65
Water	0.00	79.7	100.00	539
Sulfur	119	9.1	445	78
Lead	327	5.9	1750	208

where

Q = total heat supplied or removed
m = mass of the substance
L = latent heat of fusion or vaporization

Table 2.5 lists the normal melting point, the latent heat of fusion, the normal boiling point, and the latent heat of vaporization for several substances.

Measurement of Specific Heat and Latent Heat

The measurement of specific heat and latent heat is often carried out with a device called a calorimeter. The word "calorimeter" comes from the Greek word "calor," meaning heat, and the Latin word "meter," meaning measure. Thus, a calorimeter literally is a device for measuring heat.

There are many types of calorimeters; the most common design is a metallic can containing water. The can contains a measured quantity of water and has a cover through which a thermometer and a stirrer are inserted. An insulating jacket, or pocket of air, surrounds the can to reduce heat losses. The basic structure is sketched in Figure 2.13.

The thermometer is used to measure temperature of the water before and after

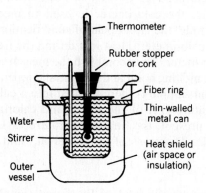

FIGURE 2.13 Calorimeter.

Thermometer

Rubber stopper or cork

Fiber ring

Thin-walled metal can

Water

Stirrer

Heat shield (air space or insulation)

Outer vessel

the unknown quantity of heat is introduced into the calorimeter. The amount of heat may be found from the measured rise in temperature of the water.

Measuring Specific Heat   To determine an unknown specific heat of a material, Equation 2.23 may be used to express the heat lost by the sample and the heat gained by the calorimeter. This yields for the heat lost

$$Q_L = m_s c_s \Delta T = m_s c_s (T_s - T_2) \tag{2.25}$$

where

Q_L = heat lost by the sample
m_s = mass of the sample
T_s = original temperature of the sample
T_2 = final temperature of sample, water, and calorimeter

The heat gained by the water and calorimeter is

$$Q_G = m_w (T_2 - T_1) + m_c c_c (T_2 - T_1) \tag{2.26}$$

where

Q_G = heat gained
m_w = mass of water in the calorimeter
T_1 = initial temperature of the water and calorimeter
m_c = mass of calorimeter
c_c = specific heat of the calorimeter

Setting the heat lost equal to the heat gained and factoring out the common term $(T_2 - T_1)$ in the expression for heat gained gives

$$m_s c_s (T_s - T_2) = (m_w + m_c c_c)(T_2 - T_1) \tag{2.27}$$

Solving the equation for the unknown specific heat yields

$$c = \frac{(m_w + m_c c_c)(T_2 - T_1)}{m_s (T_s - T_2)} \tag{2.28}$$

The quantity $m_c c_c$ is called the "water equivalent" of the calorimeter. This quantity is usually specified by the manufacturer or marked on the calorimeter. The water equivalent is added to the mass of the water.

In a similar way, the calorimeter is used to measure the latent heat of a substance, for example, the latent heat of solidification when a substance such as lead solidifies in a calorimeter. Heat lost during the freezing process is gained by the water and calorimeter. To accomplish the measurement, the unknown sample is heated above the melting point and the initial temperature (T_s) and mass (m_s) are recorded. The liquid sample is then inserted into a calorimeter containing water. The initial temperature (T_1) of the water and calorimeter is also measured. If sufficient water is present, the final temperature (T_2) will be below the freezing point of the sample. The mass of the water m_w and final temperature are also recorded.

Latent Heat   If the specific heat of the liquid sample and solidified sample is known, the latent heat of solidification (equal to the latent heat of fusion) can be found by **setting heat lost equal to heat gained.** For this case, the heat lost by the sample can be expressed as

$$Q_L = m_s c_{sl}(T_s - T_f) + m_s L_F + m_s c_{ss}(T_f - T_2) \qquad (2.29)$$

where

Q_L = heat lost by the sample
c_{sl} = specific heat of the liquid sample
L_F = latent heat of fusion
T_f = normal melting temperature
c_{ss} = specific heat of the solid sample

The corresponding heat gained by the water and calorimeter is

$$Q_G = (m_w + m_c c_c)(T_2 - T_1) \qquad (2.30)$$

If heat lost is set equal to heat gained, one obtains

$$m_s c_{sl}(T_s - T_f) + m_s L_F + m_s c_{ss}(T_f - T_2) = (m_w + m_c c_c)(T_2 - T_1) \qquad (2.31)$$

Since the only unknown is the latent heat of fusion, one can solve the equation for L_F. This gives

$$L_F = \frac{(m_w + m_c c_c)(T_2 - T_1) - m_s c_{sl}(T_s - T_f) - m_s c_{ss}(T_f - T_2)}{m_s} \qquad (2.32)$$

Example 2.17 combines all the above elements and illustrates how a calorimeter problem is solved.

EXAMPLE 2.17   Heat lost equals heat gained in a calorimeter
. . . .

Given: A 100-g sample of liquid lead at an initial temperature of 500°C has been inserted into a calorimeter containing 200 g of water at 5°C. The calorimeter has a water equivalent of 150 cal/C°. The melting point of lead is 327°C and its heat of fusion is 5.86 cal/g. The specific heat of liquid lead is 0.035 cal/g·C° and of solid lead is 0.031 cal/g·C°.

Find: The final temperature of the water.

Solution: **1.** Heat energy lost by the lead in cooling from 500°C to the melting point (327°C):

$$m_{\substack{\text{liquid} \\ \text{lead}}}\, c_{\substack{\text{liquid} \\ \text{lead}}}\,(T_i - T_f) = (100\text{ g})(0.035\text{ cal/g·C°})(500°C - 327°C)$$

$$\text{Loss (1)} = 605\text{ cal}$$

2. Heat energy lost by the lead while freezing (no change in temperature of lead occurs; that is, T remains at 327°C):

$$H_F = L_F m_s$$

$$H_F = (5.86\text{ cal/g})(100\text{ g})$$

Loss (2) = 586 cal

3. Heat energy lost by solid lead in cooling from the freezing point (327°C) to the final temperature (T_2):

$$m_{\substack{\text{solid} \\ \text{lead}}} \, c_{\substack{\text{solid} \\ \text{lead}}} (T_f - T_2) = (100 \text{ g})(0.031 \text{ cal/g·C°})(327 - T_2)$$

$$\text{Loss (3)} = (3.1 \text{ cal/C°})(327 - T_2)$$

4. Heat energy gained by water:

$$m_{\text{water}} \, c_{\text{water}} (T_2 - T_1) = (200 \text{ g})(1 \text{ cal/g·C°})(T_1 - 5°C)$$

$$\text{Gain (1)} = (200 \text{ cal/C°})(T_2 - 5°C)$$

5. Heat gain by calorimeter:

$$(\text{water equivalent})(T_2 - T_1) = (150 \text{ cal/C°})(T_2 - 5°C)$$

$$\text{Gain (2)} = (150 \text{ cal/C°})(T_2 - 5°C)$$

Since heat gain must equal heat loss, these separate factors can be added together to give an energy-balanced equation.

Heat energy gain = Heat energy loss

$$200 \text{ cal/C°} (T_2 - 5°C) + 150 \text{ cal/C°} (T_2 - 5°C)$$

$$= 605 \text{ cal} + 586 \text{ cal} + 3.1 \text{ cal/C°} (327 - T_2)$$

$$350 (T_2 - 5°C) = 1191 + 3.1 (327 - T_2)$$

$$353.1 \, T_2 = 1191 + 1014 + 1750$$

$$353.1 \, T_2 = 3955$$

$$T_2 = 11.2°C$$

Final temperature of the water is 11.2°C.

Notice that in Example 2.17 the final temperature T was found. In actual practice, the equilibrium temperature can be measured and the final equation can be used to determine the latent heat for lead.

In Example 2.17, the specific heat was assumed to be a constant, independent of temperature. This assumption is not strictly true, because specific heat does vary slightly as the temperature changes. However, at ordinary temperatures and over reasonably small temperature ranges, specific heat can be considered as constant. The error introduced by this assumption will not be great.

Another source of error involves the heat exchanged between the calorimeter and the surroundings. Therefore, a quality calorimeter should be well-insulated to minimize this loss, since heat losses may seriously affect the measurements. To reduce this error, start the experiment with the initial temperature of the water in the calorimeter below the temperature of the surroundings and end the experiment with the water temperature above the temperature of the surroundings. In such a case, the heat lost to the surroundings will approximate the heat gained from the surroundings, and the errors may almost cancel each other.

Efficiency of Systems

When the work was calculated for the various energy systems, frictional losses were neglected. For example, the work done in moving a body up an inclined plane must also include the work done against the frictional force. Since the work done against frictional forces is converted into heat, it cannot be recovered from the stored energy. For any system, the efficiency is defined as

$$\eta = \frac{W_{out}}{W_{in}} (100) \qquad (2.33)$$

where

η = percentage efficiency
W_{out} = total work output or stored
W_{in} = total work input

Examples 2.18 and 2.19 illustrate the calculation of efficiency for several systems.

EXAMPLE 2.18   Efficiency of an inclined plane
. . . .

Given: The inclined plane given earlier in Example 2.5. The frictional force is 500 N. The applied force to move the body is 1000 N.

Find: The efficiency of the inclined plane.

Solution: The work done in moving the body up the inclined plane (W_{in}) is

$$W_{in} = (F_c + F_f)$$

where

F_f = frictional force
F_c = constant applied force
W_{in} = (1000 N + 500 N)(30 m) = 45,000 J

The work stored in the body (work done against gravity) is

$$W_{out} = 30,000 \text{ J} \qquad (3000 \text{ N} \cdot 10 \text{ m} = 30,000 \text{ N} \cdot \text{m})$$

The efficiency is

$$\eta = \frac{W_{out}}{W_{in}} (100) = \frac{(30,000 \text{ J})}{(45,000 \text{ J})} (100) = 67\%$$

EXAMPLE 2.19   Efficiency of a water pump
. . . .

Given: A water pump connected to a 24-V supply for 30 sec lifts 50 kg of water to an average height of 10 m. The current in the pump circuit is 10 A.

Find: The efficiency of the system.

Solution: The amount of electrical work done (W_{in}) is

$$W_{in} = (\Delta V)(\Delta Q) = (\Delta V)It \quad \text{(Recall that } I = \Delta Q/t.)$$

$$W_{in} = (24 \text{ V})(10 \text{ A})(30 \text{ sec}) \quad (1 \text{ V·A·sec} = 1 \text{ J})$$

$$W_{in} = 7200 \text{ J}$$

The amount of energy stored in the water (W_{out}) is

$$W_{out} = (\Delta p)(\Delta V) = \rho_m g \Delta h \Delta V$$

$$m = \rho_m \Delta V \quad \text{(from definition of density)}$$

$$W_{out} = mg\Delta h$$

$$W_{out} = (50 \text{ kg})(9.8 \text{ m/sec}^2)(10 \text{ m}) = 4900 \text{ J}$$

The efficiency is

$$\eta = \frac{W_{out}}{W_{in}}(100) = \frac{(4900 \text{ J})}{(7200 \text{ J})}(100) = 68\%$$

SECTION 2.4   EXERCISES

1. State the general equation for work and explain how it applies in each of the following energy systems:

    a. Mechanical rotational   d. Electrical
    b. Mechanical translational   e. Magnetic
    c. Fluid   f. Thermal

2. A gas is in a cylinder with a piston. If the gas is compressed by doing 2000 J of work and the heat transferred from the gas is 500 J, what is the increase in internal energy of the gas? Will the temperature of the gas change?

3. A beaker of water is heated from 25°C to 75°C by the addition of 90 Btu of heat. Determine the mass of water heated.

4. Define and give examples of sensible heat and latent heat.

5. What is the specific heat of a substance if it requires 1500 cal to raise the temperature of 400 g of the substance 10 C°?

6. How many calories of heat are required to raise the temperature of 500 g of copper from 29°C to 95°C? (See Table 2.4.)

7. Show that 1150 cal is required to change 10 g of ice at −7°C to water at 32°C. Use 0.5 cal/g·C° as the specific heat of ice. Assume no heat is lost to the environment. Show that just using the equation $H = mc\Delta T$ yields a value of thermal energy less than 1150 cal. How is the rest of the heat energy used?

8. A flywheel of radius 1 m is submerged in water and brought to rest in 2 revolutions by a braking force of 100 N applied at the outer edge of the flywheel. Assuming that all the energy lost by the flywheel goes into heating the water, what will the final temperature of the water be if there are 200 g of water and the initial temperature of the water is 20°C? Recall that 1 cal = 4.184 J.

9. A water pump connected to a 24-V supply for 1 min lifts 100 kg of water to an average height of 15 m. The current in the pump circuit is 15 A. What is the efficiency of the pump?

10. An electrical heating coil is inserted into a calorimeter. Ten amperes of current at 12 V are drawn by the coil for 30 sec. The calorimeter has 300 g of water, initially at a temperature of 20°C. The water equivalent of the calorimeter and coil is 120 cal/C°. What is the final water temperature in the calorimeter after the heater is turned on?

S U M M A R Y

Work is done in a physical system whenever a forcelike quantity causes a change (displacementlike quantity) within the system. Energy is the capacity to do work and is conserved within a closed system. When a system is in equilibrium, it contains energy, but no net work is being done, because no displacement is occurring as the result of net forcelike quantities within the system. When a system is not in equilibrium, displacement of some kind occurs as the result of net forcelike quantities, and energy is converted to work or work is converted to energy within the system. Work and energy are equivalent and are measured in the same units. Table 2.6 shows the general work equation and the form it takes in mechanical, fluid, and electromagnetic systems. The equation for heat energy required for a temperature increase also is included, although it is not derived from the general work equation.

Table 2.6 Work as a unifying concept

Energy System	Work = Forcelike Quantity × Displacementlike Quantity
Mechanical	
Translational	Work = Average Force × Displacement (in direction of force)
Rotational	Work = Average Torque × Angular Displacement
Fluid	Work = Average Pressure Difference × Volume Change
Electrical	Work = Average Potential Difference × Charge Transferred
Magnetic	Work = Average Magnetomotive Force × Magnetic Flux Established
Thermal	Heat Energy = Temperature Change × Heat Capacity

Rate

If a car travels 120 miles in 3 hours, it travels an average of 40 miles every hour. The car's average speed is 40 miles per hour. Forty miles per hour is referred to as speed or the rate of motion of the car. Both distance traveled and time required are used to determine rate of motion. Rate is a quantity that describes how rapidly an occurrence takes place.

This chapter presents the concept of rate as it applies to mechanical, fluid, electromagnetic, and thermal systems. In each case, rate expresses the ratio of a displacementlike quantity to an elapsed time.

- In a linear mechanical system, velocity (or speed) is the ratio of distance traveled to elapsed time.

- In a rotational mechanical system, angular velocity (or speed) is the ratio of angle of rotation traveled to elapsed time.

- In a fluid system, volume flow rate is the ratio of volume of fluid moved to elapsed time; mass flow rate is the ratio of fluid mass moved to elapsed time.

- In an electrical system, current is the ratio of charge transferred to elapsed time.

- In a magnetic system, the induced voltage is proportional to the change in magnetic flux to the elapsed time.

- In a thermal system, heat flow rate is the ratio of thermal energy transferred to elapsed time.

Chapter 1 described forcelike quantities as those physical quantities that cause motion in each of the four energy systems. Chapter 2 described work done in an energy system in terms of changes that occur in that system as the result of the action of the forcelike quantities present. A useful description of the effects of forcelike quantities in a system and work done in that system requires that the time necessary for the changes to occur be considered.

Movement is an occurrence common to every energy system. The **rate** of movement in each system can be determined by **dividing** the **displacementlike quantity** by the **elapsed time** in which the displacement takes place. The **unifying formula** used for calculating rate is

$$\text{Rate} = \frac{\text{Change in Displacementlike Quantity}}{\text{Elapsed Time}} \tag{3.1}$$

This chapter discusses the application of this general rate equation in mechanical, fluid, electromagnetic, and thermal systems. It presents the correct mathematical formulation and units for rate equations in the four different energy systems.

3.1 .... OBJECTIVES ....

RATE IN MECHANICAL SYSTEMS

Upon completion of this section, the student should be able to

- Define the following rates and, where applicable, express their basic units in the SI and English systems of units:
  Average speed and velocity
  Instantaneous velocity
  Average acceleration
  Instantaneous acceleration
  Average angular velocity and acceleration
  Instantaneous angular velocity and acceleration

- In a linear mechanical system with constant acceleration, given all the quantities except one in each of the following groups, determine the unknown quantity:
  Final position, initial position, elapsed time, initial velocity, and acceleration
  Initial velocity, final velocity, elapsed time, and acceleration

Initial velocity, final velocity, initial position, final position, and acceleration
Mass, force, and acceleration

- In a rotational mechanical system with constant angular acceleration, given all the quantities except one in each of the following groups, determine the unknown quantity:

  Initial angular displacement, final angular displacement, elapsed time, initial angular velocity, and angular acceleration

  Initial angular velocity, final angular velocity, elapsed time, and angular acceleration

  Initial angular velocity, final angular velocity, initial angular displacement, final angular displacement, and angular acceleration

- Calculate the radial acceleration and centripetal force for uniform circular motion.

DISCUSSION

Rate in Translational Systems

Speed   A fundamental rate in a translational system is the average speed. Average speed can be calculated by dividing the total path traveled by the time required to move along the path. This definition for average speed is expressed mathematically

$$v_{av} = \frac{\Delta s}{\Delta t} \tag{3.2}$$

where

$v_{av} = \bar{v}$ = average speed
Δs = total path length traveled
Δt = corresponding elapsed time

The average speed does not have a unique direction. Thus, it is a **scalar** quantity.

Velocity   It is often more useful to calculate a quantity called the average velocity between two points. The average velocity is defined by the equation

$$\mathbf{v}_{av} = \frac{\Delta \mathbf{d}}{\Delta t} \tag{3.3}$$

where

$\mathbf{v}_{av} = \bar{\mathbf{v}}$ = average velocity
$\Delta \mathbf{d}$ = vector displacement between the two points
Δt = corresponding time

The **vector displacement** is the **straight line** drawn from the initial starting point to some final point, as illustrated in Figure 3.1.

Another possible path is also shown in Figure 3.1. From this figure, one can see that the path length will always be greater than or equal to the magnitude of $\Delta \mathbf{d}$.

FIGURE 3.1 Vector
displacement Δ**d**.

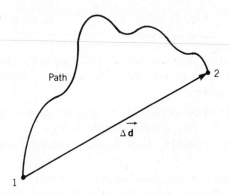

FIGURE 3.1 Vector
displacement Δ**d**.

Thus, the average speed will be greater than or equal to the magnitude of the average velocity. The average velocity is a vector quantity that has the same direction as the displacement vector Δ**d.**

Example 3.1 illustrates the use of Equations 3.2 and 3.3.

EXAMPLE 3.1   Speed and velocity of an automobile
. . . .

Given: An automobile travels 250 ft east and then 500 ft north in 30 sec.

Find: **a.** The average speed.

              b. The average velocity.

Solution: Draw a diagram of the movement. Notice that the displacement Δ**d** is the vector sum of the two displacement vectors, 250 ft east and 500 ft north.

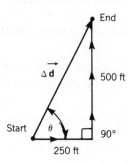

a. The average speed is the path length divided by the time.

$$v_{av} = \frac{\Delta s}{\Delta t} = \frac{250 \text{ ft} + 500 \text{ ft}}{30 \text{ sec}} = 25 \text{ ft/sec}$$

b. The average velocity is the displacement vector Δ**d** divided by the elapsed time.

$$\mathbf{v}_{av} = \frac{\Delta \mathbf{d}}{\Delta t}$$

$$\Delta \mathbf{d} = \sqrt{(250)^2 + (500)^2} = 559 \text{ ft}$$

$$\tan \theta = \frac{500}{250}$$

$$\theta = 63.4°$$

$$|\mathbf{v}_{av}| = \frac{559}{30} = 18.6 \text{ ft/sec} \quad \text{(magnitude of the velocity)}$$

$$\mathbf{v}_{av} = 18.6 \text{ ft/sec at } 63.4° \text{ north of east}$$

Path length and displacement are measured in units such as feet, miles, or meters and time is measured in units such as seconds, minutes, or hours. Common units for speed are summarized in Table 3.1.

Example 3.2 illustrates the use of Equation 3.2 to determine elapsed time if the speed and distance are known.

EXAMPLE 3.2   Time to run a mile
. . . .

Given: A runner can average 13.34 mi/hr.

Find: The time required to run one mile.

Solution: From Equation 3.2

$$\Delta t = \frac{\Delta s}{v_{av}}$$

$$\Delta t = \frac{1 \text{ mi}}{13.34 \text{ mi/hr}} = 0.0750 \text{ hr}$$

$$\Delta t = 0.0750 \text{ hr} \times 60 \text{ min/hr}$$

$$\Delta t = 4.5 \text{ min}$$

TABLE 3.1   Common units for speed

System of Measure	Common Units for Speed
English	ft/sec or mi/hr (44 ft/sec = 30 mi/hr)
SI	m/sec or km/hr (0.278 m/sec = 1 km/hr)

Linear Motion   When an object moves in a straight line, it is said to have linear motion. The object may stop or reverse its direction, but its motion always occurs along the same straight line. For this case, it is advantageous to plot the magnitude of the displacement vector **d** as a function of time as shown in Figure 3.2. This curve can be used to define the instantaneous speed of a body in linear motion by calculating the average speed for smaller and smaller time intervals as described in the following paragraph.

The magnitude of the average velocity between points 1 and 2 is

$$\bar{v}_2 = \frac{\Delta d_2}{\Delta t_2} = \text{slope of line 1-2}$$

Note: In accordance with common custom, $v_{av} = \bar{v}$. Thus $(v_2)_{av} = \bar{v}_2$.

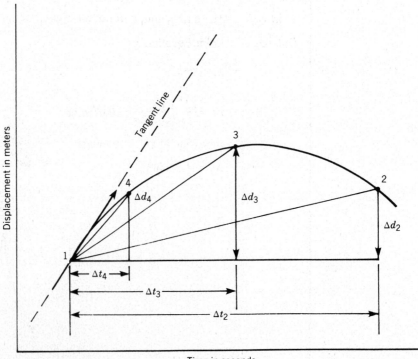

FIGURE 3.2   Displacement *d* versus time *t*.

This may be considered as an approximation to the magnitude of the instantaneous velocity at point 1. If a better approximation is wanted, then point 3 may be used. This yields

$$\bar{v}_3 = \frac{\Delta d_3}{\Delta t_3} = \text{slope of line 1--3}$$

A still better approximation is obtained if point 4 is used to give

$$\bar{v}_4 = \frac{\Delta d_4}{\Delta t} = \text{slope of line 1--4}$$

One can see that the line drawn between the first point and a different point on the curved path approaches a tangent line as the different points approach the first point. The corresponding speed approaches the slope of the tangent line. If one uses the slope of the tangent line to calculate the speed, it is called the instantaneous speed at the point where the tangent line intersects the curve. If the speed is constant, the curve is a straight line. In that case, the average speed is equal to the instantaneous speed.

When the speed is not constant, it is necessary to introduce a quantity called the acceleration. The average acceleration for linear motion is

$$\bar{a} = \frac{v_f - v_i}{\Delta t} = \frac{\Delta v}{\Delta t} \tag{3.4}$$

where

$\bar{a}$ = average linear acceleration
v_i = initial speed
v_f = final speed
Δt = time in which Δv takes place

The units of acceleration are velocity units (m/sec, ft/sec, mi/hr) divided by time units. The most commonly used acceleration units are ft/sec² and m/sec². An acceleration of 1 m/sec² increases the velocity by 1 m/sec in 1 sec.

Example 3.3 shows the use of Equation 3.4 in determining acceleration.

EXAMPLE 3.3 Acceleration of a car
. . . .

Given: A driver increases the speed of a car uniformly from 35 miles per hour to 55 miles per hour in 15 sec.

Find: The acceleration of the car.

Solution: $\bar{a} = \dfrac{v_f - v_i}{\Delta t}$

$\bar{a} = \dfrac{55 \text{ mi/hr} - 35 \text{ mi/hr}}{15 \text{ sec}}$

$\bar{a} = \dfrac{20 \text{ mi/hr}}{15 \text{ sec}}$

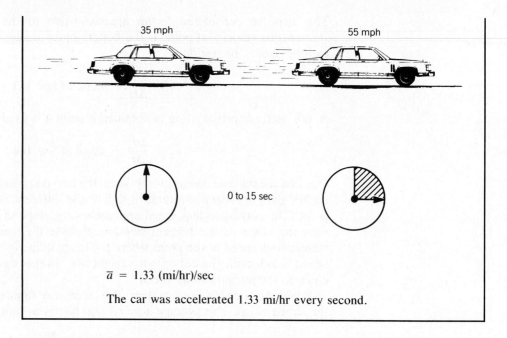

35 mph   55 mph

0 to 15 sec

$$\overline{a} = 1.33 \ (\text{mi/hr})/\text{sec}$$

The car was accelerated 1.33 mi/hr every second.

EXAMPLE 3.4   Alternate solution for example 3.3
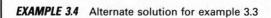

Given: A driver increases the speed of a car uniformly from 35 miles per hour to 55 miles per hour in 15 sec.

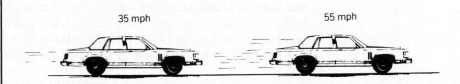

35 mph   55 mph

          0 to 15 sec

Find: The acceleration of the car.

Solution: $\overline{a} = \dfrac{v_\text{f} - v_\text{i}}{\Delta t}$

$$\overline{a} = \frac{(55 - 35) \text{ mi/hr} \left(\dfrac{44 \text{ ft/sec}}{30 \text{ mi/hr}}\right)}{15 \text{ sec}}$$

$$\overline{a} = \frac{29.3 \text{ ft/sec}}{15 \text{ sec}}$$

$$\overline{a} = 1.96 \text{ (ft/sec)/sec} = 1.96 \text{ ft/sec}^2$$

The units ft/sec $\div$ sec $=$ ft/sec^2

As was done in the case of speed, an instantaneous acceleration can be defined. Consider the plot of speed versus time shown in Figure 3.3.

The average acceleration between points 1 and 2 is

$$\overline{a}_2 = \frac{v_2 - v_1}{t_2 - t_1} = \frac{\Delta v_2}{\Delta t_2} = \text{slope of line 1-2}$$

The value of $\overline{a}_2$ is a poor approximation of the instantaneous acceleration. However, the average acceleration between points 1 and 3 is a better approximation:

$$\overline{a}_3 = \frac{v_3 - v_1}{t_3 - t_1} = \frac{\Delta v_3}{\Delta t_3} = \text{slope of line 1-3}$$

Further decrease of the elapsed time gives an approximation that is even better:

$$\overline{a}_4 = \frac{v_4 - v_1}{t_4 - t_1} = \frac{\Delta v_4}{\Delta t_4} = \text{slope of line 1-4}$$

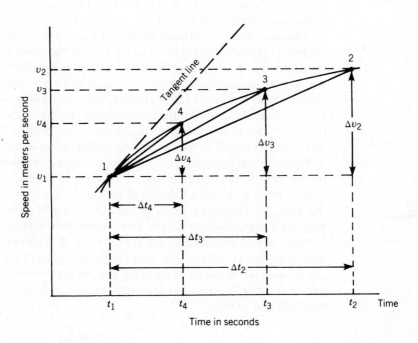

FIGURE 3.3 Speed versus time.

FIGURE 3.4 The effect of constant and nonuniform acceleration.

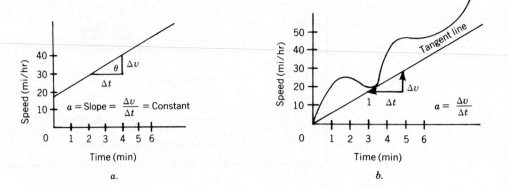

a.

b.

Thus, one can see that the approximation approaches the slope of the tangent line as a point is taken closer and closer to the initial point. The slope of the tangent line is the instantaneous linear acceleration.

Acceleration is also illustrated in Figure 3.4, where speed is plotted versus time. A constant linear acceleration is shown in Figure 3.4a. The straight line has a constant slope that is equal to the linear acceleration. A nonuniform acceleration is illustrated in Figure 3.4b. The determination of the instantaneous acceleration at any point requires construction of a tangent line at the point and measuring its slope as the figure illustrates.

The concepts of speed and acceleration are illustrated further in Figure 3.5, where the changing speed of a car during a short trip is shown. Section A of the graph shows the acceleration of the car as it goes from a stop to a speed of approximately 10 mi/hr. Since the graph rises very rapidly the rate of increase in speed (acceleration) is large. The car is in low gear. Section B represents a slowing down (deceleration) of the car as it changes from low gear to second gear. Section C represents the acceleration as the car increases its speed while in second gear. Section D represents another deceleration as the gears are shifted from second to third. Section E at first represents an increasing speed or an acceleration. But notice that acceleration during section E does not remain constant. The graph clearly shows a period of high acceleration followed by a period of lower acceleration. Section F shows a rapid deceleration, which is due to a braking action from the driver. Section G shows no acceleration, because the speed is constant during this time. Section H represents a final deceleration, which brings the car to a stop.

Figure 3.5 includes both accelerations and decelerations. "Acceleration" is the term generally applied to a rate of increase in speed, and "deceleration" is the term applied to a rate of decrease in speed. If a moving object is decelerating, the value for change of speed in Equation 3.4 will be negative, giving an overall negative value for acceleration. Thus, a deceleration is a "negative acceleration."

From a plot of acceleration versus time, the velocity can be determined by measuring the corresponding area under the curve. This problem is similar to the one of calculating the work from the area under the curve of the forcelike quantity versus the displacementlike quantity. From the definition of acceleration (Equation 3.4) one obtains

$$\Delta v_n = \overline{a}_n \Delta t_n \qquad (3.5)$$

FIGURE 3.5 Plot of speed versus time for a short trip.

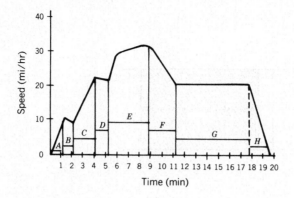

where

Δt_n = a very small change in time
Δv_n = corresponding change in speed
$\overline{a}_n$ = average acceleration

The time axis, as shown in Figure 3.6, is divided into a large number of time intervals (Δt_n's). For each Δt_n, the area of the rectangle $a_n \Delta t_n$ is approximately equal to the area under the curve. Since each small area approximates Δv_n, one may write

$$v_f - v_i = \Sigma \Delta v_n = \Sigma a_n \Delta t_n = \Sigma(\text{small areas}) = \text{area under curve}$$

where

v_f = final velocity
v_i = initial velocity

FIGURE 3.6 Determination of change in velocity from a plot of acceleration versus time.

FIGURE 3.7 Acceleration versus time for a constant acceleration.

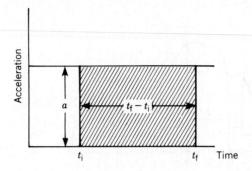

For constant acceleration, as shown in Figure 3.7, the area is a rectangle. Thus, the total change in velocity between the times t_f and t_i is

$$v_f - v_i = \text{area} = a\,(t_f - t_i)$$
$$v_f = v_i + a\,(t_f - t_i) \tag{3.6}$$

For many applications, it is convenient to let the initial time, t_i, be zero. Equation 3.6 then becomes

$$v_f = v_i + at \tag{3.7}$$

where

$t = $ final time

Examples 3.5, 3.6, and 3.7 show the application of Equation 3.7 in solving problems.

EXAMPLE 3.5   Acceleration of an automobile
. . . .

Given: An automobile accelerates from 15 m/sec to 18 m/sec in 10 sec.

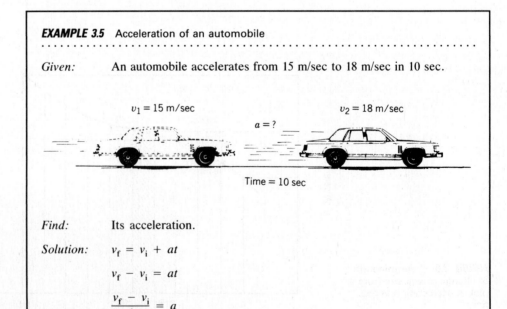

Find: Its acceleration.

Solution: $v_f = v_i + at$

$v_f - v_i = at$

$\dfrac{v_f - v_i}{t} = a$

$$a = \frac{v_f - v_i}{t}$$

$$a = \frac{18 \text{ m/sec} - 15 \text{ m/sec}}{10 \text{ sec}} = \frac{3 \text{ m/sec}}{10 \text{ sec}}$$

$$a = 0.3 \text{ m/sec}^2$$

EXAMPLE 3.6   Final speed of a rocket
. . . .

Given: A rocket traveling at 8000 m/sec accelerates at 10 m/sec^2 for 3 min.

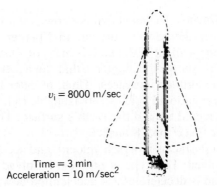

$v_i = 8000$ m/sec

Time = 3 min
Acceleration = 10 m/sec^2

$v_f = ?$

Find: Its new speed.

Solution: $v_f = v_i + at$

$v_f = 8000 \text{ m/sec} + (10 \text{ m/sec}^2)(180 \text{ sec})$

$v_f = 8000 \text{ m/sec} + 1800 \text{ m/sec}$

$v_f = 9800 \text{ m/sec}$

After 3 min the speed is 9800 m/sec.

EXAMPLE 3.7   Time required to reach a given speed
. . . .

Given: An object is accelerating at 0.5 m/sec^2.

Find: Time required to produce a speed of 27 m/sec, beginning from rest.

Solution: $v_f = v_i + at$

$v_f - v_i = at$

$\dfrac{v_f - v_i}{a} = t$

$t = \dfrac{v_f - v_i}{a}$

$t = \dfrac{27 \text{ m/sec} - 0 \text{ m/sec}}{0.5 \text{ m/sec}^2} = \dfrac{27 \text{ m/sec}}{0.5 \text{ m/sec}^2}$

$t = 54 \text{ sec}$

The object has accelerated to 27 m/sec in 54 sec.

Gravity   The most familiar constant acceleration is the acceleration due to gravity near the earth's surface. In Chapter 1, it was shown that the acceleration is independent of the body's mass or shape if air resistance can be neglected. It was also shown in Chapter 1 that the acceleration due to gravity slowly decreases as the altitude increases. Thus, in calculating the velocity of bodies in the earth's gravitational field, it is reasonable to assume that the acceleration is constant and directed toward the earth's surface. The value used in solving these problems is 32 ft/sec^2 or 9.8 m/sec^2.

Problems in which velocity and acceleration are in opposite directions must be solved. This type of problem is solved in Example 3.8. A ball is thrown upward and is decelerated (or accelerated downward) by the pull of gravity until it stops its upward motion and begins to fall with increasing speed toward the earth's surface.

EXAMPLE 3.8   Speed of a ball thrown upward
. . . .

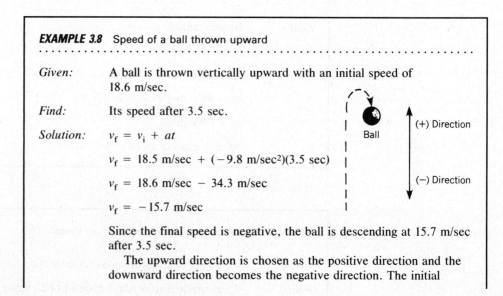

Given: A ball is thrown vertically upward with an initial speed of 18.6 m/sec.

Find: Its speed after 3.5 sec.

Solution: $v_f = v_i + at$

$v_f = 18.5 \text{ m/sec} + (-9.8 \text{ m/sec}^2)(3.5 \text{ sec})$

$v_f = 18.6 \text{ m/sec} - 34.3 \text{ m/sec}$

$v_f = -15.7 \text{ m/sec}$

Since the final speed is negative, the ball is descending at 15.7 m/sec after 3.5 sec.

The upward direction is chosen as the positive direction and the downward direction becomes the negative direction. The initial speed is upward (positive) and the acceleration downward (negative). Since the final speed is negative, the ball travels downward. The direction chosen as positive is unimportant because the same sign convention is maintained throughout the solution.

Determination of Position When the speed as a function of time for the linear motion of a body and its initial speed and position are known, it becomes possible to determine the position of the body at any other time. Consider the plot of velocity versus time shown in Figure 3.8. The time axis is again divided into many very small time intervals (Δt_n's). The area of each corresponding rectangle is $v_n \Delta t_n$, where v_n is the instantaneous speed. From the definition of instantaneous speed,

$$\Delta s_n = v_n \Delta t_n \tag{3.8}$$

where

Δt_n = a very small interval of time
Δs_n = corresponding small change in distance
v_n = instantaneous speed

It can be seen that each small area represents a small change in displacement. The area also approaches the area under the curve, if Δt_n is very small. The total displacement becomes

$$s_f - s_i = \Sigma \Delta s_n = \Sigma v_n \Delta t_n = \Sigma(\text{small areas}) = \text{area under the curve}$$

where

s_f = final displacement
s_i = initial displacement from the origin

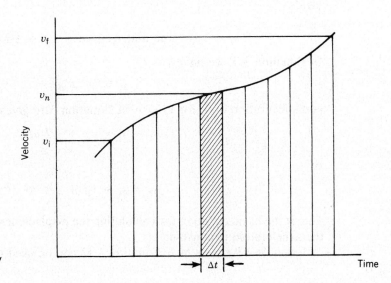

FIGURE 3.8 Plot of velocity versus time.

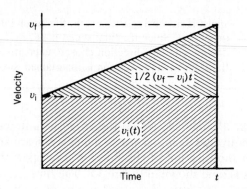

If the velocity is constant, one has the area of a rectangle as in the case for constant acceleration; and

$$s_f - s_i = v\,(t_f - t_i)$$

If the initial time is set equal to zero, then the equation reduces to

$$s_f = s_i + vt \tag{3.9}$$

where

t = final time

Equation 3.9 has the same form as the equation for the velocity when the acceleration is constant (Equation 3.7).

A more useful case occurs when the acceleration is constant. Using Equation 3.7 to plot the velocity as a function of time yields the curve shown in Figure 3.9, where the slope is the acceleration. The distance traveled is the area under the curve. The total area is the area of the rectangle plus the area of the triangle. This yields

$$s_f - s_i = \text{area under curve} = v_i t + 1/2\,(v_f - v_i)\,t \tag{3.10}$$

In Equation 3.7, we have

$$v_f - v_i = at$$

and substitution of at for $v_f - v_i$ in Equation 3.10 gives

$$s_f - s_i = v_i t + 1/2\,at^2 \tag{3.11}$$

or

$$s_f = s_i + v_i t + 1/2\,at^2 \tag{3.12}$$

This is the basic equation for calculating the displacement for linear motion when the acceleration is constant.

Example 3.9 shows how Equation 3.12 can be used to find a displacement.

EXAMPLE 3.9 Displacement at constant acceleration
. . . .

Given: An object with an initial position of 10 m, an initial speed of
             2 m/sec, and an acceleration of 1 m/sec² moves in a straight line.

Find: Its position after 5 sec.

Solution: $s_f = s_i + v_i t + 1/2\ at^2$

             $s_f = 10\text{ m} + (2\text{ m/sec})(5\text{ sec}) + 1/2\ (1\text{ m/sec}^2)(5\text{ sec})^2$

             $s_f = 10\text{ m} + 10\text{ m} + 12.5\text{ m}$

             $s_f = 32.5\text{ m}$

There are three terms in Equation 3.12. The first term, s_i, is the initial position. The second term, $v_i t$, is the displacement during time, t, that would occur if the initial speed, v_i, remained constant. The third term, $1/2\ at^2$, is the additional displacement due to a speed change from a constant acceleration, a. Displacement at any time, t, is the sum of these three terms at that time. The graph of the $v_i t$ term is a straight line, and that of the $1/2\ at^2$ term is curved (parabolic). The curve representing the sum is also curved (again parabolic).

In Figure 3.11, sample calculations are shown for the terms in Equation 3.12. Each term is calculated at the end of each second over a 5-sec span. The body moves with an initial velocity of 2 m/sec and a constant acceleration of 1 m/sec².

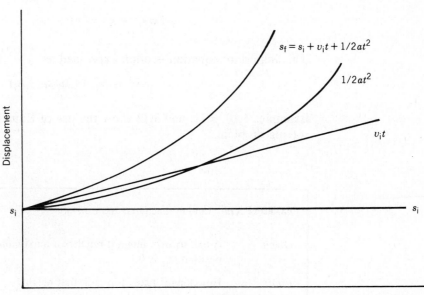

FIGURE 3.10 Plot of displacement versus time.

FIGURE 3.11 Calculation of displacement for various times.

t	s_i	$v_i t$	$1/2\ at^2$	s_f
0	10	0	0	10
1	10	2	0.5	12.5
2	10	4	2	16
3	10	6	4.5	20.5
4	10	8	8	26
5	10	10	12.5	32.5

$$s_i = 10 \text{ m}$$

$$v_i = 2 \text{ m/sec}$$

$$a = 1 \text{ m/sec}^2$$

Adding like terms in Equation 3.10 gives

$$s_f - s_i = \left(\frac{v_f + v_i}{2}\right) t \qquad (3.13)$$

Equation 3.7 can be solved for t to yield

$$t = \frac{v_f - v_i}{a} \qquad (3.14)$$

Substituting this expression for t in Equation 3.13 gives

$$s_f - s_i = \frac{(v_f + v_i)(v_f - v_i)}{2a}$$

$$2a(s_f - s_i) = v_f^2 - v_i^2 \qquad (3.15)$$

This important equation is often expressed as

$$v_f^2 = v_i^2 + 2a(s_f - s_i) \qquad (3.16)$$

Examples 3.10, 3.11, and 3.12 show the use of Equations 3.12, 3.14, and 3.15 in solving problems.

EXAMPLE 3.10   Time to reach maximum height
....

Given: A ball thrown upward reaches a maximum height of 21 m. The initial position, s_i, is 0.

Find: How long it takes to reach that height.

Solution: **a.** Known:

$$s_f = 21 \text{ m}$$

$$v_f = 0 \text{ m/sec}$$

$$a = -9.8 \text{ m/sec}^2 \qquad \text{(Upward was chosen as the positive direction.)}$$

$$v_f^2 = v_i^2 + 2\, as_f$$

$$v_i^2 = v_f^2 - 2\, as_f$$

$$v_i^2 = (0 \text{ m/sec})^2 - 2\,(-9.8 \text{ m/sec}^2)(21 \text{ m})$$

$$v_i^2 = 411.6 \text{ m}^2/\text{sec}^2$$

$$v_i = 20.29 \text{ m/sec}$$

The ball is thrown upward at an initial velocity of 20.29 m/sec.

$$v_f = v_i + at$$

$$t = \frac{v_f - v_i}{a}$$

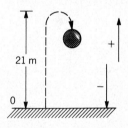

$$t = \frac{0 \text{ m/sec} - 20.29 \text{ m/sec}}{-9.8 \text{ m/sec}^2}$$

$$t = 2.07 \text{ sec}$$

It takes 2.07 sec to reach a height of 21 m.

b. Assume that the ball requires the same time to travel the 21 m in a downward direction as it does going upward. Therefore, starting at the top, $v_i = 0$ m/sec. The initial position is 21 m.

$$s_f = s_i + v_i t + 1/2\, at^2$$

$$0 = 21 \text{ m} + 0.0 \text{ m} + 1/2\,(-9.8 \text{ m/sec}^2)t^2$$

$$t^2 = \frac{(-21 \text{ m})(2)}{-9.8 \text{ m/sec}^2}$$

$$t = \sqrt{4.29 \text{ sec}^2}$$

$$t = 2.07 \text{ sec}$$

EXAMPLE 3.11   Acceleration and speed of an automobile
. . . .

Given: An automobile undergoes a constant acceleration from an initial speed of 25 m/sec for a period of 10 sec and travels 500 m in that time. The initial position is at the origin ($s_i = 0$).

Find: **a.** The acceleration of the car.
             b. The final speed.

Solution: **a.** $s_f = v_i t + 1/2\, at^2$;   solve for a

$$a = \frac{2s_f}{t^2} - \frac{2v_i}{t}$$

$$a = \frac{2(500 \text{ m})}{(10 \text{ sec})^2} - \frac{2(25 \text{ m/sec})}{10 \text{ sec}}$$

$$a = \frac{1000 \text{ m}}{100 \text{ sec}^2} - \frac{50 \text{ m/sec}}{10 \text{ sec}}$$

$$a = 10 \text{ m/sec}^2 - 5 \text{ m/sec}^2$$

$$a = 5 \text{ m/sec}^2$$

b. $v_f = v_i + at$

$$v_f = 25 \text{ m/sec} + (5 \text{ m/sec}^2)(10 \text{ sec})$$

$$v_f = 25 \text{ m/sec} + 50 \text{ m/sec}$$

$$v_f = 75 \text{ m/sec}$$

EXAMPLE 3.12   Final speed of automobile
. . . .

Given: An automobile with an initial speed of 10 m/sec experiences an acceleration of 1.1 m/sec² for a distance of 20 m.

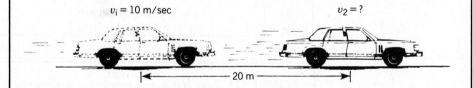

$v_i = 10$ m/sec $v_2 = ?$

|←————— 20 m —————→|

Find: Its final speed.

Solution: $v_f^2 = v_i^2 + 2a(s_f - s_i)$

$$v_f^2 = (10 \text{ m/sec})^2 + 2(1.1 \text{ m/sec}^2)(20 \text{ m} - 0 \text{ m})$$

$$v_f^2 = 100 \text{ m}^2/\text{sec}^2 + 44 \text{ m}^2/\text{sec}^2$$

$$v_f^2 = 144 \text{ m}^2/\text{sec}^2$$

$$v_f^2 = \sqrt{144 \text{ m}^2/\text{sec}^2}$$

$$v_f = 12 \text{ m/sec}$$

The car reaches a final speed of 12 m/sec.

Acceleration from Newton's Law As stated in Chapter 1, acceleration is produced by an unbalanced force according to Newton's second law of motion, Newton's second law may be expressed as

$$\mathbf{a} = \frac{\mathbf{F}}{m} \tag{3.17}$$

where

$\mathbf{a}$ = vector acceleration
$\mathbf{F}$ = applied force
m = mass of the body

Equation 3.17 shows that the acceleration is in the same direction as the applied force, its magnitude is directly proportional to the magnitude of the force, and it is inversely proportional to the mass.

To apply Newton's second law, force and mass must be in correct units. For example, when mass is measured in kilograms and force is measured in newtons, the acceleration will be m/sec². If the mass is measured in slugs and the force in pounds, the acceleration will be in ft/sec². The use of Newton's second law is illustrated in Examples 3.13, 3.14, and 3.15.

EXAMPLE 3.13 Acceleration of a mass
. . . .

Given: An unbalanced force of 35 N is applied to a 7-kg object.

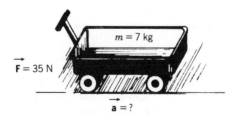

Find: The acceleration produced (neglect friction.)

Solution: $\mathbf{a} = \dfrac{\mathbf{F}}{m}$

$\mathbf{a} = \dfrac{35 \text{ N}}{7 \text{ kg}} = \dfrac{35 \text{ kg·m/sec}^2}{7 \text{ kg}}$

$\mathbf{a} = 5 \text{ m/sec}^2$

EXAMPLE 3.14   Truck deceleration
. . . .

Given: A 5000-kg truck has a speed of 20 m/sec.

Find: The braking force required to stop in 60 sec.

Solution: First, find the rate of deceleration.

$$\mathbf{a} = \frac{v_f - v_i}{t}$$

$$\mathbf{a} = \frac{0 - 20 \text{ m/sec}}{60 \text{ sec}}$$

$$\mathbf{a} = \frac{-20 \text{ m/sec } (1/\text{sec})}{60 \text{ sec}/1 \ (1/\text{sec})}$$

$$\mathbf{a} = -0.33 \text{ m/sec}^2$$

Now, find the force required.

$$\mathbf{F} = m\mathbf{a}$$

$$\mathbf{F} = (5000 \text{ kg})(-0.33 \text{ m/sec}^2)$$

$$\mathbf{F} = -1650 \text{ kg·m/sec}^2 = -1650 \text{ N}$$

A force of 1650 N stops the truck in the indicated time. The negative sign tells that the braking force was applied in the opposite direction from the motion of the truck.

EXAMPLE 3.15   Towing an automobile
. . . .

Given: A 3700-lb automobile is being towed on a level road by a tow truck accelerating at 3.8 ft/sec².

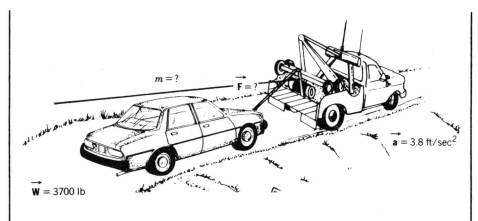

Find: Tension in the tow line (force required). Assume zero friction in turning the automobile wheels.

Solution: $m = \dfrac{3700 \text{ lb}}{32 \text{ ft/sec}^2}$

$m = 115.6 \text{ lb·sec}^2/\text{ft} = 115.6 \text{ slugs}$ (Recall that lb·sec^2/ft = slugs.)

Now, find the tension on the rope.

$\mathbf{F} = m\mathbf{a}$

$\mathbf{F} = (115.6 \text{ slugs})(3.8 \text{ ft/sec}^2)$

$\mathbf{F} = (115.6 \text{ lb·sec}^2/\text{ft})(3.8 \text{ ft/sec}^2)$

$\mathbf{F} = 439 \text{ lb}$

Tension in the tow line is 439 lb.

Motion in a Plane

Constant Acceleration   The simplest motion in a plane occurs when the acceleration is constant in magnitude and direction. To analyze this problem, consider the displacement vector, **d,** drawn from the origin of a rectangular coordinate system as shown in Figure 3.12. The displacement vector can be expressed as (definition of vector addition) a sum of component vectors:

$$\mathbf{d} = \mathbf{x} + \mathbf{y} \tag{3.18}$$

Movement along the path causes the direction as well as the magnitude of **d** to change. However, only the magnitude of its components, **x** and **y,** changes with time. By definition, the velocity of the body is

$$\mathbf{v} = \frac{\Delta \mathbf{d}}{\Delta t} = \frac{\Delta \mathbf{x}}{\Delta t} + \frac{\Delta \mathbf{y}}{\Delta t}$$

$$\mathbf{v} = \mathbf{v}_x + \mathbf{v}_y \tag{3.19}$$

FIGURE 3.12 Illustration of displacement vector, **d**₁.

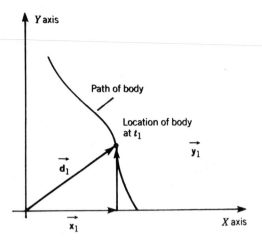

where

  v = total velocity
  $\mathbf{v}_x$ = velocity in the x direction
  $\mathbf{v}_y$ = velocity in the y direction

The definition of acceleration gives

$$\mathbf{a} = \frac{\Delta\mathbf{v}}{\Delta t} = \frac{\Delta\mathbf{v}_x}{\Delta t} + \frac{\Delta\mathbf{v}_y}{\Delta t}$$

$$\mathbf{a} = \mathbf{a}_x + \mathbf{a}_y \tag{3.20}$$

where

  a = total acceleration
  $\mathbf{a}_x$ = acceleration in the x direction
  $\mathbf{a}_y$ = acceleration in the y direction

Since the directions a_x, a_y, v_x, v_y, x, and y do not change, they may be treated as scalars as was done for linear motion. Constant values for the scalar accelerations a_x and a_y give the total acceleration a, a constant value. The equations for linear motion with constant acceleration then apply for the x direction and the y direction. This yields

$$x_f = x_i + v_{ix}t + 1/2\, a_x t^2 \tag{3.21}$$

$$v_{fx} = v_{ix} + a_x t \tag{3.22}$$

$$2\, a_x(x_f - x_i) = v_{fx}^2 + v_{ix}^2 \tag{3.23}$$

$$y_f = y_i + v_{iy}t + 1/2\, a_y t^2 \tag{3.24}$$

$$v_{fy} = v_{iy} + a_y t \tag{3.25}$$

$$2\, a_y(y_f - y_i) = v_{fy}^2 - v_{iy}^2 \tag{3.26}$$

where

v_{ix} = initial velocity in the direction of the x axis
v_{iy} = initial velocity in the direction of the y axis
v_{fx} = final velocity in the direction of the x axis
v_{fy} = final velocity in the direction of the y axis

The initial position (at $t = 0$) is indicated by the coordinates (x_i, y_i) and the final position by the coordinates (x_f, y_f).

The application of Equations 3.21 through 3.26 to practical problems is illustrated in Examples 3.16 and 3.17.

EXAMPLE 3.16   Motion of a body in a gravitational field
. . . .

Given: The initial velocity of a body in the x direction is 100 m/sec as shown in the sketch. The initial velocity in the y direction is 0.0 m/sec. The body falls a distance of 10 m in the y direction with a constant acceleration of -9.8 m/sec². The acceleration in the x direction is 0.0 m/sec².

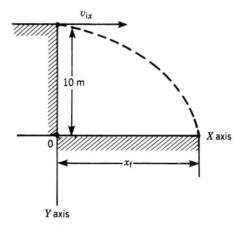

Find: The distance traveled in the x direction.

Solution: Find the time required to fall 10 m.

$$y_f = y_i + v_{iy}t + 1/2\, a_y t^2$$

$$0.0\ m = 10\ m + (0.0\ m)t + 1/2\,(-9.8\ m/sec^2)t^2$$

$$t^2 = \frac{(2)(-10\ m)}{-9.8\ m/sec^2} = 2.04\ sec^2$$

$$t = 1.43\ sec$$

Find the distance traveled in the x direction in 1.43 sec.

$$x_f = x_i + v_{ix}t + 1/2\ a_xt^2$$

$$x_f = 0.0\ \text{m} + (100\ \text{m/sec})(1.43\ \text{sec}) + 1/2\ (0.0\ \text{m})(1.43\ \text{sec})$$

$$x_f = 0.0\ \text{m} + 143\ \text{m} + 0.0\ \text{m}$$

$$x_f = 143\ \text{m}$$

If x_f is measured, v_{ix} can be calculated. This provides an easy way to measure velocity in the laboratory.

EXAMPLE 3.17   Motion of an electron in an electric field
. . . .

Given: An electron passes through a parallel-plate capacitor as shown in the following sketch. The initial velocity of 10^5 m/sec is parallel to the plates. The voltage applied is 1000 V, the length of the plates is 0.05 m, and separation is 0.02 m. Assume no fringing of the electric field at the ends of the plates.

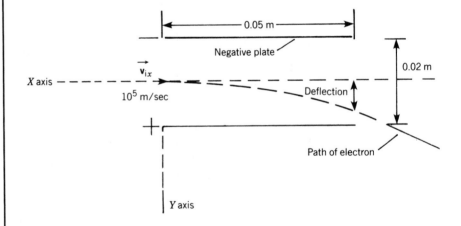

Find: The deflection of the beam at the end of the plates and the direction of the velocity.

Solution: The force on the electron is (Chapter 1)

$$\mathbf{F}_y = \mathbf{E}q_e \qquad (q_e \text{ is the electronic charge})$$

Newton's second law is

$$\mathbf{F}_y = m_e\mathbf{a}_y \qquad (m_e \text{ is the electronic mass})$$

Therefore,

$$m_e\mathbf{a}_y = \mathbf{E}q_e$$

$$\mathbf{a}_y = \mathbf{E}\left(\frac{q_e}{m_e}\right)$$

The definition of voltage in Chapter 2, $V = E \, \Delta l$, yields $E = V/\Delta l$. Substituting for E gives (q_e/m_e is 1.76×10^5 C/kg)

$$a_y = \frac{V}{\Delta l}\left(\frac{q_e}{m_e}\right) = \left(\frac{1000 \text{ V}}{0.02 \text{ m}}\right)(1.76 \times 10^5 \text{ C/}kg)$$

$$a_y = 8.8 \times 10^9 \text{ m/sec}^2$$

Thus,

$$y_f = y_i + v_{iy} + 1/2 \, a_y t^2$$

$$y_f = 0 + 0 + 1/2 \, (8.8 \times 10^9 \text{ m/sec}^2) t^2$$

$$y_f = (4.4 \times 10^9 \text{ m/sec}^2) t^2$$

The value of t can be found from the equation

$$x_f = x_i + v_{ix} t + 1/2 \, a_x t^2$$

$$0.05 \text{ m} = 0 + (10^5 \text{ m/sec}) t + 0$$

$$t = \frac{(5 \times 10^{-2} \text{ m})}{(10^5 \text{ m/sec})} = 5 \times 10^{-7} \text{ sec}$$

Thus,

$$y_f = (4.4 \times 10^9 \text{ m/sec}^2)(5 \times 10^{-7} \text{ sec})^2$$

$$y_f = 1.1 \times 10^{-3} \text{ m}$$

Thus, the displacement at the end of the plate is 1.1 mm. The velocity in the x direction is always 10^5 m/sec and the velocity in the y direction is

$$v_{fy} = v_{iy} + a_y t$$

$$v_{fy} = 0 + (8.8 \times 10^9 \text{ m/sec}^2)(5 \times 10^{-7} \text{ sec})$$

$$v_{fy} = 4.4 \times 10^3 \text{ m/sec}$$

From the definition of vector addition

$$\tan \theta = \frac{v_y}{v_x} = \frac{4.4 \times 10^3}{10^5}$$

$$\theta = \text{Arctan} \frac{4.4 \times 10^3}{10^5} = 2.5°$$

In practice, one has a beam of electrons that is deflected in a cathode-ray tube (see Chapter 1). Following the deflection, the beam strikes a screen that emits light where the beam intersects the screen. In Example 3.17, the deflection on the screen would be much larger than 1.1 mm. Suppose the screen is 30 cm from the end of the plate. Then from Figure 3.13 the deflection on the screen becomes

$$d = l_1 \tan \theta + y_f = 30 \text{ cm} \tan 2.5° + 0.11 \text{ cm}$$

$$d = 1.42 \text{ cm}$$

FIGURE 3.13 Deflection of electron beam in a cathode-ray tube.

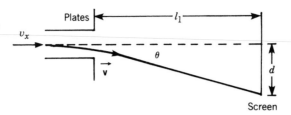

Mechanical Rotational Rates

Average Angular Velocity   Figure 3.14 shows a rigid body that rotates about a fixed axis perpendicular to the plane of the figure and passing through point Q. Any point on the body at any time can be specified by the distance, r, from the point Q and the angle θ measured from the fixed reference line. For a point, P, in a rigid body, r will be constant and only θ will vary with time. The coordinates r and θ (polar coordinates) are analogous to the rectangular coordinates x and y. As defined in Chapter 1, the value of θ in radians is

$$\theta = \frac{s}{r} \tag{3.27}$$

where

   s = arc length of the circle traced out by point P measured from the reference line

Angles measured in the counterclockwise direction are positive and angles measured in the clockwise direction are negative. Since the body is rigid, the change in the angle for any point is the same for all points, that is,

$$\Delta\theta = \theta_2 - \theta_1$$

is the same for all points.

The general rate equation

$$\text{Rate} = \frac{\text{Change in Displacement Quantity}}{\text{Elapsed Time}}$$

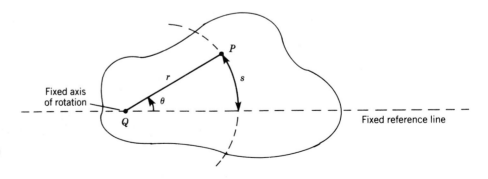

FIGURE 3.14 Rotation of a rigid body.

can be used to define the average angular velocity

$$\overline{\omega} = \frac{\Delta\theta}{\Delta t} \qquad (3.28)$$

where

θ = displacementlike quantity

Since $\Delta\theta$ is the same for all points on the rigid body, all points of a rigid body have the same average angular velocity.

The average angular velocity can be related to the speed because, from the definition of the radian,

$$\Delta\theta = \frac{\Delta s}{r} \qquad (3.29)$$

Substituting this value for $\Delta\theta$ in Equation 3.28 gives

$$\overline{\omega} = \frac{1}{r}\frac{\Delta s}{\Delta t} = \frac{\overline{v}}{r} \qquad (3.30)$$

or

$$\overline{v} = r\overline{\omega}$$

The average speed of each point is proportional to the distance from the fixed axis of rotation.

In Equation 3.30, the angular velocity must be expressed in radians per unit of time, usually seconds. It is also common to express the rotational rate in revolutions per minute (rpm). The conversion factor from rpm to rad/sec is

$$1 \text{ rpm} = \frac{2\pi}{60} \text{ rad/sec}$$

since one revolution is one trip around a circle or 2π radians, and one minute is 60 seconds.

Examples 3.18, 3.19, and 3.20 illustrate the application of the above relationships in solving problems.

EXAMPLE 3.18   Units for angular velocity
. . . .

Given: A phonograph turntable has an angular velocity of 45 rpm.

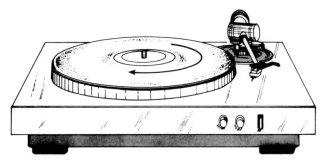

Find: The angular velocity in rad/sec.

Solution: Angular velocity $= \left(45 \dfrac{\text{rev}}{\text{min}}\right)\left(\dfrac{2\pi \ \text{rad}}{1 \ \text{rev}}\right)\left(\dfrac{1 \ \text{min}}{60 \ \text{sec}}\right)$

Angular velocity $= 4.7$ rad/sec
The angular velocity is 4.7 rad/sec.

EXAMPLE 3.19   Angular velocity
· ·

Given: A grindstone rotates 600 revolutions in 5 min.

Find: The average angular velocity.

Solution: $\overline{\omega} = \dfrac{\Delta\theta}{\Delta t}$

$\overline{\omega} = \dfrac{600 \ \text{rev}}{5 \ \text{min}}$

$\overline{\omega} = 120$ rpm

$\overline{\omega} = 120 \left(\dfrac{\text{rev}}{\text{min}}\right)\left(\dfrac{2\pi \ \text{rad}}{\text{rev}}\right)\left(\dfrac{1 \ \text{min}}{60 \ \text{sec}}\right)$

$\overline{\omega} = 120 \dfrac{2\pi \ \text{rad}}{60 \ \text{sec}}$

$\overline{\omega} = 4\pi$ rad/sec

$\overline{\omega} = 12.56$ rad/sec

EXAMPLE 3.20   Rotation of automobile tires
· ·

Given: An automobile with tires of 0.35-m radius moves at a speed of 18.0
              m/sec and travels a distance of 1.6 km. The speed of the tire tread
              must equal the translational speed of the car, in this case 18.0 m/sec.

Find: **a.** The average angular velocity of the tires in rad/sec and in rpm.
              b. The number of revolutions made by the tires.

Solution: **a.** The average angular velocity in rad/sec is found from Equation
                      3.30.

$\overline{\omega} = \dfrac{\overline{v}}{r} = \dfrac{\text{speed of tire tread}}{\text{radius of tire}}$

$\overline{\omega} = \dfrac{18.0 \ \text{m/sec}}{0.35 \ \text{m}}$

$$\overline{\omega} = 51.4 \text{ rad/sec}$$

This result is converted to rpm.

$$\overline{\omega} = (51.4)(60/2\pi)$$

$$\overline{\omega} = 490.8 \text{ rpm}$$

b. The angular displacement θ is found from Equation 3.29.

$$\Delta\theta = \frac{\Delta s}{r}$$

$$\Delta\theta = (1.6 \times 10^3 \text{ m})/(0.35 \text{ m})$$

$$\Delta\theta = 4.57 \times 10^3 \text{ rad}$$

There are 2π radians in one revolution. The number of revolutions made in 4.57×10^3 radians is then $4.57 \times 10^3/2\pi$, or about 727.

Instantaneous Angular Velocity The instantaneous angular velocity can be obtained from a plot of the angle θ of a point on the body versus the time. This is illustrated in Figure 3.15. The instantaneous angular velocity at a point in time is equal to the slope of a tangent line drawn at that point. This result is obtained in the same way that it was obtained for linear motion. Equation 3.30 is also valid for instantaneous values:

$$\omega = \frac{v}{r} \tag{3.31}$$

where

ω = instantaneous angular velocity
v = instantaneous speed

Average Angular Acceleration If the angular velocity is not constant but changes from an initial value to a final value during a time interval, the rotating body undergoes an angular acceleration. The average angular acceleration is defined as the change in the angular velocity divided by the elapsed time, or

FIGURE 3.15 Plot of angular displacement versus time.

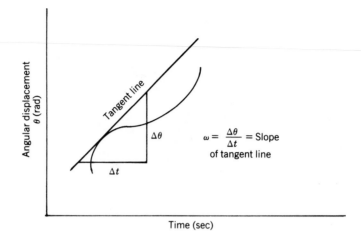

$$\bar{\alpha} = \frac{(\omega_2 - \omega_1)}{t_2 - t_1} = \frac{\Delta\omega}{\Delta t}$$

where

$\bar{\alpha}$ = average angular acceleration
ω_1 = initial angular velocity
ω_2 = final angular velocity after time interval Δt

Instantaneous Angular Acceleration   The instantaneous angular acceleration is obtained from a plot of angular velocity versus time. As in the case for linear acceleration, the instantaneous angular acceleration at a point is equal to the slope of a tangent line at the point as illustrated in Figure 3.16a. If the plot is a straight line, the acceleration is a constant as shown in Figure 3.16b.

FIGURE 3.16 Plots of angular velocity versus time for variable and constant angular acceleration.

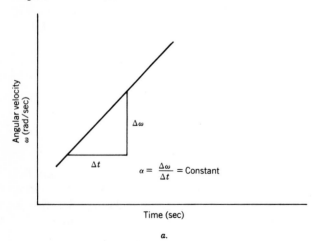

a.

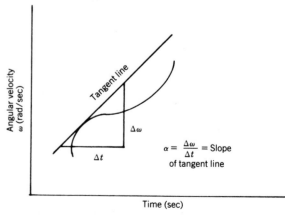

b.

Constant Angular Acceleration When α is constant, the instantaneous angular acceleration becomes

$$\alpha = \frac{\omega_f - \omega_i}{t} \tag{3.32}$$

where

ω_f = final angular velocity
ω_i = initial angular velocity
t = elapsed time (initial time is 0)

Equation 3.32 can be solved for ω_f to yield

$$\omega_f = \omega_i + \alpha t \tag{3.33}$$

Equation 3.33 is analogous to the equation for the final velocity in linear motion:

$$v_f = v_i + at$$

Following the sign convention for angles, clockwise rotation is negative and counterclockwise rotation is positive. An increase in the absolute value of ω when the body is rotating in the clockwise direction gives a negative angular acceleration, and an increase in the absolute value of ω when the body is rotating in the counterclockwise direction gives a positive angular acceleration.

Example 3.21 demonstrates the use of Equation 3.33 in solving a problem.

EXAMPLE 3.21   Angular acceleration of an airplane propeller

· ·

Given: An airplane propeller accelerates from rest at a constant angular acceleration of 420.0 rad/sec^2 for 0.5 sec.

Find: Its final angular velocity.

Solution: The initial angular velocity ω_1 is zero. From Equation 3.33

$$\omega_f = \omega_i + \alpha t$$

$$\omega_f = 0 + (420.0 \text{ rad/sec}^2)(0.5 \text{ sec})$$

$$\omega_f = 210.0 \text{ rad/sec}$$

or in rpm

$$\omega_f = 210.0 \times 60/2\pi$$

$$\omega_f = 2005 \text{ rpm}$$

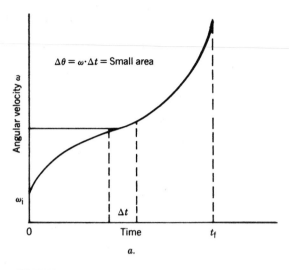

$$\Delta\theta = \omega\cdot\Delta t = \text{Small area}$$

a.

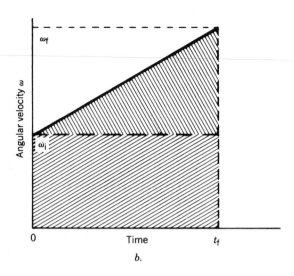

b.

FIGURE 3.17 Plots of angular velocity versus time.

Just as the total linear distance traveled is equal to the area under the velocity versus time curve, the total angular displacement is equal to the area under the angular velocity versus time curve. This is shown in Figure 3.17.

In Figure 3.17b, α is constant and the area under the curve is equal to $\theta_f - \theta_i$ (angular displacement):

$$\theta_f - \theta_i = \omega_i t + 1/2\,(\omega_f - \omega_i)t \tag{3.34}$$

From Equation 3.32, $\omega_f - \omega_i = \alpha t$. Thus, by substitution,

$$\theta_f - \theta_i = \omega_i t + 1/2\,\alpha t^2$$

or

$$\theta_f = \theta_i + \omega_i t + 1/2\,\alpha t^2 \tag{3.35}$$

This rotational equation is the counterpart to the expression $s_f = s_i + v_i t + 1/2\,at^2$ for linear motion with constant acceleration. Its use is demonstrated in Example 3.22.

EXAMPLE 3.22   Rotation of a grindstone
. . . .

Given: The angular acceleration of a grindstone is 2.5 rad/sec². The grindstone starts from rest and attains its final velocity after 4.0 sec. The initial angular position is taken as zero.

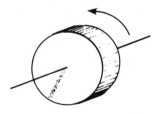

Find: **a.** Its angular displacement.
          b. Its final angular velocity after 4.0 sec.

Solution: **a.** Equation 3.35 is used to find θ_f; the initial angular velocity, ω_i, is zero.

$$\theta_f = \theta_i + \omega_i t + 1/2\ \alpha t^2$$

$$\theta_f = 0 + 0 + 1/2\ (2.5\ \text{rad/sec}^2)(4.0\ \text{sec})^2$$

$$\theta = 20.0\ \text{rad}$$

b. The final angular velocity is found from Equation 3.33.

$$\omega_f = \omega_i + \alpha t$$

$$\omega_f = 0 + (2.5\ \text{rad/sec}^2)(4.0\ \text{sec})$$

$$\omega_f = 10\ \text{rad/sec}$$

If like terms are combined in Equation 3.34, one obtains

$$\theta_f - \theta_i = \frac{(\omega_f + \omega_i)}{2}\ t \tag{3.36}$$

Solving Equation 3.32 for t gives

$$t = \frac{\omega_f - \omega_i}{\alpha}$$

Substituting this value for t in Equation 3.36 yields

$$\theta_f - \theta_i = \frac{(\omega_f + \omega_i)(\omega_f - \omega_i)}{2\alpha} \tag{3.37}$$

Multiplying both sides of Equation 3.37 by 2α gives

$$2\alpha\ (\theta_f - \theta_i) = \omega_f^2 - \omega_i^2 \tag{3.38}$$

This equation has the same form as the analogous equation $2a\ (s_f - s_i) = v_f^2 - v_i^2$ in linear motion with constant acceleration. Its use is illustrated in Example 3.23.

EXAMPLE 3.23   Rotation of a flywheel
. . . .

Given: A wheel initially rotating at 4 rev/sec decelerates at a rate of 2 rev/sec^2.

Find: The number of revolutions it makes before stopping.

Solution: In this problem, the units can be left in rev/sec without converting them to rad/sec. The final angular velocity ω_f is zero.

$$\omega_f = \omega_i^2 + 2\alpha\ (\theta_f - \theta_i)$$

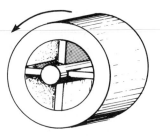

$$0 = (4 \text{ rev/sec})^2 + 2 (-2 \text{ rev/sec}^2)(\theta_f - \theta_i)$$

$$\theta_f - \theta_i = \frac{16 \text{ rev}^2/\text{sec}^2}{4 \text{ rev/sec}^2} = 4 \text{ rev}$$

The parallels between linear motion with constant acceleration and rotational motion about a fixed axis with constant angular acceleration are summarized in Table 3.2.

Example 3.24 shows how two of the equations describing rotation are solved for the problem of a flywheel.

EXAMPLE 3.24   Rotation of a flywheel
. . . .

Given: The angular acceleration of a flywheel is 2 rad/sec². It rotates through 50 rad in a 5-sec time interval sometime after it started from rest.

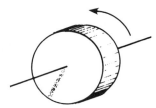

Find: How long the flywheel had been turning just before the beginning of the 5-sec interval.

Solution: Let $t = 0$ at the beginning of the 5-sec interval. The angular velocity at the beginning of the 5-sec interval is found as follows.

$$\theta_f - \theta_i = \omega_i t + 1/2 \, \alpha t^2$$

$$50 \text{ rad} = \omega_i \, (5 \text{ sec}) + 1/2 \, (2 \text{ rad/sec}^2)(5 \text{ sec})^2$$

$$\omega_i = \frac{(50 \text{ rad} - 25 \text{ rad})}{5 \text{ sec}}$$

$$\omega_i = 5 \text{ rad/sec}$$

This value is the initial angular velocity at $t = 0$, which corresponds to the beginning of the 5-sec interval. The initial angular velocity at $t = 0$ is 5 rad/sec. Since the flywheel started from rest, the corresponding final velocity is 0.0 rad/sec.

$$\omega_f = \omega_i + \alpha t$$

$$0 = 5 \text{ rad/sec} + (2 \text{ rad/sec}^2)(t)$$

$$t = \frac{(-5 \text{ rad/sec})}{(2 \text{ rad/sec}^2)}$$

$$t = -2.5 \text{ sec}$$

Thus, the flywheel started spinning from rest 2.5 sec before it turned through 50 rad in the next 5 sec.

Radial Acceleration and Centripetal Force

An important motion in a plane is that of a body traveling in a circular path at a constant speed. Although the speed is constant and equal to the magnitude of the velocity, the vector acceleration is not zero because the velocity is constantly changing direction. Thus

$$\mathbf{a} = \frac{\Delta \mathbf{v}}{\Delta t}$$

will not be zero even when Δt is very small.

Direction of the Velocity   Consider a body traveling in a plane along any path as illustrated in Figure 3.18. The average velocity between points 1 and 2 is

$$(\mathbf{v}_2)_{av} = \frac{\Delta \mathbf{d}_2}{t_2 - t_1}$$

The direction of the average velocity is in the direction of d_2. If a shorter time t_3 is taken

TABLE 3.2   Analogy between linear translation and rotation

Concept	Translation	Rotation
Displacement	s	θ
Average velocity	$\bar{v} = \dfrac{\Delta s}{\Delta t}$	$\bar{\omega} = \dfrac{\Delta \theta}{\Delta t}$
Acceleration	$\bar{a} = \dfrac{(v_f - v_i)}{t}$	$\bar{\alpha} = \dfrac{(\omega_f - \omega_i)}{t}$
Constant acceleration	$v_f = v_i + at$	$\omega_f = \omega_i + \alpha t$
	$s_f = s_i + v_i t + 1/2\, at^2$	$\theta_f = \theta_i + \omega_i t + 1/2\, \alpha t^2$
	$v_f^2 = v_i^2 + 2a(s_f - s_i)$	$\omega_f^2 = \omega_i^2 + 2\alpha(\theta_f - \theta_i)$

FIGURE 3.18 Motion of a body in a plane.

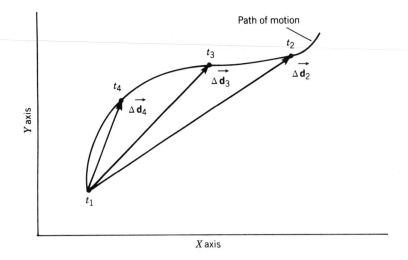

$$(\mathbf{v}_3)_{\text{av}} = \frac{\Delta \mathbf{d}_3}{t_3 - t_1}$$

If the time is again made less,

$$(\mathbf{v}_4)_{\text{av}} = \frac{\Delta \mathbf{d}_4}{t_4 - t_1}$$

One can see from Figure 3.18 that the direction of the $\Delta \mathbf{d}$'s approaches the direction of a tangent to the curve as the change in time becomes smaller. However, as the change in time becomes small, the average velocity approaches the instantaneous velocity. Thus, the instantaneous velocity is in the direction of a tangent line to the curve. Also, as $\Delta \mathbf{d}$ becomes smaller, the magnitude of $\Delta \mathbf{d}$ (its length) approaches the path length between the two points. Thus, the magnitude of the instantaneous velocity is equal to the instantaneous speed.

Radial Acceleration   To calculate the acceleration when the path is a circle, consider the diagram shown in Figure 3.19. Since a tangent line is always perpen-

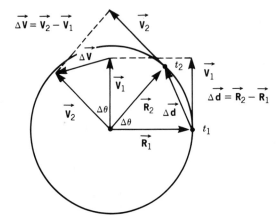

FIGURE 3.19 Acceleration of a body traveling in a circular path.

dicular to the radius, the velocity will always be perpendicular to the radius. If the body moves with constant speed, the magnitude of the velocity vectors remains constant. For similar triangles, one has

$$\frac{\Delta d}{R} = \frac{\Delta v}{v}$$

$$\Delta v = \frac{v}{R} \Delta d \tag{3.39}$$

If both sides of Equation 3.39 are divided by Δt, the equation becomes

$$\frac{\Delta v}{\Delta t} = \frac{v}{R} \frac{\Delta d}{\Delta t} \tag{3.40}$$

But, from the definition of velocity and acceleration,

$$a = \frac{\Delta v}{\Delta t}$$

$$v = \frac{\Delta d}{\Delta t}$$

Thus, by substituting these expressions in Equation 3.40, one obtains

$$a = \frac{vv}{R} \tag{3.41}$$

If the time interval, Δt, is made very small (approaches 0), then **a** becomes the instantaneous acceleration and **v** becomes the instantaneous velocity. Thus, Equation 3.41 may be written as

$$a_{c} = \frac{v^2}{R} \tag{3.42}$$

where

a_{c} = instantaneous centripetal acceleration

Equation 3.42 gives the magnitude of the instantaneous acceleration of a body moving in a circular path with constant speed (magnitude of the velocity is constant). To specify the acceleration, the direction must also be known. Consider the effect on the direction of Δv as Δt (and $\Delta \theta$) becomes smaller. The direction of Δv, which is the direction of the acceleration, approaches a line parallel to the initial radius drawn from the center of the circle to point 1. To visualize this, imagine that $\Delta \theta$ in Figure 3.19 approaches 0. It is also important to notice that Δv points toward the center rather than away from the center. The instantaneous acceleration is clearly a vector pointing toward the center of the circular path with a magnitude of v^2/R.

The acceleration can also be expressed in terms of the angular velocity, ω. From Equation 3.31 the tangential velocity is

$$v = \omega R \tag{3.43}$$

The magnitude of the centripetal acceleration can then be expressed in terms of the angular velocity:

$$a_c = \frac{v^2}{R} = \omega^2 R \qquad (3.44)$$

Since the centripetal acceleration is directed toward the center in the opposite direction of the radius vector, one may write the vector equation as

$$\mathbf{a_c} = -\omega^2 \mathbf{R} \qquad (3.45)$$

Centripetal Force   A body traveling in a circular path requires a force acting on the body. According to Newton's second law of motion the force is

$$\mathbf{F_c} = m\mathbf{a_c} = -m\omega^2\,\mathbf{R} \qquad (3.46)$$

or

$$\mathbf{F_c} = \frac{-mv^2\,\mathbf{R}}{R^2} \qquad (3.47)$$

Thus, the force is constant in magnitude and directed toward the center of the circular path. Without this force, the body would travel in a straight line with constant velocity. The force necessary to produce the circular motion with constant speed is called the "centripetal force" (centripetal means "center-seeking"). The corresponding acceleration is called the "centripetal" or "radial" acceleration.

Example 3.25 illustrates the application of Equations 3.42 and 3.47 to a problem.

EXAMPLE 3.25   Centripetal acceleration and force

. . . .

Given: A 3000-lb car travels at constant speed on a circular track having a radius of 200 ft. The car requires 15 sec to complete one lap.

Find: **a.** The tangential velocity.
             b. The centripetal acceleration.
             c. The centripetal force.

Solution: **a.** The tangential velocity can be found by use of this equation.

$$v = \frac{2\pi r}{t}$$

$$v = \frac{2\pi\,(200\ \text{ft})}{15\ \text{sec}}$$

$$v = 83.8\ \text{ft/sec}$$

b. Equation 3.42 gives the centripetal acceleration.

$$a_c = \frac{v^2}{R}$$

$$a_c = \frac{(83.8\ \text{ft/sec})^2}{200\ \text{ft}}$$

$$a_c = 35.1 \text{ ft/sec}^2$$

directed toward the center of the circular path.

c. The centripetal force is given by Equation 3.46.

$$F_c = ma_c = (w/g)\, a_c$$

$$F_c = \left(\frac{3000 \text{ lb}}{32 \text{ ft/sec}^2} \right)(35.1 \text{ ft/sec}^2)$$

$$F_c = 3291 \text{ lb}$$

directed toward the center of the circular path.

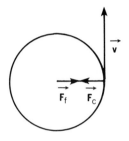

FIGURE 3.20 Forces acting on a mass on a string traveling in a circular path.

Consider the example of a mass traveling in a circular path on the end of a string, as illustrated in Figure 3.20. The centripetal force F_c acts inward on the mass toward the center of the circle. Newton's third law—"For every action there is an equal and opposite reaction"—indicates that there is an outward-acting force F_f, called the "centrifugal force," which acts outward upon the center support. These two forces, centripetal (inward) and centrifugal (outward), are equal in magnitude, opposite in direction, and act upon different objects. When the string is held by a hand, the centrifugal force acts outward on the mass.

Consider a car traveling in a circular path on a level road as shown in Figure 3.21. The centripetal acceleration and force act inward toward the center upon the tire of the car. In this case, it is a force caused by friction between the tires and road. The centrifugal force is the frictional force produced by the tires acting outward upon the road.

When a car travels in a circular path, a person in the car seems to experience an outward force, that is, a force tending to throw him or her outward from the center of the circle. Actually, the seat exerts an inward force upon the person, causing him to travel in a circular path. The person tends to travel in a straight line, but the centripetal force acting inward (provided by the seat) causes him to travel along an arc.

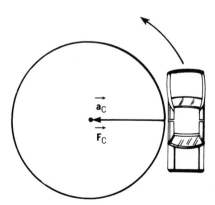

FIGURE 3.21 Car traveling a circular path.

FIGURE 3.22   Particles in a
centrifuge.

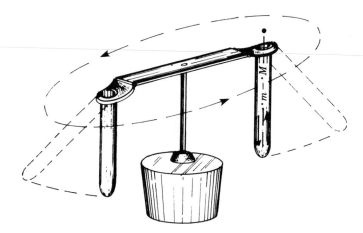

Consider two particles m and M, which have the same size and which are suspended in a uniform fluid in a centrifuge (Figure 3.22). The density of M is greater than the density of m. As the centrifuge spins, both experience the same centripetal acceleration, $a = -\omega^2 R$, since the radius and the angular velocities are the same. The centripetal force upon M, necessary to hold M in place, is given by

$$F_M = -M\omega^2 R \tag{3.48}$$

The centripetal force for M must be greater than that for m, since $M > m$, as indicated by

$$F_m = -m\omega^2 R \tag{3.49}$$

However, the fluid exerts the same fluid force on both masses, since they are the same size. Because of this, the masses will separate, with the larger mass M moving farther outward in the tube than the smaller mass m. The centrifuge uses this principle to separate particles of varying densities. The particles of greater density move toward the bottom (away from the center) of the tube.

If the same particles are placed on ends of identical springs and swung in a circle, the one with the greater mass will stretch its spring more because it requires a greater restoring (centripetal) force. The larger mass will rotate with a greater radius, as in the centrifuge.

Another application of centripetal force in circular motion is illustrated in Figure 3.23. The fluid enters the pump at the center inlet and is given a spinning

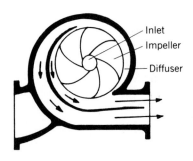

— Inlet
— Impeller
— Diffuser

FIGURE 3.23   Centrifugal
pump.

motion. Its natural tendency is to move in a straight line, tangent to its circular motion, unless an inward centripetal force keeps it in a circular path. Since there is no string or spring tied to each particle, there is no inward force. Thus, the fluid moves outward along lines tangent to its circular motion at each point. Because the fluid moves outward, it is called a "centrifugal pump." Centrifugal means "center-fleeing."

SECTION **3.1** *EXERCISES*

1. Define the following rates and express their basic units in both the SI and the English systems of units:

   Average velocity

   Average speed

   Instantaneous linear velocity

   Average acceleration

   Instantaneous linear acceleration

   Average angular velocity

   Instantaneous angular velocity

2. Calculate an airplane's average speed if it travels 3200 km in 3.5 hr.

3. If a car traveling at 55 mi/hr stops in 16.3 sec, calculate the car's average deceleration in ft/sec².

4. If a driver averages 55 mi/hr, how much time will be required to make a 250-mi trip?

5. A force of 200 N accelerates an object with a mass of 100 kg. If the object is initially at rest, determine its speed after 3 sec.

6. A flywheel initially at rest requires 25 sec to reach an angular velocity of 1750 rpm. Calculate the average angular acceleration of the flywheel.

7. A motor rotates at the rate of 5000 rpm. How much time is required for one complete revolution?

8. A ball is dropped from a window. How far will it fall in 3 sec?

9. A rock is dropped over a cliff. What is its velocity after falling 5.7 m?

10. An arrow was shot vertically upward. As it fell back to the ground, it had a velocity of 8.6 m/sec at a height of 5.1 m.

    a. What was the initial velocity of the arrow?

    b. How high did it go?

    c. How much time was required for the arrow to fall from its maximum height to the ground?

11. A car traveling at 30 m/sec stops with a constant deceleration applied for 10 sec. How far did it travel in that 10 sec?

12. A rock is thrown vertically upward. If it takes the rock a total of 5 sec to fall back to the earth, what is the maximum height the rock reaches?

13. Find the linear speed of a point on a record at the needle position

    a. at the beginning of the recording, where the radius is 15 cm, and

    b. at the end of the recording, where the radius is 7.4 cm, for a phonograph speed of 78 rpm.

14. An automobile has wheels of radius 38 cm and travels at 25 m/sec. Find:

    a. the angular speed of the wheels about the axle

    b. the angular deceleration if the wheels are brought to a stop after 30 turns

c. the distance traveled during braking.

(**Hint:** The tread speed of the tire is equal to the translation speed of the automobile.)

15. The angular speed of the flywheel of a steam engine is 160 rpm. The wheel comes to rest in 25 hr when the steam is turned off. Find the number of revolutions made by the wheel before it comes to rest.

16. If the angular acceleration of a flywheel is 5 rad/sec² and the initial angular velocity of the flywheel is 10 rad/sec, what is the time required for the flywheel to make 10 revolutions?

17. Through what angular distance does a flywheel turn during the time it takes the flywheel to accelerate from 10 rad/sec to 100 rad/sec if the flywheel undergoes a constant acceleration of 14 rad/sec²?

18. If the speed of a point on a flywheel is v, then the angular velocity of that point is v/r, where r is the radius out to that point. Suppose a car accelerates around a curve at 100 mi/hr², and just before entering the curve the car's speed is 30 mi/hr. What is the angular speed of the car as it leaves the curve if it takes the car 10 sec to move through the length of the curve? The curve has a radius of 50 ft.

19. If a block of mass 1 kg undergoes a constant acceleration and, starting from rest, travels 10 m in 1 sec, find:

a. the speed of the block at the end of the first second

b. the acceleration of the block

c. The force necessary to produce this acceleration.

20. How long will it take a car traveling 50 m/sec to overtake a car traveling 20 m/sec if the first car is initially 2000 m behind the second car?

21. A merry-go-round at a park is set into motion in such a manner that it completes one revolution each 3 sec. A child weighing 50 lb sits on the edge of the merry-go-round a distance of 6 ft from its center. Calculate:

a. the tangential velocity of the child

b. the centripetal acceleration

c. the centripetal force acting upon the child.

22. A motorcyclist loops-the-loop as shown in the figure below. What is the minimum

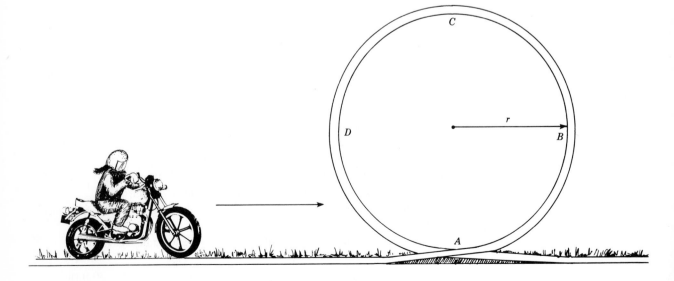

velocity he must have at C in order not to lose contact with the surface? (**Hint:** At point C, $F = mg = mv^2/r$.)

23. How long will it take a car to complete 3 laps around a circular track if the radius of the track is 100 m and the car (mass 2000 kg) experiences a constant centripetal force of 72,000 N?

24. A boy of mass 40 kg sits 20 m from the center of a merry-go-round. If the merry-go-round starts from rest and undergoes an angular acceleration of 1 rad/sec², how many revolutions will the merry-go-round make before the boy experiences a centripetal force of 100 N?

3.2

RATE IN FLUID SYSTEMS

OBJECTIVES

Upon completion of this section, the student should be able to

• Define volume and mass flow rates and express them in the SI and English systems of units.

• In a fluid system, given all the quantities except one in each of the following groups, determine the third quantity:

   Volume of fluid moved, elapsed time, and volume flow rate
   Mass of fluid moved, elapsed time, and mass flow rate

• Describe in a qualitative way

   Laminar flow
   Turbulent flow
   Developing flow
   Developed flow

• Define the Reynolds number and give the transition range from laminar to turbulent flow.

• Apply the principle of conservation of mass for incompressible flow.

DISCUSSION

We have examined speed in a translational system and angular velocity in a rotational system. Both of these quantities have many applications in mechanical energy systems. Let us now examine the volume flow rate and mass flow rate in fluid systems.

Flow Rate

Definition of Flow Rate   The fluid flow rate is an important quantity in heat transfer, hydraulic, and pneumatic systems. The volume flow rates or mass flow rates through pipes or ducts are required to evaluate these systems. The volume flow rate, the volume of fluid flowing through a cross section per unit of time, is

$$Q_v = \frac{\Delta V}{\Delta t} = \frac{\text{Change in Displacementlike Quantity}}{\text{Elapsed Time}} \qquad (3.50)$$

FIGURE 3.24 Volume flow rate as a velocity.

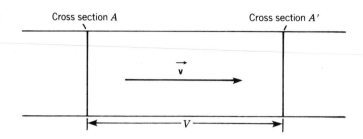

where

Q_V = volume flow rate
Δt = an element of time
ΔV = corresponding volume displaced in Δt

Notice that the displacementlike quantity in this case is the change in volume as it is in the equation for work, $\Delta W = P\Delta V$. The volume flow rate is the product of the average velocity of the fluid and the area of the corresponding cross section of the pipe or duct. This is similar to the case for the flow of electric charge and is illustrated in Figure 3.24. In a unit of time, all the fluid in the volume between the area A and A' goes through the area at A'. Thus, the volume flow rate is

$$Q_V = \bar{v}A \tag{3.51}$$

where

$\bar{v}$ = average velocity of the fluid

It is assumed in Equation 3.51 that the fluid has not been compressed or expanded during the flow. Therefore, one must be careful when measuring the volume flow rates of gases since gases are compressible.

Units of volume flow rate are usually expressed in m³/sec, ft³/sec, gal/min, liters/sec, or cm³/sec. Example 3.26 illustrates an application of Equation 3.50.

EXAMPLE 3.26 Volume flow rate

. . . .

Given: A swimming pool holds 10,000 gal; 2 hr are needed to fill the pool.

Find: The volume flow rate of water through the hose.

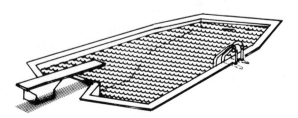

Solution: $Q_V = \dfrac{\Delta V}{\Delta t}$

$Q_V = \dfrac{10{,}000 \text{ gal}}{(2 \text{ hr})\left(\dfrac{60 \text{ min}}{1 \text{ hr}}\right)}$

$Q_V = 83 \text{ gal/min}$

In the case of gases, it is desirable to consider the mass flow rate rather than the volume flow rate. Total mass flow rate is defined as the fluid mass flowing through a cross section of the pipe or duct per unit time and is not affected by compressibility. Equation 3.52 defines the mass flow rate:

$$Q_m = \frac{\Delta m}{\Delta t} \qquad (3.52)$$

where

Q_m = mass that passed through the cross section
Δt = corresponding time

However, from the definition of density ($\Delta m = \rho \Delta V$), Equation 3.52 can also be expressed as

$$Q_m = \frac{\Delta m}{\Delta t} = \rho\,\frac{\Delta V}{\Delta t} = \rho Q_V \qquad (3.53)$$

which relates the volume flow rate to the mass flow rate. The mass flow rate is also given in terms of the average velocity by substituting the expression for Q_V given by Equation 3.51 into Equation 3.53. This yields

$$Q_m = \rho v A \qquad (3.54)$$

Example 3.27 shows how the mass flow rate can be calculated for a given situation.

EXAMPLE 3.27   Mass flow rate
. . . .

Given: 180 kg of ammonia gas flow through the cooling coils of a refrigerated truck in 6 min.

Find: The mass flow rate of the gas.

Solution: $Q_m = \dfrac{\Delta m}{\Delta t}$

$Q_m = \dfrac{180,000 \text{ g}}{6 \text{ min}}$

$Q_m = 30,000 \text{ g/min} = 30 \text{ kg/min}$

The ammonia gas is flowing at a rate of 30,000 g/min or 30 kg/min.

Reynolds Number   There are two distinctly different types of fluid flow. Reynolds injected a fine stream of colored liquid at the entrance of a tube through which water was flowing. When the velocity of the water was small, the colored liquid was visible as a straight line for the length of the tube. This showed that small elements of water moved in parallel straight lines. This type of flow is called laminar or streamline flow. As the velocity was increased, a point was reached where the line of colored liquid would become wavy for a short distance and then break up into vortices. Beyond the vortices, the color would become diffused. This type of flow, where there exists a random churning action, is called turbulent flow. It is found that the transition from laminar to turbulent flow depends on the value of a quantity called the Reynolds number. For a tube, this dimensionless quantity is

$$\text{Re} = \frac{\rho v D}{\mu} \tag{3.55}$$

where

ρ = density of the fluid
D = tube diameter
v = average velocity
μ = dynamic viscosity

The dynamic viscosity is discussed in Chapter 5. It has been found experimentally that a transition from laminar to turbulent flow usually occurs when the Reynolds number lies in the approximate range

$$2000 < \text{Re} < 4000 \tag{3.56}$$

The application of Equations 3.55 and 3.56 is shown in Example 3.28.

EXAMPLE 3.28   Calculation of Reynolds number
. . . .

Given: The diameter of a smooth pipe is 0.4 ft. Water flows through the
                  pipe with an average velocity of 1000 ft/hr. The density is
                  1.95 slug/ft^3 and the viscosity is 0.074 slug/ft·hr.

Find: **a.**   The Reynolds number.

                b.   Determine if the flow is laminar or turbulent.

Solution: **a.**   The Reynolds number is

$$Re = \frac{\rho v D}{\mu}$$

$$Re = \frac{(1.95 \text{ slug/ft}^3)(1000 \text{ ft/hr}) \left(\dfrac{1 \text{ hr}}{3600 \text{ sec}}\right)(0.4 \text{ ft})}{0.074 \text{ slug/ft·hr} \cdot \left(\dfrac{1 \text{ hr}}{3600 \text{ sec}}\right)}$$

$$Re = 10,500$$

b.   Since Re > 4000, the flow is probably turbulent.

Velocity in a Tube   The velocity in a tube is a function of the radius. The maximum velocity occurs at the axis of the tube and the velocity is zero at the wall. This is illustrated for laminar flow in Figure 3.25. At the entrance cross section of the tube, the velocity profile has a flat distribution that quickly changes to zero at the wall. As the flow progresses down the tube, the streamlines are slowed down near the wall due to friction until the final velocity profile is developed at a distance *l*. At this point, the velocity profile becomes constant. The final velocity profile is a parabola expressed as

$$v = 2v \left(1 - \frac{r^2}{r_o{}^2}\right) \tag{3.57}$$

where

r_o = radius of the tube
v = average velocity
r = any radial distance in the tube

FIGURE 3.25   Velocity of a fluid in a tube as a function of radius.

Equation 3.57 shows that the maximum velocity along the axis ($r = 0$) is equal to twice the average velocity.

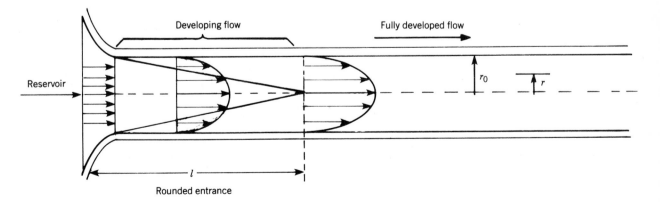

Developing flow   Fully developed flow

Reservoir

Rounded entrance

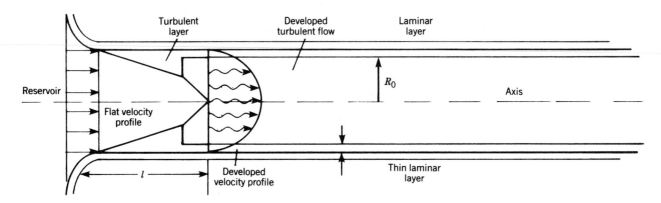

FIGURE 3.26 Velocity profile for turbulent flow.

The velocity profiles for turbulent flow are illustrated in Figure 3.26. Fully developed turbulent flow occurs after some distance, *l*, down the tube. Beyond this distance the velocity profile is constant. The velocity profile in the developed region rises more sharply near the wall and is flatter in the central region than the profile for developed laminar flow. The average velocity is approximately 0.8 times the maximum velocity at the axis. The laminar layer shown in the developed region is only a few thousandths of an inch thick.

Conservation of Mass   If the density of the fluid remains constant, the mass flow into a junction must equal the mass flow out of a junction. This is a form of the equation of continuity or the conservation of mass and can be expressed as

$$\Sigma \ Q_m = 0 \tag{3.58}$$

where flow into a junction is positive and flow out of a junction is negative.

The application of the conservation of mass for an incompressible fluid (ρ_m = constant) is given in Example 3.29.

EXAMPLE 3.29   Conservation of mass

· ·

Given: The flow system shown in the sketch.

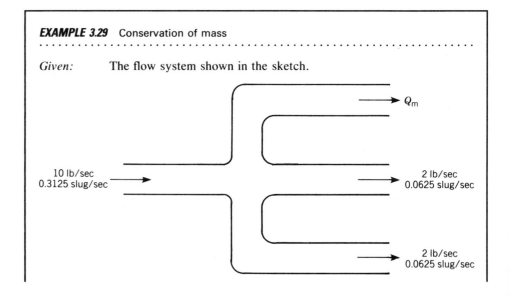

Find: The mass flow rate Q_m.

Solution: Flow rate into junction + flow rate out of junction = 0

                 0.3125 slug/sec − 0.0625 slug/sec − 0.0625 slug/sec + Q_m = 0

                 Q_m = −0.1875 slug/sec

SECTION *3.2*   EXERCISES

1. Define volume flow rate and mass flow rate. Express their basic units in the SI and the English systems of units.

2. What is the mass flow rate in kg/sec through a hose if 20 l of water flow out of the end of the hose in 10 sec? The density of water is 1000 kg/m³. (**Note:** l is the symbol for liter.)

3. A gas pump moves gasoline from an underground storage tank to the tank in a car at the rate of 15 l/min. Calculate the time required to fill a 20-gal tank with this pump.

4. In Example 3.28, what is the maximum velocity of the water if the flow is to remain laminar (assume laminar flow if the Reynolds number is less than 2000).

5. Water flows into and out of the junction shown in the sketch. Find the unknown flow rate in slugs/min.

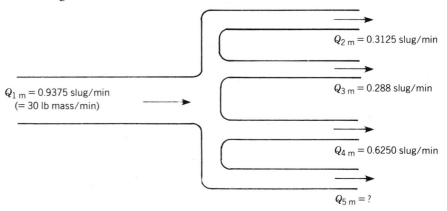

$Q_{1\,m}$ = 0.9375 slug/min
(= 30 lb mass/min)

$Q_{2\,m}$ = 0.3125 slug/min

$Q_{3\,m}$ = 0.288 slug/min

$Q_{4\,m}$ = 0.6250 slug/min

$Q_{5\,m}$ = ?

6. If the volume flow rate of water is 0.5 m³/sec, what is the mass flow rate in lb/min?

3.3 .... OBJECTIVES ....

RATE IN Upon completion of this section, the student should be able to
ELECTROMAGNETIC
SYSTEMS   • Define electric current and induced voltage, giving their basic units in the SI and English systems of units.

                 • Given two of the following quantities in an electrical system, determine the third:
                 Charge transferred
                 Elapsed time
                 Current

- Describe in a qualitative way direct and alternating current.

- Apply the principle of conservation of charge.

- Calculate the induced voltage in a rotating loop in a magnetic field and in a current loop in a changing magnetic field.

- Describe the operation of an electrical transformer.

- Define the following terms and give their units:
  Temporal frequency
  Period
  Spatial frequency
  Spatial separation

DISCUSSION

The rates in mechanical (velocity) and fluid systems (volume flow rate and mass flow rate) have been examined. Let us now turn to the electromagnetic system, where the rates are current and induced voltage.

Electrical Systems

Current   The most basic ratelike quantity that needs to be understood in relation to electrical systems is the current. Current is defined as the rate of movement of the displacementlike quantity, charge, from one point to another. Direct current in an electrical circuit is the rate at which charge moves past a cross section in the circuit when a constant electrical potential is applied. Alternating current is the rate of charge flow back and forth through a cross section in the circuit when the applied voltage changes direction periodically.

Some of the electrons in an electrical conductor are free to move in the material. These electrons are in constant motion with a total energy that depends on the temperature. As the electrons move through the material they interact with the fixed atoms, which causes a change in their direction. These interactions result in random velocities. These free electrons will acquire a net velocity when influenced by an electric field due to an applied voltage. This velocity will be in the opposite direction of the electric field. The source of additional electrons at the negative terminal is called the cathode and the sink for electrons at the positive terminal is called the anode.

As long as an electrical potential difference exists, the net drift of electrons through the conductor will continue. The conducting material itself has no net change in charge. For every electron that enters the conductor from the negative terminal, one electron leaves the conductor at the positive terminal.

To indicate the rate at which current flows through a conductor, a meter must count the net electrons that pass through a cross section in a given time interval. This process can be visualized by imagining a plane surface within the conductor perpendicular to the electron flow (perpendicular to the conducting wire) as shown in Figure 3.27.

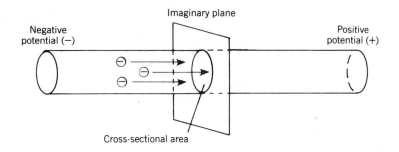

FIGURE 3.27 Electrons flowing through a cross section.

When a net charge, ΔQ, passes through the cross section in a time t, the current is defined by

$$I = \frac{\Delta Q}{\Delta t} \tag{3.59}$$

where

I = current
ΔQ = net charge passing through the cross section
Δt = corresponding change in time

Thus, the current through a conductor is the net charge per unit time that passes through a plane perpendicular to the applied electric field.

If the charge is measured in coulombs and the time is measured in seconds, the resulting current is in amperes. An ampere is defined as one coulomb per second. Examples 3.30 and 3.31 show the use of Equation 3.59 for solving problems.

EXAMPLE 3.30 Current calculation
....

Given: 2.4 C (coulombs) passing through a copper wire in 0.6 sec.

Find: The current.

Solution: $I = \dfrac{\Delta Q}{\Delta t}$

$I = \dfrac{2.4 \text{ C}}{0.6 \text{ sec}}$

$I = 4.0 \text{ C/sec} = 4.0 \text{ A}$

The current is 4.0 A.

EXAMPLE 3.31 Charge calculation
....

Given: A current of 62.5 A flows in 16 msec.

Find: How much charge passes through the conductor.

Solution: $I = \dfrac{\Delta Q}{\Delta t}$

$\Delta Q = I \times \Delta t$

$\Delta Q = (62.5 \text{ A})(16 \text{ msec})$ (msec is equal to 10^{-3} sec)

$\Delta Q = (62.5 \text{ C/sec})(16 \times 10^{-3} \text{ sec})$

$\Delta Q = 1.0 \text{ C}$

A charge of 1.0 C passes through the conductor.

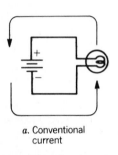

a. Conventional
current

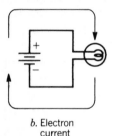

b. Electron
current

FIGURE 3.28 Direction of conventional current and electron current.

The current has been described in terms of electron flow. In metal conductors, the charge carriers are electrons (negative carriers), which move from a negative potential toward a more positive potential.

In some semiconductor materials, in ionized gases, and in many solutions, charge is transferred by positive-charge carriers. Positive carriers result from the removal of an electron at the positive terminal. Such positive-charge carriers move toward the negative terminal, where they receive an electron and become neutral. The result is the same as with electron flow: one unit of charge is transferred through the conductor.

The direction of current flow is confusing because at the time electricity was discovered, the positive charge was thought to be the moving particle. Some texts used in the study of electricity use the term ''conventional current direction'' rather than electron flow direction. The direction of conventional current flow is from positive to negative, and the charge carriers are assumed to be positive. The direction of electron flow is from negative to positive, and the charge carriers are negative electrons. In some situations, positive-charge particles (ions) and electrons flow at the same time. For this case, the total current is the sum of the two currents.

Both conventional current and electron current are shown in Figure 3.28. The magnitudes in both directions are the same. Only the directions are different.

Alternating Current   Direct current is the rate of charge flow resulting from a constant voltage difference across an electrical conductor. If the voltage changes, the current also changes.

Alternating current (ac) flows first in one direction and then in the other and results in a circuit when the polarity of the voltage source alternates as shown in Figure 3.29. At time t_0 in this figure, the applied voltage is zero; no current flows as is indicated by t_0 in Figure 3.30. At t_1 a maximum positive voltage is applied to one end of the conductor, while the other end remains at 0 V. This voltage difference produces the electron flow shown at t_1 of Figure 3.30. At t_2 both voltage and current are again zero, and at t_3 an applied voltage of $-V_{\max}$ causes a current flow in the opposite direction. At t_4 the voltage and current are both zero again. This waveform repeats at a frequency of 60 cycles per second in ordinary household circuits.

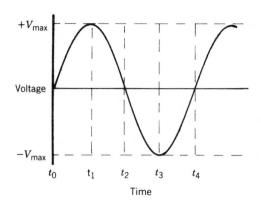

FIGURE 3.29 Plot of voltage versus time.

FIGURE 3.30 Electron flow at different times. (right)

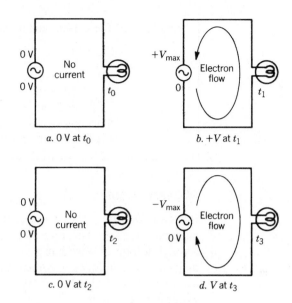

Power companies generate alternating current because it can be easily transformed to high voltages and back to low voltages. This is done because the energy lost in the transmission lines is proportional to the current squared, whereas the energy transmitted is proportional to the product of voltage and current. Keeping the current as low as possible reduces the line loss. The same energy can be transmitted by using a high voltage.

An important aspect of current flow is the principle of continuity, which is based on the conservation of charge. This requires, if there is no storage of charge, that the current flowing in any junction must equal the total current flowing out of the junction. If the current flowing into the junction is given a positive sign and the current flowing out a negative sign, then

$$\Sigma I_n = 0 \qquad (3.60)$$

The application of Equation 3.60 is illustrated in Example 3.32.

EXAMPLE 3.32 Principle of conservation of charge

. . . .

Given: The circuit shown in the sketch.

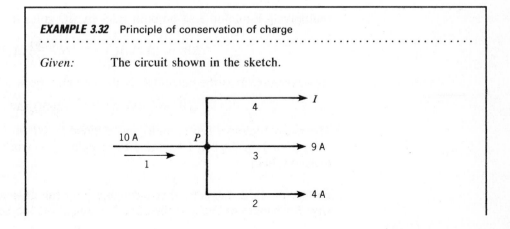

Find: The current and its direction in wire #4.

Solution: $\Sigma I_n = 0$

$10 - I - 9 - 4 = 0$

$I = -3$ A

Since the sign is negative, the assumed direction of I is not correct. I is flowing into point P.

Magnetic Systems

Induced Voltage The displacementlike quantity in a magnetic system is the magnetic flux, ϕ. The rate of change of the magnetic flux with respect to time through a loop of wire has a very important physical significance. The changing magnetic flux induces a voltage in the loop of wire. This phenomenon was first discovered by Michael Faraday. If the loop has a number of turns, it can be shown experimentally that

$$\Delta V = -N \frac{\Delta \phi}{\Delta t} \tag{3.61}$$

where
  ΔV = induced voltage in volts across the coil
  Δt = a very small interval of time in seconds
  $\Delta \phi$ = corresponding change in the flux through the loop in webers
  N = number of turns of wire in the loop

For an electric system, it was shown in Chapter 2 that a small amount of work, ΔW, can be expressed as

$$\Delta W = (\Delta V)(\Delta Q) \tag{3.62}$$

where
  ΔV = voltage difference between two points
  ΔQ = charge transported between the two points

Multiplying Equation 3.61 on both sides by ΔQ gives the work done:

$$\Delta W = (\Delta V)(\Delta Q) = -N \frac{(\Delta \phi)}{\Delta t} (\Delta Q) \tag{3.63}$$

However, $\Delta Q/\Delta t$ is the current I_c in the coil and the work becomes

$$\Delta W = -NI_c\Delta \phi = -(\mathrm{mmf})_c(\Delta \phi) \tag{3.64}$$

The magnetomotive force, mmf, was defined as NI_c in Chapter 2. This justifies using Equation 3.64 in Chapter 2 to calculate the work done in establishing a magnetic field, ϕ.

Generator The magnetic flux through a loop can change because the flux density, B, changes or because the effective area of the loop changes. The latter case

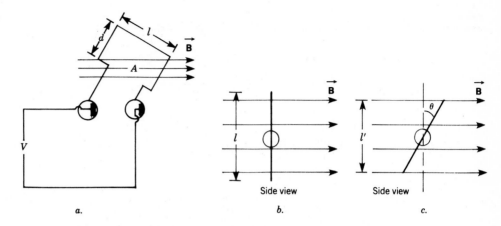

FIGURE 3.31 Rotation of a wire loop through a magnetic field.

is illustrated in Figure 3.31. As the loop turns the total magnetic flux through the area of the loop will change. To calculate the magnetic flux through the area of the loop when B is perpendicular to the plane of the loop, one multiplies the flux density B times the area of the loop as expressed in

$$\phi = (B)(l)(d) = BA \tag{3.65}$$

where

   A = area of the loop

However, when the loop has rotated through the angle θ, the flux through the loop becomes

$$\phi = B(l \cos \theta)d = (B)(l)(d) \cos \theta$$

$$\phi = BA \cos \theta \tag{3.66}$$

Substitution of Equation 3.66 into Equation 3.61 gives the induced voltage (if there are N turns in the loop) as

$$V = -N\frac{\Delta\phi}{\Delta t} = -NBA \frac{\Delta(\cos \theta)}{\Delta t} \tag{3.67}$$

If the coil rotates with a constant angular velocity, ω, then $\theta = \omega t$. With this condition, Equation 3.67 becomes

$$V = -NBA \frac{\Delta(\cos \omega t)}{\Delta t} \tag{3.68}$$

In calculus, it is shown for very small Δt's that

$$\frac{\Delta(\cos \omega t)}{\Delta t} = -\omega \sin \omega t \tag{3.69}$$

Using Equation 3.69 to substitute in Equation 3.68 yields

$$V = NBA \, \omega \sin \omega t \tag{3.70}$$

If slip rings are used as illustrated in Figure 3.31, the rotating loop becomes an

FIGURE 3.32 Electric transformer.

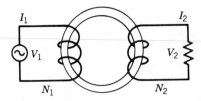

alternating current generator with an angular frequency of ω and an amplitude of $NBA\omega$.

Transformers   Equation 3.61 can also be used to explain the operation of electric transformers. An electric transformer consists of two coils wound on a common core as illustrated in Figure 3.32. An alternating current flowing in coil 1 produces a magnetic flux in the core that changes with time. This changing flux passes through coil 1 and coil 2 and induces a voltage in both coils. The voltage induced in coil 2 causes a current flow in coil 2, which produces an additional flux change in the core that modifies the induced voltages in coil 1 and coil 2. Since the same total flux, ϕ, passes through both coils, the induced voltages from Equation 3.61 become

$$V_2 = -N_2 \frac{\Delta\phi}{\Delta t}$$ (3.71)

and

$$V_1 = -N_1 \frac{\Delta\phi}{\Delta t}$$ (3.72)

where
  V_1 = primary coil voltage
  N_1 = number of turns in the primary
  V_2 = secondary coil voltage
  N_2 = number of turns in the secondary

Dividing Equation 3.71 by Equation 3.72 yields

$$\frac{V_2}{V_1} = \frac{N_1}{N_2}$$

$$V_2 = \frac{N_1}{N_2} V_1$$ (3.73)

Inductors   If only one coil is wound on a core, the magnetic device is called an inductor. From Equation 3.61 the induced voltage in the coil is

$$V = -N \frac{\Delta\phi}{\Delta t} = -N \left(\frac{\Delta\phi}{\Delta I}\right)\left(\frac{\Delta I}{\Delta t}\right)$$ (3.74)

The quantity $N \Delta\phi/\Delta I$ is called the inductance, and it is usually given the symbol L. The unit of inductance is the henry (H):

$$1 \ H \ = \ 1 \ \frac{Wb \cdot turn}{A}$$

$$L = N \frac{\Delta\phi}{\Delta I} \tag{3.75}$$

For the coil in Example 2.14, the magnetic flux is

$$\phi = BA = \left(\frac{\mu NA}{l}\right) I \tag{3.76}$$

where

A = area of the coil
l = average distance around the coil
I = current through the coil
μ = magnetic permeability

For a change in ϕ, Equation 3.76 gives

$$\Delta\phi = \frac{\mu NA}{l} \Delta I$$

$$\frac{\Delta\phi}{\Delta I} = \frac{\mu NA}{l}$$

By substitution, the value of the inductance from Equation 3.75 becomes

$$L = \frac{\mu N^2 A}{l} \ H \tag{3.77}$$

To calculate the inductance of other coils, it is necessary to find the mathematical relationship between the magnetic flux and the current in the coil. In general, this is not easy to do, but L is often a constant and can be measured.

The self-induced voltage of an inductor is usually written as

$$V = -L \frac{\Delta I}{\Delta t} \tag{3.78}$$

where L has been substituted for $N\Delta\phi/\Delta I$ in Equation 3.74.

Applications of Equations 3.70, 3.73, and 3.78 are given in Examples 3.33, 3.34, and 3.35.

EXAMPLE 3.33   Voltage of simple generator
. . . .

Given: A simple coil of 100 turns with an area of 0.01 m² rotates in a magnetic field of 0.1 Wb/m² at an angular velocity of 377 rad/sec.

Find: **a.**   The voltage as a function of time.
            b.   The maximum voltage.

Solution: The voltage as a function of time is given by Equation 3.70.

            a. $V = \omega NBA \sin \omega t$

$$V = \left(\frac{377}{\text{sec}}\right)(100 \text{ turns})(0.1 \text{ Wb/m}^2)(0.01 \text{ m}^2) \sin 377t$$

$$V = 37.7 \sin 377t \text{ Wb·turns/sec} = (37.7 \sin 377t) \text{ V}$$

b. The maximum voltage occurs when $\sin 377t = 1$.

$$V_{\text{max}} = 37.7 \text{ V}$$

EXAMPLE 3.34   Transformer voltage
. . . .

Given: The primary coil of a transformer has 200 turns and the secondary has 100 turns. The input voltage is 100 V.

Find: The output voltage.

Solution: From Equation 3.73

$$V_2 = \frac{N_2}{N_1} V_1$$

$$V_2 = \left(\frac{100}{200}\right)(100 \ V)$$

$$V_2 = 50 \ V$$

EXAMPLE 3.35   Inductance of a coil
. . . .

Given: The current in a coil is changed at the rate of -10 A/sec. The induced voltage is 2 V.

Find: The self-inductance of the coil.

Solution: From Equation 3.78

$$V = -L\frac{\Delta I}{\Delta t}$$

Solve the equation for L.

$$L = \frac{-V}{\Delta I/\Delta t}$$

$$L = \frac{-2 \text{ V}}{-10 \text{ A/sec}}$$

$$L = 0.2 \frac{\text{V·sec}}{\text{A}} = 0.2 \text{ H}$$

Frequency

Temporal Frequency   Temporal frequency describes events that repeat after a specific elapsed time. Events that can be described in terms of temporal frequencies include water dripping from a faucet, the ticking of a clock, or the motion of a planet around a star. Mathematically, temporal frequency can be expressed by

$$f = \frac{\Delta n}{\Delta t} \tag{3.79}$$

where

f = temporal frequency
Δn = number of repetitions
Δt = time in seconds

Example 3.36 illustrates how the equation for frequency is used to solve a problem. (**Note:** From this point the term "frequency" will be used to refer to temporal frequencies.)

EXAMPLE 3.36   Frequency of an electrical wave
· ·

Given: A commercial generator produces 240 electrical waves in 4 sec and operates for 1 hr.

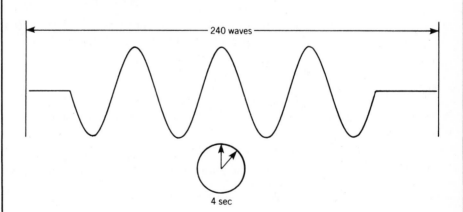

Find: **a.**   The frequency of the waves.
            b.   The number of waves produced in this time.

Solution: **a.**   Use Equation 3.79.

$$f = \frac{\Delta n}{\Delta t}$$

$$f = \frac{240 \text{ waves}}{4 \text{ sec}}$$

$$f = 60 \text{ waves/sec}$$

b. $f = \dfrac{\Delta n}{\Delta t}$

$\Delta n = f\,\Delta t$ (multiplying both sides by time)

$\Delta n = (60 \text{ waves/sec})(1 \text{ hr})$

$\Delta n = (60 \text{ waves/sec})(1 \text{ hr}) \left(\dfrac{60 \text{ min}}{1 \text{ hr}}\right)\left(\dfrac{60 \text{ sec}}{1 \text{ min}}\right)$

$\Delta n = 216{,}000 \text{ waves}$

Example 3.36 shows how to calculate the frequency of a repetitive event and demonstrates how a knowledge of the frequency allows the calculation of the number of events occurring in a given time.

The standard units of frequency are the same in the mks, cgs, SI, and English measuring systems. Frequency is measured in hertz (Hz), where

$$1 \text{ Hz} = 1 \text{ cycle/sec} = 1 \text{ event/sec}$$

"Cycles" or "events" cannot be expressed in terms of units of mass, displacement, or time. Therefore, in a strict sense, frequency is measured in reciprocal time units:

$$1 \text{ Hz} = 1/\text{sec} = 1 \text{ sec}^{-1}$$

These units immediately lead to Equation 3.80, another form for the definition of frequency, which is equivalent to Equation 3.79:

$$f = \frac{1}{T} \tag{3.80}$$

or

$$T = \frac{1}{f} \tag{3.81}$$

where T is the period (time) for one event. The use of Equation 3.81 is illustrated in Example 3.37.

EXAMPLE 3.37   Period of vibration of a tuning fork
. . . .

Given: A tuning fork vibrates at a frequency of 800 Hz.

Find: The period of vibration of the tuning fork.

Solution: From Equation 3.81

$$T = \frac{1}{f}$$

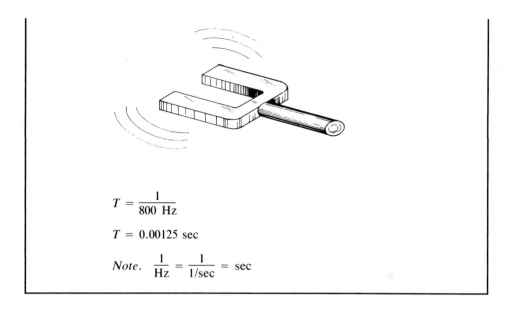

$$T = \frac{1}{800\ \text{Hz}}$$

$$T = 0.00125\ \text{sec}$$

Note. $\dfrac{1}{\text{Hz}} = \dfrac{1}{1/\text{sec}} = \text{sec}$

By far, the most common application of the frequency concept is the measurement of time. In this application, time is measured by counting repetitive phenomena. If the frequency of the repetitive event is constant, then the period of the motion defines a basic unit of time. The smallest unit of time that can be measured by any repetitive event is equal to the period of the event. Therefore, the higher the frequency of any repetitive phenomenon, the smaller the basic unit of time that it can measure. Thus, higher frequencies allow more precise time measurements.

Many modern clocks use the frequency of alternating current to determine their basic time unit. The frequency of this electrical voltage is 60 Hz (50 Hz in most foreign countries).

The natural vibrations of a quartz crystal provide a finer division of the second. Typical natural frequencies for these crystals range from 1 to 10 megahertz (1 MHz = 10^6 Hz), allowing basic time units of 10^{-6} to 10^{-7} sec, respectively.

The ultimate in timekeeping precision is presently the atomic clock. This device uses the frequency of electromagnetic radiation emitted from certain atoms or molecules (typically cesium or ammonia) for a time standard. The frequencies involved are generally in the range from 1 to 10 gigahertz (1 GHz = 10^9 Hz). One such clock, operated by the National Bureau of Standards, is accurate to within a few seconds over one million years.

Spatial Frequency Any object or event that repeats after a given distance can be characterized by a spatial frequency. The definition of spatial frequency can be stated in a form similar to Equation 3.81, as indicated by

$$k = \frac{\Delta n}{\Delta d} \tag{3.82}$$

FIGURE 3.33   Wavelength.

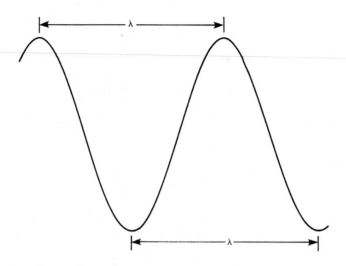

where

   k = spatial frequency
   Δn = number of spatial events that occur in distance d
   Δd = distance

The spatial frequency also can be expressed in terms of the spatial separation, that is, the distance between two consecutive objects or events. Spatial separation is defined by

$$k = \frac{1}{d} \qquad (3.83)$$

Notice the similarity between Equations 3.82 and 3.83. In Equation 3.83 the units of spatial frequency are reciprocal distance units; spatial frequency, therefore, has different units in the three systems of units. One class of phenomena that is readily described by spatial frequencies is wave motion. The spatial separation of waves (for instance, the waves in Figure 3.33) is the distance between two consecutive peaks (or two consecutive troughs) and is called the "wavelength." Example 3.38 shows how the spatial frequency of waves is calculated.

EXAMPLE 3.38   Spatial frequency of water waves
. . . .

Given: Ocean waves have peaks every 10 m.

Find: The spatial frequency of these waves.

Solution: Wavelength = spatial separation = 10 m . Then

$$k = \frac{1}{d}$$

$$k = \frac{1}{10 \text{ m}}$$

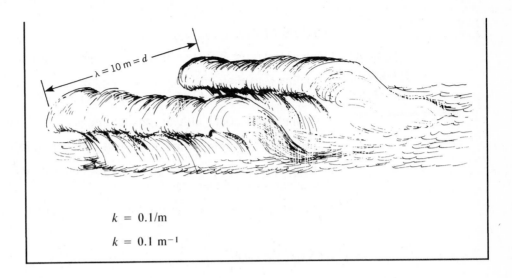

$$k = 0.1/m$$
$$k = 0.1\ m^{-1}$$

SECTION 3.3 EXERCISES

1. Define electric current and induced voltage. Express their basic units in both the SI and the English systems of units.

2. If 1 min is required for 10 C of charge to flow through a conductor, what is the current?

3. How much time is required for a current of 25 mA to transfer a charge of 0.5 C?

4. In the circuit to the left, indicate:

    a. the direction of electron flow with a solid line

    b. the direction of conventional current flow with a dashed line.

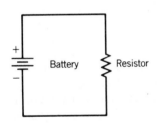

5. Draw a graph showing alternating current as a function of time in an ac circuit.

6. Recalling that $W = Q\Delta V$, find the current required to do 50 J of work in 40 sec if the operating voltage is 24 V.

7. How many electrons flow through a cross section of a wire in 3 min if the current through the wire is 3A? (**Hint:** 1 electron = 1.602×10^{-19} C.)

8. The rate of change of the total flux through a coil of 100 turns is 10.0 Wb/sec. What is the induced voltage?

9. The rate of change of current through an inductor is 1.00 A/sec. The inductance of the coil is 1.0 H. What is the induced voltage in the coil?

10. A transformer has 100 turns on the primary coil and 1500 turns on the secondary coil. If the primary voltage is 100 V, what is the secondary voltage?

11. Define:
Temporal frequency
Period
Spatial frequency

12. A train of identical boxcars travels at constant speed past an observer. The observer sees a car go by every 4 sec. With what temporal frequency do the cars pass the observer? If each car is 20 m long, what is the spatial frequency of the boxcars?

13. Find the period for waves with the following frequencies:

    a. 50 Hz   **b.** 50 kHz   **c.** 50 MHz.

3.4 ....

**RATE IN
THERMAL
SYSTEMS**

OBJECTIVES ....

Upon completion of this section, the student should be able to

- Define heat flow rate and cooling rate or heating rate, expressing their basic units in the SI and English systems of units.

- Name three modes of heat transfer and give an equation for each mode.

- Calculate heat transfer rates for heat conduction in solids.

- Calculate a heat transfer coefficient for convective heat transfer in a pipe for laminar and turbulent flow.

- Estimate the radiant heat transfer for simple geometric conditions.

DISCUSSION

The study of rate will be concluded by an examination of heat flow and the rate of cooling or heating in a thermal system.

Thermal Systems

Heat transfer is the flow of energy due to a temperature difference. There are three modes of heat transfer: conduction, convection, and radiation. The mechanisms for these three modes of energy transfer will be explained qualitatively and the corresponding rate of heat transfer will be discussed.

Conduction

Mechanisms   Heat conduction occurs in gases, liquids, and solids. In gases, the energy transfer is due to the molecular motion. The molecular energy is directly proportional to the temperature. All the molecules are in random motion, and they collide with each other and exchange energy. The net effect of these processes is a diffusion (movement) of molecules from a region of high temperature to a region of low temperature, with a corresponding diffusion of molecules from a low-temperature region to a high-temperature region. Although there can be a continuous flow of energy by diffusion if the temperature difference is maintained, there is no continuous net flow of mass.

The mechanism for thermal conduction in liquids is similar to the mechanism in gases. The analysis of heat conduction in liquids from a molecular point of view, however, is much more complex because of the very high collision rates and strong forces of interaction between molecules.

Thermal energy is also conducted in solids although the atoms are not free to move through a solid. The atoms vibrate about their equilibrium points. Their energy of vibration increases with the temperature. The atoms at a region of higher temperature transfer part of their vibrational energy to their neighbors. As long as a temperature difference exists, heat energy is transported through the solid by this process. If the solid is a conductor, there are electrons that are free to move throughout the solid. These electrons also transport energy in a way similar to that of molecules in a gas. The energy transported by the electrons is usually greater than the energy transported by the vibrating atoms. Thus, electrical conductors are also good heat conductors. If the circuit is not closed to form a thermocouple, there will not be a continuous net flow of charge.

Conduction Heat Flow Rate   The heat flow rate in conduction is the amount of heat energy flowing through a surface per unit of time. This may be expressed mathematically by the equation

$$q_c = \frac{\Delta Q}{\Delta t} \tag{3.84}$$

where

q_c = heat flow rate by conduction
Δt = an element of time
ΔQ = heat energy transferred in Δt by conduction

It can be shown experimentally (Fourier's law) that the heat flow rate is equal to the product of three factors: the area perpendicular to the direction of heat flow, the difference in temperature divided by its corresponding distance, and a proportionality factor, k, called the thermal conductivity of the material. Mathematically the expression is

$$q_c = -kA \frac{\Delta T}{\Delta l} \tag{3.85}$$

where

A = area perpendicular to the heat flow
k = thermal conductivity
ΔT = change in temperature
Δl = corresponding distance between ΔT

The negative sign is required because heat flows from a region of high temperature to a region of low temperature. Thus, $-\Delta T = -(T_2 - T_1)$ will be positive if T_1 is greater than T_2. Table 3.3 lists values for the thermal conductivity of several common materials.

In Equation 3.85, one can see a similarity between temperature difference ΔT and the other forcelike quantities. Each forcelike quantity is associated with the

TABLE 3.3   Material thermal conductivities

Material	Conductivity (Btu × inch/hr × ft²·F°)
Plywood	0.80
Glass fiber insulation	0.25
Concrete	12.0
Brick	9.0

flow of a corresponding quantity. In fluid systems it is the pressure difference and mass flow rate, and in electrical systems it is the voltage difference and the current flow. For thermal conduction, it is the temperature difference and the flow of heat energy.

Examples 3.39 and 3.40 use Equation 3.85 to solve problems in conductive heat transfer.

EXAMPLE 3.39   Heat flow through a brick wall
· ·

Given: An old, broken 5 × 8 ft window is removed and replaced with a 4-inch-thick brick wall. On a particular winter day, the outside temperature is 10°F and the inside temperature is 72°F.

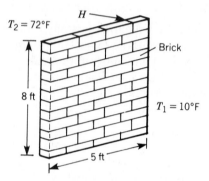

Find: The heat flow through the bricked-up opening.

Solution: From Table 3.3, the thermal conductivity of 4-inch brick is

$$k = \frac{9 \text{ Btu} \times \text{inch}}{\text{hr} \times \text{ft}^2\cdot\text{F}°}$$

$$A = 5 \text{ ft} \times 8 \text{ ft} = 40 \text{ ft}^2$$

$$\Delta l = 4 \text{ inch}$$

$$(T_2 - T_1) = 10°\text{F} - 72°\text{F} = -62 \text{ F}°$$

$$H = \frac{-kA \ (T_2 - T_1)}{\Delta l}$$

$$H = \frac{-9 \text{ BTU} \times \text{inch}}{\text{hr} \times \text{ft}^2\cdot\text{F}°} \times \frac{40 \text{ ft}^2 \times -62 \text{ F}°}{4 \text{ inch}} = 5580 \text{ Btu/hr}$$

EXAMPLE 3.40   Heat flow through concrete and plywood
· ·

Given: All conditions being equal.

Find: The thickness of concrete that will have the same heat flow as 1 ft² of 1-inch plywood.

Solution:

$$q_{cc} = \frac{-k_c A (T_2 - T_1)}{(\Delta l)_c}$$

$$q_{cc} = \frac{-12 A (T_2 - T_1)}{(\Delta l)_c}$$

$$q_{cp} = \frac{-k_p A (T_2 - T_1)}{(\Delta l)_p}$$

$$q_{cp} = \frac{-0.80 A (T_2 - T_1)}{(\Delta l)_p}$$

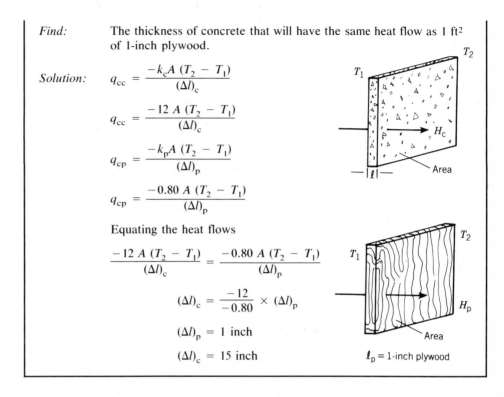

Equating the heat flows

$$\frac{-12 A (T_2 - T_1)}{(\Delta l)_c} = \frac{-0.80 A (T_2 - T_1)}{(\Delta l)_p}$$

$$(\Delta l)_c = \frac{-12}{-0.80} \times (\Delta l)_p$$

$$(\Delta l)_p = 1 \text{ inch}$$

$$(\Delta l)_c = 15 \text{ inch}$$

$\ell_p = 1$-inch plywood

Convection When heat transfer is associated with the flow of a fluid over a surface, the calculation of heat transfer rates becomes very difficult. The effect of the fluid flow on the energy transfer rate (convection) has been experienced by everyone. A hot object will cool faster if placed in front of a fan. An object in water will cool faster if the water is flowing. In both of these examples convection plays an important part in the rate of energy transfer from the body to the fluid.

As an example, consider the flow of water over a smooth plane surface. Because of frictional forces (viscous forces), the velocity will be zero at the surface and increase to the free-stream velocity at some distance from the surface as shown in Figure 3.34. In addition, there will be a temperature distribution in the water near the surface. The temperature will drop from the wall temperature to the free-stream temperature at some distance from the wall. The temperature distribution will be affected by the velocity distribution. The effect of the flow will be to reduce the temperature of the fluid near the wall. This will cause $\Delta T/\Delta l$ near the wall to be greater with flow. The velocity of the fluid at the wall is zero, and heat will be conducted in the fluid away from the wall. The heat transfer rate in convection at the wall is expressed as

$$q_c = -k_F A \left[\frac{\Delta T}{\Delta l} \right]_{wall} \tag{3.86}$$

where

k_F = thermal conductivity of the fluid

$[\Delta T/\Delta l]_{wall}$ = change in temperature for a very small change in distance in the fluid evaluated at the wall

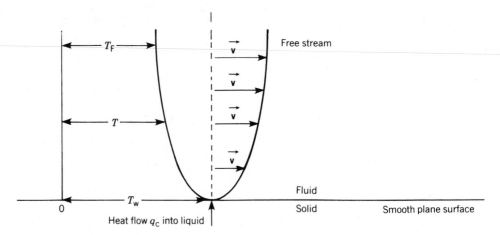

FIGURE 3.34 The effect of fluid flow on heat flow rate.

Free stream

T_F

T

T_w

0

Heat flow q_c into liquid

Fluid

Solid

Smooth plane surface

Since $[\Delta T/\Delta l]_{wall}$ is greater with flow, the energy transfer increases. The heat transfer, however, is still a conduction process.

For almost all real problems, it is very difficult to determine $[\Delta T/\Delta l]_{wall}$, so a heat transfer coefficient is used to calculate the heat transfer rate. The heat transfer coefficient is usually based on experimental data. The equation for the convective heat transfer to the fluid is written as

$$q_F = hA \, (T_w - T_F) \tag{3.87}$$

where

q_F = heat transfer rate to the fluid
A = area of the surface
T_w = wall temperature
T_F = free-stream temperature (or bulk temperature)
h = convective heat transfer coefficient

If q_c is equated to q_F, since both represent the same heat transfer rate, one obtains

$$-k_F \left[\frac{\Delta T}{\Delta l}\right]_{wall} = h \, (T_w - T_F) \tag{3.88}$$

$$h = \frac{-k_F \left[\dfrac{\Delta T}{\Delta l}\right]_{wall}}{T_w - T_F} \tag{3.89}$$

The common area, A, cancels in Equation 3.88. The major difficulty in using Equation 3.89 to calculate values for h is the problem of determining $[\Delta T/\Delta l]_{wall}$.

A simplified mathematical analysis for developed, laminar flow in a pipe with a constant heat flux at the wall yields

$$h = \frac{4.36 \, k_F}{D} \tag{3.90}$$

where

D = diameter of the pipe
k_F = thermal conductivity of the fluid

The corresponding heat transfer rate from Equation 3.87 is

$$q_F = hA \, (T_w - T_b) = \left[\frac{4.36 \, k_F}{D}\right](A)(T_w - T_b) \tag{3.91}$$

where

T_b = average temperature across the pipe
T_w = wall temperature

Better values of h for laminar and turbulent flow can be obtained from empirical data presented in handbooks on heat transfer. The use of Equation 3.88 is illustrated in Example 3.41.

EXAMPLE 3.41   Approximate heat transfer for laminar flow in a pipe
. . . .

Given: Heat is transferred from a pipe to flowing water at a constant heat flux such that the difference between the wall temperature and average water temperature is 20 F°. The diameter of the pipe is 0.10 ft and the average velocity of the water is 60 ft/hr. The water temperature enters the pipe at 70°F.

Find: The heat transferred per hour for 1 ft of pipe. The dynamic viscosity, μ, is 0.074 slug/ft·hr.

Solution: Determine if the flow is laminar or turbulent. The Reynolds number is

$$Re = \frac{\rho v D}{\mu}$$

At 70°F

$\rho = 1.95$ slug/ft³   (mass in slugs, mass density of water)

$\mu = 0.074$ slug/ft·hr

$$Re = \frac{(1.95 \text{ slug/ft}^3)(60 \text{ ft/hr})(0.10 \text{ ft})}{0.074 \text{ slug/ft·hr}}$$

$Re = 158$

Since $Re < 2000$, the flow is laminar. Therefore, Equation 3.88 applies.

$$h = 4.36 \frac{k}{D}$$

At 70°F

$k_F = 3.49$ Btu/hr·ft·F°

$$h = (4.36) \frac{3.49}{0.1 \text{ ft}} \frac{\text{Btu}}{\text{hr·ft·F°}}$$

$h = 152$ Btu/hr·ft²·F°

From Equation 3.87

$$q_F = hA (T_w - T_b)$$

For a 1-ft length

$$q_F = (152 \text{ Btu/hr} \cdot \text{ft}^2 \cdot \text{F}°)(0.10 \text{ ft})(\pi)(1.0 \text{ ft})(20 \text{ F}°)$$

$$q_F = 9.55 \times 10^2 \text{ Btu/hr}   \quad (1\text{-ft length})$$

For fully developed turbulent flow in smooth pipes, the empirical relationship given by Equation 3.92 can be used to approximate h if the heat flux is constant:

$$h = 0.023 \text{ Re}^{0.8} \text{ Pr}^n \frac{k_F}{D} \tag{3.92}$$

where

$n = 0.4$ for heating a fluid
$n = 0.3$ for cooling a fluid
Re = Reynolds number
Pr = Prandtl number

The Prandtl number is defined as

$$\text{Pr} = \frac{c_p \mu}{k_F} \tag{3.93}$$

where

c_p = specific heat at constant pressure
μ = dynamic viscosity
k_F = thermal conductivity

Since the Reynolds number increases with the velocity, h also increases with the velocity of the fluid in turbulent flow. Example 3.42 illustrates the use of Equation 3.92.

EXAMPLE 3.42   Approximate heat transfer for turbulent flow
. . . .

Given: Heat is transferred from a pipe to flowing water at a constant heat flux so that the difference between the wall temperature and the average water temperature is 20 F°. The diameter of the pipe is 0.10 ft and the average velocity of the water is 2000 ft/hr. The water enters the pipe at 70°F.

Find: The heat transferred per hour for 1 ft of pipe.

Solution: Determine if the flow is laminar or turbulent.

$$\text{Re} = \frac{\rho v D}{\mu}$$

At 70°F

$\rho = 1.95$ slug/ft³

$\mu = 0.074$ slug/ft·hr

$\text{Re} = \dfrac{(1.95 \text{ slug/ft}^3)(2000 \text{ ft/hr})(0.10 \text{ ft})}{0.074 \text{ slug/ft·hr}}$

$\text{Re} = 5270$

Since Re > 4000, the flow is turbulent.
The Prandtl number is

$\text{Pr} = \dfrac{c_p \mu}{k}$

For 70°F

$c_p = 31.616$ Btu/slug·F°

$k = 0.349$ Btu/hr·ft·F°

$\text{Pr} = \dfrac{(31.616 \text{ Btu/slug·F°})(0.071 \text{ slug/ft·hr})}{(0.349 \text{ Btu/hr·ft·F°})}$

$\text{Pr} = 6.43$

From Equation 3.92

$h = 0.023 \ \text{Re}^{0.8} \ \text{Pr}^{0.4} \dfrac{k}{D}$

$h = (0.023)(5270)^{0.8} (6.43)^{0.4} \dfrac{(0.349 \text{ Btu/hr·ft·F°})}{0.10 \text{ ft}}$

$h = (0.023)(949)(2.11)(3.49 \text{ Btu/hr·ft}^2\text{·F°})$

$h = 161$ Btu/hr·ft²·F°

From Equation 3.87

$q_F = hA \ (T_w - T_F)$

$q_F = (161 \text{ Btu/hr·f·ft}^2\text{·F°})(0.10 \text{ ft})(\pi)(1 \text{ ft})(20 \text{ F°})$

$q_F = 1.01 \times 10^3$ Btu/hr   (1-ft length)

In Example 3.42, 2000 ft/hr corresponds to 0.56 ft/sec. Thus, it would be possible to increase the velocity significantly. This would increase the value of q_F since h depends on $(v)^{0.8}$. For example, by doubling v, q_F increased by a factor of 1.74.

Radiation   The third mode of heat transfer is by radiation. Radiant heat energy can be transmitted through a vacuum since, like light, it is electromagnetic radiation. In heat transfer, one usually limits the transfer of radiant energy to that which is due to a temperature difference. It is found that, when two bodies exchange radiant heat energy, the net energy exchange is

$$q_R = G_{12} \, \epsilon \, \sigma \, A_1 \, (T_1^4 - T_2^4) \tag{3.94}$$

where

q_R = radiant heat transfer rate
A_1 = area of the surface of body 1
T_1 = temperature at the surface of body 1 in °R
σ = a constant equal to 0.171×10^{-8} Btu/hr·ft²·R°⁴
ϵ = a constant that depends on the nature of the surface called the emissivity
G_{12} = a geometric factor
T_2 = temperature at the surface of body 2 in °R

If ϵ is one, the maximum value, the body is called a blackbody. A blackbody absorbs all the radiant energy incident on it. For surfaces with an ϵ less than one, Equation 3.94 becomes an approximation. If it is a good approximation, the body is called a gray body. The geometric factor, G_{12}, is usually difficult to calculate. This shape factor can be found in heat transfer handbooks for a number of different conditions. The use of Equation 3.94 is demonstrated in Example 3.43.

EXAMPLE 3.43   Radiant heat transfer between parallel disks
· ·

Given: The parallel disks shown in the sketch. The value of ϵ for each surface is 1.00. The temperature of surface 1 is 700°R and the temperature of surface 2 is 600°R.

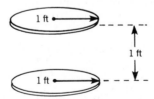

Find: The radiant heat transfer rate from surface 1 to surface 2.

Solution: From a handbook, the geometric factor is 0.37. The equation for q_R is (Equation 3.94)

$$q_R = G_{12} \, \epsilon \, \sigma \, A_1 \, (T_1^4 - T_2^4)$$

$$q_R = (0.37)(1)(0.171 \times 10^{-8} \text{ Btu/hr·ft}^2\text{·R}°^4)(\pi)(1.0 \text{ ft})^2$$
$$[(700°R)^4 - (600°R)^4]$$

$$q_R = (2.0 \times 10^{-9} \text{ Btu/hr·R}°^4)(2.4 \times 10^{11}°R^4 - 1.3 \times 10^{11}°R^4)$$

$$q_R = 220 \text{ Btu/hr}$$

If the emissivity of the two bodies is different, then Equation 3.94 becomes

$$q_r = \sigma\, A_1 G_{12}\, (\epsilon_1 T_1^4 - \epsilon_2 T_2^4) \qquad (3.95)$$

In addition to emission and absorption by the surfaces, part of the incident radiant energy may be reflected. Problems that include reflection are too complex to be discussed here. However, radiant heat transfer will be discussed in greater detail in Chapter 13.

Heating and Cooling   The rate at which a body changes temperature is called the heating rate or cooling rate. The rate is defined by the equation

$$R = \frac{\Delta T}{\Delta t} = \frac{T_2 - T_1}{t_2 - t_1} \qquad (3.96)$$

where

R = heating or cooling rate
T_2 = temperature at time t_2
T_1 = temperature at time t_1

Since $t_2 - t_1$ is always positive, the sign of R is determined by $T_2 - T_1$. If T_2 is greater than T_1, the body is heating and the rate is positive. If T_2 is less than T_1, the body is cooling and the rate is negative.

The units of the heating or cooling rate are units of temperature per unit of time such as C°/sec or F°/hr.

The cooling or heating of a body is due to a combination of conduction, convection, and radiation. A change in the heat content changes the internal energy of the body. Since the temperature of the body is a measure of its internal energy, the temperature changes as the heat content changes. This can be expressed mathematically by using the definition of specific heat given in Chapter 2:

$$\Delta Q = mc\, \Delta T \qquad (3.97)$$

where

m = body's mass
c = specific heat of the solid or liquid

Dividing Equation 3.97 by the corresponding time, Δt, yields

$$\frac{\Delta Q}{\Delta t} = mc\, \frac{\Delta T}{\Delta t}$$

or

$$R = \frac{\Delta T}{\Delta t} = \frac{1}{mc}\, \frac{\Delta Q}{\Delta t} \qquad (3.98)$$

To illustrate the various modes of heat transfer, consider the small disk of total area A_D enclosed in a chamber at temperature T_1 as shown in Figure 3.35. The chamber has a window that allows a laser beam to be absorbed by one side of the disk. The temperature of the chamber is T_2. The rate at which energy is absorbed from the laser beam is equal to the radiant beam power, q_L, since the disk is a blackbody. The disk is supported by a fiber, and the energy lost by conduction

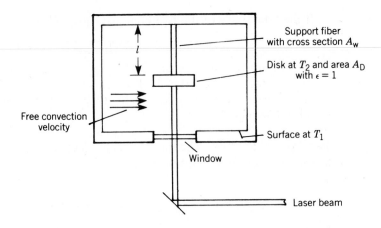

down the fiber is (Equation 3.85)

$$q_c = \frac{-kA_w}{l}(T_2 - T_1) \tag{3.99}$$

The convection loss due to the velocity of air over the surface may be expressed
as (Equation 3.87)

$$q_F = h\,A_D\,(T_2 - T_1) \tag{3.100}$$

An approximation to the radiation heat loss is given by Equation 3.94:

$$q_R = \sigma\,G_{12}\,\epsilon A_D\,(T_2^4 - T_1^4) \tag{3.101}$$

For small temperature differences, Equation 3.102 may be used to simplify the
equation for q_R:

$$T_2^4 - T_1^4 = (T_2^2 + T_1^2)(T_2^2 - T_1^2) = (T_2^2 + T_1^2)(T_2 + T_1)(T_2 - T_1) \tag{3.102}$$

Since the temperature difference will be small, one may approximate the above
expression by letting $T_2 = T_1$ in the first two factors on the right. This yields

$$T_2^4 - T_1^4 \approx (2T_1^2)(2T_1)(T_2 - T_1) = 4T_1^3(T_2 - T_1) \tag{3.103}$$

Substituting the expression $T_2^4 - T_1^4$ given by Equation 3.103 into Equation 3.101
yields

$$q_R = \sigma\,G_{12}\,\epsilon A_D(4\,T_1^3)(T_2 - T_1) \tag{3.104}$$

Thus, if $T_2 - T_1$ is small, the temperature difference acts as a forcelike quantity
for radiative heat transfer.

By combining the various modes of heat transfer rates, one has

$$\frac{\Delta Q}{\Delta t} = q_L - q_c - q_F - q_R \tag{3.105}$$

$$[\text{transfer rate}] = \begin{bmatrix} \text{laser power} \\ \text{absorbed} \end{bmatrix} - \begin{bmatrix} \text{conduction} \\ \text{loss rate} \end{bmatrix} - \begin{bmatrix} \text{convection} \\ \text{loss rate} \end{bmatrix} - \begin{bmatrix} \text{radiation} \\ \text{loss rate} \end{bmatrix}$$

The corresponding rate of temperature change can be determined from Equation
3.98:

$$\frac{\Delta T}{\Delta t} = \frac{1}{mc} \frac{\Delta Q}{\Delta t} = \frac{1}{mc} [q_L - q_c - q_F - q_R] \qquad (3.106)$$

Substituting in Equation 3.106 the expressions for q_c, q_F, and q_R from Equations 3.99, 3.100, and 3.104 yields

$$\frac{\Delta T}{\Delta t} = \frac{1}{mc} [q_L - \left(\frac{kA_w}{l} + h A_D + 4T_1^3 \sigma G_{12} \epsilon A_D\right)(T_2 - T_1)] \qquad (3.107)$$

where $(T_2 - T_1)$ has been factored out of the last three terms. When the body comes to equilibrium, $\Delta T/\Delta t$ will be zero. If Equation 3.107 i solved for $T_2 - T_1$ when equilibrium is established, one obtains

$$T_2 - T_1 = \frac{q_L}{\dfrac{kA_w}{l} + h_c A_D + 4T_1^3 \sigma G_{12} \epsilon A_D} \qquad (3.108)$$

A thermocouple is usually attached to the disk. The output voltage is proportional to the temperature difference $(T_2 - T_1)$ and is a direct measurement of the laser power. To make the detector more sensitive, the chamber can be evacuated. This eliminates the convection heat loss. In addition, the conduction losses are reduced by using material with low thermal conductivity to support the disk. The radiation loss is minimized by keeping the area of the disk as small as possible relative to the laser beams that are to be measured. Example 3.44 illustrates the use of Equation 3.108.

EXAMPLE 3.44   Temperature rise of a small disk
. . . .

Given: A small blackbody disk with a total area of 0.003 ft² is in an evacuated chamber. Conduction heat losses can be neglected. The temperature of the chamber is 530°R. A 10-mW laser beam strikes the disk; the geometric factor is approximately 1 and the emissivity ϵ is 1.

Find: The rise in temperature of the disk.

Solution: Convection can be neglected since the chamber is evacuated. It is given that conduction can be neglected. Equation 3.108 then becomes

$$T_2 - T_1 = \frac{q_L}{4T_1^3 \sigma \epsilon A_D}$$

$$\sigma = 0.171 \times 10^{-8} \text{ Btu/hr·ft}^2\text{·°R}^4$$

$$A_D = 0.003 \text{ ft}^2$$

$$T_1 = 530°R$$

The laser power is given in milliwatts.

$$10mW = 0.010 \text{ W}$$

$$1 \text{ W} = 1 \text{ J/sec} = 1/4.18 \text{ cal/sec}$$

$$0.010 \text{ W} = \frac{0.010}{4.18} \text{ cal/sec} = 0.0024 \text{ cal/sec}$$

$$1 \text{ cal/sec} = \frac{1}{252} \text{ Btu/sec}$$

$$0.0024 \text{ cal/sec} = \frac{0.0024}{252} \text{ Btu/sec} = 9.5 \times 10^{-6} \text{ Btu/sec}$$

$$1 \text{ Btu/sec} = 3600 \text{ Btu/hr}$$

$$9.5 \times 10^{-6} \text{ Btu/sec} = (9.5 \times 10^{-6})(3600) \text{ Btu/hr} = 0.0342 \text{ Btu/hr}$$

The laser power is 0.0342 Btu/hr.

$$T_2 - T_1 = \frac{0.0342 \text{ Btu/hr}}{(4)(530)^{3\circ}\text{R}^3 (0.171 \times 10^{-8} \text{ Btu/hr·ft}^2\text{·R}^{\circ 4})(1)(0.003 \text{ ft}^2)}$$

$$T_2 - T_1 = 11.2 \text{ R}^\circ = 11.2 \text{ F}^\circ$$

SECTION 3.4 EXERCISES

1. Define heat flow rate and express its basic unit in both the SI and English systems of units.

2. What thickness of fiberglass insulation is required for a heat flow rate of 5 Btu/hr through 1 ft² of insulation if the temperature difference is 50 F°?

3. What is the thermal conductivity of a type of wood if a wall made of the wood is 6 inch thick and heat flows through the wall at a rate of 200 Btu/hr when the temperature difference on opposite sides of the wall is 30 F° and the wall has an area of 40 ft²?

4. If the heat flow through one outside wall (area 60 ft²) is 5000 Btu/hr on a cold day, what is the heat flow out of another outside wall of the house if the second wall is made of the same material and is the same thickness as the first but has an area of only 40 ft²?

5. Heat is transferred from a pipe to water flowing through the pipe. The temperature difference between the pipe wall and the average water temperature remains 30 F°. If the heat transfer coefficient is 60 Btu/hr·ft²·F° and the inner diameter is 0.2 ft, what is the amount of heat transferred in 2 hr if the pipe is 10 ft long?

6. If the flow in Problem 5 is turbulent, what would the heat transfer coefficient be if the flow velocity is doubled and all other factors remain the same?

7. A 0.5-m-diameter sphere is inside a 1.0-m-diameter sphere. Both spheres have the same center and are blackbodies. If the inner sphere is maintained at a temperature of 600°R and the outer sphere is maintained at 600°R, how much heat is transferred by radiation in 2 hr? The geometric factor for the inner sphere is 1.0. Use the inner sphere area.

8. State the mathematical formula for the heating or cooling rate. What are the appropriate units? How does one determine if the rate is to be interpreted as heating or cooling?

9. State three methods by which heat energy is transferred to produce heating or cooling.

10. Suppose that one places a bowl of gelatin in the refrigerator to cool. The initial temperature of the gelatin is 200°F. After 5 min its temperature is 150°F; after 10 min

110°F; after 15 min 80°F; after 20 min 60°F; after 25 min 50°F; after 30 min 44°F; after 35 min 40°F; and after 40 min 38°F. Plot the cooling rate versus time.

11. Suppose that one places a steel ingot in a high-temperature furnace. In the early stages of the heating, the heating rate is nearly constant. If after the first 10 sec the temperature of the ingot goes from 70°F (room temperature) to 200°F, what is the temperature of the ingot after 20 sec?

S U M M A R Y

The concept of rate is important in all energy systems. Any change in a system takes place in a certain elapsed time. Rate is the ratio of change in a physical quantity to the time required for that change. This chapter has dealt with the basic rates in four energy systems—mechanical, fluid, electromagnetic, and thermal. Many other rates commonly are used in science and technology. One rate is so important that it is considered to be a separate unifying concept. Power is the rate of energy transfer and will be discussed in detail in Chapter 7.

Table 3.4 summarizes the basic rates presented in this chapter.

Table 3.4 Rate as a unifying concept

Energy System	$\text{Rate} = \dfrac{\text{Displacementlike Quantity}}{\text{Elapsed Time}}$	
Mechanical Translational	$\text{Velocity} = \dfrac{\text{Displacement}}{\text{Elapsed Time}}$	
Rotational	$\text{Angular Velocity} = \dfrac{\text{Angular Displacement}}{\text{Elapsed Time}}$	
Fluid	$\text{Volume Flow Rate} = \dfrac{\text{Volume Displaced}}{\text{Elapsed Time}}$	
	$\text{Mass Flow Rate} = \dfrac{\text{Mass Displaced}}{\text{Elapsed Time}}$	
Electrical	$\text{Current} = \dfrac{\text{Charge Transferred}}{\text{Elapsed Time}}$	
Magnetic	$\text{Induced Voltage} = -N\left(\dfrac{\text{Change in Magnetic Flux}}{\text{Elapsed Time}}\right)$	
Thermal	$\text{Heat Flow Rate} = \dfrac{\text{Heat Energy Transferred}}{\text{Elapsed Time}}$	

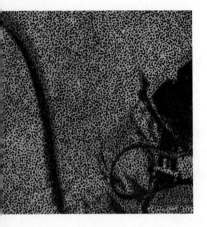

Momentum

INTRODUCTION

Moving objects and fluids possess a very important physical quantity called "momentum." Linear momentum, a vector quantity, depends on the mass and the linear velocity of the object. Angular momentum, also a vector quantity, depends on the moment of inertia and the angular velocity of a rotating mass. Fluid momentum depends on the velocity and density of the fluid.

This chapter explains the mathemtical relationships involved with linear, angular, and fluid momentum. Elastic and inelastic collisions are considered, as well as the law of conservation of momentum. The forces resulting from the change in the direction of a fluid's velocity are also studied.

The law of conservation of momentum is an important unifying principle in physics. Whenever momentum is transferred from one object to another, or from one system to another, total momentum is always conserved. The concept of momentum is most important in mechanical and fluid systems.

4.1................ OBJECTIVES...

MOMENTUM IN MECHANICAL SYSTEMS

Upon completion of this section, the student should be able to

- Define the following terms, state the appropriate units in the mks (SI) and the cgs systems, and give the equation for each:

  Linear momentum
  Angular momentum
  Impulse
  Angular impulse
  Moment of inertia

- Given two of the following quantities, determine the third:

  Mass of an object
  Velocity of the object
  Momentum of the object

- Explain each of the following concepts in a short paragraph:

  Conservation of linear momentum in mechanical systems
  Conservation of angular momentum in mechanical systems

- Given two of the following quantities, determine the third:

  Moment of inertia of an object
  Angular velocity of the object
  Angular momentum of the object

- Given all the following quantities except one describing a linear collision, determine the unknown quantity:

  Mass of first object
  Initial velocity of first object
  Final velocity of first object
  Mass of second object
  Initial velocity of second object
  Final velocity of second object

- State the right-hand rule for determining the direction of a rotational vector.

- Label a diagram of a spinning gyroscope with the following vectors:

  Angular velocity, ω
  Angular momentum, **L**
  Force, **F,** *acting on the gyroscope*
  Torque, τ*, acting on the gyroscope*
  Angular momentum change, Δ**L**
  Angular velocity of precession, Ω

- Given the magnitudes of two of the following vectors, find the magnitude of the third:

  Angular momentum, **L**
  Torque, τ
  Angular velocity of precession, Ω

- List the major parts of a gyrocompass and explain the function of each.

- List the two methods for stabilizing a gyrocompass and explain the effect of each on the magnitude of the torque, τ, or the angular momentum, **L.**

DISCUSSION

A large ocean liner has a tremendous mass. When such a liner approaches a dock (Figure 4.1), the crew reduces the ship's velocity to a very low value. As the enormous ship makes contact with the dock, a huge force must be absorbed by the dock in order to stop the ship's motion completely.

A bullet fired from a gun at high velocity has a tremendous impact on a target, although the mass of the bullet may be only a few grams. The huge force exerted by the bullet on the target can be seen by the destructive power of even a small-caliber bullet fired at a glass bottle (Figure 4.2). If the bottle is filled with water, the velocity of the bullet is reduced more upon impact and the destruction of the target is even greater. Rapidly moving objects are difficult to stop.

These examples illustrate the concept of momentum. Momentum is an important physical quantity related to the motion of objects. Both a small object moving with a high velocity and a large object moving with a low velocity possess relatively large amounts of momentum and exert large forces on other objects when stopped. Of course, a large object moving with a high velocity possesses even more momentum. A knowledge of both mass and velocity is necessary in determining the momentum of a body.

FIGURE 4.1 Ocean liner approaching dock.

FIGURE 4.2 Bullet shattering a bottle.

Consideration of momentum for moving objects is divided into two classifications.

1. Linear momentum if the object is moving in a straight line.
2. Angular momentum if the object is rotating.

Linear Momentum

Definition of Momentum   Linear momentum of a single object of mass m moving with velocity v is defined by

$$\mathbf{p} = m\mathbf{v} \tag{4.1}$$

where
   $\mathbf{p}$ = linear momentum
   m = mass
   $\mathbf{v}$ = velocity

To indicate that they are vectors **p** and **v** are set in boldface type.)

Since velocity is a vector quantity (one that has both magnitude and direction), momentum also must be a vector quantity if Equation 4.1 is to hold true. The direction of the momentum vector is the same as that of the velocity vector, but its magnitude and units are different from those of the velocity vector. The vector nature of momentum will be considered more thoroughly later in this chapter. As indicated in Equation 4.1, vector quantities are set in bold face type when the direction, as well as the magnitude, is important in the problem. In momentum problems, both magnitude and direction are important.

Units for momentum are consistent with Equation 4.1. If mks units are used for mass (kg) and velocity (m/sec), then the momentum units obtained are kg·m/sec. However, if the cgs system is used, mass in grams times velocity in cm/sec yields momentum in units of g·cm/sec. Examples 4.1 and 4.2 illustrate the calculation of momentum for two greatly different masses.

EXAMPLE 4.1   Linear momentum of large mass
. . . .

Given: An automobile whose mass is 2500 kg is moving with a velocity of 70 km/hr. (Assume that the car is moving in a straight line.)

Find: Its linear momentum.

Solution: The momentum of the automobile will be in the same direction as its velocity.

$$\mathbf{p} = m\mathbf{v}$$

$$\mathbf{p} = (2500 \text{ kg})\left(\frac{70{,}000 \text{ m}}{\text{hr}}\right)\left(\frac{1 \text{ hr}}{3600 \text{ sec}}\right)$$

$$\mathbf{p} = 48{,}600 \text{ kg·m/sec} \quad \text{(in the direction of the velocity)}$$

EXAMPLE 4.2   Linear momentum of small mass
. . . .

Given: An electron whose mass is 9.1×10^{-28} g is moving in an electrical wire at an effective velocity of approximately 0.1 cm/sec.

Find: Its momentum.

Solution: $\mathbf{p} = m\mathbf{v}$

$\mathbf{p} = (9.1 \times 10^{-28} \text{ g})(1 \times 10^{-1} \text{ cm/sec})$

$\mathbf{p} = 9.1 \times 10^{-29} \text{ g·cm/sec}$ (a small momentum!) (in the direction of the velocity)

Impulse and Momentum Change   When a force is applied to a moving object, a velocity change is produced. The velocity may either increase or decrease, depending on the direction in which the force is applied to the object. A change in the velocity requires a corresponding change in the momentum of the object, because momentum depends on velocity, as indicated in Equation 4.1. The amount of change in momentum is equal to the product of the force and the elapsed time during which the force acts. This relationship is expressed by

$$\Delta\mathbf{p} = \mathbf{F}\Delta t \tag{4.2}$$

where

$\Delta\mathbf{p}$ = change in momentum
 $\mathbf{F}$ = average force acting
Δt = elapsed time during which the force acts

The term applied to the product of force and time is "impulse."

The concepts of impulse and momentum change are understood better when analyzed with the following short derivation. The derivation shows that Equation 4.2 is just another way of expressing Newton's second law of motion.

$$\mathbf{F} = m\mathbf{a} \quad \text{(Newton's second law)}$$

$$\mathbf{F} = m\frac{\Delta\mathbf{v}}{\Delta t} \quad \left(\text{since } \mathbf{a} = \frac{\Delta\mathbf{v}}{\Delta t} = \frac{\text{change in velocity}}{\text{time elapsed}}\right)$$

$$\mathbf{F}\Delta t = m\frac{\Delta\mathbf{v}}{\Delta t}\Delta t \quad \text{(multiplying both sides by } \Delta t)$$

$$\mathbf{F}\Delta t = m\Delta\mathbf{v} = m\mathbf{v}_2 - m\mathbf{v}_1 = \mathbf{p}_2 - \mathbf{p}_1 = \Delta\mathbf{p}$$

Impulse = Change in momentum

From this derivation, impulse is shown to be equivalent to the change in momentum produced when a force is applied to an object. Consider the case of a baseball being struck by a bat. The baseball is given its initial momentum and velocity by the pitcher. The direction of the quantities is, of course, toward the batter. As the bat makes contact with the ball, an enormous force is applied to the ball for the short time during which the two objects are in contact. This force is large enough to cause a considerable change in the momentum and velocity of the ball. The direction of the ball is changed as well as its speed. The impulse, or the product of the average force applied and the time during which the bat and the ball are in contact, is equal to the change in momentum of the ball. Examples 4.3 and 4.4 illustrate the use of Equation 4.2 in solving problems.

EXAMPLE 4.3   Calculation of change in velocity
. . . .

Given: A constant force of 10 N is applied to a 25-kg mass for a time of 2 sec.

Find: **a.** Impulse.
                b. Change in velocity.

Solution: **a.** Impulse is defined as $F\Delta t$.

$$F\Delta t = (10\ N)(20\ sec)$$

$$F\Delta t = 20\ N\cdot sec \quad \text{(in the direction of the force)}$$

b. $F\Delta t = m\Delta v$

$$\Delta v = \frac{F\Delta t}{m}$$

$$\Delta v = \frac{20\ N\cdot sec}{25\ kg}$$

$$\Delta v = 0.8\ \frac{\frac{kg\cdot m}{sec^2}\cdot sec}{kg}$$

$$\Delta v = 0.8\ m/sec \quad \text{(in the direction of the force)}$$

Thus, if the mass is initially stationary, after 2 sec it has a velocity of 0.8 m/sec, in the direction of the force.

Question.   Is the result of $\Delta v = 0.8$ m/sec valid regardless of the initial velocity of the mass?

EXAMPLE 4.4   Calculation of duration of force
. . . .

Given: A constant force of $+5$ N is applied to a 25-kg mass and changes its velocity by $+10$ m/sec.

Find: The duration of the applied force.

Solution: $F\Delta t = m\Delta v$

$$\Delta t = \frac{m\Delta v}{F}$$

$$\Delta t = \frac{(25 \text{ kg})(10 \text{ m/sec})}{5 \text{ kg·m/sec}^2}$$

$$\Delta t = 50 \text{ sec}$$

Figure 4.3 represents a constant force F_1 applied to an object and the velocity of that object, as functions of time. The impulse is the area $F_1 \Delta t_1$. This impulse causes a momentum change of $m\Delta v$, where m is the mass of the object and Δv is the change in its velocity. The acceleration is the slope of the velocity versus time curve, $\Delta v/\Delta t$.

Figure 4.4 shows a larger force F_2 applied for a shorter period Δt_2 to produce the same total impulse and the same net change in momentum ($m\Delta v$). Equal areas (shaded) under the force versus time curves represent the same impulse and so cause the same change in momentum. For example, an applied force F is increased a hundred times (to $100F$) and the time of application Δt is decreased to

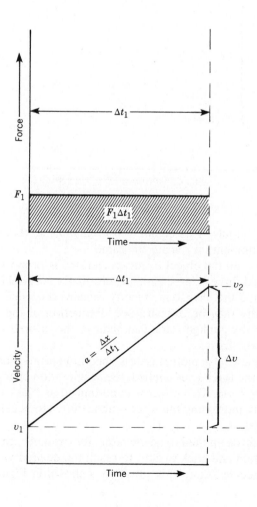

FIGURE 4.3 Small force over a long time.

FIGURE 4.4 Large force over a short time.

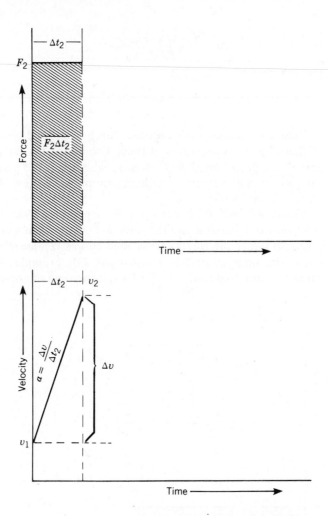

$\Delta t/100$ (the impulse $F\Delta t$ equals $100\ F\ \Delta t/100$), so that both the impulse and the change in momentum remain the same.

The stress on the object being accelerated is proportional to the applied force. If the applied force is large enough deformation may occur, although Equation 4.2 still holds and the change in velocity remains the same. A machine brought up to speed rapidly may be overstressed. Reduction of the force and increase in the duration of the applied force can achieve the desired change in momentum, but with reduced stress.

A machine whose motion is described in Figure 4.3 has an initial velocity v_1. At the instant the force F_1 is applied, the acceleration changes, almost abruptly, from zero to $a = F_1/m$. This effect is experienced as "jerk," which can be eliminated by gradually increasing the force—and, thus, the acceleration—from zero to the desired level.

The net force applied to an elevator, for example, changes smoothly from zero to a maximum and back to zero to reach the desired velocity, without the abrupt change in acceleration. This situation is shown in Figure 4.5. The impulse is the

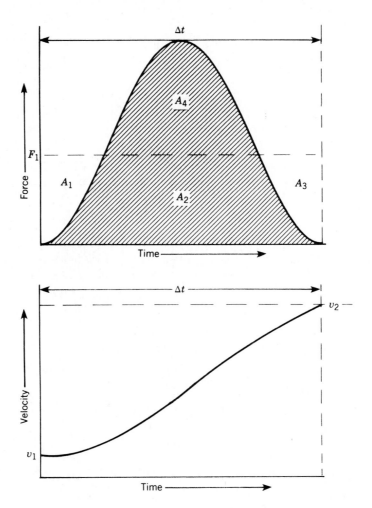

FIGURE 4.5 Plots of a gradually changing force versus time and the corresponding velocity versus time.

same as in Figures 4.3 and 4.4, but in the curve described in Figure 4.5, there is no abrupt jerk in the motion. The impulse depicted in Figure 4.5 is the sum of the areas A_2 and A_4. The impulse associated with a constant force F_1 (dashed line in Figure 4.5) is the sum of the areas A_1, A_2, and A_3. These impulses are equal if $A_1 + A_3 = A_4$. The average force, F, then produces the same impulse as the variable force.

Example 4.5 demonstrates how to determine the change in the momentum of an object from an impulse graph.

EXAMPLE 4.5 Impulse graph
. . . .

Given: A 10-kg object initially at rest is subjected to the net force described in the graph shown.

Find: The final velocity of the mass.

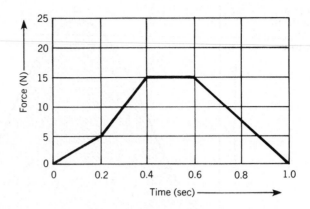

Solution: The area of one square of the graph is 5 N × 0.20 sec = 1 N·sec. The area under the impulse curve is approximately 8.5 squares.

Impulse = 8.5 squares × 1 N·sec/square

$$\mathbf{F}\Delta t = 8.5 \text{ N·sec}$$

$$\mathbf{F}\Delta t = m\Delta\mathbf{v}$$

$$\Delta\mathbf{v} = \frac{F\Delta t}{m}$$

$$\Delta\mathbf{v} = \frac{8.5 \text{ kg·m/sec}}{10 \text{ kg}}$$

$$\Delta\mathbf{v} = 0.85 \text{ m/sec}$$

$$\mathbf{v}_f = \mathbf{v}_0 + \Delta\mathbf{v} = 0 + 0.85 = 0.85 \text{ m/sec} \qquad \text{(in the direction of the average force)}$$

In an automotive engine, a cam working against a valve spring smoothly changes the acceleration of the valve. If the acceleration is too large or abrupt, the valve may "float" and not remain in contact with the cam, causing damage to the engine. Use of the proper cam shape and the maintenance of engine speed below the design limit decrease the applied force (and acceleration) and increase the duration of acceleration.

Newton's third law states that for every action, there is an equal and opposite reaction; therefore, if object *A* imparts an impulse to *B*, then *B* imparts an equal and opposite impulse to *A*. Deformation of both objects may then occur if the stress produced exceeds their elastic limits. Care must be taken to avoid this in all mechanical systems. Aircraft wings are subject to variable impulse loadings during flight, because of turbulent air or landings. Little can be done about the magnitude of the impulse (net change in momentum) during these situations. However, the force element of the impulse can be reduced by allowing the wings to flex, thereby increasing the duration of the impulse. This controlled deformation serves as a shock absorber. Aircraft landing gears include a shock-absorbing system that, for a given impulse, increases the duration Δt of the applied force and reduces the magnitude of the applied force. The elasticity of the shroud lines of a parachute also serves this shock-absorbing function and makes the sudden opening of the chute and reduction of momentum more tolerable.

In crash situations, final momentum is reduced to zero. The impulse experienced by the occupants, therefore, is equal to their initial momentum. The crash time must be lengthened to reduce the force applied to the occupants. This lengthening of time can be realized, to some extent, by designing the vehicle to deform more gradually, in a controlled manner. (Incidentally, inflatable air bags between the driver and the automobile frame are designed to do just that.)

Conservation of Momentum   According to Newton's third law, for every force acting on a body there is an equal but opposite force acting on a different body. Thus, the total impulse and corresponding momentum of the "system" are zero. Mathematically this can be shown for two interacting bodies by writing, for the system as a whole,

$$\mathbf{F}_a \Delta t_a + \mathbf{F}_b \Delta t_b = m_a \Delta \mathbf{v}_a + m_b \Delta \mathbf{v}_b \tag{4.3}$$

where

$\mathbf{F}_a$ = force exerted on body a by body b
$\mathbf{F}_b$ = force exerted on body b by body a

Then by Newton's third law,

$$\mathbf{F}_b = -\mathbf{F}_a$$

Since both forces act for the same time,

$$\Delta t_a = \Delta t_b$$

Thus, the left-hand side of Equation 4.3 is 0. Equation 4.3 then becomes

$$m_a \Delta \mathbf{v}_a + m_b \Delta \mathbf{v}_b = 0 \tag{4.4}$$

Equation 4.4 states that the total change in momentum is 0. Thus, the momentum before the interaction must equal the momentum after the interaction.

Since the total momentum of an isolated system does not change, we say it is conserved. An isolated system is one in which there are no external, unbalanced forces acting on the system. For example, the forward momentum of a baseball does not change noticeably from the time that it leaves the pitcher's hand until it is struck by the bat because the only force acting upon it along the direction of motion is air resistance, which is relatively small. A large momentum is given to a golf ball on a drive. Because the golf ball is in the air for a longer time than the baseball, much of its momentum gradually is transferred to the molecules in the air. However, the momentum gained by the molecules in the air is equal to the momentum lost by the ball. In this case, the isolated system must include both the golf ball and the surrounding air. Conservation of momentum applies to all isolated systems, whether the system be colliding atoms, billiard balls, automobiles, or galaxies of stars.

FIGURE 4.6 Boater stepping from boat.

Consider an isolated system of boat and boater. The boater steps from the boat to a dock, as illustrated in Figure 4.6. The boat and boater are initially at rest. As the boater steps forward from the boat, a force is applied to the boat, as well as to the boater. The boater pushes back on the boat, but the boat pushes forward on the boater with an equal force. These forces cause the boater of mass m_1 to move forward with velocity v_1 at the same time that the boat of mass m_2 moves backward with velocity v_2. Because momentum is conserved, the product m_1v_1 (the boater's momentum) is equal to the product m_2v_2 (the momentum of the boat). The initial momentum of boater and boat was zero before the boater stepped from the boat to the dock. After the boater steps from the boat, the total momentum must still be zero, since momentum is conserved. But the boater's momentum when stepping from the boat is m_1v_1, whereas the boat's momentum is m_2v_2. The sum of $m_1v_1 + m_2v_2$ must remain zero. Thus $m_1v_1 = -m_2v_2$. The minus sign indicates that the boater and boat move away from each other in opposite directions, each with the same numerical value of momentum. The total momentum of boater and boat in this system is zero before, during, and after the boater steps from boat to dock, and momentum is conserved.

An important and interesting observation to be made from the above example is that a boat with a much larger mass will "recoil" with a lower velocity. Thus, the velocity of the boat backward will be much greater for a lightweight canoe than for a heavy runabout. To step from a heavy boat is much easier than from a lightweight boat. Conservation of momentum in this example can be expressed as

$$m_1\mathbf{v}_1 = -m_2\mathbf{v}_2 \tag{4.5}$$

where

m_1 = mass of boater
m_2 = mass of boat
$\mathbf{v}_1$ = velocity of boater
$\mathbf{v}_2$ = velocity of boat

An important application of the conservation of momentum occurs in the firing of a rifle. When the rifle fires, the expanding gases push on both the bullet and the rifle with equal force. The force produces a much larger velocity on the small bullet than on the massive rifle. As a result, the bullet leaves the barrel of the rifle with a very large velocity. The massive rifle, however, recoils with a much lower velocity. A relation equivalent to Equation 4.5 also applies to this case, since momentum is conserved. Equation 4.6 summarizes the results, and Example 4.6 shows how recoil velocity for a rifle can be calculated from

$$m_2 \mathbf{v}_2 = -m_1 \mathbf{v}_1 \tag{4.6}$$

where

m_1 = mass of bullet
m_2 = mass of rifle
$\mathbf{v}_1$ = velocity of bullet
$\mathbf{v}_2$ = velocity (recoil) of rifle

EXAMPLE 4.6   Conservation of momentum and rifle recoil velocity
. . . .

Given: A bullet of mass 5 g is fired from a rifle whose mass is 6 kg. The velocity of the bullet is 675 m/sec when it leaves the rifle barrel.

Find: The recoil velocity of the rifle.

Solution: Both rifle and bullet considered separately—but as part of the same isolated system—are at rest before firing. Thus, the initial momentum of bullet and rifle is zero. Since the momentum is conserved, the final momentum of the bullet and rifle is zero.

Using Equation 4.6,

$$m_2 \mathbf{v}_2 = -m_1 \mathbf{v}_1$$

The negative sign is an important reminder that the recoil velocity of the rifle is in a direction opposite to the velocity of the bullet. The magnitudes of $m_2 v_2$ and $m_1 v_1$ are equal, so we can write

$$|m_2 \mathbf{v}_2| = |m_1 \mathbf{v}_1|$$

The magnitude of the recoil velocity of the rifle is given by v_2, which is

$$v_2 = \left| \frac{m_1 \mathbf{v}_1}{m^2} \right|$$

Substitution of the numerical values yields

$$v_2 = \frac{5 \text{ g} \times 675 \text{ m/sec}}{6000 \text{ g}}$$

$$v_2 = 0.56 \text{ m/sec}$$

The rifle kicks *back* with a recoil velocity of 0.56 m/sec.

Another common application of the law of conservation of momentum involves the collision of two or more objects. Each object may have momentum before the collision, and each may have momentum after the collision. When Equation 4.4 is generalized to account for the collision of two or more objects, Equation 4.7 is obtained:

$$(m_a \mathbf{v}_{a_1} + m_b \mathbf{v}_{b_1} + m_c \mathbf{v}_{c_1} + \cdots) = (m_a \mathbf{v}_{a_2} + m_b \mathbf{v}_{b_2} + m_c \mathbf{v}_{c_2} + \cdots) \quad (4.7)$$

Momentum before the Interaction = Momentum after the Interaction

Equation 4.7 can also be written as

$$(\mathbf{P}_a + \mathbf{P}_b + \mathbf{P}_c + \cdots) = (\mathbf{p}_a + \mathbf{p}_b + \mathbf{p}_c + \cdots) \quad (4.8)$$

where

P = momentum of object(s) before collision

p = momentum of object(s) after collision

Equation 4.8 is used to solve a problem in Example 4.7, which considers a bowling ball colliding with a stationary pin in such a manner that all motion takes place in a straight line.

EXAMPLE 4.7   Conservation of momentum
. . . .

Given: A 7-kg bowling ball moving with a velocity of 2 m/sec collides head-on with a bowling pin. Immediately after the collision, the 1.0-kg pin moves with a velocity of 6 m/sec.

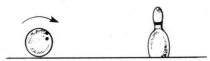

Find: The velocity of the ball immediately after the collision.

Solution: Equation 4.8 applies, since momentum must be considered:

$$(\mathbf{P}_{ball} + \mathbf{P}_{pin}) = (\mathbf{p}_{ball} + \mathbf{p}_{pin})$$

Momentum before = Momentum after

$\mathbf{P}_{ball}$ = mass of ball × velocity before collision

      = (7 kg)(2 m/sec) = 14 kg·m/sec

$\mathbf{P}_{pin} = 0$,   since the pin has zero velocity before the collision

$\mathbf{p}_{ball}$ = (7 kg)(unknown velocity, v_2)

$\mathbf{p}_{pin}$ = (1.0 kg)(6 m/sec) = 6 kg·m/sec

Thus, use of the equation above yields

$$(14 \text{ kg·m/sec} + 0) = [(7 \text{ kg})(v_2) + 6 \text{ kg·m/sec}]$$
Before   After

Solve for v_2:

$$(7 \text{ kg})(v_2) = 8 \text{ kg·m/sec}$$

$$v_2 = 1.1 \text{ m/sec}$$

The bowling ball thus continues straight ahead after the collision, with an *initial* speed of 1.1 m/sec. **Note:** Friction of the ball against the floor and the spin on the ball will alter this speed as the ball continues rolling. Friction represents an external force; therefore, this system is not totally isolated *after* collision.

Momentum is conserved in all collisions. When a certain type of collision occurs, in which all the kinetic energy possessed by the objects remains as kinetic energy after the collision, it is called an "elastic collision." When some, or all, of the kinetic energy is changed to other forms of energy during the collision, an "inelastic collision" results. (Kinetic energy can be changed to heat energy as a result of friction or can be used in changing the shapes of the objects on collision.) Of course, the total amount of all forms of energy is conserved in both types of collisions. Perfectly elastic collisions are not common except in collisions between atoms and subatomic particles. Some collisions, like those between billiard balls, are almost elastic. Most other real-world situations involve inelastic collision.

Example 4.8 illustrates the application of Equation 4.7 to a collision in two dimensions. For this particular case, energy is conserved.

EXAMPLE 4.8 Conservation of momentum in a two-dimensional collision
· ·

Given: Two spheres collide as shown in the diagram. The initial velocity of sphere a is 10 m/sec and the initial velocity of sphere b is zero. The mass of sphere a is 1.0 kg and the mass of sphere b is 1.0 kg.

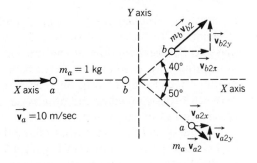

Before collision   After collision

Find: The magnitude of the velocities v_{b_2} and v_{a_2} after the collision.

Solution: The vector momentum before the collision must equal the total vector momentum after the collision. Thus, the sum of the momentum components in the x direction before the collision must equal the sum of the components in the x direction after the collision. Using subscript 1 before and subscript 2 after collision, we write

$$m_a v_{a_1 x} = m_a v_{a_2 x} + m_b v_{b_2 x}$$

$$(1)(10) = (1)\ v_{a_2} \cos 50° + (1)\ v_{b_2} \cos 40°$$

Also the sum of the components in the y direction before the collision must equal the sum of the components in the y direction after the collision.

$$0 = m_a v_{a_2 y} + m_b v_{b_2 y}$$

$$0 = -1.0\ v_{a_2} \sin 50° + 1.0\ v_{b_2} \sin 40°$$

$$v_{a_2} = 0.839\ v_{b_2}$$

Substitute for v_{a_2} in the x-direction equation.

$$10 = 0.839\ v_{b_2} \cos 50° + v_{b_2} \cos 40°$$

$$10 = 0.539\ v_{b_2} + 0.766\ v_{b_2}$$

$$10 = 1.31\ v_{b_2}$$

$$v_{b_2} = 7.66 \text{ m/sec}$$

Use the relationship between v_{a_2} and v_{b_2} to find v_{a_2}.

$$v_{a_2} = (0.839)(7.66)$$

$$v_{a_2} = 6.43 \text{ m/sec}$$

In Chapter 6, it will be shown that this is an elastic collision, that is, kinetic energy is also conserved in the collision.

Angular Momentum

Definition of Angular Momentum   Angular momentum is very similar to linear momentum. In practice, it often refers to objects rotating about a fixed axis. A massive flywheel rotating at the end of an engine crankshaft has large angular momentum. It is used to smooth the rotation of an engine between the firing of the pistons. Because of its large angular momentum, it tends to keep rotating evenly between firings, avoiding the otherwise jerky motion that would accompany the spaced firings of the pistons.

To understand the importance of angular momentum, consider a body traveling in a circular path as shown in Figure 4.7a. The force acting on the body can be resolved into two components, one component F_r directed toward the center of the circle and one component F_t tangent to the circle. According to Newton's second law

$$\mathbf{F}_t = m\mathbf{a}_t \tag{4.9}$$

FIGURE 4.7 Forces
affecting rotational motion.

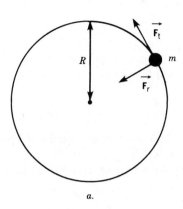

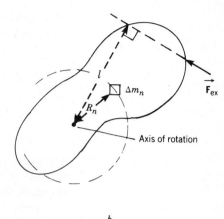

a.

b.

and

$$F_r = ma_r = m\omega^2 R \tag{4.10}$$

where

a_t = tangential acceleration
a_r = radial acceleration (centripetal acceleration)

If both sides of Equation 4.9 are multiplied by R, the radius of the circle, the equation becomes

$$RF_t = mRa_t \tag{4.11}$$

The left-hand side of Equation 4.11 is the total torque about the center of the circle. No torque about the center is produced by $\mathbf{F}_r$ since the moment arm is zero. From Equations 3.30 and 3.32, in Chapter 3, one has

$$a_t = R\alpha \tag{4.12}$$

Substituting this value for a_t in Equation 4.11 yields

$$\tau = mR^2\alpha \tag{4.13}$$

where

τ = torque
m = mass
R = radius of the circle
α = angular acceleration

The quantity mR^2 is called the moment of inertia of the mass about the point of rotation and is usually given the symbol I. Thus,

$$\tau = I\alpha \tag{4.14}$$

The above derivation can be extended easily to a rigid body rotating about a fixed axis as shown in Figure 4.7b. Each small element of mass in the body travels in a circular path. For each small element of mass Δm_n, at a distance R_n from the axis, one may write

$$\tau_n = \Delta m_n R_n^2 \alpha \tag{4.15}$$

Since the body is rigid, each element of mass must have the same angular acceleration. If the equation is summed over all the elements of mass, Equation 4.15 becomes

$$F_{ex}l + \Sigma\tau_n = \alpha\Sigma\Delta m_n R_n^2 \tag{4.16}$$

For every internal force there is another equal but opposite internal force. Thus the sum of the internal torques, $\Sigma\tau_n$, is zero. The force F_{ex} is an externally applied resultant force producing the rotation, and l is its moment arm. Thus, Equation 4.16 reduces to

$$\tau_{ex} = \alpha\Sigma\Delta m_n R_n^2 \tag{4.17}$$

where

$$\tau_{ex} = \text{externally applied torque}$$
$$\Sigma\Delta m R_n^2 = \text{body's moment of inertia, } I$$

Equation 4.17 is usually written as

$$\tau = I\alpha \tag{4.18}$$

where it is understood that τ is the externally applied torque. Since the definition of α is $\Delta\omega/\Delta t$, Equation 4.18 may be written as

$$\tau = I\frac{\Delta\omega}{\Delta t} = \frac{I\omega_2 - I\omega_1}{t_2 - t_1} \tag{4.19}$$

The quantity

$$L = I\omega \tag{4.20}$$

is called the angular momentum. Equation 4.19 becomes by substitution

$$\tau = \frac{L_2 - L_1}{t_2 - t_1} = \frac{\Delta L}{\Delta t} \tag{4.21}$$

Thus net torque equals the rate of change of angular momentum. This is analogous to the equation for linear momentum

$$\mathbf{F} = \frac{\mathbf{p}_2 - \mathbf{p}_1}{t_2 - t_1} = \frac{\Delta\mathbf{p}}{\Delta t}$$

The torque, τ, is the forcelike quantity in Equation 4.21.

The moment of inertia, $I = \Sigma\Delta m_n R_n^2$, is a quantity analogous to the mass in a translational system. The location of the axis of rotation greatly affects the moment of inertia. If a small particle of mass, such as a grain of sand, is spinning about an axis a distance R from the grain of sand, the moment of inertia of the sand is equal to mR^2, where m is the mass of the sand particle. For large objects, consisting of many mass particles at different distances from the axis, the overall moment of inertia must be calculated. It varies according to the shape of the objects and the location of the rotation axis. A few examples of moment of inertia

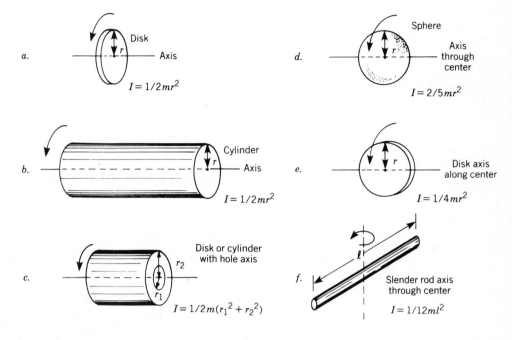

FIGURE 4.8 Moments of inertia for several regularly shaped bodies.

a. Disk — Axis — $I = 1/2 mr^2$

b. Cylinder — Axis — $I = 1/2 mr^2$

c. Disk or cylinder with hole axis — $I = 1/2 m(r_1{}^2 + r_2{}^2)$

d. Sphere — Axis through center — $I = 2/5 mr^2$

e. Disk axis along center — $I = 1/4 mr^2$

f. Slender rod axis through center — $I = 1/12 ml^2$

of regular bodies are given in Figure 4.8. Notice that units of I are always mass-times-distance-squared.

Units for angular momentum can be determined by substitution into Equation 4.20:

$$L = I\omega \rightarrow (\text{kg·m}^2)(\text{rad/sec}) \rightarrow \text{kg·m}^2/\text{sec}$$

Thus, in the mks system, the angular momentum has units of kg·m²/sec. In the cgs system

$$L = I\omega \rightarrow (\text{g·cm}^2)(\text{rad/sec}) \rightarrow \text{g·cm}^2/\text{sec}$$

Note: As discussed in Chapter 3, the radian is a dimensionless unit and can be omitted in expression of units.

Example 4.9 indicates how angular momentum is calculated for a flywheel.

EXAMPLE 4.9 Angular momentum of a flywheel
. . . .

Given: A 30-cm (radius) flywheel with a mass of 12 kg rotating about an axis through its center at 500 rev/min (rpm).

Find: Its angular momentum.

Solution: $L = I\omega$

$I = 1/2 \, mR^2$,   since a flywheel is a disk rotating as shown in Figure 4.8*a*

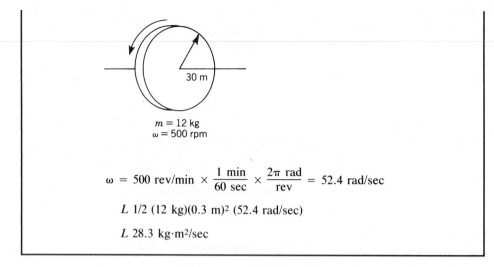

$$\omega = 500 \text{ rev/min} \times \frac{1 \text{ min}}{60 \text{ sec}} \times \frac{2\pi \text{ rad}}{\text{rev}} = 52.4 \text{ rad/sec}$$

$$L \; 1/2 \; (12 \text{ kg})(0.3 \text{ m})^2 \; (52.4 \text{ rad/sec})$$

$$L \; 28.3 \text{ kg·m}^2/\text{sec}$$

Recall that linear momentum can be changed when a force is applied for a given time (impulse). Similarly, angular momentum can be changed when a torque is applied for a given time. Angular impulse is the product of torque (forcelike quantity) and the time during which the torque acts. Equation 4.21 can be written as

$$\Delta L = \tau \Delta t \qquad (4.22)$$

or

$$I \Delta \omega = \tau \Delta t \qquad (4.23)$$

where

$\quad \Delta L$ = change in angular momentum (angular impulse)
$\quad \tau$ = torque
$\quad \Delta t$ = elapsed time
$\quad \Delta \omega$ = change in angular velocity
$\quad I$ = moment of inertia

Relating this to the flywheel on the engine crankshaft, each time a piston fires, an angular impulse—a torque acting for a short time—acts on the engine shaft to increase its angular momentum. Since the moment of inertia of the crankshaft remains fixed, angular velocity is the quantity that increases. Angular velocity is transformed by the differential to increase the linear speed of the car. In Example 4.10, the amount of angular impulse produced by a certain braking torque, when applied to a rotating flywheel, is examined.

EXAMPLE 4.10   Angular impulse and change in angular momentum
· ·

Given: A braking torque of 5 N·m is applied to the flywheel in Example 4.9

for 2.5 sec. Initial angular momentum, as worked out in Example 4.9, is 28.3 kg·m²/sec.

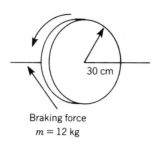

Braking force
$m = 12$ kg

Find: **a.** The angular impulse produced.

b. The angular momentum of the flywheel after the braking action.

Solution: **a.** $\Delta L = \tau \Delta t$

$\Delta L = (5 \text{ N·m})(2.5 \text{ sec})$

$\Delta L = 12.5 \text{ N·m·sec}$

The units for angular impulse appear to cause trouble at first, because they are not the same as those for angular momentum. A short dimensional analysis is necessary to verify that the units are equivalent.

Substitution:

$12.5 \text{ N (m·sec)} = 12.5 \text{ (kg·m/sec}^2)(\text{m·sec})$

$= 12.5 \text{ kg·m}^2/\text{sec}$

b. Thus, the change in angular momentum of the flywheel is 12.5 kg·m²/sec. The angular momentum of the flywheel after the braking action then is found from

Original Value − Change = Final Value

$28.3 \text{ kg·m}^2/\text{sec} − 12.5 \text{ kg·m}^2/\text{sec} = 15.8 \text{ kg·m}^2/\text{sec}$

The law of conservation of angular momentum is as important as the laws of conservation of linear momentum and the law of conservation of energy. It simply states that "the angular momentum of a system remains constant, provided no net external torques are applied"; that is, angular momentum is conserved in an "isolated" system.

An observation that can be made from Equation 4.20 ($L = I\omega$) is that several "hidden variables" are present. Both mass and radius of rotation are included in the moment of inertia term I. These facts are explained further in Examples 4.11 and 4.12.

EXAMPLE 4.11   Angular momentum and ice skaters

. . . .

Given: Skaters practicing pirouettes extend their arms to slow down and hold their arms tight to their body to increase their angular speed.

Find: The reason.

Solution: A skater rotating on smooth ice will conserve angular momentum like all other isolated systems. Thus, angular momentum while arms are extended must equal the angular momentum while arms are tucked in. When arms are extended, the effective radius of rotation of the skater, and therefore I, becomes larger; thus the angular velocity must be less. In other words, the product $I\omega$ must remain constant; as I increases with a larger radius, ω must decrease. The same principle is used by platform divers and ballet dancers.

EXAMPLE 4.12   Merry-go-rounds and angular momentum

. . . .

Given: A merry-go-round turning with negligible friction, and with no one pushing, will slow down when a heavy person gets on.

Find: The reason.

Solution: The angular momentum of the merry-go-round before and after the additional person gets on must remain the same. When the heavy person gets on the merry-go-round, the mass of the system increases. This increased mass causes the value for I to increase in the equation $I\omega$, thus ω must decrease to keep the product unchanged. The product must remain unchanged if angular momentum is to be conserved.

Gyroscopes

Definition of Vectors Before the gyroscopic effect can be discussed, a framework within which the pertinent rotational quantities can be treated as vectors must be developed. This requires a system for determining both the magnitude and the direction of such vector quantities as torque, angular velocity, and angular momentum.

Equations for determining the magnitudes of these quantities have been previously presented. The magnitude of the angular velocity of the flywheel in Figure 4.9, for instance, can be calculated from the equation

$$\omega = \frac{\Delta\theta}{\Delta t} \tag{4.24}$$

where

ω = average angular velocity in rad/sec
$\Delta\theta$ = angle through which the rotating body turns in time Δt, measured in rad
Δt = time during which the rotating body advances through angle $\Delta\theta$

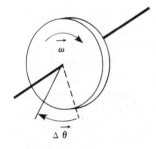

FIGURE 4.9 Rotating flywheel.

The vector nature of angular velocity is evident from the fact that the flywheel can rotate in either of two directions, clockwise or counterclockwise. The vector representing ω must then point in one direction to signify clockwise rotation and a different direction to signify counterclockwise rotation. The standard means of representing this velocity vector is illustrated in Figure 4.10. The angular velocity vector of the clockwise-rotating flywheel lies along the axis of rotation pointing

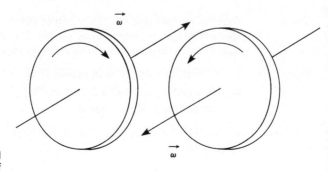

FIGURE 4.10 Clockwise and counterclockwise rotation of a flywheel.

a.   *b.*

FIGURE 4.11   Right-hand rule.

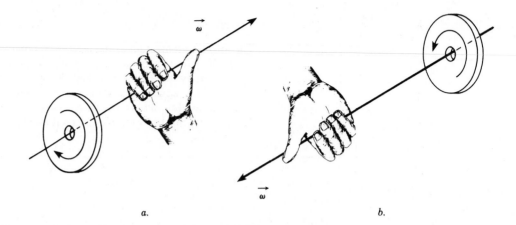

a.   b.

away from the viewer (Figure 4.10*a*). The angular velocity vector of the counterclockwise-rotating flywheel lies along the axis of rotation but points toward the viewer (Figure 4.10*b*).

In general, the vector associated with any rotational quantity lies along the axis of rotation in a direction determined by the right-hand rule. Wrap the fingers of the right hand around the axis of rotation in the direction of rotation. The extended thumb points in the direction of the rotational vector. Figure 4.11 indicates how to find the direction of ω for the rotating flywheel using the right-hand rule.

This is a convenient point at which to introduce two new symbols to represent vectors aligned perpendicular to the plane of the page. The symbol ⊙ represents a vector pointing straight up out of the page toward the reader. The symbol ⊗ represents a vector pointing straight down into the page, away from the reader. One can think of them as the tip and the tail of an arrow.

Example 4.13 shows how the basic vectors **ω**, **L** and **τ** can be found for a spinning gyroscope supported at one end.

EXAMPLE 4.13   Angular velocity, angular momentum, and torque on a spinning gyroscope
. . . .

Given: A gyroscope of mass 100 g, moment of inertia 900 g·cm², and angular speed 3000 rpm is supported as shown. The distance, *r*, from the end of the gyroscope to its center of gravity is 4.50 cm.

Find: **a.** Its angular velocity **ω** in rad/sec.

              b. Its angular momentum **L** in g·cm²/sec.

              c. The torque **τ** acting upon it.

Solution: **a.** To find the magnitude of **ω**, convert the units from rpm to rad/sec.

                     ω = 3000 rpm

Sketch 1

$$\omega = (3000 \text{ rev/min})(2\pi \text{ rad/rev})\left(\frac{1 \text{ min}}{60 \text{ sec}}\right)$$

$$\omega = 314 \text{ rad/sec}$$

The direction of **ω** is found by wrapping the fingers of the right hand around the spin axis in the direction of rotation, as shown. The angular velocity now can be represented as

$$\xrightarrow{\quad \omega \quad}, \quad \text{where } |\omega| = 314 \text{ rad/sec}$$

b. Magnitude of **L**:
Angular Momentum = Moment of Inertia × Angular Velocity

$$L = I\omega \tag{4.20}$$

$$L = (900 \text{ g·cm}^2)(314 \text{ rad/sec})$$

$$L = 2.83 \times 10^5 \text{ g·cm}^2/\text{sec}$$

Direction of **L**:
The equation $\mathbf{L} = I\boldsymbol{\omega}$ is a vector equation and indicates that the vector **L** equals the scalar I times the vector **ω**. Thus, **L** and **ω** have the same direction. The vectors **ω** and **L** are shown as

$$\xrightarrow{\quad \omega \quad} \qquad \xrightarrow{\quad L \quad} |L| = 2.83 \times 10^5 \text{ g·cm}^2/\text{sec}$$

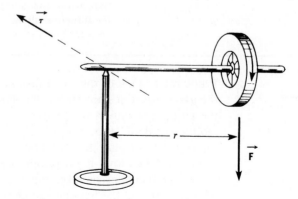

Sketch 2

    c. Magnitude of τ:

Torque = Force × Lever Arm

$$\tau = Fr$$

The force on the gyroscope is its weight, so

Force = Mass × Acceleration of Gravity

$$\mathbf{F} = m\mathbf{g}$$

$$F = (100 \text{ g}) (980 \text{ cm/sec}^2)$$

$$F = 9.80 \times 10^4 \text{ dynes}$$

Then

$$\tau = (9.80 \times 10^4 \text{ dynes})(4.50 \text{ cm})$$

$$\tau = 4.41 \times 10^5 \text{ dyne cm}$$

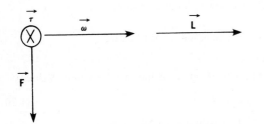
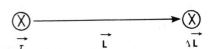

Sketch 3

Direction of τ:
The force, **F**, acting on the moment arm, **r**, as shown in the preceding sketch, tends to cause a rotation of the gyroscope assembly in a clockwise direction—as viewed by the reader. The product of **F** and r is equal to the torque. Thus, when the right-hand rule is applied to the figure, the torque vector τ has the direction along the dashed line, as shown, away from the viewer. The vectors $\boldsymbol{\omega}$, τ, **L**, and **F** are shown in sketch 3.

Precession One of the most fascinating aspects of the gyroscope effect is that the gyroscope does not fall off the support. Instead, the free, unsupported end of the gyroscope moves in a horizontal circle with the support at its center. This motion is called ''precession.'' The rate at which the spin axis precesses around the support is called the ''angular velocity of precession,'' $\boldsymbol{\Omega}$.

Figure 4.12 shows a gyroscope precessing around its supported end.

To understand precession, one may begin with the relation between the angular impulse $\tau \Delta t$ and the angular momentum change $\Delta \mathbf{L}$. This already has been presented as Equation 4.22. In vector form, Equation 4.22 becomes

FIGURE 4.12 Precession of a gyroscope.

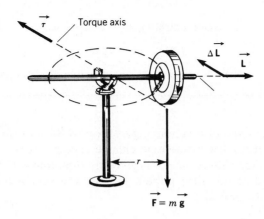

$$\Delta \mathbf{L} = \boldsymbol{\tau} \Delta t \qquad (4.25)$$

where

$\Delta \mathbf{L}$ = change in angular momentum (angular impulse)

$\boldsymbol{\tau}$ = torque

Δt = elasped time

Equation 4.25 states that the effect of a torque $\boldsymbol{\tau}$ acting on the gyroscope for a length of time equal to Δt is to change the angular momentum of the gyroscope by an amount $\Delta \mathbf{L}$. Furthermore, the angular momentum changes in the direction of the applied torque. Example 4.14 shows how $\Delta \mathbf{L}$ is calculated for the gyroscope of Figure 4.12.

EXAMPLE 4.14 ΔL for a gyroscope
. . . .

Given: The gyroscope of Figure 4.2 has a mass, m, of 0.2 kg and a lever arm, r, of 0.05 m.

Find: Its angular momentum change ΔL after an elapsed time of 0.03 sec.

Solution: **a.** Calculate the magnitude of the torque acting.

        $\boldsymbol{\tau} = \mathbf{F}r$

        $\boldsymbol{\tau} = mgr$

        $\boldsymbol{\tau} = (0.2 \text{ kg})(9.8 \text{ m/sec}^2)(0.05 \text{ m})$

        $\boldsymbol{\tau} = 0.098 \text{ kg·m}^2/\text{sec}^2$

        The direction of $\boldsymbol{\tau}$ is found by applying the right-hand rule around the torque axis.

    b. Find the magnitude of $\Delta \mathbf{L}$.

        $\Delta L = \boldsymbol{\tau} \Delta t$

$$\Delta L = (0.098 \text{ kg·m}^2/\text{sec}^2)(0.03 \text{ sec})$$

$$\Delta L = 2.94 \times 10^{-3} \text{ kg·m}^2/\text{sec}$$

From Equation 4.25, the direction of τ and ΔL is the same.

Figure 4.13 shows a spinning gyroscope with its weight, torque, angular momentum, and angular momentum change vectors shown. The angular momentum change ΔL after a short time Δt is given by Equation 4.25. The angular momentum L at the end of this time period must be the vector sum of the initial angular momentum L_0 and ΔL.

$$L = L_0 + \Delta L \tag{4.26}$$

Figure 4.14 shows these vectors added graphically. The final angular momentum L has approximately the same magnitude as L_0, but is displaced through angle $\Delta\phi$ from L_0. Since ΔL is perpendicular to L_0, the vector L_0 rotates but does not change in magnitude. Since L always lies along the spin axis, the spin axis rotates, or precesses, in the direction of ΔL. Notice that the torque vector rotates with ΔL, so that they remain parallel.

Figure 4.14 can be used to estimate the magnitude of the angular velocity of precession Ω, which is just the rate of change of angle $\Delta\phi$. From Figure 4.14,

FIGURE 4.13 Angular momentum change in precession of a gyroscope.

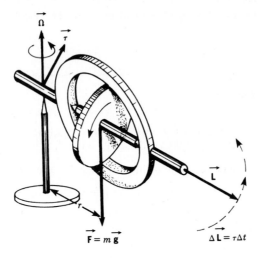

FIGURE 4.14 Angular displacement of the angular momentum vector.

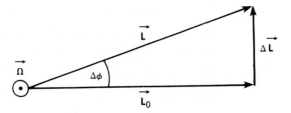

$$\tan (\Delta\phi) = \frac{\Delta L}{L_0}$$

Since $\Delta\phi$ is a very small angle,

$$\tan (\Delta\phi) \simeq \Delta\phi \qquad \text{(in radians)}$$

Then

$$\Delta\phi = \frac{\Delta L}{L_0}$$

$$\Delta\phi = \frac{\tau \Delta t}{L_0} \qquad \text{(since } \Delta L = \tau\Delta t)$$

$$\frac{\Delta\phi}{\Delta t} = \frac{\tau}{L_0} \qquad \text{(both sides are divided by } \Delta t)$$

$$\Omega = \frac{\tau}{L_0} \qquad \left(\text{since } \Omega = \frac{\Delta\phi}{\Delta t}\right) \qquad\qquad (4.27)$$

Example 4.15 shows how to find the magnitude and direction of Ω for a precessing gyroscope.

EXAMPLE 4.15   Angular velocity of precession for a gyroscope
. . . .

Given: The gyroscope of Example 4.13   ($m = 100$ g, $I = 900$ g·cm^2, $r = 4.50$ cm) spins clockwise at the rate of 3000 rpm.

Find: Its angular velocity of precession Ω.

Solution: **a.** Find τ:

$$\tau = Fr$$

$$\tau = mgr \qquad \text{(since } F = mg, \text{ the gyroscope's weight)}$$

$$\tau = (100 \text{ g})(980 \text{ cm/sec}^2)(4.50 \text{ cm})$$

$$\tau = 4.41 \times 10^5 \text{ g·cm}^2/\text{sec}^2$$

The direction of τ is found by applying the right-hand rule around the torque axis.

b. Find **L**:

$$L = I\omega \qquad \text{(Equation 4.20)}$$

$$L = (900 \text{ g·cm}^2)(3000 \text{ rpm})$$

$$L = (900 \text{ g·cm}^2)(3000 \text{ rev/min})(2\pi \text{ rad/rev})\left(\frac{1 \text{ min}}{60 \text{ sec}}\right)$$

$$L = 2.83 \times 10^5 \text{ g·cm}^2/\text{sec}$$

The direction of **L** is found by applying the right-hand rule around the spin axis.

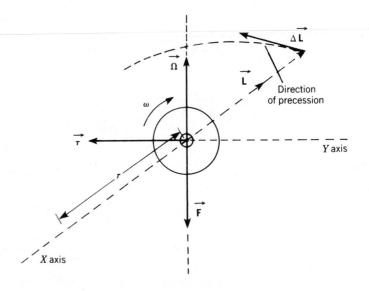

c. Calculate Ω:

$$\Omega = \frac{\tau}{L}$$

$$\Omega = \frac{4.41 \times 10^5 \text{ g·cm}^2/\text{sec}}{2.83 \times 10^5 \text{ g·cm}^2/\text{sec}}$$

$$\Omega = (1.56 \text{ g·cm}^2/\text{sec})(\text{sec}/\text{g·cm}^2)$$

$$\Omega = 1.56 \frac{1}{\text{sec}}$$

$$\Omega = 1.56 \text{ rad/sec}$$

The direction of precession at each instant is the same as the direction of $\Delta\mathbf{L}$. Equation 4.25 indicates that $\Delta\mathbf{L}$ and τ point in the same direction; so the gyroscope must precess in the direction of τ. Application of the right-hand rule around the precession axis indicates that the vector Ω points up. The gyroscope precesses along the dashed arc shown.

Gyroscopic Stability   Examination of Equation 4.27 shows that as the torque on the gyroscope becomes smaller, Ω also decreases. If the torque is reduced to zero in a model such as shown in Figure 4.16, the gyroscope stops precessing, and the spin axis remains rigidly fixed in space. This aspect of the gyroscopic effect is called "gyroscopic stability."

A bicycle or motorcycle being ridden demonstrates gyroscopic stability and precession. The spinning wheels of the cycle are gyroscopes. With the rider's weight positioned in the same vertical plane as the wheels (as in Figure 4.15*a*), there is no torque exerted on the wheels. The precessional velocity is zero and

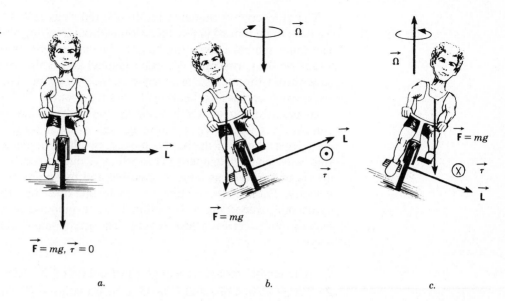

FIGURE 4.15 Gyroscopic effect applied to a bicycle.

a.   *b.*   *c.*

gyroscopic stability keeps the bicycle upright. This is why the act of riding a bicycle is a very simple task, but balancing on a stationary cycle is very difficult. When the rider leans to either side, his or her weight exerts a torque that acts on the gyroscopes (wheels). The wheels precess, and the bicycle turns right or left (Figure 4.15*b*). When we learn to ride a bicycle, we do this without thinking.

Gyroscopic stability allows the use of the gyroscope as a direction reference. A gyroscope mounted in such a manner that there is no torque acting on it maintains its spin axis in the same direction indefinitely. This phenomenon is the basic operating principle of the gyrocompass diagrammed in Figure 4.16.

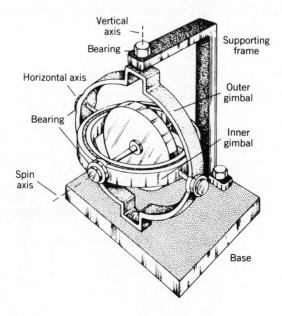

FIGURE 4.16 Gyrocompass.

The gyroscope is mounted inside pivoted rings called "gimbals," which allow the base to be moved in any direction without exerting a torque on the gyroscope. The inner gimbal allows the base to be turned left or right without exerting a torque on the gyroscope. The outer gimball allows the base to be tilted forward or backward without exerting a torque on the gyroscope. The gyroscope and gimbals are pivoted on jeweled bearings to reduce friction.

In practice, no matter how finely the bearings are ground, they cannot be completely frictionless. Friction causes a small torque to be exerted on the gyroscope whenever the base is moved, resulting in precession, which rotates the spin axis out of alignment. This effect is minimized by making the amount of inertia of the rotor as large as possible and by spinning it at a very high angular velocity. Each of these steps serves to increase the angular momentum of the gyroscope. According to Equation 4.27, an increase in L causes Ω to decrease, making the precession less severe. The gyrocompass is thus stabilized in two ways:

1. The gimbal mounting makes τ in Equation 4.27 very small.

2. Large values of I and ω make L in Equation 4.27 very large.

Gyroscopic Effect   In addition to the navigational uses of the gyrocompass, the gyroscopic direction finder has many applications. It is used in road mapping to measure accurately the curves in highways. It can measure curves and grades of sections of track in railroad surveying. Gyroscopes also are attached to oil-drilling equipment to determine and to control the direction of drilling.

The gyroscopic effect often intrudes detrimentally on the operation of commonplace devices. The rotating flywheel of an engine is a gyroscope with a very large amount of inertia. The effects of this large gyroscope are seen in the flight characteristics of single-engine aircraft. Figure 4.17a shows an airplane in level flight. The propeller and engine flywheel rotate clockwise (as viewed from the pilot's seat). The angular momentum **L** is found by application of the right-hand

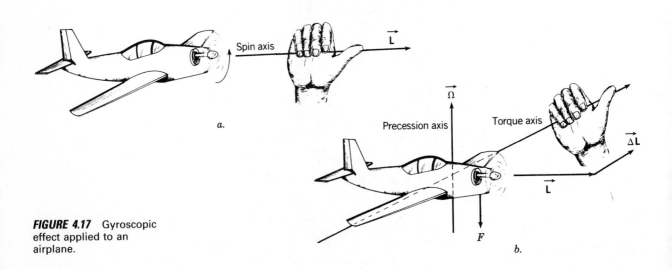

FIGURE 4.17 Gyroscopic effect applied to an airplane.

rule around the spin axis. Figure 4.17*b* shows the vectors associated with the aircraft as it goes into a dive. The force **F** pulls the nose down. The torque τ is found by applying the right-hand rule around the torque axis, which causes precession around the precession axis in the direction of Δ**L**. Thus, the aircraft veers left when the nose is dropped.

The effect is most noticeable when the angular momentum of the engine is small, as readily seen from Equation 4.27, since **L** appears in the denominator. The moment of inertia of the flywheel is constant, so the angular momentum can be changed only by changing the angular velocity of the engine. Therefore, the effect is most pronounced at low engine speeds (small angular momentum).

In twin-engine aircraft, the gyroscopic effect is nullified if the engines are designed to rotate in opposite directions. If the left engine rotates clockwise, the right engine rotates counterclockwise. This difference in direction causes the precessional motion of the engines to be equal in magnitude and opposite in direction, so that there is no net precession.

SECTION *4.1* *EXERCISES*

1. Define the following terms. State both the mks and cgs units for each, and give the equation for each.

Linear momentum

Moment of inertia

Angular velocity

Angular momentum

Impulse

Angular impulse

2. What is the momentum of a 250-lb object moving with a velocity of 5 ft/sec? (**Hint:** Be sure to convert 250-lb weight to mass in slugs.)

3. If a rifle were lighter than the bullet it fired, which would be safer—the shooter or the target? Explain your answer in terms of the law of conservation of momentum.

4. A proton (mass = 1.67×10^{-27} kg) moves with a velocity of 5×10^6 m/sec. Upon colliding with a stationary particle of unknown mass, the proton rebounds along its original path with a velocity of 3×10^6 m/sec. The collision sends the unknown particle forward with a velocity of 1.5×10^6 m/sec. What is the mass of the unknown particle?

5. A projectile with a mass of 0.5 kg and a velocity of 700 m/sec collides with a 20-kg mass initially at rest. The projectile imbeds itself within the mass. What is the velocity of the 20-kg mass and projectile after the collision? (Remember that momentum before equals momentum after!)

6. The radius of a solid shaft is 10 cm and its mass is 5 kg. It is rotating with an angular velocity of 1800 rpm about an axis of rotation through its center, as shown in Figure 4.8*b*.

    a. What is the angular momentum of the shaft?

    b. What angular impulse is needed to increase the angular velocity to 2400 rpm?

    c. What constant torque is needed to increase the angular velocity from 1800 rpm to 2400 rpm in 30 sec?

7. An Olympic high diver changes his speed of rotation while in flight by tucking himself into a small ball or by straightening himself out. Explain why this is possible in light of the conservation of angular momentum.

8. State the right-hand rule for determining the direction of a rotational vector.

9. The rotor of a gyroscope spins as shown in the diagram. Label the diagram with the following vectors:

Angular velocity, ω

Angular momentum, $\mathbf{L}$

Force (weight), $\mathbf{F}$

Torque, τ

Angular momentum change, $\Delta\mathbf{L}$

Angular velocity of precession, Ω

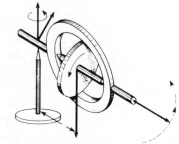

10. A spinning gyroscope precesses at the rate of 0.3 rad/sec and has angular momentum of 5000 rpm. Find the torque acting upon the gyroscope. The moment of inertia $I = 900$ g·cm².

11. List the major parts of a gyrocompass and explain the function of each.

12. List the two methods of stabilizing a gyrocompass and explain the effect of each on the magnitude of τ or $\mathbf{L}$.

4.2

MOMENTUM IN FLUID SYSTEMS

OBJECTIVES

Upon completion of this section, the student should be able to

- Explain each of the following in a short paragraph:
  Conservation of linear momentum in a fluid system
  Conservation of angular momentum in a fluid system

- Calculate the reaction force of a free jet on a stationary surface.

- Calculate the reaction force on a bend in a pipe due to a momentum change in the fluid flow.

- Calculate the reaction force of a jet on a reservoir.

DISCUSSION

Momentum and its conservation are important concepts in fluid systems as well as in mechanical systems. We will examine these properties as they apply in fluid systems.

Fluid Momentum

Momentum Equation   A flowing fluid has momentum. The local momentum is usually expressed in terms of the local velocity and density, which give the momentum per unit volume. This may be expressed as

$$\mathbf{p}_v = \rho_m \mathbf{v} \tag{4.28}$$

where

$\mathbf{p}_v$ = momentum per unit volume
ρ_m = local mass density
$\mathbf{v}$ = local fluid velocity

An important property is the momentum flow rate. The momentum that flows through a unit area perpendicular to the velocity per second is

$$\frac{\Delta \mathbf{p}_v}{\Delta t} = (\rho_m \mathbf{v})v \tag{4.29}$$

If the momentum flowing into a system is different from the momentum flowing out of the system, there must be a corresponding force according to Newton's second law of motion:

$$\mathbf{F} = \frac{\Delta \mathbf{p}}{\Delta t} \tag{4.30}$$

If the velocity is constant across the entrance area and exit area, a simple expression can be obtained for the force. The momentum flow into the system is

$$\frac{\Delta \mathbf{p}_1}{\Delta t} = (\rho_{m_1} \mathbf{v}_1)v_1 A_1 \tag{4.31}$$

where

A_1 = entrance area

The momentum flowing out of the system is

$$\frac{\Delta \mathbf{p}_2}{\Delta t} = (\rho_{m_2} \mathbf{v}_2)v_2 A_2$$

where

A_2 = exit area

Thus, the total change in momentum in the system per unit time is

$$\frac{\Delta \mathbf{p}}{\Delta t} = (\rho_{m_2} \mathbf{v}_2)v_2 A_2 - (\rho_{m_1} \mathbf{v}_1)v_1 A_2 \tag{4.32}$$

It is important to note that $\rho_m v A$ is the mass flow rate as shown in Chapter 3. For steady flow, the mass flowing into the system per unit time must equal the mass flowing out of the system per unit time. Thus,

$$\rho_{m_1} v_1 A_1 = \rho_{m_2} v_2 A_2 = \frac{\Delta m}{\Delta t}$$

where

$\Delta m / \Delta t$ = mass flow rate

Substituting $\Delta m/\Delta t$ for the mass flow rate in Equation 4.32 yields

$$\frac{\Delta \mathbf{p}}{\Delta t} = \left(\frac{\Delta m}{\Delta t}\right)(\mathbf{v}_2 - \mathbf{v}_1) = \frac{\Delta m}{\Delta t}\,\Delta \mathbf{v}$$

Since $F = \Delta p/\Delta t$, one has

$$\mathbf{F} = \frac{\Delta m}{\Delta t}\,\Delta \mathbf{v} \qquad (4.33)$$

Force on Pipes and Surfaces   The resultant force exerted on a pipe or surface by a flowing fluid can be due to static, frictional, and dynamic forces. By Newton's third law, an equal but opposite force is exerted on the fluid. To analyze these forces, it is necessary to consider the change in the momentum of the fluid as well as the static pressure and frictional forces.

To simplify the analysis, consider the force exerted on a fluid jet by a stationary surface as shown in Figure 4.18. Since the jet is open to the atmosphere, the pressure throughout the jet is essentially atmospheric pressure. Thus, the resultant force exerted by the surface will be the force that causes the change in momentum per unit time. This force, according to Equation 4.33, becomes

$$\mathbf{F}_R = \frac{\Delta m}{\Delta t}\,\Delta \mathbf{v} = Q_m \Delta \mathbf{v} \qquad (4.34)$$

where

   Q_m = mass flow rate, $\Delta m/\Delta t$

Equation 4.34 can also be expressed in terms of x and y components. For the velocities, one has

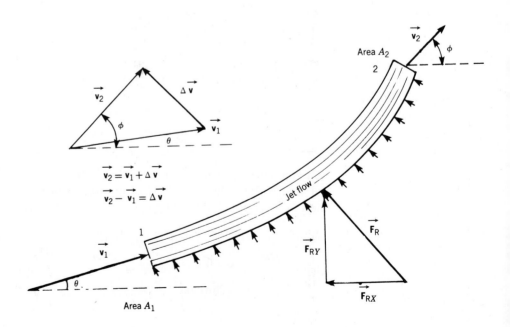

FIGURE 4.18 Forces exerted on a fluid jet.

$$v_{1x} = v_1 \cos \theta; \quad v_{2x} = v_2 \cos \phi$$

$$\Delta v_x = v_2 \cos \phi - v_1 \cos \theta$$

and

$$v_{1y} = v_1 \sin \theta; \quad v_{2y} = v_2 \sin \phi$$

$$\Delta v_y = v_2 \sin \phi - v_1 \sin \theta$$

$$v_{1x} = v_1 \cos \theta; \quad v_{2x} = v_2 \cos \phi$$

$$\Delta v_x = v_2 \cos \phi - v_1 \cos \theta$$

and

$$v_{1y} = v_1 \sin \theta; \quad v_{2y} = v_2 \sin \phi$$

$$\Delta v_y = v_2 \sin \phi - v_1 \sin \theta$$

The corresponding force components become

$$-F_{Rx} = Q_m \Delta v_x = Q_m(v_2 \cos \phi - v_1 \cos \theta) \tag{4.35}$$

$$F_{Ry} = Q_m \Delta v_y = Q_m(v_2 \sin \phi - v_1 \sin \theta) \tag{4.36}$$

The application of Equations 4.35 and 4.36 is illustrated in Example 4.16.

EXAMPLE 4.16   Force on a stationary body
. . . .

Given: The stationary surface shown in the sketch. The initial velocity of
the water jet, v_1, is 50 ft/sec, and the velocity v_2, due to friction
losses, is 40 ft/sec. The initial diameter of the water jet is 0.2 ft.

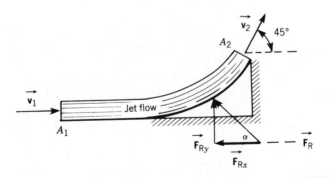

Find: The force on the surface.

Solution: The force on the surface will be equal in magnitude but in the opposite direction of F_R.

$$Q_m = \rho_m v_1 A_1 = (\rho_w/g) \, v_1 A_1$$

$$Q_m = \left(\frac{62.4 \text{ lb/ft}^3}{32.2 \text{ ft/sec}^2}\right)(50 \text{ ft/sec})\left(\frac{\pi (0.2)^2 \text{ ft}^2}{4}\right)$$

$Q_m = 3.04$ slugs/sec

From Equations 4.35 and 4.36, the components of force are

$-F_{Rx} = Q_m(v_2 \cos \phi - v_1 \cos \theta)$

$-F_{Rx} = (3.04 \text{ slugs/sec})(40 \text{ ft/sec } \cos 45° - 50 \text{ ft/sec } \cos 0°)$

$F_{Rx} = 66.0$ lb

$F_{Ry} = (3.04 \text{ slugs/sec})(40 \text{ ft/sec } \sin 45° - 50 \text{ ft/sec } \sin 0°)$

$F_{Rx} = 86.0$ lb

To obtain the magnitude of $\mathbf{F}_R$

$F_R = \sqrt{(F_{Rx})^2 + (F_{Ry})^2}$

$F_R = \sqrt{(66.0 \text{ lb})^2 + (86.0 \text{ lb})^2}$

$F_R = 108$ lb

To obtain the direction of $\mathbf{F}_R$

$\tan \alpha = \dfrac{F_y}{F_x}$

$\tan \alpha = \dfrac{86.0}{66.0}$

$\alpha = 52.5°$

The force on the surface will be in the opposite direction.

When the flow is in a pipe, one must also consider the forces due to the static pressure at the ends of the two areas as illustrated in Figure 4.19. As in the case for jet flow, the total force must be equal to the momentum change in the volume between A_1 and A_2. However, one has additional forces at A_1 and A_2 due to the static pressure. These forces are

$$\mathbf{F}_1 = \mathbf{p}_1 A_1 \tag{4.37}$$

$$\mathbf{F}_2 = \mathbf{p}_2 A_2 \tag{4.38}$$

where

$\mathbf{p}_1 =$ static pressure at A_1
$\mathbf{p}_2 =$ static pressure at A_2

If the flow is turbulent, it is a good approximation to assume that the velocity across the pipe is constant since the velocity profile will tend to be flat. For this case, the equations for jet flow can be modified by adding the pressure forces to them. The x and y components due to the pressure forces are

FIGURE 4.19 Forces exerted on a fluid flow in a pipe.

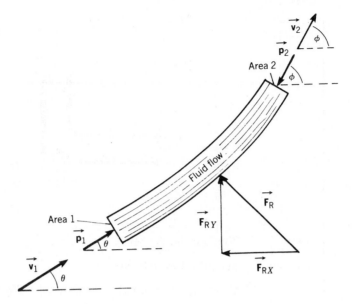

$$F_{1x} = p_1 A_1 \cos \theta; \qquad F_{2x} = p_2 A_2 \cos \phi$$

$$F_{1y} = p_1 A_1 \sin \theta; \qquad F_{2y} = p_2 A_2 \sin \phi$$

Thus, adding these forces to the corresponding components of F_R in Equations 4.35 and 4.36 yields

$$F_{1x} + F_{2x} - F_{Rx} = Q_m(v_2 \cos \phi - v_1 \cos \theta)$$

$$p_1 A_1 \cos \theta - p_2 A_2 \cos \phi - F_{Rx} = Q_m(v_2 \cos \phi - v_1 \cos \theta) \qquad (4.39)$$

$$F_{1y} + F_{2y} + F_{Ry} = Q_m(v_2 \sin \phi - v_1 \sin \theta)$$

$$p_1 A_1 \sin \theta - p_2 A_2 \sin \phi + F_{Ry} = Q_m(v_2 \sin \phi - v_1 \sin \theta) \qquad (4.40)$$

Stated in words, the sum of the forces in the x direction is equal to the change in momentum in the x direction, the sum of the forces in the y direction is equal to the change in momentum in the y direction.

The above derivations for turbulent flow assume that the velocity distribution across the areas is constant. For laminar flow in a pipe, the velocity distribution is a parabolic. However, the average velocity can be used in Equations 4.39 and 4.40 for laminar flow, if the right-hand sides are multiplied by the factor of 4/3.

An application of Equations 4.39 and 4.40 is given in Example 4.17.

EXAMPLE 4.17   Force on a pipe bend
. . . .

Given: Water under a static pressure of 5000 lb/ft² with a velocity of 20
                  ft/sec flows through a right-angle bend. The diameter of the pipe is
                  1.0 ft. There is no significant pressure drop and the flow is turbulent.

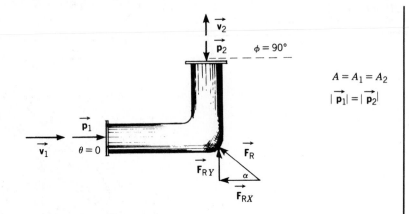

$$A = A_1 = A_2$$
$$|\vec{p_1}| = |\vec{p_2}|$$

Find: The force on the bend.

Solution: The mass flow rate is

$$Q_m = \rho_m v_1 A = \rho_m v_2 A \qquad (|\mathbf{v}_1| = |\mathbf{v}_2| = |v|)$$

$$Q_m = \left(\frac{62.4 \text{ lb/ft}^3}{32.2 \text{ ft/sec}^2}\right)(20 \text{ ft/sec})\left(\frac{\pi \text{ ft}^2}{4}\right)$$

$$Q_m = 30.4 \text{ slug/sec}$$

The area is $(\pi)(1 \text{ ft}^2)/4 = 0.785 \text{ ft}^2$.
From Equations 4.39 and 4.40   ($\theta = 0°$ and $\phi = 90°$)

$$p_1 A_1 \cos \theta - p_2 A_2 \cos \phi - F_{Rx} = Q_m(v_2 \cos \phi - v_1 \cos \theta)$$

$$(5000 \text{ lb/ft}^2)(0.785 \text{ ft}^2) - 0 - F_{Rx} = 30.4 \text{ slug/sec } (0 - 20) \text{ ft/sec}$$

$$3925 - F_{Rx} = -608$$

$$F_{Rx} = +4533 \text{ lb}$$

$$p_1 A_1 \sin \theta - p_2 A_2 \sin \phi + F_{Ry} = Q_m (v_2 \sin \phi - v_1 \sin \theta)$$

$$0 - (5000 \text{ lb/ft}^2)(0.785) + F_{Ry} = 30.4 \text{ slug/sec } (20 - 0) \text{ ft/sec}$$

$$-3925 + F_{Ry} = 608$$

$$F_{Ry} = 4533 \text{ lb}$$

The resultant force is

$$F_R = \sqrt{(4533 \text{ ft})^2 + 4533 \text{ ft}^2}$$

$$F_R = 6410 \text{ lb}$$

$$\tan \alpha = \frac{4533}{4533}$$

$$\alpha = 45°$$

The force on the pipe is in the opposite direction.

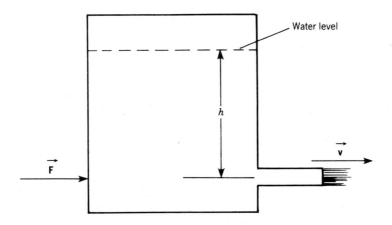

FIGURE 4.20 Jet of fluid from a reservoir.

Reaction of a Jet A jet of fluid flowing from a large reservoir will exert a reaction force on the reservoir. If the reservoir is large, the velocity of the fluid in the reservoir may be neglected. The reservoir and jet are shown in Figure 4.20. The change in velocity is just the velocity of the jet. The force required to change the momentum can be determined from Equation 4.33.

$$\mathbf{F} = \left(\frac{\Delta m}{\Delta t}\right)\mathbf{v} = Q_m \mathbf{v} \tag{4.41}$$

The reaction force on the reservoir due to the momentum change will be in the opposite direction. In Chapter 6, it will be shown that the jet velocity is $\sqrt{2gh}$ if friction can be neglected. Q_m in Equation 4.41 may be expressed as

$$Q_m = \rho_m v A \tag{4.42}$$

where

A = cross-sectional area of the jet

The force required to change the momentum may then be written as

$$F = \rho_m A\, v^2$$

Substituting $v = \sqrt{2\,gh}$ yields

$$F = 2\rho_m ghA = 2\rho_w hA \tag{4.43}$$

It is interesting to note that $2\rho_w h$ in Equation 4.43 is just two times the liquid pressure at a distance h below the surface of the liquid. The use of Equation 4.43 is illustrated in Example 4.18.

EXAMPLE 4.18 Reaction of a jet on a stationary reservoir
. . . .

Given: An orifice in the vertical side of a tank is 25 ft below the surface of the water. The diameter of the orifice is 0.2 ft. Assume no friction.

Find: The reaction on the tank.

Solution: Use Equation 4.43.

$$F = -2\rho_w hA$$

$$F = (-2)(62.4 \ \text{lb/ft}^3)(25 \ \text{ft}) \frac{(\pi)(0.2 \ \text{ft})^2}{4}$$

$$F = -98 \ \text{lb}$$

The force exerted by a stream on a moving object requires an analysis of relative velocities. This analysis is required for the study of turbines, an application that will be covered in Chapter 9.

SECTION *4.2*   *EXERCISES*

1. A free jet flows over the surface as shown in the diagram. The initial velocity, v_1, of the jet is 40 ft/sec in the horizontal direction and the velocity v_2 is 35 ft/sec in the vertical direction. The initial diameter of the jet is 0.15 ft. Find the force on the surface. Calculate the force on the surface if the velocity v_2 is 40 ft/sec in the vertical direction. In real situations, the magnitude of the velocity v_2 will be less than the magnitude of v_1 because of friction.

2. Water under a static pressure of 1000 lb/ft^2 with a velocity of 10 ft/sec flows through a right-angle bend with no significant pressure drop. The diameter of the pipe is 0.5 ft. If the flow is turbulent, find the force on the bend.

4.3

MOMENTUM IN ELECTROMAGNETIC SYSTEMS

OBJECTIVE

Upon completion of this section, the student should be able to

• Calculate the momentum flow of a laser beam of known energy flow.

DISCUSSION

Momentum is as important in electromagnetic systems as it is in mechanical or fluid systems. However, the development of it is very difficult, and for this reason the treatment presented here is limited in depth.

Electromagnetic Radiation

Momentum Traveling electromagnetic waves have momentum. The development of the electromagnetic equations for momentum are beyond the scope of this text. However, the problem can be approached in a simpler way. Electromagnetic radiation, such as radiant heat energy, light, x rays, and γ rays, can be considered to be small bundles of energy called photons that always travel at the speed of light. It was found experimentally (photoelectric and Compton effects) that each photon has an energy equal to the frequency of the radiation times a constant called Planck's constant. In addition, according to the theory of relativity, its energy must also equal its mass times the velocity of light squared. Mathematically this may be expressed as

$$E_p = mc^2 = h\nu \tag{4.44}$$

where

E_p = energy of a single photon
m = photon's relativistic mass
c = speed of light
h = Planck's constant (6.62×10^{-34} J·sec)
ν = frequency of the radiation

Since a photon always travels at the speed of light, its momentum becomes

$$p_p = mc \tag{4.45}$$

However, from Equation 4.44

$$m = \frac{h\nu}{c^2}$$

Thus, the momentum of a photon is

$$p_p = \left(\frac{h\nu}{c^2}\right)c = \frac{h\nu}{c} \tag{4.46}$$

Since $h\nu = E_p$, Equation 4.46 may be expressed as

$$p_p = \frac{E_p}{c} \tag{4.47}$$

The energy passing through a cross-sectional area of a laser beam per unit time, q_v, may be expressed as

$$q_v = \frac{E_{p_2} - E_{p_1}}{\Delta t} \tag{4.48}$$

where

$E_{p_2} - E_{p_1}$ = energy transferred through the area
Δt = corresponding time

However, from Equation 4.47, $E_p = cp_p$. Thus, by substitution,

$$q_v = \frac{cp_{p_2} - cp_{p_1}}{\Delta t} = cq_p$$

or

$$q_v = c\left[\frac{p_{p_2} - p_{p_1}}{\Delta t}\right] = cq_p \tag{4.49}$$

where

q_p = momentum passing through a cross-sectional area of the beam per unit time

Solving Equation 4.49 for q_p gives

$$q_p = \frac{q_v}{c} \tag{4.50}$$

Since the speed of light is very large (3×10^8 m/sec), q_p will be small.

The energy flow, q_v, can be easily measured in a laser beam with a system similar to the one illustrated in Example 3.42. The corresponding momentum flow can then be obtained by using Equation 4.50. If the beam strikes a surface and is reflected back, the change in momentum per unit of time at the surface is $2q_p$. According to Newton's second law, the corresponding force is

$$F = 2q_p \tag{4.51}$$

The force due to the photon rate of change in momentum is so small that it is usually neglected. This is illustrated in Example 4.19.

EXAMPLE 4.19 Force due to a laser beam
. . . .

Given: A 5000-W laser beam strikes a surface and is reflected. The speed of light is 3×10^8 m/sec.

Find: The force on the surface.

Solution: A watt is a joule per second. From Equations 4.51 and 4.50

$$F_R = 2\,q_p = q_p = 2\,q_v/c$$

$$F_R = \left(\frac{2}{3 \times 10^8 \text{ m/sec}}\right)(5000 \text{ N·m/sec})$$

$$F_R = 3.33 \times 10^{-5} \text{ N}$$

$$F_R = (3.3 \times 10^{-5} \text{ N})\left(\frac{1 \text{ lb}}{4.45 \text{ N}}\right)$$

$$F_R = 7.5 \times 10^{-6} \text{ lb}$$

SECTION 4.3 EXERCISES

1. The energy flow in a laser beam is 100 J per sec. Find the corresponding momentum flow in newtons. If the beam is absorbed by a surface, what is the force on the surface?

S U M M A R Y

Momentum is an important property in mechanical, fluid, and electromagnetic systems. Momentum changes in all three systems require the action of a force. The momentum is always conserved for an isolated system.

Table 4.1 summarizes the momentum in the systems examined in this chapter.

TABLE 4.1 Momentum as a unifying concept

Energy System	Momentum = Mass × Rate
Mechanical	
Translational	Momentum = Mass × Velocity
Rotational	Angular Momentum = Moment of Inertia × Angular Velocity
Fluid	Momentum Flow Rate = Mass Flow Rate × Velocity
Electromagnetic	Momentum = $\dfrac{\text{Energy}}{\text{Velocity of Light}}$

....

Resistance

....

....

....

....

INTRODUCTION

Resistance is an opposition to motion. It acts to oppose the forcelike quantities in all energy systems, thus causing a reduction of rates. If no forcelike quantities are present to counteract the effects of resistance, motion in the system will die out gradually and eventually will cease altogether. Although some systems maintain motion without assistance for a longer time than others, motion ultimately ceases in all instances. Resistance appears in different forms in different energy systems, but in each case creates a forcelike quantity that opposes motion within the system.

If a moving system is to be maintained in motion at a constant rate, a forcelike quantity must act continuously on the system and, in acting on a moving system, must do work. The work done against friction in all energy systems is converted to heat energy. This process is taken advantage of in systems in which the conversion of other forms of energy to heat energy is desirable. In other cases, the energy converted to heat is wasted and cannot be recovered. Often, this heat energy causes thermal problems in a system, and steps must be taken to remove it. A stream of water used to cool the edge of a glass-cutting wheel is an example.

The following are examples of the effects of resistance in the energy systems discussed in this chapter:

- *In mechanical systems, resistance appears as friction between any two surfaces in contact. The effect of friction is to slow the motion of moving objects and to prevent the motion of stationary objects. Lubricants are used to reduce frictional forces and, thus, to reduce the force necessary to maintain motion and the amount of energy converted to heat.*

- *In fluid systems, resistance appears as drag (fluid friction) forces on solid objects moving through fluids and as fluid resistance that limits fluid flow rates through pipes. Much of the energy used by an automobile appears as work done against the fluid friction of the air.*

- *In electrical systems, electrical resistance opposes current in conductors. In electrical power distribution systems, this resistance must be minimized to reduce energy losses. In heating devices, such as toasters and electric heaters, resistance is employed to convert electrical energy into heat energy.*

- *In magnetic systems, the reluctance of the magnetic path determines the magnetic flux for a given magnetomotive force. In this respect, the material acts as a resistance. However, no energy is converted into heat by the magnetic flux. When the flux changes, heat can be generated as a result of the friction that opposes the turning about of the magnetic domains in a magnetic material. In addition, a changing field will induce eddy currents in a conductor that cause resistance heating. All these effects, for example, occur in electric transformers and reduce their efficiency.*

- *In thermal systems, resistance appears as opposition to heat flow. Materials, such as metals, that transmit heat readily have low thermal resistance. Thermal insulators are materials having high resistances to heat flow. In this system, resistance acts to reduce the rate of motion only. Since only heat energy is present in purely thermal systems, no other kinds of energy are available for conversion into heat.*

All these points bring to mind a fascinating question, not of vital concern at this stage of study, but we pause for a moment to ask it. "Is a completely frictionless system possible?" If such a system could be developed or found, no energy waste would occur as a result of resistance effects; in other words, the system would be 100% efficient. In fact, sometimes such an ideal situation can be approached very closely. For example, a rocket ship moving through the near vacuum of outer space encounters very little resistance to its motion and coasts for great distances without additional energy input. In electrical systems, an electrical current can be maintained for a very long time without additional energy input in the "resistanceless" circuit made of a superconductor. The fact remains, however, that a perfectly frictionless system is an ideal.

In summary, a moving system involves some degree of resistance to motion. To maintain motion, additional energy input to the system is required. Where the heat produced by this additional energy is not desirable, it represents a waste of energy, and an effort is made to reduce the resistance of the system. There are cases, however, when the heat produced is taken advantage of and put to some useful purpose, for example, with industrial heat exchangers.

The remainder of this chapter discusses in greater detail the types of resistance introduced above.

5.1

RESISTANCE IN MECHANICAL SYSTEMS

OBJECTIVES

Upon completion of this section, the student should be able to

- Calculate the magnitudes of starting and frictional forces, given the mass or weight of the object, the coefficients of friction, and the angle of incline.

- Describe the differences between sliding and rolling friction.

DISCUSSION

The resistance that retards motion is friction. The effect of friction, which occurs as one surface moves across another, is examined in this section.

Dry Friction

Bodies in contact and in relative motion with respect to one another, that is, rubbing against each other, experience a resistance to the motion, usually called "dry friction." This term is applied to distinguish this type of friction from "fluid friction," the friction that a body experiences as it moves through a fluid. Fluid friction is caused by the fluid's viscosity, which will be considered later.

A simple and effective way for the student to understand this discussion is to think of a block of wood being dragged along the top surface of a level table. A string is tied to the side of the block and the block is being pulled in a horizontal direction (Figure 5.1). If a spring scale is introduced between the hand and the

FIGURE 5.1 Friction between block and table.

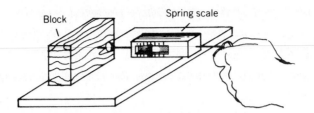

string, the scale will register the force required to drag the block along the table. The force that must be applied to do nothing more than slide the block along at constant speed is a measure of the frictional force resisting the motion. This frictional force originates at the point where the block surface and table surface are in actual contact.

Dry friction, therefore, depends on the types of surfaces involved. Wood dragged along a smooth tile floor involves less friction than wood dragged against a wooden floor, and so on.

How the lubrication of surfaces reduces frictional force should be obvious now. If the table top is covered with a film of liquid or oil, for example, the film can act to hold the block and table surfaces apart enough to reduce collisions of their irregular surface bumps (Figure 5.2).

Apart from the types of surfaces involved, what other factors might influence the frictional force? Obviously, the weight (w) of the block makes a difference. A heavy crate is more difficult to drag than a light crate of the same size and shape. More precisely, the frictional force really depends on the force pushing the two surfaces together. This force usually is called the ''normal force'' (N), because its direction is perpendicular, or ''normal,'' to the surface area. This bit of precision is important, because weight and normal force are not always identical. Example 5.1 shows a case in which weight and normal force are the same and a case in which they are different. Example 5.1 also shows the importance of treating forces as vectors and of being able to resolve such vectors into components.

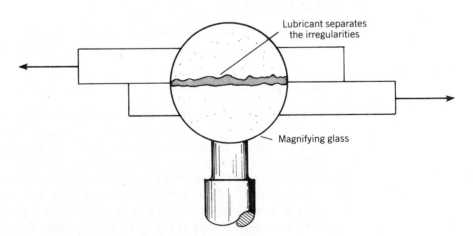

FIGURE 5.2 The effect of a lubricant on sliding surfaces.

EXAMPLE 5.1   Normal force problem

. . . .

Given: Two crates weigh 100 lb each. One rests on a horizontal surface and
                the other on a surface 30° from horizontal.

Find: The normal force on each crate.

Solution: **a.**   In the first case, the weight of the crate acts perpendicularly to
                         the horizontal floor. Thus, weight and normal force are identical.
                         $w = 100$ lb. $N = 100$ lb.

                b.   In this case, since weight continues to act vertically downward,
                         its component perpendicular to the surface is less than the
                         weight and can be found from the vector triangle.

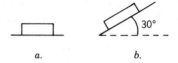

a.   *b.*

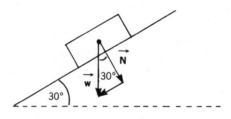

$$N = w \cos 30°$$

$$N = (100)(\cos 30°)$$

$$N = (100)(0.866)$$

$$N = 86.6 \text{ lb}$$

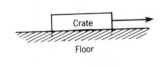

a.

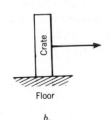

b.

FIGURE 5.3 Effect of
contact area.

What factors, other than the types of surfaces and the normal force pressing
them together, are involved in determining the frictional force? Surprisingly,
nothing else matters much. For example, one might guess that the amount of
surface area itself makes a significant difference. Consider a crate being dragged
along the floor, first on its large side and then on its small side (Figure 5.3). One
side (Figure 5.3*a*) has a large surface area and the other (Figure 5.3*b*) a small
surface area in contact with the floor. Of course, the weight of the crate does not
change. One finds there is no significant difference in the frictional forces acting.
This may seem rather remarkable at first, because our understanding of the source
of friction would lead us to believe that the larger surface area in contact with the

TABLE 5.1   Coefficients of sliding friction (μ)[a]

Wood on wood	0.3
Wood on stone	0.4
Metal on metal	0.15
Metal on wood	0.5
Leather on wood	0.4
Rubber on dry concrete	0.7
Rubber on wet concrete	0.5

[a] Values are approximate.

floor would have more friction. There are more surface irregularities that can become locked together. However, we have overlooked the fact that in this case there is less weight directly above the irregularities pressing them into the floor. Apparently, these two effects cancel each other. Thus, dry friction is relatively independent of the contact surface area.

Consider next whether the speed with which the block is dragged along makes a difference, Again, experiment shows that speed is not important—unless speeds are reached at which air friction itself is no longer negligible. Dry friction is found to be relatively independent of velocity.

The conclusion can be made that dry-frictional forces depend mainly on two factors:

1. Nature of the sliding surfaces.
2. Normal force pressing them together.

Experiment shows that the frictional force is directly proportional to the normal force. Therefore, if the normal force is doubled, the frictional force is doubled. All this information is summarized in

$$F_f = \mu N \tag{5.1}$$

where

F_f = frictional force
N = normal force
μ = "coefficient of friction"

The coefficient of friction μ depends on the nature of the two surfaces involved. It is a pure number without units, being simply the ratio of two forces, F_f/N. Some approximate values for the coefficient of friction for common materials are given in Table 5.1. Example 5.2 illustrates the usefulness of Equation 5.1.

EXAMPLE 5.2   Dry sliding friction
. . . .

Given: The 100-lb crate of Example 5.1 has a coefficient of friction between crate and floor of 0.25.

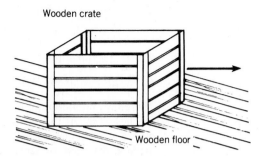

Wooden crate

Wooden floor

Find: What force must be applied to overcome friction and to keep the
                 crate sliding at constant speed in each case.

Solution: **a.** $N = 100$ lb, $\mu = 0.25$

                         $F_f = \mu N = (0.25)(100) = 25.0$ lb

                 b. $N = 86.6$ lb, $\mu = 0.25$

                         $F_f = \mu N = (0.25)(86.6) = 21.65$ lb

There is a small complication in this simple theory. A larger force is required to
start a box sliding than to keep it sliding, that is, more effort is required to unlock
the two surfaces when they are stationary than to keep them sliding when motion
has begun, which means we must distinguish between coefficients of "starting
friction" and coefficients of "sliding friction." The first is called a static friction
coefficient, μ_s, and the second a kinetic friction coefficient, μ_k. For the same two
surfaces, μ_s will be slightly greater than μ_k. Consequences of the difference in μ_s
and μ_k are illustrated in Example 5.3.

EXAMPLE 5.3   Static and kinetic dry friction
. . . .

Given: A metal box weighing 140 lb rests on a horizontal wooden floor. The
                 coefficient of static friction for these surfaces is 0.55 and the coeffi-
                 cient of kinetic friction is 0.50. A pull rope is attached.

Find: What happens as horizontal pull on the rope is increased.

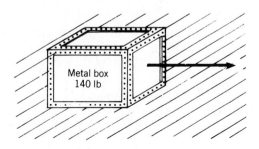

Metal box
140 lb

Solution: Static frictional force is

$$F_{fs} = \mu_s N = (0.55)(140) = 77 \text{ lb}$$

Kinetic frictional force is

$$F_{fk} = \mu_k N = (0.50)(140) = 70 \text{ lb}$$

Therefore, no motion of the box occurs until pull on the rope reaches 77 lb. At this point, the box begins to slide. Since pull required to keep the box sliding is only 70 lb, the box will begin to increase in speed, or accelerate, unless the pull is reduced to 70 lb. If the pull is reduced below 70 lb, however, the box will come to a stop.

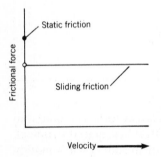

FIGURE 5.4   Plot of frictional force versus velocity.

Sliding friction, remember, is not dependent on the velocity of the surfaces involved. Figure 5.4 is a graph of frictional force versus velocity. Static friction is the maximum possible value of frictional force that can exist between stationary surfaces for a given normal force. If a force applied parallel to the surface exceeds the magnitude of the static friction, motion results. Once motion has begun, the frictional force drops to the sliding friction value and remains constant as long as the surfaces are in motion.

Rolling Friction

Figure 5.5 shows how the effect of friction is reduced when a round object rolls across a flat surface. The surface irregularities and friction still are present, but

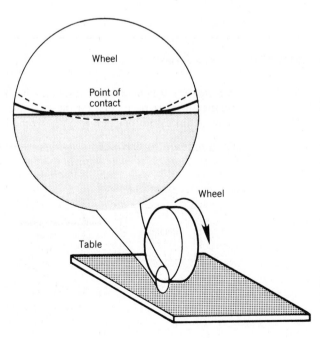

FIGURE 5.5   Rolling friction.

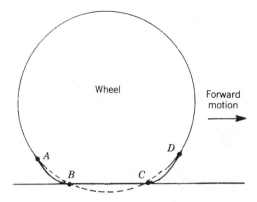

FIGURE 5.6 Point deformations.

their effect is minimized because the two surfaces do not slide. At the point of contact, they are motionless with respect to one another. The forward surface of the round object moves downward, causing the point of contact to move forward. At the same time, the rear surface is lifted away from the irregular surface of the flat object. Some friction still is present as the two surfaces make and break contact, but its effect is small compared with what is properly called "rolling friction."

Figure 5.6 illustrates the origin of rolling friction. In a real system, surfaces of the objects are not perfectly rigid and some surface deformation occurs around the point of contact. At points A and D, the round object bulges slightly. From point C to point B, it is slightly flattened, and the plane surface on which it rests is indented. When the object rolls, both surfaces change shape as the point of contact moves forward. Internal resistance of objects to this change in shape produces rolling friction. The magnitude of the frictional force produced depends on the normal force between the objects and the rigidity of their surfaces. Rigid surfaces (such as steel on steel) undergo less change in shape, producing smaller frictional forces than flexible surfaces, such as rubber on wood.

Figure 5.7 shows the force when a round object rolls across a plane surface. The applied force is horizontal and is shown acting through the center of mass (usually applied by an axle of a wheel). The resulting reaction force, R, of the supporting surface acts at point A in front of the vertical line OB. The gravitational force, w, transmitted by the axle acts downward through point O. If it is assumed that the body moves with constant velocity, the sum of these three forces must equal zero. Since there are only three forces in equilibrium, they must meet in the common point O. If this were not true, an unbalanced torque would exist and the body would not be in equilibrium. The three concurrent forces are shown in Figure 5.8. For equilibrium, $\Sigma F_x = 0$ and $\Sigma F_y = 0$. The condition $\Sigma F_x = 0$ gives

$$F_f = R \sin \theta \qquad (5.2)$$

The condition $\Sigma F_y = 0$ gives

$$w = R \cos \theta$$

or

$$R = \frac{w}{\cos \theta} \qquad (5.3)$$

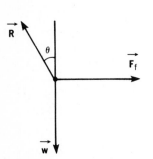

FIGURE 5.7 Forces required to roll a round object across a plane.

FIGURE 5.8 Forces acting on a rolling wheel.

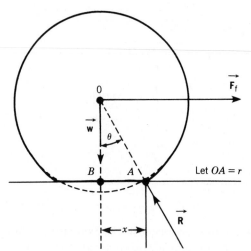

Substituting the value for R from Equation 5.3 into Equation 5.2 gives

$$F_f = w \frac{\sin \theta}{\cos \theta}$$

$$F_f = w \tan \theta \qquad (5.4)$$

However, from the space diagram in Figure 5.7,

$$\tan \theta = x/r$$

Substituting this expression for $\tan \theta$ in Equation 5.4 yields

$$F_f = \frac{wx}{r} \qquad (5.5)$$

The actual frictional force is equal in magnitude to F_f, but it is in the opposite direction.

Experiments have shown that the distance x is essentially constant for varying loads and for varying radii as long as the loads are within reasonable limits. The distance x is called the coefficient of rolling friction. Several values for x are given in Table 5.2.

The application of Equation 5.5 to a problem is given in Example 5.4.

TABLE 5.2   Coefficients of rolling friction

Wheel Material	Plane Material	x (cm)
Cast iron	Cast iron	≈ 0.018
Steel	Steel	≈ 0.014

EXAMPLE 5.4 Calculation of static and kinetic friction
· ·

Given: A 12-lb wheel is rolled across a wooden floor. A 2-oz force is re-
              quired to start it rolling and a 1.5-oz force is required to keep it
              rolling at a constant velocity. The radius of the wheel is 0.5 ft.

Find: The coefficients of static and kinetic rolling friction.

Solution: Static friction is

$$F_f = \frac{wx}{r}$$

$$x_s = \frac{Fr}{w}$$

$$x_s = \frac{(2 \text{ oz})\left(\frac{1 \text{ lb}}{16 \text{ oz}}\right)(0.5 \text{ ft})(12 \text{ inch/ft})}{12 \text{ lb}}$$

$$x_s = 0.063 \text{ inch}$$

Kinetic friction is

$$x_k = \frac{(1.5 \text{ oz})\left(\frac{1 \text{ lb}}{16 \text{ oz}}\right)(0.5 \text{ ft})(12 \text{ inch/ft})}{12 \text{ lb}}$$

$$x_k = 0.047 \text{ inch}$$

Ball bearings   Cylindrical
                    roller bearings

a.   *b.*

FIGURE 5.9 Ball bearings
and roller bearings.

In most practical cases, rolling friction is much smaller than sliding friction. In addition, the difference between the static and kinetic coefficients of rolling friction is less than the corresponding difference for sliding friction.

In many mechanical systems, sliding friction exerted on shafts rolling in plain bearings can be reduced drastically when roller or ball bearings are installed, thereby replacing the sliding friction by a lesser rolling friction. In sliding friction, resistance to motion is reduced if the sliding parts are separated with a thin film of lubricating oil. Use of rolling friction bearings, however, simplifies both the lubrication and the wear problems. In Figure 5.9, two common types of rolling friction bearings are shown, one with ball bearings and the other with roller (cylindrical) bearings.

SECTION **5.1** *EXERCISES*

1. A 60-lb block can be pulled along at constant speed by a 15-lb force acting at an angle of 20° above the horizontal.

   a. How much of the 15 lb is lifting the block? How much is dragging it horizontally?

   b. What is the normal force, that is, the resultant force, pressing block and floor together?

c. What is the coefficient of kinetic friction?

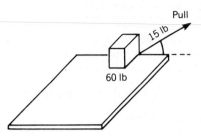

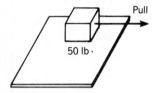

2. Only 12.5 lb are required to pull a 50-lb crate at constant speed, but 15 lb are required to get it started. What are the coefficients of static and kinetic friction?

3. Two blocks are arranged on a horizontal surface as shown. What must be the acceleration of block *B* if block *A* is to start to slide on the top of block *B*? The coefficient of static friction between blocks *A* and *B* is 0.3. The mass of block *A* is 2 kg and the mass of block *B* is 8 kg.

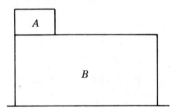

4. What force is required to push a 200-kg crate up a plane inclined at 20° if the coefficient of sliding friction between the crate and the inclined plane is 0.2?

5. What is the acceleration of a 40-kg wooden crate that is pushed horizontally across a stone floor by a horizontal force of 400 N? (See Table 5.1).

6. A block rests on an inclined plane. The coefficient of static friction between the block and the plane is 0.2. What is the maximum angle of inclination of the plane for which the block will not slide down the plane?

7. What is the difference in the magnitudes of the coefficients of static and kinetic friction between a block and a table if the same force that is required to just start the block in motion accelerates the block 0.1 m/sec² once the block starts moving?

8. A 20-lb wheel is rolled across a horizontal plane. The wheel and plane are made from steel. If the radius of the wheel is 1.0 ft, what force is required to keep the wheel rolling at a constant velocity?

5.2 . . . .

RESISTANCE IN FLUID SYSTEMS

OBJECTIVES . . . .

Upon completion of this section, the student should be able to

• Given two of the following quantities in laminar flow, determine the third:
  Forcelike quantity
  Rate
  Resistance

- Calculate the total resistance for pipes in series and in parallel.
- Define the coefficient of viscosity.
- Describe the difference between laminar and turbulent shear stress.
- Describe the difference between laminar and turbulent flow in pipes.
- Determine the effect of turbulence and pipe roughness on fluid flow.

DISCUSSION

Resistance to flow in fluid systems is due to the viscosity of the fluid. The treatment that follows examines the effect of various factors on the resistance to flow.

Viscosity—Laminar Flow

The important property of a fluid that causes frictional losses is called the viscosity. It is a measure of the fluid's resistance to an angular deformation. To define the property called viscosity, consider two parallel plates a small distance apart. The space between the plates is filled with a fluid. The lower plate remains stationary and the upper plate moves with a velocity such that the plates remain parallel. This is shown in Figure 5.10. The fluid at the surfaces of the plate has the same velocity as the plate with which it is in contact. For laminar conditions, the velocity of the fluid will increase linearly with distance, from zero at the stationary plate to the velocity (v) at the moving plate. Experimental data for many fluids show that

$$F_f = \frac{\mu A \Delta v}{\Delta y} \tag{5.6}$$

where

F_f = force required to keep the upper plate moving with constant velocity
A = area of the upper plate
Δv = difference in velocity between the upper and lower plates
Δy = small distance between the plates
μ = a constant called the viscosity

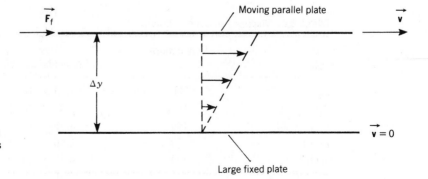

FIGURE 5.10 Behavior of a fluid between two plates as one moves across the other.

If Equation 5.7 is solved for the viscosity, one obtains

$$\mu = \frac{F_f \Delta y}{A \Delta v} \tag{5.7}$$

Therefore, the units of viscosity are

$$\text{lb·ft/(ft}^2\text{·ft/sec)} = \text{lb·sec/ft}^2 \quad \text{(English system)}$$

$$\text{N·m/(m}^2\text{·m/sec)} = \text{N·sec/m}^2 \quad \text{(international system)}$$

Fluids that satisfy the relationship given by Equation 5.7 are called Newtonian fluids.

Viscosity is also expressed in other units. A common unit for viscosity is called the poise (1 centipoise = 0.01 poise). The basic units for the poise are

$$1 \text{ poise} = 1 \text{ dyne·sec/cm}^2$$

Water at 68°F has a viscosity of approximately 1.0 centipoise. In many engineering applications, the English gravitational system is also used. For this system, the units of viscosity are

$$\text{poundal·ft/(ft}^2\text{·ft/sec)} = \text{poundal·sec/ft}^2$$

The viscosity of several fluids is given in Table 5.3. When the force is measured in poundals, the mass is measured in pounds. To change the viscosity in poundal·sec/ft^2 to lb·sec/ft^2, divide by 32 poundal/lb.

If one uses the simple kinetic theory of gases and analyzes the drag forces on the parallel layers of fluid between the two plates shown in Figure 5.10, one obtains the following equation for the viscosity $\mu(T)$ of a gas at temperature T:

$$\mu(T) = \sqrt{\left(\frac{M_m kT}{3A^2}\right)} \tag{5.8}$$

where

$\mu(T)$ = viscosity of the gas
M_m = mass of a molecule
k = Boltzmann's constant
T = absolute temperature
A = collisional cross-sectional area for the gas molecules

TABLE 5.3   Viscosity of several fluids

Fluid	Temperature (°F)	Viscosity (poundal·sec/ft²)	Viscosity (lb·sec/ft²)
Air	200	1.53×10^{-5}	4.78×10^{-7}
Air	100	1.44×10^{-5}	4.50×10^{-7}
Engine oil	140	4.33×10^{-2}	1.35×10^{-3}
Freon	50	1.86×10^{-4}	5.81×10^{-6}
Water	100	4.58×10^{-4}	1.43×10^{-5}

The simple kinetic theory indicates that the viscosity of a gas depends on the temperature, but is independent of the pressure. If it is assumed that the collisional cross section is independent of temperature, Equation 5.8 may be used to estimate the viscosity at any temperature provided it is known at one temperature. For the known temperature Equation 5.8 may be written as

$$\mu(T_1) = \sqrt{\left(\frac{M_m k T_1}{3A^2}\right)} \tag{5.9}$$

where

T_1 = temperature at which $\mu(T)$ is known

Dividing Equation 5.8 by Equation 5.9 and solving for $\mu(T)$ gives

$$\mu(T) = \sqrt{\left(\frac{T}{T_1}\right)} \mu(T_1) \tag{5.10}$$

An application of Equation 5.10 is given in Example 5.5.

EXAMPLE 5.5   Viscosity of hydrogen
....

Given: The viscosity of hydrogen at 299°K is 1.88×10^{-7} lb·sec/ft².

Find: The viscosity at 500°K.

Solution: Equation 5.10 may be used.

$$\mu(T) = \sqrt{\left(\frac{T}{T_1}\right)} \mu(T_1)$$

$$\mu(500) = \sqrt{\left(\frac{500°K}{299°K}\right)} \ (1.88 \times 10^{-7} \ \text{lb·sec/ft}^2)$$

$$\mu(500) = 2.43 \times 10^{-7} \ \text{lb·sec/ft}^2$$

The measured value of $\mu(500)$ is 2.66×10^{-7} lb·sec/ft².

Turbulent Shear

For laminar flow, Equation 5.6 may be written as a force per unit area:

$$\tau_1 = \frac{F_f}{A} = \mu \frac{\Delta v}{\Delta y} \tag{5.11}$$

The force per unit area, τ_1, due to the viscous force is called the shear stress. This stress is caused by the random molecular motions transporting a net momentum across an area parallel to the directed flow. In turbulent flow, however, many small eddies are formed in the flow. These eddies grow in size and disappear as they merge into adjacent eddies. Thus, there is a mixing of clumps of molecules that have a mixing length similar to the mean free path of molecules. These clumps of molecules transfer larger amounts of net momentum across an area parallel to the directed flow, and a larger shear stress is obtained.

By analogy, a shear stress for turbulent flow may be defined as

$$\tau_T = \eta \frac{\Delta v}{\Delta y} \qquad (5.12)$$

However, η is not constant but depends on the level of turbulence. Its magnitude varies from zero to values thousands of times greater than μ. The total shear stress then becomes

$$\tau = \tau_l + \tau_T = \mu \frac{\Delta v}{\Delta y} + \eta \frac{\Delta v}{\Delta y} \qquad (5.13)$$

At a smooth wall, however, there can be no movement of the clumps in a direction perpendicular to the wall. Thus the only momentum transfer is due to normal molecular motion. For this reason, the shear stress at the wall is

$$\tau_{\text{wall}} = \tau_l = \mu \left(\frac{\Delta v}{\Delta y} \right)_{\text{wall}} \qquad (5.14)$$

However, the turbulence does cause the value of $\Delta v/\Delta y$ near the wall to increase, producing a greater value of the shear stress near the wall.

Motion of a Body through a Fluid

Ideal Fluid An ideal fluid is an incompressible fluid with no viscosity. For this case, Bernoulli's equation, which will be developed in Chapter 6, applies. Using Bernoulli's equation, it can be shown that there is no net drag force on a body moving through an ideal fluid with constant velocity. This situation is known as "D'Alembert's paradox." Thus, the drag that a body actually experiences when moving through a real fluid with constant velocity is due to the properties of the real fluid, the geometry of the body, and its surface roughness.

Real Fluids In real fluids a moving object encounters resistive forces, or fluid friction, in its motion. A minimal amount of energy loss occurs when the fluid is simply separated and then comes together again after the passage of the object, without the creation of eddies and other kinds of swirling motion of the fluid. This simpler motion of the fluid is accomplished by the streamlining of the shape of the object. Unlike the case of dry friction, the shape of the object now is all-important. Much is learned about streamlining from the natural shapes of birds and fish as they move effortlessly through air and water. Their shapes may be compared with current designs of automobiles and especially of jet airplanes. Turbulence (swirling) is more difficult to avoid at high velocities, therefore, streamlining is especially important in the design of vehicles that must move at relatively high speeds through air or water. In contrast, recall the cumbersome shapes of weather satellites. Although they move in orbit at tremendous speeds, they move through a near vacuum—that is, the total absence of air—so that streamlining is not important.

At higher speeds, more turbulence is created, therefore, more energy is dissipated because of fluid friction in heating the fluid. This additional energy is

FIGURE 5.11 Sky diver.

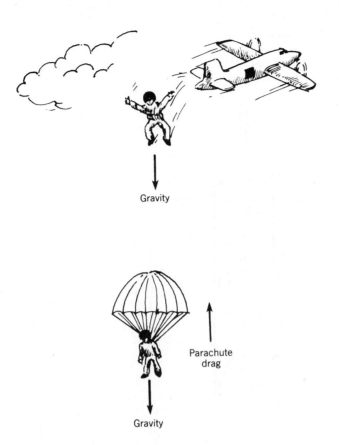

wasted. The operation of an automobile at high speed costs much more than it does at low speed, as evidenced by the lower figure for miles per gallon of gasoline in operating an automobile against air friction at higher speeds. Or equivalently, recall the difference in effort required in walking through water and in running through it at the beach.

The sport of skydiving (Figure 5.11) illustrates the increase of fluid friction with velocity. A sky diver who falls a great distance before opening the parachute quickly reaches a steady velocity of fall—a "terminal velocity" of about 130 miles per hour. From the moment of jumping, the diver's speed increases because of weight (that is, the pull of the earth). At the same time, the velocity and the frictional force of air are increasing. When this upward force of air (drag) becomes equal to the downward force (caused by the diver's weight), these forces cancel each other, and the diver maintains a constant or terminal velocity. The sky diver releases the parachute by pulling on the rip cord. The parachute is anything but streamlined in shape, and it increases air drag to such an extent that the sky diver continues to fall at a much smaller terminal velocity of only a few feet per second.

Unlike dry friction, fluid friction depends on velocity of the motion of a body through the fluid. If the body is well streamlined and if the fluid moves past the body without turbulence, the fluid flow is called "streamline" or "laminar" flow. In this case, frictional forces are very nearly proportional to the velocity, that is, frictional force doubles when the velocity doubles. As velocity is increased

FIGURE 5.12 Thought experiment.

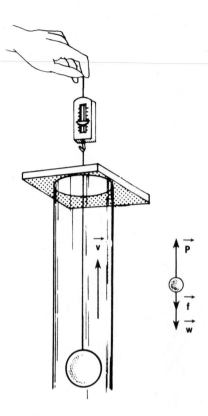

further, however, turbulence is encountered, and then the frictional forces increase more rapidly with velocity. For example, air drag on an airplane wing is proportional to the square of the velocity: frictional forces become four times as great when the speed doubles. At even higher speeds, the drag forces may vary as the cube or higher powers of the velocity. Velocity is an important factor, therefore, in the determination of fluid friction, but the exact equation depends very much on conditions and must be determined experimentally.

Fluid friction presents a way of thinking about the resistance of a system that will be valid in other systems as well. The following paragraphs present an experiment and a graphical analysis of the results. The relationships examined will be applied to other energy systems later in this chapter. Although these experiments can be done in principle, some of the measurements would be difficult to make because of the choice of materials tested. The materials will be ignored, since the principles are more important.

Imagine taking data from the apparatus in Figure 5.12 and plotting it in the following simple experiment, with the objective of determining the frictional force on a solid body moving through a specified fluid. A cylinder is filled with a fluid, water. A metal sphere suspended on a string is raised at constant speed, encountering fluid frictional drag. The spring scale registers pull of the string, P, upward on the sphere. Pulling downward on the sphere are its weight, w, and the frictional drag, f. When weight and measured force, P, are known, the frictional drag can be determined from $f = P - w$. (To be more exact, a buoyant force also forces the sphere upward. The buoyant force equals the weight of the fluid that the small

FIGURE 5.13 Plots of frictional force versus velocity.

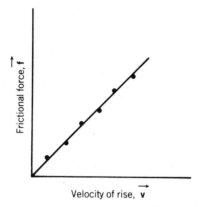

sphere displaces.) Data are taken in this way for values of f corresponding to different velocities of rise, v. The plotting of all these points gives a graph that looks something like that shown in Figure 5.13. Each point represents a corresponding pair of values of v and f. The set of points appears to determine a straight line drawn through the points.

Next, imagine that the experiment is repeated, but that the cylinder is filled with molasses, and then with alcohol. If all these graphs are plotted together, something like Figure 5.14 would result. Each straight line has a different inclination, or slope, whose significance will be investigated.

At a given velocity of rise, such as v_1, the graph shows that molasses presents the greatest frictional force, water the next largest, and alcohol the least. These values are shown on the graphs as f_m, f_w, and f_a, respectively. Therefore, the line with the greater slope indicates a fluid that offers greater fluid friction. A quick glance at the curves then shows that molasses is more viscous than water and water more viscous than alcohol (as might be expected); therefore, when forces are plotted against velocities in this way, slopes of the graphs are a measure of resistance of the system. If the velocities used in this experiment were large enough to create turbulence, the graphs would be curved rather than straight. When force is proportional to velocity, the graph is a straight line; when force becomes proportional to the second or third powers of velocity, the curves deviate from straight lines.

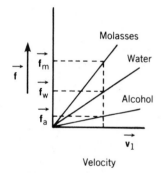

FIGURE 5.14 Plots of frictional force versus velocity for molasses, water, and alcohol.

Friction in Pipes

Laminar and Turbulent Flow For laminar flow, the pressure drop for a given length of horizontal pipe is directly proportional to the velocity. However, when the flow changes from laminar to turbulent flow, there is an abrupt increase in the dependence on velocity. After the transition region is passed, the pressure drop is proportional to v^n, where n has a range from 1.72 and 2.0. The value of n is lower for pipes with smooth walls and increases with wall roughness.

Frictional Force For a horizontal pipe of uniform diameter, the net force corresponding to a pressure difference must be balanced by the frictional force at the

FIGURE 5.15   Frictional force in a horizontal pipe.

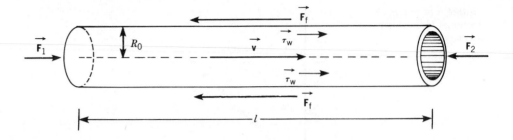

FIGURE 5.15   Frictional force in a horizontal pipe.

wall. This is illustrated in Figure 5.15. From the definition of pressure, the net force is

$$\mathbf{F}_1 - \mathbf{F}_2 = p_1 A - p_2 A = (p_1 - p_2)A$$

where

   A = area of the pipe's cross section

This force, $(p_1 - p_2)A$, is equal to the magnitude of the frictional force for steady flow. The frictional force, F_f, is equal to the wetted surface area of the pipe times the force per unit area, τ_w. The frictional force is given by

$$\mathbf{F}_f = -\tau_w\,(2\pi\,R_o l) \tag{5.15}$$

where

   $\mathbf{F}_f$ is in the negative direction
   R_o = radius of the pipe
    l = length of the pipe
   τ_w = shear stress at the wall

Setting $\Sigma F_x = 0$ gives

$$(p_1 - p_2)(\pi\,R_o^2) - \tau_w(2\pi\,R_o\,l) = 0$$

or

$$(p_1 - p_2) = 2\,\tau_w l/R_o \tag{5.16}$$

or

$$\tau_w = \frac{(p_1 - p_2)R_o}{2l}$$

Laminar Flow   The equation for the velocity profile for laminar flow given in Chapter 3 is

$$v = 2\bar{v}\left[1 - \left(\frac{R}{R_o}\right)^2\right] \tag{5.17}$$

where

   $\bar{v}$ = average velocity
   R = any radial distance

At the wall, $R = R_o$ and $v = 0$. At any other radius near the wall, it is convenient to express R as

$$R = R_o - \Delta R$$

where

ΔR = a very small positive number

The velocity at $R_o - \Delta R$ is determined by substituting $R_o - \Delta R$ into Equation 5.17. This gives

$$v = 2\bar{v}\left[1 - \left(\frac{R_o - \Delta R}{R_o}\right)^2\right]$$

$$v = 2\bar{v}\left[1 - \left(1 - \frac{\Delta R}{R_o}\right)^2\right]$$

$$v = 2\bar{v}\left[1 - 1 + \frac{2\Delta R}{R_o} - \left(\frac{\Delta R}{R_o}\right)^2\right]$$

$$v = \frac{4\bar{v}\Delta R}{R_o} - 2\bar{v}\left(\frac{\Delta R}{R_o}\right)^2$$

For a radius very near the wall, $\Delta R/R_o$ is much smaller than one. Therefore, $(\Delta R/R_o)^2$ will be much smaller than $\Delta R/R_o$. For example, $(0.001)^2 = 0.000001$ is much less than 0.001. Thus, to a good approximation

$$v = \frac{4\bar{v}(\Delta R)}{R_o} \tag{5.18}$$

The change in velocity, Δv, from $R_o - \Delta R$ to R_o at the wall is equal to v, since $v = 0$ at the wall. Thus $\Delta v = v$. Substituting Δv for v in Equation 5.18 yields

$$\Delta v = \frac{4\bar{v}\Delta R}{R_o}$$

or

$$\left(\frac{\Delta v}{\Delta R}\right)_{\text{wall}} = \frac{4\bar{v}}{R_o} \tag{5.19}$$

Substituting this change in velocity with respect to distance in Equation 5.14 yields

$$\tau_w = (\mu)\left(\frac{4\bar{v}}{R_o}\right) \tag{5.20}$$

If this expression for τ_w is substituted in Equation 5.16, the final equation for laminar flow is obtained:

$$p_1 - p_2 = \frac{8\bar{v}\mu l}{R_o^2} \tag{5.21}$$

Equation 5.21 can be solved for $\bar{v}$ to give

$$\bar{v} = \frac{(p_1 - p_2)R_o^2}{8\,\mu l} \tag{5.22}$$

In Chapter 3, the volume flow rate was shown to be

$$Q_V = A\bar{v} = \pi R_o^2 \bar{v}$$

Thus, by using Equation 5.22 to substitute for v, Q_V becomes

$$Q_V = \frac{\pi R_o^4 (p_1 - p_2)}{8\,\mu l} \qquad (5.23)$$

For laminar flow, the viscosity is constant for a given temperature. For this reason, it is possible to define a constant resistance that relates the ratelike quantity Q_V to the forcelike quantity $\Delta p = p_1 - p_2$. The resistance is defined by the equation

$$R_F = \frac{\Delta p}{Q_V} \qquad (5.24)$$

Solving Equation 5.23 for $\Delta p / Q_V$ gives

$$R_F = \frac{\Delta p}{Q_V} = \frac{8\,\mu l}{\pi R_o^4} \qquad (5.25)$$

The resistance is directly proportional to the viscosity and the length and inversely proportional to the radius of the pipe to the fourth power. Applications of Equations 5.24 and 5.25 are given in Examples 5.6 and 5.7.

EXAMPLE 5.6   Pipe resistance for laminar flow
. . . .

Given: A smooth horizontal pipe has a length of 400 ft and a radius of 0.05 ft.

Find: The flow resistance for the laminar flow of water at 100°F.

Solution: The resistance is given by Equation 5.25.

$$R_F = \frac{8\,\mu l}{\pi R_o^4}$$

The viscosity of water at 100°F $= 1.43 \times 10^{-5}$ lb·sec/ft^2.

$$R_F = \frac{(8)(1.43 \times 10^{-5} \text{ lb·sec/ft}^2)(400 \text{ ft})}{(3.14)(0.05 \text{ ft})^4}$$

$$R_F = 2.33 \times 10^3 \text{ lb·sec/ft}^5$$

The units of R_F in the English system are lb·sec/ft^5.

EXAMPLE 5.7   Volume flow rate for laminar flow
. . . .

Given: The pipe in Example 5.6.

Find: The volume flow rate. The pressure difference across the pipe is 5.0 lb/ft².

Solution: From Equation 5.24

$$R_F = \frac{\Delta p}{Q_V} = 2.33 \times 10^3 \text{ lb·sec/ft}^5$$

$$Q_V = \frac{\Delta p}{R_F}$$

$$Q_V = \frac{5.0 \text{ lb/ft}^2}{2.33 \times 10^3 \text{ lb·sec/ft}^5}$$

$$Q_V = 2.15 \times 10^{-3} \text{ ft}^3/\text{sec}$$

Question. Is the flow laminar?

The average flow velocity can be estimated from the flow rate.

$$Q_V = \bar{v}A = \bar{v}\,\pi R_o^2$$

$$\bar{v} = \frac{Q_V}{\pi R_o^2}$$

$$\bar{v} = \frac{2.15 \times 10^{-3} \text{ ft}^3/\text{sec}}{(3.14)(0.05 \text{ ft})^2}$$

$$\bar{v} = 0.274 \text{ ft/sec}$$

The Reynolds number, Re, is

$$\text{Re} = \frac{vD_o\rho_m}{\mu}, \quad \text{where } \rho_m \text{ is mass density and } D_o \text{ the pipe diameter}$$

For water at 100°F, $\rho_m = 1.94$ slug/ft³.

$$\text{Re} = \frac{(0.274 \text{ ft/sec})(0.10)(1.94 \text{ slug/ft}^3)}{1.43 \times 10^{-5} \text{ lb·sec/ft}^2}$$

$$\text{Re} = 3720$$

Thus, the Reynolds number is in the critical range and the flow may be turbulent. To ensure laminar flow, a smaller pressure drop must be used.

Turbulent Flow   For real pipe flow, the flow is usually turbulent and the walls are not perfectly smooth. The velocity profile is nearly flat except near the walls. The calculation of the frictional force, however, is much more difficult because it depends on the Reynolds number and wall roughness. If the flow rate is given for a particular pipe, the resistance can be estimated and the pressure drop calculated. If the pressure drop is given, however, the flow rate must be determined by a trial and error process because the velocity must be known before the resistance can be calculated.

From a dimensional analysis, it can be shown that the shear stress for turbulent

TABLE 5.4   Average roughness (e)

Pipe Material	e (ft)
Drawn tubing (glass, brass, etc.)	5×10^{-6}
Steel	1.5×10^{-4}
Cast iron	$\approx 8.5 \times 10^{-4}$
Galvanized iron	5×10^{-4}
Concrete	$\approx 5 \times 10^{-3}$

flow has the form given by

$$\tau_w = \frac{f \rho_m \bar{v}^2}{8} \tag{5.26}$$

where

τ_w = wall shear stress
f = a friction factor with no dimensions
ρ_m = mass density
$\bar{v}$ = average velocity

The friction factor, f, has been estimated from theory and experimental measurements. Equation 5.27 gives an approximate fit to these results:

$$f = (5.5 \times 10^{-3}) \left[1 + \left(2 \times 10^4 \frac{e}{D_o} + \frac{10^6}{\text{Re}} \right)^{1/3} \right] \tag{5.27}$$

where

e = an average wall roughness size in feet
D_o = diameter of the pipe in feet
Re = Reynolds number

Several values for e are given in Table 5.4.

To solve a problem when the flow rate is given, first find the flow velocity and calculate the Reynolds number. Use a table to find e and calculate e/D_o. With these numbers, f can be calculated. The values of f and v can then be used to determine τ_w. The pressure drop is then obtained by using Equation 5.16. An application is shown in Example 5.8.

EXAMPLE 5.8   Pressure drop in a pipe
. . . .

Given: Water is flowing through a 1-inch-diameter, galvanized iron pipe at a rate of 58.8 gal/min. The length of the pipe is 200 ft. The temperature is 100°F.

Find: The pressure drop ($p_1 - p_2$).

Solution: Change gal/min to ft³/sec.

$$(58.8 \text{ gal/min}) \left(\frac{1 \text{ min}}{60 \text{ sec}}\right)\left(\frac{1 \text{ ft}^3}{7.48 \text{ gal}}\right) = 0.131 \text{ ft}^3/\text{sec}$$

Find the average velocity in ft/sec.

$$Q_V = \bar{v}A = \bar{v} \, \pi R_o^2$$

$$\bar{v} = \frac{Q_V}{\pi R_o^2}$$

0.5-inch radius = 0.0417 ft.

$$\bar{v} = \frac{0.131 \text{ ft}^3/\text{sec}}{(3.14)(0.0417 \text{ ft})^2}$$

$$\bar{v} = 24 \text{ ft/sec}$$

Calculate the Reynolds number.

$$\text{Re} = \frac{\bar{v}\rho_m D_o}{\mu}$$

$$\text{Re} = \frac{(24 \text{ ft/sec})(1.94 \text{ slug/ft}^3)(0.0833 \text{ ft})}{1.43 \times 10^{-5} \text{ lb·sec/ft}^2}$$

$$\text{Re} = 2.71 \times 10^5 \qquad \text{(The flow is turbulent.)}$$

Calculate e/D_o. From Table 5.4, $e = 5 \times 10^{-4}$ ft.

$$\frac{e}{D_o} = \frac{5 \times 10^{-4} \text{ ft}}{0.0833 \text{ ft}} = 6 \times 10^{-3}$$

Calculate the friction factor.

$$f = (5.5 \times 10^{-3})\left[1 + \left(2 \times 10^4 \frac{e}{D_o} + \frac{10^6}{\text{Re}}\right)^{1/3}\right] \qquad \text{(Equation 5.27)}$$

$$f = (5.5 \times 10^{-3})\left[1 + \left\{(2 \times 10^4)(6 \times 10^{-3}) + \frac{10^6}{2.71 \times 10^5}\right\}^{1/3}\right]$$

$$f = (5.5 \times 10^{-3})[1 + (120 + 3.7)^{1/3}]$$

$$f = 3.29 \times 10^{-2}$$

Calculate τ_w.

$$\tau_w = \frac{f\rho_m \bar{v}^2}{8} \qquad \text{(Equation 5.26)}$$

$$\tau_w = \frac{(3.29 \times 10^{-2})(1.94 \text{ slug/ft}^3)(24 \text{ ft/sec})^2}{8}$$

$$\tau_w = 4.60 \text{ slug·ft/sec}^2\text{·ft}^2 = 4.60 \text{ lb/ft}^2$$

Calculate Δp.

$$\Delta p = \frac{2\tau_w l}{R_o} \qquad \text{(Equation 5.16)}$$

$$\Delta p = p_1 - p_2 = \frac{(2)(4.60 \text{ lb/ft}^2)(200 \text{ ft})}{0.0417 \text{ ft}}$$

$$p_1 - p_2 = 4.41 \times 10^4 \text{ lb/ft}^2$$

$$p_1 - p_2 = 306 \text{ psi}$$

In Example 5.8, the friction factor is almost independent of the Reynolds number. When this occurs, the friction factor is essentially constant. This simplifies calculating the flow rate if the pressure drop is given. To handle this case, Equation 5.26 can be solved for $\bar{v}^2$. This yields

$$\bar{v}^2 = \frac{8\tau_w}{f\rho_m} \tag{5.28}$$

Equation 5.16 can then be used to substitute for τ_w in Equation 5.28. This gives

$$\bar{v}^2 = \left(\frac{8}{f\rho_m}\right)\left[\frac{(p_1 - p_2)}{2l} R_o\right] \tag{5.29}$$

The flow rate, Q_V, is

$$Q_V = \bar{v}A \tag{5.30}$$

where

A = area of the pipe (πR_o^2)

Using the value of v given by Equation 5.29 to substitute in Equation 5.30 yields

$$Q_V = \sqrt{\left(\frac{8}{f\rho_m} \frac{(p_1 - p_2)}{2l} R_o\right)}\pi R_o^2 \tag{5.31}$$

Thus, for turbulent flow, the flow rate depends on $\sqrt{\Delta p}$ rather than Δp. The application of Equation 5.31 is illustrated and checked in Example 5.9.

EXAMPLE 5.9   Flow rate in a pipe
. . . .

Given: A 0.05-ft-radius, 100-ft-long, steel pipe has a pressure drop of 40 psi.

Find: The flow rate if the temperature is 100°F.

Solution: First assume that the friction factor depends only on the roughness factor.
Calculate e/D_o.

$$\frac{e}{D_o} = \frac{1.5 \times 10^{-4} \text{ ft}}{0.1 \text{ ft}}$$

$$\frac{e}{D_o} = 1.5 \times 10^{-3}$$

Calculate the estimated value of f.

$$f \approx 5.5 \times 10^{-3} \left\{ 1 + \left[(2 \times 10^4) \frac{e}{D_o} \right]^{1/3} \right\}$$

$$f = 5.5 \times 10^{-3} \{ 1 + [(2 \times 10^4)(1.5 \times 10^{-3})]^{1/3} \}$$

$$f \approx 2.26 \times 10^{-2}$$

Calculate the estimated value of Q_V.

$$Q_V = \sqrt{\left(\frac{8}{f\rho_m} \frac{(p_1 - p_2)}{2l} R_o \right) \pi R_o^2}$$

$$Q_V = \sqrt{\left(\frac{8}{(2.26 \times 10^{-2})(1.94 \text{ slug/ft}^3)} \frac{5760 \text{ lb/ft}^2}{(2)(100 \text{ ft})} 0.05 \text{ ft} \right) (3.14)(0.05 \text{ ft})^2}$$

Note. 40 psi = 40 psi $\times$ 144 inch^2/ft^2 = 5760 lb/ft^2

$$Q_V = 0.127 \text{ ft}^3/\text{sec}$$

Question. How does the Reynolds number affect f?
First, calculate the average velocity.

$$v = \frac{Q_V}{A}$$

$$v = \frac{0.127 \text{ ft}^3/\text{sec}}{(3.14)(0.05 \text{ ft})^2}$$

$$v = 16.2 \text{ ft/sec}$$

Calculate the Reynolds number.

$$\text{Re} = \frac{v\rho_m D_o}{\mu}$$

$$\text{Re} = \frac{(16.2 \text{ ft/sec})(1.94 \text{ slug/ft}^3)(0.1 \text{ ft})}{1.43 \times 10^{-5}}$$

$$\text{Re} = 2.20 \times 10^5$$

Calculate f using the Reynolds number.

$$f = 5.5 \times 10^{-3} \left\{ 1 + \left[(2 \times 10^4) \frac{e}{D_o} + \frac{10^6}{\text{Re}} \right]^{1/3} \right\}$$

$$f = 5.5 \times 10^{-3} \left\{ 1 + \left[(2 \times 10^4)(1.5 \times 10^{-3}) + \frac{10^6}{2.20 \times 10^5} \right]^{1/3} \right\}$$

$$f = 2.34 \times 10^{-2}$$

This value of f is not too different from the value of f used to calculate Q_V. If a closer answer for Q_V is desired, Q_V can be calculated again using $f = 2.34 \times 10^{-2}$. This would be necessary if the two f's were not nearly the same value.

Other Losses   Other pressure losses occur in a flow system because of local disturbances in the flow. These losses may be caused by elbows, valves, or anything that projects into the flow. There also can be pressure losses due to a sudden contraction or enlargement in the diameter. For very long pipes, these losses are usually negligible. However, they may be major losses for very short pipes.

Pipes in Series

General Equations   The total pressure difference for pipes connected in series is the sum of the pressure differences measured across each pipe. In addition, the flow rate is the same through each pipe. For *n* pipes, these conditions expressed mathematically are

$$\Delta p = (\Delta p)_1 + (\Delta p)_2 + \cdots + (\Delta p)_n \qquad (5.32)$$

and

$$Q_V = Q_{V_1} = Q_{V_2} = \cdots = Q_{V_n} \qquad (5.33)$$

where

Q_V = common volume flow rate

For each pipe, volume flow rate may be expressed as

$$Q_V = \bar{v}_m A_m \qquad (5.34)$$

where

A_m = area of the *m*th pipe
v_m = average velocity in the *m*th pipe

The continuity condition given by Equation 5.33 may also be written in the form of Equation 5.35, by using Equation 5.34 to substitute for the individual Q_{V_m}.

$$Q_V = \bar{v}_1 A_1 = \bar{v}_2 A_2 = \cdots = \bar{v}_n A_n \qquad (5.35)$$

Equations 5.32, 5.33, and 5.35 apply to laminar and turbulent flow.

Laminar Flow   For laminar flow, the flow resistance for each pipe is a constant at a given temperature. According to Equation 5.24, one may write for each pipe

$$Q_V = Q_{V_m} = \frac{(\Delta p)_m}{R_m}$$

or

$$(\Delta p)_m = R_m Q_V \qquad (5.36)$$

Substituting the corresponding values for each Δp_m given by Equation 5.36 in Equation 5.32 yields

$$\Delta p = R_1 Q_V + R_2 Q_V + \cdots + R_n Q_V$$

or

$$\Delta p = Q_V(R_1 + R_2 + \cdots + R_n) \tag{5.37}$$

or

$$\Delta p = Q_V R_T \tag{5.38}$$

where the total pipe resistance, R_T, is the sum of the individual resistances,

$$R_T = R_1 + R_2 + \cdots + R_n \tag{5.39}$$

Turbulent Flow   For turbulent flow, the problem is more difficult and a trial and error method may be necessary. (See Example 5.9 for the trial and error method.) For turbulent flow, Equation 5.31 applies to each pipe in series. The equation is

$$Q_{V_m} = \sqrt{\left[\left(\frac{8}{f\rho_m}\right)\left(\frac{p_1 - p_2}{2l}\right)R_{om}\right]\pi R_{om}^2}$$

where the subscript m refers to the mth pipe. By squaring both sides, this equation may also be written as

$$Q_V^2 = \frac{(\Delta p)_m}{R_m} \tag{5.40}$$

where

R_m is the resistance, which is a function of the velocity given by

$$R_m = \frac{2f\rho_m \, l}{8 \, R_{om}^5 \, \pi^2}$$

If f is dominated by the roughness factor e/D_o, then R_m is essentially constant and the problem is simplified. To obtain the total resistance, Equation 5.40 is first solved for $(\Delta p)_m$:

$$(\Delta p)_m = R_m Q_V^2 \tag{5.41}$$

Substituting these values in Equation 5.32 gives

$$\Delta p = Q_V^2[R_1 + R_2 + \cdots + R_n] \tag{5.42}$$

Equation 5.42 may be written as

$$\Delta p = Q_V^2 R_T \tag{5.43}$$

where

$$R_T = R_1 + R_2 + \cdots + R_n \tag{5.44}$$

Equation 5.44 gives the total resistance for pipes in series when the flow is turbulent. Equation 5.43 may then be written as

$$Q_V^2 = \frac{(\Delta p)}{R_T} \tag{5.45}$$

where

Δp = total pressure drop
R_T = total resistance

Pipes in Parallel

General Equations   When pipes are connected in parallel, the pressure drop is the same across each pipe. In addition, the total flow rate is now the sum of the individual flow rates. Expressed mathematically, these conditions for n pipes are

$$\Delta p = (\Delta p)_1 = (\Delta p)_2 = \cdots = (\Delta p)_n \tag{5.46}$$

and

$$Q_V = Q_1 + Q_2 + \cdots + Q_n \tag{5.47}$$

Substituting for the individual flow rates the equivalent values $A_m \bar{v}_m$ in Equation 5.47 yields

$$Q_V = A_1 \bar{v}_1 + A_2 \bar{v}_2 + \cdots + A_n \bar{v}_n \tag{5.48}$$

Equations 5.46, 5.47, and 5.48 may be used for turbulent or laminar flow.

Laminar Flow   Since the flow resistance of each pipe is a constant, one may write

$$Q_{V_m} = \frac{\Delta p}{R_m} \tag{5.49}$$

as in the case for series flow. Substituting these equivalent expressions for ΔQ_{V_m} in Equation 5.47 gives

$$Q_V = \frac{\Delta p}{R_1} + \frac{\Delta p}{R_2} + \cdots + \frac{\Delta p}{R_n}$$

or

$$Q_V = \Delta p \left[\frac{1}{R_1} + \frac{1}{R_2} + \cdots + \frac{1}{R_n} \right]$$

or

$$Q_V = \Delta p \left[\frac{1}{R_T} \right] \tag{5.50}$$

where

$$\frac{1}{R_T} = \frac{1}{R_1} + \frac{1}{R_2} + \cdots + \frac{1}{R_n} \tag{5.51}$$

The quantity R_T is the total effective resistance of the parallel pipes. An application of Equations 5.49, 5.50, and 5.51 is illustrated in Example 5.10.

EXAMPLE 5.10   Pipes in parallel for laminar flow
. . . .

Given: Three smooth-walled pipes are connected in parallel. If the flow is laminar, the resistances of the pipes are 1.0×10^3, 3×10^3, and 2×10^3 lb·sec/ft^5.

Find: **a.** The total effective resistance.

b. The total flow rate if the pressure difference is 1.0 lb/ft².

Solution: **a.** The total resistance is determined from

$$\frac{1}{R_T} = \frac{1}{R_1} + \frac{1}{R_2} + \frac{1}{R_3}$$

$$\frac{1}{R_T} = \frac{1}{10^3} + \frac{1}{3 \times 10^3} + \frac{1}{2 \times 10^3}$$

$$\frac{1}{R_T} = 1 \times 10^{-3} + 0.33 \times 10^{-3} + 0.50 \times 10^{-3}$$

$$\frac{1}{R_T} = 1.83 \times 10^{-3}$$

$$R_T = 5.45 \times 10^2 \text{ lb·sec/ft}^5$$

b. The total flow rate is

$$Q_V = \frac{\Delta p}{R}$$

$$Q_V = \frac{1.0 \text{ lb/ft}^2}{5.46 \times 10^2 \text{ lb·sec/ft}^5}$$

$$Q_V = 1.83 \times 10^{-3} \text{ ft}^3/\text{sec}$$

Turbulent Flow   For turbulent flow, the square of the flow rate in each pipe, according to Equation 5.40, is

$$Q_{V_m}^2 = \frac{(\Delta p)_m}{R_m}$$

where the subscript refers to the mth pipe in parallel. Taking the square root of both sides of the equation and noting that all the $(\Delta p)_m$ are the same for pipes in parallel gives

$$Q_{V_m} = \sqrt{\left(\frac{\Delta p}{R_m}\right)} \tag{5.52}$$

The total flow rate, according to Equation 5.47, is the sum of the individual flow rates. Substituting the individual flow rates given by Equation 5.52 into Equation 5.47 yields

$$Q_V = \sqrt{\left(\frac{\Delta p}{R_1}\right)} + \sqrt{\left(\frac{\Delta p}{R_2}\right)} + \cdots + \sqrt{\left(\frac{\Delta p}{R_n}\right)}$$

or

$$Q_V = \left[\frac{1}{\sqrt{R_1}} + \frac{1}{\sqrt{R_2}} + \cdots + \frac{1}{\sqrt{R_n}}\right] \sqrt{\Delta p} \tag{5.53}$$

Equation 5.53 may be written as

$$Q_V = \sqrt{\left(\frac{\Delta p}{R_T}\right)} \tag{5.54}$$

where

$$\frac{1}{\sqrt{R_T}} = \frac{1}{\sqrt{R_1}} + \frac{1}{\sqrt{R_2}} + \cdots + \frac{1}{\sqrt{R_n}}$$

If Equation 5.54 is squared, one obtains

$$Q_V^2 = \frac{\Delta p}{R_T}$$

where

R_T = total resistance of the pipes in parallel

Notice that Equation 5.54 is different from the equation for pipes in parallel for laminar flow.

SECTION 5.2   EXERCISES

1. Describe differences between laminar and turbulent fluid flow and give examples of the importance of streamlining.

2. A garden hose has a pressure drop of 5 psi along its length when the water flow rate is 0.5 ft³/min. If the hose diameter is 0.10 ft and the temperature is 100°F, determine:
   a. the fluid resistance of the pipe
   b. the Reynolds number
   c. the nature of flow, laminar or turbulent.

3. The resistance of a hydraulic line is 1.5 (lb/inch²)(gal/min). Determine the flow rate when the pressure drop is 4.5 psi.

4. Draw a graph showing resistance as the ratio of a forcelike quantity to a rate in the following energy systems for laminar fluid flow.

5. State the fundamental difference between sliding friction and resistance in a fluid system.

6. What is the mass of water deposited in a swimming pool by a hose with a pressure drop of 10 psi across its length and a fluid resistance of 0.3 (lb/inch²)/(ft³/min) in a period of 1 hr if the flow is turbulent?

7. Water flows out of a pipeline that is connected to a tank as shown. If the pressure at the outlet at the end of the pipe is 1.90×10^4 N/m² when the height of water in the tank is 10 m, what is the fluid resistance of the pipe if the volume flow rate is 0.02 m³/min and the flow is laminar?

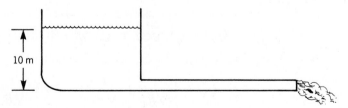

10 m

8. An experimentalist is modeling resistance to turbulent flow by using the relationship

$$R_F = \frac{\Delta p}{(Q_H)^n}$$

where n is an integer. Find the most probable value of n given the following data. Assume Δp is constant throughout the experiment.

Trial	$R_F \left(\dfrac{lb/inch^2}{gal/min} \right)$	Q_H (gal/min)
1	0.14	27
2	0.20	24
3	0.46	18

9. State how the laminar fluid flow rate through a pipe changes as each of the following quantities is changed.
  - **a.** Pressure differential
  - **b.** Pipe diameter
  - **c.** Pipe length
  - **d.** Fluid viscosity

10. Draw graphs showing the laminar fluid flow rate as a function of the following quantities.
  - **a.** Pressure differential
  - **b.** Pipe diameter
  - **c.** Pipe length
  - **d.** Fluid viscosity

11. Draw a typical curve of pressure differential versus fluid flow rate on the graph shown to the left.

12. The Reynolds number of a particular oil is 100 when the oil is flowing at a rate of 45 gal/min through a 1.5-inch-diameter pipe. What is the pressure differential between the ends of the pipe under these conditions if the pipe is 6 ft long? The specific gravity of the oil is 0.87.

13. What is the "fluid resistance" in a length of pipe if the flow is turbulent, the flow rate is 100 gal/min, and the pressure differential is 26 psi?

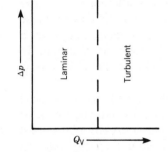

5.3

RESISTANCE IN ELECTROMAGNETIC SYSTEMS

OBJECTIVES

Upon completion of this section, the student should be able to

- Given two of the following quantities in electrical or magnetic circuits, determine the third:
  Forcelike quantity
  Rate
  Resistance

- Calculate the total resistance for resistances in series and parallel.

- Understand the effect of collision frequency and electron number density on electrical resistance.

- Describe the nature of conductors, semiconductors, and insulators.

- Describe a current–voltage characteristic plot of a junction diode.
- Describe a transistor.
- Describe the current–voltage characteristic plot for an electric arc.
- Describe a laser power supply.
- Describe the difference between a linear and a nonlinear system.
- Describe a magnetization curve.

DISCUSSION

Resistance in electrical and magnetic systems is an important quantity. In the following treatment, several similarities will be seen when compared to the resistance in fluid systems. These are especially notable for the parallel and series arrangements.

Electrical Systems

Ohm's Law An electric current in a wire is a drift of electrons in the wire. For direct current, this motion is a steady progress of electrons from one end to the other. For alternating current, there is a drift of electrons back and forth at the frequency (usually 60 cycles per second) of the electrical system. Of course, these electrons undergo collisions and random motions in the process, but when a current exists, the electrons experience a net drift through the material. In contrast, when no current exists in the wire, the motion of the electrons is random, and the effect is simply no net motion at all.

Because these electrons are "free" to wander through the bulk of the material without being captured by any one atom in particular, they contribute to both heat and electrical conduction. Good conductors of heat (metals) usually make good electrical conductors, and vice versa. Materials differ widely in their capability to conduct electricity—from copper, near the good-conductor end of the spectrum, to hard rubber, near the good-insulator end. Even liquids (salt water, for example) make rather good electrical conductors. In this case, individual electrons do not carry the current, but charged atoms or ions do the job.

An experiment is now described that will enable electrical resistance to be presented in a manner analogous to the fluid resistance in laminar flow in pipes previously discussed. In Chapter 1, the forcelike quantity in the electrical system that produces charge transport, or electrical current, is described as a potential difference ΔV, in volts. The ratelike quantity is the rate at which charge is transferred through the material, which is precisely what is meant by electrical current. The most common unit of current is the number of coulombs of charge moved per second, which is called amperes of current.

Consider a copper element (a piece of copper of a particular size and shape) included in an electrical circuit (Figure 5.16) made up of a variable voltage source and two measuring instruments—a voltmeter to measure potential difference across the copper element and an ammeter to measure charge flow rate or current

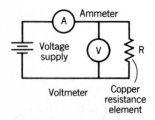

FIGURE 5.16 Experiment to measure current and potential difference.

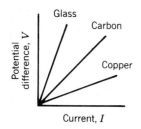

FIGURE 5.17 Plots of current versus potential difference.

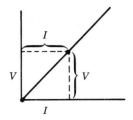

FIGURE 5.18 Slope of current versus potential difference.

through the element. The symbols used represent a battery and a resistor, R, for the copper element. Simply vary the potential difference, ΔV, and read the corresponding value of current, I, off the ammeter. If these data are plotted as shown, a straight-line graph results for copper.

Data are taken for a sample of glass and for a sample of carbon, both cut to the same size and shape as the copper, to enable comparison of the results. By now, these results should be meaningful. Plotted in Figure 5.17 is the forcelike quantity, ΔV, against the ratelike quantity, I. The slope of these lines is again a measure of the resistance. Evidently, glass offers more resistance to electrical conduction than does carbon. Copper turns out to be the best conductor of the three, as expected.

The slope of these straight lines is found, as usual, by taking an interval, ΔV, between any two points on the line and dividing by the corresponding interval, ΔI. If one point on the interval is chosen as the origin, shown in Figure 5.18, then slope equals V/I. As discussed previously, the slope is a measure of resistance of the system. Here electrical resistance, R, is defined as the ratio of V/I, as in

$$R = \frac{\Delta V}{I} \qquad (5.55)$$

where

R = resistance
ΔV = voltage across the resistor
I = current through the resistor

If ΔV is in volts and I is in amperes, R is defined in terms of a unit called the "ohm." One ohm (1 Ω) of resistance is the resistance of an element in which one ampere of current results when one volt of potential difference exists across it. Briefly, an ohm is one volt per ampere. The name "ohm" was chosen in honor of the discoverer of this basic law, the German scientist Georg Simon Ohm. This law is referred to as Ohm's law. Examples 5.11 and 5.12 illustrate the use of Ohm's law to calculate resistance and current in electrical circuits.

EXAMPLE 5.11 Ohm's law
. . . .

Given: The circuit described above, with a current of 2 A when the voltage is 10 V.

Find: The resistance of the resistive element.

Solution: The voltage, current, and resistance are related simply through Ohm's law, $\Delta V = IR$, or

$$R = \Delta V/I = \frac{(10\ \text{V})}{(2\ \text{A})} = 5\ \Omega$$

EXAMPLE 5.12   Ohm's law
....

Given: Voltage is reduced to 4 V.

Find: The current through the same resistor.

Solution: Ohm's law is applied again. This time, current is the unknown, so
                  solving for *I*,

$$I = \Delta V/R = \frac{(4\ \text{V})}{(5\ \Omega)} = 0.8\ \text{A}$$

The equations of motion developed in Chapter 3 can be used to develop a simple theory for the flow of electric charge in a conducting medium. The force on a charged particle due to an electric field is

$$\mathbf{F} = \mathbf{E}q \tag{5.56}$$

where

  $\mathbf{E}$ = field strength
  q = particle's charge

If the particle is free to move, such as a free electron in a metallic conductor, it will accelerate in the direction of the electric field. From Newton's second law, the acceleration is

$$\mathbf{a} = \frac{\mathbf{F}}{m} = \frac{\mathbf{E}q}{m} \tag{5.57}$$

where

  m = mass of the particle

If the particle were completely free, it would continue to accelerate across the conductor. This does not occur because the charged particles suffer collisions with atoms or molecules. In addition to their acceleration, the particles move about in the conductor in a random fashion. As they move about they collide with atoms a number of times each second. The number of collisions a charged particle makes per second is called the collision frequency, f_e; the average time between collisions is then $1/f_e$. Thus, on the average, a charged particle will only accelerate for a time equal to $1/f_e$. When a collision occurs, the particle tends to lose the directed velocity it gained because of its acceleration. After the collision it starts to accelerate again. The directed velocity at the time of a collision, on the average, is

$$\mathbf{v}_d = \mathbf{a}t = \frac{\mathbf{a}}{f_e} \tag{5.58}$$

Since the directed velocity starts from 0 and the acceleration is constant, the average directed velocity between collisions is

$$\overline{\mathbf{v}}_d = \frac{\mathbf{a}}{2f_e}$$

Substituting for a from Equation 5.57 yields

$$\overline{\mathbf{v}}_d = \frac{\mathbf{E}q}{2f_e m} \tag{5.59}$$

where

$\mathbf{E}$ = field strength
q = particle's charge
f_e = collision frequency
m = particle's mass
$\overline{\mathbf{v}}_d$ = average directed drift velocity

It is now advantageous to define a quantity called the current density. The current density is the total charge passing through a unit area perpendicular to the directed or drift velocity per unit time. Since the total current is the total charge passing through a cross-sectional area of the conductor,

$$I = jA \tag{5.60}$$

where

I = total current
j = current density
A = cross-sectional area of the conductor

To express j in terms of the properties of the conductor and the voltage across the conductor, it is necessary to consider the number of charged particles per unit volume that are free to move under the action of the field. On the average each charged particle has a drift velocity given by Equation 5.59. In one second, each of these particles will drift in the direction of the field a distance equal to v_d. Thus, as shown in Figure 5.19, the number passing through a unit cross section per unit time is nv_d. Since each particle has a charge, q, the current density becomes

$$j = qn\overline{\mathbf{v}}_d \tag{5.61}$$

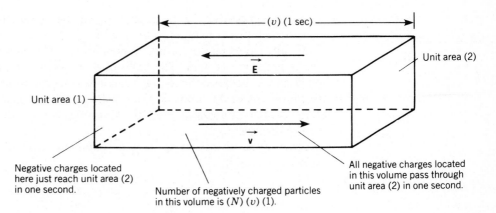

FIGURE 5.19 Illustration of current density.

where

n = number of charged particles per unit volume

Substituting the value of j given in Equation 5.61 into Equation 5.60 yields

$$I = qn\bar{v}_{d}A \tag{5.62}$$

Substituting the value of v_d from Equation 5.59 gives

$$I = \frac{nq^2AE}{2f_e m} \tag{5.63}$$

From the definition of voltage given in Chapter 2

$$\Delta V = E\Delta l$$

or

$$E = \frac{\Delta V}{\Delta l}$$

where

ΔV = voltage across the conductor
Δl = length of the conductor

Thus, by substituting this expression for E in Equation 5.63, one obtains

$$I = \left(\frac{nq^2}{2f_e m}\right)\left(\frac{A}{\Delta l}\right)(\Delta V) \tag{5.64}$$

Comparing Equation 5.64 with Ohm's law (Equation 5.55) shows that this simple theory gives

$$R = \left(\frac{2f_e m}{nq^2}\right)\left(\frac{\Delta l}{A}\right) \tag{5.65}$$

Thus, the resistance should be directly proportional to the length of the conductor and inversely proportional to the area. The quantity

$$\rho = \frac{2f_e m}{nq^2} \tag{5.66}$$

is called the **resistivity** and is a property of the conducting material.

In a metallic conductor, the number of free electrons per unit volume is essentially constant for ordinary conditions. The mass of an electron and its charge are also constant. The collisional frequency, f_e, however, increases with temperature, because the average random velocity of the electrons increases and the vibrations of the atoms in the crystal lattice increase in amplitude. Thus, the effective area for collisions with the vibrating atoms in the crystal lattice increases with temperature.

Direct experiments using metallic conductors of different lengths and areas yield the equation

$$R = \rho \frac{\Delta l}{A} \tag{5.67}$$

where ρ depends on the temperature of the material. Thus, the experimental data are in conformity with the simple theory. The resistivity for several materials is given in Table 5.5. An application of Equation 5.67 is given in Example 5.13.

EXAMPLE 5.13   Resistance of a copper wire
. . . .

Given: A copper wire at 20°C with a diameter of 0.1 mm and a length of 50 m.

Find: The resistance. (Refer to Table 5.5 for resistivity.)

Solution: $R = \dfrac{\rho \Delta l}{A} = \dfrac{\rho \Delta l}{\pi r^2}$ (Recall that $A = \pi r^2$.)

$R = \dfrac{(1.72 \times 10^{-8}\ \Omega \cdot m)(50\ m)}{(3.14)(5 \times 10^{-5}\ m)^2} = \dfrac{86 \times 10^{-8}\ \Omega \cdot m^2}{(3.14)(25 \times 10^{-10}\ m^2)}$

$R = 109\ \Omega$

Resistance of the wire at 20°C is 109 Ω.

Effect of Temperature on Resistance   The change in resistance of conductors with temperature can be investigated experimentally. Since the measured change in resistance with temperature is not a strong function of temperature (almost linear), Equation 5.68 is a good approximation as long as the temperature change is within reasonable bounds.

$$R = R_o[1 + \alpha_o (T - T_o)] \qquad (5.68)$$

where

R = resistance at any temperature T
R_o = resistance at temperature T_o, often taken to be 20°C
α_o = temperature coefficient of resistance at the reference temperature, T_o

Values for α_o are given in Table 5.5 based on a reference temperature of 20°C. In the use of Equation 5.68, it is important to use the α_o associated with T_o since α_o changes a little with T_o. The values of α_o in Table 5.5 are for a T_o of 20°C.

The simple theory for electrical conduction can be extended to estimate α_o. Space limitations, however, do not permit this to be done here. For pure materials, the simple theory gives

$$\alpha_o = \frac{0.005}{C°}$$

for $T_o = 20$°C. This is in reasonable agreement with the measured values given in Table 5.5.

The most common conductors for which these equations give useful results are wires. The accepted standard for specifying the size of wire is the American Wire

TABLE 5.5 Resistivity and temperature coefficients of various materials at 20°C

Material	Resistivity (ρ) ($\Omega \cdot m$)	Temperature Coefficient of Resistance (α)(per $C°$)
Silver	1.47×10^{-8}	0.0038
Copper	1.72×10^{-8}	0.00393
Aluminum	2.63×10^{-8}	0.0039
Tungsten	5.51×10^{-8}	0.005
Constantan	4.41×10^{-7}	<0.00001
Nichrome	1.50×10^{-6}	0.00017

TABLE 5.6 American Wire Gage values at 20°C

Gage No.	Diameter (mm)	Cross-sectional Area (mm²)
8	3.264	8.367
14	1.628	2.082
20	0.8118	0.5176
24	0.5106	0.2048
26	0.4049	0.1288
36	0.1270	0.01267
38	0.1007	0.007964
40	0.07987	0.005010

Gage. Table 5.6 is given to aid in solving problems using this nomenclature.

Examples 5.14 and 5.15 illustrate the use of Equations 5.67 and 5.68 to solve problems.

EXAMPLE 5.14 Resistance of a number-24 copper wire

Given: Number 24 copper wire.

Find: The length having a resistance of 1 Ω at 5°C.

Solution:
$$R_T = \rho \frac{\Delta l}{A} p[1 + \alpha_o (T - 20°C)]$$

$$\Delta l = \frac{R_T A}{\rho [1 + \alpha_o (T - 20°C)]}$$

$$\Delta l = \frac{(1\ \Omega)(2.048 \times 10^{-7}\ m^2)}{(1.72 \times 10^{-8}\ \Omega \cdot m)[1 + (3.93 \times 10^{-3}\ C°^{-1})(5°C - 20°C)]}$$

$$\Delta l = \frac{2.048 \times 10^{-7} \ \Omega \cdot m^2}{(1.72 \times 10^{-8} \ \Omega \cdot m)(0.941)}$$

$$\Delta l = 12.7 \ m$$

The required length of copper wire is 12.7 m.

EXAMPLE 5.15   Resistance of nichrome wire
. . . .

Given: A piece of nichrome wire with a resistance of 25 Ω at 20°C.

Find: The temperature at which it will have a resistance of 28 Ω.

Solution: $R_T = R_o \ [1 + \alpha_o \ (T - T_o)]$

Solving for T,

$$T = T_o + \frac{(R_T/R_o - 1)}{\alpha_o}$$

$$T = 20°C + \frac{(28 \ \Omega/25 \ \Omega - 1)}{1.7 \times 10^{-4} \ C°^{-1}}$$

$$T = 20°C + 706 \ C°$$

$$T = 726°C$$

Thus, the nichrome wire increases its resistance from 25 Ω to 28 Ω when temperature increases by 706 C° above room temperature of 20°C.

Conductors, Semiconductors, and Insulators   Based on their resistance to current flow, materials are classified as conductors, semiconductors, or insulators. A material's classification is determined by its atomic structure.

An atom is composed of a positively charged nucleus, or core, surrounded by negatively charged electrons. The electrons orbit the nucleus in a definite pattern of shells and subshells. Figure 5.20 shows the first four electron shells of an atom. It also shows that shell 2 has two subshells, shell 3 has three subshells, and shell 4 has four subshells.

The energy of an electron is related to its distance from the nucleus, with the closest electron being the least energetic (or most stable). In the formation of an atom, electrons naturally fill in the lower energy levels first. Helium, for example, has two electrons, both of which lie in the first shell. Copper has 29 electrons. Its first three shells are filled, with one electron left over. This remaining electron occupies the *s* subshell of the fourth shell ($2 + 8 + 18 + 1 = 29$).

The most stable configuration for any atom is to have its subshells filled. The attempt of an atom to have filled subshells causes bonding between atoms.

FIGURE 5.20 Classical view of an atom showing the first four shells and the corresponding subshells.

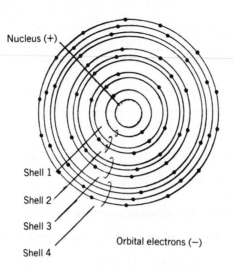

Nucleus (+)

Shell 1

Shell 2

Shell 3

Orbital electrons (−)

Shell 4

SHELL	SUBSHELL	ELECTRONS	TOTAL
1	*s*	2	2
2	*s*	2	8
	p	6	
3	*s*	2	18
	p	6	
	d	10	
4	*s*	2	32
	p	6	
	d	10	
	f	14	

Figure 5.21 shows an atom of sodium (Na) with 11 electrons and an atom of chlorine (Cl) with 17. Ten of the sodium electrons fill the first three subshells of sodium, leaving one extra electron. Chlorine, on the other hand, needs one more electron to fill its fifth subshell. The sodium atom gives up its electron to the chlorine atom, as shown in Figure 5.21.

Because the atoms are neutral (have no net charge) to begin with, losing one electron gives the sodium atom a charge of +1, and gaining one electron gives the chlorine atom a charge of −1. The attraction of unlike charges holds the two charged atoms, or ions, together. This attraction is called "ionic bonding."

A second way in which atoms may fill their outer subshells is shown in Figure 5.22, in which the structure of polyethylene molecules is illustrated. There, each carbon atom (C) with six electrons needs four additional electrons to fill its outer subshell, and each hydrogen atom (H) with one electron needs one additional electron. All these subshells are filled by sharing electrons. Each carbon atom

FIGURE 5.21 Classical view of ionic bonding in sodium chloride.

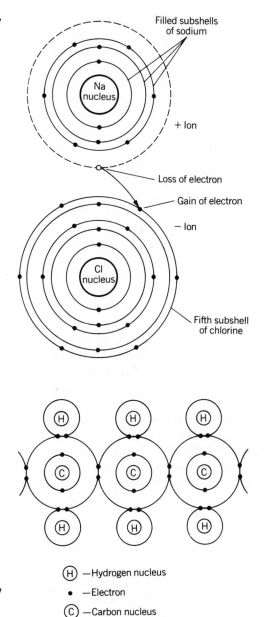

Filled subshells of sodium

Na nucleus

+ Ion

Loss of electron

Gain of electron

− Ion

Cl nucleus

Fifth subshell of chlorine

H —Hydrogen nucleus

• —Electron

C —Carbon nucleus

FIGURE 5.22 Classical view of covalent bonding in polyethylene.

shares one electron from each of two hydrogen atoms and two adjacent carbon atoms. This sharing of electrons is called "covalent bonding."

A third method of bonding occurs in metals. Figure 5.23 shows two copper atoms. The single electron in the $4s$ subshell of each atom is very loosely bound to the atom. In a solid piece of copper, the attraction between this electron and the nucleus is too weak to hold the electron in its orbit. The electron does not really "belong" to any one copper nucleus; therefore, it is free to move throughout the piece of copper. The copper actually consists of an array of positive copper ions

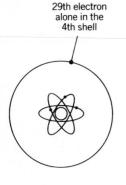

29th electron
alone in the
4th shell

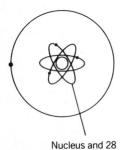

Nucleus and 28
tightly bound
electrons of
a copper atom

FIGURE 5.23 Classical view of metallic bonding in copper.

suspended in a "cloud" of free electrons. The repulsive forces between the positive copper ions cause them to have fixed positions within the electron cloud. The attractive force between the positive ions and the negative electrons holds the copper atoms together and is called "metallic bonding."

Insulators are materials whose outer electrons are bound tightly to the atoms and cannot serve as charge carriers. The bond may be either ionic or covalent. Materials commonly employed as electrical insulators are usually covalent, because this type of bonding is not broken down as easily as ionic bonding.

Conductors are materials whose outer electrons are free to move throughout the material. As shown by Equation 5.64, how well a material conducts electricity depends directly on the number of free electrons per unit volume, n, and inversely on the collisional frequency, f_e. In the case of conductors, n is essentially constant as the temperature increases. However, the collisional frequency increases because of the increased random velocity of the free electrons and the increased collisional cross section caused by the large amplitude of the vibrating atoms in the crystal lattice. For semiconductors, however, n increases with temperature, a fact that more than offsets the effect of the increased collisional frequency of a free electron. Thus, the resistance of a semiconductor actually decreases with an increase in temperature. Figure 5.24 illustrates the effect of temperature on the resistance of metals and semiconductors for large temperature changes. How an increase in the temperature can cause an increase in the number of free electrons per unit volume is discussed in the following paragraphs.

Semiconductors are materials that are ordinarily rather good insulators but can be made to conduct by supplying a small amount of extra energy, for example, by raising the temperature.

Some of the best-known semiconductors are materials in which the outer subshell of atoms has four electrons and needs another four electrons to fill the subshell completely. These semiconductors include carbon (6 electrons), silicon (14 electrons), and germanium (32 electrons). Such materials form a crystalline structure through covalent bonding, as shown in Figure 5.25.

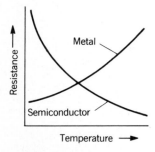

FIGURE 5.24 Plots of resistance versus temperature for metals and semiconductors.

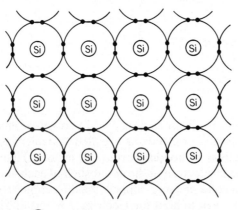

FIGURE 5.25 Bonding in a silicon crystal.

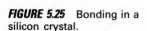

(Si) — Silicon nucleus + first 10 electrons

•   —Valence electron

FIGURE 5.26 Relative resistivities of several materials.

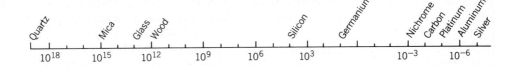

Because all the outer electrons are involved in covalent bonds, there are very few free electrons. These loosely bound, or so-called "valence," electrons, however, are barely attached to the atoms. Some of them can be moved by adding only a small amount of extra energy. For this reason, the resistance of semiconductors is high at low temperature but decreases rapidly as the temperature rises.

Figure 5.26 gives relative resistivities of several insulators, semiconductors, and conductors.

Conduction electron

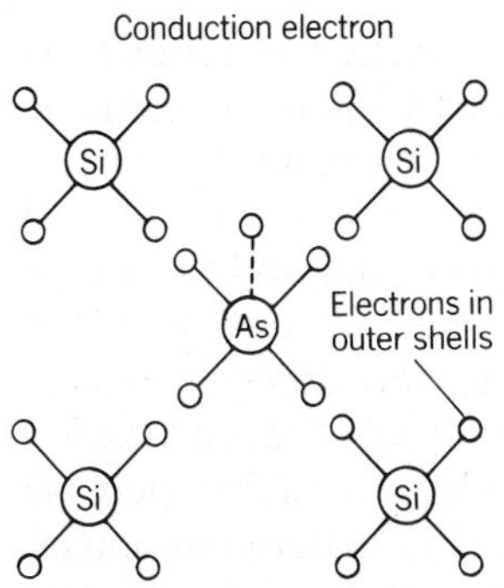

FIGURE 5.27 Silicon doped with arsenic to produce an n-type extrinsic material.

Extrinsic Semiconductors "Intrinsic" (or pure) germanium or silicon can have its natural resistance altered by the addition of "impurities" in a process called "doping." Doped germanium or silicon is then called "extrinsic." As mentioned before, germanium or silicon atoms normally have four electrons in their outer orbit. These atoms are locked to neighboring atoms in the crystalline structure of the material. Elements such as phosphorous, arsenic, and antimony have five electrons in their outer orbit. If germanium or silicon is doped with any of these three impurities, then, proportional to the amount of doping, the impurity atoms can replace some of the germanium or silicon atoms in the lattice. At each place in the crystalline structure where this process happens, four of the five electrons of the impurity atom lock into the crystalline structure and, as before, leave the fifth electron available for conduction, as shown in Figure 5.27. The heavier the doping, the greater the number of "fifth" electrons available for current flow, and the lower the resistance of the extrinsic material. This new, doped, extrinsic material is called "n-type" because electrons (negative) are available as the main, or majority, carrier of current.

Boron, aluminum, and gallium are examples of elements that have three electrons in their outer orbit. When these elements are used to dope germanium or silicon, a "hole" is created where each impurity atom locks into the crystalline structure. The crystalline structure can accommodate four electrons per atom, but only three are present. The space at which the fourth electron is missing is called a hole, as shown in Figure 5.28. A hole has the property of a positive charge (absence of a negative charge); it can move or flow in the same manner as an electron moves, although holes flow in the opposite direction from electrons because they are positive.

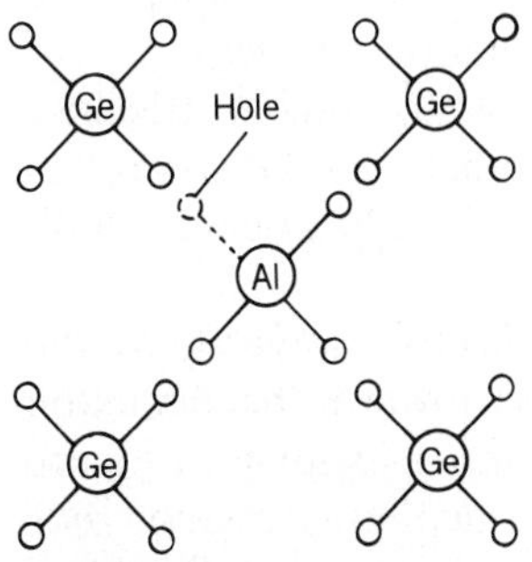

The idea that a hole (being "nothing") can move may seem mystifying at first, but it is a simple concept. In Figure 5.29, a piece of "p-type" material doped with

FIGURE 5.28 Germanium doped with aluminum to produce a p-type extrinsic material.

FIGURE 5.29 The movement of a hole during current flow.

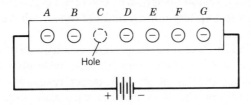

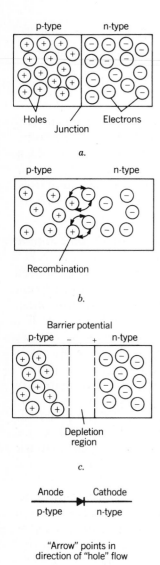

p-type   n-type

Holes   Electrons
        Junction

a.

p-type   n-type

Recombination

b.

Barrier potential
p-type   −   +   n-type

Depletion
region

c.

Anode   Cathode

p-type   n-type

"Arrow" points in
direction of "hole" flow

d.

FIGURE 5.30 Depletion at
the junction of a pn
junction diode.

impurity atoms to produce holes is connected to an external battery to produce current flow. The electrons, being negative charges, are attracted to the positive terminal of the battery and will move to the left. The holes should move to the right. Electron D moves to the left and replaces the hole at C. The hole, therefore, is now to the right, at the previous location of D. Next, electron E fills the hole at D, and the hole has moved another position to the right, to E. As electrons move to the left, holes move to the right. Holes can be represented as stepping stones that allow electrons to travel more easily across the material.

p-Type and n-type materials, by themselves, are nothing spectacular. They are merely conductors whose resistance is controlled by the amount of doping. In n-type material, electrons are the majority carriers (of current), whereas in p-type material, holes are the majority carriers. The combination of these two materials in a special way results in a device with properties of voltage-controlled resistance.

When n-type material is joined to p-type material, a junction is formed, as shown in Figure 5.30a. Immediately, the holes at one side of the junction combine with the electrons at the other side. Where this happens, the majority carriers have been "depleted," and a "barrier potential" is created. The p-type material has become slightly negative because it has gained some electrons and the n-type material has become slightly positive because it has lost some electrons, as shown in Figures 5.30b and 5.30c. Recombination may continue as more holes recombine with electrons, but the barrier potential created by the first level of recombination slows down the process. The depletion region becomes so large, and the opposing barrier potential becomes so great, the holes from the p-type material no longer can cross over the depletion region to recombine with the electrons from the n-type material. Recombination then ceases.

The device so formed is called a pn junction diode. The barrier potential created in a germanium diode is about 0.3 V; the barrier potential for a silicon diode is about 0.7 V. The resistance of the p and n materials themselves is a function of the percentage of doping and is generally very low (a few ohms). The resistance of the depletion region, because there are no majority carriers, is very high, and is inversely proportional to the width of the depletion region. Because the width of this region, and hence the resistance of the diode, can be controlled by an external, or "bias," dc voltage, the diode is called a "voltage-controlled" resistor.

When a dc voltage is applied in the forward direction (plus to p, minus to n), the holes and electrons are pushed by the battery voltage toward the depletion region—electrons toward electrons and holes toward holes, as shown in Figure 5.31a. Since the charges repel, this process shrinks the depletion region, thus lowering the resistance of the diode. When the applied voltage exceeds the 0.3 V for germanium or 0.7 V for silicon, the barrier potential is overcome completely and the depletion region disappears. The resistance of the diode at this point and beyond is very low (a few ohms), and the diode behaves like a short circuit or a closed switch.

In Figure 5.31b, a reverse-bias voltage causes holes and electrons to be attracted toward the battery terminals, thus widening the depletion region and causing the diode to exhibit an even larger resistance than in its normal state. Under this condition, the diode exhibits very high resistance and behaves like an open circuit or like an open switch.

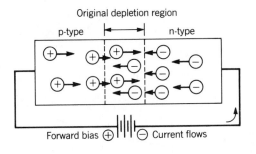

a.

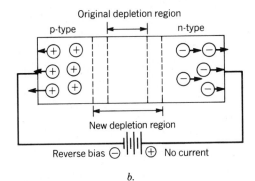

b.

Figure 5.32 shows the current–voltage relationship of a pn junction diode for various amounts of forward and reverse bias. Resistance at any point on the curve can be found simply by dividing the voltage by the current. As forward bias is applied, starting from zero volts, the depletion region is still large and the diode resistance therefore high. Diode resistance at this point is predominantly the junction resistance, or the resistance of the depletion region. As the forward-bias voltage is increased toward the barrier potential (known as the "knee voltage"), the junction resistance decreases because the depletion region narrows. When the knee voltage is reached, the depletion region disappears; then the curve takes a sharp turn upward, indicating low resistance. Ideally, the diode is considered to have a resistance of zero ohms when operated beyond its barrier potential, or to appear to be a short circuit (indicated by the vertical dotted line above the knee voltage.) However, practical p or n material has some resistance, called bulk resistance, of a few ohms, shown by the sloped line labeled bulk resistance. Any device has a power limitation (such as a $\frac{1}{4}$-W resistor) and diodes are no exception. If too much forward current is forced through the diode, the I^2R will exceed the power rating of the diode and burnout will occur.

As reverse bias is applied, the diode exhibits some "reverse" resistance, because the widened depletion region has a very high, but measurable, resistance, on the order of megaohms. As reverse bias continues to be applied, a point is reached at which breakdown occurs. This is called the "zener point," and is the point at which the applied voltage is so great that it forces carriers across the depletion region. This situation is similar to "arc-over" in a switch. With the depletion region broken down, the resistance of the diode drops drastically, and the diode conducts heavily, as the sharp, downward turn of the curve shows.

FIGURE 5.32   Plot of current versus voltage for a pn junction diode.

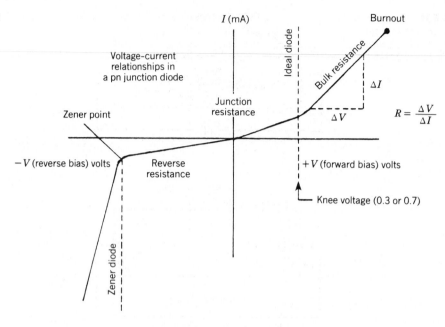

This condition is not necessarily bad, as long as the burnout point is not exceeded. Special diodes, called zener diodes, are built specifically to operate in this zener region and are useful as voltage regulators. The dashed line labeled "zener diode" on the curve shows that in this region of operation the current flow through the diode can vary. The voltage across the diode, however, remains almost constant, making it valuable as a voltage regulator.

Transistors   The fact that a pn junction diode is a voltage-controlled resistor whose resistance is a function of the bias voltage applied to it provides a clue about how a transistor operates. Essentially, a transistor is nothing more than two diodes constructed back-to-back.

Figure 5.33 shows the two possible variations of transistor construction. Figure 5.33*a* shows a pn junction (emitter-base) connected to an np junction (base-collector), with the base (n) being common to both. The reverse is shown in Figure 5.33*b*. The terminal with the arrow is called the "emitter." The arrow points in the direction of conventional current flow (hole flow), which is the opposite direction of electron flow. In normal operation of the transistor, when properly biased, majority carriers (either holes for pnp-type or electrons for npn-type) are "emitted" by the emitter and "collected" by the collector. The amount of this current flow is regulated by the voltage-controlled resistance properties of the transistor.

The biasing of a transistor is important to its proper operation. Figure 5.34 illustrates how an npn transistor is connected to a battery supply for correct operation. The emitter-base junction is always forward-biased, therefore, an npn transistor will have its emitter connected to the minus terminal of the battery and the base connected to the plus terminal. The resistance of the corresponding diode is low, and current flows very easily in the emitter-base circuit loop. The base-

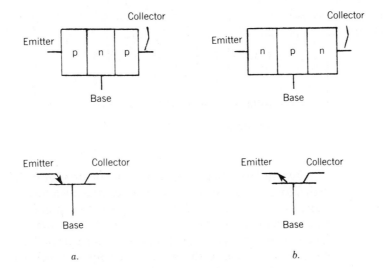

FIGURE 5.33 pnp and npn transistors.

a.   *b.*

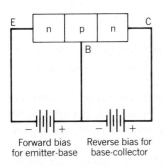

Forward bias for emitter-base   Reverse bias for base-collector

FIGURE 5.34 Normal bias of an npn transistor.

collector junction is always reverse-biased. As shown in Figure 5.34, the collector, which is n-type material, is connected to the plus terminal of a battery and the base is connected to the minus terminal.

The resistance of this base-collector loop is thus very high and, theoretically, no current flows in this loop. There is, however, one important requirement in transistor construction that makes the transistor work. This same requirement explains why transistor action cannot be achieved merely by placing two diodes back-to-back. In transistor construction, the base region is very thin. The depletion regions of each diode practically touch each other. When these majority carriers enter the base region, they "punch through" the base and end up in the collector region, where they are attracted by the plus voltage on the collector terminal. The percentage of electrons traveling through the base to the collector (punching through) is very large, about 90 to 99 percent.

The significance of this punch-through action is this: Normally the resistance of the base-collector junction is very high, and no current would flow in this circuit. Because of the punch-through, the resistance of the base-collector junction is altered, and current flow in the collector circuit is inversely proportional to the altered resistance. In effect, the resistance of the collector (output) circuit is being controlled by the resistance of the emitter (input) circuit. The resistance of the input circuit is transferred to the output circuit; hence, the origin of the name "tran-sistor" (meaning "TRANsfer-reSISTOR").

Gaseous Conductors Figure 5.35 depicts a gas-filled tube. A voltage is maintained across the tube by connecting each electrode to one of the terminals of a battery or power supply. The electrode connected to the positive terminal is called the "anode." The electrode connected to the negative terminal is termed the "cathode." The resistor R_B is called the ballast resistor.

Under ordinary circumstances, the atoms of the gas are electrically neutral and, therefore, are not attracted to either the anode or the cathode. No conduction occurs along the tube and the resistance of the gas is said to be "infinite."

FIGURE 5.35 Gas-filled
tube as a circuit element.

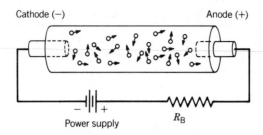

When an electron is separated from an atom or molecule, the process is called ionization. The atom or molecule is called a positive ion. If a gas is ionized to form many free electrons and positive ions, the ionized gas (called a plasma) can conduct an electric current according to Equation 5.64 (in general n and f_e are functions of position). For this case, both electron flow and ion flow contribute to the current. Because the mass of the ion is much larger than the mass of the electron, it can be shown that the current is primarily due to electron flow in most instances. According to Equation 5.64, the current will be directly proportional to the number of free electrons per unit volume and inversely proportional to the collisional frequency.

The initial ionization or breakdown of the gas is pressure dependent. It is a very complex phenomenon and will not be discussed here. However, the ionization process in a developed discharge can be explained in simple terms.

The nature of the ionization process in high-pressure discharges is different from the ionization process in low-pressure discharges. In the high-pressure case, the discharge is called an electric-arc discharge. The plasma is heated by the electric current to temperatures in the range of approximately 9000 to 17,000°K. The high temperature causes the particles to move with large random velocities. Some of the random velocities are large enough to cause atoms or molecules to become ionized when collisions occur. This process is called thermal ionization. Equations derived for thermodynamic equilibrium can usually be applied to arc plasmas. A major loss of energy from arcs occurs in the form of radiation, and arc discharges are often used as high-intensity lamps. The electric power source can be a dc, ac, or radio-frequency supply.

If the average current increases, the arc—or gas—temperature increases, so the number of free electrons increases. This tends to reduce the resistivity according to Equation 5.66. If the gas pressure is constant (arc is open to the atmosphere or else the confined gas is maintained at a constant pressure) the collisional frequency also decreases. That occurs because an increase in temperature at constant pressure "thins out" the number of particles. Fewer particles mean a lower collision frequency. This also, according to Equation 5.66, means a decrease in resistivity. Thus, the total arc resistance decreases with increasing current. The voltage across an arc as a function of current is illustrated in Figure 5.36.

Because the resistance decreases with an increase in the current, the arc is not stable. To stabilize the discharge, a resistor is placed in series with the arc. The corresponding load line is shown in Figure 5.36. This line represents all the possible values of current and voltage that can be delivered to the arc. Mathe-

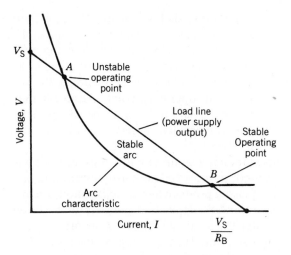

FIGURE 5.36 Plot of voltage versus current for an arc across a gas-filled tube.

matically the equation of the line is

$$V = V_S - IR_B \tag{5.69}$$

where

V = voltage available to the discharge
I = current
V_S = open circuit voltage of the supply
R_B = ballast resistor

Equation 5.69 states simply that the output voltage is equal to the power supply voltage minus the voltage drop in the ballast resistor. Two possible arc operating points are shown in the diagram. As indicated, only one (point B) is a stable operating point.

At low pressures, the time between collisions is greater. A free electron can then gain a larger final velocity between collisions under the action of the electric field. At sufficiently low pressures, many electrons gain enough energy from the field to cause ionization when they collide with an atom or molecule. In the low-pressure region, the average random energy of the electrons becomes much greater than the average random energy of the ions and molecules, and no unique temperature exists. Thus, the equations for thermodynamic equilibrium do not apply. In some cases, however, two temperatures can be defined: an electron temperature, T_e, and a heavy particle temperature, T_a. For example, in the low-pressure discharge that occurs in fluorescent lamps, $T_e \gg T_a$. Although the electron temperature is very high, the tube is relatively cool to the touch, since $T_a \ll T_e$.

The heating of the electrons by the electric field can cause additional ionization that reduces the resistivity of the plasma. Thus, as the current increases, the resistance of the low-pressure discharge also decreases. The V–I characteristic curve is similar in shape to an arc V–I characteristic curve. The resistivity of a low-pressure discharge plasma, however, is much greater than the resistivity of an arc plasma.

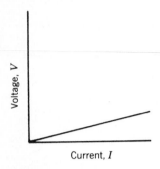

FIGURE 5.37 Voltage versus current.

For many materials, such as metals, the resistivity does not increase rapidly with temperature (a weak function of temperature). For this case, the V–I characteristic is almost a straight line as shown in Figure 5.37. When the V–I characteristic is a straight line with a positive slope, the system is called a linear system. A gaseous discharge, as shown in Figure 5.36, is a nonlinear system with a negative slope at an operating point. Although the resistance is positive, any V–I negative slope is said to have a "negative resistance" (that is, $\Delta V/\Delta I$ has a negative value). The negative slope indicates that the discharge must be stabilized with a ballast resistor.

An important application of a low-pressure discharge is the steady-state, electrically excited, gas laser. The nonequilibrium excited states of atoms or molecules in the plasma produce light amplification. Figure 5.38 is a block diagram of a typical laser power supply that consists of a starter, a running supply, and a current limiter (also called a ballast resistor).

The purpose of the starter is to generate a dc pulse of sufficiently high voltage to cause ionization of the gas in the laser tube. For this reason, the voltage of the pulse provided by the starter must exceed the breakdown voltage of the gas. Figure 5.39 is the V–I characteristic of a laser tube containing a mixture of helium and neon gases. For this helium–neon laser, the starter-voltage pulse must be in excess of 3400 V.

Once ionization occurs, however, the high voltage provided by the starter is excessively large for maintaining the discharge. Hence, gas laser power supplies are designed in such a way that the starter element is deenergized when ionization occurs in the tube. The discharge then is maintained by the lower-voltage running supply.

The current limiter (ballast resistor) is used to limit the current to the desired value. To determine its value for a given operating current, Equation 5.69 can be used. It is important, however, to be sure that the operating point is a **stable** operating point. This can be checked by drawing the load line on the same graph as the V–I characteristic to obtain the two points of intersection. It should be noted that the desired stable operating point may not be possible if the no-load output voltage is too low. Example 5.16 illustrates the determination of a ballast resistance for a given discharge characteristic curve.

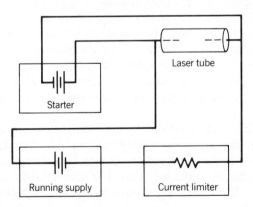

FIGURE 5.38 Gas laser power supply.

FIGURE 5.39 Plot of voltage versus current for a helium–neon laser.

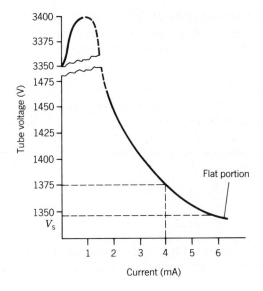

EXAMPLE 5.16 Ballast resistance for a helium-neon laser

. . . .

Given: The helium-neon laser described by the *V–I* characteristic of Figure 5.38 is to be operated from a 3000-V running supply at a current level of 4 mA.

Find: The ballast resistance required.

Solution: $R_B = \dfrac{V - V_t}{I}$

Use the *V–I* characteristic (Figure 5.39) to find V_t when

$I = 4$ mA

$V_t = 1375$ V

Then

$$R_B = \frac{3000 \text{ V} - 1375 \text{ V}}{4 \times 10^{-3} \text{ A}} \quad \text{(using } I \text{ in amps)}$$

$R_B = 406{,}250 \ \Omega \approx 406 \text{ k}\Omega$

The ballast resistance required to stabilize the current at 4 mA is approximately 406 kΩ. Is this a stable operating point?

The voltage, V_S, corresponding to the flat portion of the curve shown in Figure 5.39 is called the "sustaining voltage." It is the smallest voltage that can maintain the discharge. The current can still be increased by increasing the power supply voltage or by reducing the resistance of the ballast resistor. In the flat region, however, the tube voltage no longer decreases as the current increases. If the current is made too great, burnout will occur.

Resistances in Series   Electrical resistances can be connected in series in a manner similar to that used for pipes connected in series. Because mass is conserved, the mass flow rate in steady flow is the same in each pipe. In the electrical case, the charge is conserved. Thus, the current (charge flow rate) is the same in each resistor. As in the case for pipes, this condition can be expressed mathematically as

$$I = I_1 = I_2 = \cdots = I_n \tag{5.70}$$

where

$$I = \text{value of the common current through each resistor}$$
$$I_2, I_2 \text{ etc.} = \text{current through each resistor}$$

In fluids, the total pressure drop (forcelike quality) is equal to the sum of the individual pipe pressure drops when the pipes are connected in series. In the electrical case, the total work per unit charge (voltage drop) is equal to the individual voltage drops when the resistors are connected in series. The voltage drops are forcelike quantities. Expressed mathematically,

$$\Delta V = \Delta V_1 + \Delta V_2 + \cdots + \Delta V_n \tag{5.71}$$

where

$$\Delta V = \text{total voltage drop}$$
$$\Delta V_1, \Delta V_2, \text{ etc.} = \text{voltage drop across each individual resistor}$$

From Ohm's law, the voltage drop across each resistor can be expressed as

$$\Delta V_m = IR_m \tag{5.72}$$

where

$$\Delta V_m = \text{voltage drop across the } m\text{th resistor}$$
$$R_m = \text{resistance of the } m\text{th resistor}$$
$$I = \text{common current}$$

Substituting this expression for the individual voltage drops in Equation 5.71 yields

$$\Delta V = IR_1 + IR_2 + \cdots + IR_n$$

or

$$\Delta V = I(R_1 + R_2 + \cdots + R_n) \tag{5.73}$$

The circuit acts as if there is a single resistance given by

$$R_T = R_1 + R_2 + \cdots + R_n \tag{5.74}$$

Thus, Equation 5.73 becomes

$$\Delta V = IR_T \tag{5.75}$$

According to Equation 5.74, the total resistance of a number of resistors in series is equal to the sum of the individual resistances. Equations 5.70, 5.74, and 5.75 are used to solve a problem in Example 5.17.

EXAMPLE 5.17 Series resistors

. . . .

Given: The current through ten 12-V lamps connected in series is 0.50 A.

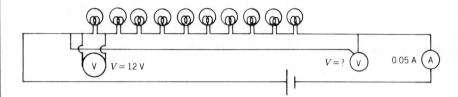

Find:

a. The resistance of each bulb.

b. The total resistance.

c. The total voltage.

Solution:

a. Use Ohm's law.

$$R = \frac{V}{I}$$

$$R = \frac{12 \text{ V}}{0.50 \text{ A}}$$

$R = 24 \ \Omega$, the resistance of each bulb

b. The total resistance is

$$R_T = R_1 + R_2 + \cdots + R_n$$

$$R_T = (24 \ \Omega)(10)$$

$$R_T = 240 \ \Omega$$

c. The total voltage is

$$\Delta V = IR_T$$

$$\Delta V = (0.50 \text{ A})(240 \ \Omega)$$

$$\Delta V = 120 \text{ V}$$

or, using Equation 5.71,

$$\Delta V = (12 \text{ V})(10) = 120 \text{ V}$$

Resistances in Parallel When resistors are connected in parallel, the conservation of charge requires that

$$I = I_1 + I_2 + \cdots + I_n \tag{5.76}$$

where

$$I = \text{total current}$$
$$I_1, I_2, \text{etc.} = \text{individual currents through each resistor}$$

Equation 5.76 states that the total current flowing into a junction must equal the current flowing out of the junction. This was illustrated in Chapter 3 for the mass flow rates of a fluid in pipes where mass is conserved. Flow rate and current are the rates.

For resistors connected in parallel, the voltage drop across each resistor is the same, just as the pressure drop across pipes in parallel is the same for each pipe. Stated mathematically,

$$\Delta V = \Delta V_1 = \Delta V_2 = \cdots = \Delta V_n \tag{5.77}$$

where

$$\Delta V = \text{common voltage across the parallel resistors}$$
$$\Delta V_1, \Delta V_2, \text{etc.} = \text{voltage across each resistor}$$

From Ohm's law, the current through each resistor is

$$I_m = \frac{\Delta V}{R_m} \tag{5.78}$$

where

$$\Delta V = \text{common voltage}$$
$$I_m = \text{current through the } m\text{th resistor}$$
$$R_m = \text{resistance of the } m\text{th resistor}$$

Substituting this expression in Equation 5.76 and factoring out ΔV yields

$$I = \Delta V \left[\frac{1}{R_1} + \frac{1}{R_2} + \cdots + \frac{1}{R_n} \right] \tag{5.79}$$

or

$$I = \frac{\Delta V}{R_T} \tag{5.80}$$

Thus, the parallel circuit acts as if the total effective resistance R_T is given by

$$\frac{1}{R_T} = \frac{1}{R_1} + \frac{1}{R_2} + \cdots + \frac{1}{R_n} \tag{5.81}$$

An application of Equations 5.76 and 5.81 is given in Example 5.18.

EXAMPLE 5.18   Resistors in parallel
. . . .

Given: The voltage across each resistor shown in the sketch is 12 V.

Find: **a.**   The total resistance.

              b.   The total current.

Solution: **a.**   The total resistance is given in this case by the equation

$$\frac{1}{R_T} = \frac{1}{R_1} + \frac{1}{R_2} + \cdots + \frac{1}{R_n}$$

$$\frac{1}{R_T} = \frac{1}{10\ \Omega} + \frac{1}{20\ \Omega} + \frac{1}{30\ \Omega}$$

$$\frac{1}{R_T} = \frac{0.100 + 0.050 + 0.033}{\Omega}$$

$$\frac{1}{R_T} = \frac{0.183}{\Omega}$$

$$R_T = 5.45\ \Omega$$

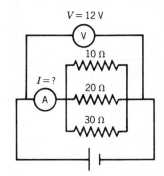

b.   The total current is

$$I = \frac{\Delta V}{R_T}$$

$$I = \frac{12\ \text{V}}{5.45\ \Omega}$$

$$I = 2.20\ \text{A}$$

Series-Parallel Circuits   Practical electrical circuits often consist of combinations of series and parallel circuits. Because such a great number of combinations of series-parallel circuits are possible, there is no definite procedure that can be followed in solving them. The method depends on the arrangement. When possible, however, the problem should be solved by replacing each parallel branch with an equivalent series branch. This technique is demonstrated in Example 5.19.

EXAMPLE 5.19   Reducing a resistance network
. . . .

Given: The circuit shown in sketch 1.

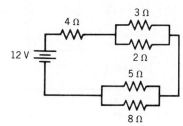

Sketch 1

Find: The total current.

Solution: The 2-Ω and 3-Ω resistors in the upper arm of the circuit form a parallel network. By use of Equation 5.81, they can be reduced to a single equivalent resistance as follows:

$$\frac{1}{R_T} = 1/2 + 1/3 = 3/6 = 2/6 = 5/6$$

$$R_T = 6/5 \ \Omega$$

$$R_T = 1.20 \ \Omega$$

In the same way, the 5-Ω and 8-Ω resistors in the lower arm form a parallel network that can be simplified as

$$\frac{1}{R_T} = 1/5 + 1/8 = 8/40 + 5/40 = 13/40$$

$$R_T = 40/13 \ \Omega$$

$$R_T = 3.08 \ \Omega$$

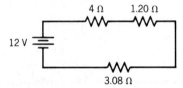

Sketch 2

The equivalent circuit now has this appearance. As the circuit is now reduced to one with three resistances in series, these can be combined into one equivalent resistance by use of Equation 5.74 as follows:

$$R_T = 4 + 1.20 + 3.08$$

$$R_T = 8.28 \ \Omega$$

The circuit in its simplest form then looks like this. Using Ohm's law, the total current in the circuit can be found.

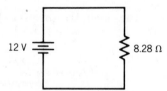

Sketch 3

$$I = \frac{V}{R_T}$$

$$I = \frac{12 \ V}{8.28 \ \Omega}$$

$$I = 1.45 \ A$$

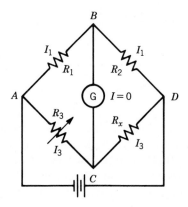

FIGURE 5.40 Wheatstone bridge.

Measurement of Electrical Resistance There are many ways to measure electrical resistance. Only one method, the Wheatstone bridge, will be discussed here. This method is accurate and can be used for a wide range of resistances. The circuit diagram is shown in Figure 5.40. A very sensitive galvanometer is connected across points B and C, and the unknown resistance is connected across points C and D. The resistance R_3 is adjusted until no current flows through the galvanometer. Thus, the voltage at B is equal to the voltage at C, and the current through R_2 is equal to the current through R_1. Also, the current through R_3 is equal to the current through R_x. From Ohm's law, under these conditions, one obtains

$$I_3 R_x = I_1 R_2 \tag{5.82}$$

$$I_3 R_3 = I_1 R_1 \tag{5.83}$$

Dividing Equation 5.82 by Equation 5.83 yields

$$\frac{R_x}{R_3} = \frac{R_2}{R_1}$$

$$R_x = R_3 \left(\frac{R_2}{R_1}\right) \tag{5.84}$$

The resistances R_3, R_2, and R_1 are standard resistances usually known to 0.1 percent or less. An application of Equation 5.84 is shown in Example 5.20.

EXAMPLE 5.20 Wheatstone bridge
. . . .

Given: A Wheatstone bridge is used to measure an unknown resistance. The ratio of R_2 to R_1 (R_2/R_1) is set to 10. The value of the variable resistance, when the galvanometer reads zero, is 10,000 Ω.

Find: The unknown resistance.

Solution: The working equation is

$$R_x = R_3 \ (R_2/R_1)$$

$$R_x = (10,000\ \Omega)(10)$$

$$R_x = 1.000 \times 10^5\ \Omega$$

Magnetic Systems

Magnetic Flux It is possible to analyze magnetic circuits in a way similar to the method used for electric circuits. The magnetic field lines form closed loops so that the total magnetic flux, ϕ, is conserved. Therefore, the magnetic flux is analogous to the mass flow rate or electric current. If the magnetic induction, B, is constant across a material, then the magnetic flux may be expressed as

$$\phi = BA \tag{5.85}$$

where

ϕ = magnetic flux in webers
A = cross-sectional area in m^2
B = magnetic induction in webers/m^2

B in this case is analogous to the current density, j, in the equation

$$I = jA$$

Magnetic Scalar Potential The magnetic scalar potential was defined in Chapter 1 (Equation 1.49) as

$$\Delta\Omega = \frac{B}{\mu} \Delta l \cos\theta \tag{5.86}$$

where

$\Delta\Omega$ = change in the potential
μ = permeability in henrys/meter
Δl = a small displacement in meters
θ = angle between $\Delta \mathbf{l}$ and $\mathbf{B}$

The magnetic scalar potential is analogous to the voltage difference in an electric system.

Reluctance Consider the closed magnetic path shown in Figure 5.41. The cross-sectional area and permeability are constant throughout the magnetic path. According to Equation 1.54 in Chapter 1,

$$\Sigma\Delta\Omega = \Sigma H\Delta l \cos\theta = NI \tag{5.87}$$

where

NI = magnetomotive force in ampere-turns (mmf)
H = magnetic field strength in ampere-turns/meter

From Equation 1.52,

$$\mathbf{H} = \mathbf{B}/\mu \tag{5.88}$$

FIGURE 5.41 Closed magnetic path.

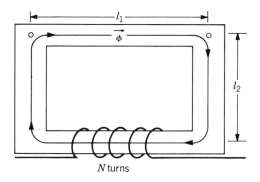

N turns

As an approximation, Equation 5.85 yields

$$B = \phi/A$$

Substituting this expression for B into Equation 5.88 gives

$$H = \frac{\phi}{\mu A} \qquad (5.89)$$

When Equation 5.89 is substituted into Equation 5.87, one obtains

$$\Sigma \Delta \Omega = \frac{\phi}{\mu A} \Sigma \, \Delta l = \text{mmf} \qquad (5.90)$$

where the common constant, $\phi/\mu A$, has been factored out of the sum and Δl is taken in the direction of **H** ($\cos \theta = 1$). The $\Sigma \Delta l$ is the total path length, $l = 2 \, l_1 + 2 \, l_2$. Thus, Equation 5.90 becomes

$$\phi \left[\frac{l}{\mu A} \right] = \text{mmf} \qquad (5.91)$$

where l is the total path length. This equation is analogous to Ohm's law

$$IR = \Delta V$$

where

$$R = \frac{\rho l}{A} \qquad (5.92)$$

The resistivity, ρ, is thus analogous to $1/\mu$. The quantity $R = l/\mu A$ is called the reluctance of the magnetic path. An application of Equations 5.91 and 5.92 is given in Example 5.21.

EXAMPLE 5.21 Magnetic Circuit
. . . .

Given: The value of l_1 and l_2 in Figure 5.41 is 0.20 m. The cross-sectional area of the cast iron material is 10^{-4} m². A magnetomotive force of 10 A-turn produces a magnetic flux of 1.12×10^{-5} Wb.

Find: The permeability of the material.

Solution: The reluctance is

$$R = \frac{\text{mmf}}{\phi}$$

$$R = \frac{10 \text{ A-turn}}{1.12 \times 10^{-5} \text{ } Wb}$$

$$R = 8.93 \times 10^5 \frac{\text{A-turn}}{Wb}$$

The reluctance also is

$$R = \frac{l}{\mu A}$$

$$\mu = \frac{l}{RA}$$

$$\mu = \frac{(0.80 \text{ m})}{\left(8.93 \times 10^5 \frac{\text{A-turn}}{Wb}\right)(10^{-4} \text{ m}^2)}$$

$$\mu = 8.96 \times 10^{-3} \frac{Wb}{\text{A-turn·m}}$$

Magnetization Curves   The value of the permeability, μ, for magnetic materials is not a constant. Thus, the magnetic circuit is a nonlinear system. This makes the solution of magnetic circuits more difficult. Before a solution can be obtained, it is necessary to determine how **B** varies with **H**. An experimentally determined curve of **B** versus **H**, whose slope at any point is the permeability, is called a magnetization curve. A magnetization curve for cast iron is shown in Figure 5.42. As **H** increases, the number of magnetic domains aligned in the direction of **H** increases, which in turn increases the value of **B**. When all the magnetic domains are essentially aligned in the direction of **H**, **B** increases at the same rate as it would in air. The material is then said to be saturated.

Magnetic Materials in Series   When magnetic materials are in series, the total flux, ϕ, through each material is the same as in the case of the current through resistors in series. This is expressed mathematically as

$$\phi = \phi_1 = \phi_2 = \cdots = \phi_n \tag{5.93}$$

where

$$\phi = \text{total magnetic flux}$$
$$\phi_1, \phi_2, \text{ etc.} = \text{flux through each material}$$

In addition, the magnetomotive force must equal the sum of the differences in the magnetic scalar potential across each material. This may be expressed as

$$NI = (\Delta\Omega)_1 + (\Delta\Omega)_2 + \cdots + (\Delta\Omega)_n \tag{5.94}$$

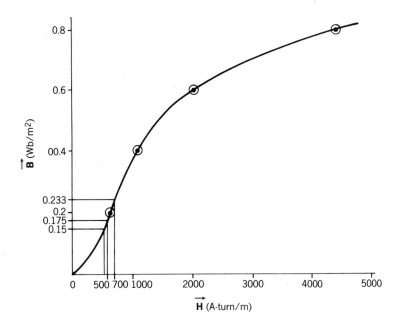

FIGURE 5.42 Plot of magnetic induction versus magnetic field strength for cast iron.

where

$$NI = \text{magnetomotive force (mmf)}$$
$$(\Delta\Omega)_1, (\Delta\Omega)_2, \text{ etc.} = \text{individual changes in the magnetic scalar potential}$$

For each material (Equation 5.90)

$$(\Delta\Omega)_m = H_m(\Delta l)_m \tag{5.95}$$

where $(\Delta l)_m$ is taken in the direction of H so that $\cos\theta = 1$ and m refers to the mth material. If the magnetic flux is given, the value of B_m in the mth material can be calculated from Equation 5.85:

$$B_m = \frac{\phi}{A_m} \tag{5.96}$$

Once the value of B_m has been determined, the magnetization curve can be used to find H_m. Equation 5.95 can then be used to calculate each $(\Delta\Omega)_m$. The required magnetomotive force is then the sum of the $(\Delta\Omega)_m$'s (Equation 5.94). The solution of a problem where the flux is given for materials in series is presented in Example 5.22.

EXAMPLE 5.22 Determination of magnetomotive force
. . . .

Given: The magnetic circuit shown in the sketch. The core material is cast iron with a 2×10^{-4}-m^2 cross-sectional area.

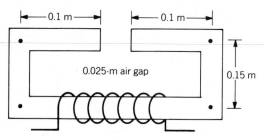

Find: The magnetomotive force required to give a magnetic flux of 3.0×10^{-5} Wb.

Solution: Assume no fringing in the air gap. The value of B in the air gap is

$$B_1 = \frac{\phi}{A_1}$$

$$B_1 = \frac{3.0 \times 10^{-5}\ Wb}{2 \times 10^{-4}\ m^2}$$

$$B_1 = 0.15\ Wb/m^2$$

The value of H in the air gap is, using

$$\mu = 1.257 \times 10^{-6}\ Wb/m \cdot A\text{-turn for air}$$

$$H_1 = \frac{B_1}{\mu_1}$$

$$H_1 = \frac{(0.15\ Wb/m^2)}{(1.257 \times 10^{-6}\ Wb/m \cdot A\text{-turn})}$$

$$H_1 = 1.19 \times 10^5\ A\text{-turn/m}$$

The scalar change in potential across the gap is

$$(\Delta \Omega)_1 = H_1(\Delta l)_1$$

$$(\Delta \Omega)_1 = (1.19 \times 10^5\ A\text{-turn/m})\ (0.025\ m)$$

$$(\Delta \Omega)_1 = 2.98 \times 10^3\ A\text{-turn}$$

The value of **B** in the cast iron is

$$B_2 = \frac{\phi}{A_2} \qquad (A_2 = A_1)$$

$$B_2 = 0.15\ Wb/m^2$$

Use Figure 5.41 to obtain H.

$$H = 500\ A\text{-turn/m}$$

Calculate $(\Delta \Omega)_2$ for the cast iron:

$$(\Delta \Omega)_2 = H_2\ (\Delta l)_2$$

$$(\Delta l)_2 = (2)\ (0.15\ m) + 2(0.1\ m) + (0.225\ m)$$

$$(\Delta l)_2 = 0.725 \text{ m}$$

$$(\Delta\Omega)_2 = (500 \text{ A-turn/m}) (0.725 \text{ m})$$

$$(\Delta\Omega)_2 = 363 \text{ A-turn}$$

The total magnetomotive force is

$$\text{mmf} = (\Delta\Omega)_1 + (\Delta\Omega)_2$$

$$\text{mmf} = 2980 \text{ A-turn} + 363 \text{ A-turn}$$

$$\text{mmf} = 3343 \text{ A-turn}$$

Magnetic Materials in Parallel For parallel magnetic circuits, the total flux is the sum of the individual fluxes, as in the case of electric currents:

$$\phi = \phi_1 + \phi_2 + \cdots + \phi_n \tag{5.97}$$

where

$$\phi = \text{total magnetic flux}$$
$$\phi_1, \phi_2, \text{ etc.} = \text{fluxes in each parallel branch}$$

As in the case for electric circuits, the change in the scalar potential is the same across each parallel branch:

$$\Delta\Omega = (\Delta\Omega)_1 = (\Delta\Omega)_2 = \cdots = (\Delta\Omega)_n \tag{5.98}$$

Equations 5.95 and 5.96 also apply.

In real applications, a magnetic circuit is usually a combination of a series-parallel magnetic circuit. A typical problem is illustrated in Example 5.23.

EXAMPLE 5.23   Series-parallel magnetic circuit

Given: The magnetic circuit shown in the sketch. The material is cast iron. The cross-sectional area of the central core is 3×10^{-4} m² and the cross-sectional area of each identical parallel branch is 2×10^{-4} m². The magnetic path in the core is 0.06 m and the magnetic path in each branch is 0.15 m.

Find: The magnetomotive force required to establish a total flux of 7.0×10^{-5} Wb.

Solution: Find the flux density in the core.

$$B_c = \frac{\phi}{A_c}$$

$$B_c = \frac{7.0 \times 10^{-5} \text{ Wb}}{3.0 \times 10^{-4} \text{ m}^2}$$

$B_c = 0.233$ Wb/m^2

Use Figure 5.41 to obtain **H**.

$H_c = 700$ A-turn/m

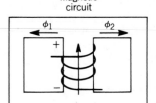

Magnetic
circuit

Find the change in scalar potential across the core.

$(\Delta\Omega)_c = H_c l_c$

$(\Delta\Omega)_c = (700$ A-turn/m$)$ $(0.06$ m$)$

$(\Delta\Omega)_c = 42$ A-turn

The scalar potential across the parallel branches is the same for each branch.

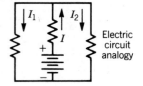

Electric
circuit
analogy

$$(\Delta\Omega)_1 = (\Delta\Omega)_2$$

$$H_1 l_1 = H_2 l_2$$

$$H_1(0.15 \text{ m}) = H_2(0.15 \text{ m})$$

$$H_1 = H_2$$

If $H_1 = H_2$, $B_1 = B_2$ since $B = \mu H$.
The total flux is

$$\phi = \phi_1 + \phi_2 = B_1 A_1 + B_2 A_2 \qquad (A_1 = A2, B_1 = B_2)$$

$$\phi = 2 B_1 A_1$$

$$B_1 = B_2 = \frac{\phi}{2A_1}$$

$$B_1 = B_2 = \frac{7.0 \times 10^{-5} \text{ Wb}}{(2) (2 \times 10^{-4} \text{ m}^2)}$$

$$B_1 = 0.175 \text{ Wb/m}^2$$

From Figure 5.41

$H = 560$ A-turn/m

The scalar potential across the parallel branches is

$(\Delta\Omega)_1 = H_1 l_1$

$(\Delta\Omega)_1 = (560$ A-turn/m$)$ $(0.15$ m$)$

$(\Delta\Omega)_1 = (\Delta\Omega)_2 = 84$ A-turn

The magnetomotive force is

$\text{mmf} = (\Delta\Omega)_c + (\Delta\Omega)_1$

$\text{mmf} = 42$ A-turn $+ 84$ A-turn

$\text{mmf} = 126$ A-turn

SECTION *5.3* *EXERCISES*

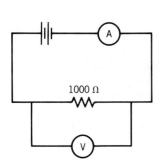

1. Draw a graph showing resistance as the ratio of a forcelike quantity to a rate in an electrical system.

2. State the fundamental difference between sliding friction and resistance in an electrical system.

3. A 1000-Ω resistor is placed in the circuit shown in the figure to the left. Assume that the resistances of the battery and ammeter are negligible.

    a. What voltage is required to produce a reading of 1 mA by the ammeter A?

    b. What current is required to produce a reading of 6 V by the voltmeter V?

4. What a 100-W light bulb is plugged into a 100-V circuit, the current is 0.91 A. What is the resistance of the light bulb?

5. A voltmeter placed across the series arrangement of resistors shown reads 6 V. What is the current flowing through the resistors?

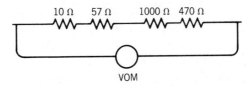

6. In a series circuit, the potential drop across a 100-Ω resistor is 2 V. What will be the potential drop across a 1-kΩ resistor that is in this circuit? (**Hint:** In a series circuit the same current flows through all the circuit elements.)

7. Find the value of the unknown resistor in the circuit represented in the sketch if the total current flowing out of the battery is 0.7 A.

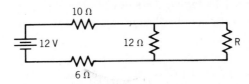

8. How many 10-Ω appliances connected in series are necessary to reduce the current flowing in a circuit to 10 mA if the applied voltage is 12 V?

9. If a current of 1 A flows through a circuit in which there is only one 10-Ω resistor, what current will flow if the circuit is expanded to include four 10-Ω resistors connected in parallel and the applied voltage remains the same? (Figure to the left.)

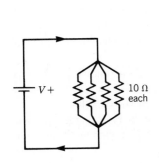

10. Define:

Ionic bonding

Covalent bonding

Metallic bonding

Insulator

Conductor

Semiconductor

11. Explain why the resistance of a copper wire increases when it is heated.

12. Explain why the resistance of a carbon rod decreases when it is heated.

13. Explain why copper is a better conductor than carbon.

14. Explain why glass is a poor conductor.

15. Why are materials with ionic bonding seldom employed as electrical insulators in practical circuits?

16. What is the resistivity of a material if a 10-m wire made of the material has a resistance of 10 Ω and the radius of the wire is 5 mm?

17. If a wire has a resistivity of 5×10^{-7} $\Omega \cdot$m, what length of the wire has a resistance of 1 Ω if the cross-sectional area of the wire is 2.9×10^{-5} m^2?

18. It is known that the resistance of a conductor increases linearly with increasing temperature. It is determined experimentally that the resistance of a certain length of a particular conductor at 20°C is 1 Ω and at 80°C is 1.3 Ω. Write an equation that gives the resistance of the conductor as a function of temperature. What should the resistance of the wire be at 100°C?

19. The resistance of a semiconducting material as a function of temperature can be expressed as

$$R = R_S \left[1 + k \left(\frac{1}{T_S} - \frac{1}{T} \right) \right]$$

where k is a constant that depends on the material and R_S is the resistance of the semiconducting material at some reference temperature T_S. (All temperatures are expressed in Kelvin degrees.) If the resistance of a semiconductor is 100 kΩ at 20°C, find the resistance at 150°C if $k = 3$ K°.

20. If a potential difference of 100 V is applied across the ends of a 10-m length of wire of radius 2 mm, and the resulting current is 1×10^{-8} A, should the wire be classified as a conductor, semiconductor, or insulator?

21. Draw and label a diagram showing the parts of an npn transistor and the proper connections for bias batteries.

22. Draw diagrams showing the movement of electrons and holes in a diode for
   a. forward bias
   b. reverse bias.

23. Explain how the resistance of the base-collector circuit of an npn transistor is controlled by current in the emitter-base circuit.

24. Describe the behavior of the voltage across a gas laser tube relative to the current through it when the tube is operating in the so-called negative resistance region.

25. List the three elements of a gas laser power supply and briefly explain the function of each.

26. A laser described in the V–I characteristic graph at the top of the following page is to be operated from a 2500-V power supply. The desired operation current is 3.5 mA. Find the ballast resistance necessary to stabilize the current at that value.

27. Determine the breakdown voltage and sustaining voltage for the laser that has the V–I characteristic shown in the graph. Label the negative resistance region on the graph at the top of the following page.

28. If the ballast resistance necessary to stabilize an operating current of 5 mA is 400 kΩ, find the supply voltage. Assume that the V–I characteristics of the laser are described in Figure 5.28.

29. If the tube voltage is related to the current by the relationship $V_t = 4/I$ in the negative resistance region, what is the tube voltage when the current is 3.5 mA?

30. Find the tube voltage if the ballast resistance is 275 kΩ, the supply voltage is 3200 V, and the current is 4.3 mA.

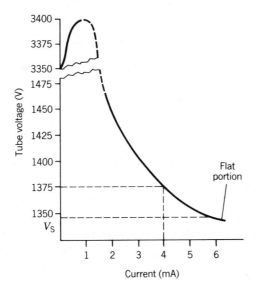

31. The reluctance of a magnetic path is 1.5×10^6 A-turn/Wb and the magnetomotive force is 20 A-turn. What is the magnetic flux?

32. A magnetic circuit is composed of two magnetic materials in series. The reluctance of the first material is 0.5×10^6 A-turn/Wb and the second material has a reluctance of 0.75×10^6 A-turn/Wb. The magnetomotive force is 30 A-turn. What is the magnetic flux?

33. The cross-sectional area of a magnetic circuit is 4.00×10^{-4} m². The circuit has a 0.01-m air gap, and the total length of the magnetic material is 0.40 m. If the magnetic material is cast iron and the flux is 7.0×10^{-5} Wb, what is the magnetomotive force?

34. Define the terms
   a. intrinsic material
   b. p-type material
   c. n-type material
   d. depletion region.

35. On the curve in the accompanying graph identify the following regions:
   a.   **b.**   **c.**   **d.**   **e.**

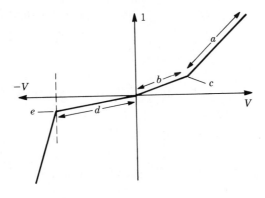

5.4 .... OBJECTIVES ....

RESISTANCE IN THERMAL SYSTEMS

Upon completion of this section, the student should be able to

- Given two of the following quantities in heat transfer systems, determine the third:

  Forcelike quantity

  Rate

  Resistance

- Calculate the overall heat transfer coefficient for plane surfaces.

DISCUSSION

Resistance to a flow of heat energy is of interest in a number of applications. It is a property that provides insulation for refrigerators and buildings. In this section, several factors that contribute to this property will be examined.

Thermal Resistance

In thermal systems, there is a resistance to the flow of thermal energy. In Chapter 3, the equation for thermal conduction was defined as

$$Q_H = \left(\frac{kA}{\Delta l}\right)(\Delta T) \tag{5.99}$$

where

Q_H = heat flow rate in cal/sec or Btu/hr
k = thermal conductivity in cal/m·sec·C° or Btu/hr·ft·F°
A = area of the conductor in m² or ft²
ΔT = temperature difference across the conductor in C° or F°
Δl = length of the conductor in m or ft

In Equation 5.99, Q_H is the rate analogous to the fluid flow rate, electric current, or magnetic flux. The temperature drop, ΔT, is analogous to the pressure drop, Δp, the voltage drop, ΔV, or the magnetomotive force. The analogy is clear in a review of the relations

$$Q_V = \left(\frac{\pi R_o^4}{8\mu l}\right)\Delta p \qquad \text{(laminar fluid flow)}$$

$$R = \frac{8\mu l}{\pi R_o^4}$$

$$I = \left(\frac{A}{\rho l}\right)\Delta V \qquad \text{(electric current flow)}$$

$$R = \frac{\rho l}{A}$$

$$\phi = \left(\frac{\mu A}{l}\right) \text{mmf} \qquad \text{(magnetic flux)}$$

TABLE 5.7 Thermal conductivity at 70°F

Material	Thermal Conductivity (Btu/hr·ft·F°)
Window glass	≈ 0.45
Yellow pine	0.085
Gypsum plaster	0.28
Common brick	0.40
Face brick	0.76
Aluminum	118
Iron	42
Copper	223
Silver	235

$$R = \frac{l}{\mu A}$$

For many materials, the thermal conductivity is a weak function of the temperature and is essentially independent of Q_H. Thus, in many cases, the heat conduction is an approximately linear process that simplifies calculations.

By analogy, the thermal resistance becomes

$$R = \frac{\Delta l}{kA} \qquad (5.100)$$

and Equation 5.99 can be written as

$$Q_H = \frac{\Delta T}{R} \qquad (5.101)$$

The thermal resistance is usually not known. However, values for the thermal conductivity of various materials are given in handbooks. Equation 5.100, therefore, is usually used to calculate the thermal resistance. Table 5.7 lists the thermal conductivity for several materials at 70°F.

An application of Equations 5.100 and 5.101 is illustrated in Example 5.24.

EXAMPLE 5.24 Heat conducted through a brick wall
. . . .

Given: The surface temperature on one side of a brick wall is 10°F and the surface temperature on the other side is 70°F. The dimensions of the wall are shown in the sketch.

Find: The heat flow rate through the wall.

Solution: The thermal resistance is

$$R = \frac{\Delta l}{kA} \qquad (k = 0.76 \text{ Btu/hr·ft·F°})$$

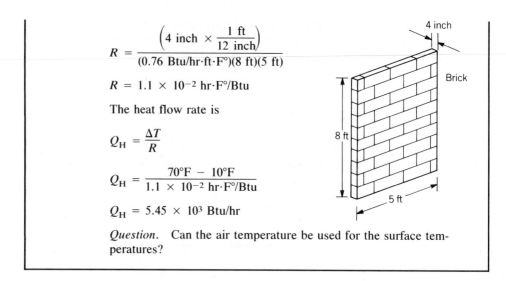

$$R = \frac{\left(4 \text{ inch } \times \dfrac{1 \text{ ft}}{12 \text{ inch}}\right)}{(0.76 \text{ Btu/hr·ft·F°})(8 \text{ ft})(5 \text{ ft})}$$

$$R = 1.1 \times 10^{-2} \text{ hr·F°/Btu}$$

The heat flow rate is

$$Q_H = \frac{\Delta T}{R}$$

$$Q_H = \frac{70°F - 10°F}{1.1 \times 10^{-2} \text{ hr·F°/Btu}}$$

$$Q_H = 5.45 \times 10^3 \text{ Btu/hr}$$

Question. Can the air temperature be used for the surface temperatures?

Heat Resistances in Series Materials can be placed in layers as shown in Figure 5.43. In the steady state (time independent), the heat flow rate is the same through each slab just as the current is the same through series resistors or the mass flow rate is the same through pipes in series. Thus, one may write

$$Q_H = Q_{H_1} = Q_{H_2} = \cdots = Q_{H_n} \tag{5.102}$$

In addition, the sum of the temperature differences must equal the total temperature difference. This was also the case for the voltage differences and the pressure differences when resistors and pipes were connected in series. Mathematically this can be expressed as

$$\Delta T = (\Delta T)_1 + (\Delta T)_2 + \cdots + (\Delta T)_n \tag{5.103}$$

Since the heat flow rate (analogous to Ohm's law) is $Q_H = \Delta T_m/R_m$ for the mth slab, one may write

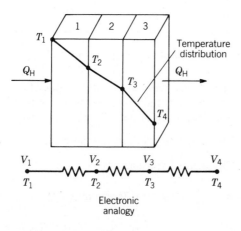

FIGURE 5.43 Thermal resistances in series.

$$(\Delta T)_m = Q_H R_m \tag{5.104}$$

where

Q_H = common heat flow rate

If these values for the $(\Delta T)_m$ are substituted in Equation 5.103, one obtains

$$\Delta T = Q_H R_1 + Q_H R_2 + \cdots + Q_H R_n$$

$$\Delta T = Q_H(R_1 + R_2 + \cdots + R_n) \tag{5.105}$$

$$\Delta T = Q_H R_T \tag{5.106}$$

Equation 5.105 states that the total resistance for materials in series is equal to the sum of the individual resistances:

$$R_T = R_1 + R_2 + \cdots + R_N \tag{5.107}$$

An application of Equation 5.107 is shown in Example 5.25.

EXAMPLE 5.25   Heat conduction for slabs in series
. . . .

Given: Three slabs are layered as shown in Figure 5.43. Slab #1 is made from gypsum plaster and is 0.05 ft thick. Slab #2 is made from yellow pine and is 0.025 ft thick. The third slab is made from face brick and is 0.33 ft thick. The area of each slab is 20 ft².

Find: The heat transfer rate if the total temperature difference is 70 F°.

Solution: Find each resistance.

$$R_m = \frac{(\Delta l)_m}{k_m A_m}$$

$$R_1 = \frac{0.05 \text{ ft}}{(0.28 \text{ Btu/hr·ft·F°})(20 \text{ ft}^2)}$$

$$R_1 = 8.93 \times 10^{-3} \text{ hr·F°/Btu}$$

$$R_2 = \frac{0.025 \text{ ft}}{(0.085 \text{ Btu/hr·ft·F°})(20 \text{ ft}^2)}$$

$$R_2 = 14.7 \times 10^{-3} \text{ hr·F°/Btu}$$

$$R_3 = \frac{0.33 \text{ ft}}{(0.76 \text{ Btu/hr·ft·F°})(20 \text{ ft}^2)}$$

$$R_3 = 21.7 \times 10^{-3} \text{ hr·F°/Btu}$$

Find the total resistance.

$$R_T = R_1 + R_2 + R_3$$

$$R_T = (8.9 \times 10^{-3} + 14.7 \times 10^{-3} + 21.7 \times 10^{-3}) \text{ hr·F°/Btu}$$

$$R_T = 45.3 \times 10^{-3} \text{ hr·F°/Btu}$$

Find the heat transfer rate.

$$Q_H = \frac{\Delta T}{R_T}$$

$$Q_H = \frac{70 \ F°}{45.3 \times 10^{-3} \ hr \cdot F°/Btu}$$

$$Q_H = 1.55 \times 10^3 \ Btu/hr$$

Overall Heat Resistance   Consider the slab shown in Figure 5.44. The left-hand side is exposed to a hot flowing fluid and the right-hand side is exposed to a cooler fluid. Heat is transferred from the hot fluid to the slab by convection. The heat is then transferred through the slab by conduction, and it is transferred to the cold fluid by convection.

In Chapter 3, the heat transfer coefficient, *h,* for convection was defined as

$$Q_H = h_m A_m (T_B - T_w) \tag{5.108}$$

where

h_m = heat transfer coefficient for convection
T_B = free-stream temperature or bulk temperature
T_w = wall temperature
A_m = area of the surface

Equation 5.108 may also be written as

$$Q_H = \frac{(\Delta T)_m}{R_m}$$

where

$\Delta T = T_B - T_w$
$R_m = 1/h_m A_m$

The resistance R_m is the convective heat transfer resistance. Since the resistances are in series as shown in Figure 5.43,

$$R_T = R_1 + R_2 + R_3 \tag{5.109}$$

where

$R_1 = 1/h_1 A$
$R_2 = \Delta l/kA$
$R_3 = 1/h_2 A$

Thus, the value of Q_H becomes

$$Q_H = \frac{\Delta T}{R_T} = \frac{T_4 - T_1}{R_T} \tag{5.110}$$

Substituting the individual resistances into Equation 5.109 yields

FIGURE 5.44 Overall heat resistance of a slab.

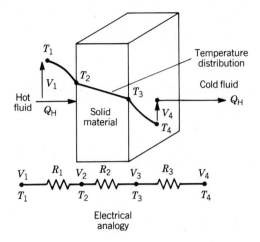

$$R_T = \frac{1}{h_1 A} + \frac{\Delta l}{kA} + \frac{1}{h_2 A}$$

$$R_T = \frac{1}{A}\left[\frac{1}{h_1} + \frac{\Delta l}{k} + \frac{1}{h_2}\right] \tag{5.111}$$

If this expression for R_T is substituted in Equation 5.110, the result is

$$Q_H = \frac{A\Delta T}{\left(\dfrac{1}{h_1} + \dfrac{\Delta l}{k} + \dfrac{1}{h_2}\right)} \tag{5.112}$$

It is common practice to express Q_H in the form

$$Q_H = h_o A \Delta T \tag{5.113}$$

where

h_o = overall heat transfer coefficient

By comparing Equation 5.113 with Equation 5.112, one can see that

$$h_o = \frac{1}{\left(\dfrac{1}{h_1} + \dfrac{\Delta l}{k} + \dfrac{1}{h_2}\right)} \tag{5.114}$$

An application of Equations 5.109 and 5.110 is given in Example 5.26.

EXAMPLE 5.26   The overall heat transfer resistance
. . . .

Given: Air flows over both sides of a slab as shown in Figure 5.44. The convection heat transfer coefficient on the left-hand side is 2.0 Btu/hr·ft²·F°. The temperature of the air on the left-hand side is 70°F. The slab is 0.02 ft thick. The convective heat transfer on the right-hand side is 1.5 Btu/hr·ft²·F° and the air temperature is 30°F. The slab area is 18 ft² and it is made from window glass.

Find: The heat transfer rate.

Solution: The convective heat transfer resistance is

$$R_m = \frac{1}{h_m A_m}$$

$$R_1 = \frac{1}{(2.0 \text{ Btu/hr·ft²·F°}) \, (18 \text{ ft}^2)}$$

$$R_1 = 27.8 \times 10^{-3} \text{ hr·F°/Btu}$$

$$R_3 = \frac{1}{(1.5 \text{ Btu/hr·ft²·F°}) \, (18 \text{ ft}^2)}$$

$$R_3 = 37.0 \times 10^{-3} \text{ hr·F°/Btu}$$

The conductive heat transfer resistance is

$$R = \frac{\Delta l}{kA}$$

$$R_2 = \frac{(0.02 \text{ ft})}{(0.45 \text{ Btu/hr·ft·F°}) \, (18 \text{ ft}^2)}$$

$$R_2 = 2.5 \times 10^{-3} \text{ hr·F°/Btu}$$

The total resistance is

$$R_T = R_1 + R_2 + R_3$$

$$R_T = (27.8 \times 10^{-3} + 2.5 \times 10^{-3} + 37.0 \times 10^{-3}) \text{ hr·F°/Btu}$$

$$R_T = 67.3 \times 10^{-3} \text{ hr·F°/Btu}$$

The heat flow rate is

$$Q_H = \frac{\Delta T}{R_T}$$

$$Q_H = \frac{(70°F - 30°F)}{(67.3 \times 10^{-3} \text{ hr·F°/Btu})}$$

$$Q_H = 594 \text{ Btu/hr}$$

SECTION 5.4   EXERCISES

1. Draw a graph showing resistance as the ratio of a forcelike quantity to a rate in a thermal system.

2. State the fundamental difference between sliding friction and resistance in a thermal system.

3. What is the thermal resistance of a calorimeter if the calorimeter's contents are maintained at a constant temperature of 40°C by electrical energy supplied at a constant rate of 5 cal/hr to the calorimeter? The temperature of the room in which the calorimeter is kept is 22°C.

4. Three slabs are placed in series as shown in Figure 5.43. The first slab is made from iron and is 1 ft thick. The second slab is made from aluminum and is 1 ft thick. The third slab is made from copper and is 1 ft thick. The area of each slab is 10 m². The surface temperature of the iron slab is maintained at 100°F and the surface temperature of the copper slab is maintained at 0°F. What is the heat transfer rate through the slab? What is the temperature difference across each slab? Give an equivalent electric circuit.

5. What is the thermal conductivity of a type of wood if a wall made of the wood is 6 inch thick and heat flows through the wall at a rate of 200 Btu/hr when the temperature difference on opposite sides of the wall is 30 F° and the wall has an area of 40 ft²?

6. A refrigerator must remove 2050 Btu/hr from its interior to maintain an internal temperature of 35°F when the outside temperature is 95°F. Determine the following quantities:

    a. The thermal resistance of the refrigerator box.

    b. The thermal energy that must be removed each hour if the outside temperature is only 80°F (assume the resistance remains constant).

7. If the heat flow through one outside wall (area 60 ft²) is 5000 Btu/hr on a cold day, what is the heat flow out of another outside wall of the house if the second wall is made of the same material and is the same thickness as the first but as an area of only 40 ft²?

8. The inside wall temperature of a house is maintained at 72°F. If the outside wall temperature is 40°F, the heat flow through a particular wall is 1000 Btu/hr. What is the outside wall temperature if the heat flow through the same wall is 2000 Btu/hr?

S U M M A R Y

In different physical systems, an important phenomenon shows up as a resistance to motion with a corresponding loss of energy.

In the case of sliding friction, resistance is independent of velocity (rate). In fluid, electromagnetic, and thermal systems, the resistance can be analyzed and determined graphically by plotting the forcelike quantity (y axis) against the ratelike quantity (x axis) for the particular system (Figure 5.45). The resistance of the system is then the slope of the curve. Usually, the curve is a straight line of constant slope passing through the origin. For this case, the slope can be written simply as the ratio of a forcelike quantity to a ratelike quantity:

FIGURE 5.45 Plot of forcelike quality versus rate. (left)

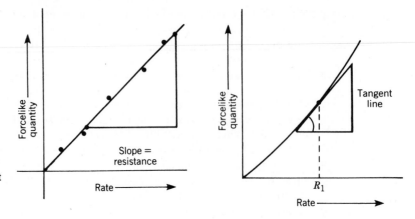

FIGURE 5.46 Resistance of a system when operating at rate R_1 is the slope of the tangent line. (right)

$$\text{Resistance} = \frac{\text{Forcelike Quantity}}{\text{Rate}}$$

When the graph of the forcelike quantity versus the rate is not a straight line, the system is said to be nonlinear. This is illustrated in Figure 5.46. This occurs in turbulent flow, electric discharges, and magnetic circuits. In turbulent pipe flow, the forcelike quantity, Δp, is proportional to the square of the ratelike quantity, Q_V. Also, the flow resistance depends on the Reynolds number. Both of these effects cause the flow system to be nonlinear. Although the magnetic flux, ϕ, is directly proportional to the magnetomotive force, NI, the reluctance of the path changes as ϕ changes if the magnetization curve is not a straight line. Thus, a plot of NI versus ϕ will not be a straight line.

In electric discharges, the resistance changes with the current because of changes in the electron and ion densities. As the current increases the resistance decreases. Thus, an electric discharge is a nonlinear system. Nonlinear systems are more difficult to analyze than linear systems. Resistance of the system can be found for any rate if the slope of the curve at the corresponding point (Figure 5.46) is found. Since the slope of a curve at a point on the curve is simply the slope of the tangent line to the curve at the point, resistance can always be found.

Table 5.8 contains a summary of the resistance in those systems in which resistance is dependent on rate.

TABLE 5.8 Generalized resistance

System	Linear	Forcelike Quantity	Rate	Resistance[a]	Generalized Law	Series Case	Parallel Case
Laminar flow	Yes	Pressure difference Δp	Volume flow rate Q_V	$\dfrac{8\mu\Delta l}{AR_o^2}$	$R=\dfrac{\Delta p}{Q_V}$	$R_T = \Sigma\, R_n$	$\dfrac{1}{R_T}=\Sigma\dfrac{1}{R_n}$
Turbulent flow	No	Pressure difference Δp	Volume flow rate Q_V	$\dfrac{f\rho_m\Delta l}{2A^2R_o}$	$R=\dfrac{\Delta p}{Q_V^2}$	$R_T = \Sigma\, R_n$	$\dfrac{1}{R_T}=\Sigma\dfrac{1}{R_n}$
Electrical conductors	Yes	Voltage difference ΔV	Electric current I	$\dfrac{\rho\Delta l}{A}$	$R=\dfrac{\Delta V}{I}$	$R_T = \Sigma\, R_n$	$\dfrac{1}{R_T}=\Sigma\dfrac{1}{R_n}$
Gaseous conductors	No	Voltage difference ΔV	Electric current I	$\dfrac{2f_e m\Delta l}{nq^2 A}$	$R=\dfrac{\Delta V}{I}$	—	—
Magnetic circuits	No	Magnetomotive force mmf	Magnetic flux ϕ	$\dfrac{\Delta l}{\mu A}$	$R=\dfrac{\text{mmf}}{\phi}$	$R_T = \Sigma\, R_n$	$\dfrac{1}{R_T}=\Sigma\dfrac{1}{R_n}$
Thermal	Yes	Temperature difference ΔT	Heat flow rate Q_H	$\dfrac{\Delta l}{kA}$	$R=\dfrac{\Delta T}{Q_H}$	$R_T = \Sigma\, R_n$	$\dfrac{1}{R_T}=\Sigma\dfrac{1}{R_n}^{\,b}$

[a] $A = \pi R^2 =$ cross-sectional area.

[b] Parallel conductors not in contact at the sides.

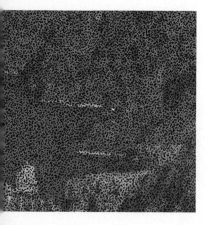

C H A P T E R 6

Potential
and
Kinetic
Energy

INTRODUCTION

Chapter 2, "Work," described work as the product of a displacementlike quantity and a forcelike quantity. Energy was defined as "the ability to do work." Energy can be contained in a system in a variety of forms. Today, most of the world's energy needs are supplied by the burning of fossil fuel, which converts stored chemical energy to thermal energy. Solar energy is electromagnetic radiation emitted by the sun. This chapter discusses two important forms of energy in technology and demonstrates the unification of physics. Kinetic energy is the energy of a moving mass; it is understood most easily in mechanical and fluid systems. Potential energy is the energy that exists because of position or potential difference. Potential energy involves no motion, but has the capacity to produce motion at some later time. The principle of the conservation of energy states, "Energy can be neither created nor destroyed." The total energy in any closed system remains constant, although it may change in form. This chapter describes potential and kinetic energy and discusses the conservation of energy in mechanical, fluid, electromagnetic, and thermal systems.

The fundamental definition of work in a mechanical system as presented in Chapter 2 is the magnitude of the force times the magnitude of a small

displacement times the angle between the two vectors. This concept of work is generalized to include other energy systems and is expressed as

$$\text{Work} = \frac{\text{Forcelike}}{\text{Quantity}} \times \frac{\text{Displacementlike}}{\text{Quantity}}$$

The work done on a system is converted to other forms of energy. This energy may be energy as a result of motion, which is called kinetic energy, or energy as a result of position, which is called potential energy. If the kinetic energy is composed of many bodies having random motions, the work done is transformed into internal energy. The inverse process also can occur. That is, if there exists kinetic, potential, or internal energy, work can be done. For this reason, energy often is defined as the ability to do work. If the energy is due to the random motion of bodies such as gas molecules, it is not possible to convert all their energy to work. Thus, part of their energy is not available to do work. This is a very important factor in the design of heat engines.

When work is done, energy is transferred. The energy may change form, as when frictional forces change mechanical energy into thermal energy, or may retain the same form, as when a moving body strikes a stationary one causing it to move. If all possible forms of energy are considered for an isolated system, the total energy of the system is conserved. This is a fundamental law of the physical sciences.

6.1

POTENTIAL AND KINETIC ENERGY IN MECHANICAL SYSTEMS

OBJECTIVES

Upon completion of this section, the student should be able to

- Define potential energy, kinetic energy, and conservation of energy for a mechanical system.

- Given any two of the quantities in the following groups, determine the third:
  Mass, velocity, and kinetic energy
  Mass, height, and potential energy
  Spring constant, spring displacement, and potential energy
  Moment of inertia, angular velocity, and kinetic energy

- Explain how the law of conservation of energy applies to an elastic collision and an inelastic collision.

- Given the masses and velocities of two objects before a totally inelastic collision, determine their common velocity after the collision and the loss of kinetic energy.

DISCUSSION

In mechanical systems, the kinetic and potential energy can arise in a number of modes. The following discussion begins with the translational mode.

Translational

Kinetic Energy   If a constant force is applied to a body at rest and there are no frictional forces, the body will accelerate according to Newton's second law. The velocity will increase as long as the force is applied. Because the body moves in the direction of the applied force, positive work is done. The total work done can be related to the final motion of the body. When the force is removed, the body moves with constant velocity. If this body interacts with another system it can do work on that system, but the velocity will decrease. The energy associated with the body's motion is called its kinetic energy.

To calculate the kinetic energy, consider Newton's second law:

$$\mathbf{F} = m \frac{\Delta \mathbf{v}}{\Delta t} \tag{6.1}$$

where

    $\mathbf{F}$ = applied force
    m = body's mass
    $\Delta \mathbf{v}$ = change in the body's velocity
    Δt = a small element of time

If both sides of Equation 6.1 are multiplied by a small element of the body's displacement, one obtains

$$F\Delta s = \frac{m\Delta s}{\Delta t} \Delta v \tag{6.2}$$

where

    F = magnitude of the applied force
    Δv = magnitude of the velocity change
    Δs = small displacement of the body in time Δt

It is assumed that the motion is linear, so that the angle between $\mathbf{F}$ and $\Delta \mathbf{s}$ is 0. Under the condition, a small element of work is

$$\Delta W = F\Delta s$$

Substituting this expression for $F\Delta s$ in Equation 6.2 and noting that

$$\frac{\Delta s}{\Delta t} = v$$

one obtains

$$\Delta W = (mv)\Delta v \tag{6.3}$$

To obtain the total work done on the body, one must add all the small elements of work ΔW. To do this, one can plot mv versus v as shown in Figure 6.1. From the

FIGURE 6.1 Plot of mass times velocity versus velocity.

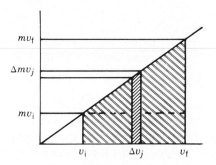

graph, one can see that ΔW is equal to the shaded area, $(mv)\Delta v$, which is also approximately equal to the area under the curve. By making Δv smaller, the two areas approach each other. Thus, to find the sum of all the ΔW's between the initial velocity and the final velocity, it is only necessary to calculate the corresponding area under the curve. This yields

$$W = \Sigma\Delta W = \Sigma mv_i\Delta v_i = \text{area under the curve}$$

$$W = (\text{area of rectangle}) + (\text{area of triangle})$$

$$W = mv_i(v_f - v_i) + 1/2(mv_f - mv_i)(v_f - v_i)$$

$$W = (v_f - v_i)\left(mv_i + \frac{mv_f}{2} - \frac{mv_i}{2}\right)$$

$$W = 1/2\, m\, (v_f - v_i)(v_f + v_i)$$

$$W = 1/2\, m\, (v_f^2 - v_i^2) = 1/2\, mv_f^2 - 1/2\, mv_i^2 \qquad (6.4)$$

If the initial velocity is zero, Equation 6.4 becomes

$$W = 1/2\, mv_f^2 \qquad (6.5)$$

This equation states that the work done on the body results in a kinetic energy of one-half the mass times the velocity squared, whereas Equation 6.4 states that the work done is equal to the change in the kinetic energy. The kinetic energy will be designated as E_k. Thus

$$E_k = 1/2\, mv^2 \qquad (6.6)$$

The sample problem in Example 6.1 illustrates the use of Equation 6.6.

EXAMPLE 6.1   Kinetic energy of a moving ball
. . . .

Given: A 250-g ball moves with a velocity of 15 m/sec.

Find: Amount of kinetic energy it possesses.

Solution: From Equation 6.6

$$E_k = 1/2\ (0.25\ \text{kg})(15\ \text{m/sec})^2$$

$E_k = 28.1$ J

Note: When only units are substituted into the equation, kinetic energy is shown to have the same units as work (i.e., joules).

$E_k \sim (kg)(m/sec)^2$

$E_k \sim kg \cdot m/sec^2$ (m)   (Recall that $kg \cdot m/sec^2$ = newton (N).)

$E_k \sim N \cdot m$

$E_k \sim J$

Potential Energy   When work is done to lift an object, the work is added to the potential energy of the object. To lift the box to the shelf in Figure 6.2, the woman must exert a force equal to the weight of the box. It was shown in Chapter 2 that the work done by the woman is equal to *wh* (where *w* = weight and *h* = height from floor to shelf). This value is the exact amount of increase in potential energy of the box after it is placed on the shelf. When the box is placed on the shelf, the gravitational potential energy of the box with respect to the floor is

$$E_p = wh \qquad (6.7)$$

where

E_p = potential energy
w = weight
h = height of shelf from floor

Because the weight of the object is related to its mass by the gravitational acceleration according to $w = mg$, Equation 6.7 often is given in the form of

$$E_p = mgh \qquad (6.8)$$

where

g = gravitational acceleration

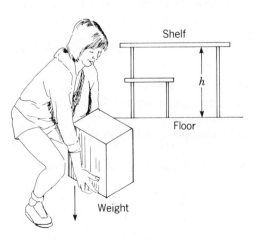

FIGURE 6.2 The work done on the box increases the potential energy of the box.

FIGURE 6.3 Potential
energy stored in a spring.

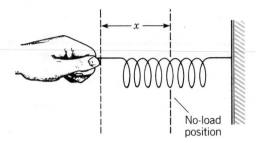

FIGURE 6.3 Potential
energy stored in a spring.

No-load
position

Gravitational potential energy is an important and useful concept. There are means of storing energy other than the raising of weight; for example, a stretched or compressed spring waiting to be released, as in Figure 6.3, possesses potential energy.

It was shown in Chapter 2, Example, 2.2, that the work necessary to stretch a spring x number of units beyond its normal length is

$$E_p = 1/2\ kx^2 \tag{6.9}$$

where

   k = spring constant (depends on the mechanical stiffness of the spring)
   x = distance spring is stretched or compressed beyond its no-load length

The work put into the spring is stored by the stretched spring as potential energy ($E_p = 1/2\ kx^2$). The value of k is just the magnitude of the force required for unit stretch of the spring; for example, 5 lb/inch. The value of k for a given spring is constant, provided the spring is not stretched beyond its elastic limit (i.e., the point at which the spring will not return to the no-load position if force is removed).

Conservation of Mechanical Energy   If frictional forces are neglected, the sum of the kinetic and potential energies is constant. This can be shown for a body in the earth's gravitational field by referring to Equation 3.15 in Chapter 3. The value of a is negative because the acceleration due to gravity is in the negative direction. Equation 3.15 then becomes

$$v_f^2 - v_i^2 = -2\ g\ (h_f - h_i) \tag{6.10}$$

where

   g = 9.8 m/sec^2 or 32 ft/sec^2
   v_f = velocity at height h_f above the earth's surface
   v_i = initial velocity
   h_i = initial height above the earth's surface

If Equation 6.10 is multiplied by 1/2 m, one obtains

$$1/2\ mv_f^2 - 1/2\ mv_i^2 = -mgh_f + mgh_i$$

or

$$1/2\ mv_f^2 + mgh_f = mv_i^2 + mgh_i$$

From the definitions of kinetic and potential energies, one obtains

$$E_{kf} + E_{pf} = E_{ki} + E_{pi}$$

Thus, the sum of the kinetic energy and potential energy is always equal to the initial kinetic energy plus the initial potential energy. This shows that the mechanical energy is conserved in the gravitation field near the earth's surface.

This result can be generalized for linear motion if the only force acting on the body can be written as

$$F_s = \frac{-\Delta E_p}{\Delta s} \qquad (6.11)$$

where

Δs = a very small linear displacement

If Equation 6.11 is true, the force F_s is called a conservative force. In addition to the validity of Equation 6.11, the work done by a **conservative force** is independent of the path for a given displacement and the energy ($F_s\Delta s$) is totally recoverable. For a force to be conservative, friction must not have an effect.

For a conservative force, one may write, from the definition of work,

$$\Delta W = F_s\Delta s = (mv)\Delta v$$

$$\Delta W = \left(\frac{-\Delta E_p}{\Delta s}\right)\Delta s = -\Delta E_p$$

or

$$-\Delta E_p = mv\Delta v$$

If all the small elements are summed as before, one obtains

$$-E_{pf} + E_{pi} = \frac{m}{2} v_f^2 - \frac{m}{2} v_i^2$$

$$E_{pi} + \frac{m}{2} v_i^2 = \frac{m}{2} v_f^2 + E_{pf} \qquad (6.12)$$

This shows that the total mechanical energy is conserved if Equation 6.11 is true. For this case, the force F_s is a conservative force.

From the above discussions the force of gravity acting on a body should be equal to

$$F_h = -\frac{\Delta E_p}{\Delta h}$$

Because $E_p = mgh$, $\Delta E_p = mg\Delta h$. By substitution, one obtains

$$F_h = -\frac{\Delta E_p}{\Delta h} = -mg \qquad (6.13)$$

Thus, the gravitational force is a conservative force. It also can be shown that the force of a spring is a conservative force.

Frictional forces are not conservative forces. One cannot assign a unique potential function for frictional forces because the work done against a frictional force depends on the total path length as was illustrated in Example 2.3. Frictional forces always oppose the motion and eventually cause moving objects to slow down and stop. Although friction cannot be completely eliminated in a mechanical system, it is often neglected to obtain approximate solutions to problems. An application of the concept of the conservation of mechanical energy is given in Examples 6.2 and 6.3.

EXAMPLE 6.2   Potential energy of stored water
. . . .

Given: Two tons of water are at the top of a 200-ft waterfall, ready to go over the edge.

Find: **a.** The amount of potential energy that the water possesses with respect to the bottom of the falls.

        b. The amount of potential and kinetic energy the water has when it has fallen halfway down.

        c. The amount of kinetic energy the water has when it has reached the bottom.

Solution: **a.** At the top of the falls

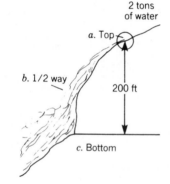

$$E_p = mgh - wh$$

$$E_p = (400 \text{ lb})(200 \text{ ft})$$

$$E_p = 800,000 \text{ ft·lb}$$

$$E_k = 0$$

        b. At the halfway point

$$E_p = (4000 \text{ lb})(100 \text{ ft})$$

$$E_p = 400,000 \text{ ft·lb}$$

$$E_k + E_p = 800,000 \text{ ft·lb}$$

$$E_k = 400,000 \text{ ft·lb}$$

        c. At the bottom

$$E_p = 0$$

$$E_k = 800,000 \text{ ft·lb}$$

We have assumed that the sum of $E_k + E_p$ is the same throughout.

EXAMPLE 6.3   Maximum velocity of a simple pendulum
. . . .

Given: The pendulum shown in the sketch. The velocity at point A is zero.

Find: The maximum velocity of the bob.

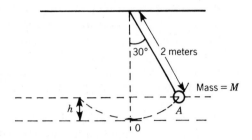

Solution: The maximum velocity occurs at point 0. The initial energy is all potential energy.

$$E = E_{pi} = mgh$$

$$h = (2 \text{ m}) - (2 \text{ m}) \cos 30°$$

$$h = 0.268 \text{ m}$$

$$E_{pi} = (m)(9.8 \text{ m/sec}^2)(0.268 \text{ m})$$

$$E = E_{pi} = (m)(2.63 \text{ m}^2/\text{sec}^2)$$

The energy at point 0 is all kinetic energy.

$$E = 1/2 \ mv_f^2$$

Because energy is conserved

$$1/2 \ mv_f^2 = m(2.63 \text{ m}^2/\text{sec}^2)$$

$$v_f^2 = (2)(2.63 \text{ m}^2/\text{sec}^2)$$

$$v_f = 2.29 \text{ m/sec}$$

The fact that the sum of potential and kinetic energy remains unchanged, even while there is a transformation of one form of energy to the other, is a demonstration of the principle of the conservation of energy. This principle is one of the cornerstones of science and technology.

Consider the changes in kinetic and potential energy of a roller coaster car as it travels on the track illustrated in Figure 6.4.

From point *A* up to *B*, energy must be supplied to the car by an electric winch. Assume the winch is released at *B*. At point *B*, energy supplied by the electric winch has been stored in the car as potential energy. This energy is equal to *mgh*, as given by Equation 6.8. As the car coasts down the first hill past point *C* toward

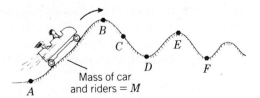

FIGURE 6.4 Energy changes of a roller coaster.

point D, its energy is changing rapidly from potential to kinetic. As kinetic energy increases, so does the car's velocity (speed), according to Equation 6.6 ($E_k = 1/2\ mv^2$). The car always is moving faster at low points on the track (points D and F) and slower at high points (B and E). At any point the sum total of its kinetic and potential energy is equal to the original amount of energy supplied to the car during its ascent from A to B. Friction, if considered, gradually steals some of the energy away.

A spring with an attached mass bouncing from its end is another example of a system with changing kinetic and potential energy (Figure 6.5).

Work is done on the system in Figure 6.5 when the spring (with mass attached) is stretched from position a to b, a distance equal to x. The amount of energy necessary to do this work is given by Equation 6.9 ($E_p = 1/2\ kx^2$). When the spring is released, it moves upward through the rest position. At the rest position, the displacement x is zero, so that potential energy $1/2\ kx^2$ equals zero at this point. However, as the mass passes the rest point, it is moving with its greatest speed, and thus has it greatest kinetic energy. As the mass rises above its rest position, its speed is reduced by its gravitational potential energy and spring potential energy are increased. This process continues until the highest point is reached, where its velocity and kinetic energy are zero, but its gravitational potential energy and spring potential energy are maximum. At any point during the cycle, total energy of the mass is

(potential energy stored in spring) + (kinetic energy)
+ (gravitational potential energy),

or

$$1/2\ kx^2\ +\ 1/2\ mv^2\ +\ mgh$$

This process would continue forever if friction did not dissipate the energy gradually and reduce the vertical oscillation.

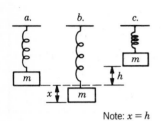

a. = Rest position
b. = Stretched position
c. = Uppermost position

FIGURE 6.5   Mass bouncing on a spring.

Collisions   It was shown that a moving body has momentum and energy. When two such bodies collide, the total momentum always is conserved, but the total kinetic energy after the collision is usually less than the total kinetic energy before the collision. The lost kinetic energy is converted into heat.

In a collision, the initial momenta and energies are usually known, but the final momenta and energies are desired. However, more information is needed before the final momenta and energies can be determined. If energy is conserved or if the energy loss is known, three equations can be written. The total momenta in the x direction before the collision must equal the total momenta in the x direction after the collision, and the total momenta in the y direction before the collision must equal the total momenta in the y direction after the collision. In addition, one can write an energy equation. However, when two bodies collide, there are four unknowns: two components of velocity in the x direction and two components of velocity in the y direction. Thus, additional information is required. The additional information usually is given as an angle between the two velocity vectors after the collision.

For an elastic collision, the two conservation equations are

$$m_a \mathbf{v}_{a_1} + m_b \mathbf{v}_{b_1} = m_a \mathbf{v}_{a_2} + m_b \mathbf{v}_{b_2} \tag{6.14}$$

$$\frac{m_a v_{a_1}^2}{2} + \frac{m_b v_{b_1}^2}{2} = \frac{m_a v_{a_2}^2}{2} + \frac{m_b v_{b_2}^2}{2} \tag{6.15}$$

where

$\mathbf{v}_{a_1}$ and $\mathbf{v}_{b_1}$ = vector velocities before the collision
$\mathbf{v}_{a_2}$ and $\mathbf{v}_{b_2}$ = vector velocities after the collision
m_b = mass of body b
m_a = mass of body a

The applications of Equations 6.14 and 6.15 are demonstrated in Examples 6.4 and 6.5.

EXAMPLE 6.4   Head-on collision of two spheres of equal mass
. . . .

Given: Two bodies sustain a head-on collision. The initial velocity of body a is 10 m/sec and body b is at rest. The mass of body a is equal to the mass of body b. Energy is conserved.

Find: The velocities of body a and body b after the collision.

Solution: For a head-on collision, all velocities lie along a straight line.

Momentum Equation

$$m_a\, v_{a_1} + m_b v_{b_1} = m_a v_{a_2} + m_b v_{b_2}$$

Since the masses are the same

$$v_{a_1} + v_{b_1} = v_{a_2} + v_{b_2}$$

or

$$v_{a_1} - v_{a_2} = v_{b_2} - v_{b_1} \tag{1}$$

Kinetic Energy Equation, Energy Conserved

$$1/2\, m_a\, v_{a_1}^2 + 1/2\, m_b v_{b_1}^2 = 1/2\, m_a v_{a_2}^2 + 1/2\, m_b v_{b_2}^2$$

Since the masses are the same

$$v_{a_1}^2 + v_{b_1}^2 = v_{a_2}^2 + v_{b_2}^2$$

or

$$v_{a_1}^2 - v_{a_2}^2 = v_{b_2}^2 - v_{b_1}^2$$

or

$$(v_{a_1} - v_{a_2})(v_{a_1} + v_{a_2}) = (v_{b_2} - v_{b_1})(v_{b_2} + v_{b_1}) \tag{2}$$

Since $v_{a_1} - v_{a_2} = v_{b_2} - v_{b_1}$ (Equation 1), then $v_{a_1} + v_{a_2} = v_{b_2} + v_{b_1}$. Two simultaneous equations occur:

$$v_{a_1} - v_{a_2} = v_{b_2} - v_{b_1}$$

and

$$v_{a_1} + v_{a_2} = v_{b_2} + v_{b_1}$$

Adding the two equations gives

$$2v_{a_1} = 2v_{b_2}$$

$$v_{a_1} = v_{b_2}$$

Subtracting the two equations gives

$$-2v_{a_2} = -2v_{b_1}$$

$$v_{a_2} = v_{b_1}$$

Because $v_{a_1} = 10$ m/sec and $v_{b_1} = 0$

$$v_{b_2} = v_{a_1} = 10 \text{ m/sec}$$

$$v_{a_2} = v_{b_1} = 0$$

Thus, the two spheres just exchange momentum and energy.

EXAMPLE 6.5   Elastic collision of two equal masses in two dimensions
. . . .

Given: Two equal masses collide. The initial velocity of the struck mass is
0. Energy is conserved.

Find: The angle between the two final velocity vectors.

Solution: *Momentum is conserved*

$$m_a \mathbf{v}_{a_1} = m_a \mathbf{v}_{a_2} + m_b \mathbf{v}_{b_2}$$

Because $m_a = m_b$

$$\mathbf{v}_{a_1} = \mathbf{v}_{a_2} + \mathbf{v}_{b_2}$$

θ is the angle between $\mathbf{v}_{a_2}$ and $\mathbf{v}_{b_2}$

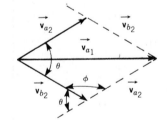

Energy is conserved

$$1/2\ m_a v_{a_1}^2 = 1/2\ m_a v_{a_2}^2 + 1/2\ m_b v_{b_2}^2$$

Because $m_a = m_b$

$$v_{a_1}^2 = v_{a_2}^2 + v_{b_2}^2$$

However, $v_{a_1}^2 = v_{a_2}^2 + v_{b_2}^2$ is a statement of the Pythagorean theorem. Therefore, angle ϕ is a right angle. If angle ϕ is a right angle,
then angle θ is a right angle. Thus, the angle between the two final
velocities is a right angle. If energy is not conserved, will the angle
be a right angle?

Totally inelastic collisions are those where the colliding bodies stick together. In this type of collision, a large fraction of the kinetic energy is lost. Example 6.6 illustrates the loss of kinetic energy in a totally inelastic collision.

EXAMPLE 6.6   Kinetic energy in a totally inelastic collision
. . . .

Given: The two bodies depicted here collide head-on and stick together.

Find: The kinetic energy before and after the collision.

$$v_1 = 10 \text{ m/sec} \qquad v_2 = 4 \text{ m/sec}$$

$$m_1 = 1 \text{ kg} \qquad m_2 = 2 \text{ kg}$$

Solution: The kinetic energy before the collision is the sum of the individual kinetic energies.

$$E_k = 1/2 m_a v_{a_1}^2 + 1/2 m_b v_{b_1}^2$$

$$E_k = (0.5)(1 \text{ kg})(10 \text{ m/sec})^2 + (0.5)(2 \text{ kg})(4 \text{ m/sec})^2$$

$$E_k = 50 \text{ kg·m}^2/\text{sec}^2 + 16 \text{ kg·m}^2/\text{sec}^2$$

$$E_k = 66 \text{ kg·m}^2/\text{sec}^2$$

$$E_k = 66 \text{ J}$$

Before the kinetic energy after the collision can be calculated, the final velocity v_f of the two bodies first must be determined. Substitution of $v_f = v_{a_2} = v_{b_2}$ into the conservation of momentum equation gives

$$m_a v_{a_1} + m_b v_{b_1} = (m_a + m_b) v_f$$

$$v_f = \frac{m_a v_{a_1} + m_b v_{b_1}}{m_a + m_b}$$

$$v_f = \frac{(1 \text{ kg})(10 \text{ m/sec}) + (2 \text{ kg})(-4 \text{ m/sec})}{(1 \text{ kg} + 2 \text{ kg})} \qquad \text{(Right is chosen as positive.)}$$

$$v_f = \frac{(10 \text{ kg·m/sec} - 8 \text{ kg·m/sec})}{3 \text{ kg}}$$

$$v_f = 0.67 \text{ m/sec}$$

$$E_k = 1/2 \, mv^2 \qquad \text{(where } m = m_1 + m_2 = 3 \text{ kg)}$$

$$E_k = (0.5)(3 \text{ kg})(0.67 \text{ m/sec})^2$$

$$E_k = 0.67 \text{ kg·m}^2/\text{sec}^2$$

$$E_k = 0.67 \text{ J}$$

Kinetic energy lost $= 66 \text{ J} - 0.67 \text{ J} = 65.33 \text{ J}$

FIGURE 6.6   Ballistic
pendulum.

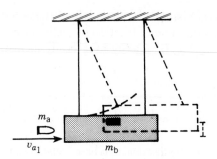

FIGURE 6.6   Ballistic
pendulum.

Because the total energy in any system is constant according to the law of conservation of energy, this decrease in kinetic energy must be accompanied by an equivalent increase in other forms of energy. This equivalent increase of energy usually appears as either a temperature increase (increased kinetic energy of molecules) or a deformation of the object.

Ballistic Pendulum   The ballistic pendulum is a device that can measure the velocity of a projectile. The projectile makes a totally inelastic collision with a stationary body of much greater mass. The momentum of the system immediately after the collision is the same as the momentum of the projectile before the collision. This same momentum can be determined if the kinetic energy after the collision is known.

Figure 6.6 depicts a bullet of mass m_a striking a stationary pendulum of mass m_b. The velocity of the bullet before the collision is v_{a_1}. The velocity of the pendulum and bullet after the collision is v_f. The center of mass of the pendulum swings upward through a vertical height h. The kinetic energy of the system immediately after the collision is

$$E_k = \frac{1}{2}(m_a + m_b)v_f^2$$

This value is equal to the potential energy of the system at the top of the pendulum swing

$$E_p = (m_a + m_b)gh$$

Thus,

$$E_k = E_p$$

$$\frac{1}{2}(m_a + m_b)v_2^2 = (m_a + m_b)gh$$

$$v_2^2 = 2gh$$

$$v_2 = \sqrt{2gh} \tag{6.16}$$

From the conservation of momentum

$$m_a v_{a_1} = (m_a + m_b)v_2$$

$$v_{a_1} = \frac{(m_a + m_b)v_2}{m_a} \tag{6.17}$$

Substitution of Equation 6.16 into Equation 6.17 yields

$$v_{a_1} = \frac{m_a + m_b}{m_a} \sqrt{2\,gh} \tag{6.18}$$

Thus, the initial velocity of the bullet can be determined from the masses and the height of the pendulum swing. The use of Equation 6.18 is illustrated in Example 6.7.

EXAMPLE 6.7 The ballistic pendulum
. . . .

Given: A bullet having a mass of 20 g strikes a ballistic pendulum having a mass of 1.5 kg. The pendulum swings through a vertical height of 5 cm.

Find: The velocity of the bullet before the collision.

$$v_{a_1} = \frac{m_a + m_b}{m_a} \sqrt{2\,gh}$$

$$v_{a_1} = \frac{1500 \text{ g} + 20 \text{ g}}{20 \text{ g}} \sqrt{2\,(9.8 \text{ m/sec}^2)(0.05 \text{ m})}$$

$$v_{a_1} = (76) \sqrt{0.98 \text{ m}^2/\text{sec}^2}$$

$$v_{a_1} = (76)\,(0.99 \text{ m/sec})$$

$$v_{a_1} = 75.24 \text{ m/sec}$$

Rotational Systems

Kinetic Energy Chapter 4, "Momentum," introduced the concept of moment of inertia and suggested that it is the quantity in a rotational system that corresponds to mass in a linear system. Recall that the moment of inertia (I) varies with the shape of the object and the position of its axis of rotation. Formulas for determining the moment of inertia of various objects are summarized in Chapter 4.

For a rotational system an element of work done on a system rotating about a fixed axis is

$$\Delta W = \tau \Delta\theta \tag{6.19}$$

where

τ = torque (forcelike quantity)
$\Delta\theta$ = small angular displacement (displacementlike quantity)

In addition, it was shown that

$$\tau = I \frac{\Delta\omega}{\Delta t} \tag{6.20}$$

where

$\Delta\omega$ = change in the angular velocity

Multiplying both sides of Equation 6.20 by $\Delta\theta$ and substituting into Equation 6.19 yields

$$\Delta W = \tau\Delta\theta = I\frac{(\Delta\omega)(\Delta\theta)}{\Delta t} = I\omega\Delta\omega \tag{6.21}$$

Equation 6.21 has the same form as Equation 6.3, where v corresponds to ω and m to I. Thus, by analogy, the work done on a rotating body becomes

$$W = \frac{1}{2}I\omega_f^2 - \frac{1}{2}I\omega_i^2 \tag{6.22}$$

where

W = work done on or by the rotating system
ω_f = final angular velocity
ω_i = initial angular velocity
I = moment of inertia

In the derivation, it is assumed that there are no frictional forces. The quantity $1/2\ I\omega^2$ is called the rotational kinetic energy and is given the symbol E_{kR}. Thus

$$E_{kR} = 1/2\ I\omega^2 \tag{6.23}$$

Two problems that illustrate the application of Equation 6.23 are given in Examples 6.8 and 6.9.

EXAMPLE 6.8   Energy used to start a chain saw
. . . .

Given: A manual starting system on a chain saw requires a force of 180 N to pull the cord. The homeowner pulls the rope a total of 0.8 m when attempting to start the chain saw. The flywheel around which the rope is wound has a radius of 12 cm and a mass of 0.7 kg. The flywheel has the shape of a solid cylindrical disk.

Find: a. The moment of inertia of the flywheel.
              b. The work put into pulling the cord.

   c. The kinetic energy and angular velocity given the flywheel, assuming no-load conditions.

Solution: **a.** $I = 1/2\ mr^2$

$I = 1/2\ (0.7\ \text{kg})(0.12\ \text{m})^2$

$I = 0.005\ \text{kg·m}^2$

   b. Recall from Chapter 4 that

$W = \tau\theta$

$\tau = Fl$

$\tau = (180\ \text{N})(0.12\ \text{m})$

$\tau = 21.6\ \text{N·m}$

$\theta = \dfrac{s}{r}$

$\theta = \dfrac{0.8\ \text{m}}{0.12\ \text{m}}$

$\theta = 6.67\ \text{rad}$

$W = (21.6\ \text{N·m})(6.67\ \text{rad})$

$W = 144\ \text{J}$

or alternately

$W = Fx$

$W = (180\ \text{N})(0.8\ \text{m})$

$W = 144\ \text{N·m}$

$W = 144\ \text{J}$

   c. $E_{kR} = 1/2\ I\omega^2$

If no friction is present, all the work is transferred into the kinetic energy of the flywheel. Thus, the kinetic energy = 144 J.

$144\ \text{J} = 1/2\ (0.005\ \text{kg·m}^2)\omega^2$

$\omega^2 = \dfrac{2\ (144)}{0.005}\ \dfrac{\text{rad}^2}{\text{sec}^2}$

$\omega^2 = 57{,}600\ \text{rad}^2/\text{sec}^2$

$\omega = 240\ \text{rad/sec}$

An object can be rotating at the same time that it is moving forward. If this is the case, the object has both rotational kinetic energy and translational kinetic energy. Total kinetic energy of the object is found by determining each energy separately and then adding the results, as illustrated in Example 6.9.

EXAMPLE 6.9   Kinetic energy of a bowling ball

. . . .

Given: A 7.2-kg bowling ball with a radius of 12 cm is thrown with a velocity of 4 m/sec.

4 m/sec ⟶

Find: Total kinetic energy of the ball before it strikes the pins.

Solution: Linear kinetic energy equals

$$E_k = 1/2 \; mv^2$$

$$E_k = 1/2 \; (7.2 \text{ kg})(4 \text{ m/sec})^2$$

$$E_k = 57.6 \text{ J}$$

Recall that

$$\omega = v/r$$

$$\omega = \frac{4 \text{ m/sec}}{0.12 \text{ m}}$$

$$\omega = 33.3 \text{ rad/sec}$$

Rotational kinetic energy equals

$$E_{kR} = 1/2 \; I\omega^2$$

$$I = 2/5 \; mr^2 \text{ for a sphere}$$

$$I = 2/5 \; (7.2 \text{ kg})(0.12 \text{ m})^2$$

$$I = 0.0415 \text{ kg·m}^2$$

$$E_{kR} = 1/2 \; (0.0415 \text{ kg·m}^2)(33.3 \text{ rad/sec})^2$$

$$E_{kR} = 23.0 \text{ J}$$

$$E_k \text{ total} = E_k + E_{kR}$$

$$E_k \text{ total} = 57.6 \text{ J} + 23.0 \text{ J}$$

$$E_k \text{ total} = 80.6 \text{ J}$$

FIGURE 6.7 Flywheel in an automotive engine.

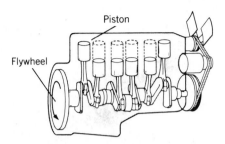

Applications In many mechanical devices, flywheels are used to smooth the motion of the reciprocating parts. Figure 6.7 shows the flywheel of a six-cylinder internal combustion engine. In this system, part of the kinetic energy of the flywheel is used to force the piston upward to compress the air–fuel mixture in the cylinder. When the mixture burns, some of the energy produced is stored in the flywheel to be used in the next compression stroke.

The friction motors of toy cars are excellent examples of pure kinetic energy drives. The axle is connected to a flywheel by a series of gears. When the toy is pushed forward, the flywheel is caused to spin at a high rate. When the toy is released, the stored kinetic energy is transmitted to the axle by the same gear system to maintain the forward motion.

In recent years, several investigations have considered storage of large amounts of rotational kinetic energy in flywheels as a solution to practical transportation problems. Flywheels weighing as much as 3000 lb and smaller wheels with rotational rates as high as 12,000 rpm have been installed in vehicles. In some systems, direct mechanical coupling of the flywheel to the drive wheels has been employed but most use the electrical conversion system shown in Figure 6.8. In this system, the flywheel drives a generator that supplies electrical power to a motor that powers the vehicle.

Figure 6.9 shows a bus equipped with a kinetic energy drive. This drive stores 25 MJ and gives the bus a range of several miles. A similar system can run a small automobile for seven to nine miles at speeds of 30 miles per hour. Combinations of kinetic energy and electrical drives offer both lower energy utilization and lower pollution for urban transportation.

FIGURE 6.8 Flywheel used in an electrical conversion system.

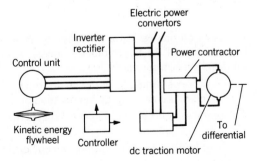

FIGURE 6.9 Bus using
kinetic energy drive.

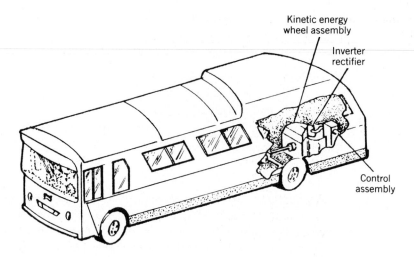

Kinetic energy
wheel assembly

Inverter
rectifier

Control
assembly

SECTION **6.1**   *EXERCISES*

1. Define potential energy, kinetic energy, and conservation of energy, using examples from mechanical systems.

2. An automobile weighing 2800 lb is traveling at 55 mi/hr. What is its kinetic energy?

3. What would be the speed of the automobile in Problem 2 if its kinetic energy were doubled?

4. An 1800-lb elevator is lifted a vertical distance of 70 ft by an electric motor. Calculate potential energy of the elevator at the 70-ft height.

5. A spring requires a force of 25 N to stretch it 15 cm. What is the spring constant (k)? How much energy is stored in the spring when it is stretched 45 cm?

6. A solid wheel (disk) with a mass of 5 kg has a rolling speed of 20 m/sec. Determine its total kinetic energy.

7. Discuss conservation of energy as it applies to the mechanical system.

8. A 1.0-kg mass is fixed to the end of a spring that is on a horizontal, frictionless table as shown in the figure below. If the spring is stretched 10 cm beyond its rest point and then released, how fast will the mass be moving as it passes through the rest point? Assume that the spring has a spring constant of 100 N/m and that there are no sources of dissipative friction.

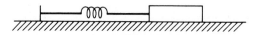

9. A 40-kg mass rests on one end of a teeter-totter. A 300-kg mass is dropped from a height of 10 m. If 40% of the larger block's energy is transferred to the small mass, how high does the small mass rise?

10. A 400-kg mass is dropped from a height of 40 m. Roughly sketch the block's potential energy versus time and kinetic energy versus time on the same graph.

11. A solid wheel's total kinetic energy is 3000 J. What is the wheel's speed if its mass is 7 kg?

12. What is the ratio of translational to rotational kinetic energy of a solid wheel moving at a constant translational speed?

13. A 0.1-kg ball is dropped from a height of 20 m. At the instant before the ball hits the ground, what is the speed of the ball?

14. Explain the difference between the conservation of linear momentum and the conservation of energy in collisions.

15. What happens to the kinetic energy in each of the following collisions?
  a. Totally elastic
  b. Totally inelastic
  c. Partially inelastic

16. A 1000-kg automobile traveling to the right at 10 m/sec has a totally inelastic collision with a 1250-kg automobile traveling to the left at 8 m/sec. How much kinetic energy is converted to heat in the collision?

17. A bullet having a mass of 9 g is fired into a ballistic pendulum having a mass of 2 kg. The pendulum swings through a vertical distance of 3.2 cm. Determine:
  a. initial velocity of the bullet
  b. initial kinetic energy of the bullet
  c. velocity of the pendulum after the collision
  d. kinetic energy converted to heat in the collision.

6.2

POTENTIAL AND KINETIC ENERGY IN FLUID SYSTEMS

OBJECTIVES

Upon completion of this section, the student should be able to

- Given Bernoulli's equation and the height of liquid in a tank, determine the exit velocity at the bottom of the tank if there is no fluid friction.

- Define the ideal gas law.

- Describe the following gas laws:
  Boyle's law
  Charles' law
  Gay-Lussac's law

- Understand the pressure and temperature of an ideal gas in terms of the molecular motion.

- Understand the concept of the mean-free path of a molecule.

- Define the concept of an ideal gas.

- Calculate the fluid velocity in a horizontal pipe using the energy equation for incompressible flow for a venturi flowmeter inserted into the pipe.

- Use the pitot tube equation to calculate fluid speeds.

DISCUSSION

The sum of potential and kinetic energy in mechanical systems has been shown to be constant in cases where friction and other nonconservative forces do not have an effect. Other cases have been examined where the nonconservative forces enter, such as inelastic collisions. The discussion now turns to fluid systems.

Gases

A fluid can be either a liquid or gas. If the fluid is a liquid, equations for the conservation of mass, momentum, and energy are greatly simplified because a liquid can be treated as an incompressible fluid to a very good approximation. Gases, on the other hand, are easily compressed. This compressibility can cause the solution of flow problems to be very difficult.

Nature of a Gas   Gases at rest have important properties associated with their compressibility. These properties arise from the nature of a gas. A gas is composed of a very large number of molecules that are free to move about in their container. Each molecule has a translational kinetic energy, $1/2\ mv^2$. If the molecule is composed of several atoms, it also can have a rotational kinetic energy, $1/2\ I\omega^2$. The atoms in the molecule can also vibrate; vibrational energy occurs at high temperatures. In addition, the molecules attract each other, resulting in potential energy. The sum of the kinetic and potential energies is called the internal energy of the gas.

The molecular motion within a gas is illustrated by Figure 6.10, where molecules are shown striking the container and its movable, airtight top with such force that a weight upon the top is suspended or balanced there. The combined force of the top and the weight compresses the gas until balanced by the upward force exerted by the gas.

Ideal Gas Law   For most gases under ordinary conditions, it has been shown experimentally that pressure times volume divided by the absolute temperature is equal to the mass divided by the mass of a molecule times a universal constant. This is expressed mathematically as

$$\frac{PV}{T} = \left(\frac{m}{m_{\mathrm{m}}}\right) k \tag{6.24}$$

where

P = absolute gas pressure
V = gas volume
T = absolute temperature of the gas
m = mass of the gas

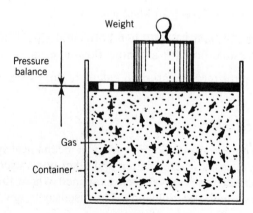

FIGURE 6.10   Force from a compressed gas.

m_m = mass of a single molecule
k = Boltzmann's constant

Gases that obey Equation 6.24 are called ideal gases. As real gases are compressed, they tend to deviate from this law.

Equation 6.24 also may be written as

$$\frac{P}{T} = \left(\frac{m}{V}\right)\left(\frac{k}{m_m}\right)$$

However, m/V was defined as the mass density ρ_m. Thus, Equation 6.24 also takes the form

$$\frac{P}{T} = \rho_m \left(\frac{k}{m_m}\right) \tag{6.25}$$

However, the density of the gas may be written as

$$\rho_m = n m_m \tag{6.26}$$

where

m_m = mass of a single molecule
n = number of molecules per unit volume

Using the expression in Equation 6.25 gives

$$\frac{P}{T} = nk$$

or

$$P = nkT \tag{6.27}$$

The quantity k, called Boltzmann's constant, is one of the basic constants of nature. The experimentally determined value of k is 1.38×10^{-23} J/molecule·K°. An application of Equation 6.27 is given in Example 6.10.

EXAMPLE 6.10   Molecular number density of a gas
. . . .

Given: The temperature of a gas is 273°K and its pressure is 1.01×10^5 N/m² (one atmosphere).

Find: The number of molecules per cubic meter.

Solution: For any gas under ordinary conditions

$$P = nkT$$

$$n = P/kT$$

$$n = \frac{(1.01 \times 10^5 \text{ N/m}^2)}{(1.38 \times 10^{-23} \text{ N·m/mol·K°})(273°K)}$$

$$n = 2.68 \times 10^{25} \text{ mol/m}^3$$

FIGURE 6.11 Path of a
molecule of gas.

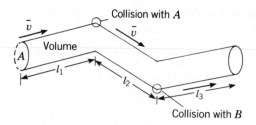

Mean-Free Path   As the gas particles move they collide with each other and develop a distribution of velocities. However, there is an average velocity $\bar{v}$. On the average, one molecule will travel a distance v in one second. The molecule has a cross-sectional area associated with the probability of a collision. This area sweeps out a volume in one second equal to

$$V = A\cdot\bar{v} \qquad\qquad (6.28)$$

where

  A = cross-sectional area for collisions
  $\bar{v}$ = average velocity
  V = volume swept out in one second

Every time a collision occurs, the direction of the velocity will change as shown in Figure 6.11. The number of collisions per second is determined by the volume swept out per second times the number of molecules per unit volume. Every molecule in this volume causes a collision. This may be expressed mathematically, by using Equation 6.28 for the volume per second, as

$$f = (A\bar{v})n \qquad\qquad (6.29)$$

where

  n = number of molecules per unit volume
  f = collisions per second (collision frequency)

From Chapter 3, the average time between collisions is

$$\bar{t} = \frac{1}{f} = \frac{1}{A\bar{v}n} \qquad\qquad (6.30)$$

The average distance between collisions is equal to the average velocity $\bar{v}$ times the average time between collisions $\bar{t}$:

$$\bar{l} = \bar{v}\bar{t} = \frac{1}{An} \qquad\qquad (6.31)$$

where

  $\bar{l}$ = mean-free path

It is found experimentally that the area A is of the order of 3×10^{-19} m² for a number of molecules. An application of Equation 6.31 is given in Example 6.11.

EXAMPLE 6.11   Calculation of mean-free paths
. . . .

Given: A chamber initially at 273°K and atmospheric pressure is pumped down to 10^{-7} atm at 273°K.

Find: **a.** The initial mean-free path.
 b. The final number density.
 c. The final mean-free path.

Solution: **a.** From Example 6.10, the initial number density is 2.68×10^{25} molecules/m³. The equation for the mean-free path is

$$\bar{l} = \frac{1}{nA}$$

An approximate value of A is 3×10^{-19} m².

$$\bar{l} = \frac{1}{(2.68 \times 10^{25} \text{ mol/m}^3)(3 \times 10^{-19} \text{ m}^2)}$$

$$\bar{l} = 1.24 \times 10^{-7} \text{ m}$$

 b. The number density is

$$n = \frac{P}{kT}$$

At the same temperature

$$\frac{n_2}{n_1} = \frac{P_2}{kT_2} \times \frac{kT_1}{P_1}$$

$$n_2 = \frac{P_2}{P_1} n_1$$

Let $n_2 = 2.68 \times 10^{25}$ molecules/m³ (1 atm).

$$P_1 = 1 \text{ atm}$$

$$n_2 = \frac{10^{-7} \text{ atm}}{1 \text{ atm}} (2.68 \times 10^{25} \text{ mol/m}^3)$$

$$n_2 = 2.68 \times 10^{18} \text{ mol/m}^3$$

 c. The mean-free path is

$$\bar{l} = \frac{1}{nA}$$

$$\frac{\bar{l}_2}{\bar{l}_1} = \frac{n_1 A}{n_2 A}$$

$$\bar{l}_2 = \frac{n_1}{n_2} \bar{l}_1$$

From part b

$$\frac{n_1}{n_2} = \frac{P_1}{P_2}$$

$$\bar{l}_2 = \frac{P_1}{P_2}\bar{l}_1$$

$$\bar{l}_2 = \left(\frac{1 \text{ atm}}{10^{-7} \text{ atm}}\right)(1.24 \times 10^{-7} \text{ m})$$

$$\bar{l}_2 = 1.24 \text{ m}$$

At 10^{-7} atm, the mean-free path is of the same order as the size of the container (vacuum chamber).

Boyle's Law   If the temperature and mass of a gas remain constant, Equation 6.24 can be written as

$$PV = C_2 \tag{6.32}$$

where

C_2 = a constant

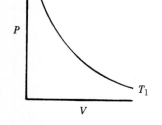

FIGURE 6.12   Boyle's law.

As displayed in Figure 6.12, decreases in volume are accompanied by an increase in pressure, or a decrease in pressure results in an increase in volume when the temperature and mass are held constant. This relationship between pressure and volume was first stated as a law in 1662 by Robert Boyle, after whom it was named. Equation 6.32 is utilized in solving the problem in Example 6.12.

EXAMPLE 6.12   Surge tank pressure and Boyle's law
. . . .

Given: An 85-gal surge tank is installed in a water system, as illustrated in the drawing. The tank is 63 inch high and 20 inch in diameter. Pressure switches in the surge tank start the pump at a pressure of 20 psig (gage) and stop the pump when the pressure reaches 40 psig. The space above the water (called "ullage") is filled with air. The temperature of the air remains essentially constant during this process. The water is 32 inch high when the pressure is 20 psig.

Find: The water height when the pressure is 40 psig.

Solution: Symbols used in the calculations:

psia = pounds per square inch absolute

V_T = total tank volume

V_L = volume of water at h_1

V_H = volume of water at h_2

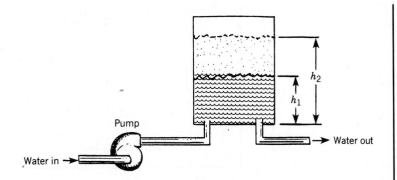

V_1 = volume of air when water volume is V_L

V_2 = volume of air when water volume is V_H

$V_T = \pi/4\ d^2 h_T = \pi/4\ (20\ \text{inch})^2\ (63\ \text{inch})$

$V_T = 19{,}792\ \text{inch}^3$

$V_L = \pi/4\ d^2 h_1$

$V_L = \pi/4\ (20\ \text{inch})^2\ (32\ \text{inch})$

$V_L = 10{,}053\ \text{inch}^3$

$V_1 = V_T - V_L$

$V_1 = 19{,}792 - 10{,}053 = 9739\ \text{inch}^3$

$P_1 = 20\ \text{psig} + 14.7\ \text{psia} = 34.7\ \text{psia}$

$P_2 = 40\ \text{psig} + 14.7\ \text{psia} = 54.7\ \text{psia}$

From Equation 6.32

$P_1 V_1 = P_2 V_2 = C_2$

or

$$\frac{P_1}{P_2} = \frac{V_2}{V_1}$$

Therefore,

$$V_2 = V_1 = \frac{P_1}{P_2}$$

So

$$V_2 = (9739\ \text{inch}^3) \left(\frac{34.7\ \text{psia}}{54.7\ \text{psia}} \right)$$

$V_2 = 6178\ \text{inch}^3$

and

$V_H = 19{,}792\ \text{inch}^3 - 6178\ \text{inch}^3$

$$V_H = 13,614 \text{ inch}^3$$

Then,

$$h_2 = \frac{4V_H}{\pi D^2}$$

$$h_2 = \frac{4 \,(13,614 \text{ inch}^3)}{\pi \,(20 \text{ inch})^2}$$

$$h_2 = 43.3 \text{ inch}$$

Charles' Law   If the pressure and mass of a gas remain constant, Equation 6.24 becomes

$$V/T = C_3 \tag{6.33}$$

where

C_3 = a constant

Increases in temperature cause an increase in volume when the pressure and mass are maintained at the same value. This relationship is named after its discoverer, Jacques Charles. Figure 6.13 indicates how volume is plotted against temperature using Charles' law.

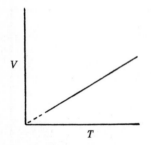

FIGURE 6.13   Charles' law.

Gay-Lussac's Law   Gay-Lussac's law is for a constant volume process, as plotted in Figure 6.14. When the volume and mass remain constant, Equation 6.24 takes the form

$$P/T = C_4 \tag{6.34}$$

where

C_4 = a constant

An increase in the temperature of a gas when the volume and mass are held constant causes an increase in pressure.

Equation 6.34 is used to solve a problem in Example 6.13.

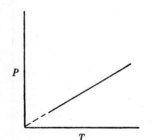

FIGURE 6.14   Gay-Lussac's law.

EXAMPLE 6.13   Tire pressure and Gay-Lussac's law
. . . .

Given: An automobile tire with a volume of 1500 inch³ is pressurized to 25 psig at a temperature of 70°F.

Find: Pressure in the tire after the car has been driven several hours and air temperature in the tire has increased to 170°F.

Solution: P_1 = 25 psig + 14.7 psia = 39.7 psia

$$T_1 = 70°F + 460 = 530°R \qquad \text{(Converting to an absolute temperature scale)}$$

$$T_2 = 170°F + 460 = 630°R$$

Assume that the volume of the tire remains essentially constant. Equation 6.34 is used as follows:

$$\frac{P_1}{T_2} = \frac{P_2}{T_2}$$

or

$$P_2 = \frac{P_1 T_2}{T_1}$$

$$P_2 = \frac{(39.7 \text{ psia}) (630°R)}{(530°R)}$$

$$P_2 = 47.2 \text{ psia}$$

or

$$P_2 = 47.2 \text{ psia} - 14.7 \text{ psia}$$

$$P_2 = 32.5 \text{ psig}$$

Adiabatic Process In addition to the constant temperature, constant pressure, and constant volume processes, there are many other processes where the ideal gas law can be used to approximate the behavior of real gases. When there is no heat transferred to or from the gas, this is called an adiabatic process. Recalling Equation 2.20, in Chapter 2,

$$\Delta Q = \Delta W + \Delta U \qquad (6.35)$$

for an adiabatic process $\Delta Q = 0$, and the work done by the gas or on the gas ΔW is equal to the increase or decrease in the internal energy ΔU of the gas. It can be shown that the equation of an adiabatic curve for an ideal gas of constant mass is

$$PV^\gamma = C_5 \qquad (6.36)$$

where

$\gamma = c_p/c_v$
$C_5 = $ a constant
$c_p = $ specific heat at constant pressure
$c_v = $ specific heat at constant volume

The value of γ is approximately 1.67 for monatomic gases, 1.40 for diatomic gases, and 1.30 for polyatomic gases.

Real Gases The concept of an ideal, imaginary gas is useful because its behavior can be predicted. Behavior of an actual gas also is predicted from ideal gas laws to the extent that the gas conforms to the requirements of an ideal gas. At tem-

FIGURE 6.15 Plots of pressure versus volume at various temperatures.

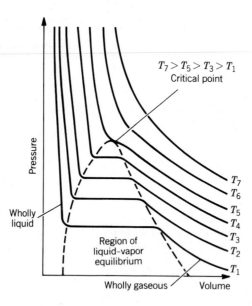

peratures much higher than the boiling point of the gas and at normal or low pressures, most real gases come very close to satisfying the conditions of an ideal gas. For most gases, these conditions are satisfied at 70°F and normal atmospheric pressure. However, steam must be "superheated" to several hundred degrees before its behavior approximates that of an ideal gas.

Thomas Andrews investigated the pressure–volume relationships of CO_2 at Queen's College, Belfast, between 1861 and 1869. He observed that as he lowered the temperature, the smooth curves began developing irregularities that eventually led to a change in volume without a change in pressure, as shown in Figure 6.15. Andrews accurately concluded that the gas actually was changing to a liquid under these conditions. The highest temperature at which the gas can liquefy is the "critical temperature"; the pressure at the critical point (Figure 6.15) is the "critical pressure."

The "critical point" indicates the point at which to expect the failure of the ideal gas law. There is considerable deviation from the ideal gas law for temperatures and pressures near the critical point. Several more sophisticated equations have been developed to represent the behavior of a gas that does not obey the ideal gas law. These equations are empirical and can only approximate the behavior of gases under conditions far removed from their liquefaction points. Such equations are beyond the scope of this chapter.

Caution must be exercised in the application of the laws expressed in Equations 6.32 through 6.36 to assure that the required conditions are not violated. For example, because compression is so rapid in an internal combustion engine, there is no opportunity for an appreciable heat transfer. As a result, the temperature of gas in the cylinder is not constant. Therefore, the adiabatic process should be used in this case. Example 6.14 compares the results obtained assuming a constant temperature process with the adiabatic process for the rapid compression of a gas in a cylinder.

EXAMPLE 6.14 Internal combustion engine pressure

Given: An eight-cylinder, internal combustion engine has a compression ratio of 10 and a displacement of 350 inch³. Assume that the cylinder is filled with air at a pressure of 0 psig (atmospheric pressure) and no ignition occurs.

Find: Cylinder pressure when the piston is at top dead center

    a. assuming a constant temperature process (Equation 6.32)

    b. assuming no heat loss (Equation 6.36).

Solution: Because the engine has a compression ratio of 10, $V_2/V_1 = 1/10$; that is, the gas volume after compression is one-tenth the initial gas volume, assuming that no ignition has occurred.

    a. From Equation 6.32

$$P_1 V_1 = P_2 V_2$$

or

$$\frac{P_1}{P_2} = \frac{V_2}{V_1}$$

Therefore

$$\frac{P_1}{P_2} = \frac{1}{10}$$

and

$$P_2 = 10\, P_1$$

Because $P_2 = 0$ psig $+ 14.7$ psia $= 14.7$ psia, then

$$P_2 = 10\,(14.7\ \text{psia}) = 147\ \text{psia}$$

or

$$P_2 = 147\ \text{psia} - 14.7\ \text{psia} = 132.3\ \text{psig}$$

    b. From Equation 6.36

$$P_1 V_1 \gamma = P_2 V_2 \gamma$$

or

$$\frac{P_1}{P_2} = \left(\frac{V_2}{V_1}\right)^{\gamma}$$

Therefore

$$\frac{P_1}{P_2} = \left(\frac{1}{10}\right)^{1.4}$$

$$\frac{P_1}{P_2} = 0.0398$$

Then

$$P_2 = \frac{P_1}{0.0398} = \frac{14.7 \text{ psia}}{0.0398}$$

$$P_2 = 369.3 \text{ psia}$$

or

$$P_2 = 369.3 \text{ psia} - 14.7 \text{ psia}$$

$$P_2 = 354.6 \text{ psia}$$

A comparison of the two calculated pressures (132.3 psig and 354.6 psig) reveals a considerable difference, which could be an important factor, depending on how these results are used. If the initial temperature (T_1) and the temperature following compression (T_2) had been measured, then Equation 6.24 could have been used.

Kinetic Theory   The pressure of a gas can be related to the molecular motion. Consider a container in the form of a cubic meter filled with molecules having an average velocity v. Because the cube has six sides, one-sixth of the total number of molecules will be moving toward one wall on the average. Of these molecules, only those that lie a distance equal to or less than the mean-free path will strike the wall before colliding with another molecule on the average. This situation is illustrated in Figure 6.16. The time it takes a molecule located $\overline{l}$ away from the wall to reach the wall is, from the definition of velocity,

$$\Delta t = \frac{\overline{l}}{\overline{v}} \tag{6.37}$$

During this time, all the molecules traveling toward the wall that are l or less from the wall will hit the wall. This number of molecules Δn is equal to the volume

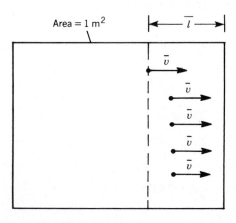

FIGURE 6.16 Gas molecules available for collision with a wall.

(l) (1 m^2) times the number of molecules per unit volume n. Thus,

$$\Delta n = 1/6 \; n \cdot l \tag{6.38}$$

Dividing Δn by Δt gives the number of molecules hitting the wall per unit time that do not suffer a collision on the average. This yields

$$\frac{\Delta n}{\Delta t} = \frac{1}{6} \frac{nl}{\Delta t}$$

Substituting for Δt from Equation 6.37 gives

$$\frac{\Delta n}{\Delta t} = \frac{1}{6} \frac{nl}{l/v} = \frac{1}{6} \, nv \tag{6.39}$$

The average momentum per molecule is

$$p = m_{\mathrm{m}} v$$

where

p = average momentum of a molecule
m_{m} = mass of a molecule

When a molecule strikes the wall, it reverses its velocity. The corresponding change in momentum is

$$\Delta p = 2 \, m_{\mathrm{m}} v$$

The total change in momentum per second becomes the change in momentum of one molecule times the number of molecules hitting the wall per second:

$$\frac{\Delta p}{\Delta t} = (2 \, m_{\mathrm{m}} v) \, (1/6 \, nv)$$

or

$$\frac{\Delta p}{\Delta t} = 1/3 \; n m_{\mathrm{m}} v^2 \tag{6.40}$$

According to Newton's second law, force is equal to the time rate of change in momentum. Thus, Equation 6.40 can be written as

$$F = 1/3 \; n \, m_{\mathrm{m}} v^2 \tag{6.41}$$

Because the area of the wall is one square meter, the force in Equation 6.41 is also the pressure. Thus

$$P = \frac{1}{3} n \, m_{\mathrm{m}} v^2 = \frac{2}{3} n \left(\frac{m_{\mathrm{m}} v^2}{2} \right) \tag{6.42}$$

Equation 6.42 states that the pressure of a gas is equal to two-thirds the kinetic energy per unit volume.

From Equation 6.27, the pressure of an ideal gas is

$$P = nkT \tag{6.43}$$

Setting the expression for pressure given by Equation 6.42 to the pressure given by 6.43 yields

$$nkT = 2/3 \ nE_k$$

or

$$3/2 \ kT = E_k \tag{6.44}$$

where

$$E_k = 1/2 \ m_m v^2 \qquad \text{(the average kinetic energy of a molecule)}$$

Thus, the temperature is a measure of the average linear momentum of a molecule of the gas.

If the gas consists of monatomic particles such as He, Ar, Kr, Ne, and Xe, then the energy of a gas at ordinary conditions consists primarily of linear kinetic energy. Equation 6.44 then can be used to estimate the specific heat at constant volume. The total energy per unit volume becomes

$$E_T = nE_k = 3/2 \ nkT \tag{6.45}$$

where

n = number of atoms per unit volume
E_T = total kinetic energy per unit volume

The mass per unit volume is (from the definition of mass density)

$$\rho_m = n m_m$$

$$n = \frac{\rho_m}{m_m} \tag{6.46}$$

where

m_m = mass of a single atom

Substituting the value of n given by Equation 6.46 into Equation 6.45 gives

$$E_T = \left(\frac{3}{2}\right) \left(\frac{\rho_m}{m_m}\right) (kT)$$

or

$$\frac{E_T}{\rho_m} = \frac{3}{2} \frac{kT}{m_m} \tag{6.47}$$

The value of E_T/ρ_m is the energy per unit volume divided by the mass per unit volume. This ratio, therefore, is the energy per unit mass, E_m. Thus

$$E_m = 3/2 \frac{kT}{m_m} \tag{6.48}$$

If the temperature changes at constant volume

$$\Delta E_m = 3/2 \frac{k \Delta T}{m_m} \tag{6.49}$$

where

ΔE_m = change in energy per unit mass
ΔT = corresponding change in temperature

The specific heat is defined as the change in energy per unit mass divided by the corresponding change in temperature. Dividing both sides of Equation 6.49 by ΔT yields

$$c_V = \frac{\Delta E_m}{\Delta T} = 3/2 \, \frac{k}{m_m} \tag{6.50}$$

where

c_V = specific heat at constant volume

For atoms composed of two atoms (diatomic molecules) and molecules composed of many atoms (polyatomic molecules), part of the molecular energy is kinetic energy of rotation. For these molecules, Equation 6.50 does not give the correct value for the specific heat. However, near room temperature

$$c_V \approx 5/2 \, \frac{kT}{m_m} \qquad \text{(diatomic and linear molecules)} \tag{6.51}$$

$$c_V \approx 3 \, \frac{kT}{m_m} \qquad \text{(nonlinear polyatomic molecules)} \tag{6.52}$$

Flow Equations

The conservation of mass, the conservation of energy, and the conservation of momentum provide the basic laws for fluid flow. In addition, an equation of state, such as the ideal gas law, provides additional information when the fluid is compressible. To simplify the development of the flow equations, it will be assumed that the flow is independent of time. Flow that is independent of time is called steady flow.

Conservation of Mass To obtain an equation for the conservation of mass in steady flow, consider the two cross-sectional areas shown in Figure 6.17.

Figure 6.17 shows a pipe. The cross-sectional area of the pipe at section a is different from the cross-sectional area at section b. However, for steady flow, the mass that flows through section a in a small time Δt must equal the mass that flows through section b in the same time if there are no sources or sinks between the two sections. From Figure 6.17, the mass Δm that flows through a in time Δt is equal to

$$\Delta m = \rho_{ma} \, (\Delta l_a) A_a = \rho_{ma} \Delta V_a \tag{6.53}$$

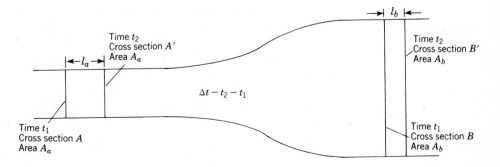

FIGURE 6.17 Pipe with tw different cross-sectional areas.

where

ρ_{ma} = density in section a

ΔV_a = element of volume that flows through section a in time Δt

In a similar way, an element of mass equal to Δm must flow through section b in Δt:

$$\Delta m = \rho_{mb}(\Delta l_b)A_b = \rho_{mb}\Delta V_b \tag{6.54}$$

where

ρ_{mb} = density in section b

ΔV_b = element of volume that flows through section b in time Δt

Setting Equation 6.53 equal to Equation 6.54 yields

$$\Delta m = \rho_{ma}\Delta V_a = \rho_{mb}\Delta V_b \tag{6.55}$$

Equation 6.55 is an expression for the conservation of mass.

Conservation of Energy   During fluid flow, total energy must be conserved. The work done on the fluid by a pump plus the heat transferred to or from the liquid must equal the change in the fluid's energy. The energy change of the fluid depends on the pressure change, the kinetic energy change, the potential energy change, and the internal energy change between two cross sections of the flow. This may be explained mathematically as

$$\Delta E = \Delta W + \Delta Q \tag{6.56}$$

where

ΔW = work done on the fluid by a pump

ΔQ = heat transferred

ΔE = change in the energy of the fluid between the sections where work is done and heat is transferred

If heat is added to the fluid, it will be given a positive sign; if heat is transferred from the fluid to the surroundings, it will be given a negative sign.

To develop the expressions for the terms in the energy equation, consider Figure 6.18. Assume the pipe has changing cross sections that do not all lie at the same level. According to Equation 6.55 (the conservation of mass), equal masses will flow through sections a and b in a small time Δt.

Consider the work done by the fluid pressure at sections a and b. According to Equation 2.14 the pressure work is equal to

$$\Delta W_P = P_b\Delta V_b - P_a\Delta V_a \tag{6.57}$$

The negative sign in this equation occurs because the pressure force is in the opposite direction of the assumed motion. In Equation 6.57, P_a is the pressure at section a and P_b is the pressure at section b.

The change in the linear kinetic energy in time Δt between the two sections will be equal to the kinetic energy of Δm at section b minus the kinetic energy of Δm at section a. This may be expressed mathematically as (Equation 6.6)

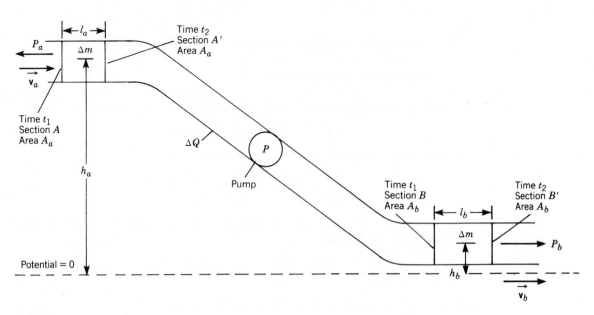

FIGURE 6.18 Generalized steady flow in a pipe.

$$\Delta E_{\mathrm{k}} = \frac{\alpha_b \, (\Delta m)}{2} \, v_b^2 - \frac{\alpha_a \, (\Delta m)}{2} \, v_a^2 \tag{6.58}$$

where

v_b = average velocity across section b

v_a = average velocity across section a

Because the velocities are not constant across their corresponding cross sections, correction factors, α_b and α_a, are needed to obtain the correct kinetic energy. If the flow is turbulent, the velocity profiles are nearly constant; α_b and α_a may be set equal to one for a good approximation. However, for laminar flow in a circular pipe, the value of α_b and α_a is 2.

Because the height changes, there is also a change in the potential energy of the mass Δm. This may be expressed as

$$\Delta E_{\mathrm{p}} = g(\Delta m)h_b - g(\Delta m)h_a \tag{6.59}$$

where

h_a = height of the element Δm at section a

h_b = height of the element Δm at section b

The distances h_a and h_b are measured from the reference plane to the center of mass of the Δm's.

Because of friction or heat transferred to or from the surroundings, the temperature at section a may be different from the temperature at section b.

The corresponding change in the internal energy of the fluid is

$$\Delta U = c_{\mathrm{V}}(\Delta m)(T_b - T_a) \tag{6.60}$$

where

c_V = specific heat at constant volume
T_b = temperature of Δm at section b
T_a = temperature of Δm at section a

The total change in the fluid's energy ΔE between sections a and b becomes

$$\Delta E = \Delta W_p + \Delta E_k + \Delta E_p + \Delta U \tag{6.61}$$

If Equation 6.61 is divided by Δm, one obtains

$$\frac{\Delta E}{\Delta m} = \frac{\Delta W_p}{\Delta m} + \frac{\Delta E_k}{\Delta m} + \frac{\Delta E_p}{\Delta m} + \frac{\Delta U}{\Delta m} \tag{6.62}$$

If Equations 6.57 through 6.60 are substituted for the quantities on the right in Equation 6.62, one obtains

$$\frac{\Delta E}{\Delta m} = P_b \frac{\Delta V_b}{\Delta m} - P_a \frac{\Delta V_a}{\Delta m} + \frac{v_b^2}{2} - \frac{v_a^2}{2} + g\,(h_b - h_a) + c_V\,(T_b - T_a) \tag{6.63}$$

where α_a and α_b are assumed to be 1. From Equation 6.55

$$\Delta m = \rho_{ma}\Delta V_a = \rho_{mb}\Delta V_b$$

Using this relationship to substitute for Δm in Equation 6.63 yields

$$\frac{\Delta E}{\Delta m} = \frac{P_b}{\rho_{mb}} - \frac{P_a}{\rho_{ma}} + \frac{v_b^2}{2} - \frac{v_a^2}{2} + g\,(h_b - h_a) + c_V\,(T_b - T_a) \tag{6.64}$$

Total Energy Change per Unit Mass	=	Pressure Work per Unit Mass	+	Kinetic Energy Change per Unit Mass	+	Potential Energy Change per Unit Mass	+	Internal Energy Change per Unit Mass

Each term in Equation 6.64 expresses an energy change per unit mass. By dividing Equation 6.56 by Δm, one also obtains an equivalent expression for the total energy change per unit mass:

$$\frac{\Delta E}{\Delta m} = \frac{\Delta W}{\Delta m} + \frac{\Delta Q}{\Delta m} \tag{6.65}$$

Total Energy Change per Unit Mass	=	Pump Work per Unit Mass	+	Heat Transferred per Unit Mass

By setting Equation 6.65 equal to Equation 6.64, the general energy equation for fluid flow in a pipe is obtained.

Gases   Because gas densities are very small, terms associated with changes in potential energy are usually neglected. Two interesting cases occur for flowing gases. They are isothermal flow (constant temperature) and adiabatic flow (no heat transfer). To develop the equations for these two cases, it is desirable to use the ideal gas law. If Equation 6.25 is multiplied by T and divided by ρ_m, one obtains

$$\frac{P}{\rho_m} = \frac{k}{m_m} T$$

Substituting this expression in Equation 6.64 yields

$$\frac{\Delta E}{\Delta m} = \frac{k}{m_\mathrm{m}}(T_b - T_a) + \frac{v_b^2 - v_a^2}{2} + c_\mathrm{V}(T_b - T_a) \qquad (6.66)$$

where the potential energy term has been dropped.

If the temperature is constant, Equation 6.66 reduces to

$$\frac{\Delta E}{\Delta m} = \frac{\Delta W}{\Delta m} + \frac{\Delta Q}{\Delta m} = \frac{v_b^2 - v_a^2}{2} \qquad \text{(isothermal gas)} \qquad (6.67)$$

For adiabatic flow, $\Delta Q/\Delta m = 0$. For this case, Equation 6.66 becomes

$$\frac{\Delta E}{\Delta m} = \frac{\Delta W}{\Delta m} = \frac{k}{m_\mathrm{m}} + c_\mathrm{V}(T_b - T_a) + \frac{v_b^2 - v_a^2}{2} \qquad \text{(adiabatic flow)} \qquad (6.68)$$

The equation for adiabatic flow also can be expressed in terms of the pressure by using the adiabatic gas law. When this is done one obtains the equation

$$\frac{\Delta W}{\Delta m} = \left[1 + \frac{c_\mathrm{V} m_\mathrm{m}}{k}\right]\left[\frac{1}{\rho_{am}}\right]\left[P_b\left(\frac{P_a}{P_b}\right)^{1/\gamma} - P_a\right] + \frac{v_b^2 - v_a^2}{2} \qquad (6.69)$$

An application of Equation 6.69 is given in Example 6.15.

EXAMPLE 6.15   Measurement of gas velocity
. . . .

Given: Two pressure measurements are made as shown in the sketch. The velocity at point b is zero. The measured pressure difference between points a and b is $P_b - P_a = 0.10$ atm. The measured total pressure at point a is 1.0 atm. The gas is nitrogen. For nitrogen, $\gamma = 1.4$ and ρ_{am} is 1.3 kg/m³.

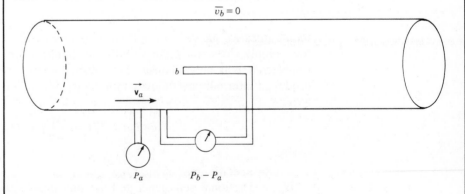

Find: The velocity at a. Assume frictionless adiabatic flow with $\alpha_a = \alpha_b = 1$.

Solution: For frictionless adiabatic flow, Equation 6.69 applies. Nitrogen is a diatomic gas. Thus, $c_\mathrm{V} = 5/2\ k/m_\mathrm{m}$. 1 atm is 1.01×10^5 N/m².

There is no pump between a and b, so $\Delta W/\Delta m = 0$. $k = 1.38 \times 10^{-23}$ J/K°.

Equation 6.68 becomes

$$\frac{v_b^2 - v_a^2}{2} + \left[1 + \frac{c_V m_m}{k} \right] \left[\frac{1}{\rho_{am}} \right] \left[P_b \left(\frac{P_a}{P_b} \right)^{1/\gamma} - P_a \right] = 0$$

Find P_b.

$$P_b - P_a = 0.10 \text{ atm}$$

$$P_b - 1 \text{ atm} = 0.10 \text{ atm}$$

$$P_b = 1.10 \text{ atm}$$

$$P_b = (1.10 \text{ atm})(1.01 \times 10^5 \text{ N/m}^2 \cdot \text{atm})$$

$$P_b = 1.11 \times 10^5 \text{ N/m}^2$$

Substitute in Equation 6.68

$$0 + \frac{v_a^2}{2} = \left[\left(1 + \frac{5}{2} \right) \left(\frac{1}{1.3 \text{ kg/m}^3} \right) \right]$$

$$\left[(1.11 \times 10^5 \text{ N/m}^2) \left(\frac{1.01}{1.11} \right)^{1/1.4} - 1.01 \times 10^5 \text{ N/m}^2 \right]$$

$$\frac{v_a^2}{2} = \left(\frac{7}{2} \right) \frac{1}{1.3 \text{ kg/m}^3} \right) (2.76 \times 10^3 \text{ N/m}^2)$$

$$\frac{v_a^2}{2} = 7.43 \times 10^3 \text{ N·m/kg} = 7.43 \times 10^3 \text{ m}^2/\text{sec}^2$$

$$v_a = \sqrt{1.48 \times 10^4 \text{ m}^2/\text{sec}^2}$$

$$v_a = 122 \text{ m/sec}$$

Incompressible Fluids   If the fluid is incompressible, then $\rho_{am} = \rho_{bm}$. Because the volume does not change with the addition or removal of heat, the heat transferred will only change the internal energy. Part of the heat energy is produced internally by fluid friction. Thus, for an incompressible fluid, one may write

$$c \, (T_b - T_a) = \frac{\Delta Q}{\Delta m} + H_{ab} \tag{6.70}$$

where

c = specific heat of the fluid
H_{ab} = frictional heat generated per unit mass between section a and b

If Equation 6.70 is substituted into Equation 6.63 and set equal to Equation 6.65, one obtains

$$\frac{v_b^2}{2} - \frac{v_a^2}{2} + g(h_b - h_a) + \frac{1}{\rho_m} (P_b - P_a) + H_{ab} = \frac{\Delta W}{\Delta m} \tag{6.71}$$

where ρ_m is the constant density of the incompressible fluid. Equation 6.71 is a general equation for incompressible flow.

For many cases the frictional heating H_{ab} can be neglected. In addition, there is often no pump between sections a and b. For these simplified conditions, Equation 6.71 becomes

$$\frac{v_b^2}{2} - \frac{v_a^2}{2} + g(h_b - h_a) + \frac{1}{\rho_m}(P_b - P_a) = 0$$

or

$$\frac{v_b^2}{2} + gh_b + \frac{P_b}{\rho_m} = \frac{v_a^2}{2} + gh_a + \frac{P_a}{\rho_m} \tag{6.72}$$

In addition, for no friction, Equation 6.70 becomes

$$c(T_b - T_a) = \frac{\Delta Q}{\Delta m} \tag{6.73}$$

Equations 6.72 and 6.73 represent the simplest expressions for the conservation of energy in flow and have a wide application. The wide application occurs because Equation 6.72 often can be applied to gaseous flow even though gases are compressible.

Equation 6.72 is Bernoulli's equation. David Bernoulli (1700–1782) applied the conservation of energy principle to incompressible fluids without friction to obtain the equation, which states that the kinetic energy per unit mass plus the potential energy per unit mass plus the pressure work per unit mass is equal to a constant at any cross section if no pump lies between the cross sections. If the equation is a good approximation for gases, the potential energy term can be dropped because gases have very small densities.

Example 6.16 gives an application of Bernoulli's equation.

EXAMPLE 6.16   Measuring pressure changes in a pipe with varying diameters
. . . .

Given: Water flows through the horizontal pipe shown, with a velocity at a of 4 m/sec and at b of 10 m/sec. The pressure gage at point a reads 250×10^3 N/m². Assume that $H_{ab} = 0$ and $\alpha_a = \alpha_b = 1$.

Find: The pressure gage reading at point b.

Solution: Because the center of the pipe stays at the same elevation, potential energy terms in Equation 6.72 are equal, and Equation 6.72 reduces to

$$\frac{P_a}{\rho_m} + \frac{v_a^2}{2} = \frac{P_b}{\rho_m} + \frac{v_b^2}{2}$$

or

$$P_a + \frac{\rho_m v_a^2}{2} = P_b + \frac{\rho_m v_b^2}{2}$$

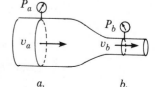

Substitution of the proper values into the equation yields (recall that ρ_m for water equals 1 g/cm³ or 1000 kg/m³)

$$250 \times 10^3 \text{ N/m}^2 + 1/2 \,(1000 \text{ kg/m}^3)(4 \text{ m/sec})^2$$

$$= P_b + 1/2 \,(1000 \text{ kg/m}^3)(10 \text{ m/sec})^2$$

$$[250 \times 10^3 + 8 \times 10^3] \text{ N/m}^2 = P_b + 50 \times 10^3 \text{ N/m}^2$$

$$[258 \times 10^3 - 50 \times 10^3] \text{ N/m}^2 = P_b$$

$$P_b = 208 \times 10^3 \text{ N/m}^2$$

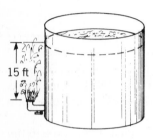

15 ft

FIGURE 6.19   Tank filled with fluid.

Consider a tank filled with fluid (Figure 6.19). When the valve at the bottom of the tank is opened, fluid escapes with velocity v. Bernoulli's equation can be applied to the cross section at the upper surface of the reservoir and the point of outflow if it is assumed that $H_{ab} = 0$.

At the upper surface, the pressure is atmospheric pressure P_a and velocity is negligible, $v_a = 0$. The right side of Equation 6.72 becomes

$$\frac{P_a}{\rho_m} + gh_a$$

At the exit pipe, the pressure again is atmospheric pressure. The height at this point is $h = 0$, and the velocity is that of the escaping fluid. Thus, the left side of Equation 6.72 is

$$\frac{P_b}{\rho_m} + \frac{v_b^2}{2}$$

Bernoulli's principle requires that left- and right-hand quantities be equal:

$$(P_a/\rho_m) + gh_a = (P_b/\rho_m) + v_b^2/2 \qquad (6.74)$$

Because $P_a = P_b$ = atmospheric pressure,

$$gh_a = v_b^2/2$$

Solving for v_b gives

$$v_b^2 = 2\,gh$$

or

$$v_b = \sqrt{2\,gh} \qquad (6.75)$$

Bernoulli's equation has allowed an approximation for the exit velocity of a fluid from a tank system. This is illustrated in Example 6.17.

EXAMPLE 6.17   Calculating efflux velocity
. . . .

Given: A tank filled with 15 ft of water is equipped with a valve at the bottom, as shown in Figure 6.19. Assume the velocity of the efflux is turbulent.

Find: **a.** Velocity of the efflux.

b. If the spigot is aimed upward, how high will the water rise?

c. If the water actually rises less than the amount predicted, how can this be accounted for?

Solution: **a.** $v = \sqrt{2\,gh}$

$v = \sqrt{2\,(32\ \text{ft/sec}^2)(15\ \text{ft})}$

$v = \sqrt{960\ \text{ft}^2/\text{sec}^2}$

$v = 31\ \text{ft/sec}$

b. If the spigot is pointed upward, the kinetic energy of the water would be turned back into potential energy and the water would rise to the original 15-ft height.

c. If the water does not actually rise 15 ft, the discrepancy can be explained in the following way: Friction within the fluid and from the fluid moving against walls of the pipe and valve has consumed a portion of the energy ($H_{ab} \neq 0$).

Fluid Measurements

Venturi Meter   The change in pressure and kinetic energy that occurs at different cross sections in fluid flow can be used to measure the fluid velocity and the mass or volume flow rate. A device designed for these measurements is called a venturi meter. A diagram of a venturi meter is shown in Figure 6.20.

If the differential pressure is measured between sections *a* and *b* along with the corresponding areas, the flow velocity can be determined from the flow equations for the conservation of mass and energy. Because $h_a = h_b$ and no pump exists

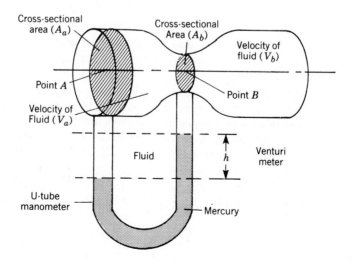

FIGURE 6.20   Venturi meter.

between sections a and b, Equation 6.71 for incompressible fluids becomes

$$\frac{v_b^2}{2} + \frac{P_b}{\rho_m} + H_{ab} = \frac{v_a^2}{2} + \frac{P_a}{\rho_m} \tag{6.76}$$

In addition, Equation 6.55 for the conservation of mass can be divided by Δt to yield ($\rho_{ma} = \rho_{mb} = \rho_m$)

$$\frac{\Delta m}{\Delta t} = \rho_m A_a \left(\frac{\Delta l_a}{\Delta t}\right) = \rho_m A_b \left(\frac{\Delta l_b}{\Delta t}\right) \tag{6.77}$$

or

$$A_a v_a = A_b v_b \tag{6.78}$$

Solving Equation 6.78 for v_b yields

$$v_b = v_a \left(\frac{A_a}{A_b}\right) = v_a \left(\frac{D_a^2}{D_b^2}\right) \tag{6.79}$$

where

D_a = diameter of section a
D_a = diameter of section b

If the expression for v_b given by Equation 6.79 is substituted into Equation 6.76, one obtains

$$\frac{v_a^2}{2}\left(\frac{D_a^4}{D_b^4}\right) - \frac{v_a^2}{2} = \frac{P_b - P_a}{\rho_m} - H_{ab} \tag{6.80}$$

Equation 6.80 can be solved for v_a. This gives

$$\frac{v_a^2}{2}\left[\frac{D_a^4}{D_b^4} - 1\right] = \frac{P_b - P_b}{\rho_m} - H_{ab}$$

$$v_a = \sqrt{\frac{\left[2\left(\dfrac{P_b - P_a}{\rho_m}\right) - 2H_{ab}\right] D_b^4}{(D_a^4 - D_b^4)\,\rho_m}} \tag{6.81}$$

It has been assumed that α_a and α_b are equal to 1, which is reasonable for turbulent flow. Because H_{ab} usually is not known, it is assumed to be equal to 0 (frictionless flow).

The pressure change between positions a and b in Figure 6.20 can be measured with a U-tube manometer. The simplified diagrams in Figure 6.21 illustrate how a U-tube manometer registers the pressure drop. In Figure 6.21a, representing a condition of no-flow in the venturi meter, the mercury in the U-tube is at the same level in each arm. In the lower drawing, under flow conditions, the pressure at position b decreases relative to position a. This drop is registered by a difference in the height of the mercury level in the two arms of the U-tube.

The pressure difference $\Delta P = P_a - P_b$ is obtained from

$$\Delta P = P_a - P_b = (\rho_{MG} - \rho_m)gh_G \tag{6.82}$$

FIGURE 6.21 Pressure difference in a venturi meter.

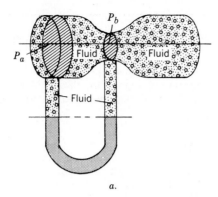

a.

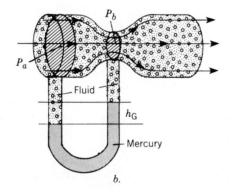

b.

where

ρ_{MG} = density of mercury (13.6 g/cm³)
ρ_m = density of liquid flowing through venturi flowmeter
h_G = height differential of mercury columns (cm)
g = acceleration due to gravity (980 cm/sec²)

The measured value of ΔP given by Equation 6.82 can be substituted into Equation 6.80 to give the velocity of the fluid at section *a*:

$$v_a = \sqrt{\left(\frac{2\,(\rho_{MG} - \rho_m)\,gh_G\,D_b^4}{(D_a^4 - D_b^4)\,\rho_m}\right)} \tag{6.83}$$

In Equation 6.83, it is assumed that $H_{ab} = 0$.

Because H_{ab} is assumed to be 0, the measured velocity obtained from Equation 6.83 will be too high. In addition, errors are introduced by assuming that α_a and α_b are equal to 1. Measurements of velocity profiles in a venturi meter show that α_a is greater than α_b. Therefore, it can be shown that the latter assumption causes the measured velocity to be too low. Thus, the errors tend to cancel each other. For water with a diameter ratio of 0.5, the error in the calculated velocity ranges from 5 to 1 percent as the Reynolds number changes from 10^4 to 10^7.

Example 6.18 illustrates use of Equation 6.83 in determining flow velocity in a pipe.

EXAMPLE 6.18   Flow measurement with a venturi flowmeter
. . . .

Given: Water flows through a pipe 6 cm in diameter into a venturi tube with a constriction of 4 cm at the throat. Mercury fills the manometer U-tube; as the steady flow proceeds, a differential height of 8 cm is established.

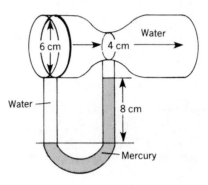

Find: Velocity of fluid in the pipe.

Solution: From Equation 6.83

$$v_a = \sqrt{\left(\frac{2\,(\rho_{MG} - \rho_m)(gh_G)(D_b^4)}{(D_a^4 - D_b^4)\rho_m} \right)}$$

$$v_a = \sqrt{\left(\frac{(2)(13.6 - 1)(980)(8)(4)^4}{[(6)^4 - (4)^4]\,(1)} \right)}$$

$$v_a = 220 \text{ cm/sec}$$

The usual purpose of the average velocity measurement is to determine the mass or volume flow rate. The mass flow rate according to Equation 6.77 is

$$Q_m = \frac{\Delta m}{\Delta t} = \rho A_a \frac{\Delta l_a}{\Delta t} = \rho_m A_a v_a$$

$$Q_m = \rho_m A_a v_a \tag{6.84}$$

where

Q_m = mass flow rate
ρ_m = mass density

If Equation 6.55 is divided by Δt, one obtains

$$Q_m = \frac{\Delta m}{\Delta t} = \rho_m \frac{\Delta V}{\Delta t} \tag{6.85}$$

as $\rho_{am} = \rho_{bm}$ for incompressible flow. By comparing Equation 6.85 with Equation 6.84, the volume flow rate becomes

$$Q_V = \frac{\Delta V}{\Delta t} = A_a v_a \tag{6.86}$$

The mass and volume flow rates for Example 6.18 are calculated in Example 6.19.

EXAMPLE 6.19   Volume and mass flow rates

. . . .

Given: Water flows through a 6-inch pipe with a velocity of 220 cm/sec.

Find: **a.** The volume flow rate.

            b. The mass flow rate.

Solution: **a.** The volume flow rate is

$$Q_V = \frac{\Delta V}{\Delta t} = A_a v_a$$

$$A = \frac{\pi}{4} D^2 = \frac{\pi}{4} (6 \text{ inch} \times 2.54 \text{ cm/inch})^2$$

$$A = 182 \text{ cm}^2$$

$$Q_V = (220 \text{ cm/sec})(182 \text{ cm}^2) = 4.00 \times 10^4 \text{ cm}^3/\text{sec}$$

            b. The mass flow rate is

$$Q_m = \frac{\Delta m}{\Delta t} = (A_a v_a) \, \rho_m$$

$$Q_m = (4.00 \times 10^4 \text{ cm}^3/\text{sec})(1 \text{ g/cm}^3)$$

$$Q_m = 4.00 \times 10^4 \text{ g/sec}$$

Pitot Tube   The pitot tube measures the velocity of a fluid by directly measuring the pressure required to stop the flow. This pressure is called the stagnation pressure. In addition, the static pressure of the flowing fluid is determined by a static tube. A pitot tube of popular design is illustrated in Figure 6.22. It is constructed of two concentric tubes. The inner tube is open to the moving fluid and is referred to as the "pitot tube." The outer tube, referred to as the "static tube," is shielded from the oncoming flow, although in its side it has small holes perpendicular to the flow of the fluid stream.

The small holes in the static tube provide a means by which static stream pressure is measured. The stagnation pressure occurs at the mouth of the instrument. The velocity of the fluid at this impact region is zero, and the region is sometimes called the "stagnation point." The speed of the fluid stream is proportional to the square root of the difference in the stagnation pressure and static pressure. Pitot tubes use a sensitive differential-pressure measuring device to measure this pressure difference, such as the manometer shown in Figure 6.22.

Equation 6.69, as illustrated in Example 6.15, may be used to determine the flow velocity of gases if the gas properties are known. However, if the measured

FIGURE 6.22　Pitot tube.

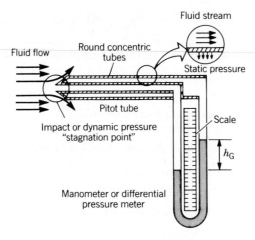

pressure difference is small, Bernoulli's equation for incompressible fluids gives essentially the same result for the velocity. For the pitot tube, $h_a = h_b$ and $v_b = 0$. Thus, Equation 6.72 (Bernoulli's equation) reduces to

$$\frac{P_a}{\rho_m} + \frac{v_a^2}{2} = \frac{P_b}{\rho_m} + 0$$

or

$$v_a^2 = \frac{2\,(P_b - P_a)}{\rho_m}$$

or

$$v_a = \sqrt{\left(\frac{2\,(P_b - P_a)}{\rho_m}\right)} \tag{6.87}$$

The difference in pressure $\Delta P = P_b - P_a$ is determined from the measurement of h_G as illustrated in Figure 6.22. As in the case of the venturi meter

$$\Delta P = (\rho_{mg} - \rho_m)\, h_G g \tag{6.88}$$

where

ρ_{mg} = density of the manometer fluid
h_G = height illustrated in Figure 6.22

Substituting this expression for ΔP in Equation 6.82 gives

$$v_a = \sqrt{\left(\frac{2(\rho_{mg} - \rho_m)\, h_G g}{\rho_m}\right)} \tag{6.89}$$

If the flow velocity is not too large, Equation 6.89 gives a very good approximation to the real velocity. This is illustrated in Example 6.20.

EXAMPLE 6.20 Velocity of fluids
. . . .

Given: The data from Example 6.15.

Find: The fluid velocity using Equation 6.87 for incompressible fluids.

Solution: Equation 6.87 is

$$v_a = \sqrt{\left(\frac{2\Delta P}{\rho_m}\right)}$$

$$v_a = \sqrt{\left(\frac{(2)(0.101 \times 10^5 \text{ N/m}^2)}{1.3 \text{ kg/m}^3}\right)}$$

$$v_a = 125 \text{ m/sec}$$

The value obtained for a compressible gas was (Example 6.15)

$$v_a = 122 \text{ m/sec}$$

As the velocity increases, the deviation will become greater. Other errors associated with pitot tube measurements are usually less than three percent.

SECTION **6.2 EXERCISES**

1. Starting with Bernoulli's equation, derive the expression for the velocity of a frictionless fluid flowing from an opening in a large tank. Use this equation to determine the velocity of water exiting through a hole 50 ft below the water level of a tank.

2. Discuss conservation of energy as it applies to the fluid system.

3. How high is a waterfall if 3000 kg of water acquire a kinetic energy of 400,000 J by falling from the top of the waterfall to the ground?

4. How high above a valve at the bottom of a tank is the water level in the tank when 10 m³ of water flow out of the tank per minute if the area of the valve opening is 0.1 m²?

5. The initial conditions of a given mass of gas are temperature = 300°K, pressure = 1 atm, and volume = 1 m³. If the pressure is changed to 2 atm and the temperature is changed to 600°K, what is the new volume of gas?

6. Describe the following gas laws using words and equations. State which variable is held constant in each law.

   Boyle's law

   Charles' law

   Gay-Lussac's law

7. If the temperature is held constant, how does the mean-free path of a molecule change with pressure?

8. The height difference registered in the two arms of a mercury manometer with water above the mercury is 17.5 cm when connected to a horizontal venturi meter. The diameters of the pipe and throat are 15 cm and 7 cm, respectively. Neglecting frictional losses, calculate the volume flow rate of discharge through the meter.

9. A horizontal pipe has a cross-sectional area of 4.0 inch^2. In the center of the pipe, there is a constriction having a cross-sectional area of 1.0 inch^2. Gasoline (weight density 42 lb/ft^3) flows with a speed of 6.0 ft/sec^2 in the larger section. $P = 10$ lb/inch^2. Calculate (a) gasoline speed in the constriction and (b) pressure in the constriction.

10. What is the density of a fluid flowing through a venturi meter with throat and pipe areas of 20 cm^2 and 100 cm^2, respectively, if the height difference in a U-tube filled with mercury is 16 cm when the velocity of fluid flow through the pipe is 2 m/sec?

11. If the height differential in a manometer connected to a venturi meter is 12 cm when water is flowing through the venturi meter, what is the height differential when a fluid with a density of 4 g/cm^3 flows through the meter at the same rate as the water?

12. A pitot tube used to measure air velocity in a wind tunnel is connected to a manometer that indicates a differential head of 0.20 inch of water. Compute air velocity in ft/sec if the density of air is 0.075 lb/ft^3.

13. The velocity of dry air is measured by a pitot tube to be 20 m/sec when the differential height in the manometer is 2 cm. What is the velocity of the air when the height differential is 7 cm?

14. What is the density of the liquid used in the pitot tube in Problem 13?

15. *Compare the velocity obtained by using Equation 6.87 with the value obtained by using the method outlined in Example 6.15, if $\Delta P = 0.3$ atm, $\gamma = 1.4$, and $\rho_{am} = 1.3$ kg/m^3. The static pressure is 1.0 atm.

16. How is the temperature of an ideal gas related to the molecular motion of the gas?

---

*Challenge question.

6.3

POTENTIAL AND KINETIC ENERGY IN ELECTROMAGNETIC SYSTEMS

OBJECTIVES

Upon completion of this section, the student should be able to

• Given any two of the quantities in the following groups, determine the third:
   Capacitance, voltage, and potential energy
   Inductance, current, and potential energy

• Calculate the equivalent capacitance of a series or parallel network of capacitors.

• Calculate the equivalent inductance for a series or parallel network of independent inductors when an ac source is applied to the network.

DISCUSSION

As the focus changes from fluid systems to electromagnetic systems, a number of similarities will be noticed between the electromagnetic system and the previous systems. Try to find the reason for these similarities!

Electrical Systems

In electrical systems, the energy equations are similar to those for mechanical systems. The primary components of electrical systems are capacitors, resistors, inductors, and sources of electrical potential. In all electrical systems, the quantity that flows (displacementlike quantity) is electrical charge. In conductors, electrons flow; in gaseous discharges, electrons, positive ions, and negative ions flow. Electrons and negative ions flow in one direction, and positive ions flow in the opposite direction. However, all moving charged particles contribute to the total current flow.

A general energy equation for any system must include all possible forms of energy transformation. For example, electrical energy can be transformed into mechanical energy (motor) or mechanical energy can be converted into electrical energy (generator). In this chapter, only the electrical energy associated with the primary circuit elements will be considered.

Capacitors   In Chapter 2, it is shown that the energy stored in a capacitor is

$$E_p = 1/2 \ CV^2 \tag{6.90}$$

where

   E_p = electrical energy stored in joules
   C = capacitance in farads
   V = voltatge in volts

From the definition of capacitance in Chapter 1, the relationship among voltage, charge, and capacitance was expressed as

$$V = Q/C \tag{6.91}$$

where

   Q = charge on the capacitor in coulombs (displacementlike quantity)

By substituting the expression for V given by Equation 6.91 into Equation 6.90, the energy becomes

$$E_p = 1/2 \ Q^2/C \tag{6.92}$$

This form of stored energy in a capacitor is analogous to the mechanical energy stored in a spring given by the equation

$$E_p = 1/2 \ kx^2 \tag{6.93}$$

Q is the displacementlike quantity. $1/C$ corresponds to the spring constant k. The application of Equations 6.92 and 6.93 is illustrated in Example 6.21.

EXAMPLE 6.21   Potential energy
. . . .

Given: A spring is compressed 0.01 m. The spring constant is 10^3 N/m.

Find: **a.**   The magnitude of the force to hold the spring.
            b.   The required charge on a 10-μF capacitor so that the stored energy is the same as the spring's energy.

                    c. The capacitor's voltage.

Solution: **a.** $F = kx$

$$F = (10^3 \text{ N/m})(0.01 \text{ m})$$

$$F = 10 \text{ N}$$

$$F = (10 \text{ N}) \left(\frac{1 \text{ lb}}{4.45 \text{ N}} \right) = 2.2 \text{ lb}$$

b. The spring's energy is

$$E_p = 1/2 \; kx^2$$

$$E_p = 1/2 \; (10^3 \text{ N/m})(0.01 \text{ m})^2$$

$$E_p = 0.05 \text{ J}$$

The charge on capacitor is

$$E_p = 1/2 \; Q^2/C$$

$$Q = \sqrt{2 \; E_p C}$$

$$Q = \sqrt{(2)(0.05 \text{ J})(10^{-5} \text{ F})}$$

$$Q = 10^{-3} \text{ C}$$

c. The capacitor's voltage is

$$V = Q/C$$

$$V = \frac{(10^{-3} \text{ C})}{10^{-5} \text{ F}}$$

$$V = 100 \text{ V}$$

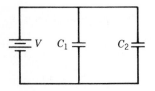

FIGURE 6.23 Capacitors in parallel.

Capacitors in Parallel Figure 6.23 is a schematic diagram of two capacitors connected in parallel across a voltage source. As can be seen from the diagram, the voltage across each capacitor is the same. The energy stored in each capacitor is given by Equation 6.90, and the total stored energy is the sum of the two energies. The charge on each capacitor is obtained by solving Equation 6.91 for Q:

$$Q = VC \qquad\qquad (6.94)$$

The charge on each capacitor can be determined from the common voltage and the individual capacitance. The total stored charge is the sum of the two charges. These ideas are illustrated in Example 6.22.

EXAMPLE 6.22   Charge and energy stored by capacitors in parallel
. . . .

Given: A 4-μF and a 5-μF capacitor are connected in parallel across a 20-V dc source.

Find: **a.** The charge stored by each capacitor.

         b. The energy stored by each capacitor.

Solution: **a.** $Q = CV$ (Equation 6.94)

4-μF capacitor:

$Q_1 = (4 \times 10^{-6} \text{ F})(20 \text{ V})$

$Q_1 = (80 \times 10^{-6} \text{ F})(\text{V})$

$Q_1 = 80 \times 10^{-6} \text{ C}$

5-μF capacitor:

$Q_2 = (5 \times 10^{-6} \text{ F})(20 \text{ V})$

$Q_2 = (100 \times 10^{-6} \text{ F})(\text{V})$

$Q_2 = 100 \times 10^{-6} \text{ C}$

         b. $E_p = 1/2 \, CV^2$ (Equation 6.90)

4-μF capacitor:

$E_{p_1} = (1/2)(4 \times 10^{-6} \text{ F})(20 \text{ V})^2$

$E_{p_1} = (800 \times 10^{-6} \text{ F})(\text{V}^2)$

$E_{p_1} = 800 \times 10^{-6} \text{ J}$

5-μF capacitor:

$E_{p_2} = (1/2)(5 \times 10^{-6} \text{ F})(20 \text{ V})^2$

$E_{p_2} = (1000 \times 10^{-6} \text{ F})(\text{V}^2)$

$E_{p_2} = 1000 \times 10^{-6} \text{ J}$

A useful concept often used in analyzing combinations of capacitors in a circuit is equivalent capacitance. The equivalent capacitance of a group of capacitors in a circuit is defined as "the capacitance of the single capacitor that can replace the group without changing the characteristics of the circuit." A circuit in which two or more capacitors are replaced by their equivalent capacitance is called an "equivalent circuit." For the case in which the original group of capacitors is connected in parallel, the equivalent capacitance C_p is easily found.

Figure 6.24*a* depicts the circuit shown in Figure 6.23 and indicates the charge and energy associated with each capacitor. Because the characteristics of the equivalent circuit must be identical to those of the original circuit, the total potential energy E_p stored by C_p must be equal to the sum of the energies stored by C_1 and C_2. Similarly, the total charge Q_T on C_p must equal the sum of the charges on C_1 and C_2 (Figure 6.24*b*). Either of these criteria can be used to determine the equivalent capacitance. Using the latter

$$Q_T = Q_1 + Q_2$$

FIGURE 6.24 Equivalent capacitance.

$$a.$$

$$b.$$

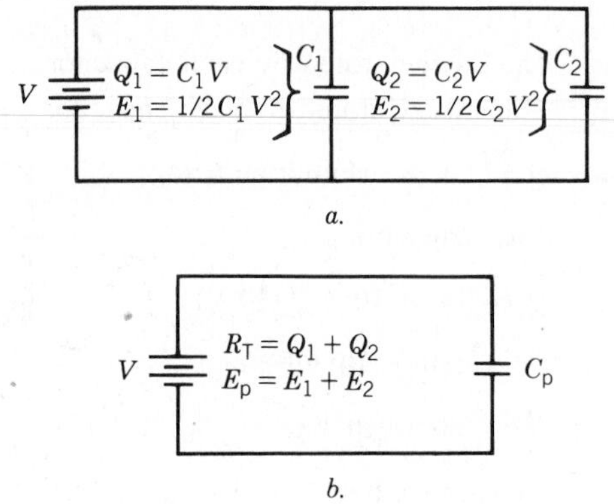

or

$$C_p V = C_1 V + C_2 V \quad \text{(substitution from Equation 6.94)}$$

or

$$C_p = C_1 + C_2 \quad \text{(division of both sides by } V)$$

As seen previously, the equivalent capacitance of C_1 and C_2 connected in parallel is exactly the sum of their individual capacitances. In general, for any number of capacitors connected in parallel, the equivalent capacitance is simply the sum of the individual capacitances, as indicated by

$$C_p = C_1 + C_2 + C_3 + \cdots \text{etc.} \tag{6.95}$$

In addition, the total charge and energy stored by the circuit can be calculated directly from Equations 6.94 and 6.90, respectively, by using the value of C_p for the capacitance. The equivalent capacitance for two capacitors in parallel is calculated in Example 6.23.

EXAMPLE 6.23 Equivalent capacitance of a parallel network
. . . .

Given:　　A 4-μF and a 5-μF capacitor are connected in parallel across a 20-V dc voltage source, as in Example 6.22.

Find:　　**a.** Their equivalent capacitance.
　　　　b. The total charge stored.
　　　　c. The total energy stored.

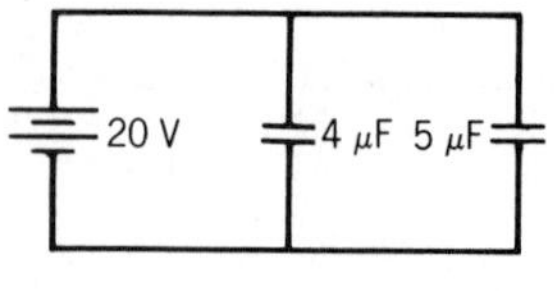

Solution:　**a.** $C_p = C_1 + C_2$

　　　　　　$C_p = 4 \ \mu F + 5 \ \mu F$

　　　　　　$C_p = 9 \ \mu F$

　　　　b. $Q_T = C_p V$

　　　　　　$Q_T = 9 \times 10^{-6} \ \text{F} \ (20 \ \text{V})$

$Q_T = 180 \times 10^{-6}$ C

$Q_T = 180 \ \mu$C

Note that this is the sum of Q_1 and Q_2 from Example 6.22.

c. $E_p = 1/2 \ C_p V^2$

$E_p = 1/2 \ (9 \times 10^{-6} \text{ F})(20 \text{ V})^2$

$E_p = 1800 \times 10^{-6}$ J

Note that this value is the sum of E_{p_1} and E_{p_2} from Example 6.22.

The circuit behaves exactly as if a single 9-μF capacitor were connected across the voltage.

Capacitors in Series   Figure 6.25a displays capacitors C_1 and C_2 connected in series across the potential difference V. In general, the voltages V_1 and V_2 across each capacitor will be different—unlike the case for capacitors in parallel; however, the charge that accumulates on the plates of each capacitor is the same.

Figure 6.25b shows the circuit as the capacitors begin to charge. The process can be visualized step-by-step as follows: A negative charge (an electron) leaves the negative terminal of the battery and travels to the nearest plate of C_2, imparting a slight, negative net charge to it (Figure 6.25b). Because the charge on

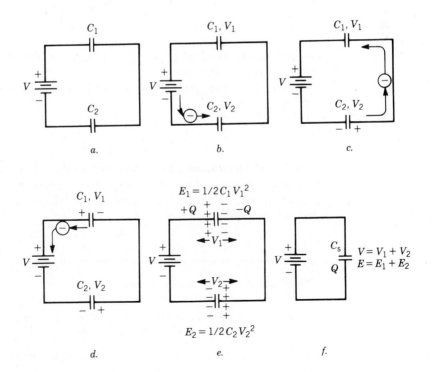

FIGURE 6.25 Behavior of capacitors in series.

one plate of a capacitor induces an equal and opposite charge on the opposite plate, an electron leaves the opposite plate of C_2 and travels to the nearest plate of C_1 (Figure 6.25c). Because the plates of each capacitor at first are electrically neutral, a positive charge of equal magnitude is left behind on C_2. Upon arriving at C_1, the electron repels another electron from the opposite plate; this electron then moves to the positive terminal of the battery (Figure 6.25d). The process continues until the capacitors are charged completely, each carrying equal charges Q (Figure 6.25e). Notice that although each capacitor carries a charge equal to Q, the battery has only supplied a total charge Q_T to the entire circuit. The net effect then is that of a single capacitor carrying a charge Q. Thus, the charge Q_T carried by the equivalent capacitor should be the same as the charge on each of the individual capacitors.

For the equivalent circuit to behave in the same manner as the series circuit, two more criteria must be met: (1) the energy stored by the equivalent circuit must be equal to the sum of the energies stored in the individual capacitors and (2) the total voltage across the equivalent circuit must be equal to the sum of the voltages across the individual capacitors. Either of these criteria (shown in the equivalent circuits of Figure 6.25f) can be used to determine the value of the series equivalent capacitance C_s. Equation 6.96 states the voltage condition mathematically:

$$V_T = V_1 + V_2 \tag{6.96}$$

Equation 6.91 gives $V = Q/C$. Substitution of the appropriate value of V and C into this result transforms Equation 6.96 into

$$\frac{Q}{C_s} = \frac{Q}{C_1} + \frac{Q}{C_2}$$

$$\frac{1}{C_s} = \frac{1}{C_1} + \frac{1}{C_2}$$

Thus, the reciprocal of the equivalent capacitance of the two capacitors connected in series is equal to the sum of the reciprocals of the individual capacitances. This relationship is true in general, no matter how many capacitors are connected in series; therefore

$$\frac{1}{C_s} = \frac{1}{C_1} + \frac{1}{C_2} + \frac{1}{C_3} + \cdots \text{etc.} \tag{6.97}$$

Once C_s is known, the charge and energy stored by the equivalent circuit can be calculated from Equations 6.94 and 6.90, respectively, using the value of C_s for the capacitance. Example 6.24 demonstrates the solution of a typical problem involving capacitors in series.

EXAMPLE 6.24   Series capacitor circuit
. . . .

Given: A 6-μF capacitor and a 4-μF capacitor are connected in series across a 12-V dc voltage source.

Find: a.   The equivalent capacitance of the circuit.

b. The total charge supplied by the battery.

c. The total energy stored in the circuit.

d. The voltage across each capacitor.

e. The energy stored in each capacitor.

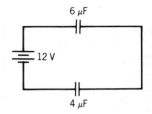

Solution:

a. $\dfrac{1}{C_s} = \dfrac{1}{C_1} + \dfrac{1}{C_2}$

$\dfrac{1}{C_s} = \dfrac{1}{6\ \mu F} + \dfrac{1}{4\ \mu F}$

$\dfrac{1}{C_s} = \dfrac{2}{12\ \mu F} + \dfrac{3}{12\ \mu F}$ (Expressed in terms of the common denominator.)

$\dfrac{1}{C_s} = \dfrac{5}{12\ \mu F}$ (The fractions are added.)

$C_s = \dfrac{12\ \mu F}{5}$ (The reciprocal of both sides is taken.)

$C_s = 2.4\ \mu F$

b. $Q_T = C_s V$

$Q_T = 2.4 \times 10^{-6}\ F\ (12\ V)$

$Q_T = 28.8 \times 10^{-6}\ C$

c. $E_p = 1/2\ C_s V^2$

$E_p = 1/2\ (2.4 \times 10^{-6}\ F)(12\ V)^2$

$E_p = 172.8 \times 10^{-6}\ J$

d. 6-μF capacitor:

$V_1 = \dfrac{Q}{C}$

$V_1 = \dfrac{28.8 \times 10^{-6}\ C}{6 \times 10^{-6}\ F}$

$V_1 = 4.8\ V$

4-μF capacitor:

$V_2 = \dfrac{28.8 \times 10^{-6}\ C}{4 \times 10^{-6}\ F}$

$V_2 = 7.2\ V$

Notice that the sum of the voltages across the individual capacitors is equal to the voltage across the equivalent capacitor (source voltage):

$4.8\ V + 7.2\ V = 12\ V$

e. $E_p = 1/2\ CV^2$

6-μF capacitor:

$$E_{p1} = 1/2\ (6 \times 10^{-6}\ \text{F})(4.8\ \text{V})^2$$

$$E_{p1} = 69.12 \times 10^{-6}\ \text{J}$$

4-μF capacitor:

$$E_{p2} = 1/2\ (4 \times 10^{-6}\ \text{F})(7.2\ \text{V})^2$$

$$E_{p2} = 103.68 \times 10^{-6}\ \text{J}$$

Notice that the sum of the energies stored by each capacitor is equal to the energy stored by the equivalent capacitor:

$$69.12 \times 10^{-6}\ \text{J} + 103.68 \times 10^{-6}\ \text{J} = 172.8 \times 10^{-6}\ \text{J}$$

Example 6.25 illustrates another application of capacitors in series.

EXAMPLE 6.25   Enhanced voltage rating of capacitors in series
. . . .

Given: A technician must construct a 12-μF equivalent capacitance network across a 48-V dc voltage source. An unlimited selection of capacitors with maximum voltage ratings of 16 V each is available.

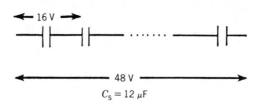

$$C_s = 12\ \mu\text{F}$$

Find: The minimum number and capacitance of identical capacitors from which the network can be constructed.

Solution: Each individual capacitor can support a maximum voltage of 16 V; therefore, a series combination is required. The minimum number of capacitors needed is given by

$$\text{number of capacitors} = \frac{\text{total voltage}}{16\ \text{V}} = \frac{48\ \text{V}}{16\ \text{V}} = 3$$

Find the capacitance of each capacitor:

$$\frac{1}{C_s} = \frac{1}{C_1} + \frac{1}{C_2} + \frac{1}{C_3} \qquad \text{(for three capacitors)}$$

$$\frac{1}{C_s} = \frac{1}{C} + \frac{1}{C} + \frac{1}{C} \qquad \text{(because } C_1 = C_2 = C_3 \text{ for identical capacitors)}$$

$$\frac{1}{12\ \mu\text{F}} = \frac{3}{C}$$

$$\frac{C}{12\ \mu F} = 3 \qquad \text{(both sides multiplied by } C)$$

$$C = 3\ (12\ \mu F) \qquad \text{(both sides multiplied by 12 } \mu F)$$

$$C = 36\ \mu F$$

Thus, three 36-μF capacitors rated at 16 V each connected in series are equivalent to a single 12-μF, 48-V capacitor.

Magnetic Systems

Inductance   The inductance of a coil is defined in Equation 3.75 as

$$L = N\frac{\Delta\phi}{\Delta I} \tag{6.98}$$

where

  L = inductance of the coil in henrys (H)
  N = number of turns
  ΔI = a very small change in current
  $\Delta\phi$ = corresponding change in magnetic flux in webers (Wb)

In Example 5.21, the relationship between the magnetic flux and the magnetomotive force (NI) is defined as

$$\phi = \frac{NI}{R} \tag{6.99}$$

where
  R = reluctance of the magnetic circuit

From Equation 6.99, one may write, if the reluctance R is constant,

$$\phi - \phi_2 = \frac{N}{R}(I_2 - I_1)$$

$$\Delta\phi = \frac{N}{R}\Delta I$$

$$\frac{\Delta\phi}{\Delta I} = \frac{N}{R} \tag{6.100}$$

Using the expression for $\Delta\phi/\Delta I$ from Equation 6.100 to substitute in Equation 6.98 yields

$$L = \frac{N^2}{R} \tag{6.101}$$

Energy Stored in an Inductor   From the results of Example 2.14 and Equation 3.75, it can be shown that the energy stored in the magnetic field of an inductor is

$$E_k = \frac{1}{2} L \left(\frac{\Delta Q^2}{\Delta t}\right)$$

or

$$E_k = 1/2 \ LI^2 \tag{6.102}$$

where

$I = \Delta Q/\Delta t$ = current through the inductor in amperes (A)

Equation 6.102 is analogous to the equation for kinetic energy:

$$E_k = \frac{1}{2} m \left(\frac{\Delta s}{\Delta t}\right)^2$$

or

$$E_k = 1/2 \ mv^2 \tag{6.103}$$

The current $\Delta Q/\Delta t$ is the ratelike quantity defined in Chapter 3. Therefore, the inductance L is analogous to the mass in a mechanical system. Equations 6.102 and 6.103 are compared in Example 6.26.

EXAMPLE 6.26   Energy in a magnetic field
. . . .

Given: An inductor has 100 turns and a current of 10 A. The reluctance is
              1.2×10^7 A-turn/Wb.

Given: **a.**   The inductance of the coil.
              b.   The energy stored in the field.
              c.   The velocity a 1-kg mass must have to have the same energy.

Solution: **a.**   The inductance is

$$L = \frac{N^2}{R}$$

$$L = \frac{(100 \text{ turns})^2}{(1.2 \times 10^7 \text{ A-turn/Wb})}$$

$$L = 8.3 \times 10^{-4} \text{ H}$$

              b.   The energy in the field is

$$E_k = 1/2 \ LI^2$$

$$E_k = (1/2)(8.3 \times 10^{-4} \text{ H})(10 \text{ A})^2$$

$$E_k = 0.0415 \text{ J}$$

              c.   The mechanical kinetic energy is

$$E_k = 1/2 \ mv^2$$

$$0.0415 \text{ J} = 1/2 \ (1 \text{ kg})(v^2)$$

$$v = \sqrt{\frac{(2)(0.0415 \text{ J})}{1 \text{ kg}}}$$

$$v = 0.288 \text{ m/sec}$$

Inductors in Series   When inductors are connected in series, the magnetic flux of each inductor may link only its own turns, or a part of the flux may link the turns of the other inductors. In the first case, the inductors act independently, and only the individual inductances need be known. In the second case, the interaction between the inductors also must be known. The interaction is expressed in terms of a quantity called the mutual inductance. Only the first case will be considered in this chapter.

Figure 6.26 shows two independent inductors connected in series. The total voltage across the inductors must equal the sum of the individual voltages:

$$V = V_1 + V_2 \tag{6.104}$$

Because the same current flows through both inductors, the voltages V_1 and V_2 (Equation 3.78) become

$$V_1 = -L_1 \frac{\Delta I}{\Delta t}$$

$$V_2 = -L_2 \frac{\Delta I}{\Delta t}$$

where $\Delta I/\Delta t$ is the common time rate of change of current through the series inductors. Substituting these expressions for V_1 and V_2 in Equation 6.104 yields

$$V = -L_1 \frac{\Delta I}{\Delta t} - L_2 \frac{\Delta I}{\Delta t}$$

or

$$V = -(L_1 + L_2) \frac{\Delta I}{\Delta t}$$

or

$$V = -L \frac{\Delta I}{\Delta t} \tag{6.105}$$

FIGURE 6.26   Inductors in series.

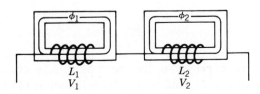

where

$$L = L_1 + L_2 \quad \text{(the total inductance)}$$

Thus, if the inductors are independent, the total inductance is the sum of the two inductances. For independent inductors in series, the general expression for the total inductance is

$$L = L_1 + L_2 + L_3 + \cdots + L_n \tag{6.106}$$

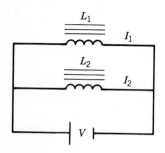

FIGURE 6.27 Inductors in parallel.

Inductors in Parallel   Two independent inductors connected in parallel are shown in Figure 6.27. For this case, the voltage across each inductor is the same. According to Equation 3.78, one may write

$$V = -L_1 \frac{\Delta I_1}{\Delta t} \tag{6.107}$$

$$V = -L_2 \frac{\Delta I_2}{\Delta t} \tag{6.108}$$

$$V = -L \frac{\Delta I}{\Delta t} \tag{6.109}$$

where

I = total current ($I = I_1 + I_2$)
L = total inductance

Solving Equations 6.107, 6.108, and 6.109 for the time rate of change of currents yields

$$\frac{\Delta I_1}{\Delta t} = \frac{V}{L_1} \tag{6.110}$$

$$\frac{\Delta I_2}{\Delta t} = \frac{V}{L_2} \tag{6.111}$$

$$\frac{\Delta I}{\Delta t} = \frac{V}{L} \tag{6.112}$$

Because charge is conserved

$$I = I_1 + I_2$$

or

$$\Delta I = \Delta I_1 + \Delta I_2$$

or

$$\frac{\Delta I}{\Delta t} = \frac{\Delta I_1}{\Delta t} + \frac{\Delta I_2}{\Delta t} \tag{6.113}$$

Substituting the expressions for the time rate of change of the currents from Equations 6.110, 6.111, and 6.112 into Equation 6.113 yields

$$\frac{V}{L} = \frac{V}{L_1} + \frac{V}{L_2}$$

or

$$\frac{1}{L} = \frac{1}{L_1} + \frac{1}{L_2} \tag{6.114}$$

Equation 6.114 states that the reciprocal of the total inductance is equal to the sum of the reciprocals of the individual inductances if they act independently. It should be noted that independent inductors follow the same rule as resistors in series and parallel. If many independent inductors are connected in parallel, then

$$\frac{1}{L} = \frac{1}{L_1} + \frac{1}{L_2} + \frac{1}{L_3} + \cdots + \frac{1}{L_n} \tag{6.115}$$

The application of Equations 6.106 and 6.115 is illustrated in Example 6.27.

EXAMPLE 6.27   Inductors in a circuit
. . . .

Given: The circuit shown in the diagram. The inductors are independent.

Find: The total inductance.

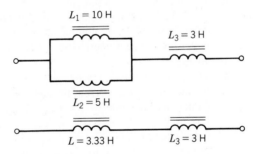

Solution: The parallel inductors L_1 and L_2 can be replaced by an equivalent inductor.

$$\frac{1}{L} = \frac{1}{L_1} + \frac{1}{L_2}$$

$$\frac{1}{L} = \frac{1}{10 \text{ H}} + \frac{1}{5 \text{ H}} = \frac{1}{10 \text{ H}} + \frac{2}{10 \text{ H}} = \frac{3}{10 \text{ H}}$$

$$L = \frac{10}{3} \text{ H}$$

$$L = 3.33 \text{ H}$$

The circuit can be replaced by two inductors in series.
For inductors in series, add the inductance.

$$L_T = L + L_2$$

$$L_T = 3.33 \text{ H} + 2.0 \text{ H}$$

$$L_T = 5.33 \text{ H} \quad \text{(total inductance of the circuit)}$$

SECTION 6.3 EXERCISES

1. An electrical capacitor of 1.5 μF is connected to a 6.0-V battery. How much potential energy is stored in the capacitor?

2. Discuss conservation of energy as it applies to the electrical system.

3. What capacitance should a capacitor have if it is to store 3×10^{-3} J of energy when it is connected to a 12-V battery?

4. How much charge is stored on the plates of a parallel-plate capacitor if the capacitor stores 5×10^{-4} J of energy when the applied voltage is 24 V?

5. What capacitance should be placed in parallel with a 10-μF capacitor if the applied voltage is 10 V when the system contains 10^{-3} J of potential energy? (**Hint:** Capacitances in parallel add.)

6. What is the total capacity of a 10-μF capacitor and a 50-μF capacitor if they are connected in series?

7. How much energy is stored in the magnetic field of an inductor if its inductance is 1 H and the current is 1.0 A?

8. Two inductors are connected in series to an ac source. The inductance of one inductor is 1.0 H and the other one has an inductance of 0.5 H. At a given instant, the current is 0.5 A. What is the total inductance? What is the energy stored in each inductor? What is the total stored energy? Can one use the total inductance to calculate the stored energy?

9. The two inductors in Problem 8 are connected in parallel. As ac source is applied to the inductors. The current through each inductor is found to depend inversely on the inductance such that $I_2/I_1 = L_2/L_1$. If the current at a given instant is 0.5 A through the 1-H inductor, what is the current through the 0.5-H inductor? What is the total inductance? What is the energy stored in each inductor? Can the total inductance be used to calculate the total stored energy at the instant of time?

10. Calculate the equivalent capacitance of each of the circuits in the following figure.

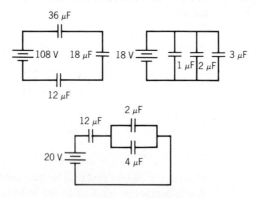

11. What capacitance should be placed in parallel with a 6-μF capacitor to store 5.0×10^{-4} J of energy in a circuit that contains only the two capacitors and a 12-V battery?

6.4................

HEAT ENERGY

....

Upon completion of this section, the student should be able to

- Define the mechanical equivalent of heat.

- Calculate the increase in a body's temperature when mechanical energy is converted into heat.

- Calculate the increase in a body's temperature when electrical energy is converted into heat.

DISCUSSION

For thermal systems, the form of energy is heat energy. Heat energy is one of the most familiar forms of energy. In the following section, the equivalence of heat energy to mechanical and electrical energy is shown.

Nature of Heat Energy

In all real systems, part of the systems' energy is converted into heat energy as changes occur. For many physical processes, the heat energy is small enough that it can be neglected without causing significant errors. For other physical processes, it may be the dominant term in the energy equation. Up to this point, heat losses have been neglected in most of the examples. Examples will be considered now that include the heat generated or transferred in the energy equation.

When heat is transferred to or from an object, there is a corresponding change of temperature. The amount of transferred heat energy can be determined with the help of

$$\Delta H = mc\Delta T \qquad (6.116)$$

where

ΔH = thermal energy added to a body to cause a temperature change ΔT
ΔT = temperature change of the body
m = mass of the body
c = constant characteristic of the body, known as the "specific heat" of the substance

Recall that by definition water requires one calorie of heat energy to raise the temperature of one gram by one Celsius degree. Specific heat of water is 1 cal/g·C°.

A specific heat such as 0.22 for aluminum means that only 0.22 cal is needed to change the temperature of one gram of aluminum one Celsius degree. Specific heats for other materials are found in standard reference handbooks.

It is important to realize that heat is the energy transferred because of a temperature difference. The heat energy transferred can cause an increase or decrease in the internal energy of a body, that is, an increase or decrease in the kinetic and potential energies of the molecules. However, the internal energy is not heat energy.

Mechanical Equivalent of Heat   Because heat is a form of energy, a relationship exists between the calorie and the joule or the British thermal unit and the foot-pound. As an illustration, consider the work done on a block as it is pulled along on a plane. Part of the work done on the block will cause an increase in the block's kinetic energy if the velocity changes, and part of the work will be converted into heat energy. The heat energy will cause the internal energy of the block and plane to increase. Suppose all the heat energy is transferred to the block and the increase in the block's temperature is recorded. The heat energy can be calculated using Equation 6.116. If the block is pulled along at constant velocity, all the work is converted into heat energy. The work done is

$$\Delta W = F \Delta s$$

where

ΔW = work done
F = constant force
Δs = distance moved

Because energy is conserved,

$$\Delta W = j \Delta H \qquad (6.117)$$

where

j = a constant which relates energy in calories to joules or energy in Btu to ft·lb.

The experiment would result in a large error for j because some of the heat energy would be transferred to the plane, some to the air, and some would be radiated to the surroundings. To minimize the heat lost, the work is done in a calorimeter as shown in Figure 6.28. The work done by the mass (m) in the earth's gravitational field is

$$\Delta W = mgh$$

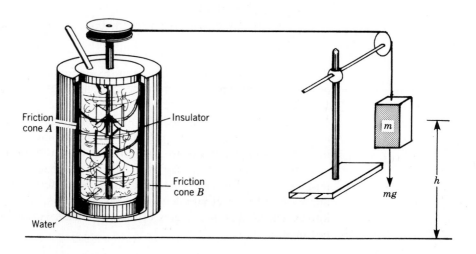

FIGURE 6.28   Calorimeter.

TABLE 6.1 Conversion table

Heat Energy	Mechanical Energy
1 cal	4.184 J
1 kcal	4184 J
1 Btu	778 ft·lb or 1054 J

The system is adjusted so that the kinetic energy gained by the mass is very small compared to ΔW. As the mass falls, friction cone A rotates and slides over friction cone B. The heat generated increases the temperature of the water and the friction cones. Friction cone B is insulated from the surroundings. The heat energy is determined by the equation

$$\Delta H = (M_m c_w + M_c c) \, \Delta T \qquad (6.118)$$

where

M_w = mass of the water
c_w = 1 cal/g·C°
M_c = mass of the sliding cones
ΔT = increase in the temperature
c = specific heat of the cones (cal/g·C°)

The corresponding work done is

$$\Delta W = mgh \qquad (6.119)$$

where

m = mass of the falling body
h = height through which m falls

From Equation 6.117, one obtains

$$\Delta W = j \, \Delta H$$

or

$$mgh = j \, (M_w \, C_w + M_c c) \, \Delta T$$

or

$$j = \frac{mgh}{(M_w \, C_w + M_c c) \, \Delta T} \qquad (6.120)$$

When the experiment is carefully done, it is found that

$$j = 4.184 \text{ J/cal}$$

Table 6.1 lists the conversion constants for several systems of units.

Examples 6.28 , 6.29, and 6.30 illustrate the equivalence of mechanical and heat energies.

EXAMPLE 6.28   Temperature increase calculation—auto brakes
· ·

Given: A 1600-kg car traveling 20 m/sec is brought to a stop. Assume 40%
of the heat developed goes into the brake drum. The specific heat of
iron is 0.11 cal/g·C° or 0.11 kcal/kg·C°.

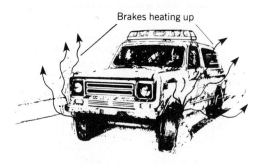

Brakes heating up

Find: The increase in temperature of the 5-kg drum.

Solution: Kinetic energy lost = $1/2\ mv^2$ = 1/2 (1600 kg) (20 m/sec)^2

$$= 320,000\ \text{J}$$

$$\text{Heat developed} = 320,000\ \text{J} \times \frac{1\ \text{kcal}}{4184\ \text{J}} = 76.5\ \text{kcal}$$

Forty percent of this heat goes into the brake drum. The other sixty
percent goes into heating the brake lining, tires, and road.

$$0.40 \times 76.5 = 30.6\ \text{kcal}$$

$$H = mc\Delta T$$

$$30.6\ \text{kcal} = (5\ \text{kg})\ (0.11\ \text{kcal/kg·C°})\ \Delta T$$

Increase in temperature = Δt = 55.6 C°

EXAMPLE 6.29   Temperature increase calculation—waterfall
· ·

Given: Assume that water is traveling at 2 m/sec as it goes over a 50-m
waterfall. After it hits the bottom its speed is zero. (Assume that all
mechanical energy is converted to heat energy.)

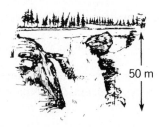

50 m

Find: Amount of temperature increase of each gram of water as it falls.

Solution: Mechanical energy = kinetic energy + potential energy

$$ME = 1/2 \, mv^2 + mgh$$

$$ME = 1/2 \, (0.001 \text{ kg}) (2 \text{ m/sec})^2 + (0.001 \text{ kg}) (9.8 \text{ m/sec}^2) (50 \text{ m})$$

$$ME = 0.002 \text{ N·m} + 0.490 \text{ N·m}$$

$$ME = 0.492 \text{ N·m or } 0.492 \text{ J}$$

$$\text{Heat developed} = 0.492 \text{ J} \times \frac{1 \text{ cal}}{4.184 \text{ J}} = 0.118 \text{ cal}$$

$$\text{Heat} = mc\Delta T$$

$$0.118 \text{ cal} = (1 \text{ g}) (1 \text{ cal/g·C°})\Delta T$$

Increase in temperature = ΔT = 0.118 C°

The temperature does not increase very much because water has a high value of specific heat.

EXAMPLE 6.30   Temperature increase calculation—lead weight
. . . .

Given: A l-g lead weight falls from rest 50 m above the ground. Lead has a specific heat of 0.031 cal/g·C°. Assume that all mechanical energy is converted to heat energy.

Find: Its increase in temperature.

Solution: Potential energy = mgh

Potential energy = $(0.001 \text{ kg})(9.8 \text{ m/sec}^2)(50 \text{ m})$
Potential energy = 0.49 J

$$\text{Heat} = 0.49 \text{ J} \times \frac{1 \text{ cal}}{4.184 \text{ J}}$$

$$\text{Heat} = 0.117 \text{ cal}$$

$$\text{Heat} = mc\Delta T$$

$$0.177 \text{ cal} = (1 \text{ g}) (0.031 \text{ cal/}g·\text{C°}) \, \Delta T$$

Increase in temperature = ΔT = 3.8 C°. This temperature is much larger than that of Example 6.29, showing that lead heats up much more than water. It has a lower value of specific heat than water.

Fluid Friction   Heat is generated in fluid flow, but can be neglected for many applications. Fluid friction produces eddies and turbulence that are eventually transformed into random internal energy. The frictional force on the flow occurs at the walls and is the result of shear stress. In fluid flow, the heat transferred to or from the surroundings in heat exchangers is often the dominant term in the energy equation. This will be considered in the chapter on power.

Electrical Heat Generation   In electrical systems at ordinary temperatures, a portion of the electrical energy always is transferred to the atoms of the conductor. This energy becomes the random kinetic and potential energy of the atoms that increase the temperature of the conductor. Eventually equilibrium is established when the heat gained per unit of time because of the electric current is equal to the heat lost by the conductor to the surroundings per unit of time. However, if the temperature at which equilibrium would occur is greater than the melting temperature of the conductor, the conductor melts.

In systems designed to convert electrical energy into mechanical energy it is desirable to minimize the electrical heating of conductors. In other electrical systems, heat may be the desired result; all the electrical energy is transformed into heat energy. In either case, it is important to understand the relationship between electrical energy and heat energy.

From Chapter 2, the work done by a source of voltage in moving a given amount of charge was expressed as

$$\text{Electrical Work} = \text{Forcelike Quantity} \times \text{Displacementlike Quantity}$$

$$\Delta W = (\Delta V)(\Delta Q) \tag{6.121}$$

or

$$\Delta W = (\Delta V)\left(\frac{\Delta Q}{\Delta t}\right)\Delta t$$

or

$$\Delta W = (\Delta V)\, I\Delta t \tag{6.122}$$

Equation 6.122 expresses the work done in terms of the current $\Delta Q/\Delta t$, the potential difference ΔV, and the time Δt. This equation gives the correct energy dissipated for resistive loads. Other loads will be discussed in the chapter on power.

The application of Equations 6.117, 6.121, and 6.122 is illustrated in Examples 6.31, 6.32, and 6.33.

EXAMPLE 6.31   Electrical energy calculation
. . . .

Given: A battery with a potential difference of 12 V is used to move 3 C of charge through a circuit.

Find: How much electrical energy is used.

Solution: $E_e = Q(\Delta V)$

$E_e = (3\ \text{C})(12\ \text{V})$

$E_e = 36\ \text{C·V} = 36\ \text{J}$

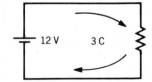

EXAMPLE 6.32   Energy and total charge calculation
. . . .

Given: A 220-Ω resistor is connected to a 12-V battery as shown. The switch is closed for 30 sec.

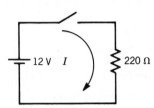

Find: **a.**   Energy converted to heat within the resistor.
              b.   The total charge that flows through the circuit.

Solution: **a.**   From Ohm's law

$$I = \Delta V/R$$

$$I = \frac{12 \text{ V}}{220 \text{ }\Omega}$$

$$I = 0.0545 \text{ A}$$

From Equation 6.122

$$\Delta W = I(\Delta V)\,(\Delta t)$$

$$\Delta W = (0.0545 \text{ C/sec})\,(12 \text{ J/C})\,(30 \text{ sec})$$

$$\Delta W = 19.62 \text{ J}$$

b.   From Equation 6.121

$$\Delta Q = \frac{\Delta W}{\Delta V}$$

$$\Delta Q = \frac{19.62 \text{ J}}{12 \text{ J/C}} \qquad \begin{array}{l} (1 \text{ A} = 1 \text{ C/sec}) \\ (1 \text{ V} = 1 \text{ J/C}) \end{array}$$

$$\Delta Q = 1.64 \text{ C}$$

or

$$Q = I\Delta t = (0.0545 \text{ C/sec})\,(30 \text{ sec}) = 1.64 \text{ C}$$

EXAMPLE 6.33   Calculating heat energy
. . . .

Given: A small heating coil is used to heat a cup of water. The voltage applied to the coil is 100 V and the current is 10 A. There are 200 g of water in the cup.

Find: The time required to raise the water temperature from 25°C to 80°C.

Solution: Total energy required is

$$\Delta H = mc\Delta T$$

$$\Delta H = (200 \text{ g}) \ (1 \text{ cal/g·C°}) \ (80°C - 25°C)$$

$$\Delta H = 11,000 \text{ cal}$$

$$\Delta W = j\Delta H$$

$$\Delta W = (4.184 \text{ J/cal}) \ (11,000 \text{ cal})$$

$$\Delta W = 4.60 \times 10^4 \text{ J}$$

The electrical energy is

$$\Delta W = (\Delta V) \ (I) \ (\Delta t)$$

$$\Delta W = (110 \text{ V}) \ (10.0 \text{ A}) \ (\Delta t)$$

$$\Delta W = 1.1 \times 10^3 \text{ J/sec} \ (\Delta t)$$

Setting the two ΔW's equal to each other

$$(1.1 \ 10^3 \text{ J/sec}) \ \Delta t = 4.60 \times 10^4 \text{ J}$$

$$\Delta t = \frac{4.60 \times 10^4}{1.1 \times 10^3} \text{ sec}$$

$$\Delta t = 41.8 \text{ sec}$$

SECTION 6.4   EXERCISES

1. Discuss conservation of energy as it applies to the thermal system.
2. How much heat must be added to a 1-kg block of aluminum to increase the temperature of the block by 20 C°?
3. How much water at 40°C could be brought to 100°C using energy produced by a rotating 5-kg solid flywheel (disk) that has a radius of 50 cm and an angular velocity of 20 rad/sec if all of the rotational energy of the flywheel went into heating the water?

S U M M A R Y

There are striking similarities between certain mathematical representations of energy that reveal analogies between physical systems. For example, the similarity of kinetic energy $1/2\ mv^2$ in the translational and $1/2\ I\omega^2$ in the rotational mechanical systems reveals the analogous roles in the two systems played by v and ω (ratelike quantities) and m and I (mass or inertial-like quantities). Also, the similarity of potential energies, $1/2\ kx^2$ and $1/2\ (1/C)\ Q^2$, in the mechanical and electrical systems reveals the analogous roles played by x and Q (displacementlike quantities) and k and I/C.

An analogy also exists between the energy stored in an inductor $1/2\ LI^2$ and the translational kinetic energy of a mass $1/2\ mv^2$. For this case, L is analogous to m and I, the ratelike quantity, is analogous to the velocity.

Energy itself is a fundamental unifying concept. In all systems, whenever work is done, there is an increase in kinetic and/or potential energy. When energy is transferred from one system to another, the total amount of energy present remains the same, including heat energy produced by friction. If friction is present, a portion of the work is represented by a frictional heat energy loss. In the absence of friction, the sum of the kinetic and potential energy remains constant, although there may be exchanges between these two forms of energy. This fact, known as the "conservation of energy" is one of the fundamental principles of science and technology.

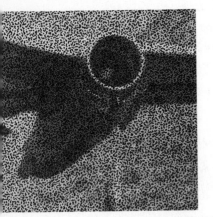

Power

INTRODUCTION

A major factor affecting our national economy is energy consumption. Energy from petroleum products is being consumed faster than new supplies are being found. Yet the high rate of energy consumption in this country has had many desirable outcomes. For instance, we have the largest gross national product in the world, which means that the United States is prosperous. The gross national product is related very closely to the rate of energy consumption. When the rate of energy consumption is high, factories approach full production. As the rate of consumption decreases, so does production. However, the high rate of energy consumption could exhaust the world's oil and coal supplies before an alternate source is available.

Thus, the rate of energy consumption is an important issue in economics and politics. As energy cost rises, it becomes a more important issue to most individuals. The rate of energy consumption, called "power," is also of great importance in science and technology. This chapter discusses the concept of power and its applications in mechanical, fluid, electromagnetic, and thermal energy systems.

Power

Definition of Power   In Chapter 2, it was shown that an element of work can be expressed as a forcelike quantity times a small change in a displacementlike quantity. If the quantities are vectors, then the product also must be multiplied by the cosine of the angle between the vectors. The **power** is the rate of doing work. Mathematically, the power may be expressed as

$$P = \frac{F\Delta D}{\Delta t} \tag{7.1}$$

where

F = forcelike quantity
ΔD = displacementlike quantity
Δt = elapsed time in doing the work

The quantity $\Delta D/\Delta t$ is the general expression for the rate discussed in Chapter 3. The power can be expressed in terms of the rate as given by

$$P = FR \tag{7.2}$$

where

R = rate

These two unifying concepts for power can be stated as

$$\text{Power} = \frac{\text{Work}}{\text{Elapsed Time}}$$

$$\text{Power} = \text{Forcelike Quantity} \times \text{Rate}$$

The application of Equations 7.1 and 7.2 to mechanical, fluid, electromagnetic, and thermal systems is developed in this chapter.

Units of Power   In all energy systems, the appropriate units of power are obtained by dividing the energy units by the time units or multiplying the rate units by the forcelike units. In the SI system, the units from Equation 7.1 are (N.m)/sec and the units from Equation 7.2 are N·(m/sec). Either method results in power units of newton-meters per second. A newton-meter (N·m) is defined as a **joule** (J) of energy; power sometimes is expressed as joules per second (J/sec). In the SI system, this unit is called a **watt** (W). When the same procedure is applied to the English system of units, the result is power units of foot-pounds per second (ft·lb/sec).

The English system of units developed over a long period of time. Whenever a convenient unit was needed, it was invented. Experiments dealing with power were done when horses were commonly used. The power of a machine was compared to that of a horse. An average horse could perform work at a rate of 550 ft·lb/sec. This unit is known as a **horsepower** and is still in use today:

$$1 \text{ hp} = 550 \text{ ft·lb/sec}$$

Thermal power is usually expressed in cal/sec in SI units and Btu/hr in the English system of units.

FIGURE 7.1 Block diagram
of power flow.

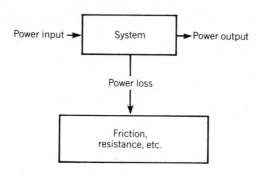

Efficiency

No system creates energy, but it is often transferred from one system to another
or within a given system. Energy cannot be transferred without some loss.
Resistance and friction, which produce heat, are the most frequent causes of
energy loss. Clearly, to operate any system, energy must be supplied to the
system at a rate that equals the rate of energy loss plus the rate at which useful
work is performed.

The rate at which energy is transferred or work is performed is defined as
power. Power input means the rate at which energy is transferred into a system,
whereas power output is the rate at which the system performs work. Figure 7.1
demonstrates this concept in general terms.

The ratio of power output to power input in any system is referred to as
efficiency. The quantity usually is expressed as a percentage and can be calculated
from

$$\eta = \frac{P_{out}}{P_{in}} (100) \tag{7.3}$$

where

η = percent efficiency
P_{out} = output power
P_{in} = input power

The efficiency of energy conversion devices is of great importance in the production and application of energy. Efficiency will be calculated using Equation 7.3 in
several examples in this chapter.

7.1

**POWER IN
MECHANICAL
SYSTEMS**

OBJECTIVES

Upon completion of this section, the student should be able to

- Define power as it applies to a mechanical system and write an equation that
  relates work, elapsed time, force, and rate to power.

- List the SI and English units used to define power for a mechanical system.

- Given any two of the following quantities in a mechanical system, determine
  the third:
  Work
  Elapsed time
  Power

- Given any two of the following quantities in a mechanical system, determine the third:
  Force
  Rate
  Power

- Define the terms
  Input power
  Output power
  Efficiency

- Given any two of the following quantities in a mechanical system, determine the third:
  Input power
  Output power
  Efficiency

DISCUSSION

In the following discussion, power is examined in translational and rotational systems. Both are widely used mechanical systems, and the results have a variety of applications.

Translational

In a mechanical system, the force and displacement are vectors. Thus, the expression for power becomes

$$P = F \frac{\Delta d}{\Delta t} \cos \theta \qquad (7.4)$$

$$P = Fv \cos \theta \qquad (7.5)$$

where

d = magnitude of the displacement vector
F = magnitude of the force vector
θ = angle between the two vectors
v = magnitude of the velocity vector

The application of Equations 7.4 and 7.5 to specific problems is shown in Examples 7.1 and 7.2.

EXAMPLE 7.1   Power of an elevator
. . . .

Given: An elevator with a mass of 1200 kg moves upward at a constant velocity of 2 m/sec.

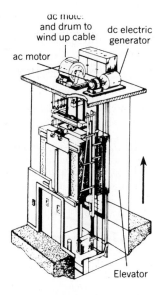

ac motor

dc motor
and drum to   dc electric
wind up cable   generator

Elevator

Find: The power that the elevator motor must supply.

Solution: Use Equation 7.5.

$$P = Fv \cos 0° = mgv \cdot 1$$

$$P = (1200 \text{ kg}) (9.8 \text{ m/sec}^2) (2 \text{ m/sec})$$

$$P = (11{,}760 \text{ N}) (2 \text{ m/sec})$$

$$P = 23{,}520 \text{ N·m/sec}$$

Use Equation 7.4. If the elevator travels at an average speed of 2 m/sec, it is displaced 2 m in 1 sec.

$$P = \frac{Fd}{t} \cos 0°$$

$$P = \frac{(11{,}760 \text{ N}) (2 \text{ m})}{(1 \text{ sec})}$$

$$P = 23{,}520 \text{ N·m/sec}$$

The power of the elevator is 23.52 kW. Note that the following are equivalent power units in the SI system:

$$W = J/\text{sec} = N·m/\text{sec} = kg·m^2/\text{sec}^3$$

EXAMPLE 7.2   Power delivered to a piston
. . . .

Given: A movable piston is forced by a connecting rod to move along a straight path. The connecting rod makes an angle of 15° with the line

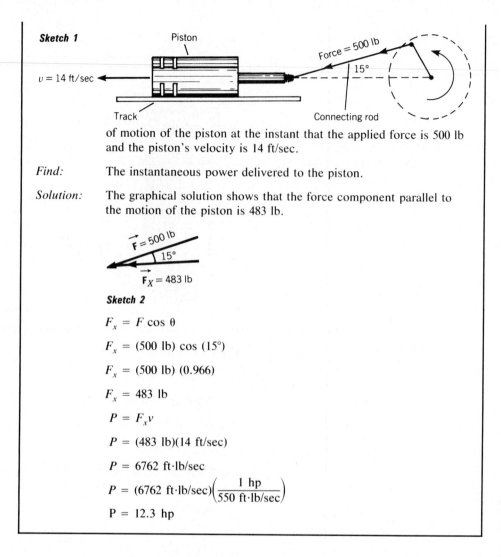

of motion of the piston at the instant that the applied force is 500 lb and the piston's velocity is 14 ft/sec.

Find: The instantaneous power delivered to the piston.

Solution: The graphical solution shows that the force component parallel to the motion of the piston is 483 lb.

Sketch 2

$$F_x = F \cos \theta$$

$$F_x = (500 \text{ lb}) \cos (15°)$$

$$F_x = (500 \text{ lb}) (0.966)$$

$$F_x = 483 \text{ lb}$$

$$P = F_x v$$

$$P = (483 \text{ lb})(14 \text{ ft/sec})$$

$$P = 6762 \text{ ft·lb/sec}$$

$$P = (6762 \text{ ft·lb/sec})\left(\frac{1 \text{ hp}}{550 \text{ ft·lb/sec}}\right)$$

$$P = 12.3 \text{ hp}$$

Rotational

In rotational mechanical systems, the equations for power are

$$P = \tau \frac{\Delta\theta}{\Delta t} \tag{7.6}$$

$$P = \tau\omega \tag{7.7}$$

where

τ = torque
$\Delta\theta$ = angle of rotation
ω = angular velocity

The units of rotational mechanical power are the same as those of linear mechanical power. Examples 7.3 and 7.4 show the use of Equations 7.6 and 7.7, respectively.

EXAMPLE 7.3 Power of a winch

Given: An electric winch is used to lift a 50-kg mass. The radius of the winch drum is 20 cm, and it turns one-half a revolution in 1 sec. The motor driving the winch has an electrical input power of 385 W.

Find:
a. Output power of winch.
b. Efficiency.

Solution:
a. Torque:

$$\tau = Fl = mgl$$

$$\tau = (50 \text{ kg})(9.8 \text{ m/sec}^2)(0.2 \text{ m})$$

$$\tau = 98 \text{ kg·m}^2/\text{sec}^2$$

$$\tau = 98 \text{ N·m}$$

Notice that N·m is a unit of torque in this case. Angle of rotation:

$$\theta = (0.5 \text{ rev}) \left(\frac{2\pi \text{ rad}}{1 \text{ rev}} \right)$$

$$\theta = 3.14 \text{ rad}$$

Power:

$$P = \frac{\tau\theta}{t}$$

$$P = \frac{(98 \text{ N·m})(3.14)}{1 \text{ sec}}$$ ("Radians" is dropped because by definition it is a dimensionless ratio.)

$$P = 307.7 \text{ W}$$

b. Efficiency:

$$\eta = \frac{P_{out}}{P_{in}} (100)$$

$$\eta = \frac{307.7 \text{ W}}{385 \text{ W}} (100)$$

$$\eta = 80\%$$

0.5 rev/sec

20 cm

50 kg

EXAMPLE 7.4 Power of a vacuum pump

Given: A vacuum pump requires a torque of 35 ft·lb and has a rotational rate of 750 rpm.

Find: The minimum power rating for a motor to drive the pump.

Solution: In Equation 7.7, ω must be expressed in units of rad/sec.

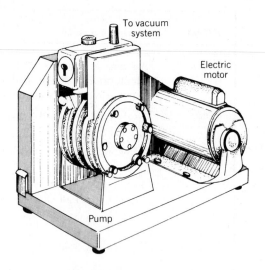

$$\omega = (750 \text{ rev/min})\left(\frac{1 \text{ min}}{60 \text{ sec}}\right)\left(\frac{2\pi \text{ rad}}{1 \text{ rev}}\right)$$

$$\omega = 2\pi \text{ rad/sec}$$

$$\omega = 78.5 \text{ rad/sec}$$

$$P = \tau\omega$$

$$P = (35 \text{ ft·lb})(78.5/\text{sec})$$

$$P = 2750 \text{ ft·lb/sec}$$

$$P = (2750 \text{ ft·lb/sec})\left(\frac{1 \text{ hp}}{550 \text{ ft·lb/sec}}\right)$$

$$P = 5 \text{ hp}$$

SECTION 7.1   EXERCISES

1. Write two equations for power for each of the following energy systems.
   a. Mechanical translational
   b. Mechanical rotational
2. A force of 250 lb is applied to push a wagon a distance of 200 ft in 2 min. Determine the horsepower required.
3. An automobile engine develops 50 hp at a rotational rate of 2000 rpm. How much torque is produced by the engine?
4. What is the mass of the load that a 1200-kg elevator carries upward with a constant velocity of 3.5 m/sec if the elevator motor produces 50,000 W of power?
5. What is the efficiency of an elevator motor that operates on 600 V and draws a current of 15 A if the motor can lift an elevator of mass 500 kg at a constant speed of 1 m/sec?

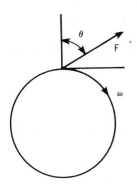

6. An electric winch is designed to lift a block of mass m at a constant speed v. The winch is driven by a motor that has an input power P. Show that the efficiency of the winch does not depend on the winch's radius.

7. A constant force of 20 N is applied to the radius of a flywheel at a distance of 1 m from the center. At what angle θ is the force applied if the flywheel rotates with an angular speed of 10 rad/sec and has an output power of 180 W? (See drawing to the left.)

8. How many horsepower are required to push a 20-lb wagon up a 20-ft incline at an angle of 30° with the horizontal in 20 sec? Assume there is no friction between the wagon and the hill. Also assume that the wagon is pushed up the hill with a constant velocity.

9. Does it take more power to lift an elevator at constant velocity than it does to lower the elevator at the same constant velocity? Explain.

10. What is the speed of a piston if a force of 20 N, at an angle of 45° with the direction of the piston's motion, drives the piston and the piston delivers 200 W of power?

11. Forty m³/sec of water falling from a height of 30 m are used to turn a flywheel. If 80% of the water's power is delivered to the flywheel and the torque on the flywheel is 20,000 N·m, what is the constant angular velocity of the flywheel?

7.2
POWER IN FLUID SYSTEMS

OBJECTIVES...

Upon completion of this section, the student should be able to

- Define power as it applies to a fluid system and write an equation that relates work, elapsed time, force, and rate to power.

- List the SI and English units used to define power for the fluid system.

- Given any two of the following quantities in a fluid system, determine the third:
  Work (or Forcelike Quantity times Displacementlike Quantity)
  Elapsed time
  Power

- Given any two of the following quantities in a fluid system, determine the third:
  Forcelike quantity
  Rate
  Power

- Given any two of the following quantities in a fluid system, determine the third:
  Input power
  Output power
  Efficiency

- Determine the horsepower required to pump water into a tank of a given height at a given rate.

DISCUSSION

The following treatment examines power in fluid systems. Fluid systems are widely used and understanding their power requirements is important.

Power

Definition of Fluid Power   For incompressible fluids, the power equations become

$$P = \frac{(\Delta p)(\Delta V)}{\Delta t} \tag{7.8}$$

$$P = (\Delta p)Q_V \tag{7.9}$$

where

Δp = pressure difference across a pump (forcelike quantity)
ΔV = volume displaced (displacementlike quantity)
Q_V = volume flow rate (rate)

Equations 7.8 and 7.9 are used to solve problems in Examples 7.5 and 7.6.

EXAMPLE 7.5   Power in a fluid system
. . . .

Given: A pump must fill a 5000-ft³ tank located 25 ft above the level of the lake. The tank is to be filled in 15 min. The weight density of water is 62.4 lb/ft³. Assume water moves with negligible speed.

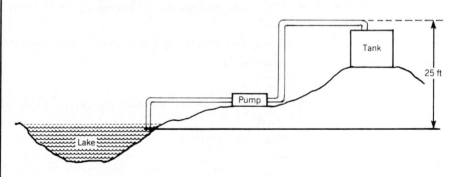

Find: The power that must be delivered by the pump.

Solution: Pressure head:
$$\Delta p = \rho_w h$$

$$\Delta p = (62.4 \text{ lb/ft}^3)(25 \text{ ft})$$

$$\Delta p = 1560 \text{ lb/ft}^2$$

Power from Equation 7.8:

$$P = \Delta p \frac{\Delta V}{\Delta t}$$

$$P = \frac{(1560 \text{ lb/ft}^2)(5000 \text{ ft}^3)}{(15 \text{ min})(60 \text{ sec/min})}$$

$$P = 8667 \text{ ft·lb/sec}$$

Volume flow rate:

$$Q_V = \frac{\Delta V}{\Delta t}$$

$$Q_V = \frac{5000 \text{ ft}^3}{900 \text{ sec}}$$

$$Q_V = 5.556 \text{ ft}^3/\text{sec}$$

Power from Equation 7.9:

$$P = Q_V \Delta p$$

$$P = (5.556 \text{ ft}^3/\text{sec})(1560 \text{ lb/ft}^3)$$

$$P = 8667 \text{ ft·lb/sec}$$

$$P = 15.8 \text{ hp}$$

EXAMPLE 7.6   Power in a hydraulic system
. . . .

Given: Water is pumped into a tank at a rate of 50 gal/min at a constant tank gage pressure of 40 psi.

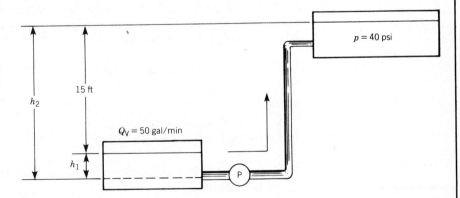

Find: The required horsepower when the difference in water levels is 15 ft.

Solution: If frictional losses are neglected, the gage pressure on the right-hand side of the pump is

$$p_b = \rho_w h_2 + p \qquad (\rho_w \text{ is weight density and } p \text{ is the tank pressure})$$

The gage pressure on the left-hand side of the pump is

$$p_a = \rho_w h_1$$

It is assumed that the velocity of the water in the tanks is zero. This is reasonable because the tanks have large diameters.
The power delivered by the pump is

$$P = (p_b - p_a)Q_V$$

$$P = [\rho_w (h_2 - h_1) + p] Q_V$$

$$P = [(62.4 \text{ lb/ft}^3)(15 \text{ ft}) + (40 \text{ lb/inch}^2)(144 \text{ inch}^2/\text{ft}^2)]$$

$$\left[(50 \text{ gal/min})\left(\frac{1 \text{ ft}^3}{7.48 \text{ gal.}}\right)\right]$$

$$P = (44,800 \text{ ft·lb/min}) \left(\frac{1 \text{ min}}{60 \text{ sec}}\right) \left(\frac{1 \text{ hp}}{550 \text{ ft·lb/sec}}\right)$$

$$P = 1.36 \text{ hp}$$

Note. 1.36 hp is an approximate value. Additional power is required to make up frictional losses in the motor, pump, and pipes.

General Power Equation   A more general equation for fluid power can be obtained from the energy equation for fluid flow developed in Chapter 6. If there are no frictional losses or heat transfer, the energy equations for incompressible fluids reduce to Bernoulli's equation (Equation 6.72):

$$\frac{v_b^2}{2} + gh_b + \frac{p_b}{\rho_m} = \frac{v_a^2}{2} + gh_a + \frac{p_a}{\rho_m} \tag{7.10}$$

Each term in Equation 7.10 is energy per unit mass. This equation can be made more general by including a source of mechanical energy per unit mass such as a pump. Equation 7.10 then becomes

$$\frac{v_b^2}{2} + gh_b + \frac{p_b}{\rho_m} = \frac{v_a^2}{2} + gh_a + \frac{p_a}{\rho_m} + \frac{\Delta W}{\Delta m} \tag{7.11}$$

where

ΔW = work done by the pump
Δm = mass moved by the pump

To change Equation 7.11 to a power equation, each term must be multiplied by the mass flow rate:

$$Q_m = \frac{\Delta m}{\Delta t} \tag{7.12}$$

where

Q_m = mass flow rate

Each term then becomes work per unit time or power. Because the density is constant in incompressible flow, it is desirable to use the volume flow rate. From the definition of density

$$\Delta m = \rho_m \Delta V \qquad (7.13)$$

Dividing Equation 7.13 by Δt gives

$$\frac{\Delta m}{\Delta t} = \rho_m \frac{\Delta V}{\Delta t}$$

$$Q_m = \rho_m Q_V \qquad (7.14)$$

where

Q_V = volume flow rate

Thus, if each term in Equation 7.11 is multiplied by $\rho_m Q_V$, the power equation is obtained. This yields

$$\frac{\rho_m Q_V v_b^2}{2} + Q_V \rho_m g h_b + Q_V p_b = \frac{\rho_m Q_V v_a^2}{2} + Q_V \rho_m g h_a + Q_V p_a + P \qquad (7.15)$$

where

$P = (\Delta W / \Delta m)(\Delta m / \Delta t) = \Delta W / \Delta t$ = pump power

If Equation 7.15 is solved for P, one obtains

$$P = Q_V(p_b - p_a) + Q_V \rho_m g \, (h_b - h_a) + \frac{Q_V \rho_m}{2} \, (v_b^2 - v_a^2) \qquad (7.16)$$

Pump Power = Pressure Power + Potential Power + Kinetic Power

It is important to remember that a and b can be any two cross sections, but the pump must lie between the two cross sections.

Equation 7.16 can be used to calculate the required pump power for the conditions given in Example 7.5. To apply Equation 7.16 to Example 7.5, section a can be taken at the lake level and section b at the end of the pipe at the top of the tank. The solution is obtained in Example 7.7. The additional required energy is the kinetic power of the fluid at the end of the pipe, which depends on the diameter of the pipe. This kinetic power is converted into heat that increases the temperature of the water in the tank.

EXAMPLE 7.7 Power in a fluid system
. . . .

Given: The data supplied in Example 7.5. In addition, the area of the pipe is 1.0 ft². Assume turbulent flow, $\alpha \approx 1$.

Find: The power that must be delivered by the pump.

Solution: From Example 7.5

$Q_V = 5.556$ ft³/sec

The power equation (no friction and no heat added) for an incompressible fluid gives

$$P = Q_V(p_b - p_a) + Q_V\rho_m g(h_b - h_a) + \frac{\rho_m}{2} Q_V(v_b^2 - v_a^2)$$

Let section a be the lake level. Then $v_a = 0$ and p_a = atmospheric pressure. Let section b be the end of the pipe at the top of the tank. Then pb = atmospheric pressure. For steady flow, the mass flow rate is constant. Thus,

$$\rho_m v_b A = \rho_m Q_V = \Delta m/\Delta t$$

$$v_b = \frac{Q_V}{A}$$

$$v_b = \frac{5.556 \text{ ft}^3/\text{sec}}{1 \text{ ft}^2}$$

$$v_b = 5.556 \text{ ft/sec}$$

The power equation becomes ($\rho mg = 62.4$ lb/ft³)

$$P = 0 + (5.556 \text{ ft}^3/\text{sec})(62.4 \text{ lb/ft}^3)(25 \text{ ft})$$

$$+ \left(\frac{62.4 \text{ lb/ft}^3}{32 \text{ ft/sec}^2}\right)\left(\frac{5.556}{2} \frac{\text{ft}^3}{\text{sec}}\right)(5.556 \text{ ft/sec} - 0 \text{ ft/sec})^2$$

$$P = 8667 \text{ ft·lb/sec} + 167 \text{ ft·lb/sec}$$

$$P = 8834 \text{ ft·lb/sec}$$

The extra power is the kinetic power of the flow out of the pipe. This will be converted into heat in the tank. How could this energy be saved?

The modern technical world depends heavily on the application of fluid power. Perhaps the most obvious applications are in modern automobiles equipped with hydraulic brakes, automatic hydraulic transmissions, and, very frequently, power brakes, power steering, and other fluid-power devices. Most modern construction and agricultural equipment uses fluid power for operation or control, particularly of remotely located components. Industrial equipment uses fluid power extensively for both operation and control, and the availability of fluid power has been one of the most important factors in the development of automation.

Some of the more spectacular uses of fluid power are in aircraft and missiles. A single aircraft similar to the Boeing 707 contains miles of hydraulic tubing plus hundreds of hydraulic devices to power components such as flaps, control surfaces, and landing gear, in addition to the complex systems within each jet engine. Hydraulic systems are widely used to control the engines of missiles. Spacecraft attitude-correction devices are wholly fluid-power units.

SECTION **7.2 EXERCISES**

1. Write two equations for power in a fluid system.

2. A 5-hp (output) water pump has an electrical input power of 5 kW. It is used to fill a 2500-ft³ tank and must raise the water to a height of 46 ft. Determine the efficiency of the pump and the time required to fill the tank. (The weight density of water is 62.4 lb/ft³.) (Assume no energy of flow or friction loss in the pipe.)

3. How high above a reservoir should a 2000-m³ tank be placed if a motor that delivers 20,000 W is to fill the tank in 40 min? (Assume no losses because of the kinetic energy of flow or friction in the pipe.)

4. Water is pumped into a 10,000-ft³ tank through a 6-inch-diameter pipe. The tank is filled in 30 min. What is the pump power delivered to the water if the water is pumped 20 ft above a reservoir as shown in the sketch? (Neglect friction in the pipe.) (p_a is atmospheric pressure.)

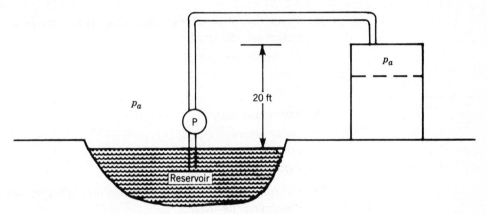

5. *Water is pumped into a 10,000-ft³ tank through a 6-inch-diameter pipe as shown in the sketch. The cross-sectional area of the tank is 666 ft². The tank is filled in 20 min. What is the initial pump power? What is the final pump power? If the kinetic energy of flow is neglected, what is the average pump power? What is the approximate energy required to fill the tank?

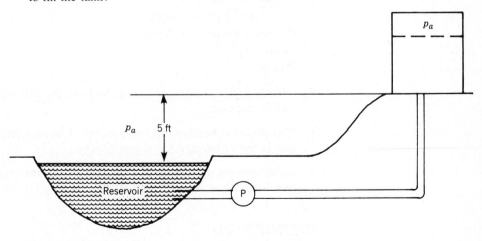

* Challenge question.

7.3

**POWER IN
ELECTROMAGNETIC
SYSTEMS**

OBJECTIVES

Upon completion of this section, the student should be able to

- Define power as it applies to an electromagnetic system and write an equation that relates work, elapsed time, force, and rate to power.

- List the SI and English units used to define power for an electromagnetic energy system.

- Given any two of the following quantities in an electromagnetic system, determine the third:
  Work (or Forcelike Quantity times Displacementlike Quantity)
  Elapsed time
  Power

- Given any two of the following quantities in an electromagnetic system, determine the third:
  Forcelike quantity
  Rate
  Power

- Define the terms
  Input power
  Output power
  Efficiency

- Given any two of the following quantities in any electromagnetic system, determine the third:
  Input power
  Output power
  Efficiency

- Define for electrical systems
  Root-mean-square current
  Root-mean-square voltage
  Phase angle
  Power factor

- Define electrical power in watts and energy consumption in watt-hours for ac and dc supplies.

- Explain what happens to the portion of input energy of a motor or generator that is not converted to output energy.

- Determine the efficiency of a dc motor from the input voltage and current and from the output torque and angular velocity.

DISCUSSION

Electrical power is consumed in a wide variety of tasks, such as opening cans or pulling large freight trains. The treatment below examines the use of electrical power. Electric power often is used in conjunction with coupled magnetic power and electrical power can be lost by a magnetic path.

Instantaneous Power

Electric Power   The forcelike quantity in electrical systems is the voltage difference between two points. The displacementlike quantity is the electrical charge transported between the two points. Equations 7.17 and 7.18 give the instantaneous electrical power:

$$P_i = (\Delta V_i)\frac{\Delta Q}{\Delta t} \tag{7.17}$$

$$P_i = (\Delta V_i)\, I_i \tag{7.18}$$

where

Δt = a very small change in time
P_i = instantaneous power in joules (J)
ΔV_i = instantaneous voltage in volts (V)
I_i = instantaneous current in amperes (A)

In electric circuits, power is dissipated in resistance loads. To express power in terms of the resistance, Ohm's law can be used:

$$\Delta V_i = I_i R \tag{7.19}$$

where

ΔV_i = instantaneous voltage across the resistance in volts (V)
R = resistance in ohms (Ω)

If Equation 7.19 is used to substitute for ΔV_i in Equation 7.14, one obtains

$$P_i = I_i^2 R \tag{7.20}$$

Average Power   For most applications, the average power is the important quantity. For constant direct currents, the average power is equal to the instantaneous power. For this case, the average power can be expressed as

$$P = (\Delta V)I \tag{7.21}$$

where

P = constant power
I = constant current
ΔV = constant voltage

For alternating currents, a circuit may include capacitors, inductors, and resistors. The power, however, is only dissipated in resistors. For alternating currents, Equation 7.22 gives the instantaneous dissipated power:

$$P_i = I_i^2 R \tag{7.22}$$

where

R = total effective resistance

For sinusoidal currents, the instantaneous current is

$$I_i = I_m \sin (2\pi f\, t) \qquad (7.23)$$

where

I_m = maximum current
f = frequency

Substituting this expression for the instantaneous current in Equation 7.22 yields

$$P_i = [I_m^2 \sin^2 (2\pi f\, t)]R \qquad (7.24)$$

It is advantageous, at this point, to use the trigonometric identity

$$\sin^2 (2\pi f\, t) = \frac{1 - \cos (4\pi f\, t)}{2} \qquad (7.25)$$

Notice that the frequency is doubled in the cosine term on the right-hand side of Equation 7.25. Thus, the power frequency is two times the current frequency. Because the current is squared to obtain the power, the power is positive or zero in a resistance load as the time changes. Substituting for the $\sin^2 (2\pi f\, t)$ given by Equation 7.25 in Equation 7.24 gives

$$P_i = \frac{I_m^2 R}{2} - \frac{I_m^2 R}{2} \cos (4\pi f\, t) \qquad (7.26)$$

The instantaneous power and corresponding current are shown in Figure 7.2.

From Figure 7.2, one can see that the average current is 0, whereas the average power is

$$P = \frac{I_m^2}{2} R \qquad (7.27)$$

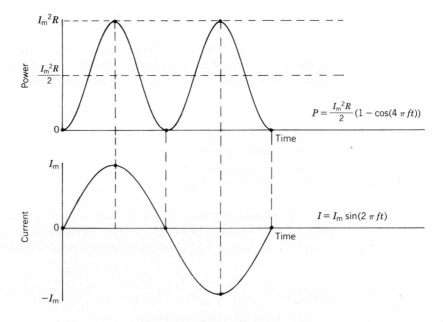

FIGURE 7.2 Power frequency and current frequency.

$$P = \left(\frac{I_m}{\sqrt{2}}\right)\left(\frac{I_m}{\sqrt{2}}\right) R \qquad (7.28)$$

where

P = average power dissipated in an alternating current circuit

$I_m/\sqrt{2}$ = definition of the root-mean-square current (rms)

In most references, the symbol I is used for the root-mean-square current for alternating-current circuits. The power equation then is expressed as

$$P = I^2 R \qquad (7.29)$$

where

I = root-mean-square current

R = total effective resistance

The average alternating current power also can be expressed as

$$P = VI \cos \alpha \qquad (7.30)$$

where

I = rms current ($I = I_m/\sqrt{2}$)

V = rms voltage ($V = V_m/\sqrt{2}$)

α = phase angle between the instantaneous current and voltage

The value of $\cos \alpha$ is called the power factor. It may lie between 0 and 1. Equation 7.30 is illustrated in Figure 7.3 for a sinusoidal voltage applied to a capacitor. It is assumed in Figure 7.3 that there are no frictional losses in the capacitor. As shown in the figure, energy is first stored in the capacitor as the stored charge increases. Because the initial charge is assumed to be zero, the voltage also will be zero because the charge on a capacitor is equal to the voltage times the capacitance. The voltage and stored charge are said to be in phase as shown in Figure 7.3. The current is maximum when there is no charge on the capacitor and is zero when the charge is a maximum. Because $I = \Delta Q/\Delta t$ for very small times, the current is the slope of the charge versus time curve; it is 90° ahead of the voltage and charge. It is said to lead the voltage by 90°. As the current goes through zero, the power becomes zero. The energy that was stored in the capacitor then is given back to the source and the power is negative.

A similar case occurs for a lossless inductor. Energy is stored in the magnetic field for one half-cycle and is given back to the source during the next half-cycle. This results in an average power of 0. The corresponding phases are shown in Figure 7.4. Because $V = -L \Delta I/\Delta t$ for small Δt, the voltage is proportional to the slope of the sinusoidal current as shown in Figure 7.4. To align the voltage and current curves, the current wave would have to be moved forward by 90°. Thus, the current lags the voltage by 90° in an inductor if there are no frictional losses. When the absolute value of the current is zero, the absolute value of the voltage is a maximum. When the current is a maximum, the voltage is zero. From the figures, one can see that the inductive power is 180° out of phase with the capacitance power.

FIGURE 7.3 Plots of power, voltage, current, and charge on a capacitor versus time for a capacitor in an ac circuit.

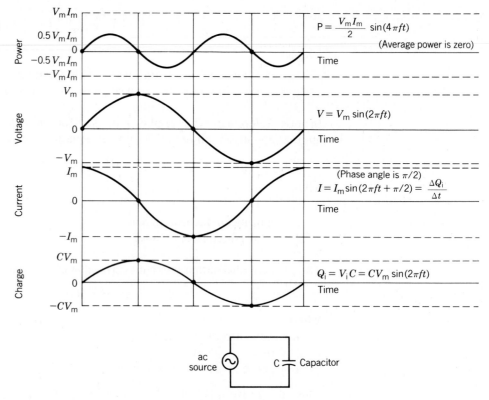

FIGURE 7.4 Plots of power, voltage, and current versus time for an inductor in an ac circuit.

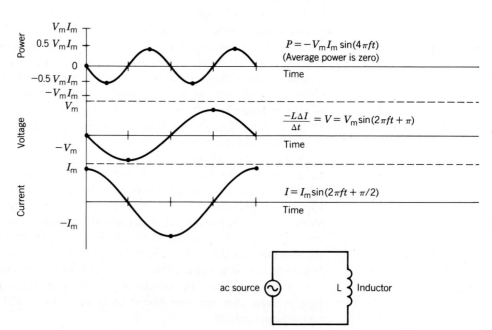

If either current contains resistance, the phase angle will be less than 90° and greater than −90°. Power will be consumed on the average as given by Equation 7.29 or Equation 7.30.

When one reads voltage or current on a calibrated ac meter, it is usually the root-mean-square value. If the waveforms are not sinusoidal, some meters may not give the correct root-mean-square values.

Applications of the power equations are given in Examples 7.8 and 7.9.

EXAMPLE 7.8 Current required by a water heater
. . . .

Given: An electric water heater operates at 240 V and is rated at 1500 W (only a resistance load).

Find: The current required.

Solution: $P = IV$

$$I = \frac{P}{V}$$

$$I = \frac{1500 \ W}{240 \ V}$$

$$I = 6.25 \ A$$

EXAMPLE 7.9 Power in an ac circuit
. . . .

Given: An inductor, capacitor, and resistor are connected in a series as shown. The root-mean-square current is 10.0 A, and the total root-mean-square voltage is 289 V.

Find: **a.** The average power delivered by the source.
            b. The power factor.

Solution: **a.** For an ac circuit, the average power is

$$P = I^2R$$

$$P = (10 \ A)^2 \ (10 \ \Omega)$$

$$P = 1000 \ W \quad \text{(average)}$$

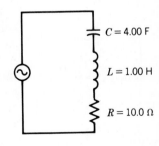

            b. The average power is also

$$P = VI \cos \alpha$$

            Thus

$$\cos \alpha = \frac{P}{VI}$$

$$\cos \alpha = \frac{1000 \text{ W}}{(289 \text{ V})(10 \text{ A})}$$

$$\cos \alpha = 0.346$$

$$\text{Power factor} = 0.346$$

Electrical Energy   The total electrical energy delivered to a load is usually measured in watt-hours or kilowatt-hours. For dc loads, the total energy may be expressed as

$$W = VIt \tag{7.31}$$

where

    W = watt-hours
    t = time of operation in hours

Equation 7.31 assumes that the load is constant. For constant ac loads, the electrical energy is

$$W = VI \text{ (PF)}t \tag{7.32}$$

where

      V = root-mean-square voltage
      I = root-mean-square current
   (PF) = power factor (cos α)

The number of watts consumed in an electrical circuit changes as various loads are switched into and out of the circuit. Total watt-hours are calculated by summing the products of watts and time elapsed for the individual time intervals.

Note that a kilowatt is 1000 watts and a kilowatt-hour is 1000 watt-hours.

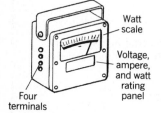

FIGURE 7.5   Wattmeter.

Wattmeters   Figure 7.5 shows a four-terminal wattmeter. The four terminals are connected in the manner shown in Figure 7.6. (**Caution:** When using the wattmeter, observe both voltage and ampere readings on the meter. Neither should be exceeded individually. Furthermore, the watt rating should not be exceeded.)

Figure 7.7 illustrates the internal workings of a single-phase wattmeter. The voltage coil is basically the meter movement and is attached to the meter needle mounted on pivots so that it rotates clockwise against a return spring (not shown). As more voltage is applied, the voltage coil tends to rotate farther against the spring. Current also passes through the two stationary current coils. These coils produce a magnetic field in which the voltage coil turns. As current increases, the magnetic-field strength increases, and the voltage coil tends to rotate further against the spring.

As either voltage or current or both are increased, there is a greater deflection of the needle or a higher reading of the wattmeter. This deflection, like power itself, is proportional to the product of volts and amps. Therefore, the meter scale can be calibrated to read watts directly.

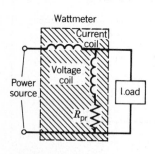

FIGURE 7.6   Schematic of wattmeter.

FIGURE 7.7 Operation of wattmeter.

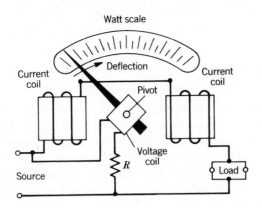

For the cases of an alternating current in phase with the voltage, the current through the moving coil changes direction. However, the current through the stationary coil also changes direction at the same time. Thus, the torque and corresponding deflection are in the same direction. When the current and voltage are in phase, the deflection will be a maximum. When the voltage and current are 90° out of phase, the average torque will be zero, indicating that no average power is being delivered to the circuit. It can be shown that the wattmeter reads the correct average power for any power factor. The wattmeter can be calibrated using direct current. This is desirable because the power can be measured using dc meters. The earth's magnetic field may cause an error when using or calibrating a wattmeter on direct current. This error can be eliminated by reversing the voltage and current terminals and averaging the two readings. This is not necessary when the meter is used to measure alternating power because the reversal automatically takes place. A sample problem is given in Example 7.10.

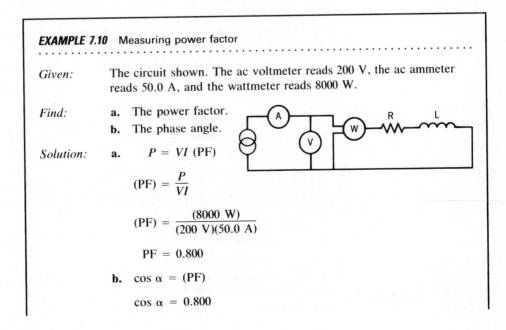

EXAMPLE 7.10 Measuring power factor
. . . .

Given: The circuit shown. The ac voltmeter reads 200 V, the ac ammeter reads 50.0 A, and the wattmeter reads 8000 W.

Find: **a.** The power factor.
  b. The phase angle.

Solution: **a.** $P = VI \, (PF)$

$$(PF) = \frac{P}{VI}$$

$$(PF) = \frac{(8000 \text{ W})}{(200 \text{ V})(50.0 \text{ A})}$$

$$PF = 0.800$$

  b. $\cos \alpha = (PF)$

$$\cos \alpha = 0.800$$

$$\alpha = \cos^{-1} 0.800$$

$$\alpha = 70°$$

Energy Meters   A kilowatt-hour meter measures the total (cumulative) energy consumption. Its output is connected to a device similar to the mileage counter (odometer) on an automobile. The meter is read monthly. The previous month's reading is subtracted to obtain the kWh usage. This method is used by utility companies to determine energy used by customers.

The direct-current (dc) kWh meter is used seldom, but it does illustrate some principles applicable to the ac kWh meter. The dc meter shown in Figure 7.8 is geared to a counter, which is calibrated so that each kWh fed through the meter to the load will cause meter rotation and indicate a count of 1 kWh. Voltage is applied to the armature coil A. Current is applied to the field coils F. In a manner similar to the wattmeter (Figure 7.7), the output rotation is directly proportional to watts (the product of volts and amps). When connected by gears to a counter, the net rotation indicates kilowatt-hours.

The alternating-current kWh meter is a two-phase induction motor designed to measure the total energy delivered to an electric load. Figure 7.9 shows the essential parts of the meter. The voltage coil has a high inductance and is connected directly across the line. The current in this coil is directly proportional to the line voltage and lags behind it by approximately 90°. A lag coil is used to cause the current to lag the voltage by 90°. The current coil consists of a few turns of wire connected in series with the load. When the power factor is zero, the line current is 90° behind the voltage. Therefore, the fluxes produced by the two coils are in phase and no torque is developed. When the power factor is 1, the two fluxes are 90° out of phase and the torque is a maximum. At any other power factor, it can be shown that the torque is proportional to $VI \cos \theta$. The armature disk rotation is stabilized by the damping magnets. The rotational speed is proportional to the watts in use. A counter, which measures the number of rotations, directly reads kWh. A typical kWh meter found attached to houses is shown in Figure 7.10.

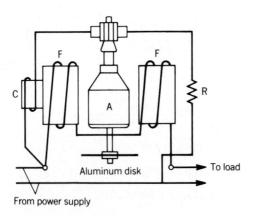

FIGURE 7.8   Operation of a dc watt-hour meter.

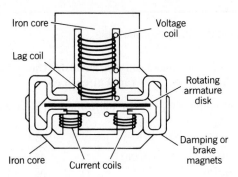

FIGURE 7.9 Operation of an ac watt-hour meter.

Residential electrical rate structures vary with time and location. A sample rate structure is shown in Table 7.1.

The sample rate structure in Table 7.1 will be used in Example 7.11 to demonstrate how an electric bill can be calculated. In the example, the total kilowatt-hour use is computed from an itemized knowledge of the power used per day for a 30-day month. In actual practice, the kWh meter (Figure 7.10) would be read monthly. Thus, the kWh meter reading would indicate 604.4 kWh used if the meter read 0000 at the beginning of the month.

TABLE 7.1 Residential electric rate structure

Fixed charge	$2.75 plus	
First	200 kWh/month	$0.0435/kWh
Next	1300 kWh/month	$0.0314/kWh
Over	1500 kWh/month	$0.0287/kWh

Plus electric adjustment factor × kWh use

Plus 4% tax

(Adjustment factor 0.00138)

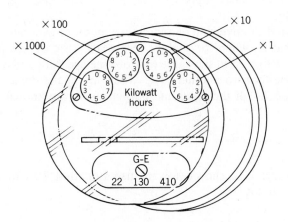

FIGURE 7.10 kWh meter.

EXAMPLE 7.11   Calculating an electric bill
. . . .

Given: Two 1-W clocks run continuously. One 1.5-kW heater runs all day.
             Four 100-W bulbs run 9 hr each day.

Find: Monthly total electric bill using Table 7.1 data below. Total kWh is

Item	No.	kW	hr/day	days	kWh
Clocks	2	0.001	24	30	1.44
Heater	1	1.5	11	30	495.00
Bulbs	4	0.1	9	30	108.00
					604.44 kWh
				Use factor	0.83 kWh
				Total	605.27 kWh

Solution: Correcting for use factor (adjustment factor, Table 7.1):
             604.44 × 0.00138 = 0.83 kWh
             Fixed cost $ 2.75
             First 200 kW × 0.0435   8.70
             Next 405.27 × 0.0314   12.73
                                            24.18
                              + 4% tax   0.97
                      Total cost = $25.15/month

Electric Motors

Efficiency   Electric motors and generators (Figure 7.11) are two applications of the same device. When this device is employed to convert mechanical energy to electrical energy, it is called a "generator." When it converts electrical energy to mechanical energy, it is called a "motor." In both cases, some energy losses occur. As a result, the output energy always is less than the input energy. Because these losses are the same for motors and generators, a discussion of loss factors in either device also applies to the other. This discussion uses the motor as an example.

When the convertion of energy in a motor or generator takes place at a uniform rate (that is, when the energy received by the device per unit of time and the energy delivered in the same unit of time are both constant), power is converted from one form to another. The efficiency of the device can be calculated from Equation 7.33 if the input and output power are known:

$$\eta = \frac{P_{\text{out}}}{P_{\text{in}}} (100) \tag{7.33}$$

Example 7.12 shows how to use Equation 7.33 to solve a problem for an electric motor.

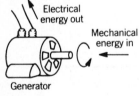

Generator

a.

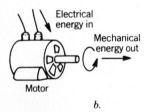

Motor

b.

FIGURE 7.11 Electric generator and electric motor.

EXAMPLE 7.12   Efficiency of an electric motor
. . . .

Given: An electric motor produces an output power of 1/2 hp when the input power is 434 W.

Find: Efficiency of the motor.

Solution: Convert output power to watts: 1 hp = 746 W.

$$P_{out} = (1/2 \text{ hp})(746 \text{ W/hp})$$

$$P_{out} = 373 \text{ W}$$

$$\eta = \frac{P_{out}}{P_{in}} (100)$$

$$\eta = \frac{373 \text{ W}}{434 \text{ W}} (100)$$

$$\eta = 86\%$$

The remainder of the energy (14% or 61 W in Example 7.12) must be accounted for in some way. This energy appears as heat energy in the motor. If input energy is converted to heat at a high rate, the motor can be damaged and burn out.

Power Losses   There are three classifications of losses in an electric motor.

- Rotational losses are those that are caused by the rotation of the armature.
- Copper losses are those that are caused by the rotation of the conductors of the motor.
- Stray-load losses are additional losses that cannot be accounted for in the two categories above, and often are difficult to analyze.

Rotational losses may be subdivided into the following classes:

- Bearing friction
- Brush friction
- Wind friction (windage)
- Hysteresis
- Eddy currents

The first three rotational losses are mechanical, that is, they are caused by the rotation of the armature. They can always be reduced by good mechanical design and lubrication. The last two are electrical in nature, but are classified as rotational losses because they also result from, and are dependent on, the rate of rotation of the armature.

Hysteresis occurs when a material is magnetized. Materials that can be magnetized, such as the iron in an armature core, are composed of many tiny magnetic

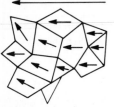

Net magnetic field in material

Magnetic domains

FIGURE 7.12 Domains in a magnetic material.

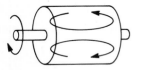

FIGURE 7.13 Eddy currents.

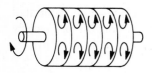

FIGURE 7.14 Reduced eddy currents.

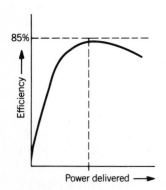

FIGURE 7.15 Efficiency versus output power.

domains, each with its own field of orientation, as in Figure 7.12. In unmagnetized iron, the direction of the magnetic fields of the domains is random. When an external magnetic field is applied, the domain fields line up to produce a net magnetic field in the material. In a motor armature, the direction of the applied field changes as the armature turns. This causes the direction of the magnetic fields of the domain to change. The directions of the domain fields always lag behind the direction of the applied field. This lag is called "hysteresis." The energy necessary to change the direction of the fields of magnetic domains eventually appears as heat in the armature. Hysteresis losses are always present in electric motors.

Eddy currents are the result of the motion of conducting parts of the armature through the magnetic fields in the motor. When the armature rotates in the magnetic field, the outer edges of the armature core move through the field at a greater velocity than the center portions. This difference in velocity causes electrical potential differences to appear within the core and creates the eddy currents shown in Figure 7.13. These eddy currents heat the material of the core by joule heating (I^2R).

Most of the eddy current flows parallel to the axis of rotation. The eddy current can never be eliminated completely, but it can be reduced greatly by the use of a laminated core composed of thin plates of conductor coated with varnish. The smaller eddy currents of such a core are shown in Figure 7.14.

Copper losses are the result of current flowing through the resistance presented by electrical conductors within the motor. These losses are dependent on the current and resistance (I^2R) and are independent of armature rotation. There are three sources of copper loss:

- Resistance of the armature windings
- Resistance of field windings
- Resistance of brush contacts

Copper losses may be reduced by using larger-diameter wire to reduce the resistance of the windings. Thus, motors with high efficiencies are larger and more expensive than lower-efficiency motors.

Stray-load losses are attributed to the lack of a uniform division of current through various paths in the windings, to short-circuit currents in coils undergoing commutation, to radiated electromagnetic energy, and to distortion of magnetic flux lines because of armature reaction. In most cases, the magnitude or exact causes of stray-load losses are impossible to determine; in modern motors, they are small. A value of 1% usually is assumed for stray-load losses in large motors, and they are often neglected completely in efficiency calculations for small motors.

The overall efficiency of electric motors varies with the type and size of the motor and the power that the motor delivers. Peak efficiencies in the range of 80 to 90% are common. The efficiency of a particular motor varies as the load varies. A typical graph for efficiency versus output power of an electric motor is shown in Figure 7.15.

Measuring Efficiency   Although determining the exact sources and magnitudes of losses in an electric motor may be complicated, experimental determinations of efficiency may be made with relative ease by the measurement of the electrical input power and the mechanical output power. For dc motors, the input power may be determined by the measurement of the input voltage and current and multiplying to obtain power. In ac motors, the power factor also must be considered. The task of measuring the input power is simplified by the use of an ac wattmeter that measures electrical power directly in watts.

The mechanical output power is the product of the angular velocity of the motor and the torque delivered. The rotational rate may be measured with a tachometer or with a stroboscope. Torque is measured by a dynamometer or with a prony brake. Mechanical power then is determined by Equation 7.7:

$$P_{out} = \tau\omega$$

where

P_{out} = mechanical output power
τ = torque
ω = angular velocity

Example 7.13 shows the use of this equation in calculating the efficiency of an electric motor.

EXAMPLE 7.13 Efficiency calculation
. . . .

Given:　　A dc electric motor has an input of 18 A at 24 V. It rotates at 1500 rpm and produces a force of 26.5 oz on a prony brake with a lever arm of 1 ft.

Find:　　Efficiency of motor.

Solution:　　Input power:

$$P_{in} = IV$$
$$P_{in} = (18 \text{ A}) (24 \text{ V})$$
$$P_{in} = 432 \text{ W}$$

Power out:
　Torque:

$$\tau = Fl$$

$$\tau = (26.5 \text{ oz})\left(\frac{1 \text{ lb}}{16 \text{ oz}}\right)(1 \text{ ft})$$

$$\tau = 1.66 \text{ ft·lb}$$

Angular velocity:

$$\omega = (1500 \text{ rev/min})\left(\frac{2\pi \text{ rad}}{1 \text{ rev}}\right)\left(\frac{1 \text{ min}}{60 \text{ sec}}\right)$$

$$\omega = 157 \text{ rad/sec}$$

$$P_{out} = \tau\omega$$

$$P_{out} = (1.66 \text{ ft·lb})(157 \text{ rad/sec})$$

$$P_{out} = (260.6 \text{ ft·lb/sec})$$

$$P_{out} = (260.6 \text{ ft·lb/sec})\left(\frac{1 \text{ hp}}{550 \text{ ft·lb/sec}}\right)$$

$$P_{out} = 0.474 \text{ hp}$$

$$P_{out} = (0.474 \text{ hp})(746 \text{ W/hp})$$

$$P_{out} = 353.6 \text{ W}$$

Efficiency:

$$\eta = \frac{P_{out}}{P_{in}}(100)$$

$$\eta = \frac{353.6 \text{ W}}{432 \text{ W}}(100)$$

$$\eta = 81.9\%$$

SECTION 7.3 EXERCISES

1. Write two equations for power in the electrical energy system.

2. Calculate the efficiency of a system in which an electric motor is used to turn a flywheel attached to a load if the motor draws 7.5 A when connected to a 240-V line. The flywheel requires a constant torque of 42 ft·lb to maintain a constant angular velocity of 250 rpm. (1 ft·lb/sec = 1.356 W.)

3. A current of 10 A flows through a 2-Ω resistor for 1 min. The conversion of electrical energy to thermal energy has an efficiency of 100%. How much heat energy is produced? (1 cal = 4.186 J.)

4. What is the resistance of a given resistor if a current of 30 mA flowing through the resistor dissipates 5×10^{-3} cal/sec?

5. How much charge can be moved through a resistor with one end at -3.0 V with respect to ground and the other at $+8.0$ V with respect to ground in 1 min if the power required is 40 W?

6. What should be the operating voltage of a motor if it is to lift 20 kg through a distance of 40 m in 200 sec? The motor draws a current of 0.5 A and is 70% efficient.

7. What is the current flowing through a 1-kΩ resistor if, when the resistor is immersed in a calorimeter filled with 100 g of water, the temperature of the water increases by 0.2 C°/sec? The calorimeter has a water equivalent of 100 cal/C°.

7.4

POWER IN THERMAL SYSTEMS

OBJECTIVES

Upon completion of this section, the student should be able to

- List the SI and English units used to define power for a thermal system.

- Given any two of the following quantities in a thermal system, determine the third:
  Work (or Forcelike Quantity times Displacementlike Quantity)
  Elapsed time
  Power

- Given any two of the following quantities in a thermal system, determine the third:
  Forcelike quantity
  Rate
  Power

- Given any two of the following quantities in a thermal system, determine the third:
  Input power
  Output power
  Efficiency

- Given the flow rate and the input and output temperatures of a heat exchanger, calculate the thermal power.

- Given a cylindrical heat exchanger, the input and output temperatures, the flow rate, and the heat transfer coefficient, calculate the required area.

DISCUSSION

Heat exchangers are used in a wide variety of applications. In general, their purpose is to dissipate unrecoverable power from a system. In this section, several factors that determine the efficiency of a heat exchanger are examined.

Thermal Power

Definition of Thermal Power The thermal power may be expressed by the equation

$$\text{Thermal Power} = \frac{\text{Heat Energy Transferred}}{\text{Time Required}}$$

$$P_Q = \frac{\Delta Q}{\Delta t} \qquad (7.34)$$

The thermal power may be expressed in cal/sec, Btu/hr, W, ft·lb/sec, or any other unit of power. If the thermal power is included in a general power equation, the units must be consistent with the units used for the other terms in the equation.

Power Equation   The general energy equation for a fluid was derived in Chapter 6. Each term in the energy equation is the energy per unit mass. It was shown that for fluid systems each term can be changed to work-per-unit-time by multiplying each term by $Q_V\rho_m$, where $Q_V\rho_m$ is the volume flow rate. For incompressible fluids, the density ρ_m remains constant. The general power equation for turbulent flow becomes

$$\frac{Q_V\rho_m}{2}[v_b^2 - v_a^2] + Q_V\rho_m g[h_b - h_a] + Q_V[p_b - p_a] \tag{7.35}$$

$$+ Q_V\rho_m c_V[T_b - T_a] = \frac{\Delta Q}{\Delta t}$$

$$\begin{bmatrix}\text{Kinetic}\\\text{power}\end{bmatrix} + \begin{bmatrix}\text{Potential}\\\text{power}\end{bmatrix} + \begin{bmatrix}\text{Pressure}\\\text{power}\end{bmatrix} + \begin{bmatrix}\text{Internal}\\\text{energy power}\end{bmatrix} = \begin{bmatrix}\text{Thermal}\\\text{power}\end{bmatrix}$$

If it is desirable, pump power also may be included in the equation. The letters a and b represent any two cross-sectional areas, but the heat source or sink represented by the heat power must lie between the two cross sections.

For many heat transfer problems, the sections a and b are taken at the inlet and outlet of a heat exchanger. For this case, h_a is usually approximately equal to h_b. Thus, the potential power term drops out. In addition, the inlet and outlet cross-sectional areas are usually the same. Under this condition for incompressible flow, $v_a = v_b$; the kinetic power term becomes 0. The power equation then reduces to

$$P_Q = \frac{\Delta Q}{\Delta t} = Q_V(p_b - p_a) + Q_V\rho_m c_V(T_b - T_a) \tag{7.36}$$

Because it was assumed that there was no pump between the two cross sections, the term $Q_V(p_b - p_a)$ represents the frictional loss in the exchanger. In most heat exchangers this term is small compared to $Q_V\rho_m(T_b - T_a)$. Thus, a good working equation for most cases becomes

$$P_Q = Q_V\rho_m c_V (T_b - T_a) \tag{7.37}$$

Equation 7.37 shows that the power of a heat exchanger is found by multiplying the specific heat of the incompressible fluid times the volume flow rate times the density times the temperature difference across the exchanger. Example 7.14 shows the use of Equation 7.37 to solve a problem.

EXAMPLE 7.14   Determining the power of a heat exchanger

Given: Heat exchanger shown in the sketch and

$$T_a = 80°C$$

$$T_b = 50°C$$

$$\Delta V/\Delta t = 10.2 \text{ l/min}$$

$$c = 1 \text{ cal/g·C° (water)}$$

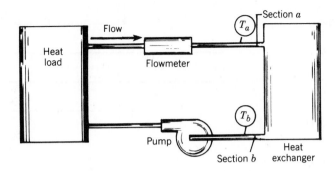

Find: The power of the heat exchanger.

Solution:

$$P_Q = Q_V \rho_m c_V (T_b - T_a)$$

Convert l/m to cm³/sec.

$$(10.2 \text{ l/m})(1000 \text{ cm}^3/\text{l})\left(\frac{1 \text{ min}}{60 \text{ sec}}\right) = 170 \text{ cm}^3/\text{sec}$$

$$P_Q = (170 \text{ cm}^3/\text{sec})(1 \text{ g/cm}^3)(1 \text{ cal/g·C°})(50°C - 80°C)$$

$$P_Q = 5.10 \times 10^3 \text{ cal/sec}$$

$$P_Q = (5.10 \times 10^3 \text{ cal/sec})(4.19 \text{ J/cal})$$

$$P_Q = 2.14 \times 10^3 \text{ W}$$

$$P_Q = 2.14 \text{ kW}$$

Cylindrical Heat Exchangers

In the chapter on resistance, the overall heat transfer coefficient was defined by Equation 5.114:

$$P_Q = \frac{\Delta Q}{\Delta t} = U_o A \, \Delta T_o \tag{7.38}$$

where

P_Q = total heat transfer rate
A = area
U_o = overall heat transfer coefficient
ΔT_o = overall temperature difference

The overall heat transfer coefficient was calculated for plane surfaces, and it can be calculated for cylindrical heat exchangers as shown in Figure 7.16. The area used for the overall heat transfer coefficient for cylindrical heat exchangers is the outer area of the inner tube.

Because the overall temperature difference ΔT_o is not a constant as one moves down the exchanger, an average overall temperature difference must be used. Typical temperature plots are shown in Figure 7.17 for cylindrical exchangers with parallel flow and counterflow.

FIGURE 7.16 Cylindrical
heat exchanger.

FIGURE 7.17 Temperature
differences in cylindrical
heat exchangers using
parallel and counterflow.

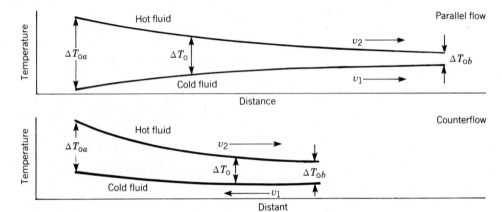

The calculation of the proper average value of ΔT_o requires the use of calculus. For the reason, only the answer will be given. The proper average is called the log-mean-temperature difference, and it is given by

$$\Delta T_o = \frac{\Delta T_{ob} - \Delta T_{oa}}{\ln \left[\dfrac{\Delta T_{ob}}{\Delta T_{oa}} \right]} \tag{7.39}$$

where

   ΔT_o = log-mean-temperature difference
   ΔT_{oa} = overall temperature difference at one end of the exchanger
   ΔT_{ob} = overall temperature difference at the other end
   ln = natural logarithm

Thus, the heat power for a cylindrical exchanger may be expressed as

$$P_Q = U_o A \, \Delta T_o \tag{7.40}$$

For other types of heat exchangers, there are plots of correction factors that can be used with Equation 7.40. Equation 7.40 can then be expressed as

$$P_Q = K U_o A \, \Delta T_o \tag{7.41}$$

where

   K = a correction factor

Example 7.15 shows the application of Equations 7.40 and 7.41 to a problem.

EXAMPLE 7.15   Heat exchanger area

· ·

Given: Water at the rate of 0.050 ft³/sec is heated from 100°F to 140°F by oil in a cylindrical heat exchanger. The oil enters the exchanger at 200°F and leaves at 150°F. The overall heat transfer coefficient is 50 Btu/hr·ft³·F°. The oil has a specific heat of 0.43 Btu/lb·F°. The fluids are used in counterflow.

Find: The required area.

Solution: The power of the heat exchanger is (use water values)

$$P_Q = Q_V \rho_m c_V (T_b - T_a)$$

$$P_Q = (0.050 \text{ ft}^3/\text{sec})(6.14 \text{ lb/ft}^3)(1 \text{ Btu/lb·F°})(140°F - 100°F)$$

$$P_Q = 123 \text{ Btu/sec}$$

$$P_Q = 4.42 \times 10^5 \text{ Btu/hr}$$

The log-mean-temperature difference for counterflow is

$$\Delta T_o = \frac{\Delta T_{ob} - \Delta T_{oa}}{\ln \left[\dfrac{\Delta T_{ob}}{\Delta T_{oa}} \right]}$$

$$\Delta T_{ob} = 200°F - 140°F = 60 \text{ F}°$$

$$\Delta T_{oa} = 150°F - 100°F = 50 \text{ F}°$$

$$\Delta T_o = \frac{(60 \text{ F}° - 50 \text{ F}°)}{\ln \left(\dfrac{60 \text{ F}°}{50 \text{ F}°} \right)}$$

$$\Delta T_o = 54.8 \text{ F}°$$

The value of P_Q from Equation 7.41 is

$$P_Q = UA\Delta T_o$$

$$A = \frac{P_Q}{U\Delta T_o}$$

$$A = \frac{4.42 \times 10^5 \text{ Btu/hr}}{(50.0 \text{ Btu/hr·ft}^2·F°)(54.8 \text{ F}°)}$$

$$A = 161 \text{ ft}^2$$

If the inlet or the exit temperatures are unknown the problem is more difficult because the log-mean-temperature difference cannot be directly calculated. However, if the flow rates, area, and overall heat transfer coefficient are known, the temperatures can be determined by a trial-and-error method.

Application of Heat Exchangers

The solid-to-air heat exchanger in Figure 7.18 consists of metal shaped to have a large surface area. Metal is chosen because it is an excellent thermal conductor. The high-temperature component to be cooled is connected directly to the heat exchanger. Heat flows into the conducting metal, raising its temperature. Forced cooler air, flowing past the hot metal surface, carries the energy away. This type of heat exchanger, often called a "heat sink," frequently is employed in the electronics industry and in air-cooled engines.

The liquid-to-air heat exchanger consists of a recirculating liquid system that moves internal energy from the high-temperature component or heat load to a large-surface-area radiator. The automobile cooling system, a common liquid-to-air heat exchanger, is shown in Figure 7.19. Forced air from the fan cools the radiator and the hot liquid in it. The cool liquid then returns to the heat source, the hot engine, to draw away more heat from the pistons. Because a liquid is used, this system is more efficient than the solid-to-air type and provides a lower operating temperature. Liquid-to-air heat exchangers are used to cool internal combustion engines. Water is used as the energy-transfer liquid, because it has a high heat capacity and is cheap, available, and relatively noncorrosive.

Liquid-to-liquid heat exchangers are even more efficient, because the "radiator" of this system is immersed in a flowing liquid, such as water. This type of heat exchanger, shown in Figure 7.20, is often found on electronic equipment requiring deionized water cooling. Figure 7.20 shows a fluid circulating between a high-temperature laser and a heat exchanger. Fluid is heated by the laser, piped to the heat exchanger, where it circulates through coils immersed in running water, then returned to the laser. At the heat exchanger, the hot fluid transfers heat energy to the flowing water.

Refrigeration systems contain two heat exchangers (Figure 7.21). The heat load of the system is located at the evaporator. A spray of cooled liquid refrigerant enters the evaporator via the expansion valve. The pressure in the evaporator is relatively low (about 30 psi). The liquid absorbs thermal energy from the walls of the evaporator, providing the cooling or refrigeration of the unit. This heat energy causes the liquid refrigerant to vaporize. The vaporized liquid flows to the compressor, where its pressure is increased to about 150 psi. The gas then moves to the condenser, where it is cooled and reliquefied, giving up heat to the outside air.

FIGURE 7.18 Heat sink. (left)

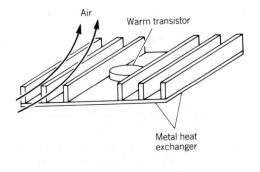

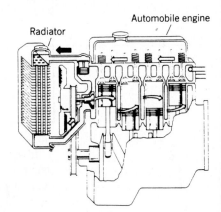

FIGURE 7.19 Automobile radiator. (right)

FIGURE 7.20 Laser cooling system.

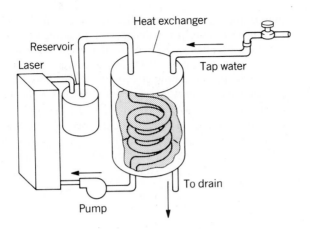

FIGURE 7.21 Refrigeration system.

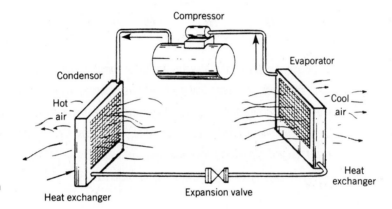

The gas condenses back to a liquid as the heat energy is transferred to the outside air. The liquid then flows back to the expansion valve to repeat the process. Thus, in the refrigeration system, one heat exchanger is used to transfer heat from the unit to the circulating refrigerant, and the second heat exchanger is used to transfer heat from the circulating refrigerant to the outside air.

Heat exchangers may be purchased. The selection will depend on the cost and the specifications. In some situations, it may be desirable to design a heat exchanger for a specific application. In any event, there are a number of factors to be evaluated. These include cost, size, pressure requirements, and heat transfer requirements. The overall heat transfer coefficient can be increased by increasing the velocity, but this requires a larger pressure drop across the exchanger. On the other hand, a smaller heat transfer coefficient requires a larger area for the same power. Thus, good judgment is needed to meet the desired requirements at a reasonable cost.

SECTION **7.4** *EXERCISES*

1. A heat exchanger using water as a coolant has an inlet temperature of 52°C and an outlet temperature of 31°C. The flow rate is 7.62 l/min. What is the power in watts?

2. *A source of constant power is located in a cylinder with copper walls. A second cylinder surrounds the first cylinder. Water passing through the coaxial cylinder is used as a coolant. Heat is transferred to the water through the copper wall of the first cylinder. The temperature of the water increases 10 C° after flowing between the coaxial cylinders, but the inner cylinder melts. Why does this happen? What can be done to increase the heat transfer? What limits the maximum heat transfer rate?

3. Water at the rate of 0.10 ft³/sec is heated from 30°C to 60°C in a cylindrical heat exchanger by oil. The oil enters the exchanger at 90°C and leaves at 80°C. The overall heat transfer coefficient is 60 Btu/hr·ft³·F°. The oil has a specific heat of 0.44 Btu/lb·F°. The flow in the exchanger is parallel. What is the required area?

* Challenge question

S U M M A R Y

In this chapter, the idea of power is applied to common problems in electrical, fluid, mechanical rotational, and mechanical translational systems. Power is the rate of energy consumption; in any given energy system, it is the product of the forcelike quantity and rate appropriate to that system. Thermal power also is defined and applied to heat exchangers. The equations for power in the different energy systems are summarized in Table 7.2.

TABLE 7.2   Summary of various energy systems

System	Forcelike Quantity	Ratelike Quantity	Power (common units)
Mechanical translational	Force (F)	Velocity (v)	$P = Fv$ (ft·lb/sec, hp, N·m/sec)
Mechanical rotational	Torque (τ)	Angular velocity (ω)	$P = \tau\omega$ (ft·lb/sec, hp, N·m/sec)
Fluid	Pressure (p)	Volume flow rate (Q_V)	$P = p\,\Delta V/\Delta t$ (ft·lb/sec, hp, N·m/sec)
Electrical	Voltage (V)	Current (I)	$P = (\Delta V)I$ (W)
Thermal	Temperature (T)	Flow rate ($\rho_m Q_V c_V$) (incompressible)	$P_Q = (\Delta T)(\rho_m Q_V c_V)$ (cal/sec, Btu/hr)

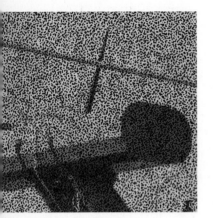

Force

Transformers

INTRODUCTION

There are times when nature seems to give something for nothing. However, the careful comparison of an output to the corresponding input shows that nature insists on the conservation of energy. More energy cannot be taken out of a system than is put into the system.

This chapter treats a class of devices called "force transformers." In such devices, the output force is different from the input force, but the output energy is never greater than the input energy. In an ideal case in which no energy is lost to frictional forces, the work output equals the work input. Force transformers are simply machines having great diversity and having a wide variety of applications, but the basic explanation of their usefulness is the same. The unified approach taken in the preceding chapters will assist in understanding force transformers in different energy systems as examples of the same physical principle.

OBJECTIVES ....

Upon completion of this section, the student should be able to

• Define in general terms
  Ideal mechanical advantage of forcelike transformers
  Actual mechanical advantage of forcelike transformers
  Efficiency of forcelike transformers

DISCUSSION

Force Transformers   The prospect of changing a flat tire without a lever jack would be a dismal one. With a jack, however, the relatively weak muscular forces of the human body are transformed, or magnified, sufficiently to lift one end of a two-ton car off the ground. Levers and screw jacks are examples of force transformers in mechanical systems.

An example of force transformers in fluid systems is a barber's chair. When the barber wishes to elevate the customer to a more convenient height, the barber makes use of a hydraulic pump. By a few effortless motions of a lever, the client is raised with the chair.

Also consider force transformers in electrical systems. Good examples of these devices are electric calculators and cassette recorders that operate from a 110-volt line. Ordinarily, these devices operate from batteries, which typically provide 6 volts. The accommodation is made by "stepping down" the 110 volts to the required 6 volts by use of a "6-volt transformer." Here, voltage, rather than mechanical force, is being transformed, but recall that in the electrical system, voltage is the "forcelike quantity." Thus, the lever/screw jack and the 6-volt transformer appear to use similar physical principles.

In each of the examples considered above, there is a coupling device, or transformer, that receives an input from some source and produces an output fed to some load, as shown schematically in Figure 8.1.

Reexamination of the three examples of force transformers given above, according to this scheme, shows that in the first example the source is an individual operating the jack. The input is the muscular force exerted by the person's arm. The coupling device is the screw jack. The load is the car, and the output is the force required to lift one end of the car off the ground. (These facts are summarized in Figure 8.2). In this case, the function of the screw jack is to multiply the input force sufficiently to lift the car with the required output force.

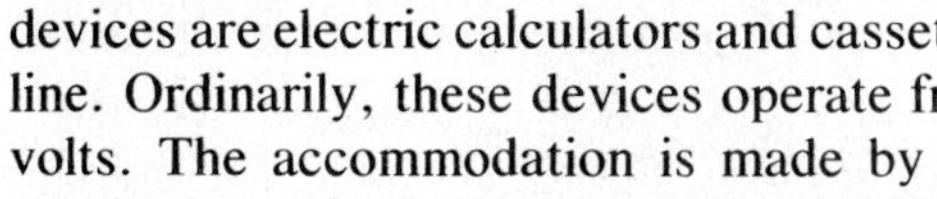

FIGURE 8.1 Generalized "forcelike transformer."

FIGURE 8.2 Force transformer used to lift a car.

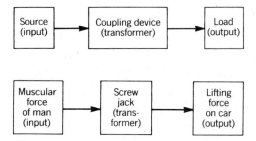

FIGURE 8.3   Barber chair.

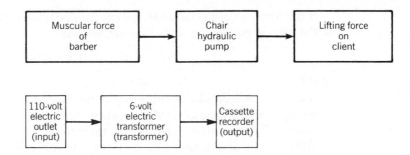

FIGURE 8.4   Six-volt transformer.

The other examples can be diagrammed in the same way (Figures 8.3 and 8.4). Energy is produced by work done at the source and is consumed as work done on the load. Recall from Chapter 2 that

$$\text{Work = Forcelike Quantity} \times \text{Displacementlike Quantity} \qquad (8.1)$$

Efficiency   The coupling device (force transformer) transfers the energy from the source to the load. Although the forcelike quantities and the displacementlike quantities at each end may have different magnitudes, the product of these quantities (which is the energy transferred from input to output) is ideally the same. In practice, the output energy will be less, because some energy will be dissipated as heat from friction in the operation of the transformer. Energy is conserved overall, but some is wasted in overcoming resistance. Reducing these losses is important in making highly efficient transformers. The percent efficiency of a transformer is defined by

$$\eta = \frac{E_{\text{out}}}{E_{\text{in}}}\,(100) \qquad (8.2)$$

where

η = percent efficiency
E_{out} = useful output energy
E_{in} = input energy
100 = multiplier to obtain percent

When useful output energy equals input energy (the ideal case), no energy is lost in the coupling device, and the formula gives 100% efficiency. In actual cases, the output energy is less, and the efficiency is below 100%. In no case can the efficiency be greater than 100%.

Mechanical Advantage and Efficiency   The actual mechanical advantage of a system is defined by

$$\text{AMA} = \frac{\text{Forcelike quantity out}}{\text{Forcelike quantity in}} = \frac{(\text{GF})_{\text{o}}}{(\text{GF})_{\text{i}}} \qquad (8.3)$$

The quantity AMA is the actual mechanical advantage of the system. Since the actual advantage depends on the frictional forces in the system, it is usually not easy to calculate. The value of AMA, however, can be determined by measuring the forcelike quantity out of the system and the forcelike quantity into the system.

If friction is neglected and the system does not store energy, then the work out of the system equals the work into the system:

$$(GF)_o (GD)_o = (GF)_i (GD)_i \qquad (8.4)$$

where

$(GF)_o$ = the forcelike quantity out of the system
$(GD)_o$ = the displacementlike quantity out of the system
$(GF)_i$ = the forcelike quantity into the system
$(GD)_i$ = the displacementlike quantity into the system

Dividing both sides of Equation 8.4 by the product $(GF)_i (GD)_o$ yields

$$IMA = \frac{(GF)_o}{(GF)_i} = \frac{(GD)_i}{(GD)_o} \qquad (8.5)$$

where

IMA = ideal mechanical advantage

The efficiency can be expressed in terms of the actual mechanical advantage and the ideal mechanical advantage. Since the percent efficiency is defined as the work out divided by the work in times one hundred,

$$\eta = \frac{(GF)_o (GD)_o}{(GF)_i (GD)_i} (100)$$

or

$$\eta = \frac{(GF)_o / (GF)_i}{(GD)_i / (GD)_o} (100)$$

or

$$\eta = \frac{AMA}{IMA} (100) \qquad (8.6)$$

Generalized Power   If Equation 8.4 is divided by the corresponding time one obtains

$$\frac{(GF)_o (GD)_o}{t} = \frac{(GF)_i (GD)_i}{t} \qquad (8.7)$$

Since the ratelike quantity is equal to the displacementlike quantity divided by the corresponding time, Equation 8.7 becomes

$$(GF)_o (GR)_o = (GF)_i (GR)_i \qquad (8.8)$$

where

$(GR)_o$ = ratelike quantity out
$(GR)_i$ = ratelike quantity in

Dividing Equation 8.8 on both sides by the product $(GF)_i (GR)_o$ yields

$$\text{IMA} = \frac{(\text{GF})_o}{(\text{GF})_i} = \frac{(\text{GR})_i}{(\text{GR})_o} \qquad (8.9)$$

Equation 8.8 states that the power into an ideal system is equal to the power out of an ideal system, whereas Equation 8.9 states that the ratio of the input ratelike quantity to the output ratelike quantity is equal to the ideal mechanical advantage.

8.1

FORCE TRANSFORMERS IN MECHANICAL SYSTEMS

OBJECTIVES

Upon completion of this section, the student should be able to

- Describe with the aid of a diagram various pulley systems and discuss how they act as force transformers.

- Calculate the ideal mechanical advantage of any pulley system.

- Describe the differential hoist and calculate its ideal mechanical advantage.

- Given the actual input and output forces of any system, calculate the actual mechanical advantage.

- Calculate the efficiency of a system from the ideal and actual mechanical advantages.

- Calculate the ideal mechanical advantage of any lever system.

- Sketch and identify the three classes of levers and give applications of each type.

- Calculate the ideal mechanical advantage of an incline plane, screw, and wedge.

- Calculate the force present when a wedge of known angle is driven into an object by a known force.

- Calculate the ideal mechanical advantage of a wheel and axle system.

- Calculate the ideal mechanical advantage in the following systems:
  Gear drive system, given number of teeth on each gear
  Belt drive system, given radius of each wheel
  Disk drive system, given radius of each disk

- Given the ideal mechanical advantage of a drive system, input torque, and angular velocity, calculate output torque and angular velocity.

DISCUSSION

Pulleys

Simple Pulley   A simple pulley is shown in Figure 8.5. A weight F_o is lifted by pulling on the cord with a force F_i. If there were no friction and no acceleration, the tension in the cord would be equal to the body's weight. Since some friction exists in the real case, F_i will be greater than F_o.

FIGURE 8.5 Simple pulley.

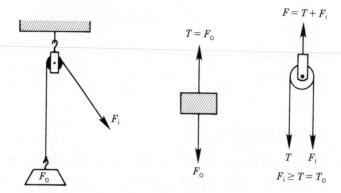

The simple pulley does not multiply the magnitude of the force since $F_i > F_o$. The system does, however, change the direction of the force. When the rope is pulled down, the weight moves up. In addition, if the cord does not stretch, the magnitude of the velocity v_i will be equal to the magnitude of the velocity v_o. Therefore, for the ideal simple pulley

$$F_o = F_i$$

$$IMA = \frac{F_o}{F_i} = 1$$

$$v_o = v_i$$

Multiple Pulleys   The system of pulleys pictured in Figure 8.6 changes the magnitude as well as the direction of the force. For the ideal case, the tension will be the same throughout the cord. Figure 8.6b shows a free-body diagram. For equilibrium, the sum of the verticle forces yields

$$F_o - 2 F_i = 0$$

$$F_o = 2 F_i$$

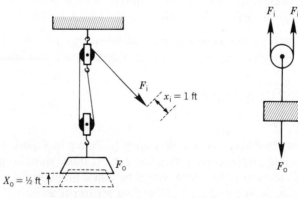

FIGURE 8.6 System of two pulleys.

a. Space diagram

b. Free-body diagram

This result can also be obtained by using the conservation of energy (input work = output work). When F_i is moved a distance of one foot the load moves up one-half foot. Thus,

$$F_o\, x_o = F_i\, x_i$$
$$(F_o)\,(1/2) = (F_i)\,(1)$$
$$F_o = 2\,F_i$$

Neither friction nor acceleration of the body is considered.

For the ideal case, the body is lifted by a force equal to one-half its weight. This force transformer changes the magnitude and direction of the input force. One pays, however, by pulling twice as far as the body rises. What is gained as a force advantage is lost as a displacement advantage; but, for the ideal case, the input work is equal to the output work.

The ideal mechanical advantage of this system is

$$\text{IMA} = \frac{F_o}{F_i} = \frac{2\,F_i}{F_i} = 2$$

Again, the actual mechanical advantage will be less than two because some friction always exists.

To determine the ideal mechanical advantage of any system of pulleys, one can take a free-body diagram that includes the load and cut through all cords that support the load. An example is shown in Figure 8.7. Calculations for different pulley systems are given in Examples 8.1, 8.2, and 8.3.

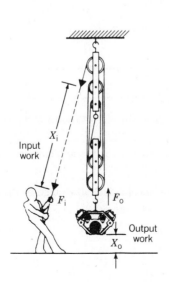

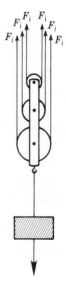

FIGURE 8.7 Multiple pulley system.

a. Space diagram

b. Free-body diagram

EXAMPLE 8.1 The pulley as a force transformer
. . . .

Given: In the pulley system shown in Figure 8.6, a load of 400 lb is lifted 5 ft. (Assume that this system is an ideal transformer.)

Find:
  a. The amount of work that is done by the source.
  b. The applied force.
  c. The distance that the input force moves.

Solution: In this case, the pulley has an ideal mechanical advantage of 2.

  a. Input work:

$$W_i = W_o$$

$$W_o = F_o\, x_o$$

$$W_o = (400 \text{ lb}) (5 \text{ ft})$$

$$W_o = 2000 \text{ ft·lb}$$

  b. Applied force:

$$F_i = \frac{F_o}{\text{IMA}}$$

$$F_i = \frac{400 \text{ lb}}{2}$$

$$F_i = 200 \text{ lb}$$

  c. Since $\text{IMA} = x_i/x_o$, the input force displacement corresponding to the load displacement of 5 ft can be found from

$$x_i = \text{IMA}\, x_o$$

$$x_i = (2) (5 \text{ ft})$$

$$x_i = 10 \text{ ft}$$

EXAMPLE 8.2 Mechanical advantage and efficiency of a pulley system
. . . .

Given: When actually measured, the applied force in Example 8.1 is found to be 225 lb.

Find:
  a. The actual mechanical advantage of the pulley.
  b. Its percent efficiency.

Solution:
  a. $\text{AMA} = \dfrac{F_o}{F_i}$

$$\text{AMA} = \frac{400 \text{ lb}}{225 \text{ lb}}$$

$$AMA = 1.78$$

b. $\eta = \dfrac{AMA}{IMA} (100)$ (from Equation 8.6)

$$\eta = \frac{1.78}{2} (100)$$

$$\eta = 89\%$$

EXAMPLE 8.3 Mechanical advantage of a block and tackle

. . . .

Given: Pulley systems *a* and *b* as shown. This combination of pulleys is called a "block and tackle."

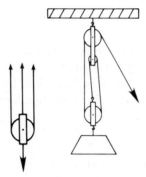 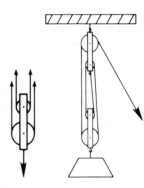

Find: The ideal mechanical advantage of the pulleys shown.

Solution: **a.** F_i must move down vertically three times as much as F_o moves up, if the cord does not stretch. Thus,

$$IMA = \frac{x_o}{x_i}$$

$$IMA = \frac{3}{1}$$

$$IMA = 3$$

              b. In this case by similar reasoning,

$$IMA = \frac{4}{1}$$

$$IMA = 4$$

Notice that IMA turns out to be simply the number of cords supporting the load, but excluding the cord directly bearing the input force.

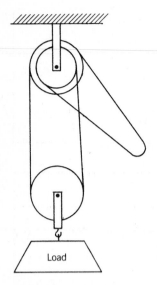

FIGURE 8.8   Chain hoist.

The Differential Hoist · Probably the most familiar differential hoist is the chain hoist used in auto repair shops and in factories (Figure 8.8).

In a differential hoist, the differential pulley at the top is composed of two parts: a large-diameter (D) wheel and a small-diameter (d) wheel. The two wheels are fixed so that they turn together. A chain is normally used on a differential hoist, and the wheels are notched so that the chain cannot slip.

To raise the load, the side of the chain that is looped over the large upper wheel is pulled. This causes both upper wheels to turn through a common angle $\Delta\theta$. In doing this the applied force moves through a distance x_i. This distance is related to the upper pulley rotation by

$$X_i = \frac{D}{2}\,\Delta\theta \qquad (8.10)$$

At the same time the other side of the chain passes over the smaller upper wheel and, as this wheel also turns through the angle $\Delta\theta$, the lower half of the chain loop moves through a distance x_s given by

$$x_s = \frac{d}{2}\,\Delta\theta \qquad (8.11)$$

Both of these movements are transmitted to the load. It tends to move up a distance x_i and down a distance x_s simultaneously. The total load movement (x_o) is half the difference in these two values because the total difference must be divided equally between the two supporting lines:

$$x_o = \frac{(x_i - x_s)}{2} \qquad (8.12)$$

With these three equations, the ratio x_i to x_o (the ideal mechanical advantage) can be determined. Dividing Equation 8.10 by Equation 8.12 yields

$$\text{IMA} = \frac{x_i}{x_o}$$

$$\text{IMA} = \frac{D\,(\Delta\theta)/2}{(x_i - x_s)/2}$$

$$\text{IMA} = \frac{D(\Delta\theta)}{(x_i - x_s)}$$

and substituting Equations 8.10 and 8.11 for x_i and x_s, respectively, on the right side gives

$$\text{IMA} = \frac{D(\Delta\theta)}{\left(\dfrac{D}{2}\right)(\Delta\theta) - \left(\dfrac{d}{2}\right)(\Delta\theta)}$$

Simplifying this expression gives

$$\text{IMA} = \frac{2D}{D - d} = \frac{x_i}{x_o} \qquad (8.13)$$

In Equation 8.13, one should notice that, if the diameters are nearly equal, the ratio x_i to x_o becomes very large.

Differential chain hoists have low efficiencies of 30 to 40% as a result of friction. If the actual mechanical advantage is measured, the efficiency can be determined from

$$\eta = \frac{AMA}{IMA} = AMA\left(\frac{D-d}{2D}\right) \tag{8.14}$$

The hand-operated chain hoist is widely used for lifting loads up to 1000 lb. If the diameters of the top wheels are nearly the same, the load will hold at any height to which it is raised; otherwise, care must be taken to keep the applied force on the chain to prevent the load from lowering (Example 8.4). This distance ratio can be advantageous when it is required to lower a load gradually to a precise position (such as an engine onto its bolt mounts) because small increments of lowering distance can be made accurately.

EXAMPLE 8.4 Differential hoist calculations

Given: The upper wheels of a chain hoist are 8 inches and 9 inches, respectively. The hoist is 80% efficient.

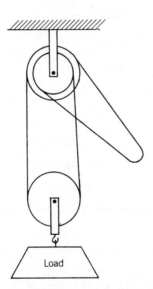

Find:
a. The maximum weight of an auto engine that can be lifted by an applied force of 50 lb.
b. How much chain must be pulled through the pulleys to raise the load by 2 ft?

Solution:
a. The ideal mechanical advantage of this hoist is, from Equation 8.13,

$$IMA = \frac{x_i}{x_o} = \frac{2D}{D-d}$$

$$IMA = \frac{2 \times 9 \text{ inch}}{9 \text{ inch} - 8 \text{ inch}}$$

$$IMA = 18$$

Since it is 80% efficient, the actual mechanical advantage is

$$AMA = \frac{(\eta)IMA}{100}$$

$$AMA = \frac{(80)(18)}{100}$$

$$AMA = 14.4$$

Thus, an applied force of 50 lb can lift an auto engine weighing

$$F_o = AMA(F_i)$$

$$F_o = 14.4 \ (50 \text{ lb})$$

$$F_o = 720 \text{ lb}$$

b. To find the distance through which the chain must be pulled, we recall from the principle of work that

$$IMA = \frac{x_i}{x_o}$$

Here

$$IMA = 18$$

$$x_o = 2 \text{ ft}$$

so

$$x_i = IMA(x_o)$$

$$x_i = (18) \ (2 \text{ ft})$$

$$x_i = 36 \text{ ft}$$

Levers

Class I Levers   The lever is one of the simplest and most common force transformers. A lever may be used to amplify either force or displacement or to change their directions.

Figure 8.9 shows a Class I lever. In a Class I lever, the fulcrum is always located between the load and the applied force. It is assumed that the lever is a rigid rod with no mass and that the forces are applied in the vertical direction. The force F_o is the load force and F_i is the input force. The force F_p is the force applied by the pivot or fulcrum.

In most applications of the lever, the system is very near equilibrium conditions (at rest or constant velocity). The conditions of equilibrium are

FIGURE 8.9 Class I lever.

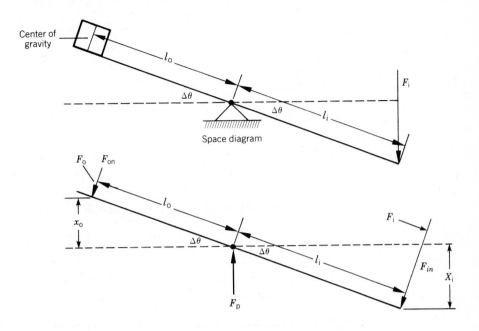

$$\Sigma\, F_v = F_i + F_o - F_p = 0$$
$$\Sigma\, \tau = F_{on}\, l_o = F_{in}\, l_i = 0 \qquad\qquad (8.15)$$

where

$\quad \Sigma\, F_v$ = sum of the vertical forces on the lever
$\quad \Sigma\, \tau$ = sum of the torques on the lever
$\quad F_{on}$ = component of F_o normal or perpendicular to l_o
$\quad F_{in}$ = component of F_i normal or perpendicular to l_i

From Figure 8.9, one can see that

$$F_{in} = F_i \cos \Delta\theta$$
$$F_{on} = F_o \cos \Delta\theta$$

Substituting these expressions into Equation 8.15 gives

$$(F_i \cos \Delta\theta)\, l_i = (F_o \cos \Delta\theta)\, l_o$$

or

$$F_i\, l_i = F_o\, l_o$$

or

$$\frac{F_o}{F_i} = \frac{l_i}{l_o} \qquad\qquad (8.16)$$

If friction is neglected, the work done by F_i equals the work out since it is assumed that there is no change in kinetic energy. This yields

$$F_o \, x_o = F_i \, x_i$$

$$\frac{F_o}{F_i} = \frac{x_i}{x_o} \tag{8.17}$$

However,

$$\sin \Delta\theta = \frac{x_o}{l_o} = \frac{x_i}{l_i}$$

Therefore,

$$\frac{l_i}{l_o} = \frac{x_i}{x_o}$$

Substituting this expression into Equation 8.5 yields

$$\text{IMA} = \frac{F_o}{F_i} = \frac{x_i}{x_o} = \frac{l_i}{l_o} \tag{8.18}$$

This last expression is consistent with Equation 8.16. Sample calculations for a Class I lever are given in Examples 8.5 and 8.6.

EXAMPLE 8.5   Class I lever arm calculation
. . . .

Given: The lever shown is used to raise a 200-lb load.

Find: **a.**   The value of F_i for equilibrium.
               b.   The height of the load if F_i moves down 0.5 ft.
               c.   The ideal mechanical advantage.

Solution: **a.** $\dfrac{F_i}{F_o} = \dfrac{l_o}{l_i}$

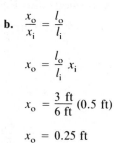

$$F_i = \frac{l_o}{l_i} F_o$$

$$F_i = \frac{3 \text{ ft}}{6 \text{ ft}} \, 200 \text{ lb}$$

$$F_i = 100 \text{ lb}$$

b. $\dfrac{x_o}{x_i} = \dfrac{l_o}{l_i}$

$$x_o = \frac{l_o}{l_i} x_i$$

$$x_o = \frac{3 \text{ ft}}{6 \text{ ft}} (0.5 \text{ ft})$$

$$x_o = 0.25 \text{ ft}$$

c. $\text{IMA} = \dfrac{l_i}{l_o}$

$$\text{IMA} = \frac{6 \text{ ft}}{3 \text{ ft}}$$

$$\text{IMA} = 2$$

EXAMPLE 8.6   Pliers as a lever
. . . .

Given: A pair of pliers

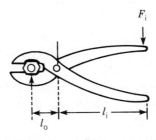

Find: The reason why a pair of pliers is effective in removing a stubborn nut.

Solution: The pair of pliers is actually a two-bar lever. If one of the handles (say, the lower one) is considered fixed, then the ideal mechanical advantage is

$$\text{IMA} = \frac{l_i}{l_o}$$

$$\text{IMA} > 1$$

Class II Lever   In a Class II lever, the fulcrum is at one end of the lever and the applied force is at the other end. The load is between the fulcrum and the applied force. In this type lever, the load force is always greater than the applied force. A Class II lever is shown in Figure 8.10.

Since work in equals work out for the ideal case,

$$F_o \, x_o = F_i \, x_i$$

$$\text{IMA} = \frac{F_o}{F_i} = \frac{x_i}{x_o}$$

From $\sin \Delta\theta$,

$$\frac{x_i}{l_i} = \frac{x_o}{l_o}$$

Thus,

FIGURE 8.10   Class II lever.

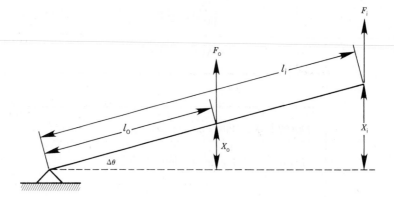

$$IMA = \frac{x_i}{x_o} = \frac{l_i}{l_o}$$

Therefore, the equations for a Class II lever have the same form as the equations for a Class I lever. Example 8.7 shows a sample calculation for a Class II lever.

EXAMPLE 8.7   Wheelbarrow load calculation
. . . .

Given: A loaded wheelbarrow weights 180 lb and its center of gravity is 2 ft from the wheel axle. The lifting force is applied 6 ft from the wheel axle.

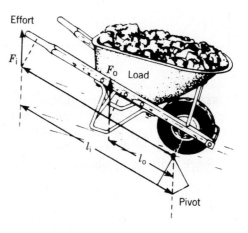

Find: **a.**   The applied force.

   b.   The distance the load rises when the applied force rises 15 inch.

Solution: **a.** $F_i = \dfrac{l_o}{l_i} F_o$

$$F_i = \frac{2\ ft}{6\ ft}\ 180\ lb$$

$$F_i = 60\ lb$$

b. $\quad x_o = \frac{l_o}{l_i}\ x_i$

$$x_o = \frac{2\ ft}{6\ ft}\ 15\ inch$$

$$x_o = 5\ inch$$

Class III Lever   In a Class III lever, the positions of the load force and applied force are exchanged as shown in Figure 8.11. The student can easily show that Equations 8.17 and 8.18 also apply to the Class III lever. Example 8.8 shows the application of the equations to a practical problem in anatomy.

EXAMPLE 8.8   Human anatomy strength calculation

. . . .

Given: The length of a person's arm from elbow to palm is 14 inch. When the biceps are shortened by 1/4 inch, the 25-lb weight in the hand is raised 1 3/4 inch.

Find: **a.**   The tension in the person's muscle.
           b.   The lever arm of the applied force.

Solution: **a.** $F_i = \dfrac{x_o}{x_i}\ F_o$

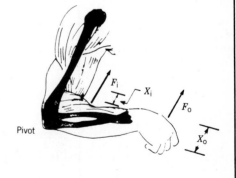

$$F_i = \frac{1.75\ inch}{0.25\ inch}\ 25\ lb$$

$$F_i = 175\ lb$$

b. $l_i = \dfrac{x_i}{x_o}\ l_o$

$$l_i = \frac{0.25\ inch}{1.75\ inch}\ 14\ inch$$

$$l_i = 2\ inch$$

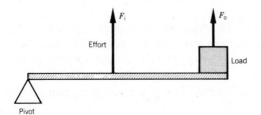

FIGURE 8.11   Class III lever.

Mechanical Advantage and Efficiency In the ideal case it is assumed that the lever has no mass, that there is no friction, and the kinetic energy is zero or constant. Because these assumptions are only approximations, the actual mechanical advantage is usually less than the ideal value. In general, the efficiency is given by Equation 8.6:

$$\eta = \frac{\text{AMA}}{\text{IMA}}\,(100)$$

Since for the lever $\text{IMA} = l_i/l_o$, the efficiency also may be written as

$$\eta = \text{AMA}\left(\frac{l_o}{l_i}\right)(100) \tag{8.19}$$

Example 8.9 illustrates the calculation of the efficiency of a lever.

EXAMPLE 8.9   Load-lifting using a crowbar
. . . .

Given: A crowbar used to lift a 72-lb load vertically with an applied force of 7.2 lb. The respective lever arms are 2.5 inch and 22 inch.

Find: The efficiency of the crowbar.

Solution: $\text{IMA} = \dfrac{l_i}{l_o}$

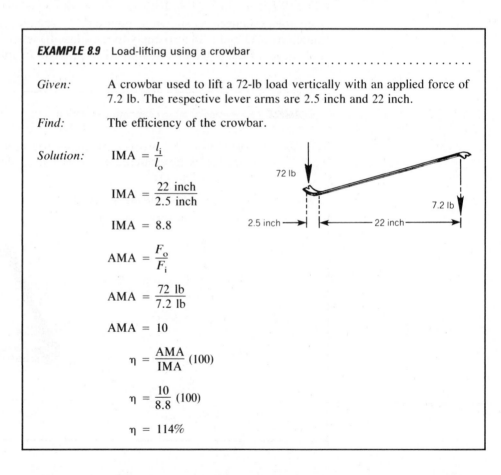

$\text{IMA} = \dfrac{22 \text{ inch}}{2.5 \text{ inch}}$

$\text{IMA} = 8.8$

$\text{AMA} = \dfrac{F_o}{F_i}$

$\text{AMA} = \dfrac{72 \text{ lb}}{7.2 \text{ lb}}$

$\text{AMA} = 10$

$\eta = \dfrac{\text{AMA}}{\text{IMA}}\,(100)$

$\eta = \dfrac{10}{8.8}\,(100)$

$\eta = 114\%$

Notice that the efficiency is greater than 100%. This is because the weight of the heavy metal bar has been neglected. The downward gravitational force acting at the center of gravity of the crowbar assists the applied force.

Application of Levers Every conscious person capable of movement uses one or

FIGURE 8.12 Applications of levers.

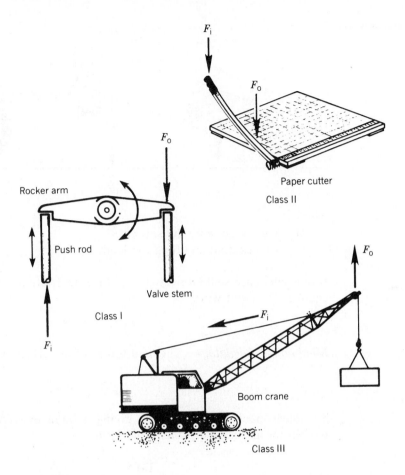

more of the lever classes each day. Examples of this simple and most common of all force transformers are abundant in contemporary life. The crowbar is one of the purest examples of the Class I lever. Other examples of Class I levers are scissors, pliers, and the rocker-arm assembly of an internal combustion engine. The paper cutter in Figure 8.12 shows one of the clearest examples of Class II levers. Class III levers are represented in the human body and machines such as the boom crane.

The Inclined Plane

Conservation of Energy Using a ramp (inclined plane) to load a crate into a truck is easier than directly lifting it (Figure 8.13). The inclined plane serves as another coupling device that enables a small force to do a big job at the expense of a displacement. In this case, the crate is rolled along the incline farther than it is lifted, that is, $x_i > x_o$. This sacrifice in displacement results in a mechanical advantage. Input work and output work are

$$W_i = F_i x_i$$

$$W_o = F_o x_o$$

FIGURE 8.13   Ramp used to
load a truck.

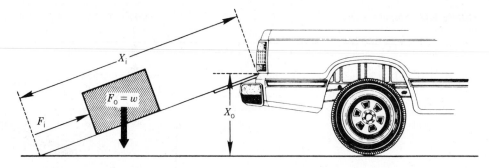

where

W_i = work done by the source
W_o = work done in lifting the load

If the rollers are well lubricated to reduce the friction, the output work is nearly equal to the input work, or

$$F_i x_i = F_o x_o$$

Mechanical Advantage   The load force is then given by

$$F_o = \frac{x_i}{x_o} F_i$$

The ideal mechanical advantage giving a force magnification of F_o/F_i can be written for an inclined plane in the form

$$\text{IMA} = \frac{x_i}{x_o} = \frac{1}{\sin \theta}$$

where

x_o = rise of incline
x_i = length of incline
θ = angle of incline

Equation 8.20 is used in Example 8.10 to solve a problem.

Notice that this equation is just another form of Equation 8.5. The ratio of incline to rise is sometimes called the grade of the incline, so that the IMA of an inclined plane is easily remembered as simply the grade of the incline (see Examples 8.10 and 8.11). This ideal mechanical advantage would not be achieved if it were necessary to slide (rather than to roll) the crate up the incline. Why?

EXAMPLE 8.10   Mechanical advantage of an inclined plane
. . . .

Given: A 300-lb safe is pushed up an incline 15 ft long and 5 ft high.
               (Assume the rollers are frictionless.)

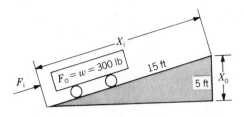

Find: **a.** The grade of the incline.

b. The IMA of the incline.

c. The input force required to roll the safe up the incline.

Solution: **a.** Grade $= \dfrac{x_i}{x_o}$

Grade $= \dfrac{15 \text{ ft}}{5 \text{ ft}}$

Grade $= 3$

b. IMA = Grade

IMA = 3

c. $F_i = \dfrac{x_o}{x_i} F_o$

$F_i = \left(\dfrac{5 \text{ ft}}{15 \text{ ft}} \right) (300 \text{ lb})$

$F_i = 100$

or

$F_i = \dfrac{F_o}{\text{IMA}}$

$F_i = \dfrac{300 \text{ lb}}{3}$

$F_i = 100 \text{ lb}$

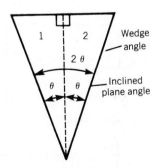

FIGURE 8.14 Wedge.

The Wedge

Mechanical Advantage and Efficiency Figure 8.14 shows a wedge. It is composed of two identical inclined planes (1 and 2) back to back. The wedge angle (2θ) is twice the angle (θ) of one of the inclined planes.

Figure 8.15 shows the forces present on a wedge. The force F_i is the total downward force on the wedge. It consists of the weight of the wedge plus the driving force applied to the wedge. Because the wedge and the forces acting on it are symmetrical about the center line, only one of the two inclined planes will be considered.

FIGURE 8.15 Forces
present on a wedge.

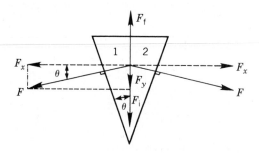

The downward force F_y on one of the inclined planes is half of F_i. This produces a normal force F_i that has F_y as a vertical component:

$$F_y = F \sin \theta$$

The horizontal component of this normal force is F_x. This is the force that does the splitting work of the wedge. It is given by

$$F_x = F \cos \theta$$

$$F_x = \left(\frac{F_y}{\sin \theta} \right) \cos \theta$$

$$F_x = \frac{F_y}{\tan \theta}$$

The total splitting force of the wedge is $2F_x$:

$$F_o = 2 F_x$$

$$F_o = 2 \frac{F_y}{\tan \theta}$$

$$F_o = \frac{F_i}{\tan \theta} \tag{8.21}$$

Example 8.11 demonstrates the use of Equation 8.21.

In Example 8.11, a vertical force of 200 lb on the wedge results in a splitting force of 746 lb on the log. The ratio of these numbers is the ideal mechanical advantage since friction is neglected.

EXAMPLE 8.11   Splitting a log
. . . .

Given: A wedge with a wedge angle of 30° used to split a log. A force of 200 lb is delivered on top of the wedge.

Find: Total splitting force applied to the log. Neglect the weight of the wedge compared to the 200-lb force.

Solution: The angle θ of each inclined plane is 30°/2 = 15°. The total splitting force is

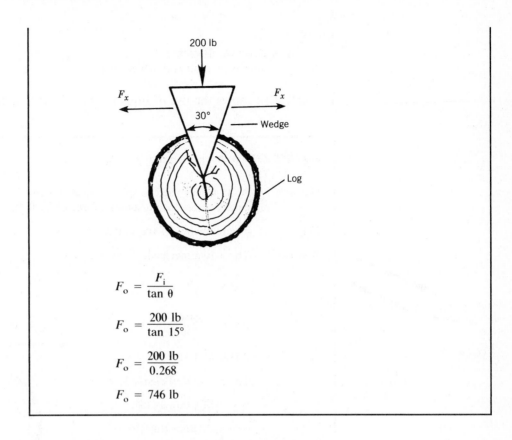

$$F_o = \frac{F_i}{\tan \theta}$$

$$F_o = \frac{200 \text{ lb}}{\tan 15°}$$

$$F_o = \frac{200 \text{ lb}}{0.268}$$

$$F_o = 746 \text{ lb}$$

The ideal mechanical advantage (no friction) is

$$\text{IMA} = \frac{F_o}{F_i}$$

Using Equation 8.21 to substitute for F_o yields

$$\text{IMA} = \frac{F_i}{F_i \tan \theta}$$

$$\text{IMA} = \frac{1}{\tan \theta}$$

$$\text{IMA} = \frac{2\,h}{b} \tag{8.22}$$

$$\tan \theta = \frac{b/2}{h} = \frac{b}{2h}$$

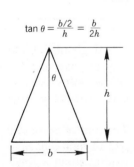

FIGURE 8.16 Dimensions of wedge used for tan θ.

The quantities h and b are shown in Figure 8.16. Thus, the ideal mechanical advantage is based on the wedge geometry.

The actual mechanical advantage will be less, and measured forces must be used to determine its value. The percent efficiency can then be expressed as

$$\eta = \frac{\text{AMA}}{\text{IMA}}\ (100)$$

$$\eta = \text{AMA} \ (\tan \theta) \ (100) \tag{8.23}$$

where

η = percent efficiency

AMA = actual mechanical advantage

Example 8.12 shows the use of Equation 8.23 in solving a problem.

EXAMPLE 8.12   Splitting a log with a wedge
. . . .

Given: A 20° wedge used to split a log. A downward force of 700 N is necessary to produce a splitting force of 2460 N.

Find: The efficiency of the wedge.

Solution: The actual mechanical advantage is

$$AMA = \frac{F_o}{F_i}$$

$$AMA = \frac{2460 \text{ N}}{700 \text{ N}}$$

$$AMA = 3.51$$

The percent efficiency is

$$\eta = AMA \ (\tan \theta)(100)$$

$$\eta = (3.51)(\tan 10)(100)$$

$$\eta = (3.51)(0.176)(100)$$

$$\eta = 62\%$$

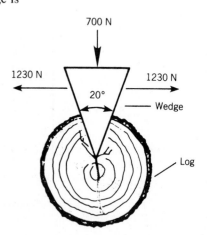

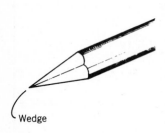

Wedge

a. Nail

Axis

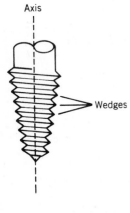

Wedges

b. Wood screw

FIGURE 8.17   Wedges used for piercing.

Applications   A wedge is used to separate two objects or split a single object into two parts. Obvious applications include cutting tools such as the axe and the knife. Other applications include the wedge-shaped bow that allows ships to travel at higher speeds. The wedge allows a small forward force to produce a large splitting force to move aside the water.

Piercing tools also employ the wedge. Figure 8.17 shows the point of a nail and a wood screw. The point of the nail is composed of two wedges at right angles to each other. The woodscrew consists of a wedge wound in a spiral about a cone. Notice that the cone-shaped piercing point is another application of the principle of the wedge.

MECHANICAL ROTATIONAL FORCE TRANSFORMERS

The Screw

Conservation of Energy   The screw is a clever adaptation of the incline plane in the rotational mechanical system. The ordinary wood screw, shown in Figure

FIGURE 8.18 Screw jack lifting the corner of a house.

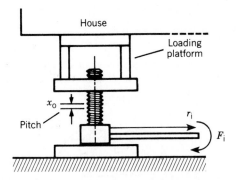

8.17*b*, is an inclined plane wrapped into a spiral around the screw axis.

To appreciate the transforming advantage of the screw, consider a type of screw jack used to raise the corner of a house (Figure 8.18). Rotation of the lever forces the loading platform on the screw upward, thus, lifting the load. Notice that this device combines two force transformers—the lever and the screw. When the handle is turned by an input force through the distance of one revolution, the load rises by a much smaller distance, equal to a linear displacement by one thread of the screw. This distance between threads of the screw usually is called the screw pitch. If the screw has five threads per linear inch, the load will be lifted one-fifth inch for each revolution of the screw.

Mechanical Advantage Equation 8.5 is used for the ideal mechanical advantage. The relative displacement of the force F_i and of the load F_o must be determined. For one revolution of the handle, the force moves through one circumference of a circle with radius r_i as shown. The corresponding displacement of the load is equal to the pitch of the screw. Therefore, the ideal mechanical advantage of a screw may be expressed using Equation 8.18 as

$$\text{IMA} = \frac{2\pi r_i}{x_o} \tag{8.24}$$

where

$2\pi r_i$ = circumference of input force circle
x_o = pitch of screw (distance between threads)

In this application of the screw, the load is usually so large that frictional forces of turning are appreciable, even when the screw is well lubricated. As a result, the actual mechanical advantage is considerably less than the ideal mechanical advantage given by Equation 8.24. As explained previously, the actual mechanical advantage can only be determined by measuring the input force F_i needed to move the load F_o. Then,

$$\text{AMA} = \frac{F_o}{F_i}$$

Equation 8.24 is used to solve problems in Examples 8.13 and 8.14.

EXAMPLE 8.13   Bench vise as a screw

. . . .

Given: The screw in a bench vise has a pitch of 0.25 inch. A worker applies a tangential force of 30 lb on the handle at a point 8 inch from the axis of the screw.

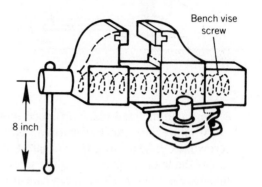

Bench vise screw

8 inch

Find: **a.**   The IMA of the vise.

   b.   The force that could be exerted by the vise jaws (under ideal conditions).

   c.   The percent efficiency of the vise. (When the force exerted by the vise jaws is measured, it is found to be 4000 lb.)

Solution: **a.** $\text{IMA} = \dfrac{2\pi r_i}{x_o}$

   $\text{IMA} = \dfrac{(2)(3.14)(8 \text{ inch})}{(0.25 \text{ inch})}$

   $\text{IMA} = 201$

   b. $F_o = \text{IMA } F_i$

   $F_o = (201)(30 \text{ lb})$

   $F_o = 6030 \text{ lb}$

   c. $\eta = \dfrac{\text{AMA}}{\text{IMA}} (100)$

   $\eta = \left(\dfrac{F_o \text{ (measured)}}{F_i}\right)\left(\dfrac{F_i}{F_o \text{ (ideal)}}\right) (100)$

   $\eta = \dfrac{F_o \text{ (measured)}}{F_o \text{ (ideal)}} (100)$

   $\eta = \dfrac{4000 \text{ lb}}{6030 \text{ lb}} (100)$

   $\eta = 66.3\%$

EXAMPLE 8.14   Screw jacks used in lifting a building
. . . .

Given: Four screw jacks are placed under the corners of a square structure weighing 25 tons. The screw jacks are identical, each with a pitch of 0.20 inch and a handle 2.5 ft long, measured from the screw axis.

Find: **a.**   The ideal mechanical advantage of each jack.

b.   The input force that would have to be applied to each jack to lift the load when all four jacks are working together, assuming 100% efficiency.

c.   The in force if the screw jacks are only 30% efficient.

Solution: **a.** $\text{IMA} = \dfrac{2\pi r_i}{x_o}$

$$\text{IMA} = \frac{(2)(3.14)(2.5\ \text{ft})(12\ \text{inch/ft})}{(0.20\ \text{inch})}$$

$$\text{IMA} = 942$$

b.   Each screw jack lifts one-fourth the total load or 25/4 tons. This value is (25/4 tons) (2000 lb/ton) = 12,500 lb.

$$F_i(\text{ideal}) = \frac{F_o}{\text{IMA}}$$

$$F_i(\text{ideal}) = \frac{12,500\ \text{lb}}{942}$$

$$F_i(\text{ideal}) = 13.3\ \text{lb}$$

c. $\eta = \dfrac{F_i(\text{ideal})}{F_i(\text{actual})}\,(100)$

$$F_i(\text{actual}) = \frac{F_i(\text{ideal})}{\eta}\,(100)$$

$$F_i(\text{actual}) = \frac{13.3\ \text{lb}}{30\%}\,(100)$$

$$F_i(\text{actual}) = 44.3\ \text{lb}$$

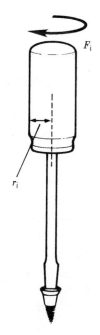

FIGURE 8.19   Screwdriver.

In using a screwdriver, the radius r_i is the radius of the driver handle, as shown in Figure 8.19.

The Wheel and Axle

Mechanical Advantage   When a wheel and axle are rigidly connected so that they turn together, another type of force transformer results. For example, Figure 8.20 shows two wheels of different radii, around which ropes are wound. One rope holds the load; the other is pulled by the input force F_i. To determine the ideal

FIGURE 8.20 Wheels of different radii and a common axle.

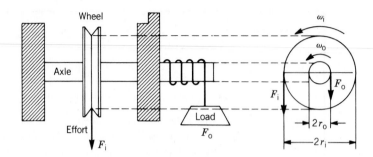

mechanical advantage of this simple device, consider one revolution of the axle. The rope applying the input force moves a distance equal to the circumference of the large wheel, $2\pi r_i$. The rope holding the weight or load moves a distance equal to the circumference of the smaller wheel, $2\pi r_o$. When Equation 8.18 is applied to this system, the ideal mechanical advantage is given by

$$IMA = \frac{x_i}{x_o}$$

$$IMA = \frac{2\pi r_i}{2\pi r_o}$$

$$IMA = \frac{r_i}{r_o} \tag{8.25}$$

where

r_i = radius of the source or drive wheel
r_o = radius of the load wheel

An application of Equation 8.25 is shown in Example 8.15

EXAMPLE 8.15   Mechanical advantage of a wheel and axle
. . . .

Given: A bucket of water weighing 30 lb is raised from a well by the "wheel and axle" shown. The turning handle has a length of 1.5 ft, and the rope holding the bucket is wrapped around a shaft of 6 inch diameter.

Find: **a.** The ideal mechanical advantage.
b. The required input force at the handle.

Solution: **a.** $IMA = \frac{r_i}{r_o}$

$$IMA = \frac{(1.5\ ft)(12\ inch/ft)}{3\ inch}$$

$$IMA = 6$$

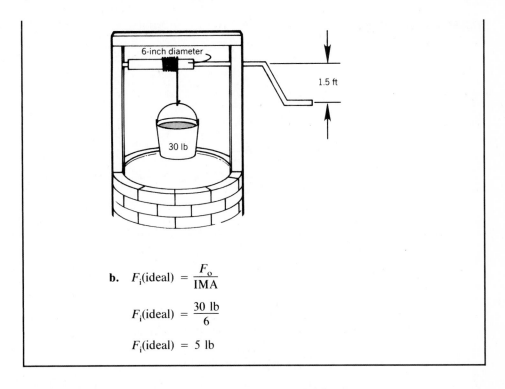

b. $F_i(\text{ideal}) = \dfrac{F_o}{\text{IMA}}$

$F_i(\text{ideal}) = \dfrac{30 \text{ lb}}{6}$

$F_i(\text{ideal}) = 5 \text{ lb}$

MECHANICAL TORQUE TRANSFORMERS

Belt Drives

Angular Velocity   Figure 8.21 shows a torque transformer called a belt drive. During a time t, a point on the belt moves from position A to position B. If there is no slipping, a point on wheel i moves through arc A_iB_i (equal to the distance AB) and a point on wheel o moves through an arc A_oB_o (also equal to the distance AB). The angular velocities ω of the two wheels are

$$\theta_i = \frac{\text{arc } A_iB_i}{r_i}$$

$$\omega_i = \frac{\theta_i}{t}$$

$$\omega_i = \frac{\text{arc } A_iB_i}{r_i \, t}$$

and

$$\theta_o = \frac{\text{arc } A_oB_o}{r_o}$$

$$\omega_o = \frac{\theta_o}{t}$$

$$\omega_o = \frac{\text{arc } A_oB_o}{r_o \, t}$$

FIGURE 8.21   Belt drive.

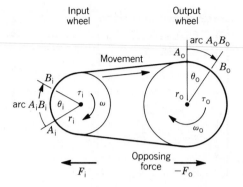

since arc A_iB_i = arc A_oB_o, dividing ω_i by ω_o results in

$$\frac{\omega_i}{\omega_o} = \frac{r_o}{r_i} \qquad (8.26)$$

where

ω_i = angular velocity of wheel i
ω_o = angular velocity of wheel o
r_i = radius of wheel i
r_o = radius of wheel o

Torque   If F_i and F_o are tangential forces at the rims of the wheels and no friction exists, the input and output torques acting on the two wheels are

$$\tau_i = F_i\, r_i \qquad (8.27)$$

and

$$\tau_o = F_o\, r_o \qquad (8.28)$$

F_i and F_o are each equal to the tension in the belt and are, therefore, equal to each other. Dividing Equation 8.27 by Equation 8.28 gives

$$\frac{\tau_i}{\tau_o} = \frac{r_i}{r_o} \qquad (8.29)$$

where

τ_i = input torque of wheel i
τ_o = output torque of wheel o

Mechanical Advantage   If there is neither friction nor energy storage in the transformer, the ideal mechanical advantage is

$$\text{IMA} = \frac{\text{Forcelike quantity out}}{\text{Forcelike quantity in}}$$

or

$$\text{IMA} = \frac{\tau_o}{\tau_i}$$

Using Equations 8.26 and 8.29, this can be written as

$$\text{IMA} = \frac{\tau_o}{\tau_i} = \frac{r_o}{r_i} = \frac{\omega_i}{\omega_o} \tag{8.30}$$

If there is an increase in torque, there must be a decrease in angular velocity. Example 8.16 shows the use of Equation 8.30 in solving a problem.

EXAMPLE 8.16   Belt drive system

. . . .

Given: A belt drive system vacuum pump has a motor wheel with a 4-inch diameter and a 16-inch diameter load wheel. The motor torque is 4 ft·lb, and the motor turns at 480 rpm.

Find: **a.**   Ideal mechanical advantage of the drive system.
         b.   Torque supplied to the load.
         c.   Angular velocity of the pump.

Solution: From Equation 8.30

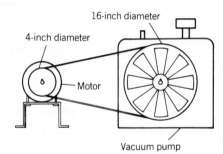

a. $\text{IMA} = \dfrac{r_o}{r_i}$

$r_i = \dfrac{4 \text{ inch}}{2}$

$r_i = 2 \text{ inch}$

$r_i = \dfrac{16 \text{ inch}}{2}$

$r_o = 8 \text{ inch}$

$\text{IMA} = \dfrac{8 \text{ inch}}{2 \text{ inch}}$

$\text{IMA} = 4$

b. $\text{IMA} = \dfrac{\tau_o}{\tau_i}$

$\tau_o = \tau_i \, \text{IMA}$

$\tau_o = (4 \text{ ft·lb})(4)$

$\tau_o = 16 \text{ ft·lb}$

c. $\text{IMA} = \dfrac{\omega_i}{\omega_o}$

$\omega_o = \dfrac{\omega_i}{\text{IMA}}$

$\omega_o = \dfrac{480 \text{ rpm}}{4}$

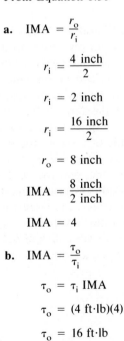

$$\omega_o = 102 \text{ rpm}$$

or

$$\frac{r_o}{r_i} = \frac{\omega_i}{\omega_o}$$

$$\omega_o = \frac{r_i}{r_o} \omega_i$$

$$\omega_o = \frac{2 \text{ inch}}{8 \text{ inch}} \; 480 \text{ rpm}$$

$$\omega_o = 120 \text{ rpm}$$

Disk Drives   Figure 8.22 shows a disk drive system where one disk drives another by friction contact. A careful comparison of this diagram with Figure 8.21 shows that the only difference is the direction of the second wheel. The disk drive in this figure reverses the direction of rotation; a belt drive does not unless the belt is crossed. All the equations for belt drives are applicable to disk drives.

Applications   Since rotational motion is a convenient method of transferring mechanical power from one point to another, drive systems are normally used with electric motors and internal combustion engines. Gear drives are common in clocks, engine lathes, and automobile transmissions. Belt drives often are employed with electric motors in pumps, blowers, and tape recorders. In an automobile, belts drive the fan, water pump, and power brakes. Disk drives are common in systems requiring accurate constant angular velocity and low torque, such as the turntable of a stereo record player.

Figure 8.23 shows a variable-diameter pulley system of a drill press. The wide v-belt rides on two cone, or variable-diameter, pulleys. Each pulley is composed on two conical sections whose separation may be varied. The pulley with the cones closest together has the largest effective diameter.

Figure 8.24 shows the disk drive system of a record player turntable. The motor shaft with a radius r_1 drives a disk of radius r_2. This disk drives the larger turntable (with radius r_3) from the inside. Example 8.17 illustrates a problem that deals with this figure.

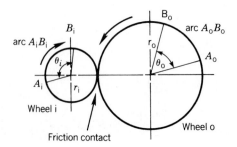

FIGURE 8.22   Disk drive.

FIGURE 8.23 Variable-diameter pulley system.

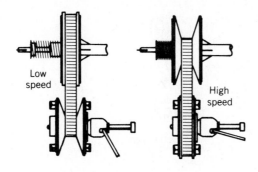

Low speed

High speed

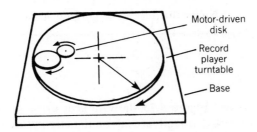

Motor-driven disk

Record player turntable

Base

FIGURE 8.24 Record turntable disk drive system.

EXAMPLE 8.17 Multidisk drive

. . . .

Given: In Figure 8.24, the motor rotates at 3600 rpm, $r_i = 0.785$ mm, $r_2 = 1.85$ cm, and $r_3 = 8.5$ cm.

Find: **a.** The ideal mechanical advantage of the disk drive system.

b. The angular velocity of the turntable.

Solution: **a.** IMA for r_1 to $r_2 = \dfrac{r_2}{r_1}$

IMA for r_1 to $r_2 = \dfrac{1.85 \text{ cm}}{0.0785 \text{ cm}}$

IMA for r_1 to $r_2 = 23.57$

IMA for r_2 to $r_3 = \dfrac{r_3}{r_2}$

IMA for r_2 to $r_3 = \dfrac{8.5 \text{ cm}}{1.85 \text{ cm}}$

IMA for r_2 to $r_3 = 4.59$

IMA for system $= [\text{IMA for } r_1 \text{ to } r_2][\text{IMA for } r_2 \text{ to } r_3]$

IMA for system $= (23.57)(4.59)$

IMA for system $= 108$

This result can also be obtained from

$$\frac{r_3}{r_1} = \left(\frac{r_2}{r_1}\right)\left(\frac{r_3}{r_2}\right)$$

b. $\text{IMA} = \dfrac{\omega_1}{\omega_3}$

$$\omega_3 = \frac{\omega_1}{\text{IMA}}$$

$$\omega_3 = \frac{3600 \text{ rpm}}{108}$$

$$\omega_3 = 33.33 \text{ rpm}$$

Notice that this is the standard speed of a long-playing record.

Gear Trains

Simple Gear Drive   Figure 8.25 shows a gear drive system identical to the disk drive system except that teeth are cut on the circumference of each disk to create a stronger, more efficient drive. The number of teeth on each gear is proportional to the circumference, and therefore the radius, of the gear

$$n_i \propto r_i \tag{8.31}$$

$$n_o \propto r_o \tag{8.32}$$

where

n_i = number of teeth on the input gear
n_o = number of matching teeth on the output gear

The proportionalities in Equations 8.31 and 8.32 yield the ratio

$$\frac{n_o}{n_i} = \frac{r_o}{r_i} \tag{8.33}$$

Equation 8.30 now becomes

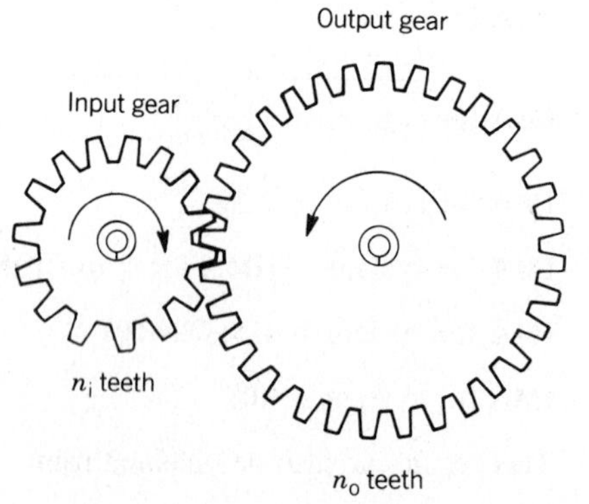

FIGURE 8.25   Gear drive system.

$$\text{IMA} = \frac{\tau_o}{\tau_i} = \frac{n_o}{n_i} = \frac{\omega_i}{\omega_o} \qquad (8.34)$$

Example 8.18 uses this relationship to solve a problem.

EXAMPLE 8.18   Gear drive in a clock
. . . .

Given: In an alarm clock, a gear with 55 teeth drives a gear with 7 teeth. The first gear has an angular velocity of 3 revolutions per day.

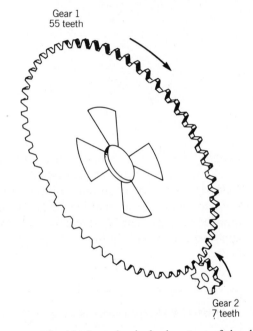

Gear 1
55 teeth

Gear 2
7 teeth

Find: **a.** The ideal mechanical advantage of the drive.

b. The angular velocity of the second gear.

Solution: **a.** $\text{IMA} = \dfrac{n_o}{n_i}$

$\text{IMA} = \dfrac{7}{55}$

$\text{IMA} = 0.1273$

b. $\text{IMA} = \dfrac{\omega_i}{\omega_o}$

$\omega_o = \dfrac{\omega_i}{\text{IMA}}$

$\omega_o = \dfrac{3 \text{ rev/day}}{0.1273}$

$\omega_o = 23.57 \text{ rev/day}$

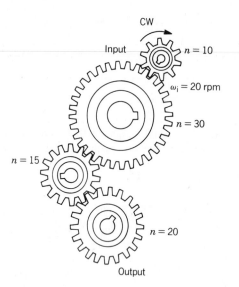

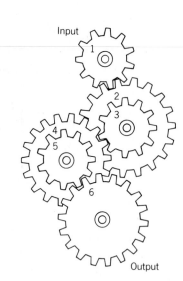

Simple Gear Train A simple gear train is one where the centers of all gears are fixed and there is only one gear per shaft. Figure 8.26 is an example of such a gear train. The direction of rotation of the output is the same as that of the input if the train contains an odd number of gears and is opposite in direction for an even number of gears. The ideal mechanical advantage of such a train is determined by applying Equation 8.34 for each pair of interlocking gears, as shown in Example 8.19.

EXAMPLE 8.19 Mechanical advantage of a simple train with more than two gears
. . . .

Given: A four-gear train has gears having the following numbers of teeth: $n_1 = 16$, $n_2 = 8$, $n_3 = 13$, $n_4 = 17$.

Find: Ideal mechanical advantage.

Solution: IMA for 1 to 2 $= \dfrac{n_2}{n_1}$

$\qquad\qquad$ IMA for 2 to 3 $= \dfrac{n_3}{n_2}$

$\qquad\qquad$ IMA for 3 to 4 $= \dfrac{n_4}{n_3}$

$\qquad\qquad$ IMA for the train $= \left(\dfrac{n_2}{n_1}\right)\left(\dfrac{n_3}{n_2}\right)\left(\dfrac{n_4}{n_3}\right)$

$\qquad\qquad$ IMA for the train $= \dfrac{n_4}{n_1}$

$\qquad\qquad$ IMA for the train $= \dfrac{17}{16}$

Example 8.19 indicates that in a simple train, the ideal mechanical advantage depends only on the number of teeth on the input gear and the number on the output gear. Intermediate gears do not affect the ideal mechanical advantage.

Compound Gear Train Figure 8.27 shows a compound gear train where some shafts carry two gears that are rigidly connected to rotate at the same rate with the same torque. The same technique used to calculate the mechanical advantage of a simple gear train can be applied to the compound train, as shown in Example 8.20.

EXAMPLE 8.20 Mechanical advantage of a compound gear train
. . . .

Given: The gear train in Figure 8.27, $n_1 = 11$, $n_2 = 18$, $n_3 = 11$, $n_4 = 18$, $n_5 = 11$, and $n_6 = 18$.

Find: Ideal mechanical advantage.

Solution: The overall ideal mechanical advantage is the product of the ideal mechanical advantage of all the gear pairs.

$$\text{IMA for 1 to 2} = \frac{n_2}{n_1}$$

$$\text{IMA for 1 to 2} = \frac{18}{11}$$

$$\text{IMA for 2 to 3} = 1 \text{ because gears are rigidly connected}$$

$$\text{IMA for 3 to 4} = \frac{n_4}{n_3}$$

$$\text{IMA for 3 to 4} = \frac{18}{11}$$

$$\text{IMA for 4 to 5} = 1 \text{ because gears are rigidly connected}$$

$$\text{IMA for 5 to 6} = \frac{n_6}{n_5}$$

$$\text{IMA for 5 to 6} = \frac{18}{11}$$

$$\text{IMA overall} = \frac{n_2}{n_1}(1)\frac{n_4}{n_2}(1)\frac{n_6}{n_5}$$

$$\text{IMA overall} = \frac{18}{11}\frac{18}{11}\frac{18}{11}$$

$$\text{IMA overall} = 4.38$$

The compound train in Figure 8.27 is composed of three simple trains—gear 1 drives gear 2, gear 3 drives gear 4, and gear 5 drives gear 6. The ideal mechanical advantage of the compound train is the product of the ideal mechanical advantages of the simple trains that it contains.

FIGURE 8.28   Reverted
train.

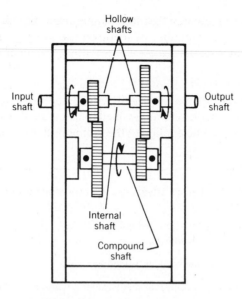

FIGURE 8.28   Reverted
train.

Compound gear trains are used in mechanical clocks and in drive trains for lathes, power tools, and automobiles. Figure 8.28 shows a special class of compound train, called a reverted train, in which the input and output shafts are coaxial. This design affords ease of changing gear ratios in a compact space. Its most familiar application is in the automobile transmission in Figure 8.29.

First gear provides the greatest mechanical advantage to give a high torque and low rotational rate for starting and low-speed operation. Second gear provides a

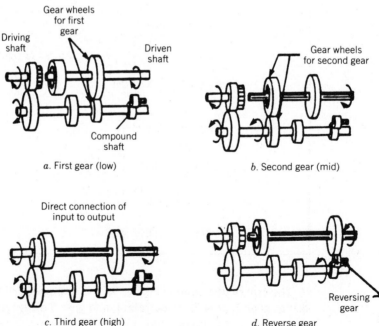

FIGURE 8.29   Automobile
transmission.

FIGURE 8.30 Planetary gear train.

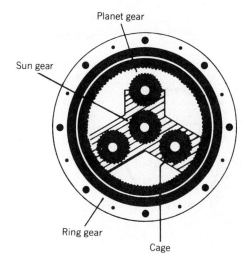

FIGURE 8.31 Planetary gear train in an automobile.

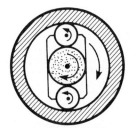

a. First gear (low)

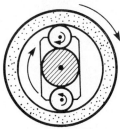

b. Second gear (mid)

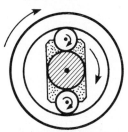

c. Third gear (high)

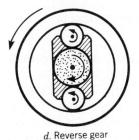

d. Reverse gear

lower mechanical advantage for intermediate speeds. In third gear the input and output shafts are directly connected to give a mechanical advantage of one for high-speed operation. In reverse, an additional gear called an idler is added to reverse the direction of rotation of the output shaft.

The Planetary Gear Train   Figure 8.30 shows a planetary gear train in which the centers of rotation of some gears are not fixed. This device consists of a central sun gear, one or more planetary gears attached to a carrier called a cage, and an outer ring gear. The sun gear, cage, and ring gear can be connected to the input or output by concentric shafts. Planetary gears are used as force transformers in bicycles and automatic transmissions for automobiles.

Figure 8.31 shows how a planetary gear train is used in an automobile. The connections are as follows:

Low gear:
  Sun gear—input
  Ring gear—locked
  Cage—output
  This arrangement gives the greatest mechanical advantage and the lowest speed.

Mild gear:
  Sun gear—locked
  Ring gear—input
  Cage—output
  This configuration gives a lower mechanical advantage and a greater speed.

High gear:
  Sun gear—locked
  Ring gear—output
  Cage—input
  This arrangement gives the least mechanical advantage and the greatest speed.

FIGURE 8.32 Automobile differential.

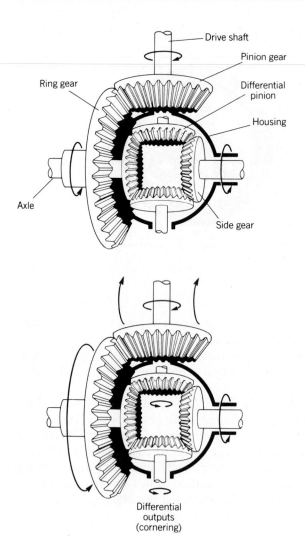

Drive shaft

Pinion gear

Differential pinion

Housing

Ring gear

Axle

Side gear

Differential outputs (cornering)

Reverse gear: Sun gear—input
                      Ring gear—output
                      Cage—locked
                      This arrangement reverses the direction of rotation of the output shaft.

The differential gear train in Figure 8.32 is used to provide two different outputs with equal torque but differing rotational rates.

The example shown is the differential train of an automobile. The ring gear is rigidly attached to the housing. Each axle is attached to a side gear. During straight travel (Figure 8.32a) the ring gear, housing, and both axles all rotate at the same rate and the differential pinions do not rotate with respect to the housing or other gears. During cornering (Figure 8.32b) one wheel must rotate faster than the other. In this case, the differential pinions rotate as shown, allowing the axles to rotate at different rates but with equal torques.

SECTION *8.1* *EXERCISES*

1. Define

Ideal mechanical advantage

Actual mechanical advantage

Efficiency

2. A wheelbarrow containing sand weighs 300 lb. If the distance from the center of gravity to the axle of the wheel is 1.5 ft, and the distance from the handle to the axle of the wheel is 4 ft, what lifting force must be applied at the handle to lift the load?

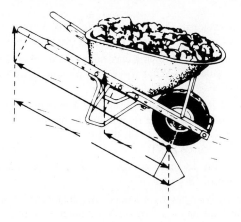

3. The total weight of skiers and equipment on a ski tow is 3000 lb. The cable of the ski tow pulls this load up an incline of 25°. What is the tension in the cable if frictional forces are negligible?

4. A wheel and axle is used to lift a 2400-lb load. The wheel diameter is 60 inches and the axle diameter is 5 inches. The operator must exert a force of 250 lb to lift the load. Determine

Ideal mechanical advantage

Actual mechanical advantage

Efficiency of the transformer

5. A screw jack is being used to lift a load of 2 tons. The screw pitch is 0.25 inch and the handle length 20 inch. In the ideal case, what force is needed at the end of the handle? If the efficiency is 20%, what force is needed?

6. What is the lever arm of the input force for a lever with an ideal mechanical advantage of 5 and a load lever arm of 1.5 ft?

7. What force must be applied to a pulley with an efficiency of 83% if the load force is 200 N and the ideal mechanical advantage is 3?

8. What is the ideal mechanical advantage of a plane inclined at an angle of 30°?

9. What is the angle of inclination of a plane if the ratio of output work to input work is 0.8 and the ratio of input force to output force is 0.4? Is there friction in this system? How can one tell?

10. What is the required input force to the source wheel of a wheel and axle if the circumference of the load wheel is 0.5 m and the radius of the drive wheel is 0.4 m, when the load is 20 N?

11. What is the efficiency of a pulley that has an ideal mechanical advantage of 2 and an input force to output force ratio of 0.7?

12. The ideal mechanical advantage of a pulley is used to predict the need for a 25-N input force. What input force is actually needed if the pulley has an efficiency of 85%?

13. What is the maximum load that can be pushed up an inclined plane that has an ideal mechanical advantage of 4 if the applied force is 55 N?

14. A lever can be used to lift a 200-kg mass when the input force is 350 N. If the ideal mechanical advantage of the lever system is 6, what is the efficiency of the system?

15. A 95-lb boy sits on a seesaw 6 ft from the pivot. How far from the pivot must an 85-lb girl sit to balance him?

16. Draw sketches to identify the three classes of levers and list two examples of each class.

17. In a nutcracker (Class II lever), a force of 32 lb is applied 6.5 inch from the fulcrum. What is the force on a nut 1.25 inch from the fulcrum?

18. In a machine for throwing clay targets, a force of 210 lb is applied to a lever 4 inches from the fulcrum. What force is experienced by the clay target 24 inches from the fulcrum?

19. A Class I lever is used to lift a 400-lb block. If the lever is 90% efficient (under all conditions), what is the minimum applied force that the lever must be able to withstand if it is to be used to lift the block? The load lever arm is 1 ft and the lever arm of the applied force is 4 ft.

20. A gear with 18 teeth drives a gear with 56 teeth. What is the ideal mechanical advantage of this system?

21. In a belt drive system for a bandsaw, the motor has a pulley with a diameter of 2.5 inch and the load has a pulley with a diameter of 14 inch. What is the ideal mechanical advantage of the system?

22. The disk drive of a turntable has three disks with diameters of 5.75, 1.15, and 0.1065 inch. What is the ideal mechanical advantage of the system?

23. The input torque of the motor in the turntable in Problem 22 is 0.42 inch-ounces and the turntable is rotating at 33 1/3 rpm. Find the output torque and the angular velocity of the motor if friction is neglected.

24. In a gear drive system, the large gear has a diameter of 30 cm and a total of 40 teeth. How many teeth does the small gear have if it has a diameter of 6 cm? What is the ideal mechanical advantage of the system?

25. A simple gear drive consists of two gears having 25 and 6 teeth, respectively. If the gear with 6 teeth develops a torque of 0.2 N·m, what torque does the other gear develop if friction is negligible?

26. A simple gear train consists of 5 gears. The ideal mechanical advantage is 6. If the first gear in the sequence has 12 teeth, what can be said about the number of teeth on the other gears?

27. A compound gear train looks like the one shown in Figure 8.27, but the number of teeth on the gears is not the same as for the gears shown. If the ideal mechanical advantage of the system is 6 and gears 1, 3, and 5 are identical, what is the ratio of the number of teeth on gear 1 to the number of teeth on gear 2?

28. A 12° wedge is driven with a total force of 90 N. What is the ideal splitting force? (The wedge angle is $2\theta = 12°$.)

29. A knife blade has a wedge angle of 8°, and a 500-g weight is set on the back of the blade. What is the ideal splitting force

    a. in newtons?

    b. in dynes?

30. A 15° wedge is used to separate two pieces of wood. The driving force of 45 lb produces a splitting force of 280 lb. What is the efficiency of the wedge?

8.2

FORCE TRANSFORMERS IN FLUID SYSTEMS

Upon completion of this section, the student should be able to

- Calculate the ideal force advantage and pressure advantage of hydraulic systems.

- Give two examples of mechanisms that apply hydraulic systems to obtain a large ideal force advantage.

- Given the diameter of two pistons in a closed hydraulic system, calculate
  Distance moved by one piston, given the distance moved by the other piston
  The output force given the input force

- Explain how a hydraulic system can be used to obtain a very high pressure.

DISCUSSION

Closed Hydraulic System   A closed hydraulic system contains a rigid tank similar to the one in Figure 8.33, completely filled with a liquid and vented to several pistons as shown. There are no leaks of liquid through the piston sides, no air space at the top of the tank, and no liquid pumped in or out during operation.

If any of the pistons are moved in or out or up or down, one or more of the other pistons would also have to move. This is because the liquid is not compressible. (Note that in an enclosed pneumatic system—air, not liquid—the air compresses easily, and the same laws do not apply.)

If piston *A* is moved inward, one piston, or a combination of the other pistons, would have to move outward, since no liquid or air could seep in, and the total volume of the system would remain the same.

Force Transformers   Consider a tank with two pistons as shown in Figure 8.34. Friction in the system will be neglected, and the movement of the pistons will be done slowly so that kinetic energy can be neglected. Under these conditions, for an incompressible fluid, the work in must equal the work out. From Chapter 2, one has

$$W = p\Delta V = pAx$$

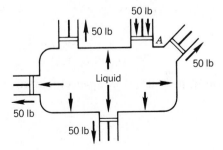

FIGURE 8.33 Closed hydraulic system.

FIGURE 8.34   Two-piston
hydraulic system.

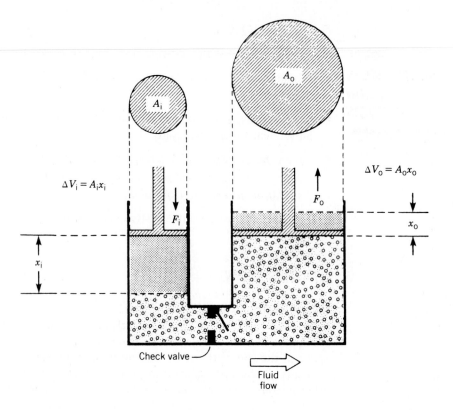

where

W = work
p = pressure
ΔV = change in volume in the cylinder
A = area of the piston
x = displacement of the piston

Thus, for the conservation of energy, one may write

$$p_o A_o x_o = p_i A_i x_i$$

Since the fluid is incompressible, the two volumes must be equal:

$$A_o x_o = A_i x_i$$

or

$$\frac{A_o}{A_i} = \frac{x_i}{x_o} \tag{8.35}$$

and

$$p_o = p_i = p \tag{8.36}$$

Using the definition of pressure (force per unit area), Equation 8.36 becomes

$$\frac{F_o}{A_o} = \frac{F_i}{A_i} \tag{8.37}$$

or

$$\frac{F_o}{F_i} = \frac{A_o}{A_i} = \frac{x_i}{x_o} \tag{8.38}$$

From the definition of the ideal mechanical advantage in Equation 8.5, one may write

$$\text{IMA} = \frac{F_o}{F_i} = \frac{A_o}{A_i} = \frac{x_i}{x_o} \tag{8.39}$$

It is important to remember, however, that the actual mechanical advantage will be less:

$$\text{AMA} = \frac{F_o \text{ (measured)}}{F_i} \tag{8.40}$$

where

F_o (measured) = the real output force

Sample calculations using Equation 8.39 are given in Examples 8.21, 8.22, and 8.23.

EXAMPLE 8.21   Calculation of a force multiplication problem

Given: An 850-lb force is applied to piston i, which has a diameter of 2.5 inch. Piston o has a diameter of 9.25 inch.

Find: Output force on piston o.

Solution: Using Equation 8.37

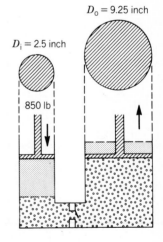

$$\frac{F_o}{A_o} = \frac{F_i}{A_i}$$

Caculate the areas of pistons i and o.

$$A_i = \frac{\pi}{4} d_i^2$$

$$A_i = \frac{(3.14)}{(4)} (2.5 \text{ inch})^2$$

$$A_i = 4.91 \text{ inch}^2$$

$$A_o = \frac{\pi}{4} d_o^2$$

$$A_o = \frac{(3.14)}{(4)} (9.25 \text{ inch})^2$$

$$A_o = 67.2 \text{ inch}^2$$

Rearranging Equation 8.37 gives

$$F_o = \frac{A_o}{A_i}$$

$$F_o = \frac{(67.2 \ \text{inch}^2)}{(4.91 \ \text{inch}^2)} \ 850 \ \text{lb}$$

$$F_o = 11,633 \ \text{lb}$$

Therefore, an 850-lb force on piston i will lift a 11,633-lb force on piston o.

EXAMPLE 8.22   Mechanical advantage of a hydraulic lift

· ·

Given: A hydraulic lift contains a load piston 9 inch in diameter and a pump piston 1 inch in diameter.

Find: **a.** IMA of the lift.

              b. The force on the pump piston required to lift a total load of 5000 lb.

Solution: **a.** $\text{IMA} = \dfrac{A_o}{A_i}$

                      $\text{IMA} = \dfrac{\pi \ r_o^2}{\pi \ r_i^2}$

                      $\text{IMA} = \dfrac{(4.5 \ \text{inch})^2}{(0.5 \ \text{inch})^2}$

                      $\text{IMA} = 81$

              b. $F_i = \dfrac{F_o}{\text{IMA}}$

                      $F_i = \dfrac{5,000 \ \text{lb}}{81}$

                      $F_i = 61.7 \ \text{lb}$

EXAMPLE 8.23   Calculation of distances forces travel

· ·

Given: The same hydraulic system as in Example 8.21 and piston i moves 6.75 inch.

Find: Distance piston o moves.

Solution: From Equation 8.39

              $$\frac{x_i}{x_o} = \frac{F_o}{F_i}$$

$$x_o = \frac{F_i}{F_o}\, x_i$$

$$x_o = \frac{850\ lb}{11{,}633\ lb}\, (6.75\ inch)$$

$$x_o = 0.493\ inch$$

Hydraulic Press or Jack Examples of force transformers that use hydraulic cylinders and a lever are the hydraulic press and the hydraulic jack. A hydraulic press is illustrated in Figure 8.35. The ideal mechanical advantage of the system will be the product of the ideal mechanical advantage of the lever and the ideal mechanical advantage of the fluid force transformer. As shown in Figure 8.36, the lever is a Class II lever and the fluid system is the same as the one shown in Figure 8.35.

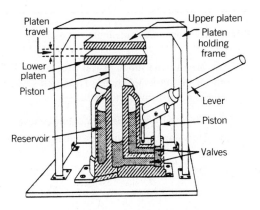

FIGURE 8.35 Hydraulic press.

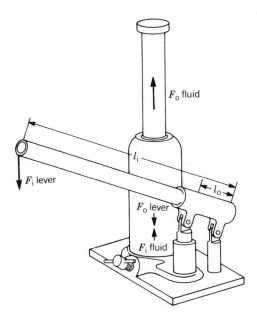

FIGURE 8.36 Mechanical advantage of a combined system.

Using Equation 8.18, which applies to all levers, and Equation 8.39 for the fluid system, one obtains

$$\text{IMA} = \left(\frac{F_o}{F_i}\right)_{\text{system}} = \left(\frac{A_o}{A_i}\right)_{\text{fluid}} \left(\frac{l_i}{l_o}\right)_{\text{lever}} \tag{8.41}$$

and

$$\text{IMA} = \left(\frac{F_o}{F_i}\right)_{\text{system}} = \left(\frac{x_i}{x_o}\right)_{\text{fluid}} \left(\frac{x_i}{x_o}\right)_{\text{lever}} \tag{8.42}$$

However, the output distance x_o for the lever is equal to the input distance x_i for the fluid system. Thus, Equation 8.42 reduces to

$$\text{IMA} = \left(\frac{x_i}{x_o}\right)_{\text{system}} \tag{8.43}$$

Example 8.24 gives an application of these equations.

EXAMPLE 8.24   Total mechanical advantage of a jack
. . . .

Given: The area of the input piston of a jack is 0.2 inch² and the area of the output piston is 3.0 inch². The input lever arm is 24 inch and the output lever arm is 2.0 inch.

Find: **a.**   The total ideal mechanical advantage.
b.   How far the input force must move to lift the load 2.0 inch.

Solution: **a.** $\text{IMA} = \left(\dfrac{A_o}{A_i}\right)_{\text{fluid}} \left(\dfrac{l_i}{l_o}\right)_{\text{lever}}$

$\text{IMA} = \dfrac{3.0 \text{ inch}^2}{0.2 \text{ inch}^2} \dfrac{24 \text{ inch}}{2.0 \text{ inch}}$

$\text{IMA} = 180$

b. $\text{IMA} = \dfrac{x_i}{x_o}$

$x_i = \text{IMA}(x_o)$

$x_i = (180)(2 \text{ inch})$

$x_i = 360 \text{ inch}$

Pressure Transformer   A hydraulic system can also be used to transform pressure. Consider the diagram shown in Figure 8.37. The pressure in system i is

$$P_i = \frac{F_i}{A_i} = \frac{F_t}{A_t} \tag{8.44}$$

where

F_i = input force at piston i

FIGURE 8.37 Pressure transformer.

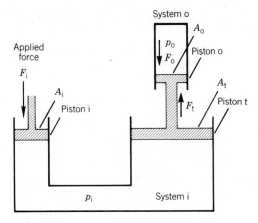

A_i = area of piston i
F_t = transmitted force at piston t
A_t = area of piston t

Since piston t is connected to piston o by a rigid rod, the force on piston o is equal to the force on piston t. Thus, the pressure in system o becomes

$$p_o = \frac{F_t}{A_o} \tag{8.45}$$

However, from Equation 8.44

$$F_t = \left(\frac{A_t}{A_i}\right) F_i$$

Substituting this value for F_t into Equation 8.45 yields

$$p_o = \left(\frac{A_t}{A_o}\right)\left(\frac{F_i}{A_i}\right) \tag{8.46}$$

Since $F_i/A_i = p_i$, Equation 8.46 may also be written as

$$p_o = \left(\frac{A_t}{A_o}\right) p_i \tag{8.47}$$

or

$$\frac{p_o}{p_i} = \frac{A_t}{A_o} \tag{8.48}$$

An application of Equations 8.47 and 8.44 is given in Example 8.25.

EXAMPLE 8.25 Pressure transformer
. . . .

Given: It is desirable to increase the pressure in a tank to 14,000 psi. The piston area in the tank is 1.0 inch². The available pressure is 1000 psi.

Find: **a.** The transmitting piston area that must be used in the 1000-psi system.

            b. The input piston area in the 1000-psi system if the input force is 500 lb.

Solution: **a.** $\dfrac{p_o}{p_i} = \dfrac{A_t}{A_o}$

                    $A_t = \left(\dfrac{p_o}{p_i}\right) A_o$

                    $A_t = \dfrac{14,000 \text{ psi}}{1,000 \text{ psi}} (1 \text{ inch}^2)$

                    $A_t = 14 \text{ inch}^2$

            b. $p_o = \left(\dfrac{A_t}{A_o}\right)\left(\dfrac{F_i}{A_i}\right)$

                    $A_i = \left(\dfrac{A_t}{A_o}\right)\left(\dfrac{F_i}{p_o}\right)$

                    $A_i = \left(\dfrac{14 \text{ inch}^2}{1 \text{ inch}^2}\right)\left(\dfrac{500 \text{ lb}}{14,000 \text{ lb/inch}^2}\right)$

                    $A_i = 0.5 \text{ inch}^2$

                    or

                    $p_i = \dfrac{F_i}{A_i}$

                    $A_i = \dfrac{F_i}{p_i}$

                    $A_i = \dfrac{500 \text{ lb}}{1000 \text{ lb/inch}^2}$

                    $A_i = 0.5 \text{ inch}^2$

The ideal pressure advantage can be expressed as

$$\text{IMA} = \frac{(\text{GF})_o}{(\text{GF})_i} = \frac{p_o}{p_i} = \frac{A_t}{A_o} \tag{8.49}$$

The actual pressure advantage is less because the effect of friction is neglected.

SECTION 8.2 *EXERCISES*

1. The two pistons of a hydraulic press have diameters of 4 inch and 64 inch. What force must be applied by the smaller piston to give a compressive force of 50 tons at the larger piston? How far does the small piston travel in moving the large piston 1 inch?

2. How much fluid is displaced on the source side of a hydraulic lift if the load piston has a cross-sectional area of 0.3 m² and is raised 5 cm?

3. What is the cross-sectional area of the source piston of the lift in Problem 2 if the lift has an ideal mechanical advantage of 12?

4. The ideal mechanical advantage of a hydraulic lift is 5. What is the ratio of the radius of the piston on the load side to that of the source piston?

5. Write a short description of the following terms and include any sketches to clarify presentation:

Closed hydraulic system

Hydraulic force transformation

Mechanical advantage

Hydraulic pressure

6. If a load of 1000 lb is moved through 21 inches on the input piston, how much load will it lift on the output piston if the input and output piston diameters are 1 and 12 inch respectively? How far will the 12-inch output piston move during this process?

7. The output piston of an ideal hydraulic system does 100 J of work by moving 20 cm. How far does the input piston move if the ratio of the area of output piston to the input piston is 3 to 1?

8. A certain thin-walled circular piston is made of a material that can withstand up to 1000 lb/ft² before it fractures. What is the maximum force this piston can withstand before it fractures if the piston's diameter is 3 inches?

8.3

VOLTAGE TRANSFORMERS IN ELECTROMAGNETIC SYSTEMS

OBJECTIVES

Upon completion of this section, the student should be able to

• For an ideal transformer with unity power factors, calculate the secondary voltage and current given the primary current, primary voltage, and turns ratio of the coils.

• Discuss the three major sources of power loss in a transformer.

• For an ideal transformer, calculate the transferred resistance from the secondary circuit to the primary circuit given the load resistance, the inductance of the primary coil, and the inductance of the secondary coil.

• Determine the voltage advantage of an ideal transformer from the turns ratio of the coils.

• Define the mutual inductance of a transformer.

DISCUSSION

Elecrtrical Transformers   In the electrical system, the forcelike quantity is voltage. The electrical transformer is a device that can increase or decrease voltage with the opposite change in the current. The source in this case is producing alternating current, and the load is some device that consumes electrical energy. In Figure 8.38, the load is shown as a simple resistor R. (Notice the symbol for the

FIGURE 8.38 Circuit
diagram with a transformer.

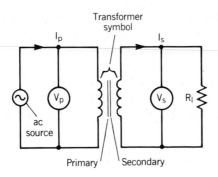

transformer used in drawing circuit diagrams.) V_p and V_s represent the input and output voltages, respectively. Voltmeters placed in the positions shown would record the appropriate voltages. The transformer itself usually is a rectangular iron core with two windings—the primary windings on the source side and the secondary windings on the load side. Figure 8.39 shows a simple drawing of a transformer in the same circuit.

As discussed in Chapter 3, the changing current through the primary and secondary windings creates a changing magnetic field through the core, which induces a voltage in each winding. From Chapter 3, the voltages across the windings are

$$V_{pi} = -N_p \frac{\Delta\phi}{\Delta t} \tag{8.50}$$

$$V_{si} = -N_s \frac{\Delta\phi}{\Delta t} \tag{8.51}$$

where

V_{pi} = instantaneous primary voltage
V_{si} = instantaneous secondary voltage
N_p = number of turns in the primary winding
N_s = number of turns in the second winding
ϕ = common magnetic flux in the core

In Equations 8.50 and 8.51, it is assumed that the entire magnetic flux passes through the core and through each coil. By dividing Equation 8.51 by Equation 8.50, one obtains

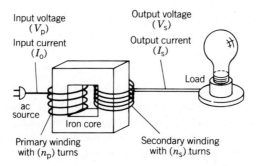

FIGURE 8.39   Transformer.

$$\frac{V_{si}}{V_{pi}} = \frac{N_s}{N_p} \tag{8.52}$$

or

$$V_{si} = \left(\frac{N_s}{N_p}\right) V_{pi} \tag{8.53}$$

In a transformer with no leakage flux, the primary voltage (forcelike quantity) is multiplied by the turns ratio to give the secondary voltage. This ratio is similar to the ideal mechanical advantage and can be expressed by

$$\text{IMA} = \frac{V_{si}}{V_{pi}} = \frac{N_s}{N_p} \tag{8.54}$$

Since Equation 8.54 is true at all times, it is also true for the average voltages and the root-mean-square voltages. Thus,

$$V_s = \left(\frac{N_s}{N_p}\right) V_p \tag{8.55}$$

where

V_s = root-mean-square secondary voltage
V_p = root-mean-square primary voltage

Since the ratio of forcelike quantities is the same as the ratio of the turns, a step-down transformer has fewer turns in the secondary coil than in the primary coil, and a step-up transformer has more turns in the secondary coil than in the primary coil. Typical transformer problems are solved in Examples 8.26 and 8.27.

EXAMPLE 8.26   Step-up transformer voltage
. . . .

Given: The transformer in the illustration has 4 turns in the primary winding and 8 turns in the secondary winding. The input voltage to the primary coil V_p is 3 V.

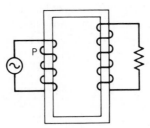

Find: Theoretical output voltage V_s from the secondary coil.

Solution: $V_s = \left(\dfrac{N_s}{N_p}\right) V_p$

$$V_s = \left(\frac{8 \text{ turns}}{4 \text{ turns}}\right) 3 \text{ V}$$

$$V_s = 6 \text{ V}$$

The transformer "steps-up" the voltage from 3 V to 6 V.

EXAMPLE 8.27   Step-up transformer winding
. . . .

Given: An electric generator delivers energy at a voltage of 15,000 V.

Find: The relative number of turns in the secondary winding if the voltage
              is to be stepped up to 300,000 V. (Assume 100% efficiency.)

Solution: $\dfrac{V_s}{V_p} = \dfrac{N_s}{N_p}$

$$N_s = \left(\frac{V_s}{V_p}\right) N_p$$

$$N_s = \left(\frac{300,000 \text{ V}}{15,000 \text{ V}}\right) N_p$$

$$N_s = (20) N_p$$

The secondary winding will need twenty times the number of turns
in the primary winding.

As long as the voltage rating of the winding insulation is not exceeded, either
winding can be used as the primary winding. A step-up transformer can be used as
a step-down transformer if its leads are reversed. For example, if the transformer
in Example 8.26 is reversed by putting 3 V into the winding with eight turns, the
voltage out of the four-turn winding is 1.5 V.

The Ideal Transformer   In real transformers there are energy losses. However, to
a good approximation in many cases, it can be assumed that these energy losses
are small. The principle of conservation of energy then gives (no energy storage)

$$\text{Power in} = \text{Power out}$$

$$V_p I_p \cos \alpha_p = V_s I_s \cos \alpha_s$$

or

$$\frac{V_p \cos \alpha_p}{V_s \cos \alpha_s} = \frac{I_s}{I_p} \tag{8.56}$$

where

$\cos \alpha_p$ = primary circuit power factor
$\cos \alpha_s$ = secondary circuit power factor

I_p = primary root-mean-square current

I_s = secondary root-mean-square current

Thus, the current ratio is equal to the inverse voltage ratio times the inverse power factor ratio.

Equation 8.56 can be related to Equation 8.55. Substituting for V_p/V_s in Equation 8.56 gives

$$\frac{I_s}{I_p} = \frac{N_p}{N_s} \frac{\cos \alpha_p}{\cos \alpha_s} \tag{8.57}$$

The current ratio is equal to the inverse turns ratio times the inverse power factor ratio. In many applications, the power factor ratio is almost unity, and transformer problems are easily estimated. Equation 8.57 is used to solve a problem in Example 8.28.

EXAMPLE 8.28 Set-up transformer current
. . . .

Given: A transformer has 500 primary turns and 3000 secondary turns. The power factors are unity ($\cos \alpha_p = 1$ and $\cos \alpha_s = 1$). The primary current is 3 A.

Find: The secondary current.

Solution: $\dfrac{I_s}{I_p} = \left(\dfrac{N_p}{N_s}\right)\dfrac{\cos \alpha_p}{\cos \alpha_s}$

$\dfrac{I_s}{I_p} = \left(\dfrac{N_p}{N_s}\right)$ (1)

$I_s = \left(\dfrac{N_p}{N_s}\right) I_p$

$I_s = \left(\dfrac{500 \text{ turns}}{3000 \text{ turns}}\right)$ (3 A)

$I_s = 0.5 \text{ A}$

The 3 A is reduced to 0.5 A in this step-up transformer.

The ideal transformer load may be a combination of inductance, capacitance, and resistance. The usual load, however, is a resistance. For this case, one may write the voltage equations for Figure 8.38 as

$$V_i = V_{pi} \tag{8.58}$$

$$V_{si} = I_{si} R_l \tag{8.59}$$

where

V_i = instantaneous voltage applied to the primary coil

R_L = resistance load in the secondary circuit

However, $V_s = (N_s/N_p)\,V_i$, and Equation 8.59 becomes

$$\left(\frac{N_s}{N_p}\right) V_i = I_{si}\,R_L \qquad\qquad (8.60)$$

Solving Equation 8.60 for I_{si} yields

$$I_{si} = \left(\frac{N_s}{N_p}\right)\left(\frac{V_i}{R_L}\right) \qquad\qquad (8.61)$$

If the power factor ratio is near unity, then from Equation 8.57

$$I_{pi} = \left(\frac{N_s}{N_p}\right) I_{si}$$

and for this simple ideal case

$$I_{pi} = \left(\frac{N_s}{N_p}\right)^2 \left(\frac{V_i}{R_L}\right) = \frac{V_i}{(N_p/N_s)^2\,R_L} \qquad\qquad (8.62)$$

It can be seen from Equation 8.62 that the primary circuit has a transferred resistance given by

$$R_t = \left(\frac{N_p}{N_s}\right)^2 R_L \qquad\qquad (8.63)$$

Applications of the equations for ideal transformers are given in Examples 8.29 and 8.30.

EXAMPLE 8.29   The ideal transformer
. . . .

Given: An ideal transformer has twice as many secondary turns as primary turns. The primary coil is powered by a 100-V source and the secondary coil is connected to a 50-Ω load. Assume that the power factor ratio is unity and neglect the coil resistance.

Find: a.   Induced voltage in the secondary coil.
             b.   Current in the secondary circuit.
             c.   Power in the secondary circuit.
             d.   Current in the primary circuit.
             e.   Is this a step-up or step-down transformer?

Solution: From Equation 8.55

   a. $V_s = \left(\dfrac{N_s}{N_p}\right) V_p$

        $V_s = (2)(100\ \text{V})$

        $V_s = 200\ \text{V}$

   b. $V_s = I_s\,R_L$ (Ohm's law)

$$I_s = \frac{V_s}{R_L}$$

$$I_s = \frac{200 \text{ V}}{50 \text{ }\Omega}$$

$$I_s = 4 \text{ A}$$

c. $P_s = I_s V_s$

$$P_s = (4 \text{ A})(200 \text{ V})$$

$$P_s = 800 \text{ W}$$

d. $P_p = P_s$ for an ideal transformer

$$I_p = \left(\frac{V_s}{V_p}\right) \left(\frac{\cos \alpha_s}{\cos \alpha_p}\right) I_s$$

$$I_p = \left(\frac{200 \text{ V}}{100 \text{ V}}\right) (1)(4 \text{ A})$$

$$I_p = 8 \text{ A}$$

or

$$P_p = I_p V_p$$

$$I_p = \frac{P_p}{V_p}$$

$$I_p = \frac{800 \text{ W}}{100 \text{ V}}$$

$$I_p = 8 \text{ A}$$

e. Since $V_s > V_p$, this is a step-up transformer.

Note. Voltage is stepped up, current is stepped down, power is conserved, and Ohm's law is obeyed.

EXAMPLE 8.30 Transferred resistance in an ideal transformer

. . . .

Given: A transformer has a load resistance of 25 Ω in the secondary circuit, 400 turns in the primary winding, and 100 turns in the secondary winding.

Find: Transferred resistance to the primary circuit.

Solution: $R_t = \left(\frac{N_p}{N_s}\right)^2 R_L$

$$R_t = \left(\frac{400}{100}\right)^2 (25 \text{ }\Omega)$$

$R_t = (16)(25\ \Omega)$

$R_t = 400\ \Omega$

Note. The load resistance of 25 Ω does not include the resistance of the winding wire, which is usually low compared to the load resistance.

The voltage equations for the primary and secondary circuits can be expressed in terms of the inductances of the coils. From Chapter 5 on resistance,

$$\phi = \frac{\text{mmf}}{R_c} \tag{8.64}$$

where

mmf = magnetomotive force
R_c = reluctance of the core

The magnetomotive force is equal to the sum of the currents times the corresponding number of turns. Thus,

$$\text{mmf} = N_p\, I_{pi} + N_s\, I_{si} \tag{8.65}$$

and

$$\phi = \frac{N_p\, I_{pi}}{R_c} + \frac{N_s\, I_s}{R_c} \tag{8.66}$$

If it is assumed that the reluctance is constant,

$$V_{pi} = N_p \frac{\Delta\phi}{\Delta t} = \frac{N_p^2}{R_c}\frac{\Delta I_{pi}}{\Delta t} + \frac{N_p N_s}{R_c}\frac{\Delta I_{si}}{\Delta t} \tag{8.67}$$

and

$$V_{si} = N_s \frac{\Delta\phi}{\Delta t} = \frac{N_p N_s}{R_c}\frac{\Delta I_{pi}}{\Delta t} + \frac{N_s^2}{R_c}\frac{\Delta I_s}{\Delta t} \tag{8.68}$$

The quantities N_p^2/R_c and N_s^2/R_c are defined as the inductances of the coils. The quantity $N_p N_s/R_c$ is called the mutual inductance. Thus, Equations 8.65 and 8.66 may be written as

$$V_{pi} = L_p \frac{\Delta I_p}{\Delta t} + (\sqrt{L_p L_s})\frac{\Delta I_{si}}{\Delta t} \tag{8.69}$$

$$V_{si} = (\sqrt{L_p L_s})\frac{\Delta I_{pi}}{\Delta t} + L_s \frac{\Delta I_{si}}{\Delta t} \tag{8.70}$$

where

$L_p = N_p^2/R_c$ = inductance of the primary coil
$L_s = N_s^2/R_c$ = inductance of the secondary coil
$\sqrt{L_p L_s} = N_p N_s/R_c$ = mutual inductance

These values for V_{pi} and V_{si} may be substituted in Equations 8.58 and 8.59 to obtain "exact" solutions for I_{pi} and I_{si} for the ideal transformer. Since the solutions require calculus, the methods used to solve the equations cannot be given here. However, for a resistance load, the root-mean-square primary current becomes

$$I_p = V_p \sqrt{\left(\frac{1}{\omega^2 L_p^2} + \frac{L_s^2}{L_p^2 R_L^2}\right)} \tag{8.71}$$

where ω is the angular frequency of the ac; for 60-cycle ac this would be $(360°)$ $(60 \text{ sec}^{-1}) = 21,600° \text{ sec}^{-1}$ or $(2\pi \text{ rad})(60 \text{ sec}^{-1}) = 377 \text{ rad·sec}^{-1}$. If $1/\omega^2 L_p^2 \ll L_s^2/L_p^2 R_L^2$, then

$$I_p = \left(\frac{V_p}{R_L}\right)\left(\frac{L_s}{L_p}\right) \tag{8.72}$$

Substituting in Equation 8.72, from the definitions of L_s and L_p, yields

$$\frac{L_s}{L_p} = \left(\frac{N_s^2}{R_c}\right)\left(\frac{R_c}{N_p^2}\right) = \left(\frac{N_s^2}{N_p^2}\right) = \left(\frac{N_s}{N_p}\right)^2 \tag{8.73}$$

$$I_p = \left(\frac{V_p}{R_L}\right)\left(\frac{N_s}{N_p}\right)^2 \tag{8.74}$$

This is the same result that is obtained by assuming that the power factor ratio is one.

The quantity ωL_p is called the inductive reactance of the primary coil. The inductive reactance must be large if $1/\omega^2 L_p^2$ is small. If the load resistance is very large ($R_L = \infty$ for an open secondary circuit) the quantity $L_s^2/L_p^2 R_L^2$ will approach zero, and $1/\omega^2 L_p^2$ will be the dominant term in Equation 8.71. The primary current for this case is

$$I_p = \frac{V_p}{\omega L_p} \tag{8.75}$$

The use of Equations 8.71, 8.72, and 8.75 are illustrated in Example 8.31.

EXAMPLE 8.31   Primary current in an ideal transformer
. . . .

Given: The primary coil inductance of an ideal transformer is 10 H. The primary coil is connected to a 110-V ac source having an angular frequency of 377 rad/sec. The inductance of the secondary coil is 5 H.

Find: **a.** The primary coil current is the secondary circuit is open.

  b. The primary coil current if the secondary circuit resistance is 100 Ω (use Equation 8.71).

  c. The primary coil current if the secondary circuit resistance is 100 Ω (use Equation 8.72).

Solution: **a.** From Equation 8.75

$$I_p = \frac{V_p}{\omega L_p}$$

$$I_p = \frac{110 \text{ V}}{(377 \text{ rad/sec})(10 \text{ H})}$$

$$I_p = 0.029 \text{ A}$$

b. $I_p = V_p \sqrt{\dfrac{1}{\omega^2 L_p^2} + \dfrac{L_s^2}{L_p^2 R_L^2}}$

$$I_p = (110 \text{ V}) \sqrt{\left(\frac{1}{(377 \text{ rad/sec})^2 (10 \text{ H})^2} + \frac{(5 \text{ H})^2}{(10 \text{ H})^2 (100 \text{ }\Omega)^2}\right)}$$

$$I_p = (110 \text{ V}) \sqrt{(7.04 \times 10^{-8}) + (2.5 \times 10^{-5})}$$

$$I_p = (110 \text{ V}) \sqrt{2.507 \times 10^{-5}}$$

$$I_p = (110 \text{ V})(5.007 \times 10^{-3})$$

$$I_p = 0.5508 \text{ A}$$

c. $I_p = \dfrac{V_p}{R_L} \dfrac{L_s}{L_p}$

$$I_p = \frac{110 \text{ V}}{100 \text{ }\Omega} \frac{5 \text{ H}}{10 \text{ H}}$$

$$I_p = 0.5500 \text{ A}$$

Note. If Equation 8.72 were used to calculate I_p for part a, the value would have been zero rather than 0.029 A. However, for parts b and c the values for I_p are essentially the same.

The Real Transformer   All the power input to a transformer is not available to the load. Some of the power is lost in the coils because of their resistance (copper losses) and some power is lost in the core (iron losses). The core losses are due to eddy currents and hysteresis.

The power lost in the coils can be calculated if the coil resistances and currents are known. For any resistance, current and voltage are in phase; and the corresponding power loss is always equal to the current squared times the resistance. Thus, the power loss due to the winding resistances is called "I^2R loss."

The changing magnetic field in the core induces a voltage in the core. The induced voltage will cause currents to flow in the core with a resulting power loss. The core currents are called "eddy currents." Eddy current losses can be reduced when the core is made from laminated sheets. These thin sheets of magnetic material are separated by insulating varnish. The lengths of the eddy current paths are reduced, and the power loss is reduced. In some transformers, the core is made from magnetic powder suspended in a ceramiclike insulating material.

Energy is required to reverse the direction of core magnetization every half-

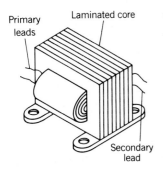

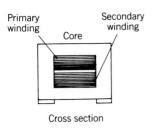

Cross section

FIGURE 8.40 Common transformer configuration.

cycle of the input frequency. This energy, called "hysteresis," is lost as heat like I^2R and eddy current losses and is controlled by the use of iron cores having low hysteresis susceptibility. Hysteresis losses are greater at higher frequencies, since core magnetization must reverse more frequently. In an effort to reduce hysteresis loss at very high frequencies, an iron core is not used. The secondary and primary coils are simply wound on a nonmagnetic, nonconducting core. However, in this case loss due to flux leakage must be considered.

Losses in a well-designed transformer do not normally exceed 2 or 3%; however, with these losses the power out of a transformer will always be less than the power in. The ratio of power out to power in is called the efficiency. In all of the discussion and examples to this point, an ideal efficiency of 100% has been assumed. For the sake of accuracy, the actual transformer efficiency should be included.

In Figure 8.38, a transformer having windings on opposite sides of the transformer core is shown. In practice, the core is shaped with a middle core piece, as shown in Figure 8.40, with the secondary coil wound on top of the primary coil.

Transformers can be wound with many secondary windings or with several connection taps on one secondary coil to suit the application.

Applications There are situations in which higher voltages are advantageous, and other situations in which low voltage is more useful. High voltage is necessary in power-transmission lines to keep I^2R losses to a minimum, and low voltage is necessary in most household appliances to keep them safe. Additionally, low voltage is needed to yield the high current necessary for welding. These voltages are produced by the proper use of step-up and step-down transformers. In addition, it may be necessary to isolate one system from another because several points are grounded. An isolation transformer is used, which has an equal number of primary and secondary turns. In this case $V_p = V_s$.

Consider the case of a power plant contracted to deliver 12,000 W of electrical power over a transmission line that has a resistance of 0.2 Ω. Consider further that the power plant unwisely decides to deliver the power at a potential of 120 V. How effective will the power transmission be?

For a unity power factor $P = IV$, the current in the transmission line is

$$I = \frac{P}{V}$$

$$I = \frac{12,000 \text{ W}}{120 \text{ V}}$$

$$I = 100 \text{ A}$$

From Ohm's law, the voltage drop in the transmission line is

$$V = IR$$

$$V = (100 \text{ A})(0.2 \text{ }\Omega)$$

$$V = 20 \text{ V}$$

This calculation shows that the voltage available to the customer is 120 V − 20 V = 100 V. In addition to the voltage drop, there is a power loss due to heating of the transmission line. At 100 A the power loss is

$$P = I^2R$$

$$P = (100 \text{ A})^2 (0.2 \ \Omega)$$

$$P = 2000 \text{ W}$$

The customer receives only 12,000 W − 2000 W = 10,000 W instead of the contracted amount.

Because a transformer was not utilized to step-up the transmission voltage, the power dropped from 12,000 W to 10,000 W and the voltage dropped from 120 V to 100 V. Both are substantial losses. Example 8.32 shows how much more efficient the power transmission can be made by use of a step-up transformer and why transformers are extremely valuable in power transmission.

EXAMPLE 8.32   Power transmission at high voltage
. . . .

Given: A power plant contracted to deliver 12,000 W of electrical power over a transmission line with a 0.2-Ω resistance. Through the use of a step-up transformer the potential can be raised from 120 V to 12,000 V.

Find: **a.**   Current transmitted.

  b.   Voltage drop in the transmission line.

  c.   Power loss in the transmission line.

  d.   Power and voltage delivered to the customer.

Solution: **a.** $P = IV$

$$I = \frac{P}{V}$$

$$I = \frac{12,000 \text{ W}}{12,000 \text{ V}}$$

$$I = 1 \text{ A}$$

  b. $V = IR$

$$V = (1 \text{ A})(0.2 \ \Omega)$$

$$V = 0.2 \text{ V}$$

  c. $P = I^2R$

$$P = (1 \text{ A})^2(0.2 \ \Omega)$$

$$P = 0.2 \text{ W}$$

  d.   Power delivered = 12,000 W − 0.02 W

  Power delivered = 11,999.8 W

  Voltage delivered:

(1) Transmitted = 12,000 V − 0.2 V

Transmitted = 11,999.8 V

(2) Stepped down = 11,999.8 V/100

Stepped down = 119.998 V

The power and voltage drop an insignificant amount.

SECTION 8.3 EXERCISES

1. A doorbell transformer has primary windings of 720 turns and secondary windings of 48 turns. The source voltage is a 120-V house circuit. What voltage is delivered to operate the bell?

2. Discuss how the resistance of a coil in a transformer dissipates power and reduces efficiency.

3. What is the primary current of an ideal transformer if the primary voltage is 18 V, the power output is 200 W, and the power factor is unity?

4. What is the input current of an ideal transformer if the ideal mechanical advantage is 6 and the output current is 0.2 A?

5. An ideal transformer with a power factor ratio of unity has N_p = 400 turns, N_s = 900 turns, V_p = 8 V, and I_p = 0.4 A. Calculate I_s and V_s.

6. An ideal transformer with a power factor ratio of one has N_p = 1000 turns, N_s = 200 turns, I_p = 0.8 A, V_p = 10 V. Calculate I_s and V_s.

7. An ideal transformer has V_p = 12 V and V_s = 3 V. Calculate the turns ratio.

8. An ideal transformer with a power factor ratio of one has I_p = 2 A and I_s = 0.1 A. Calculate the turns ratio.

9. Briefly describe the three major sources of power loss in a transformer.

10. The inductance of the primary coil of an ideal transformer is 5.0 H. If the frequency of the applied voltage is 60 Hz and the root-mean-square voltage is 120 V, what is the primary current when the secondary circuit is open? When a load resistance of 100,000 Ω is connected across the secondary coil? The inductance of the secondary coil is 1.0 H.

11. An ideal transformer has N_p = 3600 turns, N_s = 600 turns, and R_L = 50 Ω. Calculate the transferred resistance if the power factor ratio is one.

12. If an ideal transformer has N_p = 4000 turns, N_s = 2000 turns, and R_L = 50 Ω, what are the primary and secondary currents if V_p = 100 V? Assume 100% efficiency and a power factor ratio of one.

S U M M A R Y

Now that a number of force transformers, or simple machines, have been considered in detail, a few remarks to summarize their major characteristics and emphasize their basic unity may be helpful.

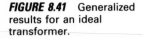

FIGURE 8.41 Generalized results for an ideal transformer.

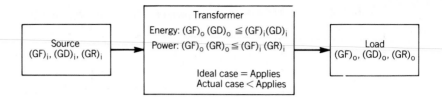

The force transformer is a coupling device between an energy source and an energy load. The transformer changes, or transforms, input values of forcelike, displacementlike, or ratelike quantities into different output values. The law of conservation of energy states that where no energy is lost through frictional effects (ideal transformer), the input energy equals the output energy. Mathematically speaking, the products of forcelike and displacementlike quantities have the same value at input and output. Additionally, where transformers operate continuously so that one can speak of the rate of work, or power, one can state that for the ideal case, the input power equals the output power, or the products of forcelike and ratelike quantities have the same values at input and output.

In the actual case, some of the energy or power input is lost because of frictional effects, so that the energy and power output is less than the corresponding input, as summarized in Figure 8.41.

A characteristic of the force transformer is that there is a mechanical advantage greater than one, $(GF)_o > (GF)_i$. This is achieved at the expense of a displacementlike quantity, $(GD)_o < (GD)_i$ or $(GR)_o < (GR)_i$. Only in this way can energy and power be conserved. The ratio $(GF)_o/(GF)_i$ is called the mechanical advantage of the transformer and has been given the symbol IMA for the ideal case and AMA for the actual case. The actual mechanical advantage is always less than the ideal mechanical advantage.

The foregoing statements apply to all energy systems, as indicated in Table 8.1, which includes all fundamental energy systems except the thermal system. A few minutes studying this table should reinforce the analogous nature of the force transformer in these different energy systems. The table also lists the IMA for each transformer.

TABLE 8.1 Summary of force transformers

System	Transformer	IMA[a]	(GF), (GD), (GR)[b]	Energy and Power Relationship[c]
Mechanical translational	Pulley	Number of supporting cords	Input: F_i, x_i, v_i	$F_o x_o \leqq F_i x_i$
	Differential hoist	$2D/(D\text{-}d)$		
	Lever	$l_i/l_o = x_i/x_o$		
	Incline plane	$x_i/x_o = 1/\sin\theta$		
	Wedge	$2h/b = 1/\tan\theta$		
Mechanical rotational	Screw	$2\pi r_i/x_o$	Input: τ_i, θ_i, ω_i	$\tau_o(\Delta\theta_o)\leqq\tau_i\,(\Delta\theta)$
	Wheel and axle	r_i/r_o	Output: τ_o, θ_o, ω_o	$\tau_o\omega_o\leqq\tau_i\,\omega_i$
	Belt drives	$r_o/r_i = \omega_i/\omega_o$		
	Gear train	$n_o/n_i = \omega_i/\omega_o$		
Fluid	Two piston	$A_o/A_i = x_i/x_o$	Input: p_i, V_i, Q_{Vi}	$p_o(\Delta V_o)\leqq p_i\,(\Delta V_i)$
			Output: p_o, V_o, Q_{Vo}	$p_oQ_{Vo}\leqq p_iQ_{Vi}$
	Hydraulic jack or press	$(A_o/A_i)\,(l_i/l_o)$		
	Pressure transformer	A_t/A_o		
Electromagnetic	Voltage transformer	N_p/V_s	Input: V_p, Q_p, I_p	$V_sQ_s\leqq V_pQ_p$
			Output: V_s, Q_s, I_s	$V_sI_s\leqq V_pI_p$ If power factor ratio is unity.

[a] IMA = Ideal mechanical advantage.
[b] (GF) = Forcelike quantity; (GD) = displacementlike quantity; (GR) = ratelike quantity.
[c] = applies for ideal case; < applies for actual case.

Energy

Convertors

INTRODUCTION

Our modern technological society depends on the production, distribution, and consumption of large amounts of energy. These processes are made possible by devices that can be classified as energy convertors. These devices accept input energy in one energy system and deliver output energy in another system. For example, an electric fan is composed of two energy-converting devices. The electric motor converts electrical energy to rotational mechanical energy, and the fan blades convert this mechanical energy to the fluid energy of moving air. Some of the input energy is lost because energy convertors are not 100% efficient.

This chapter discusses

- *A variety of energy convertors important in the conversion and application of energy.*
- *The efficiency of individual energy convertors and systems.*

The law of conservation of energy states that energy can be neither created nor destroyed. Thus, energy convertors actually do not create energy; they only

change the input energy to another form. Another consequence of energy conservation is that all input energy must appear as output energy in one form or another. Some of this output energy always appears as heat energy because of resistance in the system. If heat energy is not the desired form of the output, it represents loss in the system. The efficiency of an energy convertor is given by

$$\eta = \frac{E_{out}}{E_{in}} \text{ (100)} \tag{9.1}$$

Energy convertors are also power convertors. Since time is always involved in the energy conversion process. It's often convenient to consider input and output powers when energy convertors are in steady-state operation. Since average power is the total energy divided by the time, the average efficiency can also be written as

$$\eta = \frac{\left(\dfrac{E_{out}}{t}\right)}{\left(\dfrac{E_{in}}{t}\right)} \text{ (100)} = \frac{P_{out}}{P_{in}} \text{ (100)} \tag{9.2}$$

where

$t = $ total time
$P_{out} = $ average power out
$P_{in} = $ average power in

A wide variety of energy-conversion devices are in common use today, and most energy used in the United States is changed in form several times before its final application. The electrical energy used to operate the fan mentioned earlier originally was chemical energy contained in a fuel. The fuel was burned to produce thermal energy, which was converted to fluid energy in high-pressure steam. A turbine converted the steam to mechanical energy. Our entire society depends on our ability to convert energy from one form to another in a controlled and efficient manner. Figure 9.1 illustrates a common sequence of energy conversion.

FIGURE 9.1 Common sequence of energy conversion.

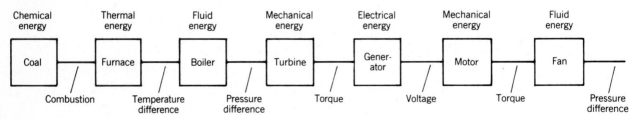

Table 9.1 is a listing of the energy convertors discussed in this chapter. Some of them have been in common use for many years, whereas others currently are in research and development. In this chapter, optical (radiant) energy is added as an additional and important form of energy. Energy convertors are discussed in the order of the form of their input energy.

TABLE 9.1 Important energy convertors

Input Energy	Output Energy				
	Mechanical	Fluid	Electrical	Thermal	Optical
Mechanical		Pump Fan	Generator		
Fluid	Turbine Windmill		Magneto-hydro-dynamic generator		
Electrical	Motor			Heating devices	Lighting devices
Thermal	Internal combustion engines	Steam generator	Thermo-electric generators		
Optical			Solar cells	Solar collectors	

9.1

MECHANICAL ENERGY TO FLUID ENERGY: THE FAN

OBJECTIVES

Upon completion of this section, the student should be able to

- Define the following terms for a fan:
  Air horsepower
  Power input
  Static efficiency
  Static pressure
  Total pressure
  Velocity pressure
  Mechanical efficiency

- Given the static pressure, velocity pressure, volume flow rate, and shaft horsepower of a fan, determine its static and mechanical efficiency.

- Given a schematic representation of an axial-flow fan and a radial-flow fan, correctly identify each and indicate the direction of airflow on both diagrams.

- Given the schematic representation of six different blade types in a centrifugal (radial) fan, correctly identify and label each of the blade types.

- Given a schematic representation showing four different inlet designs for a propeller, draw the airflow for each design. Indicate which of these four systems is most efficient.

DISCUSSION

Introduction   Fans, blowers, and compressors move air, but at greatly different pressures. Fan pressures range from a few inches of water up to about 1 psi. Blowers work to about 50 psi; compressors operate from about 35 psi and up. Fans, blowers, and compressors all basically do the same thing. When appropriate, important differences between the three will be pointed out.

A fan can be defined as a volumetric machine that operates very much like a pump. In simple terms, a fan moves quantities of gas or air from one place to another. In doing this, it overcomes resistance to flow by supplying the fluid with the energy necessary for continued motion. Physically, the essential elements of a fan are a bladed rotor (such as an impeller or propeller) and a housing in which to collect the incoming air or gas and to direct its flow. Figure 9.2 displays the essential elements of a fan.

Work   Since any device that makes something else move is doing work, a fan, blower, or compressor requires energy for operation. The amount of energy required depends on the volume of gas moved, the resistance against which the fan work is done, and the machine efficiency.

These energy relationships are clarified by the pressure–volume diagram shown in Figure 9.3 for an ideal compressor. From point 1 to point 2 in the compressor cycle, there is very little pressure change in the suction phase. During this phase, gas is sucked through the intake valve of the compressor at constant pressure until the volume at the bottom of the piston stroke is reached at point 2. Here gas is compressed, increasing the pressure. This step is shown as the compression phase in Figure 9.3 (point 2 to point 3). From point 3 to point 4, the compressor is discharging the gas at constant pressure through the discharge valve. At point 4, gas has been completely discharged. The compressor then shows a drop in pressure back to point 1, and the cycle is repeated. The work done by the compressor is equal to the area enclosed by the cycle.

Unlike blowers or compressors, which work against larger pressures and significant pressure increases, the bulk of fan work goes into moving gas at a

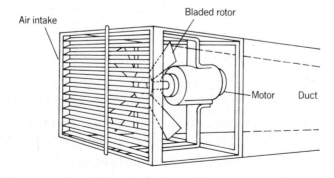

FIGURE 9.2   Elements of a fan.

FIGURE 9.3 Cycle of a compressor.

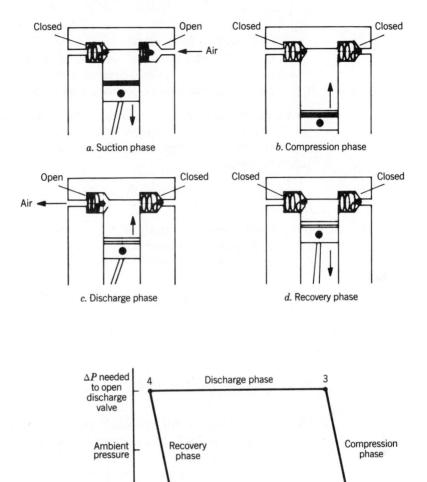

relatively low pressure. The volume remains virtually constant during the compression phase because the change in specific weight of the gas between fan inlet and discharage is negligible; in fact, for a 1-psi pressure increase, the specific weight of air changes by less than 7%. Thus, when fan power is analyzed, the density change can be thought of as zero.

If there is no change in density, the energy equation and corresponding power equation for incompressible flow can be used as an approximation. If the flow is turbulent, Equation 6.71 may be used as the energy equation. Because of the low density and approximate horizontal flow, the term $g(h_b - h_a)$ may be dropped. The energy equation then becomes

$$\frac{v_b^2}{2} - \frac{v_a^2}{2} + \frac{1}{\rho_m}\,(p_b - p_a) + H_{ab} = \frac{\Delta W}{\Delta m} \tag{9.3}$$

where

     a and b are any two cross sections

     v_a and v_b are the corresponding velocities at a and b

     p_a and p_b are the corresponding pressures at a and b

     H_{ab} = heat energy generated between a and b per unit mass

To obtain the power equation, it is necessary to multiply each term of Equation 9.3 by the mass flow rate. Thus, mass flow rate times work per unit mass gives work per unit time. Since the density is essentially constant, one obtains from the definition of density

$$\Delta m = \rho_m \Delta V \qquad (9.4)$$

where

     Δm = small amount of mass flow

     ΔV = corresponding volume flow

     ρ_m = constant mass density

Dividing both sides of Equation 9.4 by the corresponding change in time yields

$$\frac{\Delta m}{\Delta t} = \rho_m \frac{\Delta V}{\Delta t} = \rho_m Q_V$$

where

$$Q_V = \frac{\Delta V}{\Delta t} = \text{volume flow rate}$$

Multiplying the terms on the left-hand side of Equation 9.3 by $\rho_m \, Q_V$ and the terms on the right-hand side by the equivalent expression $\Delta m / \Delta t$ gives the power equation for the flow

$$Q_V\left(\frac{\rho m}{2} v_b^2 - \frac{\rho m}{2} v_a^2\right) + Q_V(p_b - p_a) + Q_V \rho_m H_{ab} = \frac{\Delta w}{\Delta t} \qquad (9.5)$$

If the fan lies between the cross sections a and b, the fan power, $\Delta w / \Delta t$, is equal to the kinetic power plus the pressure power plus the heat power between a and b. Cross section a is usually taken where the velocity is essentially zero. For a fan, this corresponds to the ambient air at atmospheric pressure. Thus, in Equation 9.5, $p_a = p_{at}$ and $v_a = 0$. Substituting these values in Equation 9.5 yields

$$\frac{Q_V \rho_m v_b^2}{2} + Q_V(p_b - p_{at}) + Q_V \rho_m H_{ab} = \frac{\Delta w}{\Delta t} \qquad (9.6)$$

If there were no heat losses, $H_{ab} = 0$, the fan power would be converted into pressure power and kinetic power of the gas. For this case, the fan would be 100% efficient in changing fan power into fluid power.

     The pressures p_b and p_{at} in Equation 9.6 are static pressures. Static pressure is the force per unit area exerted on the walls of pipes or ducts when they are parallel to the flow. Stagnation pressure is the combined effect of static pressure and pressure due to the flow when the wall is perpendicular to the flow. Both static and stagnation pressure are measured with a pitot tube.

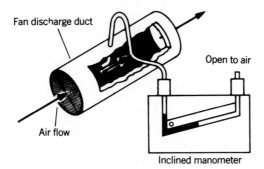

FIGURE 9.4 Measurement of stagnation pressure.

Fan discharge duct

Open to air

Air flow

Inclined manometer

Stagnation pressure minus
atmospheric pressure

To measure the stagnation pressure, the port of the measuring tube must be perpendicular to the flow as shown in Figure 9.4. Static pressure is measured when the port is parallel to the flow as shown in Figure 9.5. If both measurements are made at the same cross section, the stagnation pressure can be related to the static pressure and the velocity. It was shown in Chapter 6 that Bernoulli's equation is a good approximation for this case. If a corresponds to the port cross section, $v_a = 0$; and $p_a = p_t$, the stagnation pressure. This gives

$$\frac{p_t}{\rho_m} + \frac{0^2}{2} = \frac{p_b}{\rho_m} + \frac{v_b^2}{2} \tag{9.7}$$

Simplification yields

$$p_t = p_b + \frac{\rho_m v_b^2}{2} \tag{9.8}$$

Using this relationship to substitute into Equation 9.6 yields

$$Q_V(p_t - p_{at}) + Q_V \rho_m H_{ab} = \frac{\Delta w}{\Delta t} \tag{9.9}$$

The manometer shown in Figure 9.4 measures $p_t - p_{at}$, whereas the manometer shown in Figure 9.5 measures $p_b - p_{at}$.

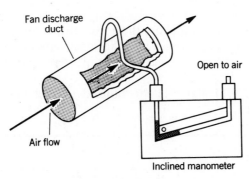

Fan discharge
duct

Open to air

Air flow

Inclined manometer

FIGURE 9.5 Measurement of static pressure.

Static pressure minus
atmospheric pressure

Air Horsepower   The quantity $Q_V(p_t - p_{at})$ in Equation 9.9 represents the total fluid power:

$$P_t = Q_V(p_t - p_{at}) \qquad (9.10)$$

where

P_t = total fluid power

This represents the fluid power gained as a result of the increased pressure and velocity from ambient conditions. If only the fluid power gained from an increase in static pressure is used, one obtains the static fluid power:

$$P_s = Q_V(p_s - p_{at}) \qquad (9.11)$$

where

P_s = static fluid power

If the pressure differences, as shown in Figures 9.4 and 9.5, are measured in inches of water and the flow rate Q_V in cubic feet per minute, Equations 9.10 and 9.11 can be used to find the total fluid horsepower and static fluid horsepower by dividing by the conversion factor 6356 (ft³/min)·inch (H₂O)/hp:

$$(\text{HP})_t = \frac{Q_V(p_t - p_{at})}{6356} \qquad (9.12)$$

$$(\text{HP})_s = \frac{Q_V(p_b - p_{at})}{6356} \qquad (9.13)$$

Efficiency   The fan efficiency can be defined in terms of the total or static horsepower and the input shaft horsepower to the fan:

$$\eta_t = \frac{(\text{HP})_t}{P_{in}} (100) \qquad (9.14)$$

$$\eta_s = \frac{(\text{HP})_s}{P_{in}} (100) \qquad (9.15)$$

where

η_t = total efficiency
η_s = static efficiency

Many small variables influence the power output efficiency of any fan; such things as pressure and temperature are the most important. For the purposes of this section, no corrections are made for temperature and pressure variations of the air or gas.

Equations 9.14 and 9.15 are utilized to find the solution to a problem in Example 9.1.

EXAMPLE 9.1   Fan efficiency
. . . .

Given: A fan used for air conditioning and heating develops an increase in static pressure of 4.5 inch (H₂O) and an increase of total pressure of

5.2 inch (H_2O). The volume flow rate is 8000 ft³/min and the shaft horsepower is 7.2 hp.

Find: **a.** The static efficiency of the fan.

                b. The total efficiency of the fan.

Solution: **a.** The static efficiency is

$$\eta_s = \frac{(HP)_s}{P_{in}}\,(100)$$

From Equation 9.13

$$(HP)_s = \frac{Q_V(p_b - p_{at})}{6356}$$

$$(HP)_s = \frac{(8000 \text{ ft}^3/\text{min})\ (4.5 \text{ inch } [H_2O])}{6356\ (\text{ft}^3/\text{min})\cdot\text{inch } (H_2O)/\text{hp}}$$

$$(HP)_s = 5.66 \text{ hp}$$

$$\eta_s = \frac{(5.66 \text{ hp})}{(7.2 \text{ hp})}\,(100)$$

$$\eta_s = 79\%$$

         b. The total efficiency is

$$\eta_t = \frac{(HP)_t}{P_{in}}\,(100)$$

From Equation 9.12

$$(HP)_t = \frac{Q_V(p_t - p_{at})}{6356}$$

$$(HP)_t = \frac{(8000 \text{ ft}^3/\text{min})\ (5.2 \text{ inch } [H_2O])}{6356\ (\text{ft}^3/\text{min})\cdot\text{inch } (H_2O)/\text{hp}}$$

$$(HP)_t = 6.54 \text{ hp}$$

$$\eta_t = \frac{(6.54 \text{ hp})}{(7.2 \text{ hp})}\,(100)$$

$$\eta_s = 91\%$$

Fan Types   There are two basic classes of fans: (1) the axial-flow design, which moves gas or air parallel to the axis of rotation, and (2) the centrifugal-flow (or radial-flow) type, which moves air or gas perpendicular to the axis of rotation. Axial-flow fans are better suited to low-resistance applications. Centrifugal-flow fans are used where higher pressures are needed.

The axial-flow fan uses the screwlike action of a multibladed rotating shaft or propeller to move air or gas in a straight-through path. The leading edge of each rotating propeller blade bites into the air or gas, which is then propelled through the fan and discharged in a helical pattern, as shown in Figure 9.6, by the blades' trailing edge.

FIGURE 9.6   Axial-flow fan.

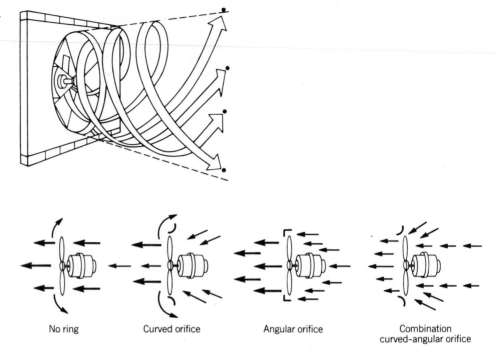

FIGURE 9.7   Airflow patterns for various inlet designs.

No ring   Curved orifice   Angular orifice   Combination curved–angular orifice

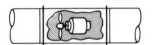

FIGURE 9.8   Tube-axial fan.

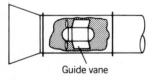

Guide vane

FIGURE 9.9   Vane-axial fan.

FIGURE 9.10   Centrifugal fan.

The efficiency of a propeller fan depends largely on its inlet design, because air recirculation around blade tips greatly reduces the efficiency of a propeller-type fan. Installation of a curved orificelike ring at the blade tips reduces the amount of recirculation and improves efficiency. An angle-shaped ring almost eliminates recirculation. Optimum efficiency is achieved with a combination curved, angle-shaped ring. The flow patterns for the air recirculating around the blade tips of these various inlet designs are illustrated in Figure 9.7.

The tube-axial design shown in Figure 9.8 is nothing more than a propeller fan enclosed in a cylinder that collects and directs airflow. In a vane-axial fan, shown in Figure 9.9 (an extension of the tube-axial design), air guide vanes on the air discharge side of the propeller straighten out the airflow pattern and increase static pressure.

Axial-flow fans can have widely differing characteristics, depending on the design, blade type, and ducting. Generally, they are used when the required outlet velocities are higher than can be produced by centrifugal fans and when pressure demands are not above actual flow limits.

Centrifugal, or radial, fans are advantageous when the air must be moved in a system in which the frictional resistance is relatively high (Figure 9.10). The air drawn into the center of the revolving wheel turns 90° and enters the space between the blades in the wheel. The bladed wheel whirls air centrifugally between each pair of blades and tosses it out peripherally at high velocity and increased static pressure. As this happens, more air is sucked in at the eye of the wheel. As the air leaves the revolving blade tips, part of its velocity is converted into additional static pressure by divergence of the scroll-shaped housing.

FIGURE 9.11 Common blade shapes used in centrifugal fans.

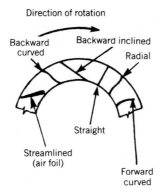

Blade shapes employed are of three basic types: forward curved, straight, and backward curved. Other configurations, including an airfoil design, are merely variations of these designs. Some commonly used shapes are given in Figure 9.11.

In general, blade type limits top fan speed. Backward-curved blade machines can operate at a higher speed than forward-curved designs. The type of blade employed also depends on space limitations, allowable noise levels, efficiency demanded by specified load conditions, and desired fan performance characteristics. Although mechanical efficiencies are much the same for each different blade type, differences in horsepower and pressure–volume relationships can be significant.

SECTION **9.1** *EXERCISES*

1. Write an equation for the following terms:
Total pressure p_t =
Static air horsepower $(HP)_s$ =
Total air horsepower $(HP)_t$ =
Static efficiency η_s =
Mechanical efficiency η_t =

2. Prepare a drawing that illustrates the airflow through a radial-flow fan and through an axial-flow fan.

3. Label the types of fan inlets illustrated below and identify the most efficient.

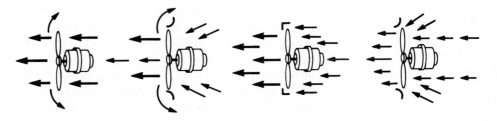

4. A fan produces an increase in static pressure of 2.5 inch (H_2O) and an increase in total pressure of 3.0 inch (H_2O). The flow rate is 3600 ft³/min. The input horsepower is 2 hp. Determine the static efficiency and total efficiency of the fan.

5. A fan with an input power of 1.0 hp produces a flow rate of 2000 ft^3/min. The increase in static pressure is 2.6 inch (H$_2$O). The total efficiency of the fan exceeds the static efficiency by 10%. Determine the total pressure.

6. How much power must be input to a fan to achieve an air horsepower of 7 hp if the mechanical efficiency of the fan is 72%?

9.2

FLUID ENERGY TO MECHANICAL ENERGY: THE TURBINE

OBJECTIVES

Upon completion of this section, the student should be able to

- Name and give the function of two major elements of a turbine.

- In one or two sentences, define the terms
  Impulse turbine
  Reaction turbine
  Staging

- Calculate the force on a moving body exerted by a fluid jet.

- Calculate the force and torque on an impulse turbine exerted by a fluid jet.

- Calculate the ideal power output of an impulse turbine as a function of the velocity of the buckets.

- Estimate the maximum ideal power of an impulse turbine when the height of the water reservoir is given.

DISCUSSION

Introduction   The turbine is a device that converts fluid energy into mechanical rotational energy. Gas turbine-powered automobiles and trucks are a reality, and gas turbine turboprop engines are in daily use in aviation. Modern ocean liners are propelled by marine turbines. Various steam turbine-equipped generating plants (including nuclear plants) produce approximately 80% of the electricity in the United States. Water-powered turbines produce nearly 20% of America's electrical power.

A turbine consists of two major elements, a high-energy working fluid and a rotor. The input energy to the turbine is provided by the high-energy fluid (combustion gases, steam, wind, or a flowing fluid such as water). The flowing medium has an initial high energy content as it enters the turbine. It flows through the turbine and causes the rotor to spin. The rotor, in effect, extracts energy from the moving fluid and converts the extracted fluid energy to a mechanical rotational form. The mechanical rotational energy is transmitted along the rotor shaft, where it is available as useful output energy. These elements of a turbine operation and energy flow are illustrated in Figure 9.12.

Equation 9.2 is used to find efficiency in Example 9.2. High efficiency is one of the distinct advantages of a turbine over reciprocating engines since the latter wastes much energy starting and stopping heavy pistons. Turbines are designed to

FIGURE 9.12 Energy flow in a turbine.

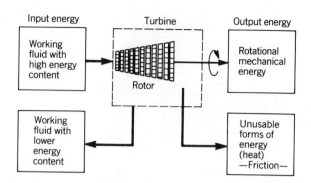

rotate smoothly at high rpm with little vibration and no change in direction of the rotor. Since there is little vibration, turbines also may be lighter in weight than other engines.

EXAMPLE 9.2   Efficiency calculations
. . . .

Given: A marine turbine that develops 50,000 hp at the rotor shaft and requires an input power of 55,555 hp.

Find: The percent efficiency.

Solution: $\eta = \dfrac{P_{out}}{P_{in}} (100)$

$\eta = \dfrac{(50,000 \text{ hp})}{(55,555 \text{ hp})} (100)$

$\eta = 90\%$

The modern water turbines used in generating electricity convert the available potential heat energy of water into electrical energy at efficiencies as high as 94%. Some of these units develop as much as 700,000 hp. The efficiency of steam turbines can be improved if the rotor is constructed with series of axial stages having alternate rows of rotating and fixed blades, as shown in Figure 9.13. Each stage extracts energy from the fluid flowing into it from the previous stage. Turbines of this type are known as impulse-reaction turbines. The term "impulse" refers to the process by which the available heat energy in steam is transformed into kinetic energy as the steam jet flows from the nozzles. This kinetic energy is transformed to work done on the rotor by impinging on a rotating row of blades. In pure impulse turbines, such as some water turbines, the working fluid undergoes little pressure change passing through a stage. The term "reaction" turbine is applied to the type of turbine that derives an additional thrust by directing the working fluid through a curved blade, thereby changing the fluid's direction. An additional thrust is gained from the change in momentum of the working fluid. Most actual turbines are a combination of impulse and reaction

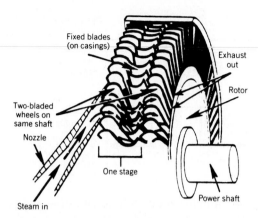

Fixed blades
(on casings)

Exhaust
out

Rotor

Two-bladed
wheels on
same shaft

Nozzle

One stage

Power shaft

Steam in

principles. Practical turbine systems for converting fluid energy to mechanical
energy in electrical generating plants often consist of two or three turbines
mounted on the same shaft. Thus, the exhaust steam from a high-pressure turbine
may be used to power another turbine designed for lower-pressure operation.

Impulse Turbine   In Chapter 4, the force exerted on a stationary surface by a fluid
jet was shown to be (Equation 4.34)

$$F_R = \frac{\Delta m}{\Delta t} \Delta \mathbf{v} = Q_m \Delta \mathbf{v} \tag{9.16}$$

where

$Q_m = \Delta m/\Delta t$ = total mass flow rate
$\Delta \mathbf{v}$ = change in velocity of the water

It was assumed that the body was not moving. For the case of an impulse turbine,
the blades or buckets are moving; this changes Equation 9.16 if a fixed coordinate
system is used. The correct equation can be obtained by using a coordinated
system attached to the body. Since the body is assumed to be moving with
constant velocity, Newton's second law of motion can be used for this coordinate
system. For this moving coordinate system, Equation 9.16 will apply since the
body is at rest in this system. Thus, it is necessary to calculate Q_m and $\Delta \mathbf{v}$ relative
to the moving coordinate system.

Consider the diagram shown in Figure 9.14. The body is moving with a constant
velocity v_{Bx} in the x direction. In addition, the incident jet has a velocity v_{Ix} in the
x direction. However, a person moving along with the body will not observe a jet
velocity equal to v_{Ix}. For example, if the body moves at a velocity equal to v_{Ix}, the
man on the body will say the velocity of the jet is zero. In general the jet velocity
observed on the body is

$$v_{Iox} = v_{Ix} - v_{Bx} \tag{9.17}$$

where

v_{Iox} = jet velocity observed from the moving body
v_{Ix} = jet velocity observed from a fixed laboratory frame of reference
v_{Bx} = velocity of the body observed from the fixed frame of reference

FIGURE 9.14 Comparison of fixed and "attached to the body" coordinate systems.

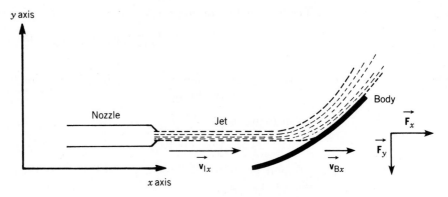

a. Velocities in a fixed coordinate system

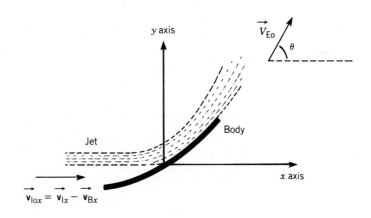

b. Coordinate system attached to the body

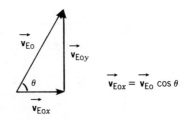

c. Vector diagram

The corresponding fluid mass striking the body per second becomes

$$\frac{\Delta m}{\Delta t} = \rho_m A \, v_{Iox} = \rho_m A (v_{Ix} - v_{Bx}) \tag{9.18}$$

where

ρ_m = mass density of the fluid
A = area of the jet

The force acting on the body in the x direction becomes (Equation 9.16)

$$F_{Rx} = \rho_m A(v_{Ix} - v_{Bx})\Delta v_{Box} \tag{9.19}$$

where

Δv_{Box} = the change in the velocity in the x direction measured by an observer on the body

From Figure 9.14,

$$\Delta v_{Box} = v_{Iox} - (-v_{Eo} \cos \theta) = v_{Iox} + v_{Eo} \cos \theta \tag{9.20}$$

Because of frictional losses, v_{Eo} will be less than v_{Iox}. It is advantageous to express v_{Eo} as

$$v_{Eo} = f v_{Iox} \tag{9.21}$$

where

f = a number less than one

Thus, Equation 9.20 may be written as

$$\Delta v_{Box} = v_{Iox}(1 + f \cos \theta) \tag{9.22}$$

Using Equation 9.17 to substitute for v_{Iox} yields

$$\Delta v_{Box} = (v_{Ix} - v_{Bx})(1 + f \cos \theta) \tag{9.23}$$

If this expression for Δv_{Box} is substituted into Equation 9.19, the force acting on the body in terms of measured velocities in the fixed laboratory reference system becomes

$$F_R = \rho_m A(v_{Ix} - v_{Bx})^2(1 + f \cos \theta) \tag{9.24}$$

An application of Equation 9.24 is given in Example 9.3.

EXAMPLE 9.3   Force exerted on a moving body by a water jet
. . . .

Given: A jet of water moving in the x direction with a diameter of 1.0 inch and a velocity of 100 ft/sec strikes a body moving in the same direction with a velocity of 50 ft/sec. The angle of the jet leaving the body with respect to the direction of the x axis is 20°. The friction loss is such that $f = 0.95$.

Find: The force acting on the body in the x direction.

Solution: All distances must be in feet, time in seconds, and mass in slugs. The density of water is

ρ_m = 1.93 slugs/ft³ at 80°F

The area of the jet is

$$A = \frac{\pi D^2}{4} = \frac{\pi}{4}\left(\frac{1.0 \text{ inch}}{12 \text{ inch/ft}}\right)^2$$

$$A = 0.00545 \text{ ft}^2$$

The force on the body in the x direction is

$$F_{Rx} = \rho_m A(v_{Ix} - v_{Bx})^2(1 + f \cos \theta)$$

$$F_{Rx} = (1.93 \text{ slug/ft}^3)(0.00545 \text{ ft}^2)(100 \text{ ft/sec} - 50 \text{ ft/sec})^2$$
$$(1 + 0.95 \cos 20)$$

$$F_{Rx} = 49.8 \text{ (slug·ft)/sec}^2 = 49.8 \text{ lb}$$

In an actual impulse turbine, many buckets are attached to a wheel as shown in Figure 9.15. To obtain the total force acting on the turbine wheel that is exerted by the jet, it is necessary to add the forces acting on different buckets. An approximate total force can be obtained by considering a large wheel with a large number of buckets such that the buckets on which the jet acts lie along a straight line. This is shown in Figure 9.16. New buckets enter the jet at position A and leave at

FIGURE 9.15 Relationship between jet and buckets in an impulse turbine.

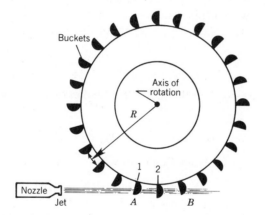

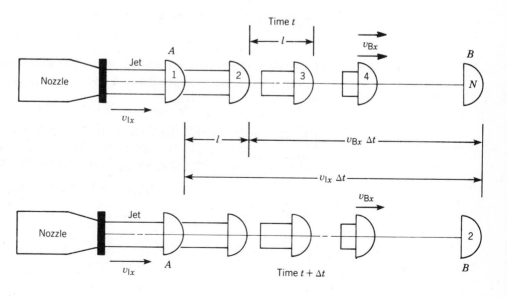

FIGURE 9.16 "Straightline" buckets interacting with a jet.

position B as the turbine wheel rotates. The distance l is the uniform distance between buckets. As a new bucket arrives at position 1, it cuts off a segment of the jet equal to l. When the left-hand side of this segement reaches bucket 2, the force on bucket 2 becomes zero. This will occur after a time Δt. During this time, the left-hand side of the segment will travel a distance equal to $v_{Ix} \Delta t$, while bucket 2 travels a distance equal to $v_{Bx} \Delta t$. As shown in Figure 9.16,

$$d = v_{Ix}\Delta t = v_{Bx}\Delta t + l \tag{9.25}$$

Equation 9.25 can be solved for Δt to yield

$$\Delta t = \frac{l}{v_{Ix} - v_{Bx}} \tag{9.26}$$

The total distance d can then be written as

$$d = v_{Ix}\Delta t = \frac{v_{Ix}l}{v_{Ix} - v_{Bx}} \tag{9.27}$$

After the distance d, no force acts on a bucket because the left-hand side of the segment of the jet that was cut off has reached the bucket in front of the segment. All buckets within this distance will have a force acting on them. Thus, the total number of buckets on which a part of the jet acts is

$$N = \frac{d}{l} = \frac{v_{Ix}}{v_{Ix} - v_{Bx}} \tag{9.28}$$

The total force is then the force on a bucket times the number of buckets, N. This yields

$$F_{Rt} = N\rho_m A(v_{Ix} - v_{Bx})^2(1 + f \cos \theta)$$

Substituting for N from Equation 9.28

$$F_{Rx} = \rho_m A(v_{Ix} - v_{Bx})(v_{Ix})(1 + f \cos \theta) \tag{9.29}$$

In Chapter 7, the power was shown to be equal to the force times the velocity of the moving body. Thus, the turbine power becomes

$$P = F_{Rt}v_{Bx} = \rho_m A(v_{Ix})(v_{Bx})(v_{Ix} - v_{Bx})(1 + f \cos \theta) \tag{9.30}$$

Equation 9.30 neglects the effects of windage and bearing losses. The maximum power (ideal power) occurs if $f = 1$ and $\theta = 0$. Equation 9.30 then becomes

$$(P_t)_{ideal} = 2\rho_m \, Av_{Ix}v_{Bx}(v_{Ix} - v_{Bx}) \tag{9.31}$$

One can easily see that the ideal power becomes zero when $v_{Bx} = 0$ and when $V_{Bx} = v_{Ix}$. It can also be shown that the maximum ideal power occurs when $v_{Bx} = v_{Ix/2}$. When this value for v_{Bx} is substituted into Equation 9.31, one obtains

$$(P_t)_{ideal} = 1/2 \, (\rho_m Av_{Ix})v_{Ix}^2 \tag{9.32}$$

when

$$v_{Bx} = \frac{v_{Ix}}{2}$$

FIGURE 9.17 Jet power, ideal power, and actual power versus bucket speed.

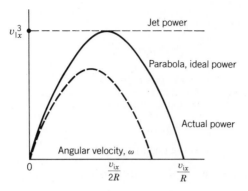

The quantity $\rho_m A v_{Ix}$ is the mass flow rate $\Delta m/\Delta t$. Thus,

$$(P_t)_{\text{ideal}} = \frac{1}{2} \frac{\Delta m}{\Delta t} v_{Ix}^2 \tag{9.33}$$

Since $1/2\ \Delta m v_{Ix}^2$ is the kinetic energy, the quantity on the right in Equation 9.33 is the jet power. A plot of the ideal power is shown in Figure 9.17 as a function of v_{Bx}. Also shown in the figure is a typical power as a function of v_{Bx} and the incident jet power. Several problems are solved in Examples 9.4 and 9.5.

EXAMPLE 9.4 The total force and torque on an impulse turbine exerted by a water jet

. . . .

Given: An impulse turbine wheel is acted upon by a water jet having a velocity of 350 ft/sec and a diameter of 6 inch. The wheel has a 5-ft diameter. The value of θ is 15° and $f = 0.95$.

Find: **a.** The force on the wheel exerted by the jet when the bucket speed is 175 ft/sec.

         b. Find the torque on the wheel exerted by the jet.

         c. Calculate the turbine power assuming no additional losses.

Solution: **a.** The force on the wheel is

$$F_{Rt} = \rho_m A(v_{Ix} - v_{Bx})(v_{Ix})(1 + f \cos \theta)$$

$$\rho_m = 1.93 \text{ slugs/ft}^3 \text{ at } 80°F$$

$$A = \frac{\pi D^2}{4} = \frac{\pi}{4}\left(\frac{6 \text{ inch}}{12 \text{ inch/ft}}\right)^2$$

$$A = 0.196 \text{ ft}^2$$

$$F_{Rt} = (1.93 \text{ slugs/ft}^3)(0.196 \text{ ft}^2)(350 \text{ ft/sec} - 175 \text{ ft/sec})$$
$$(350 \text{ ft/sec})(1 + 0.95 \cos 15)$$

$$F_{Rt} = 44{,}400 \text{ (slug·ft)/sec}^2 = 44{,}400 \text{ lb}$$

         b. The torque on the wheel is

$$\tau = F_{Rt} \cdot R$$

$$\tau = (44{,}400 \text{ lb}) \frac{(5 \text{ ft})}{2}$$

$$\tau = 111{,}000 \text{ ft·lb}$$

c. The power output is

$$P_t = F_{Rt} \cdot v_{Bx}$$

$$P_t = (44{,}400 \text{ lb})(175 \text{ ft/sec})$$

$$P_t = 7.77 \times 10^6 \text{ ft·lb/sec}$$

$$P_t = \frac{7.77 \times 10^6 \text{ ft·lb/sec}}{550 \text{ ft·lb/sec/hp}}$$

$$P_t = 1.41 \times 10^4 \text{ hp}$$

EXAMPLE 9.5   Ideal power output of an impulse turbine
. . . .

Given: The level of the water in a reservoir is 1000 ft above the nozzle of an impulse turbine. A nozzle with a 6-inch diameter is used to establish a jet.

Find:
a. Neglecting all frictional losses and any construction of the jet, calculate the maximum ideal power of the impulse turbine.
b. The estimated flow rate.

Solution:
a. The pressure at the surface of the reservoir and the jet is atmospheric pressure. The velocity at the surface of the reservoir is essentially zero. Let h_1 represent the height of the surface of the reservoir and h_2 the height of the jet. If friction is neglected, one can use Bernoulli's equation since the fluid is incompressible.

$$p_1 + g\rho h_1 + 1/2\rho v_1^2 = p_2 + g\rho h_2 + 1/2\rho v_2^2$$

$$p_1 = p_2 = pa; \qquad v_1 = 0$$

$$g\rho(h_1 - h_2) = 1/2\rho v_2^2$$

$$v_2^2 = 2g(h_2 - h_1)$$

$$v_2^2 = (2)(32.2 \text{ ft/sec}^2)(1000 \text{ ft})$$

$$v_2 = 254 \text{ ft/sec} = v_{Lx}$$

The maximum ideal power occurs when

$$v_{Bx} = \frac{v_{Lx}}{2} = 127 \text{ ft/sec}$$

The ideal power is $(v_{Bx} = v_{Lx/2})$

$$(P_t)_{ideal} = 1/2\ \rho_m A v_{Ix}^3$$

$$(P_t)_{ideal} = 1/2\ (1.93\ \text{slugs/ft}^3)\ \frac{\pi}{4}\left(\frac{6\ \text{inch}}{12\ \text{inch/ft}}\right)^2 (254\ \text{ft/sec})^3$$

$$(P_t)_{ideal} = 3.10 \times 10^6\ \text{ft·lb/sec} = 5.64 \times 10^3\ \text{hp}$$

b. The volume flow rate is

$$Q_V = A v_{Ix}$$

$$Q_V = \frac{\pi}{4}\ (0.5\ \text{ft})^2 (254\ \text{ft/sec})$$

$$Q_V = 49.9\ \text{ft}^3/\text{sec}$$

SECTION **9.2** *EXERCISES*

1. Name and give the function of the two major elements of a turbine.
2. Describe the differences between an impulse turbine and a reaction turbine.
3. Describe staging in an impulse-reaction turbine.
4. A jet of water moving in the direction of the x axis strikes a body moving in the same direction. The velocity of the object is 50 ft/sec. The angle of the jet leaving the body (θ) is 25° and f is 0.95. Find the force acting on the body in the x direction.
5. Find the force in the y direction in Problem 4. How could a bucket for an impulse turbine be designed to eliminate any force in the y direction?
6. An impulse turbine wheel is acted upon by a water jet having a velocity of 350 ft/sec and a diameter of 3 inch. The value of $\theta = 15°$ and $f = 0.90$. Find the turbine power when the velocity of the buckets is 75, 175, and 200 ft/sec (assume no additional power losses). What is the jet power?
7. Calculate the ideal power for each bucket velocity in Problem 6. When is the ideal power a maximum?
8. Plot the power calculated in Problems 6 and 7. Use at least three additional bucket velocities.

9.3

MECHANICAL ENERGY TO ELECTRICAL ENERGY: THE GENERATOR

OBJECTIVES

Upon completion of this section, the student should be able to

- Use the right-hand rule to determine the direction of the force and a moving positive charge in a magnetic field.

- Use the left-hand rule to determine the direction of the force on a moving negative charge in a magnetic field.

- Determine the direction of electron flow in a wire rotating in a magnetic field.

- Draw and label sketches showing the difference in dc and ac generators and the waveforms produced by each.

- Give the methods used to produce the magnetic field in generators.

- Calculate the force on a moving charge.

- Calculate the voltage generated in a straight conductor moving in a magnetic field.

- Calculate the voltage output of a dc generator given the angular velocity, flux through the armature, and the number of turns in series on the armature.

- Draw circuit diagrams for shunt, series, and compound generators.

- For given conditions, analyze dc, shunt, and series generators if the generated voltage is given as a function of field current.

- Discuss the power losses in dc generators.

- Show that, if mechanical frictional losses are neglected, the mechanical power input is equal to the total electrical power output.

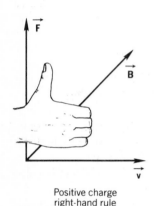

Positive charge
right-hand rule

FIGURE 9.18   Right-hand rule for defining vector directions.

DISCUSSION

Introduction   The conversion of mechanical energy into electrical energy is perhaps the most significant form of energy conversion. Today's society depends greatly on the ability to generate, transport, store, and utilize electrical energy. An electric generator is an energy convertor that changes rotational mechanical energy to electrical energy. This section presents the theory of basic dc generators.

The basic operation and design of electric generators depends on the magnetic force acting on a moving charge in a conducting medium. In Chapter 1, this force was shown to be

$$F = qvB \sin \theta \tag{9.34}$$

where

F = force on the charge in newtons
q = magnitude of the charge in coulombs
v = velocity of the charge in meters per second
B = magnitude of the magnetic field in webers per meter squared
θ = angle between the vector velocity and the direction of the magnetic field

The line of action of the vector force on the moving charge is always perpendicular to the plane formed by the velocity vector and the field vector at the position of the moving charge. The direction depends on the sign of the charge and the directions of the velocity and magnetic field. Two different cases are illustrated in Figures 9.18 and 9.19. The right-hand rule is used for positive charges and the left-hand rule for negative charges. For both cases, the fingers are curled from the velocity vector to the magnetic field vector with the thumb extended. The thumb then points in the direction of the force.

Figure 9.20 shows the application of this rule to a straight piece of wire moving through a uniform field between two magnetic poles. In Figure 9.20a the field is to the right and the wire moves upward. Magnetic force moves electrons in the wire

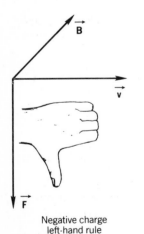

Negative charge
left-hand rule

FIGURE 9.19   Left-hand rule for defining vector directions.

FIGURE 9.20 Force acting on a charge in a conductor moving through a magnetic field.

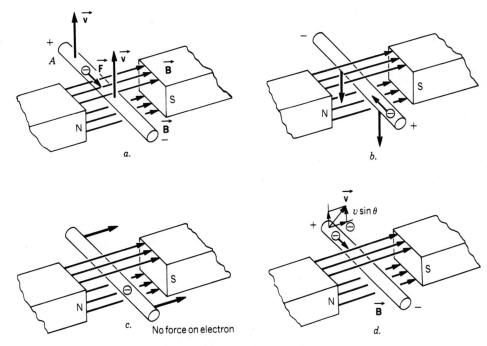

a.

b.

c. No force on electron

d.

from *A* toward *B*, resulting in a potential difference across the wire as shown. In Figure 9.20*b* the conductor moves downward, reversing the direction of force on the electron and the polarity of the voltage across the wire.

In Figure 9.20*c* the conductor moves in the same direction as the field lines. For this case the angle θ is zero. Since sin θ = 0, there will be no force on the free electrons in the conductor. Thus, no voltage is present across the conductor.

Figure 9.20*d* shows the results when the conductor moves at the angle θ to the magnetic field. For this case, the quantity *v* sin θ is the component of the velocity perpendicular to the field lines. This component produces the force on the free electrons in the conductor. The horizontal component produces no force on the electrons.

Example 9.6 illustrates the use of Equation 9.34 to calculate the force on an electron when a conductor is moved through a magnetic field.

EXAMPLE 9.6   Magnetic force on a moving electron
. . . .

Given: An electron moves with a velocity of 10 m/sec through a magnetic field of 1.5 Wb/m². The velocity vector makes an angle of 30° with the direction of the field.

Find: The force on the electron.

Solution: The force on the electron is

$$F = qvB \sin \theta$$

For an electron

$q = 1.60 \times 10^{-19}$ C

$F = (1.60 \times 10^{-19}$ C$)(10$ m/sec$)(1.5$ Wb/m$^2)(\sin 30°)$

$F = 1.2 \times 10^{-18}$ N

If the wire is a conductor, there are free electrons that can move because of the force acting on them. If the wire is attached to an external circuit, current will flow in the circuit as a result of the voltage generated by the moving wire. By definition, this voltage or electromotive force is equal to the work done by the magnetic force per unit charge. For a wire of length l moving in a magnetic field, the generated electromotive force is

$$V = \frac{W}{q} = \frac{Fl}{q} = vBl \sin \theta \qquad (9.35)$$

where

V = generated electromotive force in volts
W = work in joules
F = force in newtons on an electron of charge q as a result of the magnetic field

A sample calculation is given in Example 9.7.

As shown in Figure 9.21, the quantity $vBl \sin (90° - \theta) = vBl \cos \theta$ is the magnetic flux cut by the wire per second. In one second the wire will travel from position A to position B. This distance is equal to the magnitude of the velocity, and the area swept out by the wire in one second is equal to $l \cdot w$.

EXAMPLE 9.7   Electromotive force generated in a wire
. . . .

Given: A 20-cm long wire is moved through a magnetic field of 1.5 Wb/m^2. The velocity is 10 m/sec. The velocity vector makes an angle of 30° with the direction of the field.

Find: The generated electromotive force.

Solution: The generated voltage is

$$V = vBl \sin \theta$$

20 cm = 0.2 m

$$V = (10 \text{ m/sec})(1.5 \text{ Wb/m}^2)(0.2 \text{ m})(\sin 30°)$$

$$V = 1.5 \text{ V}$$

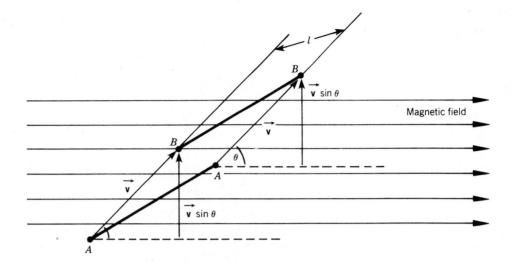

FIGURE 9.21 Conducting wire moving through a magnetic field.

The projected area perpendicular to the field lines, as shown in Figure 9.21, is $vl \sin \theta$. Thus, the total flux cut by the wire in one second is this area times the flux density B.

$$\frac{\Delta \phi}{\Delta t} = vlB \sin \theta \qquad (9.36)$$

However, $V = vlB \sin \theta$. By substitution one then obtains

$$V = \frac{\Delta \phi}{\Delta t} \qquad (9.37)$$

This equation states that the voltage generated is equal to the total flux cut by the wire per second.

Direct-Current Generators Figure 9.22 shows a dc generator. This generator consists of a square loop of wire rotating in a fixed magnetic field and a split-ring commutator.

In Figure 9.22*a*, the upper output contact (called a brush) is connected to wire segment *A*. At this time, both wire segments are moving parallel to the magnetic-field lines, and no voltage is generated. This position is point *a* on the graph in Figure 9.23. This graph shows the voltage across the generator brushes as a function of time.

Figure 9.22 shows the position of the loop a short time later. Wire segments *A* and *B* are now moving at an angle to the magnetic-field lines, and a voltage is being produced in sides *A* and *B*. Side *B* is moving downward through the field, causing a force on the electrons in the direction indicated by the arrow. Side *A* moves upward, resulting in a force in the opposite direction. These two voltages add in series to produce the output voltage shown at point *b* in Figure 9.23. Wire segment *C* contributes no voltage, because it crosses no magnetic field lines as it moves.

In Figure 9.22*c*, the two voltage-producing wire segments are moving perpendicularly to the direction of the magnetic field. This position produces the maximum voltage, indicated by point *c* in Figure 9.23.

FIGURE 9.22   A dc
generator.

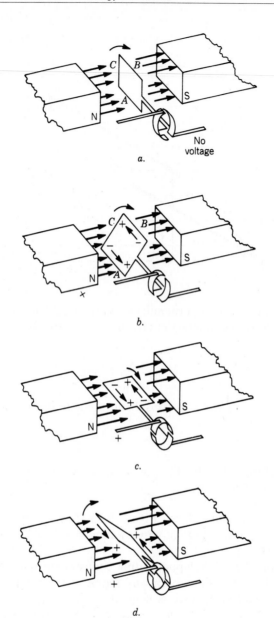

FIGURE 9.22   A dc
generator.

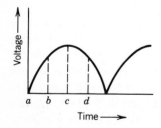

FIGURE 9.23   Plot of
voltage versus time for a dc
generator.

In Figure 9.22*d*, the motion of the wire segments is no longer perpendicular to magnetic field lines, and the output voltage has dropped to that indicated by point *d* in Figure 9.23. When the loop reaches the vertical position again, the voltage will be zero. Each brush will slip to the other segment of the commutator ring and the entire process will be repeated.

Since the amount of voltage across the wire segment depends on the strength of the magnetic field and the velocity of the segment perpendicular to the field, this voltage is small for practical fields and velocities. Generation of usable voltage requires a rotating coil of many loops, as shown schematically in Figure 9.24. As

FIGURE 9.24 A dc generator with many turns in the coil.

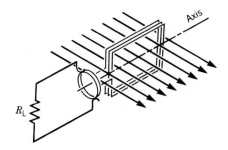

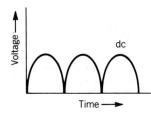

FIGURE 9.25 Plot of voltage versus time for a dc generator.

this coil rotates, a voltage is produced across each horizontal wire segment. All these voltages add in series to produce the total output voltage of the generator. This output is shown as a function of time in Figure 9.25. The amplitude (maximum voltage) of this curve is determined by three factors:

1. Number of loops in the moving coil.
2. Strength of the magnetic field.
3. Rotational rate of the coil.

When any one of these factors is increased, output voltage also increases. If the rotational rate is increased, the frequency at which the output varies is also increased.

If the same generator is equipped with a pair of slip-ring contacts as shown in Figure 9.26, an ac voltage will be produced. Such a generator is often called an "alternator" because it produces an alternating current. The generated voltage is shown in Figure 9.27.

The output voltage as a function of time for a rotating coil can be calculated by applying Equation 9.35. The magnitude of the velocity of each wire is

$$v = \omega R \tag{9.38}$$

where

ω = angular velocity
R = distance to each wire from the axis of rotation

FIGURE 9.26 An ac generator. (left)

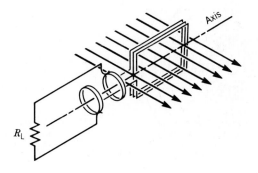

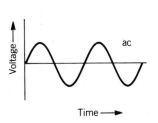

FIGURE 9.27 Plot of voltage versus time for an ac generator. (right)

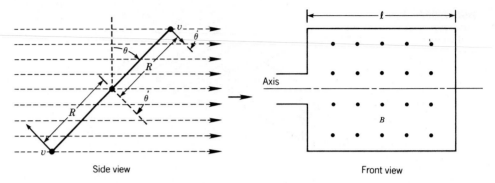

Side view   Front view

In addition, the angular displacement at any time is ωt. The front and side views of the rotating coil are shown in Figure 9.28. The voltage generated in each wire is given by Equation 9.35. Since the voltages add, each turn generates twice this voltage. The total voltage is the voltage generated in one turn times the number of turns. Thus,

$$V = 2vlBN_A \sin \theta \qquad (9.39)$$

where

N_A = number of turns in the armature coil

Substituting $v = R\omega$ and $\theta = \omega t$ into Equation 9.39 yields

$$V = 2R\omega lN_A \sin (\omega t) \qquad (9.40)$$

However, the area of the coil is $2Al$:

$$A = 2Rl \qquad (9.41)$$

where

A = area of the coil

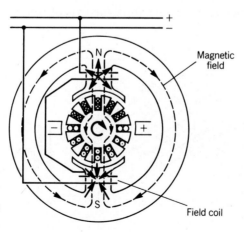

FIGURE 9.29   Electromagnets used in a generator.

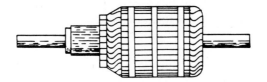

FIGURE 9.30 A multipole armature.

Substituting A for $2Rl$ into Equation 9.40 yields

$$V = N_A \omega A B \sin(\omega t) \qquad (9.42)$$

Permanent magnets are often used to provide the magnetic field in small generators. Large units require fields that cannot be obtained by permanent magnets. Figure 9.29 shows one solution to this problem. A separate dc supply is used to excite a stationary electromagnet.

To obtain higher voltages, a multiple armature, shown in Figure 9.30, is rotated through the magnetic field. The windings consist of a large number of turns connected in series between the brushes such that there are a constant number of turns per unit angle. The voltage generated in each winding depends on the angle θ. It can be shown by the use of calculus that the sum of these voltages is equal to

$$V = \frac{N_{As} \omega A B}{\pi} = \frac{N_{As} \omega \phi}{\pi} \qquad (9.43)$$

where

N_{As} = number of turns in series

$\phi = AB$ = total flux through the armature

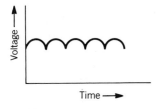

FIGURE 9.31 Output voltage of a generator with a multipole armature.

The resulting output voltage as a function of time has some variation as indicated by Figure 9.31. The output voltage for a dc generator is calculated in Example 9.8.

EXAMPLE 9.8 The electromotive force of a generator
. . . .

Given: A dc generator rotates with an angular velocity of 60 rev/sec. The total flux is 0.010 Wb, and there are 30 turns in series on the armature.

Find: The electromotive force.

Solution: The electromotive force is

$$V = \frac{\omega \phi N_{As}}{\pi}$$

60 rev/sec = $(2\pi)(60)$ rad/sec = 377 rad/sec

$$V = \frac{(377 \text{ rad/sec})(0.010 \text{ Wb})(30)}{\pi}$$

$$V = 36 \text{ V}$$

FIGURE 9.32 Plot of voltage versus field current.

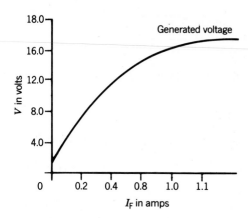

The Magnetic Field and Output Voltage   When the field coils of a dc generator are excited by a direct current, a magnetic field is established in the magnetic path of the generator. In Chapter 5, it was shown that the magnetic flux is given by

$$\phi = \frac{N_F I_F}{R_e} \tag{9.44}$$

where

ϕ = magnetic flux
N_F = number of turns in the field coils
I_F = coil current
R_e = reluctance of the magnetic path

The reluctance, however, is not constant. Thus, a plot of ϕ versus I_F is not a straight line. It has the shape of a magnetization curve, which is illustrated in Chapter 5. Since the open circuit voltage of a dc generator is

$$V = \frac{N_{As}\omega\phi}{\pi} = \frac{N_{As}\omega N_F I_F}{\pi R_e} \tag{9.45}$$

the output voltage as a function of I_F will also have the shape of a magnetization curve. There will be, however, a residual magnetic field due to previous magnetizations of the core. Thus, the curve will not start at the origin. A typical curve of the open circuit output voltage as a function of the field current is shown in Figure 9.32. To obtain this data, the field coil is excited by an external power supply.

Self-Excited Generators   If the voltage output of the generator is used to produce the magnetic field, the generator is self-excited. When the field coil is connected directly across the brushes, the generator is called a shunt generator. If the field coil is connected in series with the load, the generator is a series generator. When both coils are used, it is called a compound generator.

The circuit diagram of a shunt generator is shown in Figure 9.33. The output voltage under a load is also the voltage across the shunt coil. Thus, from Ohm's law,

$$V_o = I_F R_F \tag{9.46}$$

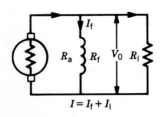

FIGURE 9.33   A schematic of a shunt generator.

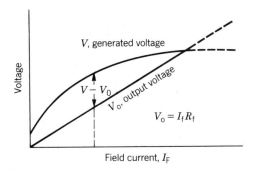

where

V_o = output voltage across the load

Plots of the generated voltage and the output voltage when a current exists are plotted in Figure 9.34 as a function of field current. The plot of generated voltage is obtained by using an external power supply to excite the field coil so that there is no voltage drop in the armature (current is zero). The difference between the generated voltage and the output voltage is equal to the voltage drop in the armature coil:

$$V - V_o = IR_A = (I_F + I_L)R_A \qquad (9.47)$$

where

R_A = armature resistance
I_L = load current

Equation 9.47 can also be written as

$$I = I_F + I_L = \frac{V - V_o}{R_A} \qquad (9.48)$$

For a given generator, plots are drawn as shown in Figure 9.35. For any given value of I_F, the value of $V - V_o$ can be determined from Figure 9.34. Equation 9.48 can then be used to determine $I_F + I_L$. By subtracting the given value of I_F from this quantity, the value of I_L is determined. Typical plots of I_l and V_o as functions of I_F are shown in Figure 9.35. When these curves are plotted, a value of load current and output voltage can be obtained for each value of field current (I_F). Thus, a plot of output voltage as a function of load current is obtained. At point C, the output current is zero and the output voltage is a maximum. This point corresponds to open circuit voltage. As the load resistance is decreased, the output current increases and the output voltage decreases until point B is obtained. This is the maximum stable output current of the generator. A voltage versus current plot is shown in Figure 9.36 for the entire curve from point C to point A. Also shown on the graph is the characteristic straight line through the origin of the voltage across a constant load resistance, R_L. The operating point is the point of intersection of the two curves. This is a stable point because an increase in the current causes the output voltage to lie below the characteristic line. Thus, the current would decrease back to point O. On the other hand, a

FIGURE 9.35 Plots of load current and output voltage versus field current. (left)

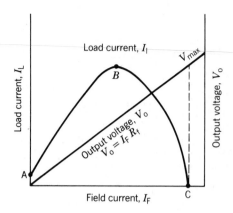

FIGURE 9.36 Plot of output voltage versus output current.

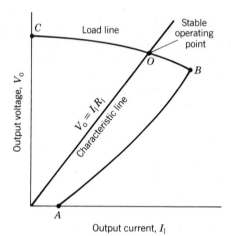

decrease in current would cause the load line to lie above the characteristic line, which would cause the current to increase back to point O. However, if point O lies between A and B, the opposite effects would occur. Thus, any operating point between points A and B on the load line is not stable. For the shunt dc generator the output voltage decreases with the load current. For many cases, this is not a desirable effect. A sample calculation for a shunt generator is given in Example 9.9.

A schematic diagram of a series generator is shown in Figure 9.37. For this case, the field current is also equal to the load current. The output voltage is equal to the generated voltage V minus the sum of the voltage drops in the armature

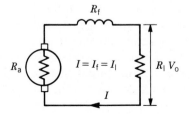

FIGURE 9.37 Schematic of a series generator.

FIGURE 9.38 Plot of voltage versus field current for a series generator.

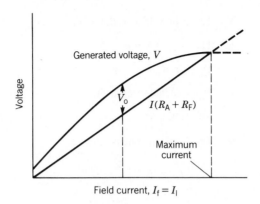

FIGURE 9.39 Plot of output voltage versus load current for a series generator.

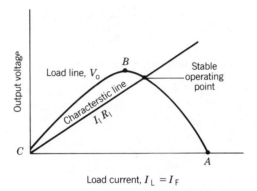

resistance and series coil resistance (Equation 9.49). In addition, there is a brush contact voltage loss of approximately two volts, which is neglected:

$$V_o = V - I(R_A + R_F) \tag{9.49}$$

The curves of V and $I(R_A + R_F)$ as functions of the field current, $I_F = I_L$, are shown in Figure 9.38. The output voltage V_o is also shown for a particular value of I_F. A plot of output voltage versus load current is shown in Figure 9.39.

EXAMPLE 9.9 Shunt generator
....

Given: When the field current of a shunt generator is 0.200 A, the generator voltage is 27.0 V. The field resistance is 100 Ω and the armature resistance is 0.15 Ω.

Find: **a.** The output voltage.
   b. The load current.
   c. The load resistance.
   d. The power output to the load resistance.
   e. The power lost in the shunt coil.

f. The power lost in the armature coil.

Solution: **a.** The output voltage is

$$V_o = I_F R_F$$

$$V_o = (0.200 \text{ A})(100 \text{ }\Omega)$$

$$V_o = 20.0 \text{ V}$$

b. The total current is

$$I_F + I_L = \frac{V - V_o}{R_A}$$

$$0.200 \text{ A} + I_L = \frac{27.0 \text{ V} - 20.0 \text{ V}}{0.150 \text{ }\Omega}$$

$$I_L = 46.7 \text{ A} - 0.200 \text{ A}$$

$$I_L = 46.5 \text{ A}$$

c. The output voltage also is

$$V_o = I_L R_L$$

$$R_L = \frac{V_o}{I_L}$$

$$R_L = \frac{(20.0 \text{ V})}{(46.5 \text{ A})}$$

$$R_L = 0.43 \text{ }\Omega$$

d. The load power is

$$P_L = V_o I_L$$

$$P_L = (20.0 \text{ V})(46.5 \text{A})$$

$$P_L = 930 \text{ W}$$

e. The shunt field power is

$$P_F = V_o I_F$$

$$P_F = (20.0 \text{ V})(0.200 \text{ A})$$

$$P_F = 4.00 \text{ W}$$

f. The power lost in the armature is

$$P_A = (I_F + I_L)^2 R_A$$

$$P_A = (46.7)^2(0.150)$$

$$P_A = 327 \text{ W}$$

Also shown in the figure is a stable operating point for a given load resistance. As the load current increases from point *C* to point *B*, the output voltage increases.

FIGURE 9.40 Schematic of a compound generator.

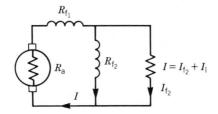

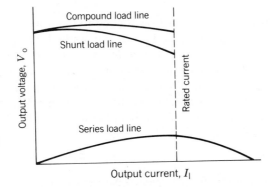

FIGURE 9.41 Plot of output voltage versus output current for a compound generator.

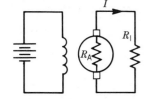

FIGURE 9.42 Schematic of a generator connected to a load.

However, as the load current increases from point *B* to point *A*, the output voltage decreases. This type of variation is usually not desirable. Thus, the series generator has few applications.

A circuit diagram of a compound generator is shown in Figure 9.40. Part of the field is due to the series coil. This will cause the voltage to increase with the load current and then decrease with current as shown in Figure 9.39. In addition, part of the field is due to the shunt coil. This will cause the voltage to decrease with an increasing load current. This effect is shown in Figure 9.36. The two curves are plotted in Figure 9.41. The fields obtained by the series and shunt windings can be adjusted to obtain an almost constant voltage up to the rated maximum load current.

Power and Efficiency   When a generator is turning with a load connected to the output, the current flowing in the load circuit is also flowing in the armature. Such current establishes a magnetic field around the armature; this field opposes the fixed magnetic field of the generator. The interaction of these two magnetic fields, one fixed and one rotating, opposes the rotation of the armature; therefore, rotational mechanical energy must be supplied to the armature to maintain its rotation. Thus, the armature of a generator rotates easily if no load is attached, but requires greater torque as the current delivered increases. This situation is required by the principle of the conservation of energy. The maximum current that may be produced by a generator is limited by the size of the wire used in the armature.

To investigate the conservation of energy in the generation of electric power, consider a generator connected to a load as shown in Figure 9.42. The field coil is excited by a separate power supply. The total electric power is

$$P_{\mathrm{E}} = VI \qquad (9.50)$$

where

P_E = total electric power generated
V = generated electromotive force
I = total current

Substituting for V from Equation 9.43 yields

$$P_E = \frac{N_{As}\omega BAI}{\pi} \tag{9.51}$$

A mechanical torque must be applied to the armature to keep it rotating at the constant angular velocity ω. Part of the applied torque overcomes the frictional torques and part is used to overcome the armature reaction torque.

To calculate the armature reaction torque, it is necessary to calculate the electrical force on each wire in the armature. The magnitude of the force on a wire conducting current in a magnetic field is given in Chapter 1 as Equation 1.47. Since the current is in the direction of the wire and the wire is perpendicular to the magnetic field, Equation 1.47 reduces to

$$F = BIl \tag{9.52}$$

The direction of this force is perpendicular to the plane formed by the direction of l and the direction of B as shown in Figure 9.43. For electron flow in the wire, the left-hand rule can be used (velocity in the direction of the wire.)

Since a loop occurs in the armature, there are two forces as shown in Figure 9.43. These two forces form a couple. The torque produced by the couple is equal

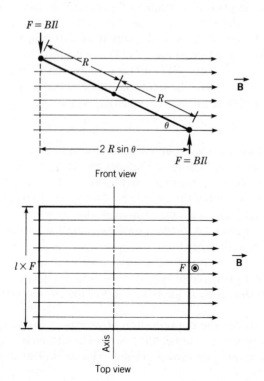

FIGURE 9.43 Forces on a loop rotating in a magnetic field.

to the magnitude of one force times the perpendicular distance between the forces. Thus,

$$\tau = (F)(2R \sin \theta) \qquad (9.53)$$

where

τ = torque on the loop

The armature, however, is composed of a large number of loops in series at different angles θ. Thus, it is necessary to sum all the torques to obtain the total torque applied to the armature as a result of the electromagnetic forces. As in the case of the generated voltage, calculus can be used to estimate this sum. This yields, since $F = BIl$ and $\phi = 2RlB$,

$$\tau_t = 2RF = \frac{(2Rl)BIN_{As}}{\pi} = \frac{\phi IN_{As}}{\pi} \qquad (9.54)$$

where

τ_t = total mechanical torque applied to the armature
N_{As} = total number of turns in a series

The corresponding mechanical power is the forcelike quantity τ_t times the ratelike quantity ω:

$$P_m = \frac{2RFN_{As}\omega}{\pi} \qquad (9.55)$$

However, $F = BIl$. Thus,

$$P_m = \frac{(2Rl)(BIN_{As}\omega)}{\pi} \qquad (9.56)$$

where

P_m = total mechanical power

The area of each coil is $2Rl$. Thus, the mechanical power may be written as

$$P_m = \frac{BAIN_{As}\omega}{\pi} = \frac{\phi IN_{As}\omega}{\pi} \qquad (9.57)$$

This equation is the same as the electric output power, and energy is conserved. There is in addition to the torque from electromagnetic forces, a torque that is required to overcome mechanical and fluid friction losses. The corresponding friction power generates heat and reduces the efficiency of the generator. Additional electrical losses occur because of the armature and field coil resistances. The useful power is the power delivered to the load. Thus, the efficiency becomes

$$\eta = \frac{P_{out}}{P_{in}}(100) = \frac{I_L^2 (100)}{P_W + I_F^2 R_F + I_A^2 R_A + I_L^2 R_L + P_B} \qquad (9.58)$$

where

P_W = power lost because of mechanical friction
P_B = brush contact power loss

A sample problem is given in Example 9.10.

EXAMPLE 9.10   Efficiency of a shunt generator
. . . .

Given: A shunt generator delivers 10 A at 100 V to a load resistance. The armature resistance is 0.40 Ω and the field resistance is 1000 Ω. The power lost because of mechanical friction is 100 W.

Find: The efficiency of the generator under the given conditions neglecting the brush contact loss.

Solution: The efficiency is given by Equation 9.58.

$$\eta = \frac{P_{out}}{P_{in}} \times 100$$

The power out is $I^2 R_L = VI$.

$$P_{out} = (10\ A)(100\ V) = 1000\ W$$

The power in is

$$P_m = P_W + I_F^2 R_F + I_A^2 R_A + VI$$

The field current is

$$I_F = \frac{V_o}{R_F} = \frac{100\ V}{1000\ \Omega} = 0.10\ A$$

The armature current is

$$I_A = I_F + I_L = 10.0\ A + 0.10\ A = 10.1\ A$$

The power in becomes

$$P_{in} = (100\ W) + (0.10\ A)^2(1000\ \Omega) + (10.1\ A)^2(0.40\ \Omega) + 1000\ W$$

$$P_{in} = 100\ W + 10\ W + 41\ W + 1000\ W$$

$$P_{in} = 1150\ W$$

The efficiency becomes

$$\eta = \frac{(1000)(100)}{1150}$$

$$\eta = 87\%$$

The most important application of dc generators arises from the fact that dc electric motors allow finer speed control than either ac motors or engines that derive their power from the combustion of chemical fuels. In locomotives and ships, diesel engines or steam turbines drive dc generators. The electrical output power is used to operate dc motors that provide propulsion.

SECTION **9.3** *EXERCISES*

1. Determine the direction of force on electrons in the wire in the illustration if the wire moves
- **a.** out of the page twoard the reader
- **b.** toward the south pole of the magnet.

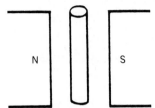

2. Indicate the direction of electron flow through the generator coil in the following diagrams. (Use the left-hand generator rule for electron flow presented in the text.)

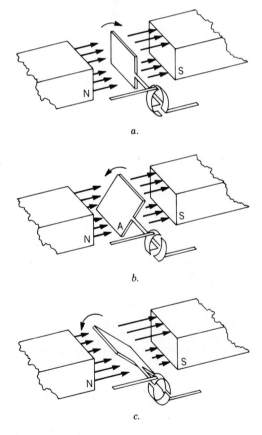

a.

b.

c.

3. List the three factors that determine the voltage produced by a generator.
4. Sketch the output waveforms from an ac generator and a dc generator.
5. State two methods of producing magnetic fields in generators.
6. Draw a circuit diagram for a shunt, a series, and a compound dc generator.

7. A proton moves through a magnetic field of 1.0 Wb/m². The magnitude of the proton's velocity is 100 m/sec. If the velocity vector makes an angle of 45° with the field, find the force on the proton.

8. Find the torque from the electromagnetic force when the armature current is 300 A in a dc generator if the generated voltage (no armature current) is 100 V when the angular velocity is 377 rad/sec. **Hint:**

$$V = \frac{\omega\phi N_{As}}{\pi}$$

$$T\omega = \frac{(AB)IN_{As}\omega}{\pi} = \frac{\phi IN_{As}\omega}{\pi}$$

$$\tau = \frac{\phi IN_{As}}{\pi}$$

9. A pulley and belt is used to drive a generator. The diameter of the pulley is 0.70 m and the armature resistance is 0.10 Ω. The armature current is 100 A and the angular velocity is 40 rad/sec. The voltage output of the shunt generator is 110 V. Find the tension in the belt. **Hint:**

$$V_o = V - I_A R_A$$

$$V = \frac{\omega\phi N_{As}}{\pi}$$

$$\tau = \frac{\phi IN_{As}}{\pi} = \frac{D}{2}T$$

where

    T = tension in the belt
    D = pulley diameter
    V_o = output voltage

10. A dc shunt generator rotates with an angular velocity of 100 rad/sec. The total flux is 0.02 Wb and there are 40 turns in series on the drum armature. Find the electromotive force.

11. In Problem 10, the armature current is 100 A and the field current is 0.5 A. The power loss because of mechanical friction is 150 W. If the field resistance is 1000 Ω, find the efficiency.

12. The output voltage of a shunt generator is 100 V when the external load current is zero and the angular velocity is 50 rad/sec. The voltage at no external load drops to 70 V when the angular velocity is decreased to 40 rad/sec. Find the percentage change in the magnetic flux and the field current. **Hint:**

$$V_o = V - I_F R_A$$

$$I_F = V_o/R_F$$

$$V = \frac{N_{As}\omega\phi}{\pi}$$

13. Calculate the input power to a motor-generator if the input voltage and current are 110 V and 1.6 A, respectively.

14. Calculate the power output of the motor-generator of Problem 13 if the output voltage and current are 110 V and 1 A. What is the efficiency of the motor-generator?

9.4

RADIANT ENERGY TO ELECTRICAL ENERGY: THE PHOTOVOLTAIC CELL

OBJECTIVES

Upon completion of this section, the student should be able to

- Describe the operation of photovoltaic devices, including energy bands in semiconductors, pn junctions, and generation of free electrical carriers by absorption of light energy near the junction leading to electrical current flow; name at least one useful photovoltaic semiconductor material.

- Describe at least two applications of photovoltaic materials.

- Calculate the voltage and current in a photovoltaic cell.

DISCUSSION

Photovoltaic cells are energy convertors made of materials such as silicon that directly convert radiant energy to electrical energy. They are often called "solar cells." Specifically, they are junctions between a p-type semiconductor and an n-type semiconductor. They are used to convert sunlight to electrical energy in satellites. However, their cost and low efficiency (approximately 10%) have made their large-scale use for electric power generation uneconomical. There are many other applications, however, where power output is not important. They are found in many systems such as cameras, burglar alarms, flame sensors, light meters, colorimeters, and night-vision devices.

To understand the operation of photovoltaic transducers, we must refer to some of the concepts related to semiconductors, as described in Chapter 5. Most materials used in photovoltaic transducers are semiconductors.

n-Type Materials As a brief review, remember that a semiconductor is a material with values of electrical conductivity intermediate between those of good metallic conductors, such as copper, and electrical insulators, such as rubber. Examples of semiconductors include materials such as silicon and germanium. The "energy band structure" of semiconductors is important. Current can be carried through a semiconductor by the motion of electrons; however, free electrons in a semiconductor are few in number compared to those in a metal, and so electrical conductivity is much lower. In addition, in a metal the free electrons have a continuous range of energy levels available to them. In a semiconductor, on the other hand, there are bands of allowed energy, separated by a band gap, a range of energies not available to the electrons (Figure 9.44). Free electrons in the conduction band are free to move under action of an electric field and contribute to electrical

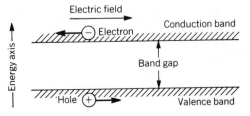

FIGURE 9.44 Relationship between conduction band, valance band, and band gap in semiconductors.

conductivity. If the dominant contribution to electrical conduction is free electrons, the material is termed an "n-type" semiconductor.

p-Type Materials   Another contribution to electrical conduction can be through missing electrons in the valence band. As electrons from neighboring sites move to fill the vacancy, the position of the missing electron may move. The movable vacancy appears similar to the motion of a positive charge through the material. The movable vacancy is called a "hole," and it acts like a positive charge carrier. Materials in which the dominant contribution to electrical conduction is the motion of the holes are termed "p-type" semiconductors. Semiconductors can be made p-type or n-type by the addition of small amounts of appropriate impurity elements.

pn-Type Junctions   A pn junction is a region in which there is an interface between p-type and n-type material, of very small dimensions, typically of the order of a few micrometers. Such a junction is called a "pn junction," which is the important junction for photovoltaic effects.

Most modern photovoltaic devices employ a pn junction in a semiconductor. The photovoltaic effect also appears at a junction between a metal and a semiconductor; in fact, many early photovoltaic devices employed such a metal-semiconductor junction. A common type is the junction between iron and the semiconductor selenium. Such devices have found great utility as photovoltaic cells and are used to turn on street lights automatically at dark. Since the applications of optical energy measurement and energy conversion usually employ semiconductor photovoltaic materials, the emphasis in this section will be on semiconductors.

The photovoltaic effect depends on distortion of the energy band structure through a pn junction (Figure 9.45). Height of the conduction band and valence band change through the junction region from n-type material to p-type material. Thus, an electron in the p-type material will move along the bottom of the conduction band, as shown. Similarly, a hole near the top of the valence band in the n-type materials will also move as shown. Thus, the junction acts very similarly to a built-in electric field, causing the motion of the electrical charge.

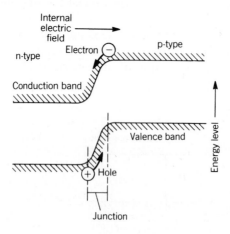

FIGURE 9.45   Band structure of a pn junction.

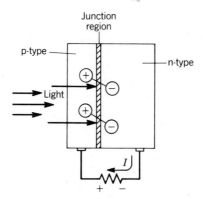

FIGURE 9.46 Light incident on a pn junction.

Thus, when p-type and n-type semiconductors form a junction, a voltage is produced across the junction that will separate holes and electrons. However, there usually are no free holes and electrons in the immediate vicinity of the junction to allow electrical current to flow. If, however, light is incident upon the junction and is absorbed, the light energy will produce free holes and electrons, which then are capable of moving under the action of the electric field (Figure 9.46).

Absorption of light energy in the vicinity of the junction thus creates current carriers, which are separated by the built-in electric field. If electrical contacts are attached at the surface on the two sides and connected through a load resistor as shown, there will be a net flow of electrical current through the load resistor and, therefore, generation of electrical power.

This generation is the photovoltaic effect direct generation of electrical energy by absorption of light energy in the vicinity of a pn junction in a semiconductor. No external voltage must be applied to produce the photovoltaic effect.

Construction and Electrical Circuits   The circuit employed for photovoltaic devices is illustrated in Figure 9.47. The light (arrows) is incident on the photodetector, shown schematically as a diode. Current I flows as shown, and the voltage is generated across the load resistor R_L. Voltage is given by Ohm's law:

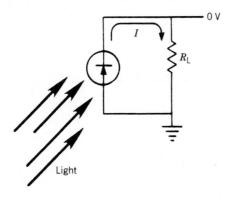

FIGURE 9.47 Schematic of a photovoltaic device in a circuit.

FIGURE 9.48   Structure of a typical silicon photovoltaic device.

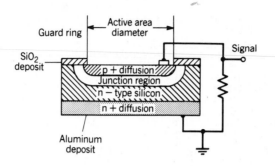

$$V = IR_{\text{L}} \qquad (9.59)$$

where

V = voltage
I = current
R_{L} = load resistor

The structure of a typical silicon photovoltaic device is displayed in Figure 9.48. A piece of n-type silicon is used as the starting material. Impurities are diffused into one side to form a p-type region and a junction. Other materials can be diffused into the opposite side to produce a heavily doped n-type region for ease in making an electrical contact. A guard-ring structure of SiO_2 is deposited on the surface to reduce stray current leakage. Electrical contacts (aluminum) are deposited on the surface. Fabrication of such devices has become a well-developed technology.

Responsivity   The response of the pn junction to light is represented by the curve in Figure 9.49. For relatively low values of the load resistor, response—that is, externally generated current—is approximately a linear function of the incident light energy. For large values of the load resistor, the response will saturate, as indicated. Response will saturate when the voltage drop across the load resistor becomes equal to the value of built-in voltage at the junction. Thus, the voltage drop has a maximum value, equal to the product of the current and the resistance. Therefore, for larger values of resistance, there is a maximum limiting current that can be generated.

The photovoltaic effect can also be used in the form of a sensor that generates a measured voltage across an open circuit. The sensor is less important as an energy

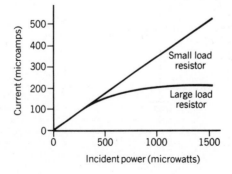

FIGURE 9.49   Response curve for a pn junction device.

convertor than some other applications. Some of these applications are discussed in Chapter 10, ''Transducers.''

Figure 9.49 is used to define responsivity, which is an important parameter of the photovoltaic detector. It is the amount of current that flows in the external circuit divided by the optical input power. It has units of current per unit power and is given by the quotient of the current divided by the power. For the example shown in Figure 9.49, responsivity is equal to 210 microamps divided by 600 microwatts, or 0.35 A/W. This is a typical value characteristic of modern photovoltaic cells based on silicon pn junctions.

Efficiency Current generated by the absorption of light that delivers power absorbed by the junction is given by

$$I = RP \tag{9.60}$$

where

R = responsivity
P = power in watts

Since electrical power generated will be I^2R_L, the efficiency (ratio of electrical power to light power) can be expressed as

$$\text{Efficiency} = \frac{I^2R_L}{P} = \frac{R^2P^2R_L}{P} = R^2PR_L \tag{9.61}$$

Notice that efficiency cannot be increased very much by an increase in R_L, because of saturation effects shown in Figure 9.49 when a large resistor is used. A calculation using Figure 9.49 is given in Example 9.11.

EXAMPLE 9.11 Calculation of current and voltage in a photovoltaic cell
. . . .

Given: Light is incident upon a photovoltaic cell that has characteristics shown in Figure 9.49. The load resistor is 100 Ω. Optical power is 500 μW.

Find: **a.** Current I
     b. Voltage V.

Solution: **a.** The resistor is a relatively small one; therefore, from the upper curve in Figure 9.49, current I is read as 175 μA at 500 μW optical power.

     b. Use Ohm's law (Equation 1):

     $V = IR_L$

     $V = (175 \times 10^{-6} \text{ A})(100 \text{ Ω})$

     $V = 1.75 \times 10^{-2}$ V

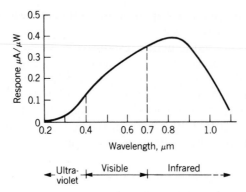

FIGURE 9.50 Spectral response of silicon photovoltaic devices.

Spectral Response   Spectral response of silicon is plotted in Figure 9.50. This graph shows the response of silicon photovoltaic devices as a function of the wavelength of the incident light. (A discussion of the concept of wavelength of light is developed in Chapter 13.) The reason for the shape of this curve is beyond the scope of this section. We simply point out that silicon detectors are most useful in the visible portion of the spectrum (0.4 to 0.7 micrometers) and in the near-infrared portion of the spectrum (0.7 to 0.9 micrometers). Other materials exhibit a strong response farther into the infrared and, therefore, have applicability as detectors for longer-wavelength infrared radiation, to which silicon detectors would not respond.

Other Semiconductor Materials   A number of different types of semiconductors have been employed, including silicon, gallium arsenide, indium phosphide, cadmium telluride, gallium phosphide, and cadmium sulfide. Some of the other materials have a higher theoretical efficiency than silicon, which has a maximum theoretical efficiency of approximately 20%. However, for large-scale applications, where low cost is a factor, silicon now is the most practical candidate. Some of the other materials can be used as detectors of infrared radiation.

APPLICATIONS

Electric Power Generation   Photovoltaic convertors have been employed for a number of years on satellite systems to provide electrical energy for satellite operation from sunlight. These are the solar cells that have been employed on many satellites and spaceprobes. Such cells have not found much application in direct generation of electricity from sunlight on the earth's surface because of the relatively high cost of the material. High cost has not been a great drawback for space applications because total areas involved are small and because the very specialized application will tolerate a higher cost per unit of electrical power. For generation of larger amounts of electrical power on the earth's surface, however, the cost will have to be reduced considerably. If the cost of silicon photovoltaic cells could be reduced to the approximate region of $0.50 per watt of electrical generating capacity, they could be competitive on the earth's surface for direct generation of electricity from sunlight.

Example 9.12 shows the calculation for the size of a photovoltaic cell unit used in a generating station.

EXAMPLE 9.12   Calculation of electrical energy using solar cells

. . . .

Given: Silicon photoelectric energy convertors are used on a satellite to generate electricity with 8% efficiency. The silicon photocells always are pointed directly at the sun. (Above the earth's atmosphere, the sun delivers radiant power per unit area of 0.14 W/cm².)

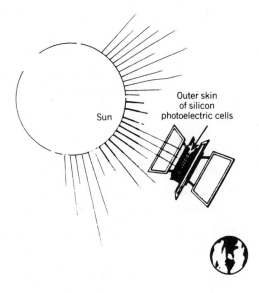

Find: Area of solar cells needed to generate 12 W of electrical power.

Solution: Required power = area × light energy × efficiency, or

$$P = AE_L\eta$$

$$12 \text{ W} = (A \text{ cm}^2)(0.14 \text{ W/cm}^2)(0.08)$$

$$A = \frac{12}{0.14 \times 0.08} = 1.07 \times 10^3 \text{ cm}^2$$

$$A = 1070 \text{ cm}^2$$

SECTION 9.4 EXERCISES

1. Describe the operation of photovoltaic devices.
2. Suppose a photovoltaic detector has a responsivity of 0.5 μA/μW and 10 mW of light energy are absorbed by the device. What is the resulting electrical current if the device is connected across a load resistor of small resistance?

3. Given the results of Problem 2, with a load resistance of 120 Ω, what is the energy efficiency of conversion of light energy into electrical energy?

4. What is the responsivity of a junction that generates 40 μA of current when absorbing 100 μW of power?

5. How large should a load resistor be if a current of 0.1 mA is to produce a voltage of 1 V?

6. How much power does sunlight deliver per square centimeter to a photovoltaic energy convertor that is 6% efficient and produces a 110-mA current across a 1000-Ω resistor if the sunlight is collected on an 800-cm² area?

9.5

ELECTRICAL ENERGY TO MECHANICAL ENERGY: THE ELECTRIC MOTOR

OBJECTIVES

Upon completion of this section, the student should be able to

- Use the left-hand motor rule to determine the direction of the force on a conductor in a magnetic field for electron flow.

- Given a diagram of a loop of wire carrying a current in a magnetic field, determine the direction of loop rotation.

- Sketch a permanent-magnet dc motor. Label and explain the function of the principal parts.

- List the three factors that determine the magnitude of the torque produced by an electric motor.

- State the difference between permanent-magnet, series, and shunt dc electric motors.

- Calculate the magnitude of the force on a straight conductor in a magnetic field.

- Calculate the torque produced by the armature of a motor given the current, magnetic field, and armature turns in series on the armature.

- Calculate the angular velocity of a motor given the applied voltage, magnetic field, current, armature current, and turns in series on the armature.

- Discuss two methods of adjusting the speed of a shunt motor.

- Discuss a method of starting a shunt motor.

- Calculate the efficiency of shunt and series motors.

- Discuss the torque and angular velocity as a function of armature current for shunt and series motors.

DISCUSSION

Introduction One of the most essential and versatile energy convertors utilized by modern technology converts electrical energy into mechanical energy. The device that accomplishes this conversion is called an "electric motor." The

FIGURE 9.51 Force on a current-carrying wire in a magnetic field.

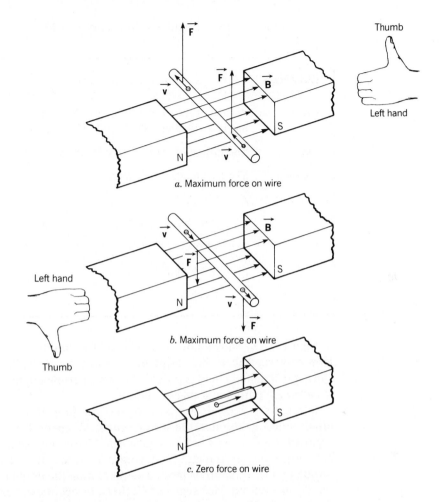

a. Maximum force on wire

Left hand

Thumb

b. Maximum force on wire

c. Zero force on wire

advantages offered by electric motors are numerous—they are mechanically simple, quiet, clean to operate, and relatively efficient.

The rotation of an electric motor is caused by the forces that act upon current-carrying wires in the motor because of the presence of a magnetic field. These forces and the resulting torque on the armature were discussed in the section on dc generators. The right- and left-hand rules used for generators can also be used for motors. The application is illustrated in Figure 9.51 for electron flow. When the velocity of the electrons is perpendicular to the field the force is a maximum, and when the electron velocity is parallel with the field the force is zero. At any angle θ between v and B, the force on the wire is (Equation 1.48)

$$F = BIl \sin \theta \tag{9.62}$$

where

F = force
B = magnetic flux density
l = length of the wire
I = current in the wire

The force on a wire is calculated in Example 9.13.

EXAMPLE 9.13   Force on a wire in a magnetic field

. . . .

Given: A straight wire is 0.5 m long and it makes an angle of 30° with a magnetic field of 0.06 Wb/m². The current in the wire is 100 A.

Find: The magnitude of the force on the wire.

Solution: The magnitude of the force is

$$F = BIl \sin \theta$$

$$F = (0.06 \text{ Wb/m}^2)(100 \text{ A})(0.5 \text{ m}) \sin 30°$$

$$F = 1.5 \text{ N}$$

$$F = (1.5 \text{ N})(0.225 \text{ lb/N})$$

$$F = 0.338 \text{ lb}$$

Figure 9.52*a* shows how a simple electric motor is constructed from a square loop of wire, called the "armature," and from a "split-ring commutator." Because the magnetic field is provided by two permanent magnets, this configuration is called a "permanent-magnet motor."

In Figure 9.52*a*, electrons flow into the loop through the upper contact, or brush, and out through the lower brush. Magnetic force is directed upward on segment *A* and downward on segment *B*. These oppositely directed forces exert a net torque on the armature, causing it to rotate in a clockwise direction. This torque on the armature is plotted as point *a* on the graph of Figure 9.53. Magnetic forces also act on wire segment *C*; these forces, however, are always parallel to the axis of rotation and, therefore, never contribute to the torque acting on the armature. For this reason, the forces on segment *C* are unimportant and will be disregarded.

A short time later the armature has rotated to the position shown in Figure 9.52*b*. Since the forces acting on *A* and *B* depend only on the current and the magnetic field strength, they have not changed. The torque acting on the armature has increased, however, because of the increase in the lever arm associated with each force. This configuration produces the maximum torque on the armature (point *b* in Figure 9.53).

In Figure 9.52*c*, segments *A* and *B* are still being pulled upward and downward, respectively. The torque is smaller because of a reduced lever arm (point *c* on Figure 9.53), and the armature continues to rotate clockwise.

Just as the loop achieves the vertical orientation shown in Figure 9.52*d*, each brush slips to the opposite segment of the commutator ring, causing the direction of electron flow to be reversed. The forces on segments *A* and *B* change direction, pulling *A* downward and pushing *B* upward. No torque acts on the loop, however, since the lever arm for each force is zero. (See point *d* in Figure 9.53.)

FIGURE 9.52   Electron flow in a permanent magnet motor.

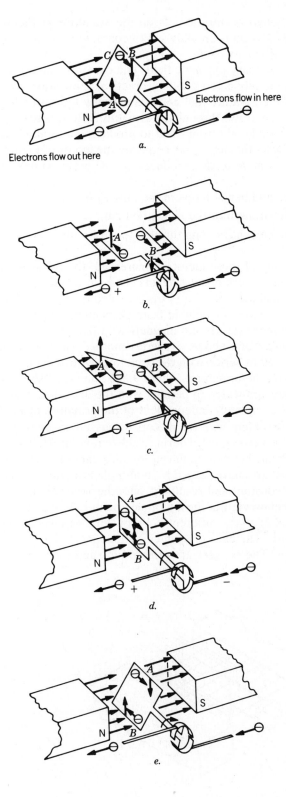

Electrons flow in here

Electrons flow out here

a.

b.

c.

d.

e.

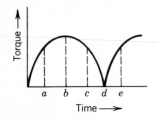

FIGURE 9.53 Plot of torque versus time for a current-carrying coil in a magnetic field.

Although no torque acts on the armature in Figure 9.52*d*, the momentum it already has built up allows it to coast into the position in Figure 9.52*e*, and the process repeats itself.

For ordinary magnetic fields and currents, the torque acting on a single loop of wire generally is too small for practical application. Torque can be made larger by use of an armature that consists of many loops of wire, as shown in Figure 9.54. As current flows through the armature, a torque acts on each single loop. Each of these individual torques add to produce the total torque on the armature. Figure 9.55 shows the net torque acting on the motor of Figure 9.54 as a function of time. The maximum torque produced by the motor is determined by three factors:

- The number of loops in the armature.
- The strength of the magnetic field.
- The current flowing through the armature.

When any of these factors is increased, the torque increases.

Usually, permanent magnets can be used to provide the magnetic field in only very small motors. More powerful electric motors require a larger magnetic field than normally is available from permanent magnets. This problem is solved by replacing the permanent magnets with field coils as in the case of a dc generator.

Figure 9.56*a* shows a "series motor," so named because the field coils are connected in series with the armature. The same current flows through the field coils and the armature. Figure 9.56*b* shows a "shunt motor," in which the field coils are connected in parallel with the armature. Current through the field coils then is completely independent of the armature current.

One problem with the motors discussed so far is that the torque acting on the armature varies greatly during a single revolution. This variation causes excessive wear on the bearings in the motor and generally causes it to "run rough." Use of a modified armature, called a "multipole armature," causes the torque to be more nearly constant and results in smoother operation.

A section view of the multipole armature is shown in Figure 9.57. This armature consists of many different coils of wire wrapped at equal intervals around the armature. The ends of each coil are connected to opposing segments of the commutator. The brushes are oriented in such a manner that they contact the com-

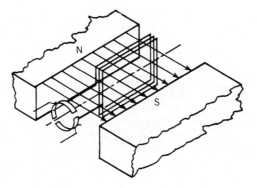

FIGURE 9.54 Many current-carrying coils in a magnetic field.

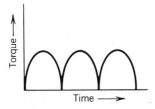

FIGURE 9.55 Plot of torque versus time for many current-carrying coils in a magnetic field.

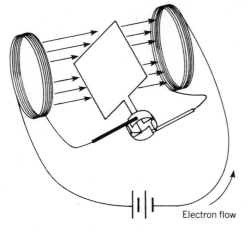

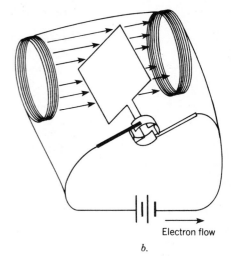

FIGURE 9.56 Diagrams of series and shunt motors.

a.   *b.*

Electron flow

Electron flow

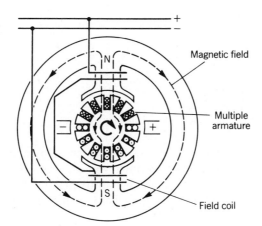

FIGURE 9.57 Multipole armature.

Magnetic field

Multiple armature

Field coil

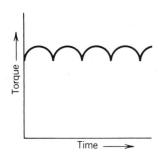

FIGURE 9.58 Plot of torque versus time for a multipole armature.

mutator segments of each coil in the maximum torque orientation. The resulting torque is shown in Figure 9.58 as a function of time.

When an electric motor is rotating, the electrons in its armature windings actually undergo two distinctly different motions through the magnetic field. One motion is that of the current flowing along the wire. This motion causes the magnetic forces that produce the torque on the armature; however, the electrons are carried around in a circular path by the spinning armature exactly as they would be in an electric generator. When the lefthand generator rule is applied for this motion, an additional force is shown to be acting on the electrons. This force pushes against the flow of electrons in the armature, resulting in a voltage being produced across the armature. This voltage is opposite in polarity to the input voltage of the motor and is called the "back emf." The back emf serves to decrease the armature current when the motor is spinning.

MOTOR ANALYSIS

Torque   When a voltage is applied to the armature of a motor, a current will flow through the armature and a torque will be produced. From Equation 9.54, the torque is

$$\tau = \frac{\phi I N_{As}}{\pi} \qquad (9.63)$$

This torque will cause the armature to rotate, and mechanical power is produced (Equation 9.57):

$$P = \omega\tau = \frac{\omega\phi I N_{As}}{\pi} \qquad (9.64)$$

However, as the armature rotates through the magnetic field, an electromotive force will be generated that opposes the applied voltage. This back electromotive force is given by Equation 9.43:

$$V_b = \frac{N_{As}\omega\phi}{\pi} \qquad (9.65)$$

The total voltage in the armature circuit becomes

$$V_{Ae} = V_a - V_b \qquad (9.66)$$

where

V_{Ae} = total voltage
V_a = applied voltage
V_b = generated back electromotive force

The armature current is given by Ohm's law. However, in addition to armature resistance, there is a brush contact voltage drop. This voltage drop is almost independent of the current, and it is approximately two volts. For this discussion, this voltage drop will be neglected. Thus, to a good approximation, the armature current is

$$I_A = \frac{V_{Ae}}{R_A} = \frac{V_a - V_b}{R_A} \qquad (9.67)$$

where

R_A = armature resistance

Substituting for V_b from Equation 9.65 yields

$$I_A = \frac{V_a - \left(\dfrac{N_{As}\omega\phi}{\pi}\right)}{R_A}$$

or

$$I_A R_A = V_a - \frac{N_{As}\omega\phi}{\pi}$$

or

$$\pi I_A R_A = \pi V_a - N_{As}\omega\phi$$

$$\omega = \frac{\pi(V_a - I_A R_A)}{N_{As}\phi}$$

(9.68)

Equation 9.68 states that the angular velocity is inversely proportional to the magnetic flux. Thus, it is very important that magnetic flux be sufficiently large to keep ω within safe limits. Since the armature resistance is small, the term $I_A R_A$ is usually small compared to V_a. Thus, the angular velocity is almost proportional to the applied armature voltage. If the armature voltage and magnetic field are held constant, the angular velocity will almost remain constant. A sample calculation is given in Example 9.14.

EXAMPLE 9.14 Angular velocity and power of a shunt dc motor
. . . .

Given: The applied voltage to a shunt dc motor is 100 V and the corresponding magnetic field is 0.015 Wb. The armature resistance is 0.15 Ω, and the number of series turns is 100.

Find:
a. The angular velocity when the armature current is 10 A and when it is 40 A.

b. The output power for the two armature currents.

Solution:
a. The angular velocity is given by Equation 9.68.

$$\omega = \frac{(V_a - I_a R_a)\pi}{N_{As}\phi}$$

For $I_A = 10$ A

$$\omega = \frac{[100\text{ V} - (10\text{ A})(0.15\text{ }\Omega)]\pi}{(100)(0.015\text{ Wb})}$$

$$\omega = 206\text{ rad/sec} = 32.8\text{ rev/sec}$$

For $I_A = 40$ A

$$\omega = \frac{[100\text{ V} - (40\text{ A})(0.15\text{ }\Omega)]\pi}{(100)(0.015\text{ Wb})}$$

$$\omega = 197\text{ rad/sec} = 31.4\text{ rev/sec}$$

b. The power output is given by Equation 9.64.

$$P = \frac{\omega N_{As}\phi I}{\pi}$$

For 10 A

$$P = \frac{(206\text{ rad/sec})(100)(0.015\text{ Wb})(10\text{ A})}{\pi}$$

$$P = 984\text{ W} = 1.32\text{ hp} \qquad (1\text{ W} = 1.341 \times 10^{-3}\text{ hp})$$

For 40 A

$$P = \frac{(197 \text{ rad/sec})(100)(0.015 \text{ Wb})(40 \text{ A})}{\pi}$$

$$P = 3760 \text{ W} = 5.04 \text{ hp}$$

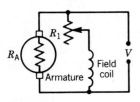

FIGURE 9.59 Circuit diagram of a field rheostat.

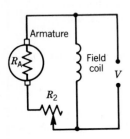

FIGURE 9.60 Circuit diagram of an armature rheostat.

Shunt dc Motor   Figures 9.59 and 9.60 show two circuit diagrams for shunt motors. By varying the field rheostat in 9.59, the current through the field coil can be adjusted. This causes the magnetic field to change, which causes the angular velocity to change (Equation 9.68). The speed can be adjusted by this method from a normal low value up to the maximum safe speed. Since the field current is small, a relatively low power rheostat can be used. In Figure 9.60, the armature rheostat can also be used to control the speed. By this method the speed can be changed from the normal value to almost zero. There is, however, a large power loss in the rheostat because of the high armature current.

The torque and angular velocity of a shunt motor as a function of armature current are shown in Figure 9.61. As previously shown, the angular velocity is nearly constant and the torque increases directly with the armature current. If the opposite direction of the angular velocity and torque are desired, they can be changed by reversing the polarity of the field. Sample calculations for a shunt motor are given in Examples 9.15 and 9.16.

EXAMPLE 9.15   Torque produced by a shunt motor
. . . .

Given: A 100-V shunt motor has an angular velocity of 30 rad/sec. The armature current is 25 A and the armature resistance is 0.5 Ω.

Find: The torque.

Solution: The angular velocity is

$$\omega = \frac{(V_a - I_A R_A)\pi}{N_{As}\phi}$$

Thus,

$$N_{As}\phi = \frac{(V_a - I_A R_A)\,\pi}{\omega}$$

$$N_{As}\phi = \frac{[100 \text{ V} - (25 \text{ A})(0.5 \text{ }\Omega)]\,\pi}{30 \text{ rad/sec}}$$

$$N_{As}\phi = 9.16 \text{ Wb·turn}$$

The torque is

$$\tau = \frac{N_{As}\phi I}{\pi}$$

$$\tau = \frac{(9.16 \text{ Wb·turn})(25 \text{ A})}{\pi}$$

$$\tau = 72.9 \text{ N·m} = 53.8 \text{ ft·lb} \qquad (1 \text{ N·m} = 0.738 \text{ ft·lb})$$

FIGURE 9.61 Plots of angular velocity and torque versus armature current.

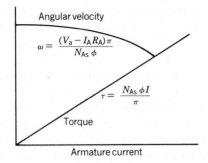

EXAMPLE 9.16 No-load angular velocity of a shunt motor
. . . .

Given: A 100-V shunt motor has an armature resistance of 0.20 Ω. The angular velocity is 377 rad/sec when the armature current is 100 A. At no external load, the armature current is 5.0 A and the magnetic flux increases by 4 percent.

Find: The no-load angular velocity.

Solution: The equation for angular velocity is

$$\omega = \frac{(V_a - I_A R_A)\,\pi}{N_{As}\,\phi}$$

Thus,

$$N_{As}\,\phi = \frac{(V_a - I_A R_A)\pi}{\omega}$$

At $I_A = 100$,

$$(N_{As}\phi)_{100} = \frac{[100\ \text{V} - (100\ \text{A})(0.20\ \Omega)]\ \pi}{377\ \text{rad/sec}}$$

$$(N_{As}\phi)_{100} = 0.667\ \text{Wb·turn}$$

At $I_A = 5$,

$$(N_{As}\phi)_5 = (N_{As}\phi)_{100} + 0.04\,(N_{As}\phi)_{100}$$

$$(N_{As}\phi)_5 = 0.667\ \text{Wb·turn} + (0.04)(0.667\ \text{Wb·turn})$$

$$(N_{As}\,\phi)_5 = 0.694\ \text{Wb·turn}$$

The angular velocity becomes

$$\omega = \frac{[100\ \text{V} - (5.0\ \text{A})(0.20\ \Omega)]\ \pi}{0.694\ \text{Wb·turn}}$$

$$\omega = 448\ \text{rad/sec}$$

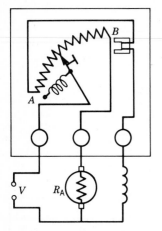

FIGURE 9.62 Schematic diagram of a starting box for a shunt motor.

Motor Starting   When the power of a dc shunt motor is greater than several horsepower, the starting current becomes too large. This occurs because the armature resistance is very low, and no back electromotive force exists when $\omega = 0$. To start the motor a protective resistance is used in series with the armature. This resistance is reduced as the angular velocity increases. This is usually accomplished with a starting box as illustrated in Figure 9.62. When the arm is in position *A*, the resistance is in series with the armature and full voltage is applied to the field coil. With full voltage applied to the field coil, the torque due to the field is a maximum. As the arm is moved from position *A* to *B*, the angular velocity increases to the maximum value. At position *B*, an electromagnet, which carries the field current, holds the arm in position *B*. If the field current is interrupted, the spring brings the arm back to position *A*.

Shunt Motor Efficiency   The power losses in a shunt motor are due to mechanical friction, armature heating, shunt coil heating, and brush contact heating. All these losses reduce the efficiency of the motor. Since energy is conserved.

$$P_{in} = P_{out} + P_{lost} = V_a(I_A + I_F) \tag{9.69}$$

or

$$P_{out} = P_{in} - P_{lost} = V_a(I_A + I_F) - I_A^2 R_A - I_F^2 R_F - P_B - P_m \tag{9.70}$$

where

V_a = input voltage
I_A = armature current
I_F = field current
R_A = armature resistance
R_F = field resistance
P_B = brush contact loss
R_m = frictional loss

All the losses can be easily estimated from measurements except the frictional loss. Part of this loss is due to the interaction of the armature with the air. This loss increases with angular velocity. Since the angular velocity does not change much for a shunt motor, this loss does not change much with the load. The frictional loss can then be estimated by measuring the power input at no load.

It has been found by experiments that the brush contact loss is approximately

$$P_B = 2I_A \tag{9.71}$$

Substituting this expression for P_B into Equation 9.70 and solving the equation for P_m yields

$$P_m = V(I_F + I_A) - I_A^2 R_A - I_F^2 R_F - 2I_A \tag{9.72}$$

P_m can be calculated at no-load conditions, and it can be used to calculate the efficiency at other loads. Example 9.17 shows the application of the equations to calculate the efficiency of a motor.

EXAMPLE 9.17 Efficiency of a shunt motor
. . . .

Given: The armature resistance of a shunt motor is 0.3 Ω and the field resistance is 300 Ω. The input voltage is 200 V. At no load the armature current is 1.3 A.

Find: The efficiency when the load requires an armature current of 15 A. Assume that the angular velocity remains constant.

Solution: The field current is

$$I_F = \left(\frac{200 \text{ V}}{300 \text{ }\Omega}\right) = 0.667 \text{ A}$$

The no-load power input is

$$(P_{in})_{no\ load} = V(I_A + I_F)$$

$$(P_{in})_{no\ load} = (200 \text{ V})(1.3 \text{ A} + 0.667 \text{ A})$$

$$(P_{in})_{no\ load} = 393 \text{ W}$$

The field loss is

$$P_F = I_F^2 R_F$$

$$P_F = (0.667 \text{ A})^2(300 \text{ }\Omega)$$

$$P_F = 133 \text{ W}$$

The armature loss is

$$P_A = (I_A)^2 R_A$$

$$P_A = (1.3 \text{ A})^2(0.3 \text{ }\Omega)$$

$$P_A = 0.5 \text{ W}$$

The brush loss is

$$P_B = 2I_A$$

$$P_B = (2)(1.3 \text{ A})$$

$$P_B = 2.6 \text{ W}$$

The frictional loss becomes

$$P_m = (P_{in})_{no\ load} - P_A - P_F - P_B$$

$$P_m = 393 \text{ W} - 0.5 \text{ W} - 133 \text{ W} - 2.6 \text{ W}$$

$$P_m = 257 \text{ W}$$

At an armature current of 15 A, the power in is

$$P_{in} = V(I_F + I_A)$$

$$P_{in} = (200 \text{ V})(0.667 \text{ A} + 15 \text{ A})$$

P_{in} = 3133 W

The armature loss is

$P_A = I_A^2 R_A$

P_A = (15 A)²(0.3 Ω)

P_A = 67.5 W

The brush contact loss is

P_B = (2)(15 A)

P_B = 30 W

The power out is

$P_{out} = P_{in} - P_A - P_F - P_B - P_m$

P_{out} = 3133 W − 67.5 W − 133 W − 30 W − 257 W

P_{out} = 2645 W

The efficiency is

$$\eta = \frac{P_{out}}{P_{in}} (100)$$

$$\eta = \frac{(2645 \text{ W})}{(3133 \text{ W})} (100)$$

$$\eta = 84.4\%$$

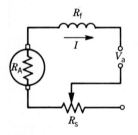

FIGURE 9.63   Schematic diagram of a series motor.

Series Motor   A circuit diagram for a series motor is shown in Figure 9.63. The same current flows through the armature, field coil, and the series rheostat. Thus, the magnetic field increases directly with the armature current. From Equation 9.44, the magnetic flux becomes

$$\phi = \frac{N_F I}{R_e} \tag{9.73}$$

As in the case of the shunt motor, the total voltage that causes the current to flow through the circuit is the applied voltage minus the back electromotive force:

$$V_e = V_a - V_b \tag{9.74}$$

where

V_e = the effective voltage for current flow

In addition, the total resistance is

$$R = R_s + R_A + R_F \tag{9.75}$$

From Ohm's law, the current becomes (neglecting the brush voltage drop of 2 V)

$$I = \frac{V}{R} = \frac{V_a - V_b}{R} \tag{9.76}$$

or

$$V_b = V_a - IR \tag{9.77}$$

Substituting for V_b from Equation 9.65 yields

$$\frac{N_{As}\omega\phi}{\pi} = (V_a - IR)$$

or

$$\omega = \frac{(V_a - IR)\pi}{N_{As}\,\phi} \tag{9.78}$$

Equation 9.73 may be used to substitute for ϕ to give

$$\omega = \frac{(V_a - IR)\pi R_e}{N_{As}N_F I} \tag{9.79}$$

Thus, the angular velocity increases as the current decreases. When the motor is started, the current will be large and the corresponding angular velocity small.

The torque produced by the series motor is given by Equation 9.63:

$$\tau = \frac{N_{As}\phi I}{\pi} \tag{9.80}$$

Using Equation 9.73 to substitute for ϕ yields

$$\tau = \frac{N_{As}N_F I^2}{\pi R_e} \tag{9.81}$$

The torque is directly proportional to the current squared. Thus, a large torque will be produced when the motor is started. Typical curves for torque and angular velocity are shown in Figure 9.64. As the current increases as a result of a mechanical load, the torque increases and the angular velocity decreases. However, the angular velocity can be adjusted by changing the resistance R_s. If Equation 9.79 is solved for I, one obtains

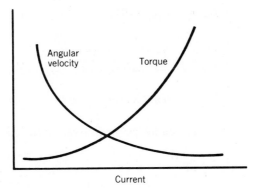

FIGURE 9.64 Plots of angular velocity and torque versus current for a series motor.

$$I = \frac{\pi V_a R_e}{N_{As} N_F \omega + P R_e \pi} \tag{9.82}$$

Substituting this value for I into Equation 9.81 gives

$$\tau = \frac{\pi N_{As} N_F V_a^2 R_e}{(N_{As} N_F \omega + R_e \pi)^2} \tag{9.83}$$

This equation shows that the torque is directly proportional to the applied voltage squared and inversely proportional to the denominator, which contains an ω^2. Thus, when the motor is started the torque will be high since ω will be zero.

Equation 9.83 can be solved for ω to yield

$$\omega = V_a \left(\frac{\pi R_e}{N_{As} N_F \tau} \right)^{1/2} - \left(\frac{\pi R R_e}{N_{As} N_F} \right) \tag{9.84}$$

If a small mechanical load is attached to the motor such that the torque is small, the angular velocity can become very large. This can destroy the motor. Thus, a sufficiently large mechanical load must be attached to the motor at all times. Sample calculations for series motors are given in Examples 9.18, 9.19, and 9.20.

EXAMPLE 9.18 Torque produced by a series motor
. . . .

Given: The current through a series motor is 25 A and the applied voltage is 200 V. The angular velocity is 200 rad/sec and the total resistance is 1.0 Ω.

Find: The torque.

Solution: The equation for the angular velocity in terms of the current is

$$\omega = \frac{(V_a - IR)\pi}{N_{As}\phi}$$

This equation can be solved for $N_{As}\phi/\pi$.

$$N_{As}\phi\omega = (V_a - IR)\pi$$

$$\frac{N_{As}\phi}{\pi} = \frac{(V_a - IR)}{\omega}$$

$$\frac{N_{As}\phi}{\pi} = \frac{200 \text{ V} - (25 \text{ A})(1.0 \text{ }\Omega)}{200 \text{ rad/sec}}$$

$$\frac{N_{As}\phi}{\pi} = 0.875 \text{ V·sec}$$

The equation for torque in terms of the current is

$$\tau = \frac{N_{As}\phi I}{\pi}$$

$$\tau = (0.875 \text{ V·sec})(25 \text{ A})$$

$$\tau = 21.9 \text{ N·m}$$

EXAMPLE 9.19   Angular velocity of a series motor
. . . .

Given: A series motor must produce a certain torque to move a car up a hill of uniform grade. When the motor runs at 10 rad/sec the applied voltage is 500 V and the current is 110 A. The total resistance is 0.3 Ω.

Find: The angular velocity if the applied voltage is reduced to 200 V. Assume that the reluctance is constant and the mechanical friction is constant.

Solution: From Example 9.18,

$$\frac{N_{As}\phi}{\pi} = \frac{(V_a - IR)}{\omega}$$

At 500 V one obtains

$$\frac{N_{As}\phi}{\pi} = \frac{500 \text{ V} - (110 \text{ A})(0.3 \text{ }\Omega)}{10 \text{ rad/sec}}$$

$$\frac{N_{As}\phi}{\pi} = 46.7 \text{ V·sec}$$

The required torque is

$$\tau = \frac{N_{As}\phi}{\pi} I$$

$$\tau = (46.7 \text{ V·sec}) (110 \text{ A})$$

$$\tau = 5137 \text{ N·m}$$

The magnetic flux is

$$\phi = \frac{N_F I}{R_e}$$

Multiply both sides by N_{As}/π.

$$\frac{N_{As}\phi}{\pi} = \frac{N_{As}N_F I}{\pi R_e}$$

Divide both sides by I.

$$\frac{N_{As}\phi}{\pi I} = \frac{N_{As}N_F}{\pi R_e}$$

$$\frac{(46.7 \text{ V·sec})}{(110 \text{ A})} = \frac{N_{As}N_F}{\pi R_e}$$

$$\frac{N_{As}N_F}{\pi R_e} = 0.425 \ \text{V·sec/A}$$

The angular velocity at 200 V can be calculated by using Equation 9.84.

$$\omega = V_a\left(\frac{\pi R_e}{N_{As}N_F} \cdot \frac{1}{\tau}\right)^{1/2} - \left(\frac{\pi R_e}{N_{As}N_F}\right) \cdot R$$

$$\frac{\pi R_e}{N_{As}N_F} = \frac{1}{(0.425 \ \text{V·sec/A})} = (2.35 \ \text{A/V·sec})$$

$$\omega = (200 \ \text{V})\left(\frac{2.35 \ \text{A/V·sec}}{5137 \ \text{N·m}}\right)^{1/2} - (2.35 \ \text{A/V·sec})(0.3 \ \Omega)$$

$$\omega = 3.57 \ \text{rad/sec}$$

EXAMPLE 9.20   Series motor
. . . .

Given: A series motor has an armature resistance of 0.09 Ω, a field resistance of 0.10 Ω, and a series resistance of 0.05 Ω. There are 50 turns in series on the armature. The magnetic flux is equal to $(3.5 \times 10^{-3})\cdot I$ Wb. The current is 100 A and the applied voltage is 200 V.

Find: a. The total torque.

             b. The angular velocity.

             c. The total power loss in the armature, field, series resistor, and brush contact.

             d. The total mechanical power output.

             e. The efficiency if the frictional loss is 800 W.

Solution: a. The torque is given by the equation

$$\tau = \frac{N_{As}\phi I}{\pi}$$

ϕ is given as

$$\phi = (3.5 \times 10^{-3})I \ \text{Wb}$$

By substitution

$$\tau = \frac{(3.5 \times 10^{-3})(N_{As})I^2}{\pi}$$

$$\tau = \frac{(3.5 \times 10^{-3} \ \text{Wb/A})(50)(100 \ \text{A})^2}{\pi}$$

$$\tau = 557 \ \text{N·m}$$

             b. The equation for ω is

$$\omega = \frac{(V_a - IR)\pi}{N_{As}\phi} \qquad \text{(neglecting the brush voltage drop)}$$

Substituting for ϕ yields

$$\omega = \frac{(V_a - IR)\pi}{N_{As}(3.5 \times 10^{-3})I}$$

The value of R is

$$R = R_A + R_F + R_S$$

$$R = 0.09\ \Omega + 0.10\ \Omega + 0.05\ \Omega$$

$$R = 0.24\ \Omega$$

The value of ω becomes

$$\omega = \frac{[(200\ \text{V}) - (100\ \text{A})(0.24\ \Omega)]\ \pi}{(50)(3.5 \times 10^{-3}\ \text{Wb/A})(100\ \text{A})}$$

$$\omega = 31.6\ \text{rad/sec}$$

c. The total power loss due to electrical resistance is

$$P_E = P_A + P_F + P_S + P_B \qquad (P_B = \text{brush contact loss})$$

$$P_E = I^2 R_A + I^2 R_F + I^2 R_S + 2I$$

$$P_E = I^2 R + 2I$$

$$P_E = (100\ \text{A})^2(0.24\ \Omega) + (2\ \text{V})(100\ \text{A})$$

$$P_E = 2600\ \text{W}$$

d. The total mechanical power output is

$$P = \omega\tau$$

$$P = (31.6\ \text{rad/sec})(557\ \text{N·m})$$

$$P = 17{,}600\ \text{W}$$

(*Note:* The calculation of ω neglects the brush contact loss. Therefore, this calculation of total mechanical power output neglects the brush contact loss.)

e. The efficiency is

$$\eta = \frac{(P_{in} - P_{lost})}{P_{in}} \times (100)$$

The power in is

$$P_{in} = V_a \cdot I$$

$$P_{in} = (200)(100) = 20{,}000\ \text{W}$$

The power lost is

$$P_{lost} = P_E + P_m \qquad (P_m = \text{frictional losses} = 800\ \text{W})$$

$$P_{\text{lost}} = 2600 \text{ W} + 800 \text{ W}$$

$$P_{\text{lost}} = 3400 \text{ W}$$

The efficiency becomes

$$\eta = \left(\frac{20{,}000 \text{ W} - 3400 \text{ W}}{20{,}000 \text{ W}}\right)(100)$$

$$\eta = 83\%$$

SECTION 9.5   EXERCISES

1. Determine the direction of the force on the current-carrying wires in the following diagrams. The direction of the electron flow is indicated in each case.

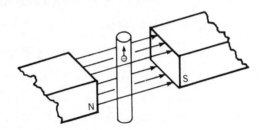

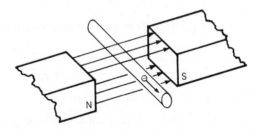

2. Indicate the direction of rotation of the current-carrying loops in the diagrams at the top of the following page. The direction of electron flow is indicated with an arrow in each case.

3. Sketch a permanent-magnet electric motor. Label the magnets, armature, commutator, and brushes and explain the function of each.

4. List the three factors that determine the maximum torque produced by a motor.

5. State the difference between the permanent-magnet, series, and shunt motors.

6. According to Figure 9.55, if the torque acting on a motor reaches its first maximum at $t = 3$ sec, when will the torque reach its second maximum?

7. Certain functions can be most easily expressed by two different functions, each of which is valid over a different subset of the domain (abscissa). Assuming the curve drawn in Figure 9.55 is the shape of a sine curve, write two expressions for torque as a function of time that describe the curve. Define the time from $t = 0$ until the time that

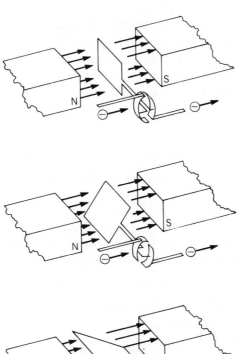

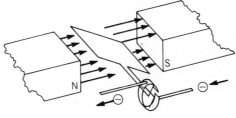

the torque first reached zero again to be one-half of a period ($T/2$) and write one expression that is good for times

$$0 < t < \frac{T}{2}, \quad T < t < \frac{3T}{2}, \quad 2\,T < t < \frac{5T}{2}, \quad \text{etc.}$$

and a second expression good for times

$$\frac{T}{2} < t < T, \quad \frac{3T}{2} < t < 2T, \quad \frac{5T}{2} < t < 3T, \quad \text{etc.}$$

8. Calculate the maximum magnitude of the force on a straight wire in a magnetic field if the magnetic flux density is 0.10 Wb/m^2 and the current is 2.0 A. What is the magnitude of the minimum force?

9. Calculate the torque of a shunt motor if the applied voltage is 500 V, the total current is 100 A, and the angular velocity is 377 rad/sec. The field resistance is 1000 Ω and the armature resistance is 0.20 Ω.

10. If the applied voltage to a dc shunt motor is 500 V when the field is 0.20 Wb, calculate the angular velocity. The armature resistance is 0.10 Ω and the number of turns in series on the armature is 90. What is the total power output?

11. The power input to a series motor is 20,000 W. The available power output is 18,500 W. What is the efficiency of the motor?

12. If the current through a series motor is 75 A, the applied voltage is 420 V, and the angular velocity is 377 rad/sec, what is the torque? The total resistance is 1.5 Ω.

13. What would happen to a dc shunt motor if the field coil became an open circuit?

14. What would happen to a dc series motor if the mechanical load became disconnected?

15. When does a dc series motor develop the greatest torque?

16. Discuss the losses in a shunt motor.

17. Discuss the losses in a series motor.

9.6

HEAT ENERGY TO FLUID ENERGY: THE BOILER

OBJECTIVES

Upon completion of this chapter, the student should be able to

• Draw and label a simple diagram of a steam generator.

• Explain in a short paragraph how thermal energy is transferred in a steam generator. Include the importance of radiation, convection in combustion gases and water, and conduction through water tubes.

• Identify three types of thermal losses in a steam generator.

• Explain how each of the following design features of modern steam generators reduces heat losses:
   Closed water system
   Thermal insulation
   Preheater (economizer)
   Superheater
   Air heater
   Stud tubes
   Purified water
   Clean combustion

• Draw and label a diagram of a two-drum steam generator showing the following parts:
   Air heater
   Economizer
   Superheater
   Water tubes
   Upper and lower drums
   Furnace

DISCUSSION

Introduction Most of the energy consumed in the United States is produced initially in the form of hot combustion gases. Internal combustion engines combine these gases and convert the resulting fluid energy directly to mechanical energy. In other cases, such as in electrical power production, a boiler is used.

The generation of steam from water in a boiler provides energy conversion from thermal energy to the potential and kinetic energy of high-temperature, high-pressure steam. All electrical generating plants that produce electrical energy

FIGURE 9.65 Simple steam generator.

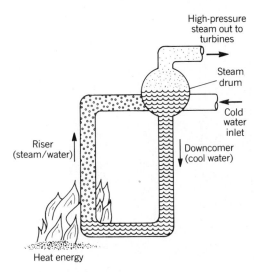

from thermal energy employ steam generators in the energy conversion process. The heat source may be the combustion of a chemical fuel such as coal, gas, or oil, or nuclear or solar energy; but in each case, the energy is transferred to high-pressure steam used to turn a turbine, which in turn drives an electric generator. This section discusses the transfer of thermal energy within the steam generator and methods to improve the efficiency of this energy transfer.

Figure 9.65 is a diagram of a simple steam generator where thermal energy is produced by the combustion of a fuel. Part of the energy from combustion is transferred to the water to produce steam. This transfer can be described by the following processes. Radiation from the internal walls of the furnace and the combustion gases is absorbed at the surface of the pipes. In addition, convection currents bring combustion gases in contact with the pipes and energy is transferred to the pipes by conduction. The outer surface of the pipes, therefore, attains a higher temperature than the inner surface, and energy is conducted from the outer surface to the inner surface of the pipes, where heat is conducted to the water in contact with the surface and steam is formed. Convection currents within the water-flow system carry the cool water down and the steam upward into a steam drum. The formation of steam at the inner surface of the pipes can reduce the heat transfer since it does not conduct heat as well as water. The convection currents tend to remove the steam from the wall, enhancing the heat transfer. If there is any deposit of material on the walls, the heat transfer will be reduced. Thus, it is advantageous to maintain a clean system. In general, the system should be designed to maximize the transfer of energy released by combustion to the water in conformity with other constraints imposed on the system.

Energy Losses   Thermal energy leaves the steam generator in three ways. Part of the thermal energy produced by the combustion process remains in the exhaust gases, which are discharged through the exhaust port. Some thermal energy escapes through the furnace and pipe walls to heat the environment around the steam generator. The remainder of the energy leaves the generator in heated steam that is used to turn turbines. Some of this energy is lost in the hot water and

FIGURE 9.66 Two-drum boiler.

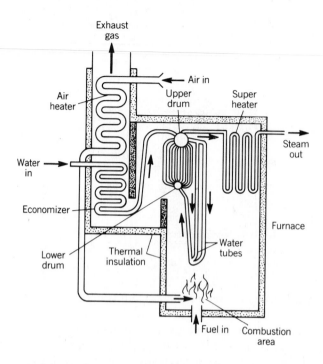

low-pressure steam discharged from the turbines. High-efficiency transfer of energy from the combustion process to the turbine requires that each of the energy losses be minimized.

Boiler Design

Figure 9.66 shows a modern two-drum boiler for steam generation. Heat losses are reduced throughout the system by thermal insulation. The efficiency of this system is increased by operating it as a closed water-flow system. The low-pressure steam leaving the turbines (not shown) is condensed to liquid water in a heat exchanger, but the temperature of water from the heat exchanger remains near the boiling point. Thus, the input water to the steam generator contains a portion of the thermal energy that left the generator earlier as steam. Less energy is required to produce steam from this hot water than from cold input water.

Air for the combustion process is also heated before it enters the combustion area. The air heater is located in the coolest part of the steam generator and absorbs thermal energy from exhaust gases as shown in Figure 9.66. This thermal energy is returned to the combustion area in the input air. Heat transfer to water occurs in three sections of the steam generator. Input water passes through a preheater called an economizer, which is located in the exhaust port just below the air heater. This process raises the temperature of the water almost to the boiling point and reclaims thermal energy that otherwise would have been lost in the exhaust gas.

The boiler consists of two large drums connected by water tubes. Heated water flows down some of these tubes by convection, and steam flows upward in others. Water tubes present a large surface area for heat transfer. In many modern

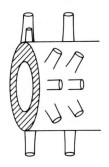

FIGURE 9.67 Stud tubes.

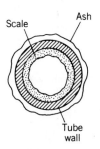

FIGURE 9.68 Scale and ash on tube wall.

boilers, the complete inner surface of the furnace is lined with water tubes that provide the necessary area for efficient heat transfer and prevent thermal damage to the walls.

Steam collects in the upper drum of the boiler and flows out the steam generator through a superheater located in the hottest part of the furnace. This heat exchanger increases the temperature of output steam to the desired level.

Heat transfer to water tubes is increased at some points in the boiler and superheater by use of stud tubes shown in Figure 9.67. These stud tubes present larger surface areas to the hot gases. Figure 9.68 shows two conditions that can reduce the efficiency of the heat transfer through water tubes. Scale formed on the tube inner surface from water impurities and ash deposits on the outer surface act as heat barriers. For these reasons, impurities must be removed from water used in the steam generator, and the combustion process must be as clean as possible.

Fuels burn with various efficiencies. For example, bituminous coal, with a heating value of 13,500 Btu/lb, burns with about 85–90% efficiency. When all energy losses are taken into account, the overall steam generator plant efficiency is about 25 to 30%.

EXAMPLE 9.21   Efficiency of a power plant

· ·

Given: A steam generator is used in conjunction with a turbine and electric generator to produce electric power.

Find: The amount of bituminous coal that will be burned per hour to produce 500 kW if the overall efficiency is 14%.

Solution: The overall efficiency is

$$\eta = \frac{P_{out}}{P_{in}} (100)$$

or

$$P_{in} = \left(\frac{P_{out}}{\eta}\right)(100)$$

$$P_{in} = \frac{(500,000 \text{ W})}{(17)} (100)$$

$$P_{in} = 2.94 \times 10^6 \text{ W}$$

$$1 \text{ W} = 3.41 \text{ Btu/hr}$$

$$P_{in} = (3.41 \text{ Btu/hr·W})(2.94 \times 10^6 \text{ W})$$

$$P_{in} = 1.00 \times 10^7 \text{ Btu/hr}$$

The energy from bituminous coal is 13,500 Btu/lb. Thus,

$$P_{in} = (13,500 \text{ Btu/lb})(Q_m)$$

where Q_m is the number of pounds per hour that are burned. Solving for Q_m gives

$$Q_m = \frac{P_{in}}{13,500 \text{ Btu/lb}}$$

$$Q_m = \frac{1.0 \times 10^7 \text{ Btu/hr}}{13,500 \text{ Btu/lb}}$$

$$Q_m = 741 \text{ lb/hr}$$

SECTION 9.6 EXERCISES

1. Draw and label a diagram of a simple steam generator.
2. Explain how each of the following contributes to heat transfer in a steam generator:
    a. Thermal radiation from the heat source.
    b. Convection currents in combustion products.
    c. Convection currents in water and steam.
    d. Conduction of heat through walls of water tubes.
3. Explain the three loss factors that lower the efficiency of steam generators using combustion as a heat source.
4. Identify the parts of the steam generator in the figure.

    A. _____

    B. _____

    C. _____

    D. _____

    E. _____

    F. _____

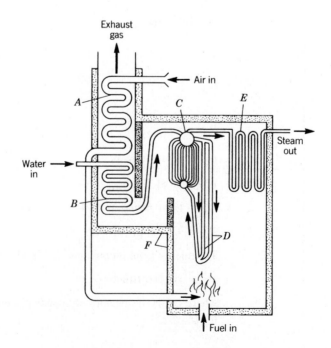

5. Explain how each of the following would effect the efficiency of the steam generator in the figure in Problem 4.
   a. Eliminate part *A*.
   b. Eliminate part *B*.
   c. Eliminate part *E*.
   d. Eliminate part *F*.
6. Explain how heat transfer is aided by the use of each of the following:
   a. Stud tubes.
   b. Purified water.
   c. Clean combustion.
7. How much energy is released from the burning of one pound of bituminous coal if the process is 87% efficient?
8. How many pounds of a fuel with an energy content of 6000 Btu/lb are necessary to produce 47,000 Btu of energy if the overall efficiency of the steam generator used to convert the energy is 20%?
9. A steam generator that is 20% efficient can produce 55,000 W of power. A device that recycles some of the heat normally lost through the exhaust port increases the efficiency to 25%. How much power can the generator now produce?
10. The burning of bituminous coal is used to produce electrical energy. If the overall process is 15% efficient and the output voltage is 110 V, how much charge can be made to flow by burning 3 lb of coal? (1 Btu = 1054 J.)
11. How efficient is a steam generator that produces 4×10^6 cal of heat energy by burning 10 lb of bituminous coal?

S U M M A R Y

Energy convertors are devices that convert energy from one form to another. For example, an electric generator converts mechanical energy into electrical energy. In the conversion process, however, part of the input energy is always converted into heat. Since the heat energy is lost in the conversion process, energy convertors are not 100% efficient. The efficiency of an energy convertor is defined as the energy out in the desired form divided by the energy in times 100. At present, the accumulative losses in energy convertors account for over half of all the energy consumed. Thus, it is important to try to maximize the efficiency of energy convertors.

The fan was discussed in this chapter as an example of an energy convertor that changes mechanical power into fluid power. The total fluid power is the volume flow rate times the quantity obtained by subtracting the atmospheric pressure from the stagnation process. The total efficiency is equal to the fluid *power divided by the mechanical power into the fan times 100. There are two classes of fans: the axial-flow design and the centrifugal-flow design. The efficiency of these fans depends on the inlet design and blade types. Axial-flow fans are better suited to low-resistance applications and centrifugal-flow fans are used in higher-pressure applications.*

The turbine is an example of an energy convertor that converts fluid energy into mechanical energy. It is one of the more efficient energy convertors, and it is used as a power source for trucks, ships, aircraft, and electric power plants. Fluids used to drive turbines include steam, hot gases, and water. The impulse turbine was analyzed, and an equation for its ideal power was developed. The maximum ideal power occurs when the bucket velocity is one-half the jet velocity. For this condition, the ideal power is equal to one-half the mass-flow rate of the jet times the velocity of the jet squared. As the bucket velocity increases from zero to one-half the jet velocity, the ideal power increases from zero to its maximum value. The ideal power then decreases and becomes zero when the bucket velocity is equal to the jet velocity. If water from a reservoir is used as a source of energy, losses will occur in the pipes, nozzle, and buckets. However, large power outputs can be obtained from an impulse turbine when the height of the reservoir is large.

The electric generator is an example of an energy convertor that converts mechanical energy into electric energy. When a conductor is moved through a magnetic field, a voltage is generated. The voltage generated is proportional to the velocity, the length, the magnetic flux density, and the sine of the angle between the velocity vector and the magnetic field vector. A large voltage can be obtained if a coil with many loops is rotated in a magnetic field. For this case, the voltages in the wires add. If slip rings are used, an alternating voltage is obtained. Direct current can be obtained by using a commutator. When current flows in the coil, a force acts on the wire that opposes the rotation. Thus, mechanical energy is required to produce electric energy. The required magnetic field can be obtained by a permanent magnet or field coils. For large generators, it is necessary to use field coils. The shunt generator has the field coil connected across the armature. The voltage output is maximum at no load, and it decreases with the load current. A series generator has the field coil in series with the armature. At no load, the output voltage is small, and it increases with the load current to a maximum voltage. As the load current increases beyond the maximum point, the output voltage decreases. For this reason, series generators are seldom used. A compound generator uses shunt and series field coils. When properly designed, the output voltage is almost constant from zero load up to the rated load. Electric power losses occur in the armature, field coils, and brushes. In addition, there are mechanical frictional losses. However, the dc generator has a relatively high efficiency.

Photovoltaic materials are semiconductor devices (p or n) that convert light energy into electrical energy. Some combinations of semiconductor materials deposited upon metal plates can generate voltage at the junctions when exposed to light radiation. Voltage is generated at junctions (barrier layers) between the two types of materials when radiant energy is absorbed. This process is termed the "photovoltaic effect." Photovoltaic cells are used as detectors of light and as generators of electric power in space. The high cost and low efficiency limit their application for power generation on the earth's surface. The efficiency for power generation is approximately 10%.

The dc electric motor is an energy convertor that changes electrical energy into mechanical energy. When current flows through a straight wire in a magnetic field, an electromagnetic force is exerted on the wire. The force is directly proportional to the current, the magnetic flux density, the length, and the sine of the angle between the direction of electron flow and the magnetic field. To obtain a large torque, a number of turns are used in series on an armature. The torque on each wire adds to give the total torque. As the armature rotates, the current through the armature coil is maintained in the proper direction by a commutator. The rotating coils also generate a voltage that is in opposition to the applied voltage. This back electromotive force limits the current through the armature. Permanent magnets may be used for small motors. For large motors, field coils must be used to establish the magnetic field. The angular velocity is almost directly proportional to the applied voltage and inversely proportional to the magnetic field. The torque is directly proportional to the current and the magnetic field. The shunt motor used a field coil in parallel with the armature. For the shunt motor, the angular velocity tends to remain constant as the mechanical load changes. The series motor has the field coil in series with the armature. The torque is directly proportional to the armature current squared. When the series motor is started the current is large. Thus, a large torque is produced. Electrical losses in motors occur in the armature, field coils, and brushes. In addition, there are mechanical frictional losses. The dc motor, however, has a relatively large efficiency.

The boiler is an example of an energy convertor that changes thermal energy into fluid energy. The thermal energy is produced by the combustion of a fuel, and the energy is transferred to the water by radiation, convection, and conduction. The boiler is designed to maximize this heat transfer. In a power plant, the steam is used as an energy source to drive a turbine that is connected to an electric generator.

Table 9.2   Important energy converters

Input Energy	Output Energy				
	Mechanical	Fluid	Electrical	Thermal	Optical
Mechanical		Pump Fan	Generator		
Fluid	Turbine Windmill		Magneto-hydro-dynamic generator		
Electrical	Motor			Heating devices	Lighting devices
Thermal	Internal combustion engines	Steam generator	Thermo-electric generators		
Optical			Solar cells	Solar collectors	

Transducers

INTRODUCTION

The operation of many technical systems requires that information be transmitted from the system to a controlling mechanism. An automobile, for example, may contain gages that indicate speed, fuel level, oil pressure, and temperature. Each of these gages is connected to a device called a "transducer." The transducer sends information about the conditions in the automobile to a gage.

Transducers accept an input signal from one energy system and convert it into an output signal in another energy system.

This chapter discusses transducers used to change information about mechanical, fluid, electrical, thermal, and optical systems into information in an electrical or mechanical system.

OBJECTIVES ....

Upon completion of this section, the student should be able to

- Define a transducer.

- Give the distinction between transducers that require external energy and those that do not.

DISCUSSION

To "transduce" means "to lead across." The function of a transducer is to move an information-carrying signal from one energy system to another. Transducers are called "sensors" because they provide input to monitoring or control systems, just as human senses provide information to the brain.

Any device that transforms a signal (as described above) can be called a "transducer." Transducers (electrical, mechanical, fluid, and thermal) can operate in all the energy systems discussed in previous chapters. For each type, the transducer function remains the same: to transform information from one system to another.

Transducers can be viewed as having an input and an output (Figure 10.1). The transducer (shown as a box) accepts some input quantity, which may represent a physical property, and produces an output, which is the response of the transducer to that property.

An electrical input (Figure 10.2) and a mechanical output illustrates this principle. In this example, the input is an electrical current, the transducer is an ammeter, and the output is the mechanical deflection of the needle on the meter.

The output varies with the input and so allows the transducer to measure the input. For example, the voltage (or current) output from a photocell is indicative of the amount of light energy incident upon the photocell. Thus, a photocell may be used as a quantitative detector of light energy.

Another aspect of a transducer is that it must also be able to transmit a signal. Again, the photocell provides an excellent example: it receives light waves and emits electrical currents. The electricity can then be sent along a wire to be displayed on a gage.

Some transducers actually are energy convertors. They can convert the energy of the input signal directly into energy of the output signal. The photocell described above is an example. When light strikes the photocell, some of the light energy is converted directly into electrical energy. Therefore, any of the energy convertors discussed in Chapter 9 can be used as transducers. One advantage to this type of transducer is that no external energy source is required.

Other transducers do require an external energy source. In most cases like this, the transducer operates by a change in resistance in response to the input. This type of transducer does not convert one type of energy directly into another. It changes its resistance in response to the input energy and, thus, changes the rate of flow of some quantity through itself. Most of these transducers operate on electrical power, but a few devices using fluid flow as an external energy source also are used in some fluid-control systems.

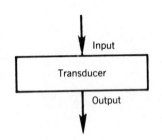

FIGURE 10.1 Basic transducer function.

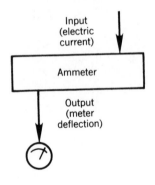

FIGURE 10.2   Ammeter as a transducer.

The remainder of this chapter discusses a variety of transducers. They are classified according to the energy systems in which the input signals originate. In all cases, the output signal is either electrical (current or voltage) or mechanical (position of an indicator).

INTRODUCTION *EXERCISES*

1. Describe the function of a transducer.
2. Describe the differences between transducers that convert input energy directly into output energy and those that do not. Give an example of each type in the following energy systems:

   Mechanical

   Fluid

   Thermal

   Optical

10.1

MECHANICAL TRANSDUCERS

OBJECTIVES

Upon completion of this section, the student should be able to

- Describe the operation of a carbon microphone.

- Describe the operation of a strain gage.

- Define the term "gage factor" and state ideal and typical values for gage factor.

- Calculate the gage factor of a wire, given the wire resistance, strain, and increase in resistance.

- Describe the piezoelectric effect.

- Given the pressure, thickness, and piezoelectric constant, calculate the open circuit output voltage of a piezoelectric crystal.

- Sketch and label the five major components of a compression accelerometer—including the critical axis—and describe briefly how it works.

- Sketch and label the three major components of a shear accelerometer and its critical axis, and describe briefly how it works.

- Sketch the voltage equivalent circuit of an accelerometer and label the signal source, capacitance, and resistance.

- Name a type of accelerometer measuring device and describe its key characteristics.

- Given two of the following three values, solve for the unknown quantity: acceleration, peak voltage, and sensitivity.

DISCUSSION

Microphone

The most common device for changing a mechanical vibration wave to an electrical wave is the microphone.

Different types of microphones employ different principles of operation. One type is the **carbon microphone,** which contains a diaphragm that moves (vibrates) in response to sound waves incident upon it (Figure 10.3). Immediately behind (and in contact with) the diaphragm is a small cup filled with tiny grains of carbon. A low-voltage electrical current passes through these carbon grains. When the diaphragm moves in response to sound waves pressing on it, the carbon grains are more closely compacted. Electrical resistance in the carbon medium changes, and electric current through the carbon varies accordingly.

Because highly compressed carbon offers less resistance than loosely packed carbon, the amount of electric current increases. If sound is weak, the diaphragm experiences light pressure, and the grains are loosely compacted. Therefore, resistance is high, and there is less electric current through the carbon.

The diaphragm and the carbon grains form a transducer, changing acoustic waves into varying electrical current. Although such a carbon microphone does not exhibit high fidelity, it does represent a good example of a transformer for sound energy into electrical energy. (Other more sophisticated microphone types can convert sound more faithfully.)

Strain Gages

Introduction   In strength of materials, four important quantities are defined. They are normal stress, linear strain, shear stress, and shear strain. For this discussion,

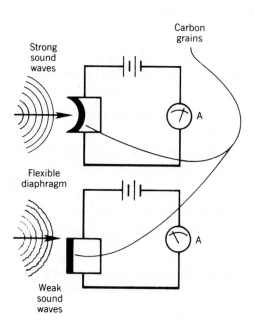

FIGURE 10.3 Varying current through carbon microphone.

FIGURE 10.4 Force on bar resulting in stress and strain.

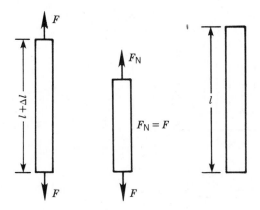

only the normal stress and strain will be considered. A simple example of normal stress and strain is illustrated in Figure 10.4. The body shown in the figure is subjected to a tension by the forces applied to its ends. Because of these forces, the body will elongate. For this case, the normal stress over the cross-sectional area shown in Figure 10.4 is defined as

$$S_N = \frac{F_N}{A} \tag{10.1}$$

where

S_N = normal stress
F_N = normal force
A = cross-sectional area

The linear strain is defined as the change in length divided by the total intitial length:

$$\epsilon_l = \frac{\Delta l}{l} \tag{10.2}$$

where

ϵ_l = linear strain
Δl = change in length
l = initial length

Elastic materials are usually used in structures and machines. For elastic materials, the stress is directly proportional to the strain (Hook's law):

$$S_N = E\epsilon_l \tag{10.3}$$

where

E = a constant called Young's modulus of elasticity

Equation 10.3 holds if the stress is not too great. If the stress on an elastic material ceases to be directly proportional to the strain, the stress has exceeded the elastic limit for that material. The stress at which the material fails is called the ultimate stress.

FIGURE 10.5 Typical wire
strain gage.

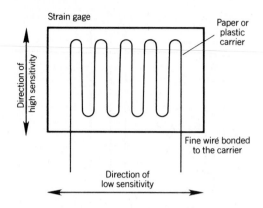

Types of Strain Gages   For many applications, it is important to measure the
linear strain. This can be accomplished by a device called a strain gage. The strain
gage consists of a material whose electrical properties change as it is stretched or
deformed. For example, a length of fine wire under tension will change its
electrical resistance as its length changes. If the wire is bonded firmly to a sample,
measurement of change in electrical resistance of the wire can be used to determine strain in the sample.

Strain gages take a number of forms: thin wires, foils, or semiconductors. A
typical form for a wire strain gage is shown in Figure 10.5. The fine wire is first
cemented to a carrier material (for example, paper or plastic). The gage then can
be cemented to the sample. When the sample is strained, strain is transmitted to
the strain gage. Measurement of electrical resistance of the wire gives the strain in
the wire and, therefore, the strain in the sample to which it is bonded.

Figure 10.6 illustrates how the strain of a wire affects its electrical resistance.
When the wire is strained, it becomes slightly longer, and the cross-sectional area
becomes smaller. Electrical resistance before the wire is strained is given by

$$R_1 = \frac{\rho l_1}{A_1} \tag{10.4}$$

where

R_1 = electrical resistance of wire before being strained
ρ = electrical resistivity of wire
l_1 = length of wire before being strained
A_1 = cross-sectional area of wire before being strained

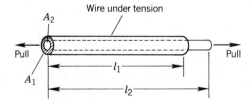

FIGURE 10.6   Effect of wire
tension on length and
cross-sectional area.

If the assumption is made that electrical resistivity is unaffected by strain, the wire's resistance after strain is applied is given by

$$R_2 = \frac{\rho l_2}{A_2} \tag{10.5}$$

where

R_2 = electrical resistance of wire after being strained
$l_2 = l_1 + \Delta l$ = length of wire after being strained
$A_2 = A_1 + \Delta A$ = cross-sectional area of wire after being strained

The change in resistance becomes

$$\Delta R = R_2 - R_1 = \frac{\rho(l_1 + \Delta l)}{A_1 + \Delta A} - \frac{\rho l_1}{A_1}$$

or

$$\Delta R = \frac{\rho A_1 \Delta l - \rho l_1 \Delta A}{A_1(A_1 + \Delta A)} \tag{10.6}$$

If ΔA is small compared to A_1, a good approximation for ΔR becomes

$$\Delta R = \frac{\rho \Delta l}{A_1} - \frac{\rho l_1 \Delta A}{A_1^2} \tag{10.7}$$

Dividing ΔR by R_1 yields

$$\frac{\Delta R}{R_1} = \frac{\Delta l}{l_1} - \frac{\Delta A}{A_1} \tag{10.8}$$

A relationship exists between the change in length per unit length and the change in area per unit area:

$$\frac{\Delta A}{A_1} = -2\nu \frac{\Delta l}{l_1} \tag{10.9}$$

where

ν = elastic constant called Poisson's ratio

Substituting this expression for $\Delta A/A_1$ into Equation 10.8 yields

$$\frac{\Delta R}{R} = \frac{\Delta l}{l_1}(1 + 2\nu) \tag{10.10}$$

Poisson's ratio ν varies slightly from one metal to another. For most metals, it is approximately 0.3, the value used in this section.

The ratio $\Delta l/l_1$ is recognized as strain ϵ_l, a dimensionless quantity given sometimes as "inches per inch" or "centimeters per centimeter." Equation 10.10 can be rewritten as shown in Equation 10.11, where the gage factor G is introduced:

$$\frac{\Delta R}{R} = \epsilon_l G \tag{10.11}$$

where

G = gage factor, $1 + 2\nu$

Equation 10.11 is used to solve a problem in Example 10.1.

The definition of the gage factor G follows from Equation 10.11. G is equal to the fractional change in resistance $\Delta R/R$ divided by strain $\epsilon_1 = \Delta l/l$. Thus, the gage factor is a fractional change in resistance per unit strain. Since G is equal to $(1 + 2\nu)$, and ν for most metals is approximately 0.3, ideally the gage factor should equal 1.6. However, Equation 10.10 was derived on the assumption that electrical resistivity ρ does not change with strain. This is not quite true. Electrical resistivity ρ does change with strain in a rather complicated manner. When this change is taken into account, the gage factor G is typically found to have values in the range from $G = 1.8$ to $G = 4.5$.

EXAMPLE 10.1 Resistance change calculation
. . . .

Given: A strain gage has a resistance of 1000 Ω; the gage material has a value of modulus of elasticity of 10^{12} dyne/cm^2 and is subjected to a stress of 10^9 dyne/cm^2. The gage factor (G) is 2.0.

Find: Change in resistance.

Solution: Strain ϵ_l is found from Hooke's law:

$$E = \frac{S_N}{\epsilon_l}$$

$$\epsilon_l = \frac{E}{S_N}$$

where

$\qquad S_N$ = normal stress
$\qquad E$ = modulus of elasticity

Using the given values,

$$\epsilon_l = \frac{10^9}{10^{12}}$$

$$\epsilon_l = 10^{-3}$$

Then, using Equation 10.11, the change in resistance ΔR is derived as

$$\frac{\Delta R}{R} = G\epsilon_l$$

Thus,

$$\Delta R = G\epsilon_l R$$

$$\Delta R = 2\,(10^{-3})(1000\ \Omega) = 2\ \Omega$$

Materials and Construction of Strain Gages   Strain gages can be made of three different types of materials:

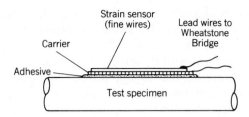

FIGURE 10.7 Typical arrangement of a bonded strain gage.

- Small-diameter wire (about 0.001 inch diameter),
- Thin sheets of metal foil (about 0.00015 inch thick), or
- Semiconductor materials (basically similar to metallic gages, but giving greater response for a given strain).

The strain gage design shown in Figure 10.5 gives relatively high sensitivity to strain along the length of the loops of wire and low sensitivity to strain perpendicular to that direction. For two-dimensional measurements, two separate gages oriented in perpendicular directions can be used.

Commercial strain gages come with the sensitive element mounted on a carrier or backing. This serves both to support the unit and to provide electrical insulation from the test specimen. The gage is bonded to the specimen by an adhesive so that the strain is transferred from the specimen to the gage (Figure 10.7). In this configuration, the device is commonly called a "bonded" strain gage. An unbonded strain gage could involve wires stretched between two components that move relative to each other. The bonded type of strain gage has become more common.

A few of the metals used in wire and foil gages are presented in Table 10.1, along with comments on useful temperature range and the recommended types of use. Situations where measurements are made of forces that are relatively constant or slowly changing are labeled static measurements. When the forces being measured are changing the measurements are called dynamic.

TABLE 10.1 Metallic materials for strain gages

Metal	Temperature Range	Recommended Type of Measurement
Copper–nickel alloy (Constantan)	−100 to +460°F	Static
Nickel–chrome alloy (Nichrome V)	to 1200°F to 1800°F	Static Dynamic
Nickel–iron alloy (dynaloy)	Limited by carrier and adhesive	Dynamic
Nickel–chrome alloy (stabiloy)	to 600°F	Static
Platinum alloy (alloy 1200)	to 1200°F to 1500°F	Static Dynamic

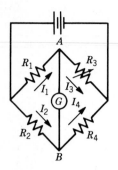

FIGURE 10.8 Schematic of Wheatstone bridge.

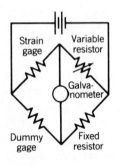

FIGURE 10.9 Wheatstone bridge with dummy strain gage to compensate for temperature effects.

Measurement of the Change in Resistance To measure the relatively small resistance changes caused by strain, a Wheatstone bridge often is used, as illustrated in Figure 10.8. Unknown resistance of the gage is R_1; resistances R_2, R_3, and R_4 are known, with R_3 being a variable resistance. To find the value of the unknown resistance R_1, the variable resistor R_3 is adjusted until the galvanometer G registers no current; that is, $I_1 = I_3$, and $I_2 = I_4$. Under this balanced condition, the four resistances must satisfy Equation 10.12 (see Chapter 5):

$$R_1 = \left(\frac{R_2}{R_4}\right) R_3 \tag{10.12}$$

R_1 can be determined by Equation 10.12 from the known values R_2, R_3, and R_4. Using commercially packaged bridge networks, the variable resistance may be adjusted simply by the rotation of dials until the meter on the instrument indicates zero current flow.

In practice, the bridge circuit may be set up as shown in Figure 10.9. A dummy gage is mounted on an unstrained sample of the same material that is being measured. The active gage is mounted on the strained material. This arrangement compensates for changes in temperature, which affect both the active gage and the dummy gage identically.

Strain gages often are used in industry for measuring dimensional changes in specimens. They offer the advantages of an easily measurable electrical output, ease of installation and operation, and low cost. In addition, they are utilized to determine the cause of strain.

Piezoelectric Pressure Transducer

Some types of crystalline material exhibit a phenomenon called the "piezoelectric effect." A mechanical pressure applied to certain crystals causes an electric charge to appear on their surfaces (Figure 10.10). (In reality, a complicated relationship exists between the direction of the applied force, the crystalline structure, and the surfaces on which the charge appears. Because this relationship is beyond the scope of this chapter, a simplified version is presented.)

When pressure is released, the charge disappears rapidly. Thus, a piezoelectric crystal is a transducer, transforming mechanical energy into electrical energy. The charge on the surface leads to a voltage in an electric circuit connected to the crystal. Thus, the crystal can be used as a pressure sensor. Quantitative measurement can be made of the pressure if the voltage is measured.

A piezoelectric crystal also can serve, in reverse, to transform electrical energy into mechanical energy. Voltage applied to a piezoelectric crystal causes it to undergo deformation and to exert mechanical pressure. This principle is commonly used for "beepers" and other alarm applications.

The piezoelectric effect is observed only in certain types of crystals, for example, Rochelle salt (sodium potassium tartrate) and tourmaline (a semiprecious gem). Significant properties of several piezoelectric crystals are summarized in Table 10.2.

The equation for the voltage that can be obtained is

$$V = kph \tag{10.13}$$

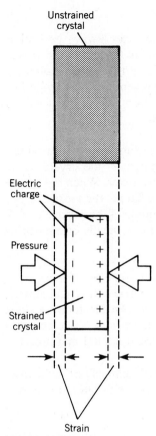

Unstrained crystal

Electric charge

Pressure

Strained crystal

Strain

FIGURE 10.10 Electric charge appearing on surface of strained piezoelectric crystal.

TABLE 10.2 Piezoelectric crystal properties

Crystal	k = Piezoelectric Constant (Voltage per Unit Applied Pressure)	Coupling Coefficient (Efficiency)
Quartz	0.055 V·m/N	9.9%
Rochelle salt (at 30°C)	0.098 V·m/N	73%
Ammonium dihydrogen phosphate	0.178 V·m/N	25%
Ethylene diamine tartrate	0.152 V·m/N	21.5%
Lithium sulfate	0.165 V·m/N	35%
Tourmaline	0.0275 V·m/N	9.2%

where

k = piezoelectric constant
p = applied pressure
h = thickness of the crystal

An application of Equation 10.13 is given in Example 10.2.

The coupling coefficients (Table 10.2) are the efficiencies for the conversion of electrical energy into mechanical energy. The values listed are obtained only with careful design, but, as Table 10.2 shows, the values can be quite high.

Piezoelectric crystals are usually employed in dynamic conditions of changing pressures or accelerations. Under static conditions, when a steady force is applied, the charge gradually leaks off the surface. Piezoelectric crystals are used, therefore, in dynamic devices, such as phonograph pickups, high-quality microphones, accelerometers, and ultrasonic submarine detectors.

EXAMPLE 10.2 Calculation of piezoelectric voltage

. . . .

Given: A pressure of 10^6 N/m² is applied to a piezoelectric crystal of ammonium dihydrogen phosphate 1 mm thick.

Find: Resulting open circuit voltage.

Solution: From Table 10.2, voltage per thickness per unit applied pressure for ammonium dihydrogen phosphate is given as 0.178 V·m/N. Voltage is given by (Equation 10.13)

$$V = kph$$

Therefore,

$$V = (0.178 \text{ V·m/N})(10^6 \text{ N/m}^2)(0.001 \text{ m})$$

$$V = 178 \text{ V}$$

10^6 N/m²

1 mm

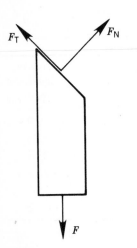

FIGURE 10.11　Normal and tangential force components.

Accelerometers

Introduction　The force acting upon the area of a body can always be broken down into a component normal to the area and a component tangent to the area as shown in Figure 10.11. The normal force, F_N, divided by the area is called the normal stress, S_N. The shear stress is defined as the tangential force divided by the area. For the accelerometers discussed in this section, the crystals will be subjected to either a normal force or a tangential force.

Compression Accelerometers　Figure 10.12 shows the components of a typical compression accelerometer. A compression is maintained on the crystal at all times by a mass to increase the sensitivity of the accelerometer. When the device experiences an acceleration along the axis shown in the figure, the mass exerts a pressure on the crystal. This pressure changes the magnitude of the surface charge on the crystal. Electrodes on the crystal surface provide a means of measuring either the charge or the voltage produced.

Shear Accelerometers　Figure 10.13 shows a shear accelerometer. The crystal is shaped like a cylinder bonded to a hollow shaft. A concentric cylindrical mass is bonded to the crystal, and the entire assembly is enclosed. An acceleration along the axis shown in the figure results in an shear force on the crystal. The electrodes are located on the inner and outer walls of the cylindrical crystal and are connected to an electrical output.

Shear accelerometers can be built with extremely small dimensions and can be mounted with a machine screw through the central hole. Most microminiature accelerometers are of this design.

Measurements　Measurement of the charge collected on the electrodes of the crystal is the most accurate measurement of acceleration. Because such measurement is difficult and expensive, however, voltage measurements usually are

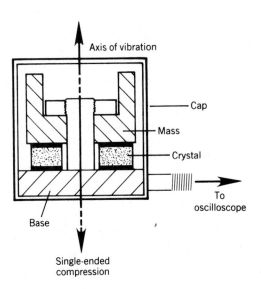

FIGURE 10.12　Compression accelerometer.

FIGURE 10.13 Shear accelerometer.

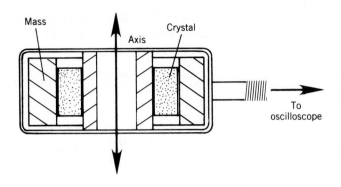

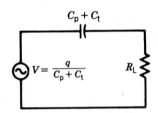

FIGURE 10.14 Voltage equivalent circuit for accelerometer measurement.

made. Figure 10.14 shows the voltage equivalent circuit of an accelerometer and voltage-measuring device. C_p is the capacitance of the accelerometer electrodes and C_t is the capacitance of the cable used to connect the accelerometer to the measuring device. R_L is the input impedance of the voltage-measuring device (a vacuum-tube voltmeter, called a VTVM, or an oscilloscope). The charge generated by the crystal is q. For the best response, $C_p + C_t$ should be as small as possible and R_L should be as large as possible.

Accelerometers usually are supplied with low-capacitance cables of known capacitance. The voltage sensitivity supplied by the manufacturer is based on the use of these cables. The accelerometer will not be calibrated correctly if other cables are used.

Figure 10.15 shows the typical deviation from the calibrated voltage sensitivity of an accelerometer for three values of load resistance. This figure indicates that good low-frequency performance requires the use of a voltage-measuring device having a very high input impedance. Thus, an oscilloscope is the most commonly used measurement device.

According to Newton's second law, the force acting on the accelerating mass is

$$\mathbf{F}_1 = m\mathbf{a}_1 \tag{10.14}$$

Since for every action there is an opposite and equal reaction, the force acting on the crystal is

$$\mathbf{F}_2 = m\mathbf{a}_2 \tag{10.15}$$

where

m = mass
$\mathbf{a}_2$ = acceleration = $-\mathbf{a}_1$

FIGURE 10.15 Typical accelerometer sensitivity for various load resistances.

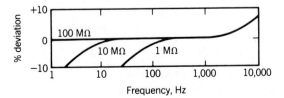

Accelerometers are designed so that the force acting on the surface of the crystal is either a normal force or a tangential force. The corresponding normal or shear stress becomes

$$S_N = \frac{F_N}{A} = \frac{ma_N}{A} \qquad \text{(normal stress)} \qquad (10.16)$$

and

$$S_s = \frac{F_T}{A} = \frac{ma_T}{A} \qquad \text{(shear stress)} \qquad (10.17)$$

where

a_N = normal acceleration
a_T = tangential acceleration

The voltage output for the two cases becomes (using Equation 10.13)

$$V_N = k_N S_N h = \frac{k_N ma_N h}{A} \qquad (10.18)$$

and

$$V_s = k_s S_s h = \frac{k_s ma_T h}{A} \qquad (10.19)$$

for a crystal of thickness h and a cross-sectional area A. Equations 10.18 and 10.19 can be solved for the acceleration to yield

$$a_N = \left(\frac{A}{k_N mh}\right) V_N = \frac{V_N}{k_N'} \qquad (10.20)$$

$$a_T = \left(\frac{A}{k_T mh}\right) V_s = \frac{V_s}{k_s'} \qquad (10.21)$$

where

$$k_N' = \frac{k_N mh}{A}$$

$$k_s' = \frac{k_s mh}{A}$$

For both cases, the acceleration is equal to the voltage divided by a constant.

The sensitivity is usually given in millivolts per g (referring to the acceleration due to gravity) with the specified external capacitance of the connecting cable provided. The resulting acceleration is measured in the number of g's. The corresponding equation may be written as

$$a = \frac{V}{k} \qquad (10.22)$$

where

V = output voltage in millivolts
k = constant in millivolts per g

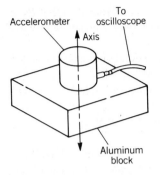

Accelerometer

To oscilloscope

Axis

Aluminum block

FIGURE 10.16 Accelerometer must be mounted firmly.

Example 10.3 shows the use of Equation 10.22 in solving a problem.

In practical applications, accelerometers must be firmly attached to the component whose acceleration is to be measured. If the accelerometer cannot be bolted directly to the component, it may be secured by a metal block held in contact with the component (Figure 10.16).

Recall from Figures 10.12 and 10.13 that accelerometers measure accelerations along one axis only. When mounted as shown in Figure 10.16, the accelerometer will respond only to those acceleration components that lie along its axis. Accelerations perpendicular to this axis cannot be detected by the device. However, accelerometer pairs can be used to measure both the magnitude and direction of accelerations.

EXAMPLE 10.3 Calculation of acceleration using acceleration readings
· ·

Given: An accelerometer with a sensitivity of 16 mV/g (g is the acceleration due to gravity) was used to produce the oscilloscope trace shown in the drawing.

Find: The acceleration.

Solution: $a = \dfrac{V}{k}$

$a = 35 \text{ mV} \div 16 \text{ mV}/g$

$a = 2.19g$

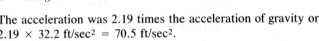

— 35 mV

The acceleration was 2.19 times the acceleration of gravity or 2.19×32.2 ft/sec^2 = 70.5 ft/sec^2.

SECTION 10.1 **EXERCISES**

1. Briefly describe the strain gage operation.

2. Define gage factor for a strain gage. State the ideal value of gage factor and list at least one typical real value.

3. List three types of materials used in strain gages. Name at least one specific metal used.

4. A strain gage has an initial resistance of 400 Ω. Under a strain of 10^{-3}, the resistance increases to 401 Ω. What is the gage factor?

5. The resistance of a wire is 10 Ω when the wire has a length of 2 m and a cross-sectional area of 0.003 m^2. What is the resistance of the wire when it is stretched to a length of 3 m and has a new cross-sectional area of 0.002 m^2? Assume that the electrical resistivity is unaffected by strain.

6. Find the fractional change in resistance ($\Delta R/R$) for a material subjected to a strain of 10^{-2}. (Take Poisson's ratio to be equal to 0.3.)

7. What is the resistance of a wire for which $G = 2$ if the wire had a resistance of 5 Ω before being subjected to a strain of 10^{-4}?

8. A Wheatstone bridge like the one shown in Figure 10.8 uses resistors such that $R_2 = 100 \ \Omega$ and $R_4 = 50 \ \Omega$. Find R_1 if the galvanometer registers no current flow when $R_3 = 200 \ \Omega$.

9. Sketch and label the five major components of a single-ended compression accelerometer, including the critical axis, and describe briefly how it works.

10. What pressure must be applied to a piezoelectric crystal of lithium sulfate 5 mm thick to produce an open circuit voltage of 200 V?

11. How much electrical energy can a quartz crystal put out if 1000 J of mechanical energy are applied to the crystal?

12. If the open circuit voltage resulting from an applied pressure of $10^7 \ N/m^2$ is 1960 V and the piezoelectric crystal is 2 mm thick, what material is the crystal probably made of? (See Table 10.2.)

13. Sketch and label the three major components of a shear-design accelerometer and its critical axis and give a brief description of how it works.

14. Sketch the voltage equivalent circuit of an accelerometer and label the signal source, capacitance, and resistance.

15. An accelerometer with a sensitivity of 15 mV/g (g is the acceleration due to gravity) produces a voltage of 0.55 V when in contact with an engine. What is the acceleration?

16. An accelerometer with a sensitivity of 12 mV/g is used to measure an acceleration of 2.6g. What voltage is produced?

17. What is the sensitivity of an accelerometer that produces a voltage of 10 mV when measuring an acceleration of 0.15g?

18. If an accelerometer has the same sensitivity as that of the one in Problem 16 and is used to measure an acceleration of 25 m/sec², what voltage is produced?

10.2

FLUID TRANSDUCERS

OBJECTIVES

Upon completion of this section, the student should be able to

• Describe the operation of a turbine flowmeter.

• Compare the operation and application of five types of pressure elements used as transducers.

• Given a specific application, select the appropriate pressure element.

• Describe the operation of an aneroid barometer.

DISCUSSION

Turbine (Fluid) Flowmeter

"Turbine flowmeters" are devices inserted into a pipe or tube to measure the rate of fluid flow. A multibladed turbine (Figure 10.17) rotates as the fluid flows past its blades. A coil of wire is placed next to the outside of the pipe. As the magnet rotates, it generates a short pulse of current in the coil by electromagnetic induction.

Each pulse represents passage of a known volume of fluid past the rotor. The pulses can be fed directly into a digital display that indicates total flow. Also, the pulses can be used to drive an electrical meter to indicate flow rate. Such devices

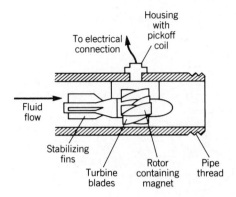

FIGURE 10.17 Turbine flowmeter measuring fluid flow.

are important in industry for measuring, recording, and controlling flow of chemicals, petroleum, cryogenic liquids, and the like. The following example shows how a known calibration can be converted into a flow rate.

EXAMPLE 10.4   Calculation of flow by flowmeter
. . . .

Given: A commercial turbine flowmeter calibrated at 1200 pulses/gal registers 13,200 pulses/min when placed into a water pipe.

Pulses per minute

Find: **a.** Flow rate.
b. Total flow (in gallons) in a 6-min period.

Solution: 13,200 pulses/min × 6 min = 79,200 pulses

Total flow = (79,200 pulses)/(1200 pulses/gal)

Total flow = 66 gal in 6 min

Flow rate = 66 gal/6 min

Flow rate = 11 gal/min

This same principle is used in a device called an "anemometer," which is used to measure wind speed.

Pressure Transducers

Bourdon Tubes   The bourdon tube is a pressure element that can be constructed with a "C"-tube, helical tube, or spiral tube (Figures 10.18 and 10.19). The tubes

FIGURE 10.18 A bourdon
C-tube pressure element.

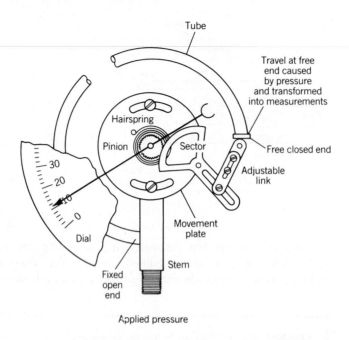

Tube

Travel at free
end caused
by pressure
and transformed
into measurements

Hairspring

Pinion   Sector   Free closed end

Adjustable
link

30
20

Movement
plate

Dial

Stem

Fixed
open
end

Applied pressure

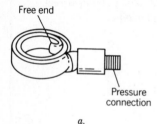

Free end

Pressure
connection

a.

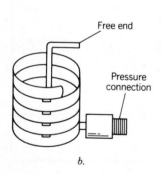

Free end

Pressure
connection

b.

FIGURE 10.19 Spiral and
helical bourdon tubes.

operate by transforming fluid work into mechanical work.

Pressure applied to the bourdon tube causes a movement of the free end. Since pressure applied through a distance represents fluid work (recall that $\Delta W = p\Delta V$), the fluid work is transformed into mechanical energy. This energy is stored as potential energy in the deformed spiral or helix. Therefore, the distortion is a direct measure of the pressure. The relationship between the pressure applied to the tube and the movement of the tube depends on the tube construction and shape. The operation is similar to a force applied to a spring and the opposition to movement by the spring. The spring moves until the force it resists is equal to the applied force.

The bourdon C-tube shown in Figure 10.18 is the most common type of pressure element, the easiest to construct, and, therefore, the least expensive. The distortion is rather small. The movement of the free end is usually amplified by use of levers and gears to cause a rotation of a shaft. When a pointer is connected to the shaft, the pointer moves along a calibrated scale.

Spiral (Figure 10.19a) and helical (Figure 10.19b) bourbon tubes operate on the same principle but have the advantage of producing more travel of the tip for a given applied pressure. The operation of a spiral element is similar to that of a "party favor," which tends to uncoil when pressure is applied (Figure 10.20). The amount of uncoiling is proportional to the pressure applied. However, the spiral element does not uncoil completely like the party favor mentioned, because a distortion of the metal would occur and the element would not return to its original shape. One end of the spiral is mounted rigidly to a frame. The movement at the free end is transmitted to a pointer that moves along a calibrated scale to indicate pressure. Table 10.3 gives the pressure range that can be measured with different spiral-element materials.

FIGURE 10.20 The effect of pressure on "party favor" is similar to the effect on bourdon tube.

TABLE 10.3 Pressure range of spiral elements

Bronze	0–10 to 0–200 psi
Beryllium–copper	0–10 to 0–200 psi
Ni-Span C alloy	0–10 to 0–200 psi
Type 316 SS	0–10 to 0–200 psi

TABLE 10.4 Pressure range of helical elements

Bronze	0–201 to 0–400 psi
Beryllium–copper	0–201 to 0–6000 psi
Ni-Span C alloy	0–201 to 0–6000 psi
Type 316 SS	0–201 to 0–6000 psi
Heavy-duty helical elements	
Type 316 SS	0–75 to 0–30,000 psi
	0–40,000 to 0–80,000 psi

The helical element (Figure 10.19*b*), instead of having coils in the same plane, is manufactured so that the coils are the same diameter. They extend a vertical distance of 1 to 5 inches and resemble a spring.

Helical elements normally are made of thicker material and so can be used for higher-pressure applications. Like the spiral elements, helical elements allow more movement at the tip than the C-tube. Sometimes they can drive an indicating device directly without the use of movement amplification by a link-and-lever system. Table 10.4 gives the pressure range for different helical element materials.

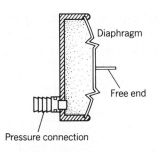

FIGURE 10.21 Diaphragm pressure element.

Diaphragm Elements For small pressure measurements, the diaphragm element may be used (Figure 10.21). The movement of the diaphragm's free end is directly proportional to the pressure. This type of energy transformer will produce a considerable movement when a small pressure is applied. Pressure gages using a diaphragm as an element can measure pressures as small as 0.30 pounds per square inch. Table 10.5 gives the pressure range for different diaphragm elements.

Bellows Elements The relationship between movement and pressure should be a direct proportionality for any type of pressure element. Ideally, equal increments of pressure should produce equal increments of movement at all points throughout the pressure range of the device. The bellows element is one of the most accurate pressure transducers because it most nearly satisfies this requirement (Figure 10.22). A bellows element in operation resembles a small accordion. When pressure is applied to a bellows element, the side of the element expands. When the pressure is removed, the element collapses to the original size. Table 10.6 gives the pressure range for different bellows element materials.

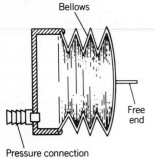

Bellows

Free end

Pressure connection

FIGURE 10.22 Bellows pressure element.

TABLE 10.5   Pressure range of diaphragm elements

| Size | Pressure Range | |
	Low	High
3-inch, Cu–Ni–Mn	0–9 inch H_2O (0–0.3 psi)	0–40 inch H_2O (0–1.4 psi)
2-inch, Cu–Ni–Mn	0–20 inch H_2O (0–0.7 psi)	0–138.7 inch H_2O (0–5 psi)

TABLE 10.6   Pressure range of bellows elements

| Type of Bellows Material | Pressure Range | |
	Low	High
Brass	0–3.6 psi (0–100 inch H_2O)	0–25 psi (0–693.3 inch H_2O)
Stainless steel (Type 316)	0–4.5 psi (0–124.8 inch H_2O)	0–29 psi (0–804.2 inch H_2O)

Pressure Switches   Pressure elements provide the driving force to operate pressure switches. For example, pressure switches can be used to start and stop water pumps and air compressors. Most pressure switches use a diaphragm or bellows, because a pressure applied to the large area of these elements can produce forces sufficient to operate the switch. For example, the switch shown in Figure 10.23 will maintain a minimum water pressure. When the pressure decreases, the bellows will contract, causing the electrical contact to close and start the motor to drive a pump. When the pump sufficiently increases the pressure, the switch will again open, stopping the pump. The adjustable spring is used to set the switching point.

Aneroid Barometer   Figure 10.24 shows a barometer used to measure atmospheric pressure. The aneroid capsule is a flexible, sealed metal can that is exhausted of air. When the external pressure changes, the height of the capsule

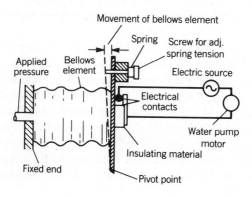

FIGURE 10.23 Bellows-driven pressure switch.

FIGURE 10.24 Aneroid barometer operation.

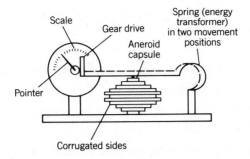

changes. This displacement is converted into a needle deflection by a gear drive. In some cases, the gear drive is connected to a variable resistor to transform the output into an electrical signal.

SECTION 10.2 *EXERCISES*

1. Discuss the operation and application of the following types of pressure elements used as transducers:
   Bourdon C-tube
   Bourdon spiral tube
   Bourdon helical tube
   Diaphragm element
   Bellows element

2. Compare the operation and application of the bourdon helical tube to that of the bellows element.

3. If a bellows moves 0.5 inch when the applied pressure is 10 psi, how much pressure is required to move the bellows 1 inch?

4. What is the total force applied to a diaphragm if the area of the diaphragm is 0.1 inch² and the fluid pressure is 0.3 psi?

5. How much fluid energy is required to change the volume of an element by 0.001 inch³ if the pressure causing the volume change is 10 psi?

6. A force of 5 N is applied to the fixed end of a bellows that has a cross-sectional area of 0.5m². What pressure is exerted on the fixed end of this bellows?

7. Convert 10 psi to units of N/m².

8. A turbine flowmeter is calibrated to give 1000 pulses/gal. If the counter registers 15,000 pulses/min, what is the flow rate?

10.3

ELECTRICAL TRANSDUCERS

OBJECTIVES..

Upon completion of this section, the student should be able to

- List three major types of meter movement and an application of each type.

- Draw and label a sketch that illustrates the major components of a stationary-magnet, moving-coil meter movement, and explain its operation.

- Draw and label diagrams of the circuits used in the following types of meters:

Voltmeter

Ammeter

Ohmmeter

- Given the meter's properties, calculate the required shunt resistance for a given maximum current.

- Given the meter characteristics, calculate the required series resistance for a given maximum voltage.

- Design an ohmmeter given the meter characteristics.

DISCUSSION

Meter Movements

Introduction   The heart of many electrical measuring instruments is the meter movement. Meter movements are transducers that use electrical and mechanical energy. Either a coil or magnet is free to rotate in a magnetic field. For either case, the moving element tends to align its magnetic field with the stationary field. This rotation, however, is opposed by a spring. A pointer attached to the rotating element indicates the value of the quantity measured. There are three major types of meter movements.

- Stationary-magnet, moving-coil mechanisms. They are the most common. They consist of a stationary magnet and a coil free to rotate through an angle. These meters provide high sensitivity and rapid response because of the small mass that must be moved.

- Stationary-coil, moving-magnet mechanisms. They are less sensitive and slower to respond, but are more rugged. They sometimes are used in the measurement of large currents.

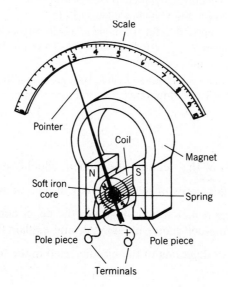

FIGURE 10.25   Stationary-magnet, moving-coil meter.

- Stationary-coil, moving-coil mechanisms. They are used in the measurement of ac quantities. The most common example is the ac wattmeter discussed in Chapter 7.

All these meter movements consist of a movable magnetic field suspended in a stationary magnetic field. The remainder of this section is a description of the operation of the most common type: stationary-magnet, moving-coil mechanisms. The other types of meter movements operate according to the same principles.

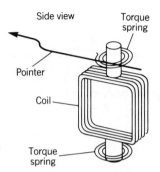

FIGURE 10.26 Movable coil with springs and pointer.

Stationary-Magnet, Moving-Coil Meters The stationary-magnet, moving-coil instrument illustrated in Figure 10.25 is used to measure single direction currents, such as those supplied by a battery, rectifier, or dc generator. A strong permanent magnet provides a fixed magnetic field. The field is concentrated and made uniform across the air gap by soft iron pole pieces and a soft iron stationary core.

A coil that rotates on jeweled bearings is located in the gap between the pole pieces and the core (Figure 10.26). This moving element must be light in weight to respond quickly to changing current and be strong enough to resist the forces of overloads.

With no current, the coil is held in the "zero" position by coil springs. When a current exists, an electromagnetic field is created by the movable coil (Figure 10.27). The coil rotates, attempting to align its magnetic field with the stationary magnetic field. However, the rotation is opposed by the coil springs. The energy of the magnetic field is converted to potential energy of the coiled springs. A pointer appropriately attached to the coil moves around a calibrated scale, indicating the value of the current through the coil.

All such mechanisms respond to the magnitude of current through the coil. The meter scale, however, may be calibrated to indicate a measurement of any quantity proportional to the coil current (for example, sound level or light intensity).

Ammeters When the direct current exceeds the current rating of the meter movement, a shunt resistor must be used. The shunt resistor has less resistance than the coil. Thus a significant portion of the current will be diverted through the shunt, and the remaining current through the meter. Figure 10.28 illustrates such an arrangement.

Recall from Chapter 3 that the charge will be conserved, so that

$$I_t = I_m + I_s \tag{10.23}$$

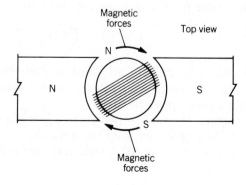

FIGURE 10.27 Magnetic forces result from coil current.

FIGURE 10.28 Meter with shunt resistor functions as ammeter.

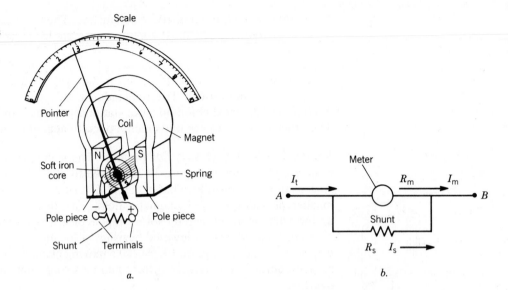

a.

b.

where

I_t = total line current
I_m = current through the meter
I_s = current through the shunt

In addition, the voltage across the meter must equal the voltage across the shunt. For this condition, Ohm's law yields

$$I_s R_s = I_m R_m \tag{10.24}$$

where

R_s = resistance of the shunt
R_m = resistance of the meter

When the maximum current that the coil can withstand, I_m, is known, one can determine the shunt resistance required to allow a measurement of total current I_t.

Solving Equation 10.24 for I_s yields

$$I_s = \frac{R_m}{R_s} I_m \tag{10.25}$$

Substituting this value for I_s into Equation 10.23 gives

$$I_t = \left(1 + \frac{R_m}{R_s}\right) I_m \tag{10.26}$$

Thus, to obtain the total current, one must multiply the meter current by $(I + R_m/R_s)$. Often R_m/R_s is much greater than 1, so that multiplying the meter current by R_m/R_s yields a good approximation.

Shunts are usually designed so that $(1 + R_m/R_s)$ is a factor of 10. The same meter scale can then be easily used for a number of different shunts. Shunts capable of carrying as much as 100 A frequently are assembled inside instrument cases. By selecting various shunts using a selector switch a wide range of current can be measured. A sample problem is given in Example 10.5.

EXAMPLE 10.5 Calculation of shunt resistance
. . . .

Given: A meter movement has a full scale reading for 1.00 mA and a resistance of 100 Ω.

Find: **a.** The required shunt resistance to extend the range to 10.0 mA.
b. The required shunt resistance to extend the range to 10.0 A.

Solution: **a.** From Equation 10.23, the shunt current is

$$I_s = I_t - I_m$$

$$I_s = 10.0 - 1.00 = 9.0 \text{ mA}$$

From Equation 10.24

$$R_s = \frac{I_m R_m}{I_s}$$

$$R_s = \frac{(1.00 \times 10^{-3})(1.00 \times 10^2)}{9.0 \times 10^{-3}}$$

$$R_s = 11.1 \ \Omega$$

b. The shunt current is

$$I_s = 10.0 - 0.001 \approx 10.0 \text{ A}$$

$$R_s = \frac{(1.00 \times 10^{-3})(1.00 \times 10^2)}{10.0}$$

$$R_s = 0.010 \ \Omega$$

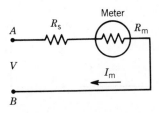

FIGURE 10.29 Meter with series resistance functions as voltmeter.

Voltmeters Stationary-magnet, moving-coil meters can be modified to measure dc voltages. For this case, the instrument coil is connected in series with a resistor to limit the current to the meter. Figure 10.29 shows the schematic of a dc voltmeter. The total resistance is

$$R = R_s + R_m \tag{10.27}$$

where

R = total resistance
R_s = series resistance
R_m = meter resistance

The corresponding voltage by Ohm's law is

$$V = I_m R = I_m(R_m + R_s) \tag{10.28}$$

where

I_m = the meter current

Thus for a maximum coil current of I_m, the series resistance R_s must be selected to measure the voltage V without meter damage. For example, a common meter current maximum is 1.00×10^{-3} A. For this meter, 1000 Ω of series resistance are needed for each volt. Another common meter gives a full scale reading for a current of 50×10^{-6} A. In this meter movement, 20,000 Ω of series resistance are needed for each volt. The higher the number of ohms per volt, the less the meter will affect a circuit. Series resistors or "voltage-multiplying resistors" may be assembled within the instrument case to provide several voltage ranges for the meter. A sample problem is given in Example 10.6.

EXAMPLE 10.6   Calculation of voltage multiplier resistance
. . . .

Given: A meter movement gives a full scale reading at 1.00 mA and has a resistance of 50.0 Ω.

Find: The required series resistance needed to give a full scale reading of 1.00 V.

Solution: From Equation 10.28

$$V = I_m (R_m + R_s)$$

or

$$R_m + R_s = \frac{V}{I_m}$$

or

$$R_s = \frac{V}{I_m} - R_m$$

$$R_s = \frac{1.00 \text{ V}}{1.00 \times 10 \text{ A}^{-3}} - 50.0 \ \Omega$$

$$R_s = 1000 \ \Omega - 50.0 \ \Omega$$

$$R_s = 950 \ \Omega$$

Ohmmeters   An ohmmeter can also be constructed from a stationary-magnet, moving-coil meter movement. A schematic is shown in Figure 10.30. The instrument coil is connected in series with a battery, series resistance, a variable resistance R_v, and the unknown resistance R_x. With points A and B shorted, the variable resistance is adjusted to give a full scale meter reading. This reading indicates zero resistance between points A and B. If a resistance R_x is connected between A and B, the meter reading will be less. The meter scale can be calibrated to read ohms directly. This scale, however, is not linear. For example, if the

FIGURE 10.30 Meter with battery and series resistance functions as ohmmeter.

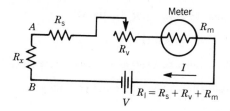

$$R_1 = R_s + R_v + R_m$$

unknown resistance is equal to the internal resistance, the current will be one-half full scale. If the external resistance is two times the internal resistance, the current will be one-third full scale.

From Figure 10.30, Ohm's law gives

$$R_x + R_I = \frac{V}{I} \tag{10.29}$$

where

R_x = unknown resistance
$R_I = R_s + R_v + R_m$, the internal resistance

When points A and B are shorted, $R_x = 0$, and the meter should read full scale. Equation 10.29 becomes

$$R_I = \frac{V}{I_m} \tag{10.30}$$

where

I_m = full scale coil current

Solving Equation 10.30 for V and substituting this expression in Equation 10.29 yields

$$R_x = \left(\frac{I_m}{I} - 1 \right) R_I \tag{10.31}$$

Thus, when $I = I_m$, $R_x = 0$, and when $I = I_m/2$, $R_x = R_I$. The required circuit voltage can be obtained from Equation 10.30. Example 10.7 illustrates a particular problem.

EXAMPLE 10.7 The ohmmeter
. . . .

Given: A meter movement has a full scale current of 1 ma and a resistance of 100 Ω. The meter is used to make an ohmmeter with a half-scale reading of 1000 Ω.

Find: The required series resistance and voltage.

Solution: From Equation 10.31, a half-scale reading occurs when the resistance is equal to R_I. Thus,

$$R_I = R_s + R_v + R_m$$

$$1000 \ \Omega = R_\mathrm{s} + R_\mathrm{v} + 100 \ \Omega$$

$$R_\mathrm{s} + R_\mathrm{v} = 900 \ \Omega$$

900 Ω must be used as the series resistance.

From Ohm's law, the required voltage is

$$V = I_\mathrm{m} R_\mathrm{I}$$

$$V = (10 \ \mathrm{A}^{-3})(10 \ \Omega^3)$$

$$V = 1 \ \mathrm{V}$$

SECTION **10.3** *EXERCISES*

1. List three types of meter movements and an application of each.

2. Identify the components of the accompanying meter movement.

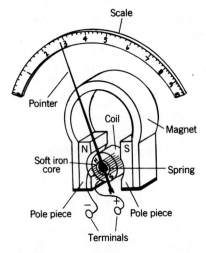

3. Explain how a stationary-magnet, moving-coil meter operates. Include a discussion of the meter movement as an energy convertor.

4. Correctly identify the following meter circuits.

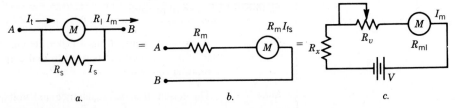

       a.   *b.*   *c.*

5. In a circuit like the one shown in Figure 10.29, find the voltage between points A and B if $I_\mathrm{t} = 0.001$ A, $R_\mathrm{s} = 10{,}000 \ \Omega$, and the voltage drop across the meter is 0.1 V.

6. Consider a circuit like the one shown in Figure 10.30. Let $V = 12$ V and $R_\mathrm{I} = 1000 \ \Omega$. If $I_\mathrm{t} = 10$ mA, what is the value of R_x?

7. Using a 1.00×10^{-3}-A and 100-Ω meter movement design a volt-ohmammeter having a voltage range of 0 to 100 V, a current range of 0 to 1.00 A, and a reading of 1000 Ω at half-scale reading. Draw a schematic for each and label each component.

10.4

THERMAL TRANSDUCERS

OBJECTIVES

Upon completion of this module, the student should be able to

*   Sketch a typical thermocouple circuit and briefly describe how it is used to measure temperature.

*   Given the temperature of the reference junction, determine the following from calibration tables for a standard thermocouple:
    Temperature of the test junction from the thermoelectric emf.
    Thermoelectric emf from the temperature of the test junction.

*   Calculate the thermoelectric emf of a nonstandard thermocouple.

*   Determine the temperature of the test junction of a thermopile from the thermoelectric voltage.

*   Calculate the thermoelectric power of a thermocouple, using calibration tables.

*   Define the terms
    Thermistor
    Self-heated mode
    Externally heated mode
    Resistance–temperature characteristic
    Voltage–current characteristic
    Current–time characteristic

*   Describe how the thermistor is used for temperature measurement, temperature control, liquid-level control, and time delay.

*   Describe in a short paragraph the theory and operation of a bimetallic strip and indicate how it can be used in a practical situation.

*   Sketch a fluid-filled thermometric system and label the pointer, pressure-sensitive element, scale, capillary tube, and fluid-filled bulb. Describe in one or two sentences the function of each.

DISCUSSION

Thermocouples

Introduction A thermocouple usually consists of junctions of two dissimilar metals *A* and *B,* as illustrated in Figure 10.31*a.* Two such junctions are used to construct a thermocouple thermometer, as represented in Figure 10.31*b.* The reference junction is maintained at a constant, known temperature T_R and the test junction is at the unknown temperature *T.*

FIGURE 10.31   Thermocouple
junction and schematic of
operation.

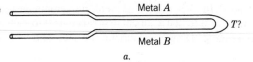

a.

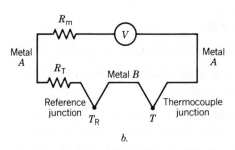

b.

A thermal emf V (also referred to as the "thermoelectric voltage" and the "thermoelectric emf") is developed across the junctions. This voltage is proportional to the difference in temperature $(T - T_R)$ between the two junctions.

Large errors in temperature measurements occur if a proper method is not used to measure the voltage. To illustrate this error, consider the schematic shown in Figure 10.31*lb*. The total resistance of the circuit consists of the meter resistance R_m and the external resistance R_T due to the wire resistance in series. The current in the circuit is

$$I = \frac{V_T}{R_m + R_T}$$
(10.32)

where

V_T = thermocouple voltage

The voltage across the meter is

$$V_m = IR_m = \left(\frac{V_T}{R_m + R_T}\right)R_m = V_T\left(\frac{R_m}{R_m + R_T}\right)$$
(10.33)

If the meter resistance R_m is much larger than the wire resistance R_T, the meter voltage will be approximately equal to the thermocouple voltage. However, millivoltmeters with ordinary meter movements have low resistances. When these meters are used, the scale is calibrated in temperature for a particular type thermocouple and resistance, R_T. It may be necessary to adjust R_T to obtain correct temperature readings. A sample problem is given in Example 10.8.

EXAMPLE 10.8   Measurement of thermocouple voltage
. . . .

Given: A millivoltmeter has a full scale reading of 10 mV and a resistance of 10.0 Ω. The meter is connected to a thermocouple with a resistance of 1.0 Ω.

Find: The error in the thermocouple voltage reading when the meter reads 5.0 mV.

Solution: From Equation 10.33

$$V_m = \frac{V_T R_m}{R_m + R_T}$$

Solving for V_T yields

$$V_T = \frac{(R_m + R_T)V_m}{R_m}$$

$$V_T = \frac{(10\ \Omega + 1.0\ \Omega)(5.0 \times 10^{-3}\ V)}{10\ \Omega}$$

$$V_T = 5.5 \times 10^{-3}\ V$$

The percent error is

$$\% \text{ error} = \left(\frac{5.5 \times 10^{-3} - 5.0 \times 10^{-3}}{5.5 \times 10^{-3}}\right)100$$

$$\% \text{ error} = 9\%$$

The problem can be eliminated by using an instrument with a very high input impedance such as a potentiometer. For this case $R_m/(R_m + R_T) \approx 1$.

A simple potentiometer is illustrated in Figure 10.32. Slide-wire potentiometers can measure unknown voltages with very small error if a good standard cell is used. With the standard cell in the circuit (switch 1 closed and switch 2 open), the sliding contact is adjusted until no current flows through the galvanometer. Under this condition, the voltage between points A and S is equal to the standard cell voltage. For a uniform wire, the resistance between A and S is proportional to the wire length:

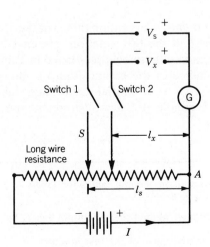

FIGURE 10.32 Schematic of a slide-wire potentiometer to measure unknown voltage.

$$R_s = kl_s \qquad (10.34)$$

where

k = constant
R_s = wire resistance between A and S
l_s = wire length between A and S

From Ohm's law, the corresponding voltage may be expressed as

$$V_s = IR_s = Ikl_s \qquad (10.35)$$

where

V_s = known standard voltage
I = current through the wire

Switch 1 is then opened and switch 2 is closed, and the slide is adjusted until no current flows through the galvanometer. From Ohm's law

$$V_x = IR_x = Ikl_x \qquad (10.36)$$

where

R_x = wire resistance between A and X
l_x = wire length between A and X

Dividing Equation 10.36 by 10.35 yields

$$\frac{V_x}{V_s} = \frac{l_x}{l_s}$$

or

$$V_x = \frac{l_x}{l_s} V_s \qquad (10.37)$$

Thus the unknown voltage is proportional to the standard voltage, the lengths of wire being determined by experiment. Potentiometers can be designed to measure voltages to an accuracy of 0.01%.

Thermocouple Calibration   Calibration tables of thermoelectric voltage as a function of temperature T of the test junction referenced to $T_R = 0°C$ or 32°F are published for the standard thermocouples listed in Table 10.7 by type and components. Table 10.8 indicates the composition of alloys commonly used in thermocouples. Table 10.9 is a partial calibration table for a chromel–alumel thermocouple and Example 10.9 demonstrates its use.

EXAMPLE 10.9   Use of thermocouple calibration tables
. . . .

Given: A chromel–alumel thermocouple has $T_R = 32°F$ and $V = 2.73$ mV.

Find: The temperature of the test junction.

Solution: The temperature in increments of 10F° is listed in the first column of Table 10.9. Succeeding columns list the thermoelectric voltage as the temperature is increased in steps of 1F°. The emf of 2.73 mV is located in the row labeled 150°F and in the column labeled 3°F. Therefore

$$T = 150°F + 3°F$$

$$T = 153°F$$

TABLE 10.7   Accepted thermocouple designations

Type	Alloy Composition
J	Iron–constantan
K	Chromel–alumel
E	Chromel–constantan
T	Copper–constantan
S	Platinum vs. platinum— 10% rhodium
R	Platinum vs. platinum— 13% rhodium
B	Platinum—6% rhodium vs. platinum—30% rhodium

TABLE 10.8   Composition of thermocouple alloys

Alloy Name	Composition
Chromel	Nickel–chromium
Alumel	Nickel–aluminum
Constantan	Copper–nickel

TABLE 10.9   Sample calibration table for type K thermocouples[a]

°F	0	1	2	3	4	5	6	7	8	9
−300	−5.51	−5.52	−5.53	−5.54	−5.54	−5.55	−5.56	−5.57	−5.58	−5.59
−290	−5.41	−5.42	−5.43	−5.44	−5.45	−5.46	−5.47	−5.48	−5.49	−5.50
−280	−5.30	−5.31	−5.32	−5.34	−5.35	−5.36	−5.37	−5.38	−5.39	−5.40
−270	−5.20	−5.21	−5.22	−5.23	−5.24	−5.25	−5.26	−5.27	−5.28	−5.29
−260	−5.08	−5.09	−5.10	−5.12	−5.13	−5.14	−5.15	−5.16	−5.17	−5.18
−250	−4.96	−4.97	−4.99	−5.00	−5.01	−5.02	−5.03	−5.04	−5.06	−5.07
−240	−4.84	−4.85	−4.86	−4.88	−4.89	−4.90	−4.91	−4.92	−4.94	−4.95
−230	−4.71	−4.72	−4.74	−4.75	−4.76	−4.77	−4.79	−4.80	−4.81	−4.82
−220	−4.58	−4.59	−4.60	−4.62	−4.63	−4.64	−4.66	−4.67	−4.68	−4.70
−210	−4.44	−4.45	−4.46	−4.48	−4.49	−4.51	−4.52	−4.53	−4.55	−4.56

continued

°F	0	1	2	3	4	5	6	7	8	9
−200	−4.29	−4.31	−4.32	−4.34	−4.35	−4.36	−4.38	−4.39	−4.41	−4.42
−190	−4.15	−4.16	−4.18	−4.19	−4.21	−4.22	−4.24	−4.25	−4.26	−4.23
−180	−4.00	−4.01	−4.03	−4.04	−4.06	−4.07	−4.09	−4.10	−4.12	−4.13
−170	−3.84	−3.86	−3.88	−3.89	−3.91	−3.92	−3.94	−3.95	−3.97	−3.98
−160	−3.69	−3.70	−3.72	−3.73	−3.75	−3.76	−3.78	−3.80	−3.81	−3.83
−150	−3.52	−3.54	−3.56	−3.57	−3.59	−3.60	−3.62	−3.64	−3.65	−3.67
−140	−3.36	−3.38	−3.39	−3.41	−3.42	−3.44	−3.46	−3.47	−3.49	−3.51
−130	−3.19	−3.20	−3.22	−3.24	−3.25	−3.27	−3.29	−3.31	−3.32	−3.34
−120	−3.01	−3.03	−3.05	−3.06	−3.08	−3.10	−3.12	−3.13	−3.15	−3.17
−110	−2.83	−2.85	−2.87	−2.89	−2.90	−2.92	−2.94	−2.96	−2.98	−2.99
−100	−2.65	−2.67	−2.69	−2.71	−2.72	−2.74	−2.76	−2.78	−2.80	−2.82
−90	−2.47	−2.49	−2.50	−2.52	−2.54	−2.56	−2.58	−2.60	−2.62	−2.63
−80	−2.28	−2.30	−2.32	−2.34	−2.36	−2.37	−2.39	−2.41	−2.43	−2.45
−70	−2.09	−2.11	−2.13	−2.15	−2.17	−2.18	−2.20	−2.22	−2.24	−2.26
−60	−1.90	−1.92	−1.94	−1.96	−1.97	−1.99	−2.01	−2.03	−2.05	−2.07
−50	−1.70	−1.72	−1.74	−1.76	−1.78	−1.80	−1.82	−1.84	−1.86	−1.88
−40	−1.50	−1.52	−1.54	−1.56	−1.58	−1.60	−1.62	−1.64	−1.66	−1.68
−30	−1.30	−1.32	−1.34	−1.36	−1.38	−1.40	−1.42	−1.44	−1.46	−1.48
−20	−1.10	−1.12	−1.14	−1.16	−1.18	−1.20	−1.22	−1.24	−1.26	−1.28
−10	−0.89	−0.91	−0.93	−0.95	−0.97	−0.99	−1.01	−1.03	−1.06	−1.08
(−)0	−0.68	−0.70	−0.72	−0.75	−0.77	−0.79	−0.81	−0.83	−0.85	−0.87
(+)0	−0.68	−0.66	−0.64	−0.82	−0.60	−0.58	−0.56	−0.54	−0.52	−0.49
10	−0.47	−0.45	−0.43	−0.41	−0.39	−0.37	−0.34	−0.32	−0.30	−0.28
20	−0.26	−0.24	−0.22	−0.19	−0.17	−0.15	−0.13	−0.11	−0.09	−0.07
30	−0.04	−0.02	0.00	0.02	0.04	0.07	0.09	0.11	0.13	0.15
40	0.18	0.20	0.22	0.24	0.26	0.29	0.31	0.33	0.35	0.37
50	0.40	0.42	0.44	0.46	0.48	0.51	0.53	0.55	0.57	0.60
60	0.62	0.64	0.66	0.68	0.71	0.73	0.75	0.77	0.80	0.82
70	0.84	0.86	0.88	0.91	0.93	0.95	0.97	1.00	1.02	1.04
80	1.06	1.09	1.11	1.13	1.15	1.18	1.20	1.22	1.24	1.27
90	1.29	1.31	1.33	1.36	1.38	1.40	1.43	1.45	1.47	1.49
100	1.52	1.54	1.56	1.58	1.61	1.63	1.65	1.68	1.70	1.72
110	1.74	1.77	1.79	1.81	1.84	1.86	1.88	1.90	1.93	1.95
120	1.97	2.00	2.02	2.04	2.06	2.09	2.11	2.13	2.16	2.18
130	3.20	2.23	2.25	2.27	2.29	2.32	2.34	2.36	2.39	2.41
140	2.43	2.46	2.48	2.50	2.52	2.55	2.57	2.59	2.62	2.64
150	2.66	2.69	2.71	2.73	2.75	2.78	2.80	2.82	2.85	2.87
160	2.89	2.92	2.94	2.96	2.98	3.01	3.03	3.05	3.06	3.10
170	3.12	3.15	3.17	3.19	3.22	3.24	3.26	3.29	3.31	3.33
180	3.36	3.38	3.40	3.43	3.45	3.47	3.49	3.52	3.54	3.56
190	3.59	3.61	3.63	3.65	3.68	3.70	3.73	3.75	3.77	3.80
200	3.52	3.84	3.87	3.89	3.91	3.94	3.96	3.98	4.01	4.03
210	4.06	4.08	4.10	4.12	4.15	4.17	4.19	4.21	4.24	4.26
220	4.28	4.31	4.33	4.35	4.38	4.40	4.42	4.44	4.47	4.49
230	4.51	4.54	4.56	4.58	4.61	4.63	4.65	4.67	4.70	4.72
240	4.74	4.77	4.79	4.81	4.83	4.86	4.88	4.90	4.92	4.95
250	4.97	4.99	5.02	5.04	5.06	5.08	5.11	5.13	5.15	5.17
260	5.20	5.22	5.24	5.26	5.29	5.31	5.33	5.35	5.38	5.40
270	5.42	5.44	5.47	5.49	5.51	5.53	5.56	5.58	5.60	5.62

continued

°F	0	1	2	3	4	5	6	7	8	9
280	5.65	5.67	5.69	5.71	5.73	5.76	5.78	5.80	5.82	5.85
290	5.87	5.90	5.91	5.93	5.94	5.96	6.00	6.02	6.05	6.07
300	6.08	6.11	6.13	6.16	6.18	6.20	6.22	6.25	6.27	6.29
310	6.31	6.33	6.36	6.38	6.40	6.42	6.45	6.47	6.49	6.51
320	6.53	6.56	6.58	6.60	6.62	6.65	6.67	6.69	6.71	6.73
330	6.76	6.78	6.80	6.82	6.84	6.87	6.89	6.91	6.93	6.96
340	6.96	7.00	7.02	7.04	7.07	7.09	7.11	7.13	7.15	7.18
350	7.20	7.22	7.24	7.26	7.29	7.31	7.33	7.35	7.38	7.40

[a] Electromotive force in absolute millivolts, temperatures in °F, and reference junctions at 32°F.

If the temperature of the reference junction is not 32°F, the temperature of the test junction still can be found from the tables referenced to 32°F. Examples 10.10 and 10.11 illustrate this procedure.

EXAMPLE 10.10   Arable reference temperature
. . . .

Given: A chromel–alumel thermocouple has $T_R = 212°F$ and $V = 3.28$ mV.

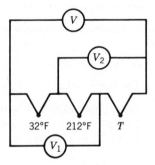

Find: Temperature T of the test junction.

Solution: The thermocouple circuit can be represented by three junctions, at 32°F, at 212°F, and at the unknown temperature T, all in series. The thermoelectric voltage is the sum of the voltage V_1 between 32°F and 212°F and the voltage V_2 between 212°F and T. From Table 10.9, $V_1 = 4.10$ mV; therefore, the sum V is

$$V = V_1 + V_2$$

$$V = 4.10 \text{ mV} + 3.28 \text{ mV}$$

$$V = 7.38 \text{ mV}$$

According to Table 10.9, the temperature corresponding to $V = 7.38$ mV is $T = 368°F$.

EXAMPLE 10.11   Single-junction thermocouple

. . . .

Given: The thermoelectric voltage across a single-junction thermocouple is 2.80 mV. Room temperature is 72°F.

Find: The temperature of the test junction.

Solution: The reference junction is essentially the connection between the thermocouple and the meter. So the reference temperature for a single-junction thermocouple is the temperature at the meter, or room temperature. Therefore

$$V = V_1 + V_2$$

$$V = 0.88 \text{ mV} + 2.80 \text{ mV}$$

$$V = 3.68 \text{ mV}$$

This value corresponds to $T = 194°F$.

To calibrate a thermocouple for use at reference temperatures other than 32°F is often desirable. Example 10.12 illustrates this procedure.

EXAMPLE 10.12   Calculation of thermoelectric voltage with known temperatures

. . . .

Given: A chromel–alumel thermocouple has $T_R = 212°F$ and $T = 150°F$.

Find: The thermoelectric voltage V_2 generated.

Solution: From Table 10.9, $V = 4.10$ mV between 32°F and 212°F, and $V = 2.66$ mV between 32°F and 150°F. (Refer back to the sketch in Example 10.10.) Therefore

$$V = V_1 + V_2$$

$$V_2 = V - V_1$$

$$V_2 = 2.66 \text{ mV} - 4.10 \text{ mV}$$

$$V_2 = -1.44 \text{ mV}$$

The sign reversal indicates that the test junction at 150°F is cooler than the reference junction at 212°F. Using a broad range of temperatures T, one can calibrate the thermocouple meter using the new reference temperature T_R.

Figure 10.33 shows two separate thermocouple schematics with test and reference junctions at the same temperatures T and T_R. One thermocouple is

FIGURE 10.33 Two thermocouple pairs of different metals.

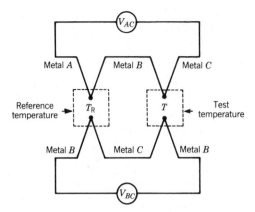

composed of metals A and C and the other of metals B and C. It may be that metal C is not available. For a thermocouple composed of metals A and B, thermoelectric emf is

$$V_{AB} = V_{AC} - V_{BC} \qquad (10.38)$$

where

V_{AB} = thermoelectric emf developed in a thermocouple of materials A and B
V_{AC} = thermoelectric emf developed in a thermocouple of materials A and C
V_{BC} = thermoelectric emf developed in a thermocouple of materials B and C

This result enables the tabulation of V as a function of T for any thermocouple made of two nonstandard thermocouple metals. Example 10.13 illustrates the calibration of such a nonstandard thermocouple.

EXAMPLE 10.13 Calibration of nonstandard thermocouple
....

Given: A cooper–constantan thermocouple with $T_R = 32°F$ and $V = 1.75$ mV at $T = 110°F$ and an iron–constantan thermocouple with $T_R = 32°F$ and $V = 2.23$ mV at $T = 110°F$.

Find: The thermoelectric emf generated across an iron–copper thermocouple at $T = 110°F$ referenced at $32°F$.

Solution: Referring to Equation 10.38, A = iron (Fe), B = copper (Cu), and C = constantan.

$$V_{Fe-Cu} = (V_{Fe-C}) - (V_{Cu-C})$$

$$V_{Fe-Cu} = 2.23 \text{ mV} - 1.75 \text{ mV}$$

$$V_{Fe-Cu} = 0.48 \text{ mV}$$

Even when thermocouple tables are unavailable, there is a way to compute the thermal emf of a thermocouple provided the temperatures to be measured are not

TABLE 10.10   Constants for lead combination thermocouples

Metal M	a ($\mu V/°C$)	b ($\mu V/°C^2$)
Aluminum (Al)	-0.47	0.003
Bismuth (Bi)	-43.7	-0.47
Copper (Cu)	2.76	0.012
Gold (Au)	2.90	0.0093
Iron (Fe)	16.6	-0.030
Nickel (Ni)	19.1	-0.030
Platinum (Pt)	-1.79	-0.035
Silver (Ag)	2.50	0.012
Steel	10.8	-0.016

higher than a few hundred degrees. The thermoelectric emf for a thermocouple consisting of lead (Pb) and any metal M referenced to 0°C can be approximated by

$$V_{M-Pb} = aT + 1/2bT^2 \tag{10.39}$$

where

V_{M-Pb} = thermoelectric voltage
a and b = constants
T = temperature of the test junction in degrees Celsius

Table 10.10 lists values for the constants a and b for several metals. Example 10.14 illustrates how the thermoelectric emf for an arbitrary thermocouple is found by using Equations 10.38 and 10.39.

EXAMPLE 10.14   Calculation of thermoelectric voltage
· ·

Given: An iron–gold thermocouple has T_R = 0°C (32°F) and T = 100°C (212°F).

Find: The thermoelectric emf V.

Solution: For iron (Fe), a = 16.6 $\mu V/°C$ and b = -0.030 $\mu V/°C^2$. For an iron–lead thermocouple

$$V_{Fe-Pb} = aT + 1/2\, bT^2$$

$$V_{Fe-Pb} = (16.6\ \mu V/°C)(100°C) + 1/2\ (-0.030\ \mu V/°C^2)(100°C)^2$$

$$V_{Fe-Pb} = 1510\ \mu V$$

$$V_{Fe-Pb} = 1.510\ mV$$

For gold (Au), a = 2.90 $\mu V/°C$ and b = 0.0093 $\mu V/°C^2$.

$$V_{Au-Pb} = aT + 1/2\, bT^2$$

$$V_{Au-Pb} = (2.90\ \mu V/^\circ C)(100^\circ C) + 1/2\ (0.0093\ \mu V/^\circ C^2)(100^\circ C)^2$$

$$V_{Au-Pb} = 336.5\ \mu V$$

$$V_{Au-Pb} = 0.3365\ mV$$

From Equation 10.38

$$V_{Fe-Au} = (V_{Fe-Pb}) - (V_{Au-Pb})$$

$$V_{Fe-Au} = 1.510\ mV - 0.3365\ mV$$

$$V_{Fe-Au} = 1.1735\ mV$$

A thermopile consists of many thermocouples connected in series. The thermoelectric voltage of the thermopile is the sum of the individual emfs and, therefore, is a larger voltage, and much easier to measure. Thermopiles can be used for detecting and measuring low-level radiant energy, such as that from stars. Example 10.15 demonstrates how the temperature is determined from thermopile voltage.

EXAMPLE 10.15   Thermopile voltage
. . . .

Given: The thermoelectric voltage across a thermopile composed of five chromel–constantan thermocouples in series is 0.1 V with reference to 0°C.

Find: The temperature of the test junction.

Solution: The thermoelectric voltage of a single thermocouple is

$$V = 0.1\ V/5$$

$$V = 0.02\ V$$

$$V = 20.0\ mV$$

The temperature corresponding to 20.0 mV in appropriate calibration tables for chromel–constantan is 287°C.

When thermocouple wires are chosen, the calibration tables should be inspected to determine which type covers the range of temperatures to be measured and which is most sensitive to changes in temperature in that range. A useful quantity in evaluating the sensitivity of a particular thermocouple is thermoelectric power, often referred to as "thermopower." It is defined as the change in thermoelectric emf per degree change in temperature. Mathematically, this relationship is stated by

$$S = \frac{\Delta V}{\Delta T} \qquad (10.40)$$

where

S = thermopower
ΔV = change in thermoelectric voltage
ΔT = change in temperature of the test junction

The larger the thermopower at a given temperature, the more sensitive is the thermocouple. Table 10.11 compares thermopowers of several types of thermocouples at 500°C (932°F) and at -100°C (-148°F). Example 10.16 demonstrates computation of thermopower.

EXAMPLE 10.16   Thermopower computation
. . . .

Given: The thermoelectric voltage V for a chromel–alumel thermocouple is 20.65 mV at 500°C and 20.69 mV at 501°C.

Find: The thermopower S.

Solution: $S = \dfrac{\Delta V}{\Delta T}$

$S = \dfrac{(20.69 \text{ mV} - 20.65 \text{ mV})}{(501°\text{C} - 500°\text{C})}$

$S = 0.04$ mV/°C

Thermistors

Introduction   The resistance of some electrical components change with temperature to such a degree that measurement of resistance can be used to determine very small temperature variations. Thermistors are such components and are used to measure temperature when connected to a bridge circuit. For example, the Wheatstone bridge circuit can be used to make accurate measurements of re-

TABLE 10.11   Thermopowers

Type	Composition	S (mV/°C) at 500°C	S (mV/°C) at −100°C
J	Iron–constantan	0.060	0.040
K	Chromel–alumel	0.040	0.030
T	Copper–constantan	—[a]	0.030
S	Platinum vs. platinum —10% rhodium	0.010	—[a]

[a] Not tabulated at this temperature.

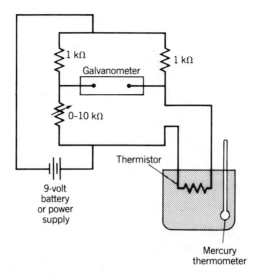

FIGURE 10.34 Wheatstone bridge used to measure thermistor resistance.

sistance (Figure 10.34). As discussed in Section 5.3, the variable resistance is adjusted to yield no deflection of the galvanometer. The thermistor resistance is then equal to the variable resistance.

Temperature Coefficient   The thermistor [therm(o) + (res)istor] is a semiconductor device. It has the unique property that its resistance varies with temperature, increasing in some types and decreasing in others. Resistance in any electrical resistor will change as the temperature varies, but the resistance change generally is small. However, the resistance of a thermistor changes substantially with temperature variations. Thus, the thermistor behaves as a resistor with a large temperature coefficient of resistance.

The temperature coefficient of resistance of a material is given by

$$\alpha = \left(\frac{\Delta R}{R_s}\right)\left(\frac{1}{\Delta T}\right) \qquad (10.41)$$

where

α = temperature coefficient of resistance
ΔR = change in resistance when temperature is changed by an amount ΔT
R_s = resistance of material at the reference temperature
ΔT = change in temperature above the reference temperature

For instance, the temperature coefficient of resistance for iron is $50 \times 10^{-4}/C°$. For nichrome it is $2 \times 10^{-4}/C°$ and for silver it is $38 \times 10^{-4}/C°$. Thus, iron increases its resistance per degree rise in temperature about 25 times as much as nichrome. However, the α values for thermistors are much higher than those for any of these metals.

Both negative-temperature-coefficient (NTC) and positive-temperature-coefficient (PTC) materials are used in the manufacture of thermistors. For negative (NTC) thermistors, the resistance decreases with temperature. For positive (PTC) thermistors, the resistance increases with temperature. Figure 10.35 shows the

FIGURE 10.35 Resistivity change with temperature for typical thermistors compared to a metallic conductor.

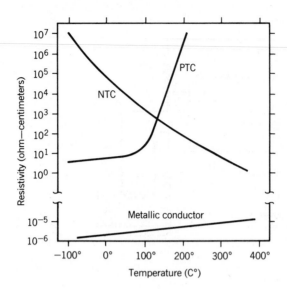

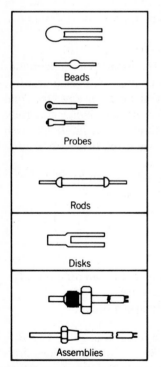

FIGURE 10.36 Common thermistor forms.

resistivity of both NTC and PTC materials over a range of temperatures as compared to that of metallic conductor (platinum).

Notice that all the numbers given above have units of 1/C°, that is, reciprocal celsius degrees. A convenient way of reading the values for the temperature coefficient is simply to say "50 × 10^{-4} per C°" for iron and "38 × 10^{-4} per C°" for silver, and so on.

The high sensitivity to temperature causes thermistors to change their resistance by as much as 6% of their normal value for each C° temperature variation. This sensitivity makes the thermistor a useful component in temperature measurement, in temperature sensing and control, and in temperature compensation applications.

Manufacturing Thermistors are semiconductor devices made of ceramic materials mixed (doped) with metallic oxides such as the oxides of manganese, nickel, cobalt, copper, iron, and uranium. After the raw materials are mixed in accurate proportions, the units are pressed or extruded to the desired shape. Thermistor elements then are sintered (heated) under carefully controlled atmospheric and temperature conditions to produce the ceramic semiconductor material. Electrode materials and lead wires are then attached. Figure 10.36 shows some common thermistor forms available today.

Modes of Operation There are two modes of thermistor operation: "self-heated" and "externally heated." Externally heated thermistors convert changes in ambient temperatures directly to corresponding changes in current. These thermistors are well suited for precision temperature measurement, temperature control, and temperature compensation, because they exhibit large changes in resistance with temperature. They are widely used for applications in the range of -100°C to over 300°C.

Self-heated NTC thermistors perform a different function. Large currents through these thermistors raise their temperature and thus decrease their resistance. Under normal operating conditions, the temperature can rise 200 to 300 C°, causing the resistance at high current to be reduced to perhaps 1/1000 of its value at low current. This mode of operation is useful in devices such as voltage regulators, microwave power meters, gas analyzers, vacuum gages, flowmeters, and automatic volume and power-level controls. Be careful circuit design, almost any desired time lag can be introduced in the response of self-heated thermistors. This makes them useful in time-delay and surge-suppression applications.

Applications   The simple circuit for temperature measurement illustrated in Figure 10.37 consists of a battery, a variable resistor, a thermistor, and a microammeter. As the temperature changes, the resistance of the thermistor changes. The current through the meter can be calibrated in terms of temperature. The variable resistor can be adjusted to compensate for battery age. In this circuit, the thermistor can be mounted a great distance from the meter, and ordinary copper wire can be used for the connections.

A simple temperature control can be made by placing a thermistor in series with a relay, a battery, and a variable resistor, as illustrated in Figure 10.38. When the current is large enough, the relay will activate. The current can be adjusted by the variable resistor to actuate the relay at the desired thermistor temperature.

If a self-heating thermistor is placed in series with a relay and a battery, the relay can be made to operate if the thermistor is suspended in air (higher temperature) and to drop out when the thermistor is immersed in cool liquid (lower temperature). This situation is shown in Figure 10.39. In air, the NTC thermistor heats up, causing the resistance to drop and permitting enough current to pull in the relay. If the thermistor is submerged in a liquid, it will be cooled, causing its resistance to increase. As a result, the current through the relay is reduced to a point at which the relay is no longer actuated. This device can be used as a liquid-level indicator or as a liquid-level control device.

If a thermistor is placed in series with a variable resistor, a battery, and a relay as shown in Figure 10.40, a variable time-delay relay can be made. When the switch is closed, current is limited by the high resistance of the thermistor. When it heats up, it permits sufficient current to close the relay. An increase in the variable resistance increases the time delay. With proper selection of components, time delays from milliseconds up to several seconds can be produced.

Other Thermal Transducers

Fluid-Filled Devices   A common temperature-measurement technique makes use

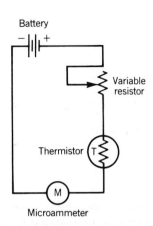

FIGURE 10.37 Simple temperature measurement using a thermistor.

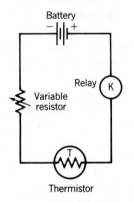

FIGURE 10.38 Simple temperature control using a thermistor.

FIGURE 10.39 Simple liquid-level control using a thermistor. (left)

FIGURE 10.40 Simple time-delay using a thermistor. (right)

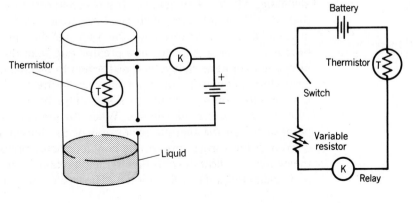

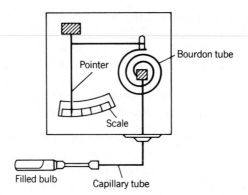

of fluid-thermal systems. An increase in temperature causes most fluids to expand. When the fluid expands in a sealed system, the motion is transmitted to a pointer or indicator by means of a link and lever system.

Fluid-filled thermal systems are categorized by the type of fluid used. The fluid may be a liquid or a gas or, in the case of mercury, a fluid metal.

All such measuring devices consist of a bulb (reservoir for the fluid), a pressure-sensitive element, and a capillary tube joining the two. An example of such a device is shown in Figure 10.41.

Liquid and mercury systems typically operate on a volumetric change of the filled material with a change in temperature. Gas systems operate on a pressure change with temperature.

The most common type of fluid system is the glass thermometer, which uses either mercury or alcohol as the fluid. Temperature changes cause a change in the volume of the fluid and the movement of the liquid along a thin cylinder of constant diameter. The change in volume is thus registered as a change in length. Thus, an empirically constructed linear scale along the cylinder can be calibrated easily in terms of temperature.

More sophisticated systems convert the movement caused by volumetric or pressure changes to a scaled pneumatic or electronic signal. This signal, in turn, is transmitted to a receiving instrument, where the signal response is interpreted as a temperature. These temperature transmitters are used in industry for control purposes, on motor vehicles for measurement of engine oil temperature, and also as temperature control devices in the home.

Bimetallic Strips   A bimetallic strip is constructed by joining together two metals that have different rates of thermal expansion. When the device is heated, the unequal thermal expansion of the two metals causes a bending of the strip, because one metal expands at a greater rate than the other. Figure 10.42 shows how a temperature change is transformed to mechanical motion in the bimetallic strip. Bimetallic strips of this construction are used most commonly in single-temperature control circuits to make or to break an electrical contact. Figure 10.43 illustrates one such device. When the temperature of the unit increases to a point at which the thermal expansion of the bimetallic strip is great enough to overcome the spring tension, the contact opens, breaking the electrical circuit to the heater. As the unit cools down, the bimetallic strip returns to its original shape and again closes the heater circuit. The cycle continues to repeat and, therefore,

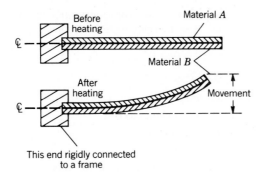

FIGURE 10.42 Effect of heating a bimetallic strip. (top left)

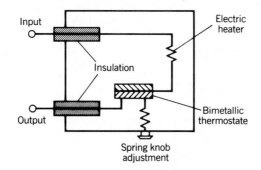

FIGURE 10.43 Bimetallic strip used as a thermostat. (bottom left)

FIGURE 10.44 Greater response obtained from spiral or helical bimetallic strips. (right)

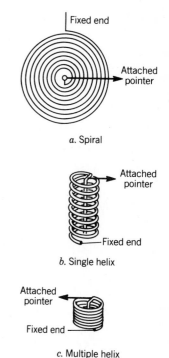

controls the temperature in the enclosure.

In some applications it is desirable to use the response of a bimetallic strip to provide temperature measurement instead of control. The bimetal element can be wound into the form of a spiral or helix (Figure 10.44). The response of the strip to temperature change is increased in this way and can be made linear over a temperature range of about 400F°. For temperature changes greater than 400F°, nonlinear scales can be empirically constructed.

SECTION 10.4 *EXERCISES*

1. Sketch a thermocouple circuit and briefly describe how it is used to measure temperature. How is the error due to the thermocouple resistance eliminated?

2. A thermoelectric emf of 6.92 mV is read across a copper–constantan thermocouple. Find the temperature of its test junction if its reference temperature is (a) 32° F or (b) 70°F. (Use thermocouple tables in the Appendix.)

3. The temperature of the test junction of a platinum-versus-platinum, 10% rhodium thermocouple is 1500°C. Find the thermoelectric voltage if the reference temperature is 0°C. (Use thermocouple tables in the Appendix.)

4. The test junction temperature of an iron–copper thermocouple is 300°C. For a reference temperature of 0°C, calculate the thermoelectric voltage (a) from calibration tables of iron–constantan and copper–constantan thermocouples and (b) from the equation for lead thermocouples. (Use thermocouple table in the Appendix.)

5. Find the temperature of the test junctions of a thermopile consisting of five iron–constantan thermocouples with the reference at 0°C, if the thermoelectric voltage is 0.1 V. (Use thermocouple tables in the Appendix.)

6. Find the thermopower of an iron–constantan thermocouple at $T = 264°C$. Use calibration tables. (Use thermocouple tables in the Appendix.)

7. Define the terms

Thermistor

Self-heated mode

Externally heated mode

Resistance–temperature characteristic

Voltage–current characteristic

Current–time characteristic

8. Describe how the thermistor is used for temperature measurement, temperature control, liquid-level control, and time delay.

9. In the accompanying circuit, how will the voltage across resistor R_2 vary if the temperature decreases? Increases? Remains the same?

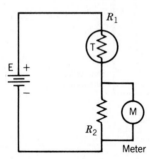

10. In the accompanying circuit R_1 and R_2 are thermistors with resistances at room temperature and identical characteristics. The only difference is that one has an NTC and the other has a PTC. How will the output voltage E_{out} change if the temperature increases? Decreases? Remains the same?

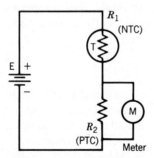

11. In a positive–temperature thermistor, how will the resistance change when the temperature increases? Decreases? Remains the same?

12. The sensitivity of a thermistor is determined by which of the following?

   a. Its rate of power dissipation.

   b. Its reaction to self-heating effects.

   c. Its ability to respond to a wide temperature range.

   d. The amount of resistance change for a given temperature variation.

13. Find the temperature coefficient of resistance of a metal wire if the resistance of the strip at room temperature (22°C) is 1 Ω and at 97°C is 1.001 Ω.

14. What length of wire will have a resistance of 10 Ω if the cross-sectional area of the wire is 0.1 cm² and the resistivity of the wire is 0.1 Ω-cm?

15. Label the parts of this fluid-filled thermal system and describe the functions of each part.

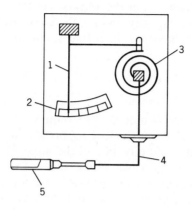

16. Briefly describe how a bimetallic strip can be used to control a thermostat.

10.5

RADIATIVE TRANSDUCERS

OBJECTIVES

Upon completion of this section, the student should be able to

- Describe the major characteristic distinguishing the two types of radiative transducers.

- Discuss four important properties to consider when selecting a radiative transducer.

- Describe the operating characteristics of the following thermal detectors with respect to light detection:
  Thermopiles
  Thermistors

- Describe the operating characteristics of the following photon detector types:
  Photoconductive detectors
  Photoemissive detectors
  Photovoltaic detectors

- Determine either the responsivity, the voltage output, the incident radiation level, or the detector surface area when given the other three quantities.

- Discuss the operation of a photomultiplier tube and estimate the output given its design.

DISCUSSION

A final class of transducers may be collectively called "radiative transducers." These devices transform the energy of light into a form used for measurement or control. Several different types of detectors are available for light measurement. Each responds differently, but most have several characteristics in common that can form a basis for comparison.

Little has been discussed thus far about the electromagnetic radiation commonly known as "light." For the purposes of this section two characteristics will be considered: the frequency (that is, the color) of light and the photon nature of light (that is, the behavior of light that simulates little packets of energy called "photons.")

Detector Types

Some applications for photodetectors require only a present/not present indication of light. Others are much more sophisticated, yielding information about the frequency, intensity, and direction of the light. The detectors that perform these tasks can be divided into two types: thermal detectors and photon detectors. Thermal detectors operate by sensing a rise in temperature of the detector surface as a result of the absorption of incident light radiation. Photon detectors utilize the interaction that occurs between the photons of the incident radiation and the electrons of the detector material. Since the latter type are very sensitive, often very sophisticated equipment is involved in their use.

Detector Characteristics

Introduction   Before discussing the different types of detectors available, the characteristics for a light detection device should be understood. These traits do not apply to all detectors, but they do represent a basis for comparison and choice.

Responsivity   Responsivity is a measure of the detector's sensitivity to radiant energy and can be calculated as

$$R = \frac{V}{HA} \tag{10.42}$$

where

R = responsivity of detector (V/W)
V = voltage output of detector (V)
H = incident radiation density (W/cm^2)
A = area of detector surface (cm^2)

Example 10.17 illustrates a typical calculation using Equation 10.42.

EXAMPLE 10.17   Responsivity calculation
....

Given: A light detector material having a responsivity of 5.0 V/W is to be used to detect a light beam of 400 mW/cm^2.

Find: The detector size required to yield a signal of 1.0 V.

Solution: Rearranging Equation 10.42, the surface area of the detector is given by

$$A = \frac{V}{HR}$$

$$A = \frac{1.0 \text{ V}}{(0.400 \text{ W/cm}^2)(5.0 \text{ V/W})}$$

$$A = 0.50 \text{ cm}^2$$

So a detector having a 0.50-cm^2 surface area is required to yield a 1.0-V reading.

Spectral Response The spectral response of a detector describes the responsivity of a detector as it varies with frequency (or energy) of light to be measured. The energy of light is proportional to its frequency. A detector that may be able to easily measure the high-frequency and high-energy ultraviolet light region may not respond to low-frequency infrared light. This characteristic is usually given for a detector in the form of a spectral response curve similar to that in Figure 10.45.

Notice that Figure 10.45 refers to the wavelength of the light rather than the frequency. The frequency and wavelength of light are related by the speed of the light:

$$c = \lambda f \tag{10.43}$$

where

c = speed of light in vacuum = 3.00×10^8 m/sec
λ = wavelength in meters
f = frequency in cycles per second (sec^{-1})

Various response curves for detectors, having either frequency or wavelength as the basis of the curve, will frequently be encountered. Example 10.18 illustrates the relationship between the frequency and wavelength of light.

EXAMPLE 10.18 Frequency and wavelength
. . . .

Given: The speed of light in a vacuum is 3.00×10^8 m/sec.

Find: The wavelength of red light from a helium–neon laser having a frequency of 4.74×10^{14} sec^{-1}.

Solution: Using Equation 10.43,

$$\lambda = \frac{c}{f}$$

$$\lambda = \frac{3.00 \times 10^8 \text{ m/sec}}{4.74 \times 10^{14} \text{ sec}^{-1}}$$

$$\lambda = 6.33 \times 10^{-7} \text{ m}$$

$$\lambda = 6330 \text{ Å} = 633 \text{ nm}$$

Since 10^{-10} m = 1 Å, this is a common unit of measure for visible light.

Response Time   It is not uncommon for light emissions to be of very short duration and hence require very fast detection devices. For applications that detect the presence or absence of a light beam (for example, an optical conveyor switch), a rapid response time is not necessary. For this reason, the response time is an important specification to consider when choosing a detector. Values can range from nanoseconds (10^{-9} sec) up to seconds.

Noise   A rather complicated but frequently encountered qualifier of light detectors involves what is commonly called "noise." Noise is the signal from the detector that is always present, even when no radiation is incident. This limits the detector's ability to yield meaningful data at very low levels of light. A fair analogy is inability of a radio or TV receiver to tune in a very weak radio or television station in the presence of electronic noise. Various indicators of a detector's ability to reduce noise have been developed. Only one will be mentioned here, the "noise equivalent power," or "NEP" as it is commonly seen in detector specifications. This represents the amount of light power that would yield

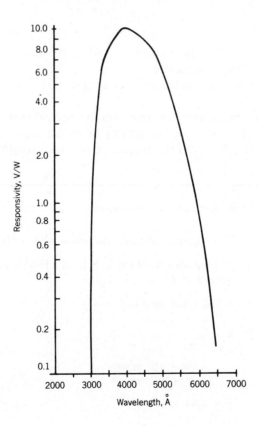

FIGURE 10.45 Typical spectral response curve; responsivity versus wavelength.

an output signal equal to the noise generated by the detector itself. These values effectively represent the minimum detectable signal level and may need to be considered when selecting a detector for a very low light level application.

Thermal Detectors

Introduction Light represents energy. One way to detect the presence of that energy is to look for the resulting temperature change of a device with incident light. Thermal detectors of light register a temperature rise of the light-sensing element in accordance with the principles outlined in Section 10.4. However, as one would expect, temperature changes from starlight in an astronomer's telescope, for example, are not very large. For this reason sophisticated techniques are used to produce very sensitive thermal transducers for light-detecting applications.

Thermopiles Obviously the light falling on a single thermocouple will produce very little temperature change and hence very little measureable output. To increase the voltage signal, several thermocouples are arranged in series to form what is known as a **thermopile.** Technological developments have improved the bimetallic junctions of the thermopiles from wire junctions to foil junctions and, more recently, from wire junctions to thin films, which are only several thousand atoms thick. This makes modern thermopiles very rugged and yet very responsive. Thermopiles are usually blackened to enable them to absorb essentially all the light spectrum from infrared to ultraviolet. This results in a generally uniform spectral response. Since thermopiles generate a voltage by the thermocouple principle, they have the advantage of being self-contained, needing no elaborate support equipment or environment. The responsivity of thermal detectors is inherently very low, since the temperature rise to be expected from low light pulses is quite low. By the same token, the response times are relatively slow, on the order of several hundred milliseconds. However, for broad-spectrum, rugged environmental applications, thermopiles are quite appropriate.

Thermistors The principle of thermistor operation covered in Section 10.4 applies here as well. The thermistors are found to be similar in performance to the thermopiles with one exception: sensitivity. Thermistors are much more sensitive to small changes in temperature and so to light intensity changes. By blackening the detector, they can be made sensitive to a broad spectral range, like the thermopiles. They suffer from the same disadvantages of thermopiles, however: relatively low responsivity and slow response times.

Pyroelectric Devices A final class of thermal detectors of light is a group of materials known as **pyroelectric** crystals. These crystals when subjected to a temperature change develop an electric charge on their surfaces. (Recall that for piezoelectric crystals, a charge is developed on the surfaces when the crystal is stressed.) Like the other thermal detectors discussed in this section, pyroelectric detectors also utilize a blackened surface and so have an essentially uniform spectral response. Their responsivity is much better than that of thermopiles and thermistors, as is their response time.

Photon Detectors

Introduction   As briefly mentioned earlier, light can be viewed as a stream of photons or packets of energy traveling at the speed of light (3.00×10^8 m/sec). Intense beams of light are composed of very, very many photons, whereas extremely faint light beams are composed of a small number of photons. In addition, a photon's energy is dependent on its frequency. Examples of relatively high energy radiation include ultraviolet light, X rays, gamma rays, and other cosmic rays. Low-energy radiations include visible light, infrared, radar, and radio and television broadcasts. All of these are types of electromagnetic radiation.

Photon detectors can be divided into three types: photoemissive, photoconductive, and photovoltaic. The latter two use various semiconductor materials. The characteristics of operation of each of the three types will be discussed.

Photoemissive Detectors   When energetic photons are incident upon certain materials in a vacuum, electrons are emitted by the material. By applying a voltage bias, the emitted electrons are attracted to a collector (called an anode) and current is created. The amount of current is proportional to the light intensity. However, the current normally developed is very small and must be amplified. Such detectors are based on the **photoelectric effect.**

The energy of an incident photon is given by

$$E = h\nu \qquad (10.44)$$

where

E = photon energy in J
h = Planck's constant = 6.63×10^{-34} J·sec
ν = frequency of light in cycles/sec, Hz, or sec^{-1}

However, it is observed that when photons of energy $E_{min} = h\nu_{min}$ or less strike the material, the current stops. That happens because the electrons are not given sufficient energy by the photons to escape the material. This energy, E_{min}, is known as the **work function** of the material and represents a minimum energy of light (and hence frequency) that can be detected. By heating the cathode material, the electrons are more energetic and more easily ejected by a photon. This lowers E_{min} slightly but still restricts the detector to a minimum frequency of operation. **Vacuum photodiodes** are photoemissive detectors that are used to measure moder-

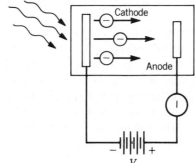

FIGURE 10.46   Incident light causes emitted electron current: the photelectric effect.

FIGURE 10.47 Schematic diagram of photomultiplier tube.

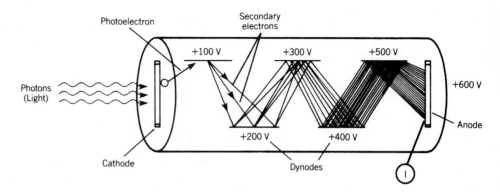

ate to high light levels. Since a voltage bias exists between the cathode material and the anode, some stray electron current will exist (noise), making it difficult to detect low light levels. For this reason, vacuum photodiodes are not generally used in low-light applications. Instead a much more efficient device is used called a **photomultiplier.** When the electrons emitted from the cathode material are attracted to a biased metal surface and strike with sufficient velocity, one or more electrons can be ejected from the metal surface. This process is called secondary emission. By arranging several metal surfaces (called dynodes) with steadily increasing voltage biases, a multiplication of emitted electrons occurs. This results in a greatly enhanced current that is much easier to detect. Figure 10.47 illustrates a schematic design for a photomultiplier tube. In Figure 10.47, three secondary electrons are shown emitted for each electron striking each of the five dynodes. In actual practice, from 2 to 10 secondary electrons per dynode can be emitted, with 10 to 16 dynodes. An idea of the electron multiplication that occurs here is shown in Example 10.19.

EXAMPLE 10.19   Photomultiplier output
. . . .

Given: A photomultiplier has 10 dynode surfaces that produce electrons by secondary emission. A single photon results in a current pulse lasting 10 nanoseconds (10×10^{-9} sec). The average number of secondary electrons produced is 5 per impact electron.

Find: **a.** The total charge collected by the anode.

            b. The average anode current.

Solution: If each collision results in 5 emitted electrons, each of which result in 5 more electrons, etc., then for all 10 dynodes, the total number of electrons n_e reaching the anode is:

            1 electron from the cathode produces 5 electrons at dynode 1.
            5 electrons from dynode 1 produce 25 electrons at dynode 2.
            25 electrons from dynode 2 produce 125 electrons at dynode 3.
            125 electrons from dynode 3 produce 625 electrons at dynode 4.
            625 electrons from dynode 4 produce 3125 electrons at dynode 5.

3125 electrons from dynode 5 produce 15,625 electrons at dynode 6.

15,625 electrons from dynode 6 produce 78,125 electrons at dynode 7.

78,125 electrons from dynode 7 produce 390,625 electrons at dynode 8.

390,625 electrons from dynode 8 produce 1,953,125 electrons at dynode 9.

1,953,125 electrons from dynode 9 produce 9,765,625 electrons at dynode 10 collected by the anode.

The charge per electron q is 1.6×10^{-19} C, so the total charge at the anode is

$$Q = n_e \, q$$

$$Q = (9.77 \times 10^6)(1.6 \times 10^{-19} \text{ C})$$

$$Q = 1.56 \times 10^{-12} \text{ C}$$

The average current is the total charge flow during a specified time span. Hence, the average current is

$$I_{\text{avg}} = \frac{Q}{T}$$

$$I_{\text{avg}} = \frac{1.56 \times 10^{-12} \text{ C}}{10 \times 10^{-9} \text{ sec}}$$

$$I_{\text{avg}} = 1.56 \times 10^{-4} \text{ C/sec} = 0.156 \text{ mA}$$

So a single photon can produce an overall current pulse of 0.156 mA!

The magnification power of the photomultiplier tube makes it a truly valuable detector for use in very low level light applications. Such applications include astronomy, luminescence, gamma rays from atomic decay, and other cases where an abundance of "light" is not available. Some drawbacks of the photomultiplier tube are their delicate and expensive nature and the amount of supporting electronics required to satisfy their high voltage supply requirements. In practice, it is usually desirable to reduce the total voltage bias to limit the continuous anode current to 1 μA or less. This limits the number of secondary emission electrons.

Photovoltaic and Photoconductive Detectors   Detectors of this type are made of semiconductor materials such as silicon, germanium, and indium. They are normally called **photodiodes** and can operate either photoconductively or photovoltaically, depending on the bias voltage applied to them. Photoconductive performance is sometimes called photoresistance. In this mode, the device's resistance changes in response to photons striking an exposed pn junction (see Section 5.3). Photovoltaic performance, on the other hand, commonly observed in "solar cells," results in a voltage, again as a result of photons striking the pn junction. (Photovoltaic cells are covered in Section 9.4) Since the light must be incident directly on the semiconductor material, the spectral response of these devices is usually somewhat limited. Germanium and indium photodiodes are usually restricted to operation in the infrared region, whereas silicon devices respond over a fairly wide region of frequency. These types of devices are probably the most common in everyday applications because of their ease of use. The variable resistance or variable voltage output is easily applied in remote controls, switching light beams, automatic light dimmers, and the like.

SECTION 10.5   EXERCISES

1. Name the two categories of radiative transducers and their distinguishing differences.
2. Name and briefly discuss four important characteristics of radiative transducers used as detectors.
3. Name and briefly discuss the operation of two types of thermal detectors used to measure light.
4. Describe the modes of operation of three types of photodetectors.
5. Describe how a photomultiplier tube is able to detect very low light levels.
6. Determine the light level one could measure with a 1-V reading from a circular cross-section detector having a responsivity of 10 V/W and a diameter of 1.0 cm.
7. Determine the voltage across a 10^6-Ω resistor of a pulse of current coming from a single photon striking the cathode of a photomultiplier tube having 10 dynode surfaces. Assume a secondary emission rate of 4 electrons per impact and a pulse duration of 10 nanoseconds.

S U M M A R Y

A transducer is a device capable of accepting an input signal containing some information from one energy system and producing an output signal containing the same information in another energy system. Some transducers are energy convertors that convert input energy to output energy. These transducers require no external energy source. Others change their resistance in response to the input and thus change the rate of flow of some quantity, usually of electric current. This second class of transducers requires an external energy source to supply the current to the transducer.

There is always a relationship of the output of a transducer to its input. For example, as temperature increases, voltage output of a thermocouple increases; as fluid flow increases, the counting rate of a turbine flowmeter increases. In other words, the output of a transducer gives a quantitative indication of its input. For this reason, transducers are used widely as measuring devices (in science, engineering, and industry) of diverse quantities (temperature, pressure, flow rate, illumination levels, and so on).

Vibrations

and

Waves

INTRODUCTION

Mechanical and electrical vibrations play an important role in technology. The periodic motion of a pendulum is used as a standard of time for clocks. The vibrations of a microphone diaphragm caused by pressure variations are used to generate equivalent electrical vibrations. Electrical vibrations are used to drive speaker cones that then generate equivalent pressure waves. Almost all mechanical and electrical systems vibrate. Sometimes these systems produce unwanted vibrations that can destroy or limit the operation of the system if they are not eliminated or reduced.

The simplest vibrations are sinusoidal in time. Sinusoidal vibrations in simple mechanical systems are analyzed in this chapter. This motion is called simple harmonic motion. Driven and free oscillators are discussed, and the phenomenon called mechanical resonance is investigated. In addition, sinusoidal electrical vibrations are related to mechanical vibrations by the forcelike quantity called voltage and the ratelike quantity called current. Electrical resonance is discussed, and methods are developed to analyze alternating current circuits.

Mechanical and electrical vibrations are often the source of waves. A wave transmits energy from the vibrating source through a medium to a receiver. For example, sound is a wave that carries energy from a mechanical vibrating source through air to a receiver (ear). Light is a wave that carries energy from a source, such as the sun, through space to a receiver (eye). In some wave motion the vibration is along the direction of propagation and is called "longitudinal wave motion." In other instances, the vibration is perpendicular to the direction of propagation and is called "transverse wave motion." Sound energy propagates as a longitudinal wave, and a wave on a string propagates as a transverse wave.

The general properties of waves are discussed in this chapter. The wave equation is introduced and it is used to explain standing waves, interference, and the beat frequency of two waves. Waves on a string and on a transmission line are analyzed. The reflection and transmission of waves are investigated at a boundary. The wave velocity and characteristic impedance of an electrical transmission line are calculated. From these discussions, the proper termination of a line can be determined to eliminate reflected waves.

11.1

VIBRATIONS IN MECHANICAL SYSTEMS

OBJECTIVES

Upon completion of this section, the student should be able to

- Gain an understanding of simple harmonic motion.
- State the classes of vibrations.
- Give an example of a complex number.
- Calculate the absolute value of a complex number.
- Add, multiply, and divide complex numbers.
- Rotate a complex number 90°, 180°, and 270°.
- State Hooke's law.
- Understand resonance in a driven mechanical oscillator.
- Understand and use the concept of mechanical impedance.
- Calculate the natural frequency of a simple mechanical oscillator.
- Use complex vectors to represent the displacement, velocity, and acceleration of simple harmonic motion.
- Calculate the displacement, mechanical impedance, and velocity of a forced oscillator with and without a friction force.
- Calculate the period of a simple and compound pendulum.

Definitions A mechanical vibration is any to-and-fro motion or oscillation of a mechanical system. Some vibrations are simple in mathematical form, such as the pendulum, whereas others are complex. In this section, only simple vibrations will be analyzed mathematically.

Some devices are specifically designed to cause vibration or shaking and are called vibrators. Foundries use special vibrators to loosen castings from molds. Vibrators are also used to stimulate blood circulation and relax muscles. All sound begins with a vibrating object; for example, when a person speaks, the vocal cords initiate the vibration that we recognize as speech sounds.

On the other hand, vibrations that weaken structures and produce wear and tear in industrial machines certainly are undesirable. In cities, trucks, subways, elevated trains, and other traffic cause considerable vibrations that are transmitted effectively throughout the steel structure of a large building. Engineers deal with this problem by using special vibration-absorbing materials. Pneumatic tires are used on automobiles in an effort to reduce undesirable vibrations created by running over rough surfaces. Automobile engines are mounted on rubber supports to reduce vibration; and more cylinders have been added to engines to smooth out engine vibration, resulting in a smoother flow of power.

Many industries deal with the problem of vibration and its effect on machinery. High-speed, rotating parts produce vibration unless they are accurately balanced. The supports of such rotating parts were once bolted securely to the machine frame or its foundation. Today the supports for high-speed, rotating parts are mounted on flexible connections, such as springs or rubber pads, thereby reducing the effect of vibration in machines.

Mechanical vibrations have various frequencies and amplitudes. Frequency is the number of complete vibrations made during a certain time period. Amplitude is the maximum distance the item travels from its normal position of equilibrium or rest. The period is the time required to make one complete vibration or oscillation and is related to frequency by

$$P = 1/\nu \tag{11.1}$$

where

P = period or time in sec
ν = frequency in Hz

Vibrations that repeat themselves at regular intervals are called periodic. Both amplitude and period are shown in Figure 11.1 for a vibration that varies as a sine wave.

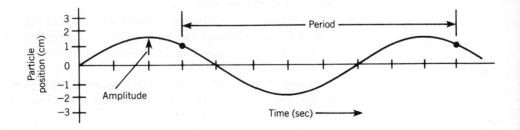

FIGURE 11.1 Periodic vibration as a sine wave.

FIGURE 11.2 Complex vibration resulting from the addition of two simple vibrations.

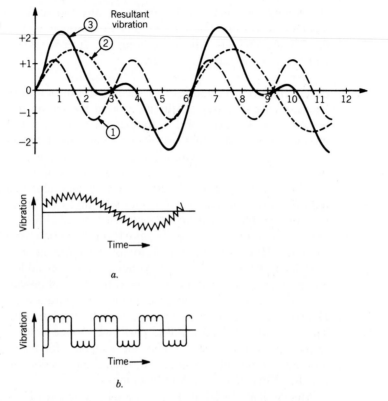

FIGURE 11.3 Examples of other types of complex vibrations.

Most vibrations are complex, meaning that they are the sum of two or more simple vibrations. Figure 11.2 illustrates the adding of vibrations 1 and 2 to obtain a resultant, or composite, vibration 3.

Figure 11.3 illustrates the nature of other types of complex vibrations.

If machinery vibrates in a regular manner, the vibrations are called "steady state." These vibrations normally are found in all machines and appliances, such as washers, dryers, and electric motors. Vibrations that are sudden and die out quickly are called "transient vibrations" or "shock vibrations." Figure 11.4 illustrates transient and steady-state vibrations.

Classes of Vibrations   Vibrations are caused either by sources external to the mechanical system or by the internal moving parts of the system. Examples of external sources are illustrated in Figures 11.5, 11.6, and 11.7. In Figure 11.5, the vibrations are caused by the uneven road surface. These vibrations can be reduced by automobile springs and shock absorber systems, which make the ride smooth and comfortable.

In Figure 11.6, an electronic instrument vibrates because of the vibrations of a ship or aircraft. Springs can be used to reduce the instrument's vibrations.

In Figure 11.7, a glass bulb is shown enclosed in a corrugated wrapper to isolate it from any sources of vibration.

Examples of vibrations caused by internal moving parts of a mechanical system are shown in Figures 11.8, 11.9, and 11.10. In Figure 11.8, an unbalanced tire causes the entire car to vibrate.

FIGURE 11.4 Transient and steady-state vibrations.

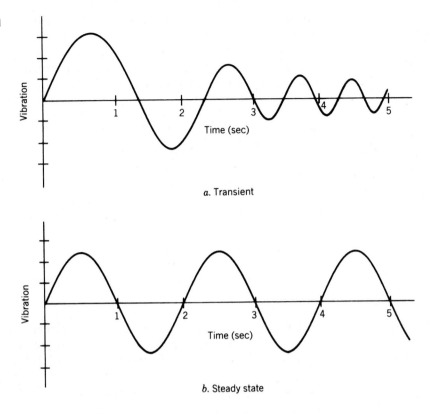

a. Transient

b. Steady state

Figure 11.9 illustrates an electromagnetic system. Vibrations can be induced in this system by rapidly changing magnetic fields. This effect can be reduced if slots are placed in the armature lamination.

Figure 11.10 shows an axle bearing that has become worn. This will cause the wheel to vibrate, particularly at high speeds. To reduce this vibration, it is usually necessary to replace the worn bearing.

Vibration Isolation Excessive vibration must be isolated or reduced for the protection of personnel and of the equipment itself. Effectiveness of vibration

FIGURE 11.5 Automobile on a rough road. (right)

FIGURE 11.6 Isolation of a sensitive instrument. (right)

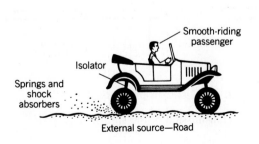

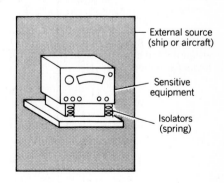

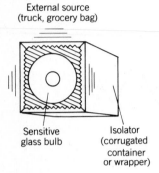

External source
(truck, grocery bag)

Sensitive
glass bulb

Isolator
(corrugated
container
or wrapper)

FIGURE 11.7 Glass bulb isolated from external vibrations.

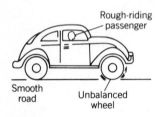

Rough-riding
passenger

Smooth
road

Unbalanced
wheel

FIGURE 11.8 Vibration from an unbalanced tire on an automobile.

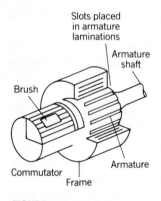

Slots placed
in armature
laminations

Armature
shaft

Brush

Commutator

Frame

Armature

FIGURE 11.9 Slots in armature and shaft to reduce vibrations.

isolation provided by a system may be evaluated by measuring the amount of vibration transmitted through an isolating medium.

Transmission of a vibration–isolation system can be measured by determining accelerations on both support structures and vibrating machinery. Transmission of the system for external vibration is given by

$$T_E = a_m/a_s \tag{11.2}$$

where

T_E = external transmission of a vibration isolation system.
a_m = acceleration of machinery
a_s = acceleration of support

For internal vibrations, transmission is defined by

$$T_I = a_m/a_s \tag{11.3}$$

where

T_I = internal transmission of vibration-isolation system

The acceleration of the vibrating source, a_s, is placed in the denominator and acceleration of the isolated structure or machinery, a_m, is placed in the numerator. In either case, the smaller acceleration is always the numerator.

Example 11.1 illustrates the use of Equation 11.3 to solve a problem.

EXAMPLE 11.1 Transmission calculation
. . . .

Given: A vacuum pump is mounted on springs (see illustration). The amplitude of trace A is 2 cm peak-to-peak and the peak-to-peak amplitude of trace B is 0.2 cm. (Note: The amplitude of the oscilloscope trace A to trace B is proportional to the acceleration of the pump and table, respectively.)

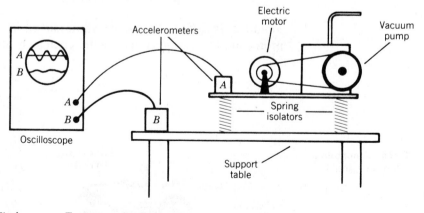

Find: T_I.

Solution: Transmission of the spring isolators is

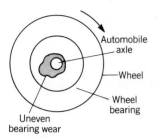

Automobile axle

Wheel

Wheel bearing

Uneven bearing wear

FIGURE 11.10 Vibration from worn axle bearing.

$$T_I = a_m/a_s$$

$$T_I = 0.2 \text{ cm}/2.0 \text{ cm}$$

$$T_I = 0.1$$

The springs reduce vibrations to 10% of their original value.

Mathematical Methods To analyze simple mechanical and electrical vibrations, it is advantageous to introduce complex numbers. It is then possible to relate the velocity to the applied force or the current to the applied voltage by defining a complex quantity called the impedance. The impedance plays the same role in simple vibrating systems that resistance plays in constant rate systems.

Ratelike Quantity = Forcelike Quantity/Impedance

Complex Numbers In addition to real numbers, which can be plotted along an x axis, there are imaginary numbers and complex numbers. The basic imaginary number is equal to $\sqrt{-1}$ and is given the symbol i. Thus $i^2 = -1$. A general imaginary number is expressed as bi, where b is any real number. Imaginary numbers can be plotted along the y axis. A complex number is the sum of a real number plus an imaginary number, and it can be expressed as

$$z = a + bi \tag{11.4}$$

where
   a and b are real numbers.

The complex number z is shown in Figure 11.11. When two complex numbers are added, one adds the real parts and the imaginary parts.

$$z_1 = a_1 + b_1 i$$

$$z_2 = a_2 + b_2 i$$

$$z_1 + z_2 = (a_1 + a_2) + (b_1 + b_2)i \tag{11.5}$$

$$|z| = \sqrt{a^2 + b^2}$$
$$a = |z| \cos \theta$$
$$b = |z| \sin \theta$$

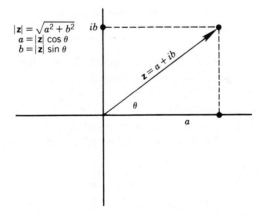

FIGURE 11.11 Graphic representation of the complex number **z**.

Thus one can see that complex numbers have vector properties since they follow the rules of vector addition. In addition,

$$z_1 - z_2 = (a_1 - a_2) + (b_1 - b_2)i \tag{11.6}$$

which is the same rule as vector subtraction. Complex numbers are usually considered vectors.

The multiplication of two complex numbers can also be defined as

$$(a_1 + b_1 i)(a_2 + b_2 i) = a_1 a_2 + (a_1 b_2 + a_2 b_1)i + b_1 b_2 i^2 \tag{11.7}$$

However, $i^2 = -1$. Thus, Equation 11.7 becomes

$$(a_1 + b_1 i)(a_2 + b_2 i) = (a_1 a_2 - b_1 b_2) + (a_1 b_2 + a_2 b_1)i \tag{11.8}$$

It is also convenient to define the complex conjugate of a complex number. It is defined as

$$z^* = a - bi \tag{11.9}$$

if

$$z = a + bi \tag{11.10}$$

where z^* is the complex conjugate of z. From the Pythagorean theorem one has

$$|z|^2 = a^2 + b^2 \tag{11.11}$$

where

$|z|$ = absolute value of z, which is the magnitude of the vector

The value $|z|^2$ can be obtained by multiplying z by z^*. For example,

$$zz^* = (a + bi)(a - bi)$$

$$zz^* = a^2 - b^2 i^2 \tag{11.12}$$

Since $i^2 = -1$, Equation 11.12 becomes

$$|z|^2 = zz^* = a^2 + b^2 \tag{11.13}$$

If the length and angle are given for a complex number, then the real and imaginary parts are easily obtained. From Figure 11.11 one obtains

$$a = |z| \cos \theta \tag{11.14}$$

$$b = |z| \sin \theta \tag{11.15}$$

which is the same method used to obtain the components of a vector.

Complex numbers can also be divided, as shown by

$$z_1/z_2 = (a_1 + b_1 i)/(a_2 + b_2 i) \tag{11.16}$$

where z_1 and z_2 are two complex numbers. To complete the division, it is advantageous to multiply the top and bottom by the complex conjugate of z_2. This yields

$$z_1/z_2 = (a_1 + b_1 i)(a_2 - b_2 i)/(a_2 + b_2 i)(a_2 - b_2 i)$$

$$z_1/z_2 = [a_1 a_2 + b_1 b_2 + (a_2 b_1 - a_1 b_2)i]/(a_2^2 + b_2^2) \tag{11.17}$$

Sample calculations with complex numbers are given in Examples 11.2 and 11.3.

EXAMPLE 11.2 Addition and subtraction of complex numbers
. . . .

Given: The two complex numbers shown in the diagram.

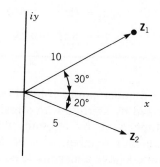

Find: **a.** The sum of the two complex numbers.

   b. The difference of the two complex numbers

Solution: **a.** First find the real and imaginary components of the numbers and then add the like components.

$$z_1 = 10 (\cos 30°) + 10 (\sin 30°)i$$

$$z_1 = 8.66 + 5.00i$$

$$z_2 = 5 (\cos 20°) + 5 (\sin 20°)i$$

$$z_2 = 4.70 + 1.71i$$

$$z_1 + z_2 = 13.36 + 6.71i$$

   b. $z_1 - z_2 = 3.96 + 3.29i$

EXAMPLE 11.3 Multiplication and division of complex numbers
. . . .

Given: Two complex numbers.

$$z_1 = 10 + 5i$$

$$z_2 = 10 + 10i$$

Find: **a.** $z_1 z_2$

   b. z_1/z_2

Solution: **a.** $z_1 z_2 = (10 + 5i)(10 + 10i)$

$$z_1 z_2 = 100 + 100i + 50i + 50i^2$$

$$z_1z_2 = (100 - 50) + (100 + 50)i$$

$$z_1z_2 = 50 + 150i$$

b. $z_1/z_2 = (10 + 5i)/(10 + 10i)$

$$z_1/z_2 = (10 + 5i)(10 - 10i)/(10 + 10i)(10 - 10i)$$

$$z_1/z_2 = (100 - 100i + 50i - 50i^2)/(200)$$

$$z_1/z_2 = (150 - 50i)/(200)$$

$$z_1/z_2 = 0.75 - 0.25i$$

A complex number can be rotated through 90° by multiplying it by i. This can be demonstrated by starting with a real number as shown in Figure 11.12.

Each multiplication by i rotates the vector by 90°. For the general case, consider a complex number, $a + bi$. Multiplying it by i yields $-b + ai$. Both complex numbers have a magnitude equal to $\sqrt{(a^2 + b^2)}$. Thus, their magnitudes are equal.

To show that the vector has been rotated by 90°, consider the complex numbers shown in Figure 11.13. If $\angle ACB$ is a right angle, then

$$(AB)^2 = (AC)^2 + (BC)^2$$

However, $(AC)^2 = a^2 + b^2$ and $(BC)^2 = a^2 + b^2$. Thus $(AB)^2 = 2a^2 + 2b^2$. Since $\angle ADB$ is a right angle, one may write $(AB)^2 = (AD)^2 + (BD)^2$. However,

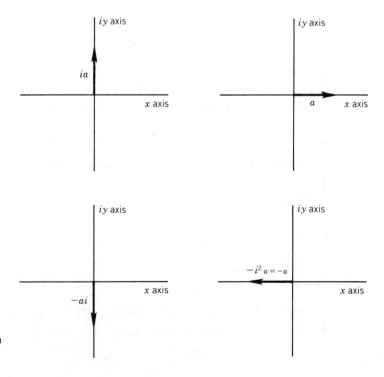

FIGURE 11.12 Rotation of a complex number by 90°.

FIGURE 11.13 Demonstration of 90° rotation.

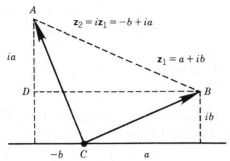

$(AD)^2 = (a - b)^2$ and $(BD)^2 = (a + b)^2$. Thus, $(AB)^2 = a^2 - 2ab + b^2 + a^2 + 2ab + b^2$ or $(AB)^2 = 2a^2 + 2b^2$. This is just the required condition to make $\angle ACB$ a right angle. Thus, multiplying a complex number z_1 by i rotates the vector by 90°.

The properties developed here for complex numbers will now be used to analyze mechanical and electrical systems.

Hooke's Law   A common force that produces mechanical vibrations is due to the deformation of elastic materials. When an elastic material is deformed, the force associated with the deformation is directly proportional to the deformation if the elastic limit is not exceeded. This is easily verified for the case of an elastic spring by direct measurements of the force and the corresponding elongation. As shown in Figure 11.14, when a mass is attached to the spring, its weight (mg) causes the spring to increase in length. If a graph of weight versus y is plotted, a straight line is obtained as shown in Figure 11.15 if the elastic limit is not exceeded. Since the line passes through the origin, the equation for the line is

FIGURE 11.14 Elongation of spring by a weight.

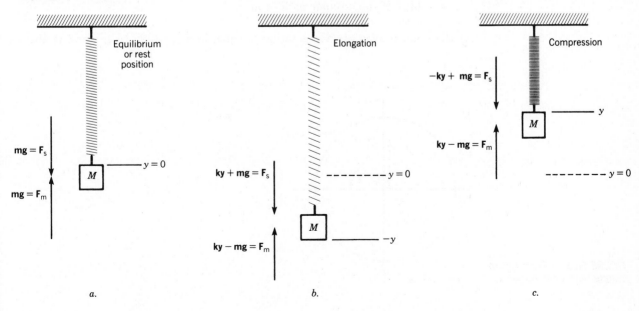

$$F_s = ky \qquad (11.18)$$

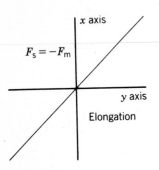

where

F_s = weight in N
y = elongation in m
k = slope in N/m

From Newton's third law, the force acting on the mass is in the opposite direction. Thus, the force acting on the mass may be written as

$$F_m = -ky \qquad (11.19)$$

FIGURE 11.15   Plot of weight versus elongation.

If the mass is displaced and released, the force will cause the mass to oscillate up and down in a motion called simple harmonic motion.

Simple Harmonic Motion   To determine the nature of simple harmonic motion, consider a point moving in a circular path with constant angular velocity as shown in Figure 11.16.

The angular displacement (θ) as a function of time is

$$\theta = \omega_0 t + \theta_0 \qquad (11.20)$$

where

θ = angular displacement in radians
ω_0 = constant angular velocity in rad/sec
t = time in sec
θ_0 = angular displacement when $t = 0$

The corresponding displacement in the y direction is

$$y = r \sin(\omega_0 t + \theta_0) \qquad (11.21)$$

where

r = radius of the circular path in m

As the point moves, its acceleration is equal in magnitude to $\omega^2 r$ and is always

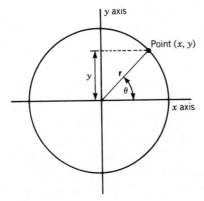

FIGURE 11.16   Illustration of simple harmonic motion.

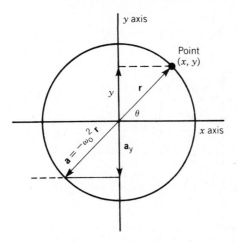

FIGURE 11.17 The y component of acceleration.

directed toward the center of the circular path. As shown in Figure 11.17, the component of acceleration in the y direction is

$$a_y = -\omega_0^2 r \sin(\omega_0 t + \theta_0) \tag{11.22}$$

From Newton's second law of motion,

$$F_y = ma_y \tag{11.23}$$

Substituting the expression for a_y from Equation 11.22 gives

$$F_y = -m\omega_0^2 r \sin(\omega_0 t + \theta_0)$$

However, from Equation 11.21,

$$y = r \sin(\omega_0 + \theta_0)$$

Thus, by substitution, one obtains

$$F_y = -m\omega_0^2 y \tag{11.24}$$

Since the force in the y direction is also equal to the spring force,

$$k = m\omega_0^2 \tag{11.25}$$

Thus, for any given spring constant k, one may solve Equation 11.25 for the required ω_0 such that the y component of r will follow the motion of the mass. This yields

$$\omega_0 = \sqrt{(k/m)} \tag{11.26}$$

The corresponding displacement y as a function of time becomes

$$y = r \sin\{[\sqrt{(k/m)}]t + \theta_0\} \tag{11.27}$$

As the mass moves, the maximum displacement occurs when $\sin\{[\sqrt{(k/m)}]t + \theta_0\} = 1$. The value r then corresponds to the maximum value of y. The quantity r is called the amplitude of the simple harmonic motion.

The magnitude of the velocity of a point moving in the circular path is $\omega_0 r$ and its direction is always perpendicular to the radius as shown in Figure 11.18. The

FIGURE 11.18   The y
component of velocity for
simple harmonic motion.

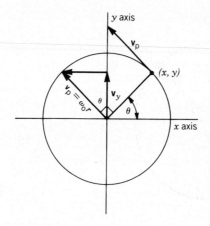

FIGURE 11.18   The y
component of velocity for
simple harmonic motion.

velocity in the y direction becomes

$$v_y = \omega_0 r \cos(\omega_0 t + \theta_0)$$

or

$$v_y = \omega_0 r \cos\{[\sqrt{(k/m)}]t + \theta_0\} \qquad (11.28)$$

The amplitude of the velocity is $\omega_0 r$. The velocity is said to be 90° out of phase with the displacement, whereas the acceleration is 180° out of phase with the displacement.

The period of the vibration is the time for one complete oscillation. This occurs when θ changes by 2π. Thus, the period may be expressed as

$$\omega_0 = \Delta\theta/\Delta t$$

or

$$\Delta t = \Delta\theta/\omega_0$$

$$P = \Delta t = 2\pi/\omega_0 \qquad (11.29)$$

where

P = period of oscillation in sec

The frequency of the oscillation is the number of oscillations per second. Thus, if it takes P seconds for one vibration there will be $1/P$ vibrations per second. Mathematically one may write

$$\nu = 1/P \qquad (11.30)$$

where

ν = frequency in oscillations per sec

Example 11.4 illustrates the use of the equations for simple harmonic motion in working a problem.

EXAMPLE 11.4 Simple harmonic motion
....

Given: A mass of 0.5 kg is attached to a spring with a spring constant of 100 N/m. The only force acting on the mass is due to the spring.

Find:
  a. The angular frequency ω_0.
  b. The period.
  c. The frequency.

Solution:
  a. The angular frequency is given by

$$\omega_0 = \sqrt{(k/m)}$$

$$\omega_0 = \sqrt{[100 \text{ N/M}/0.5 \text{ kg}]}$$

$$\omega_0 = 14.1 \text{ rad/sec}$$

  b. The period is given by

$$P = 2\pi/\omega_0$$

$$P = 2\pi \text{ rad}/(14.1 \text{ rad/sec})$$

$$P = 0.446 \text{ sec}$$

  c. The frequency is given by

$$\nu = 1/P$$

$$\nu = 1/0.446 \text{ sec}$$

$$\nu = 2.24 \text{ vibrations/sec}$$

It is advantageous to use complex vectors to represent r, v_p, and a_p as shown in Figure 11.19.

The projection of these complex vectors on the imaginary axis gives the values of y, v_y, and a_y as the system rotates at the angular velocity ω_0. This is illustrated

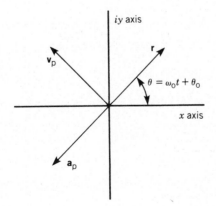

FIGURE 11.19 Complex vectors used in simple harmonic motion.

FIGURE 11.20   Projection of
the complex vectors on the
imaginary axis.

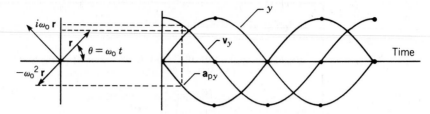

in Figure 11.20, where it is assumed that $\theta_0 = 0$. The multiplication of r by i rotates r through 90° and the multiplication of r by i^2 rotates r through 180°.

One can also draw a complex vector diagram of the forces. In solving mechanical problems it is often advantageous to consider the mass times the acceleration as a force. Newton's second law states

$$\Sigma F_A = ma \tag{11.31}$$

This may also be written as

$$\sum_{n=1}^{m} F_n - ma = 0 \tag{11.32}$$

Thus if a force equal to $-ma$ is applied in a direction opposite to the acceleration, one may write

$$\sum_{n=1}^{m+1} F_n = 0 \tag{11.33}$$

where the sum includes the term $-ma$. If a force equal to $-ma$ is applied to the body at the center of mass, then one can treat the problem as a problem in statics such that the sum of the forces must equal zero and the sum of the torques must equal zero. This is called D'Alembert's principle.

A complex vector diagram of the forces, which includes the fictitious force $-ma$, is shown in Figure 11.21. It is usually drawn at a time when the velocity vector lies along the real axis. In addition,

$$v_p = i\omega_0 r$$

and

$$a_p = -\omega_0^2 r$$

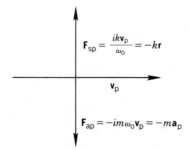

FIGURE 11.21   Complex
vector diagram of forces in
simple harmonic motion.

Expressing r in terms of the velocity yields

$$r = v_p/i\omega_0 = -iv_p/\omega_0 \qquad (11.34)$$

The acceleration may be expressed in terms of the velocity as

$$a_p = -\omega_0^2 r = iv_p\omega_0 \qquad (11.35)$$

The corresponding complex forces are

$$\text{(spring force)} \qquad F_{sp} = -kr = kiv_p/\omega_0 \qquad (11.36)$$

$$\text{(fictitious force)} \qquad F_{ap} = -ma_p = -mi\omega_0 v_p \qquad (11.37)$$

Since the sum of the forces must equal zero, one has

$$F_{sp} + F_{ap} = ikv_p - im\omega_0 v_p = 0$$

or

$$k/\omega_0 = m\omega_0$$

or

$$\omega_0 = \sqrt{(k/m)} \qquad (11.38)$$

Forced Vibrations   Often an external force in addition to the spring force is applied to the mass. In many cases this force has the mathematical form

$$F_{ey} = |F_p| \sin(\omega t + \theta_0) \qquad (11.39)$$

where

F_{ey} = applied force
$|F_p|$ = peak value or amplitude of the applied force
ω = angular frequency of the applied force

This force will cause the mass to vibrate at the applied angular frequency ω. In general, ω is different from the natural angular frequency ω_0. The rotating complex vectors now have the form shown in Figure 11.22. The corresponding complex force vectors are shown in Figure 11.23. Included in the force diagram is the fictitious force, $-ma$. Thus, $\Sigma F = 0$. This yields

$$F_p + m\omega^2 r - kr = 0$$

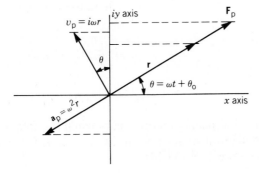

FIGURE 11.22 Rotating complex vectors for forced vibrations.

FIGURE 11.23 Complex
force vectors for forced
vibration.

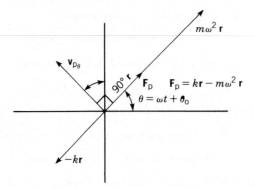

or

$$r (k - m\omega^2) = F_p$$

or

$$r = F_p/(k - m\omega^2) \tag{11.40}$$

Since the value of $v_p = i\omega r$, the value of the complex velocity becomes

$$v_p = (i\omega F_p)/(k - m\omega^2) \tag{11.41}$$

By multiplying the numerator and denominator by i and dividing by ω, one obtains

$$v_p = F_p/\{i[m\omega - (k/\omega)]\} \tag{11.42}$$

The quantity $\{i[m\omega - (k/\omega)]\}$ is called the mechanical impedance of the system. Thus, one may write

$$v_p = F_p/Z_m \tag{11.43}$$

where

$$Z_m = i[m\omega - (k/\omega)]$$

The observed displacement y and velocity v_y are the imaginary components of the complex vectors. Thus

$$y = |r|\sin(\omega t + \theta_0) \tag{11.44}$$

and

$$v_y = |v_p|\cos(\omega t + \theta_0) \tag{11.45}$$

where

$$|r| = |F_p|/(k - m\omega^2) \tag{11.46}$$

and

$$|v_p| = \omega|F_p|/(k - m\omega^2) \tag{11.47}$$

An interesting phenomenon occurs when the frequency of the driving force is equal to the natural frequency. Since $\omega_0^2 = k/m$, $k = m\omega_0^2$. Substituting this value for k in Equation 11.40 yields

$$r = F_p/(m\omega_0^2 - m\omega^2)$$

or

$$r = F_p/(m[\omega_0^2 - \omega^2])\qquad(11.48)$$

If the angular frequency of the driving force approaches the natural frequency, the complex amplitude of the motion increases without bounds. This condition is called resonance. In a real case, the complex amplitude would remain finite because there are always frictional losses.

A particular driven mechanical oscillator is analyzed in Example 11.5.

EXAMPLE 11.5 Driven mechanical oscillator with no friction
. . . .

Given: A 10-kg mass is attached to a spring. This spring constant is 150 N/m. A force having the mathematical form

$$F = (50 \sin 3t) \text{ N}$$

is applied to the mass.

Find:
  a. The natural angular frequency ω_0.
  b. The displacement as a function of time.
  c. The mechanical impedance.
  d. The velocity as a function of time.

Solution: **a.** The value of ω_0 is

$$\omega_0 = \sqrt{(k/m)}$$

$$\omega_0 = \sqrt{(150 \text{ N/m}/1.0 \text{ kg})}$$

$$\omega_0 = 12.2 \text{ rad/sec}$$

  b. The displacement amplitude is

$$|r| = |F_p|/(k - m\omega^2)$$

$$|r| = 50\text{N}/(150 \text{ N/m} - ([1.0 \text{ kg}][(3/\text{sec})2]))$$

$$|r| = (50/[150 - 9]) \text{ m}$$

$$|r| = 0.35 \text{ m}$$

The value of y as a function of time becomes

$$y = |r| \sin \omega t$$

$$y = |0.35 \text{ m}|\sin 3t$$

  c. The mechanical impedance is

$$Z_m = i[m\omega - (k/\omega)]$$

$$Z_m = (i[1.0 \text{ kg}][3 \text{ Hz}] - [150 \text{ N}/(3 \text{ Hz})])$$

$$Z_m = i(3 \text{ N·sec/m} - 50 \text{ N·sec/m})$$

$$Z_m = -47i \text{ N·sec/m}$$

d. The complex velocity is

$$v_p = F_p/Z_m$$

The square of the amplitude is

$$|v_p|^2 = v_p v_p{}^* = (F_p F_p{}^*)/(Z_m Z_m{}^*)$$

$$|v_p|^2 = (50 \text{ N})^2/(-47 \text{ N·sec/m})^2$$

$$|v_p| = 1.06 \text{ m/sec}$$

The value of $|v_p|$ can also be obtained from the equation

$$|v_p| = (\omega|F_p|)/(k - m\omega^2)$$

$$|v_p| = ([3 \text{ Hz}][50 \text{ N}]/[(150 - 9)\text{N·sec/m}])$$

$$|v_p| = 1.06 \text{ m/sec}$$

The value of v_y as a function of time becomes

$$v_y = |v_p| \cos \omega t$$

$$v_y = (1.06 \cos 3t) \text{ m/sec}$$

Forced Vibrations with Fluid Friction   If the velocity of a body is not too large, the frictional force acting on the body moving through a fluid is proportional to the velocity and it acts on the body in a direction opposing the velocity. When a significant frictional force occurs, the driving force is no longer in phase with the displacement since the frictional force is 90° out of phase with the displacement.

When friction occurs, the system of rotating complex vectors takes the form shown in Figure 11.24. As previously shown, the complex velocity leads the displacement by 90° and the complex acceleration is in the opposite direction of the displacement. The complex driving force causes all the vectors to rotate at the frequency of the sinusoidal driving force ω. If r is the complex displacement, then

$$v_p = i\omega r \tag{11.49}$$

$$a_p = -\omega^2 r \tag{11.50}$$

$$F_{fp} = -K v_p = -iK\omega r \tag{11.51}$$

where

v_p = complex velocity
a_p = complex acceleration
K = frictional constant
ω = angular frequency of the driving force
F_{fp} = complex frictional force

The angle between the complex velocity and the complex driving force is called

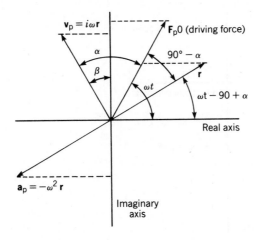

FIGURE 11.24 System of complex rotating vectors.

the phase angle, as indicated by *a* in Figure 11.24. As the vectors rotate at the angular frequency ω, the actual values of displacement, velocity, acceleration, and driving force are the projections of the corresponding complex quantities on the imaginary axis.

To analyze the vibrating system, a force diagram is drawn as shown in Figure 11.25. For this analysis, the time can be chosen so that the velocity vector lies along the real axis. The fictitious force $-ma_p$ is included so that one obtains dynamic equilibrium. The sum of all the forces must then be equal to zero. Since

$$v_p = i\omega r$$

one may write

$$r = v_p/(i\omega) = -(iv_p)/\omega \tag{11.52}$$

and

$$a_p = -\omega^2 r = i\omega v_p \tag{11.53}$$

The spring force becomes

$$F_{sp} = -kr = +(ikv_p)/\omega \tag{11.54}$$

and the fictitious force is

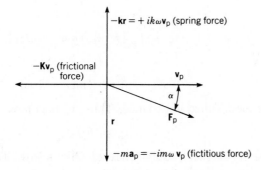

FIGURE 11.25 Force diagram of a complex vibration system.

FIGURE 11.26 Sum of
complex forces.

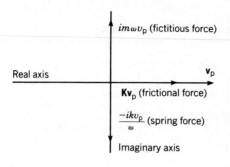

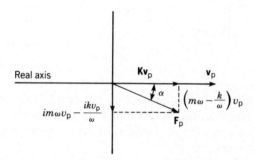

$$F_{ap} = -ma_p = -im\omega v_p \tag{11.55}$$

Since a complex vector is rotated by 90° when it is multiplied by i and rotated by $-90°$ when it is multiplied by $-i$, one obtains the force diagram shown in Figure 11.25.

Since the sum of the forces including the fictitious force must equal zero, one may write

$$F_p - Kv_p - im\omega v_p + (ikv_p)/\omega = 0$$

or

$$F_p = Kv_p + im\omega v_p - (ikv_p)\omega \tag{11.56}$$

Equation 11.56 states that the driving force is the sum of the spring force, the fictitious force, and the frictional force. This is illustrated in Figure 11.26. Equation 11.56 can be solved for v_p. This yields

$$F_p = v_p[K + im\omega - (ik)/\omega]$$

or

$$v_p = F_p/[K + i(m\omega - k/\omega)] \tag{11.57}$$

The quantity

$$Z_m = K + i(m\omega - k/\omega) \tag{11.58}$$

is called the mechanical impedance. Thus, v_p becomes

$$v_p = F_p/Z_m \tag{11.59}$$

which is the mechanical equivalence of Ohm's law. The velocity is the force divided by a resistancelike quantity Z_m.

Substituting the value for v_p given by Equation 11.59 into Equations 11.52 and 11.53 yields

$$r = -iF_p/(\omega Z_m) \qquad (11.60)$$

and

$$a_p = i\omega F_p/Z_m \qquad (11.61)$$

If the mechanical impedance and the applied force are known, then the complex velocity, displacement, and acceleration can be calculated. The phase angle α, from Figure 11.26, becomes

$$\tan \alpha = [m\omega - k/(\omega v_p)]/K v_p$$

or

$$\tan \alpha = (m\omega - k/\omega)/K \qquad (11.62)$$

The actual displacement, velocity, and acceleration as a function of time are now determined since r, v_p, and a_p are known. From Figure 11.24, the projections on the imaginary axis are

$$y = |r| \sin (\omega t - 90° + \alpha) \qquad (11.63)$$

$$v_y = |v_p| \cos (\omega t - 90° + \alpha) = |v_p|\cos \beta \qquad (11.64)$$

$$a_y = -|a_p|\sin(\omega t - 90° + \alpha) \qquad (11.65)$$

Resonance occurs when the frequency of the driving force is equal to the natural frequency ω_0. The natural frequency is

$$\omega_0 = \sqrt{(k/m)}$$

or

$$k = m\omega_0^2 \qquad (11.66)$$

Substituting this value of k into Equation 11.58 yields

$$Z_m = K + i[m\omega - (m\omega_0^2)/\omega] \qquad (11.67)$$

When $\omega = \omega_0$, the value of Z_m becomes

$$Z_m = K \qquad (11.68)$$

and

$$v_p = F_p/K \qquad (11.69)$$

Since K is a real number, v_p and F_p are in phase ($\tan \alpha = 0$). The vibrating system acts as though there is only a resistive force. Since the impedance does not go to zero, the displacement, velocity, and acceleration remain finite.

A sample problem for a forced oscillator is given in Example 11.6.

EXAMPLE 11.6 Forced simple harmonic motion with resistance
. . . .

Given: A 1.0-kg mass is attached to a spring having a spring constant of 50 N/m. A driving force $F_p = (10 \sin 10t)$ N is applied to the mass. The friction constant K is 1.0 N sec/m.

Find: **a.** The impedance.

b. The phase angle α between the velocity and the driving force.

c. The amplitude of the velocity.

d. The amplitude of the displacement.

e. The displacement amplitude if $\omega = \omega_0$ (resonance).

Solution: **a.** The impedance is

$$Z_m = K + i(m\omega - k/\omega)$$

$$Z_m = 1.0 \text{ N·sec/m} + i[(1.0 \text{ kg})(10 \text{ rad/sec}) - (50 \text{ N/m})/(10 \text{ rad/sec})]$$

$$Z_m = [1.0 + (5.0)i] \text{ N·sec/m}$$

b. The phase angle α is

$$\tan \alpha = (m\omega - k/\omega)/K$$

$$\tan \alpha = [(1.0 \text{ kg})(10.0 \text{ rad/sec}) - (50 \text{ N/m})(10 \text{ rad/sec})]/1.0 \text{ N·sec/m}$$

$$\tan \alpha = 5.0$$

$$\alpha = 78.7°$$

c. The amplitude of the velocity is

$$|v_p| = \sqrt{v_p v_p{}^*} = |F_p|/\sqrt{Z_m Z_m{}^*}$$

$$Z_m Z_m{}^* = (1.0 + (5.0)i \text{ N·sec/m})(1.0 - (5.0)i \text{ N·sec/m})$$

$$Z_m Z_m{}^* = (1.0)^2 + (25)^2 \text{ N}^2\text{·sec}^2/\text{m}^2$$

$$Z_m Z_m{}^* = 26 \text{ N}^2\text{·sec}^2/\text{m}^2$$

$$\sqrt{Z_m Z_m{}^*} = 51 \text{ N·sec/m}$$

$$|v_p| = 10 \text{ N}/(5.1 \text{ N·sec/m})$$

$$|v_p| = 1.96 \text{ m/sec}$$

d. The amplitude of the displacement is

$$|r| = |v_p|/\omega$$

$$|r| = (1.96 \text{ m/sec})/(10 \text{ rad/sec})$$

$$|r| = 0.196 \text{ m}$$

e. At resonance

$$|v_p| = |F_p|/K$$

$$|v_p| = 10 \text{ N}/(1.0 \text{ N·sec/m})$$

$$|v_p| = 10 \text{ m/sec}$$

$$|r| = |v_p|/\omega$$

$$|r| = 10 \text{ m/sec}/(10 \text{ rad/sec})$$

$$|r| = 1.0 \text{ m}$$

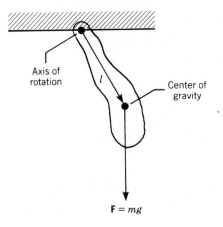

FIGURE 11.27 Physical pendulum.

F = *mg*

The Physical Pendulum   A physical pendulum is illustrated in Figure 11.27. The pendulum swings back and forth about a fixed axis of rotation. The equation for the motion is

$$\tau = I\alpha \tag{11.70}$$

where

τ = torque
I = moment of inertia
α = angular acceleration

The pendulum can have any shape and the force of gravity (*mg*) acts through the center of gravity. The gravitational force can be broken up into two components such that one is perpendicular to *l* and one is in the direction of *l*. The component of force in the direction of *l* passes through the axis of rotation and, therefore, does not contribute to the torque about the axis of rotation. The two components are illustrated in Figure 11.28. From the figure, one obtains

$$F_T = -mg \sin \theta \tag{11.71}$$

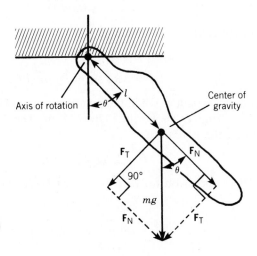

FIGURE 11.28   Forces acting on a physical pendulum.

TABLE 11.1   Sin θ and θ for small angles

θ (degrees)	θ (radians)	sin θ	Percentage Error
2.87	0.05	0.04998	0.042
5.73	0.10	0.9983	0.17
8.59	0.15	0.14944	0.37
17.2	0.30	0.2955	1.49

Thus, the total torque about the axes of rotation becomes

$$\tau = -F_T l = -mgl \sin \theta \tag{11.72}$$

Substituting this expression for the torque in Equation 11.70 yields

$$-mgl \sin \theta = I\alpha \tag{11.73}$$

The equation for θ as a function of time is greatly simplified if θ remains small. When this occurs, one can use the approximation

$$\sin \theta \approx \theta$$

This approximation is shown for small angles in Table 11.1. It is important to note that θ must be in radians for the approximation to hold. With this approximation, Equation 11.73 becomes

$$-mgl\theta = I\alpha \tag{11.74}$$

Equation 11.74 has the same mathematical form as the equation for a mass at the end of a spring:

$$-ky = ma \tag{11.75}$$

Thus, k corresponds to mgl, y corresponds to θ, m corresponds to I, and a corresponds to α. Since $\omega_o = \sqrt{(k/m)}$ for the mass at the end of a spring, the corresponding angular frequency for the pendulum becomes

$$\omega_0 = \sqrt{(mgl/I)} \tag{11.76}$$

The corresponding period and frequency are

$$P = 2\pi/\omega_0 = 2\pi/[\sqrt{(I/mgl)} \tag{11.77}$$

$$\nu = 1/P = \omega_0/2\pi = [\sqrt{(mgl/I)}]/2\pi \tag{11.78}$$

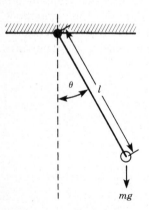

FIGURE 11.29 Special case of the physical pendulum.

An important case occurs when the pendulum consists of a small spherical bob at the end of a string as shown in Figure 11.29. The moment of inertia can be approximated by the equation

$$I = \Sigma \Delta m_i R_i^2 \approx ml^2 \tag{11.79}$$

Substituting this expression for I in Equations 11.76, 11.77, and 11.78 yields

$$\omega_0 = \sqrt{(g/l)} \tag{11.80}$$

$$P = 2\pi\sqrt{(l/g)} \tag{11.81}$$

$$\nu = \sqrt{(g/l)}/2\pi \qquad (11.82)$$

This particular pendulum is called a simple pendulum, and other pendulums are called compound pendulums. Because the pendulum swings back and forth in a definite period of time, it is often used in clocks as a timing device. Calculations for particular pendulums are given in Examples 11.7 and 11.8.

EXAMPLE 11.7 The simple pendulum

. . . .

Given: A simple pendulum.

Find: The length that will make the period one second.

Solution: The equation for a simple pendulum is

$$P = 2\pi\sqrt{(l/g)}$$

Square both sides of the equation.

$$P^2 = 4\pi^2 l/g$$

Solve the equation for l.

$$l = (P^2 g)/(4\pi^2)$$

$$l = ((1.0 \text{ sec})^2 (9.8 \text{ m/sec}))/(4\pi^2)$$

$$l = 0.248 \text{ m}$$

EXAMPLE 11.8 The compound pendulum

. . . .

Given: A pendulum is made from a slender rod and is supported at one end. The moment of inertia of a slender rod with the fixed axis of rotation at one end is $I = (ml^2)/3$.

Find: The length required to give a period of one second.

Solution: The equation for the period is

$$P = 2\pi\sqrt{(I/mgl)}$$

Substitute $(ml^2)/3$ for I.

$$P = 2\pi\sqrt{[(ml^2)/3mgl]}$$

$$P = 2\pi\sqrt{(l/3g)}$$

Square both sides of the equation.

$$P^2 = 4\pi^2 l/3g$$

Solve the equation for l.

$$l = 3P^2g/4\pi^2$$

$$l = ((3)(1.0 \text{ sec})^2(9.8 \text{ m/sec}))/(4\pi^2)$$

$$l = 0.745 \text{ m}$$

SECTION 11.1   EXERCISES

1. Distinguish between vibrations produced by internal and external sources. Give two examples of each type and prepare a labeled sketch of each.

2. In one or two sentences define each term:

    Steady-state vibration   Isolation
    Transient vibration   Frequency
    Periodic motion   Period
    Amplitude

3. A rock crusher has an acceleration of 5.5 cm/sec^2. It is spring mounted on a wooden floor, which has a measured acceleration of 3 cm/sec^2. What is the value for transmission of the vibrations?

4. Find the period of a mechanical vibration if the structure vibrating completes 3000 cycles in 6 sec.

5. What is the value for external transmission of vibration if the passenger in a car accelerates 0.1 cm/sec^2 while the wheels of the car accelerate 1.62 cm/sec^2?

6. Add and subtract the complex numbers

    $10 + 7i$

    $20 - 6i$

7. Multiply and divide the complex numbers

    $5 + 6i$

    $7 + 2i$

8. Find the absolute magnitude and angle from the real axis of the complex number

    $6 + 4i$

9. Rotate the complex number $7 - 6i$ through 90°.

10. State Hooke's law for a spring.

11. The angular frequency of a mass at the end of a spring is 2π rad/sec. The amplitude of the displacement is 0.40 m. What is the maximum velocity? What is the minimum velocity? What is the maximum acceleration? What is the minimum acceleration?

12. A 1.0-kg mass is attached to a spring. The spring constant is 20 N/m. What is the angular frequency? What is the period of the oscillation?

13. A 1.0-kg mass at the end of a spring is driven at an angular frequency of 10 rad/sec. If the spring constant is 20 N/m, what is the mechanical impedance? Assume no friction.

14. If the driving force in Problem 11 is 10 N, what is the corresponding complex velocity? What is the magnitude (amplitude) of the velocity? What is the amplitude of the displacement?

15. A 2.0-kg mass is attached to a spring with a spring constant of 30 N/m. A driving force, $F_e = 15 \sin(20t + \phi)$ N, is applied to the mass. The frictional constant is 1.5 N sec/m. Find the impedance, the amplitude of the velocity, the phase angle, and the amplitude of the displacement.

16. In Problem 13, find the amplitude of the displacement if the angular frequency of the driving force is equal to the natural frequency.

17. A simple pendulum has a length of 0.40 m. What is the period of the pendulum?

18. The moment of inertia of a compound pendulum is 0.16 kg·m² about the axis of rotation. The length to the center of mass is 0.15 m. What is the period of the pendulum?

11.2

ELECTRICAL VIBRATIONS

OBJECTIVES

Upon completion of this section, the student should be able to

- Draw an electrical circuit analogous to a simple mechanical oscillator.

- Calculate the natural frequency of an *LC* circuit.

- Calculate the current and voltage amplitudes of an *LC* circuit if the initial charge on the capacitor is given.

- Calculate the capacitive and inductive reactance as well as the total impedance of a circuit composed of resistors, capacitors, and inductors if the frequency of the voltage source is given.

- Understand resonance in electrical circuits.

- Calculate the phase angle between voltage and current in an ac circuit.

- Relate a mechanical oscillator to an equivalent electrical oscillator.

- Understand and use the concept of electrical impedance in driven electrical circuits.

DISCUSSION

Definitions   Electrical vibrations occur when the voltage and current in a circuit change periodically with time. These oscillations may also be sinusoidal or complex. However, complex oscillations can be the mathematical equivalent of a sum of sinusoidal vibrations. In this chapter, only sinusoidal oscillations will be considered, and they will be related to analogous mechanical vibrations. The oscillating current and voltage are called alternating current and voltage. For example, the alternating voltage and current may be expressed mathematically as

$$V = |V_p| \sin \omega t$$

$$I = |I_p| \sin(\omega t + \alpha)$$

where

V = voltage at any instant of time
I = current at any instant of time
$|V_p|$ = peak voltage or amplitude
$|I_p|$ = peak current or amplitude
ω = angular frequency in rad/sec
t = time in sec
α = phase angle

The voltage and current are not independent, and if one is given for a particular circuit, the other can be calculated. As in the case of the mechanical system, the period is the time for one complete oscillation and the frequency is the number of oscillations per second. One cycle per second is defined as one hertz and the frequency is usually given in hertz.

Commercially produced electricity in the United States has a frequency of 60 Hz. Thus, the theory of alternating current circuits is very important in the generation, distribution, and utilization of electric power. Other frequencies are used in various devices such as radio and television transmitters.

Natural Frequency Consider the circuit shown in Figure 11.30. As previously shown, the voltage across the capacitor is q/C and the voltage across the inductor, for very small values of Δt, is $L[\Delta I/\Delta t]$. The charge q is measured in coulombs, the capacitance C in farads, and the inductance L in henrys. In any electrical circuit, the sum of the voltages around a closed loop must equal zero. This law yields

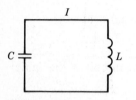

FIGURE 11.30
Schematic for an LC
(inductor–capacitor) circuit.

$$\Sigma V_n = 0 \tag{11.83}$$

$$L(\Delta I/\Delta t) + q/C = 0^1 \tag{11.84}$$

This equation has the same form as the mechanically equivalent system of a mass on the end of a spring where the sum of the forces must equal zero ($-ma_y$ included in the sum).

$$\Sigma F_n = 0$$

$$-ma_y - ky = 0$$

or

$$m(\Delta v_y/\Delta t) + ky = 0 \tag{11.85}$$

In Equation 11.85, the definition of acceleration

$$a_y = \Delta v_y/\Delta t$$

is used when Δt is very small.

The electrical equation can be obtained directly from the mechanical equation since it was shown that L corresponds to mass, $1/C$ corresponds to a spring constant, charge q corresponds to the displacement y, and the current I corresponds to the velocity v_y. In both cases, the sum of the forcelike quantity is equal to zero.

As in the case of the mechanical system, the electrical system will oscillate with a natural frequency ω_0. For the mechanical system

$$\omega_0 = \sqrt{(k/m)} \tag{11.86}$$

Thus by correspondence, ω_0 for the electrical system is

$$\omega_0 = \sqrt{(1/(LC))} \tag{11.87}$$

Also by correspondence,

[1] Note that the interval Δt for $\Delta I/\Delta t$ needs to be kept small. To be truly accurate the limit as Δt approaches zero of $\Delta I/\Delta t$ should be used; this is equivalent to dI/dt, which is the calculus expression for the slope at a point.

$$y = |v_{\mathrm{p}}|/[\omega_0 \sin(\omega_0 t + \theta_0)]$$

$$q = |I_{\mathrm{p}}|/[\omega_0 \sin(\omega_0 t + \theta_0)] \tag{11.88}$$

$$v_y = |v_{\mathrm{p}}| \cos(\omega_0 t + \theta_0)$$

$$I = |I_{\mathrm{p}}| \cos(\omega_0 t + \theta_0) \tag{11.89}$$

$$a_y = -\omega_0 |v_{\mathrm{p}}| \sin(\omega_0 t + \theta_0)$$

$$\Delta I/\Delta t = -\omega_0 |I_{\mathrm{p}}| \sin(\omega_0 t + \theta_0) \tag{11.90}$$

Since it was assumed that the circuit has no energy losses, the oscillations will continue with constant amplitude at the angular frequency ω_0. As in the mechanical case, the period is

$$P = 2\pi/\omega_0$$

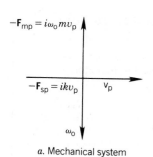

$-\mathbf{F}_{mp} = i\omega_0 m v_{\mathrm{p}}$

$-\mathbf{F}_{sp} = ikv_{\mathrm{p}}$ v_{p}

ω_0

a. Mechanical system

or

$$P = 2\pi\sqrt{(LC)} \tag{11.91}$$

and the corresponding frequency is

$$\nu = 1/P = 1/[2\pi\sqrt{(LC)}] \tag{11.92}$$

A rotating complex vector system may be used to represent the oscillations. The current, capacitor voltage, and inductor voltage are usually shown on the diagram. Since the velocity leads the displacement by 90°, the current leads the charge by 90°. Thus, the current leads the voltage across the capacitor, q/C, by 90°. The peak voltage across the capacitor, according to Equation 11.88, is

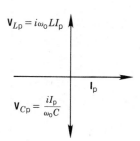

$\mathbf{V}_{Lp} = i\omega_0 L I_{\mathrm{p}}$

I_{p}

$\mathbf{V}_{Cp} = \dfrac{iI_{\mathrm{p}}}{\omega_0 C}$

b. Electrical system

FIGURE 11.31 Complex vector diagrams of mechanical and electrical systems.

$$|V_{Cp}| = |q_{\mathrm{p}}|/C = |I_{\mathrm{p}}|/C\omega_0 \tag{11.93}$$

From Equation 11.90, the peak voltage across the inductor is

$$|V_{Lp}| = L[\Delta I/\Delta t]_{\mathrm{p}} = -\omega_0 L |I_{\mathrm{p}}| \tag{11.94}$$

The complex voltage across the inductor, corresponding to ma_{p}, leads the current by 90°. The complex vector diagram is usually drawn at a time when the current lies along the real axis as shown in Figure 11.31. As the complex vectors rotate at the angular frequency ω_0, their projections on the imaginary axis give their actual values as a function of time. An equivalent system of complex vectors is also shown in Figure 11.31. Plots of the actual current, capacitor voltage, and inductor voltage are shown in Figure 11.32. As shown in Figure 11.32*b*, the sum of

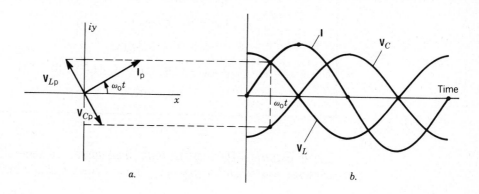

FIGURE 11.32 Phase relationships for capacitor voltage, inductor voltage, and current.

the voltages across the capacitor and inductor is always zero. Calculations for a particular *LC* circuit are given in Example 11.9.

EXAMPLE 11.9   Electrical oscillations for an *LC* circuit
. . . .

Given: A 0.25-H inductor and a 0.10-μF capacitor are connected as shown in Figure 11.30. The maximum charge on the capacitor is 1.0×10^{-4} C.

Find: **a.** The natural angular frequency ω_0, the period P, and the frequency v.
     b. The peak current I_p.
     c. The peak voltages across the capacitor and the inductor.

Solution: **a.** The angular frequency is

$$\omega_0 = 1/\sqrt{(LC)}$$

$$\omega_0 = 1/\sqrt{(0.10 \text{ H})(1.0 \times 10^{-7} \text{ F})}$$

$$\omega_0 = 1.0 \times 10^4 \text{ rad/sec}$$

The period is

$$P = (2\pi/\omega_0)$$

$$P = (2\pi/(1.0 \times 10^4 \text{ rad/sec}))$$

$$P = 6.3 \times 10^{-4} \text{ sec}$$

The frequency is

$$v = 1/P$$

$$v = 1/(6.3 \times 10^{-4} \text{ sec})$$

$$v = 1.59 \times 10^3 \text{ Hz}$$

b. The peak current is

$$|I_\text{p}| = \omega_0 |q_\text{p}|$$

$$|I_\text{p}| = (1.0 \times 10^4 \text{ rad/sec})(1.0 \times 10^{-4} \text{ C})$$

$$|I_\text{p}| = 1.0 \text{ A}$$

c. The peak voltage is

$$|V_{Cp}| = |V_{Lp}| = |q_\text{p}|/C$$

$$|V_{Cp}| = (1.0 \times 10^{-4} \text{ C})/(1.0 \times 10^{-7} \text{ F})$$

$$|V_{Cp}| = 1.0 \times 10^3 \text{ V}$$

Forced Electrical Oscillations with Resistance   In general, an electric circuit has a voltage source and resistance. Thus, a more realistic circuit is shown in Figure

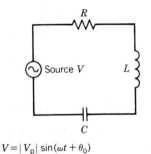

$V = |V_p|\sin(\omega t + \theta_0)$

FIGURE 11.33
Schematic of an RLC
(resistor–inductor–
capacitor) circuit.

11.33. As for the mechanical system, a common driving source has the mathematical form

$$V = |V_p|\sin(\omega t + \theta_0) \tag{11.95}$$

where

V = source voltage as a function of time
$|V_p|$ = peak voltage or amplitude
ω = angular frequency of the source
θ_0 = angle when $t = 0$

The sum of the voltages around the closed path must equal zero. This condition yields

$$|V_p|\sin(\omega t + \theta_0) - L\,\Delta I/\Delta t - q/C - IR = 0 \tag{11.96}$$

or

$$|V_p|\sin(\omega t + \theta_0) = L\Delta I/\Delta t + q/C + IR \tag{11.97}$$

The quantity IR, as previously shown, is the voltage drop across the resistor. Equation 11.97 has the same form as the driven mechanical oscillator

$$|F_p|\sin(\omega t + \theta_0) = m\Delta v_y/\Delta t + ky + v_y K \tag{11.98}$$

In Equation 11.98, K corresponds to the electrical resistance R and $|F_p|$ corresponds to $|V_p|$. Thus, a rotating complex vector system can be used to show the relationships among the electrical quantities. The mechanical and electrical vectors are compared to Figure 11.34.

The voltage across the inductor leads the current by 90°, and the voltage across the capacitor lags the current by 90°. As in the case of the mechanical system, the projections of the vectors on the imaginary axis give the voltages and current as a function of time as the vectors rotate in the counterclockwise direction with an angular velocity of ω. The angle α between V_p and I_p is called the phase angle.

Using Figure 11.34b, one may write

$$V_p = I_p R + i\omega L I_p - (iI_p\omega C) \tag{11.99}$$

or

$$V_p = I_p\{R + i[\omega L - (1/\omega C)]\} \tag{11.100}$$

The quantity ωL is called the inductive reactance and $1/[\omega C]$ is called the

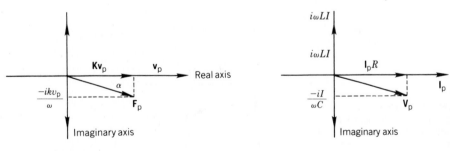

FIGURE 11.34 Comparison of complex vectors for mechanical and electrical systems.

a. Mechanical system

b. Electrical system

capacitive reactance. They are usually expressed as

$$X_L = \omega L \tag{11.101}$$

$$X_C = 1/\omega C \tag{11.102}$$

With this notation, Equation 11.100 may be expressed as

$$V_p = I_p[R + i(X_L - X_C)] \tag{11.103}$$

The quantity $R + i[\omega L - (1/\omega C)]$ is the electrical impedance of the given circuit. It relates the voltage to the current just as the mechanical impedance relates the force to the velocity. It is given the symbol Z. Thus,

$$Z = R + i(X_L - X_C) \tag{11.104}$$

and

$$V_p = I_p Z \tag{11.105}$$

This is Ohm's law for an alternating current series circuit. If both sides of Equation 11.105 are divided by $\sqrt{2}$, an equation for the root-mean-square values is obtained (see Chapter 3):

$$V_p/\sqrt{2} = (I_p Z)/\sqrt{2} \tag{11.106}$$

or

$$V_{rms} = (I_{rms})(Z) \tag{11.107}$$

The root-mean-square values are usually given the symbols $\mathbf{I}$ and $\mathbf{V}$, where $\mathbf{I}$ and $\mathbf{V}$ are complex numbers. Thus Equation 11.105 becomes

$$\mathbf{V} = \mathbf{IZ} \quad \text{(complex values)} \tag{11.108}$$

As in the mechanical case, the phase angle α is obtained from the impedance as shown in Figure 11.35. From this figure, one obtains

$$\tan \alpha = [I_p(X_L - X_C)]/(I_p R)$$

or

$$\tan \alpha = (X_L - X_C)/R \tag{11.109}$$

Resonance occurs when $X_L = X_C$. For this condition $Z = R$ and $\tan \alpha = 0$. Thus, the phase angle is zero and the applied voltage and current are in phase. If $X_L = X_C$, then $\omega L = 1/\omega C$. Solving this equation for ω yields $\omega = 1/[\sqrt{(LC)}]$. This

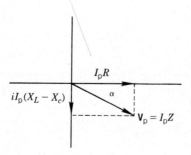

FIGURE 11.35 Complex vector diagram for electrical impedance showing phase angle.

angular frequency, however, is the natural frequency ω_0. Thus resonance occurs when $\omega = \omega_0$.

A sample problem is solved in Example 11.10 for a series alternating current circuit.

EXAMPLE 11.10   A series alternating circuit
· ·

Given: A 1.0-H inductor, a 10×10^{-6}-F-capacitor, and a 100-Ω resistor are connected in series. A voltage of the form $V = 110 \sin 377t$ is applied to the circuit.

Find: **a.** The inductive reactance.
b. The capacitive reactance.
c. The inductance.
d. The peak current.
e. The phase angle.
f. The peak voltage across the inductor.
g. The peak voltage across the capacitor.
h. The peak voltage across the resistor.
i. Draw the voltage vector diagram.

Solution: **a.** The inductive reactance is

$$X_L = \omega L$$

$$X_L = (377)(1.0)$$

$$X_L = 377 \ \Omega$$

b. The capacitive reactance is

$$X_C = 1/[\omega C]$$

$$X_C = 1/[(377)(10 \times 10^{-6} \ \text{F})]$$

$$X_C = 265 \ \Omega$$

c. The impedance is

$$Z = R + i(X_L - X_C)$$

$$Z = 100 \ \Omega + i[377 \ \Omega - 265 \ \Omega]$$

$$Z = 100 + 112i \ \Omega$$

d. The complex current is

$$I_p = V_p/Z$$

The peak current is

$$|I_p| = |V_p|/[\sqrt{(ZZ^*)}]$$

$$ZZ^* = 100^2 + 112^2 = 22{,}544$$

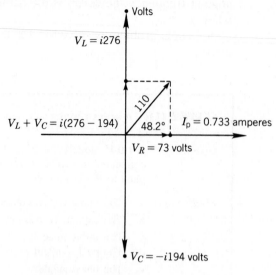

$$\sqrt{ZZ^*} = \sqrt{(22,544)}$$

$$\sqrt{ZZ^*} = 150 \ \Omega$$

$$|I_p| = (110 \ \Omega)/[150 \ \Omega]$$

$$|I_p| = 0.733 \ A$$

e. The tangent of the phase angle is

$$\tan \alpha = (X_L - X_C)/R$$

$$\tan \alpha = \{[377 \ \Omega - 265 \ \Omega]/[100 \ \Omega]\}$$

$$\tan \alpha = 1.12$$

$$\alpha = 48.2°$$

f. The peak complex voltage across this inductor is

$$V_{Lp} = i\omega L I_p = iX_L I_p$$

The peak value squared is

$$|V_{Lp}|^2 = (iX_L I_p)(-iX_L I_p^*)$$

$$|V_{Lp}|^2 = X_L^2 |I_p|^2$$

$$|V_{Lp}| = X_L |I_p|$$

$$|V_{Lp}| = 377 \ \Omega(0.733 \ A)$$

$$|V_{Lp}| = 276 \ V$$

g. The peak complex voltage across the capacitor is

$$V_{Cp} = -iI_p/\omega C = -iX_C I_p$$

The peak voltage squared is

$$|V_{Cp}|^2 = (-iX_C I_p)(iX_C I_p^*) = X_C^2 |I_p|^2$$

$$|V_{Cp}| = X_C |I_p|$$

$$|V_{Cp}| = [265 \ \Omega](0.733 \ A)$$

$$|V_{Cp}| = 194 \ V$$

h. The voltage across the resistor is

$$|V_{Rp}| = |I_p| R = (0.733 \ A)[100 \ \Omega]$$

$$|V_{Rp}| = 73.3 \ V$$

i. A vector diagram of the voltages is shown on the opposite page.

Alternating Current Circuits Using the same methods that were used for direct current circuits in Chapter 5, the student can show the following laws. For a series circuit, where the total voltage across a circuit is equal to the sum of the voltages,

$$Z_T = Z_1 + Z_2 + Z_3 + \cdots + Z_n \tag{11.110}$$

To find the total impedance of impedances in series, one adds the individual impedances.

For a parallel circuit, where the total current is the sum of the individual currents,

$$1/Z_T = 1/Z_1 + 1/Z_2 + 1/Z_3 + \cdots + 1/Z_n \tag{11.111}$$

The reciprocal of the total impedance is the sum of the reciprocals of the individual impedances. A parallel circuit is analyzed in Example 11.11.

EXAMPLE 11.11 Alternating current circuit
. . . .

Given: The electric circuit shown in the diagram.

Find: **a.** The total impedance.
               b. The peak current.
               c. The phase angle.
               d. The frequency that makes the impedance a maximum.

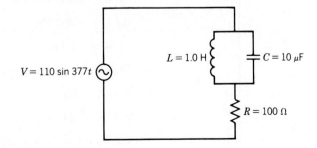

Solution: **a.**   The impedance of parallel circuit elements is

$$1/Z_P = 1/Z_1 + 1/Z_2$$

$$1/Z_P = 1/(iX_L) + 1/(-iX_C)$$

$$1/Z_P = (iX_L - iX_C)/(iX_L)(-iX_C)$$

$$1/Z_P = i(X_L - X_C)/(X_L X_C)$$

$$Z_P = (X_L X_C)/i(X_L - X_C)$$

$$Z_P = i(X_L X_C)/(X_C - X_L)$$

$$X_L = \omega L$$

$$X_L = (377 \text{ rad/sec})(1.0 \text{ H})$$

$$X_L = 377 \ \Omega$$

$$X_C = 1/[\omega C]$$

$$X_C = 1/[(377 \text{ rad/sec})(10 \times 10^{-6} \text{ F})]$$

$$X_C = 265 \ \Omega$$

$$Z_P = \{i[(377 \ \Omega)(265 \ \Omega\$]/[265 \ \Omega\text{-}377 \ \Omega]\}$$

$$Z_P = -892i \ \Omega$$

The total impedance is (impedances in series)

$$Z_T = R + Z_P$$

$$Z_T = 100 \ \Omega - 892i \ \Omega$$

$$Z_T = (100 - 892i) \ \Omega$$

b.   The peak current is

$$|I_p| = |V_p|/|Z|$$

$$|I_p| = |V_p|/[\sqrt{(ZZ^*)}]$$

$$|I_p| = 110 \text{ V}/\{\sqrt{[100^2 + 892^2]}\Omega\}$$

$$|I_p| = 0.123 \text{ A}$$

c.   The phase angle is

$$\tan \alpha = \text{Imaginary part/Real part}$$

$$\tan \alpha = -892/100$$

$$\alpha = -83.6°$$

d.   The frequency that gives the maximum impedance is

$$Z = R + Z_P$$

$$Z = R + i[X_C X_L]/[X_C - X_L]$$

The impedance increases without bounds when X_C approaches X_L.

$$X_L = \omega L$$

$$X_C = 1/[\omega C]$$

$$\omega L = 1/[\omega C]$$

$$\omega^2 = 1/LC$$

$$\omega = 1/\sqrt{(LC)}$$

$$\omega = \omega_0$$

Thus the impedance increases without bounds when ω approaches the natural frequency ω_0. This condition is called parallel resonance. In a real circuit, the inductance and capacitance will also have some resistance. The resistance will cause the impedance to remain finite when $\omega = \omega_0$.

SECTION 11.2 *EXERCISES*

1. Draw sine wave sketches showing the phase relationship between current and voltage for a resistor, a capacitor, and an inductor.

2. Draw sine wave sketches showing the phase relationship between voltage and current for an *RL* and an *RC* series circuit.

3. A 0.300-H coil is in series with a 700-Ω resistor. The frequency of the applied voltage is 1500 Hz. Calculate the impedance of the circuit and draw a vector diagram of the impedance to scale.

4. A 50-mH coil is connected in series with a 80-Ω resistor. The circuit is connected to a 60-Hz, 117-V source. Calculate the impedance, phase angle, power factor, and average power. (Review Chapter 7.)

5. What value of capacitor should be placed in the circuit in Problem 4 to produce a series resonant circuit?

6. What is the phase angle in a circuit if the total capacitance of the circuit is 20 μF, the total inductance is 30 mH, the total resistance is 130 Ω, and the circuit is connected to a 60-Hz voltage source?

7. What is the product of the inductance and the capacitance in a circuit if the resonant frequency is 60 Hz?

8. A 10-Ω resistor, a 10-μF capacitor, and a 1.0-H inductor are connected in parallel. Find the total impedance if the frequency is 60 Hz.

11.3

OBJECTIVES

WAVES Upon completion of this section, the student should be able to

- Distinguish between longitudinal and transverse waves by giving at least two examples of each type and by drawing and labeling a sketch of each.

- Define the following words associated with waves and wave motion:
  Propagation medium
  Wavelength

Frequency
Period
Displacement
Amplitude
Phase
Superposition
Standing wave
Constructive interference
Destructive interference
Beats

- Calculate the wavelength of a wave given its velocity and frequency.

- Interpret the following equations, and explain the significance of each symbol:

$$y = A_+ \sin\{[2\pi/\lambda](x - vt)]\}$$

$$y = A_+ \sin\{[\omega/v]x - [\omega t]\}$$

- Describe the superposition principle.

- Describe the wave phenomenon in each of the following energy systems:
Mechanical
Fluid
Electrical

- Calculate the resonant frequencies of a cavity given the wave velocity and length of the cavity.

- Calculate the wave velocity on a string given the mass per unit length and the tension.

- Calculate the amplitudes of the reflected and transmitted waves at the boundary between two strings of different mass per unit length.

- Calculate the wave velocity on an electrical transmission line.

- Calculate the transmitted and reflected amplitudes of voltage and current when an electrical transmission line is terminated in a given impedance.

- Calculate the characteristic impedance of two-wire and coaxial transmission lines.

- Terminate a transmission line so there is no reflected wave.

DISCUSSION

General Properties

Introduction   A wave is a disturbance in which energy is transferred from one location to another. Some well-known examples of waves are sound waves that move through the earth following an earthquake, water waves produced by the vertical motion of the surface of a lake, and light waves that can travel through a vacuum.

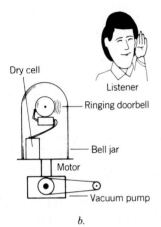

FIGURE 11.36 Sound waves require air for propagation.

Several types of waves require a material in which to travel, or propagate, themselves. This material is referred to as a "medium." Sound is an example of a wave that requires a medium through which to move. The propagation of sound waves through air is a fact of common experience. Sound also can travel through liquids and solids. Sound cannot travel through a vacuum (Figure 11.36).

Electromagnetic waves do not require a medium. Light, for example, is an electromagnetic wave that reaches the earth from the sun through the vacuum of outer space. Other waves of this type are radio waves, x rays, and infrared radiation.

The energy transmitted by a wave may be in the form of a single burst, or pulse, or it may be in the form of many identical pulses evenly spaced. Consider a person holding one end of a rope that is fastened to a wall on the other end (Figure 11.37). The person makes one up-and-down motion with an arm. This motion is transmitted to the end of the rope, which then transmits the motion gradually down the entire rope length. The result is the generation of a single pulse wave, which moves down the rope.

When the person moves an arm up and down repeatedly, the wave produced is made up of a series of identical pulses (Figure 11.38). When wave pulses are repeated at regular intervals, the wave is called a "periodic wave."

A shock wave moving away from the scene of an explosion is an example of sound energy being transferred in the form of a single pulse through the air. A musical tone coming from a vibrating guitar string is an example of a continuous wave of sound energy.

Transverse Waves "Transverse" waves are those in which the disturbance producing the wave moves perpendicularly to the direction in which the disturbance is propagated. The wave motion produced in a rope is one example.

Here a sine wave will be used as an example of transverse wave motion. Figure 11.39 shows a sine wave being produced by a swinging pendulum to which a pencil is fastened. A piece of paper is pulled under the pendulum at a constant rate of speed, while the pencil traces the path of the swinging pendulum. Motion that generates a sine wave is called "simple harmonic motion."

Displacement at any time is the distance from its center, or equilibrium posi-

FIGURE 11.37 Single pulse traveling along a rope.

FIGURE 11.38 Periodic wave traveling along a rope.

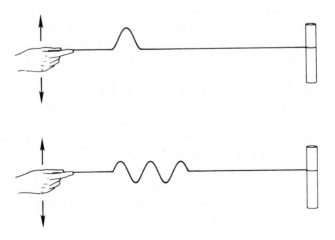

FIGURE 11.39   Sine wave produced by a swinging pendulum.

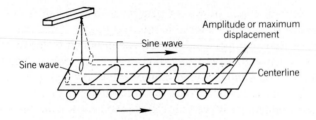

FIGURE 11.39 Sine wave produced by a swinging pendulum.

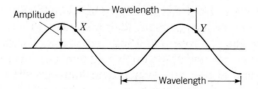

FIGURE 11.40 Amplitude and wavelength of a wave.

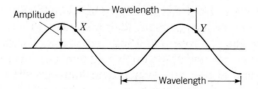

tion. The maximum displacement of the pendulum is called the "amplitude" of the wave. The energy transferred by the wave is proportional to the square of the amplitude. The wavelength λ is the distance between identical points of the wave—from crest to crest, for example. Both amplitude and wavelength are illustrated in Figure 11.40.

The pulses that make up periodic waves can have many shapes. Several pulse shapes are illustrated in Figure 11.41.

The frequency is the number of complete waves produced each second. When one wave is produced each second, the wave has a frequency of one hertz.

The time required for one complete wave pulse is known as the "period" of the wave. The period and the frequency are reciprocals of each other, as indicated by

$$P = 1/\nu \tag{11.112}$$

$$\nu = 1/P \tag{11.113}$$

where

P = period of the wave
ν = frequency of the wave

The position and motion of a given point on a wave defines the phase of the wave. Points that have the same displacement and that are moving in the same direction are in phase with each other. Thus, points x and y on the wave in Figure 11.40 are in phase.

Longitudinal Waves   A "longitudinal" wave is one in which the disturbance producing the pulses moves parallel to or along the direction in which the disturbance is propagated. If a slightly stretched spring is attached to a barely swinging pendulum in such a way that the pendulum would swing parallel to the spring (Figure 11.42), a longitudinal wave would be produced.

As illustrated in Figure 11.42, when the pendulum has moved to its maximum right-hand position, a "compression" of the spring is produced. Similarly, when it has moved to the left-hand position, an elongation, or "rarefaction," is produced.

FIGURE 11.41 Some of the possible pulse shapes of waves.

Square wave

Triangular wave (sawtooth wave)

Sound wave produced by clarinet

FIGURE 11.42 Example of a longitudinal wave.

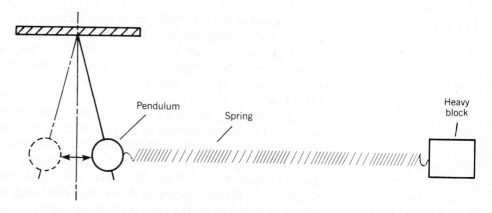

This series of compressions and rarefactions represents a periodic longitudinal wave moving from left to right in the spring. Sound waves are longitudinal waves.

A comparison can be drawn between the compression and rarefactions of a longitudinal wave and the crests and troughs of a transverse wave (Figure 11.43).

The definitions for amplitude, wavelength, frequency, and period presented for transverse waves apply equally well to longitudinal waves. For example, the wavelength λ is the distance between adjacent compressions.

Wave Velocity   A wave pulse, or disturbance, moves a distance equal to its wavelength in a time equal to the period of the wave. Since velocity is distance divided by time, the relationship expressed by Equation 11.114 holds true for all wave motion:

$$v = \lambda/P \qquad (11.114)$$

where

v = wave velocity
λ = wavelength
P = period of wave

Since

$$P = 1/v$$

the velocity becomes

$$v = v\lambda \qquad (11.115)$$

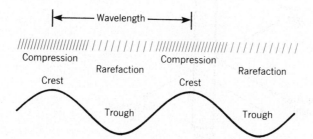

FIGURE 11.43 Comparison of transverse and longitudinal waves.

The speed of a wave is affected greatly by the medium through which the wave travels. For example, the sound of a moving train travels faster through the rails of the track than through air. Even in air, the speed of sound depends on the temperature. Other characteristics of a medium that can affect the speed of sound or light waves through that medium include density, elasticity, and index of refraction.

Waves traveling through a uniform medium do not change velocity, but waves that travel from one medium to another medium do experience velocity changes. This can be illustrated by the longitudinal wave motion in a moderately stiff and a very stiff spring connected in series. Consider a wave moving in a moderately stiff spring (Figure 11.44). When the wave pulse from the moderately stiff spring reaches the boundary between the two springs, some of the energy in the pulse will be transferred to the very stiff spring, where it will continue to travel at a slightly different velocity. The remainder of the energy will be reflected by the boundary and will bounce back in the moderately stiff spring. Thus, some of the energy is reflected and some is transmitted. The same effect occurs, but in different proportions, if the wave travels from the stiff spring into a very soft spring.

A similar effect can be seen when a beam of light waves is directed onto a flat piece of glass as shown in Figure 11.45.

When a light wave strikes a flat piece of glass, some of the light enters the glass and some is reflected. Again, some of the light passing through glass is reflected when light strikes the rear surface of the glass, and some exits the glass and passes into the air. The amounts of light reflected and transmitted will depend, among other things, on the angle of incidence. Notice that reflection and change of velocity occur when the medium changes, first from air to glass and then from glass to air.

The Wave Equation   Consider the motion of a single point of the medium as a wave passes that point. The motion of a float on a fishing line caused by a water wave (Figure 11.46*a*) is a good example. The float, moving more-or-less vertically up and down, is shown riding the crest of the wave. As the wave disturbance moves past the float, it undergoes a vertical displacement shown by the arrow (Figure 11.46*b*). Notice that the motion of the float is perpendicular (up-and-down) to the direction in which the wave moves (left-to-right). That means the wave is a transverse wave.

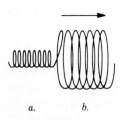

a.   b.

FIGURE 11.44   Wave at the interface of springs of different stiffness.

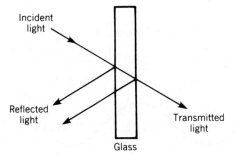

Incident light

Reflected light

Transmitted light

Glass

FIGURE 11.45   Light at an air–glass interface.

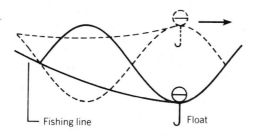

FIGURE 11.46 Fishing float riding a water wave.

Fishing line Float

a.

Displacement

Wave direction

b.

The displacement of each point along a wave can be analyzed in the same manner. If a picture is taken of a moving wave, the photograph will show that each point along the wave has a different displacement. At the instant the photograph is taken, some of the points will be displaced (Figure 11.47). However, each point will be seen to oscillate vertically with a periodic motion as time passes.

The momentary displacement of each point on the wave of Figure 11.47*a* is indicated by the arrows in Figure 11.47*b*. The direction of the arrow is the same as the direction of the displacement, and the length of the arrow indicates the amount of displacement of the point.

The mathematical expression for sinusoidal waves traveling in a positive or negative direction is given by Equations 11.116 and 11.117.

Positive direction is

$$y_+ = A_+ \sin\{[2\pi/\lambda](x - at)\} \qquad (11.116)$$

Negative direction is

$$y_- = A_- \sin\{[2\pi/\lambda](x + at)\} \qquad (11.117)$$

where

y_+ and y_- = wave displacements
A_+ and A_- = wave amplitudes
a = constant with units of velocity
λ = wavelength
t = time

Equation 11.116 is plotted in Figure 11.48 for two different times. In the time interval $t = \lambda/4a$ the wave has traveled a distance equal to $\lambda/4$ since a value of

FIGURE 11.47 Momentary displacement of points along a wave.

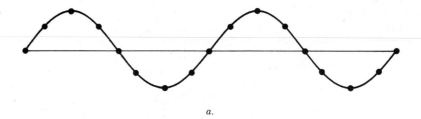

a.

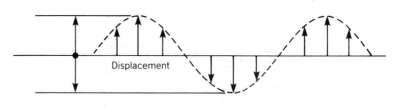

b.

FIGURE 11.48 Wave at $t = 0$ and $t = \lambda/4a$.

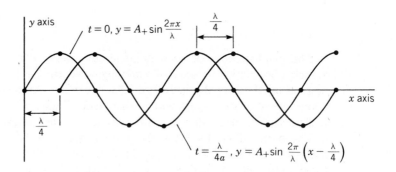

$x = \lambda/4$ will give $y = 0$. The velocity of the wave is the value of x divided by the corresponding change in time. Thus,

$$v = [\lambda/4]/[\lambda/4a] = a \qquad (11.118)$$

Substituting v for a in Equations 11.116 and 11.117 gives

$$y_+ = A_+\sin\{[2\pi/\lambda](x - vt)\} = A_+\sin\{\omega[(x/v) - t]\} \qquad (11.119)$$

$$y_- = A_-\sin\{[2\pi/\lambda](x + vt)\} = A_-\sin\{\omega[(x/v) + t]\} \qquad (11.120)$$

In Equations 11.119 and 11.120, v/v is substituted for λ and ω is substituted for $2\pi v$.

Superposition of Waves   When two or more waves traveling in the same medium overlap, a new wave is produced—a combination of the original waves. The "superposition principle" states that the resultant wave will be the sum of the two component waves. Figure 11.49 illustrates the superposition of two waves. The

FIGURE 11.49 The superposition principle.

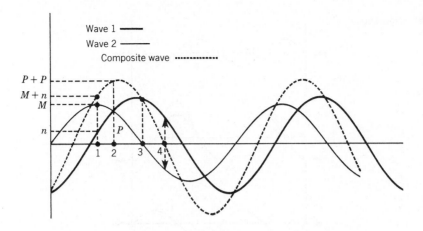

exact composite wave illustrated here exists only for the instant the component waves have the form shown.

At position 1, displacement n belongs to wave 1 and M belongs to wave 2. The two displacements are added to give the displacement $M + n$ of the composite wave. At position 2, a displacement of P is produced by both waves, so that the composite displacement at position 2 is equal to $P + P$, as shown. In position 3, no displacement is produced by wave 2. The composite displacement, therefore, is equal to the displacement of wave 1. Position 4 shows that wave 2 has a negative displacement and that wave 1 has an equal positive displacement. Here the two displacements counteract each other to give zero net displacement. A similar analysis can be conducted at any number of desired positions to verify that the composite wave is the result of the superposition of waves 1 and 2.

Standing Waves A standing wave is formed when two like waves of equal amplitude and frequency interfere with one another while traveling in opposite directions. The waves may be transverse waves, such as waves in a string, or longitudinal waves, such as sound waves. The resultant is a wave that appears to stand still in space, that is, certain regions exist wherein displacement is quite large, while other regions exist wherein there is little or no displacement. Figure 11.50 illustrates how this occurs and why the wave appears to stand still.

In Figure 11.50, two waves (dotted and solid lines) of equal amplitude and frequency travel in opposite directions. The waves are exactly out of phase at t_0, which produces a resultant represented by a straight line. At t_1 (1/4 the wave period later), the solid wave has moved 1/4 a wavelength to the right and the dotted wave has moved 1/4 a wavelength to the left. This movement places the two waves in phase and gives the maximum resultant shown. At t_2 the waves are out of phase again. At t_3 they are in phase again to produce a maximum displacement. At t_4 the period of each wave is completed and the process repeats.

As time proceeds, the standing wave shows points of maximum vibration (antinodes) separated by points of zero vibration (nodes) (A and N in Figure 11.51). At antinodal positions (A), the resultant wave moves up and down between maximum displacement limits shown (arrows in Figure 11.51). At nodal positions (N), no up and down motion occurs. Overall, there is no apparent

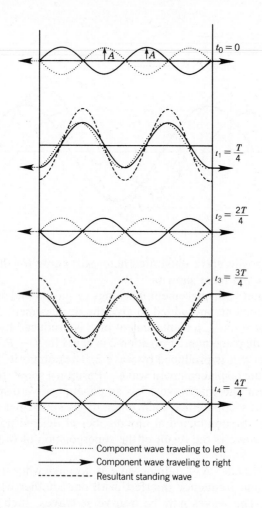

Component wave traveling to left
Component wave traveling to right
Resultant standing wave

motion of the resultant wave in a left–right direction. Hence, the resultant wave appears to stand still even though individual waves are moving to the left and right.

The value of y will be zero for all times when

$$[2\pi x]/\lambda = n\pi$$

where

$$n = 0,1,2,3,\ldots$$

or

$$x = [n\lambda]/2 \qquad \text{(nodes)} \tag{11.121}$$

These are fixed values for x. The wave does not travel, but for each value of x the medium oscillates with simple harmonic motion with an amplitude equal to $2A \sin\{[2\pi x]/\lambda\}$ as shown in Figure 11.51.

The mathematical expression for a standing wave can be obtained from Equations 11.119 and 11.120. To do this assume that $A_+ = A_- = A$ (amplitude of both waves is the same). In addition, let

FIGURE 11.51 Nodes and antinodes of a standing wave.

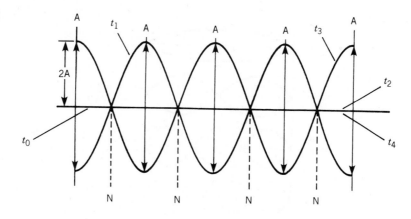

$$\{[2\pi]/\lambda\}x = \alpha$$

and

$$\{[2\pi]/\lambda\}vt = \beta$$

By the principle of superposition, the total wave is the sum of the wave traveling to the right plus the wave traveling to the left:

$$y = y_+ + y_- = A \sin[\alpha + \beta] + A \sin[\alpha - \beta] \qquad (11.122)$$

However, from trigonometry,

$$\sin[\alpha + \beta] = [\sin \alpha \cos \beta] + [\cos \alpha \sin \beta] \qquad (11.123)$$

and

$$\sin[\alpha - \beta] = [\sin \alpha \cos \beta] - [\cos \alpha \sin \beta] \qquad (11.124)$$

Using Equations 11.123 and 11.124, Equation 11.122 simplifies to

$$y = 2A [\sin \alpha \cos \beta] \qquad (11.125)$$

Substituting for α and β yields

$$y = 2A [\sin\{[2\pi]/\lambda\} \times \cos\{[2\pi]/\lambda\}vt] \qquad (11.126)$$

However, $v/\lambda = \nu$. Thus,

$$y = 2A [\sin\{[2\pi x]/\lambda\}\cos[2\pi]\nu t] \qquad (11.127)$$

The quantity $2A[\sin\{[2\pi x]/\lambda\}$ in Equation 11.127 is the amplitude of the standing wave at the position given by the value of x. At each value of x the medium undergoes simple harmonic motion given by the factor

$$\sin[2\pi \nu t] = \sin[\omega t]$$

The wave does not move and it has the form shown in Figure 11.51.

Figure 11.52 shows several possible standing waves on a string with fixed ends. At each end, the traveling waves must be reflected 180° out of phase since the ends are fixed. For standing waves to occur the length must be a multiple of half

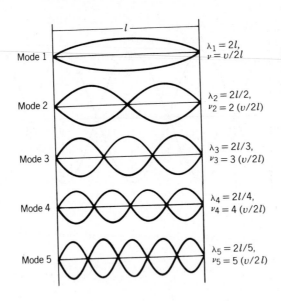

FIGURE 11.52 Allowed frequencies and wavelengths of standing waves.

wavelengths. Let $x = 0$ at one end, then $\sin\{[2\pi x]/\lambda\} = 0$ in Equation 11.127. The wave will also be zero when $\{[2\pi x]/\lambda\} = n\pi$, where n is an integer.

Since the wave must also be zero at $x = 1$, the length of the string,

$$\{[2\pi l]/\lambda_n\} = n\pi$$

or

$$\lambda_n = 2l/n \tag{11.128}$$

where

$n = 1, 2, 3, \ldots$

Equation 11.128 gives the allowed wavelengths for standing waves to occur. The corresponding allowed frequencies are

$$v_n = v/\lambda_n = nv/2l = nv_1 \tag{11.129}$$

The frequencies and wavelengths for a string are illustrated in Figure 11.52. The value of the integer n is called the mode number. The frequency difference between two successive modes can also be calculated. This frequency difference is called the mode spacing. The equation for the mode spacing is

$$\Delta v = v_{n+1} - v_n = (n + 1)v/2l - nv/2l = v/2l \tag{11.130}$$

However, from Equation 11.129, $v/2l = v$, so

$$\Delta v = v_1 \tag{11.131}$$

Equation 11.129 states that the frequency of any mode is equal to mode number times the frequency of the first mode. Equation 11.131 states that the mode spacing is equal to the frequency of the first mode. A problem using these ideas is solved in Example 11.12.

EXAMPLE 11.12 Calculation of mode spacing, frequency, and wavelength

· ·

Given: From Figure 11.52, $l = 1m$ and $v = 1000$ m/sec.

Find: **a.** Mode spacing.
              b. Frequency of third mode.
              c. Wavelength of third mode.

Solution: **a.** Mode spacing is

$$\Delta v = v/2l$$

$$\Delta v = (1000 \text{ m/sec})/2 \ (1 \text{ m})$$

$$\Delta v = 500 \text{ Hz}$$

              b. Frequency of third mode is

$$v_n = n\Delta v$$

$$v_3 = 3(500 \text{ Hz})$$

$$v_n = 1500 \text{ Hz}$$

              c. Wavelength of the third mode is

$$\lambda = v/v$$

$$\lambda = (1000 \text{ m/sec})/1500 \text{ Hz}$$

$$\lambda = 0.67 \text{ m}$$

Notice that in Figure 11.52, wavelength for the third mode is $2/3 \ l$, or 0.67 m, as determined in this example.

Resonant frequencies for a system are those frequencies for which a standing wave can exist. In Example 11.12, any multiple of 500 Hz is a resonant frequency. No other frequency can form a standing wave since it would not produce a node at both ends of the string.

Figure 11.53 shows the first four modes possible in an organ pipe with both ends open. In this case, the air executing longitudinal vibrations is free to move at

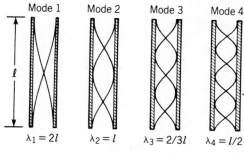

FIGURE 11.53 Modes of an organ pipe.

each end. This results in an antinode at each end of the pipe. Equations developed from Figure 11.52 hold for this example also.

Cavities   A cavity is any volume containing a medium that supports wave propagation in such a manner that some reflection of the waves occurs at the ends of the volume. In the case of the vibrating string, reflections occur at the tied ends. In the case of the organ pipes, air transmits sound waves, and open ends of the pipes actually produce partial reflection of the waves. In the optical cavity of a laser, light is reflected by a mirror at each end. In each case, reflection of waves at the ends of the cavity provides a pair of identical waves traveling in opposite directions. If the frequency of these waves is a resonant frequency of the cavity, a standing wave is formed.

Resonant cavities can be used to produce a single frequency. Most musical instruments use resonant cavities consisting of columns of air (organ pipes, trumpets, etc.) or stretched strings (pianos, violins, etc.). However, many modes are usually present, and the sound of the instrument depends on the modes present and their amplitude. The sound also depends on the method of excitation, for example, the initial shape of a plucked string.

Klystron tubes use a metallic-walled cavity to produce standing waves in the microwave region of the electromagnetic spectrum. Such tubes are employed in radar and communications. Figure 11.54 shows the elements of a klystron tube.

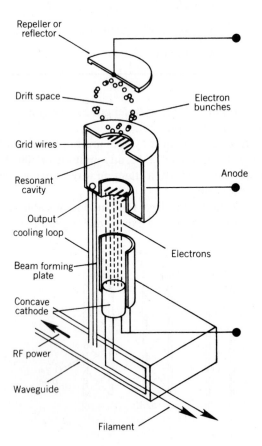

FIGURE 11.54   Diagram of a klystron tube.

Electrons from the cathode are accelerated toward the anode and pass through the grids into the drift space. They are reflected back through the upper grid by the repeller. Velocity changes of the electrons passing through the cavity produce microwave radiation. The cavity of the tube is a resonant cavity for the wavelength of microwaves to be produced, and standing waves form within it. Part of the energy is transmitted out of the tube by the output coupling loop. Wavelength of the microwaves produced can be altered by changing the shape of the resonant cavity.

Interference   Interference is the resultant cancellation or reinforcement that occurs when two or more waves are superposed. If two identical waves meet in phase (crest meeting crest and trough meeting trough), the principle of superposition predicts that a wave of twice the amplitude will result. Similarly, if the two waves meet out of phase by 180° ($\lambda/2$), then a crest of one meets a trough of the other wave, and zero amplitude results. Figure 11.55 illustrates two identical wave sources oscillating in step in such a manner as to project waves of equal wavelength and amplitude. The solid lines represent crests of the wave and the dashed lines correspond to troughs.

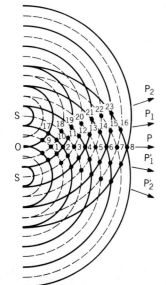

At point 1, two troughs meet and produce a double amplitude. Point 2 is a position at which two crests meet and produce a double amplitude. Similarly, waves from the two sources meet in phase at points 3, 4, 5, 6, 7, and 8. These points are said to be points of "constructive interference." Thus, constructive interference occurs along the directions of OP, OP_2, and OP'_2. "Destructive interference" occurs at points 9, 10, 11, 12, 13, 14, 15, and 16, because a wave crest from one source meets a wave trough from the other source.

Since a crest and a trough are 180° out of phase, zero displacement results at these positions. Destructive interference occurs along the directions OP_1 and OP'_1.

FIGURE 11.55   Interference pattern for two identical waves.

Constructive and destructive interference is observed when two sources generate waves of equal wavelength or frequency. If the sources produce waves of slightly different frequencies, a phenomenon known as "beats" will occur. Consider two slightly different frequency waves traveling in the same direction (Figure 11.56). Addition of the two component waves of similar frequency is shown along the lower axis. Notice that where the component waves are in phase (or nearly so), the amplitude of the composite wave is large, and where the component waves are out of phase (or nearly so), the amplitude of the composite wave is small. The result shows two "pulses," or "beats." If the pulses are not too rapid,

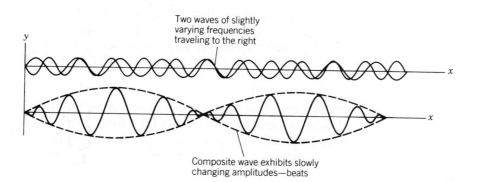

Two waves of slightly varying frequencies traveling to the right

Composite wave exhibits slowly changing amplitudes—beats

FIGURE 11.56   Pictorial representation of a beat frequency.

they can be detected separately by the human ear. Musicians listen for beats when tuning their instruments. The beat frequency turns out to be equal to the difference in frequency of the two component waves. Thus, if no beats occur, the instruments are in tune.

The beat frequency can be demonstrated mathematically. Consider two waves with equal amplitudes traveling in the positive direction with wavelength λ_1 and λ_2. Using Equation 11.119 and the substitution

$$\omega = 2\pi\nu$$

the sum of the two waves becomes

$$y = y_+ + y_- = A \sin\{[[2\pi\nu_1 x]/v] - [2\pi\nu_1 t]\} \\ + A \sin\{[[2\pi\nu_2 x]/v] - [2\pi\nu_2 t]\} \qquad (11.132)$$

Letting

$$\alpha = \{[[2\pi\nu_1 x]/v] - [2\pi\nu_1 t]\} \qquad (11.133)$$

and

$$\beta = \{[[2\pi\nu_2 x]/v] - [2\pi\nu_2 t]\}$$

one obtains

$$y = \sin\alpha + \sin\beta \qquad (11.134)$$

However, from trigonometry

$$\sin\alpha + \sin\beta = 2\cos\{1/2[\alpha - \beta]\}\sin\{1/2[\alpha + \beta]\} \qquad (11.135)$$

Substituting for α and β yields

$$y = \left\{2A \cos\left[\pi\Delta\nu\left(\frac{x}{v} - t\right)\right]\right\}\left\{\sin\left[\pi(\nu_1 + \nu_2)\left(\frac{x}{v} - t\right)\right]\right\} \qquad (11.136)$$

The first factor in Equation 11.136 acts as an amplitude, which has a slow beat frequency equal to $\Delta\nu$. For example, at a given value of x, one observes a beat frequency equal to $\Delta\nu$ because the amplitude is a maximum $\Delta\nu$ times a second. At a given time the wave has the shape shown in Figure 11.56.

Velocity of a Wave on a String   The velocity of a wave on a string can be calculated using the concept of the radius of curvature. Consider the small section of string of length ΔS shown in Figure 11.57 as a free body. The radius of curvature R is defined such that the element of length is equal to $2\theta R$. If the mass per unit length is ρ, then the mass of the element of length is $2\theta R\rho$. The force acting on this mass is $2T_y$ as shown in Figure 11.57.

The forces T_x cancel and T is the tension in the string. From Figure 11.57, $F_y = 2T \sin\theta$. However, for very small values of θ,

$$\sin\theta \approx \theta$$

Thus, the unbalanced force acting on ΔS is $2T\theta$. Instead of a wave moving to the right, one may consider the string moving to the left with velocity v. This would keep the wave stationary with respect to the observer. The element ΔS then has a centripetal acceleration equal to v^2/R. For this case, Newton's second law gives

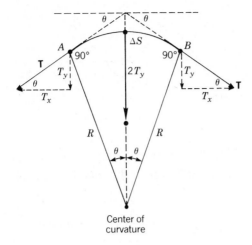

FIGURE 11.57 Free-body diagram of a vibrating string.

$$F = ma$$

or

$$2T\Delta\theta = 2R\Delta\theta\rho v^2/R \qquad (11.137)$$

or

$$T = \rho v^2 \qquad (11.138)$$

or

$$v = \sqrt{[T/\rho]} \qquad (11.139)$$

Thus, the wave velocity is equal to the square root of the tension in the string divided by the mass per unit length.

Reflected and Transmitted Waves on a String If a string is attached to a string of different density, part of an incident wave will be reflected and part will be transmitted. It is advantageous to let the value of x be zero at the point where the strings are attached as shown in Figure 11.58. The total displacement of the string on the left will be equal to the sum of the displacements due to the incident and reflected waves. At $x = 0$, the displacement for both strings is the same. Thus,

$$y_I + y_R = y_T \qquad \text{(at } x = 0)$$

where

y_I = incident wave displacement
y_R = reflected wave displacement
y_T = transmitted wave displacement

The waves on the string may be expressed as (Equations 11.119 and 11.120)

$$y_I = A_I \sin\{[\omega x/v_1] - \omega t\} \qquad (11.140)$$

$$y_R = A_R \sin\{[\omega x/v_1] + \omega t\} \qquad (11.141)$$

$$y_T = A_T \sin\{[\omega x/v_2] - \omega t\} \qquad (11.142)$$

FIGURE 11.58 Reflected, transmitted, and incident waves in a string.

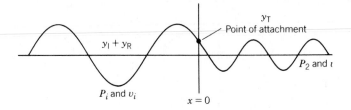

FIGURE 11.58 Reflected, transmitted, and incident waves in a string.

At $x = 0$, the equations become

$$A_I\sin[-\omega t] + A_R\sin[\omega t] = A_T\sin[-\omega t]$$

However,

$$\sin[-\omega t] = -\sin[\omega t]$$

Therefore,

$$-A_I\sin[\omega t] + A_R\sin[\omega t] = -A_T\sin[\omega t] \qquad (11.143)$$

Thus, one can see that the required condition is satisfied for all times since $\sin[\omega t]$ cancels out. The required condition may then be written as

$$-A_I + A_R = -A_T \qquad (11.144)$$

Since a string can only support a tension, the slope $\Delta y/\Delta x$ must be the same for both strings at $x = 0$. To obtain the slopes, a comparison can be made with simple harmonic motion (SHM). For SHM, we had

$$y = R\sin(\omega t + \Phi) \qquad (11.145)$$

$$\Delta y/\Delta t = \omega R\cos[\omega t + \Phi] \qquad (11.146)$$

Consider the wave equation at a particular time. Then

$$y = A \sin(\omega x/v \pm \Phi) \qquad (11.147)$$

where

Φ = a constant phase angle corresponding to the particular time

By letting t in SHM correspond to x and ω in SHM correspond to ω/x, one obtains for the slope

$$[\Delta y/\Delta x] = [\omega/v]\cos\{[\omega/v]x + \Phi\} \qquad (11.148)$$

for very small Δx. Using Equation 11.148 to set the slope on the right equal to the slope on the left at $x = 0$ for any time yields

$$[\omega/v_1]A_I\cos(-\omega t) + [\omega/v_1]A_R\cos(\omega t) = [\omega/v_2]A_T\cos(-\omega t)$$

However, $\cos[-\omega t] = \cos[\omega t]$. Using this substitution yields

$$A_I/v_1\cos[\omega t] + A_R/v_1\cos[\omega t] = A_T/v_2\cos[\omega t] \qquad (11.149)$$

This equation is satisfied for all times since $\cos[\omega t]$ will cancel, and one obtains

$$A_I/v_1 + A_R/v_1 = A_T/v_2$$

or

$$A_I + A_R = (v_1/v_2)A_T \tag{11.150}$$

Subtracting Equation 11.144 from 11.150 yields

$$2A_I = (v_1/v_2)A_T + A_T$$

or

$$2A_I = [(v_1 + v_2)/v_2]A_T$$

or

$$A_T = 2A_I v_2/(v_1 + v_2) \tag{11.151}$$

Equation 11.151 shows that $A_T = A_I$ if $v_1 = v_2$. However, for this to occur the strings must have the same mass per unit length ρ since the tension is the same. If the mass of the second string becomes very large, the velocity v_2 approaches zero. This is the case of a fixed end. If ρ_2 is very small, v_2 is much greater than v_1, and A_T approaches $2A_I$. This case approaches the condition of a free end.

If the value of A_T given by Equation 11.151 is substituted into Equation 11.144, one obtains

$$-A_I + A_R = -2A_I v_2/(v_1 + v_2)$$

or

$$A_R = -2A_I v_2/(v_1 + v_2) + A_I$$

or

$$A_R = A_I[(-2v_2 + v_1 + v_2)/(v_1 + v_2)]$$

or

$$A_R = [(v_1 - v_2)/(v_1 + v_2)]A_I \tag{11.152}$$

Equation 11.152 shows that the reflected wave amplitude is zero if $v_1 = v_2$, that is, there is no reflected wave if the mass per unit length is the same for both strings. If the mass per unit length of the second string becomes very large, then v_2 is much less than v_1 and the reflected amplitude is equal to the incident amplitude. Since the incident wave at $x = 0$ is $A_I\sin[-\omega t]$ and the reflected wave is $A_I\sin[\omega t]$, these two waves will cancel out, corresponding to a fixed end where the reflection is 180° out of phase with the incident wave. If the mass per unit length of the second string is very small, then v_2 is much greater than v_1, and A_R approaches A_I. Then the two waves add to give twice the amplitude. This corresponds to the free end.

An application of the equations for reflection and transmission of waves on a string is given in Examples 11.13 and 11.14.

EXAMPLE 11.13 Waves on a string
. . . .

Given: A string has a mass of 0.01 kg/m and it is connected to a second string that has a mass of 0.05 kg/m. The tension in the strings is 10 N.

Find: **a.**   The velocity in each string.

b.   The ratio of the amplitude of a reflected wave to an incident wave.

c.   The ratio of the amplitude of a transmitted wave to an incident wave.

Solution: **a.**   The velocity of a wave on a string is

$$v = \sqrt{[T/\rho]}$$

The velocity in the first string is

$$v_1 = \sqrt{[(10 \text{ N})/(0.01 \text{ kg/m})]}$$

$$v_1 = 31.6 \text{ m/sec}$$

The velocity of the wave in the second string is

$$v_2 = \sqrt{[(10 \text{ N})/(0.05 \text{ kg/m})]}$$

$$v_2 = 14.1 \text{ m/sec}$$

b.   From Equation 11.152

$$A_R/A_I = (v_1 - v_2)/(v_1 + v_2)$$

$$A_R/A_I = (31.6 \text{ m/sec} - 14.1 \text{ m/sec})/(31.6 \text{ m/sec} + 14.1 \text{ m/sec})$$

$$A_R/A_I = 0.383 \quad \text{(out of phase)}$$

A_R/A_I is called the reflection coefficient.

c.   From Equation 11.151

$$A_T/A_I = 2v_2/(v_1 + v_2)$$

$$A_T/A_I = [2(14.1 \text{ m/sec})]/(31.6 \text{ m/sec} + 14.1 \text{ m/sec})$$

$$A_T/A_I = 0.617 \quad \text{(in phase)}$$

A_T/A_I is called the transmission coefficient.

EXAMPLE 11.14   Reflection of light
. . . .

Given: The equations derived for a string apply to a good approximation for the reflection of light incident normal to a glass surface. The velocity of light in air is 3.0×10^9 m/sec and the velocity of light in a given glass is 2.0×10^9 m/sec.

Find: **a.**   The reflection coefficient from air to glass.

b.   The transmission coefficient from air to glass.

Solution: **a.**   The amplitude reflection coefficient is

$$E_R/E_I = (v_1 - v_2)/(v_1 + v_2)$$

where E_R and E_I are the electric fields associated with the wave.

$$\frac{E_R}{E_I} = \frac{3.0 \times 10^9 \text{ m/sec} - 2.0 \times 10^9 \text{ m/sec}}{3.0 \times 10^9 \text{ m/sec} + 2.0 \times 10^9 \text{ m/sec}}$$

$$E_R E_I = 0.2$$

The ratio of reflected power to incident power is called the reflectance and is defined as

$$P_R/P_I = (E_R^2/E_I^2)$$

$$P_R/P_I = 0.04$$

Approximately 4% of the incident power is reflected at the air–glass surface.

b. The amplitude transmission coefficient is

$$E_T/E_I = 2v_2/(v_1 + v_2)$$

$$E_T/E_I = [2(2.0 \times 10^9 \text{ m/sec})]/[(3.0 \times 10^9 \text{ m/sec}) + (2.0 \times 10^9 \text{ m/sec})]$$

$$E_T/E_I = 0.80$$

The ratio of transmitted power to incident power is called transmittance and is defined as

$$P_T/P_I = (v_1/v_2)(E_T/E_I)^2$$

$$P_T/P_I = [(3.0 \times 10^9 \text{ m/sec})/(2.0 \times 10^9 \text{ m/sec})](0.80)^2$$

$$P_T/P_I = (1.5)(0.64) \quad P_T/P_I = 0.96$$

Approximately 96% of the power is transmitted through the glass surface.

Transmission Lines A transmission line is an electrical system that transmits electromagnetic energy from one point to another. Common systems are two parallel wires and coaxial cables. Transmission lines are used to transmit power from a generating station to the consumer. In the laboratory, they are used to transmit signals from detectors and transducers to recording equipment. They are used in communications to transmit audio, radio, and television signals. In all cases, transmission lines have series inductance and resistance as well as parallel capacitance and resistance. A short section of the line may be represented by the circuit shown in Figure 11.59.

In analyzing a transmission line, it is advantageous to use the impedance per unit length. To simplify the discussion, it will be assumed that the series resistance is zero. For many applications this is a reasonable assumption. With this assumption the series impedance per unit length may be written as

$$Z_s = i\omega L_s \tag{11.153}$$

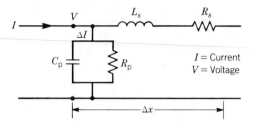

where

   ω = angular frequency of the wave
   L_s = series inductance per unit length

In addition to the series impedance, there is a parallel admittance per unit length. The admittance of a circuit is defined as the reciprocal of the impedance:

$$Y = 1/Z \qquad (11.154)$$

where

   Y = admittance
   Z = impedance

For many applications, it is reasonable to assume that the parallel resistance is very large and can therefore be neglected. The parallel impedance becomes

$$Z_p = -i/\omega C_p \qquad (11.155)$$

where

   C_p = capacitance per unit length

The corresponding admittance per unit length is

$$Y_p = [-\omega C_p]/i = i\omega C_p \qquad (11.156)$$

Consider a voltage wave traveling in the positive direction. The voltage drop over a small distance Δx is given by Ohm's law:

$$\omega V_{0I} = (I_{0I})[Z_s \Delta x]$$

or

$$[\Delta V_{0I}]/\Delta x = I_{0I}Z_s = I_{0I}[i\omega L_s] \qquad (11.157)$$

where

   V_{0I} = incident complex voltage wave
   I_{0I} = incident complex current wave

The corresponding change in current due to the parallel admittance is

$$\Delta I_{0I} = V_{0I}/Z_p = (V_{0I})[Y_p \Delta x]$$

or

$$\Delta I_{0I}/\Delta x = V_{0I}Y_p = V_{0I}[i\Delta C_p] \qquad (11.158)$$

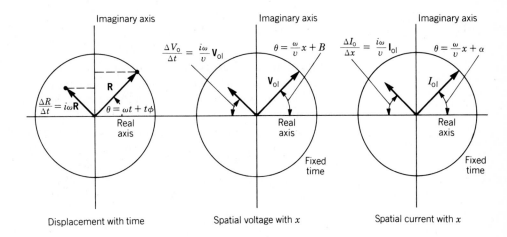

FIGURE 11.60 Complex representation of transmission line voltage and current waves.

The complex current and voltage can be related to the complex displacement in simple harmonic motion. This is illustrated in Figure 11.60, where the current and voltage are drawn for a fixed time. The waves, at a fixed time, have a sinusoidal shape in space with a spatial frequency equal to

$$\omega/v = [2\pi]/\lambda$$

For this case, t for SHM corresponds to x and the temporal frequency ω corresponds to the spatial frequency ω/v. From this correspondence, one obtains

$$\Delta V_{0\text{I}}/\Delta x = \{[i\omega]/v\}V_{0\text{I}} \tag{11.159}$$

or

$$\Delta I_{0\text{I}}/\Delta x = \{[i\omega]/v\}I_{0\text{I}} \tag{11.160}$$

Equating these values to the values of $\Delta V_{0\text{I}}/\Delta x$ and $\Delta I_{0\text{I}}/\Delta x$ given by Equations 11.157 and 11.158 yields

$$[i\omega/v]V_{0\text{I}} = i\omega L_s I_{0\text{I}}$$

or

$$V_{0\text{I}}/I_{0\text{I}} = v L_s \tag{11.161}$$

and

$$V_{0\text{I}}[i\omega C_p] = [i\omega/v]i_{0\text{I}}$$

or

$$V_{0\text{I}}/I_{0\text{I}} = 1/(v C_p) \tag{11.162}$$

Since quantities equal to the same quantity are equal to each other, Equations 11.161 and 11.162 yield

$$v L_s = 1/(v C_p)$$

or

$$v^2 = 1/(C_p L_s)$$

or

$$v = 1/[\sqrt{(C_p L_s)}] \tag{11.163}$$

Equation 11.163 gives the velocity of a sinusoidal voltage or current wave on a transmission line if the inductance per unit length and the capacitance per unit length are known. This velocity will always be less than the speed of light in a vacuum.

Substituting the line wave velocity v into Equation 11.161 gives for a positive v

$$V_{0I}/I_{0I} = L_s/v = L_s/[\sqrt{(L_s C_p)}] = \sqrt{(L_s/C_p)} = Z_0 \tag{11.164}$$

The quantity $[\sqrt{(L_s/C_p)}] = Z_0$ is called the characteristic impedance or surge impedance of the line. For a negative velocity, corresponding to a reflected wave,

$$V_{0R}/I_{0R} = L_s/(-v) = -[\sqrt{(L_s/C_p)}] = -Z_0 \tag{11.165}$$

Terminated Transmission Line   If a transmission line is terminated with a load impedance Z_L, part of the incident wave is transmitted to the load and part is reflected. However, if $Z_1 = Z_0$, there is no reflected wave. This is the desirable condition since waves reflected back and forth on a transmission line give additional signals. This is very important in communications and in the laboratory.

A terminated transmission line is shown in Figure 11.61. The voltage at the load is equal to the sum of the incident and reflected voltages. Thus, at the load

$$V_{0I} + V_{0R} = V_{0L} \tag{11.166}$$

In addition, the total current is the sum of the incident and reflected current. Thus, at the load

$$I_{0I} + I_{0R} = I_{0L} \tag{11.167}$$

From the characteristic line impedance, one may also write

$$V_{0I} = I_{0I} Z_0 \tag{11.168}$$

and

$$V_{0R} = -I_{0R} Z_0 \tag{11.169}$$

In addition, Ohm's law at the load yields

$$V_L = I_{0L} Z_L$$

Substituting these values for V_{0I}, V_{0R}, and V_L in Equation 11.166 gives

$$I_{0I} Z_0 - I_{0R} Z_0 = I_{0L} Z_L$$

or

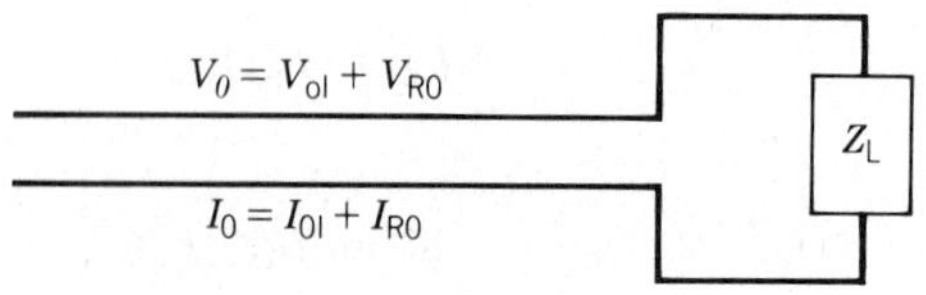

FIGURE 11.61 Schematic of a transmission line terminated at a load.

$$V_0 = V_{ol} + V_{R0}$$

$$Z_L$$

$$I_0 = I_{0I} + I_{R0}$$

$$I_{0I} - I_{0R} = I_{0L}(Z_L/Z_0) \tag{11.170}$$

Adding Equations 11.167 and 11.170 yields

$$2I_{0I} = I_{0L}[1 + (Z_L/Z_0)]$$

or

$$I_{0I} = [(Z_0 + Z_L)/2Z_0]I_{0L} \tag{11.171}$$

If I_{0I} given by Equation 11.171 is substituted into Equation 11.167, one obtains

$$[(Z_0 + Z_L)2Z_0]I_{0L} + I_{0R} = I_{0L}$$

or

$$2Z_0 I_{0R} = 2Z_0 I_{0L} - (Z_0 + Z_L)I_{0L}$$

or

$$I_{0R} = [(Z_0 - Z_L)2Z_0]I_{0L} \tag{11.172}$$

This equation is very important because it gives the required condition to eliminate a reflected wave. If the load impedance is made equal to the line impedance Z_0, then there will be no reflected wave at the load. In addition, Equation 11.171 gives $I_{0I} = I_{0L}$, if $Z_0 = Z_L$. That is, the line current is equal to the load current.

The reflection coefficient for current is defined as

$$\rho_I = I_{0R}/I_{0I} = (Z_0 - Z_L)/(Z_0 + Z_L) \tag{11.173}$$

This is obtained by dividing Equation 11.172 by Equation 11.171. If Z_L becomes large, the reflection coefficient becomes -1, and the reflected current is 180° out of phase with the incident current, resulting in essentially no current at the load. If Z_L is very small, the reflected current is in phase with the incident current, resulting in a large current at the load. Applications of the transmission line equations are given in Examples 11.15, 11.16, and 11.17.

EXAMPLE 11.15 Wave velocity and characteristic impedance of a transmission line
. . . .

Given: A transmission line has an inductance of 1×10^{-6} H/m and a capacitance of 18×10^{-12} F/m.

Find: **a.** The wave velocity.

         b. The characteristic impedance Z_0.

Solution: **a.** The wave velocity is

$$v = 1/[\sqrt{(L_s C_p)}]$$

$$v = 1/[\sqrt{(1 \times 10^{-6} \text{ H})(18 \times 10^{-12} \text{ F})}]$$

$$v = 1/(4.24 \times 10^{-9} \text{ sec/m})$$

$$v = 2.36 \times 10^8 \text{ m/sec}$$

(speed of light $= 3 \times 10^8$ m/sec)

b. The characteristic impedance is

$$Z_0 = \sqrt{L_s/C_p}$$

$$Z_0 = \sqrt{(1 \times 10^{-6} \text{ H}/18 \times 10^{-12} \text{ F})}$$

$$Z_0 = 236 \ \Omega$$

EXAMPLE 11.16   Reflection coefficient for voltage
. . . .

Given: A transmission line is terminated in an impedance Z_L. The characteristic impedance of the line is Z_0.

Find: **a.** The reflection coefficient, $\rho_V = V_{0R}/V_{0I}$.
   b. The reflection coefficient when $Z_L = 2Z_0$.

Solution: **a.** Equations 11.168 and 11.169 state

$$V_{0I} = Z_0 I_{0I}$$

and

$$V_{0R} = -Z_0 I_{0R}$$

Thus,

$$\rho_V = V_{0R}/V_{0I} = -I_{0R}/I_{0I} = \rho_I$$

From Equation 11.173

$$\rho_I = (Z_0 - Z_L)/(Z_0 + Z_L)$$

Thus,

$$\rho_V = (Z_L - Z_0)/(Z_0 + Z_L)$$

b. When $Z_L = 2Z_0$

$$\rho_V = (2Z_0 - Z_0)/(Z_0 + 2Z_0)$$

$$\rho_V = 1/3$$

EXAMPLE 11.17   Terminated transmission line
. . . .

Given: A transmission line has a characteristic impedance of 75 Ω. A 1.0-H inductor is connected to the end of the line. The frequency of the incident wave is 60 Hz.

Find: **a.** The reflection coefficient for current.
   b. The reflection coefficient for voltage.
   c. The impedance that would make the reflection coefficients zero.

Solution: **a.** The reflection coefficient for current is

$$\rho_I = (Z_0 - Z_L)/(Z_0 + Z_L)$$

$$Z_L = i\omega L$$

$$Z_L = i[2\pi](60\text{ Hz})(1.0\text{ H})$$

$$Z_L = 377i\ \Omega$$

$$\rho_I = [75\ \Omega - 377i\ \Omega]/[75\ \Omega + 377i\ \Omega]$$

$$\rho_I = [(75 - 377i)(75 - 377i)]/[(75 + 377i)(75 - 377i)]$$

$$\rho_I = [(75)^2 - (377)^2 - i(2)(75)(377)](75)^2 + (377)^2$$

$$\rho_I = (-136{,}504 - 56{,}550i)/147{,}754$$

$$\rho_I = -0.924 - 0.383i$$

b. The reflection coefficient for voltage is

$$\rho_V = -\rho_I$$

$$\rho_V = 0.924 + 0.383i$$

c. To make the reflection coefficient zero, Z_L must equal Z_0.

$$Z_L = 75\ \Omega$$

Estimation of Z_0 Two common forms of transmission lines are the two-wire transmission line and the coaxial transmission line. The two-wire line is illustrated in Figure 11.62. The quantity l is the distance between the centers of the two cylindrical wires and r is the radius of each wire. It can be shown that the characteristic impedance of this transmission line in air is

$$Z_0 = 276 \log(l/r)\ \Omega \tag{11.174}$$

The logarithm in Equation 11.174 must be taken to the base 10.

The value of Z_0 can also be calculated for a coaxial cable as shown in Figure 11.63. The medium between the two concentric conductors is usually a medium other than air. The equation for the impedance Z_0 is

$$Z_0 = 138\ \Omega/[\sqrt{\epsilon_r}]\log(r_2/r_1) \tag{11.175}$$

where

$\epsilon_r = \epsilon/\epsilon_0$ = relative permittivity of the dielectric
ϵ_0 = permittivity of air

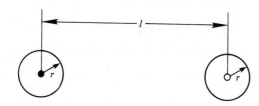

FIGURE 11.62 Geometric factors of two-line transmission line.

FIGURE 11.63 Geometric factors of coaxial transmission lines.

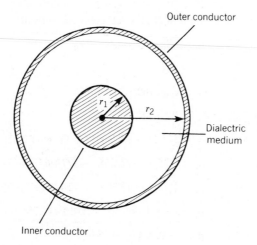

Applications of Equations 11.174 and 11.175 are given in Examples 11.18 and 11.19.

EXAMPLE 11.18   Impedance of a two-wire transmission line
....

Given: The wires of a two-wire transmission line are separated by a distance of 0.02 m, and each has a radius of 0.002 m.

Find: The impedance Z_0 of the line.

Solution: For a two-wire transmission line in air, the impedance is

$$Z_0 = 276 \log(l/r) \; \Omega$$

$$Z_0 = 276 \log(0.02 \text{ m}/0.002 \text{ m}) \; \Omega$$

$$Z_0 = 276 \; \Omega$$

EXAMPLE 11.19   Impedance of a coaxial transmission line
....

Given: A coaxial transmission line is composed of an inner conductor with a radius of 0.001 m and a concentric outer conductor with an inner radius of 0.005 m. The medium between the conductors has a relative permittivity of 2.7.

Find: The characteristic impedance of the transmission line.

Solution: The impedance of a coaxial line is

$$Z_0 = 138 \; \Omega/\sqrt{\epsilon_r} \; \log(r_2/r_1)$$

$$Z_0 = 138 \ \Omega/\sqrt{(2.7)} \ \log(0.005 \ m/0.001m)$$

$$Z_0 = (84)(0.70) \ \Omega$$

$$Z_0 = 59 \ \Omega$$

SECTION **11.3** *EXERCISES*

1. The velocity of light in vacuum is 3×10^8 m/sec. What is the frequency of a light wave with a wavelength of 5×10^{-5} cm?

2. The velocity of sound in air at sea level is 335 m/sec. What is the wavelength of a sound wave with a frequency of 800 Hz?

3. Define the term standing wave.

4. An organ pipe (open at both ends) has a length of 3.2 m. What are the frequencies of the first three modes of vibration?

5. A laser cavity with mirrors at opposite ends has a length of 25 cm. Calculate axial mode spacing in the cavity. Use 3×10^8 m/sec as the velocity of light in the cavity?

6. Predict the first five modes of vibration for an organ pipe closed at one end and open at the other. Give mode frequencies in terms of pipe length l and velocity of sound in air v.

7. If the angular frequency of a wave is 370 rad/sec, find the period of the wave.

8. Show that the equation

$$y = A \ \sin\{[2\pi/\lambda](x - vt)\}$$

is equivalent to

$$y = A \ \sin\{[\omega x/v] - \omega t\}$$

9. An organ pipe closed at both ends is 4 m long. What is the longest possible wavelength of any standing wave that would fit into the pipe? (**Hint:** For a closed organ pipe, a standing wave has node at each end of the pipe.)

10. Find the displacement of a wave at $t = 6$ sec, $x = 10$ m, if the wavelength is 3 cm and the frequency is 80 Hz. The amplitude of the wave is 0.5 cm.

11. At time t and distance x a wave has a displacement of 2 cm and a second wave has a displacement of -1 cm. What would be the displacement of a wave that is the superposition of the two waves at the same time t and distance x?

12. Two sound waves travel through air. If the frequency of one wave is 1000 Hz and the wavelength is 1 m, find the frequency of a wave traveling through air under the same condition if its wavelength is 0.7 m.

13. If the velocity of a wave along a string is 10 m/sec and the wavelength is 5 cm, find the frequency of oscillation and the period associated with the wave.

14. A string having a mass per unit length of 0.05 kg/m and tension of 4.5 N is driven by a vibrator at one end. What is the wave velocity? If the string has essentially fixed ends, what are the conditions for standing waves if the vibrator has a frequency of 60 Hz?

15. Calculate the reflection coefficient for a string having a density of 0.07 kg/m if it is connected to a string having a density of 0.10 kg/m.

16. Light travels from glass to water in a direction normal to the surface. The velocity of light in the glass is 1.9×10^8 m/sec and the velocity in the water is 2.25×10^8 m/sec. If the incident beam has a total power of 100 W, what is the transmitted power?

17. A two-wire transmission line has an inductance of 1.2×10^{-6} H/m and a capacitance of 20×10^{-12} F/m. Find the wave velocity on the line and its characteristic impedance.

18. A transmission line with a characteristic impedance of 50 Ω is terminated with a 25-Ω resistor. Find the reflection coefficient for voltage and current.

19. An oscilloscope has an input impedance of 1×10^6 Ω. A 75-Ω cable is used between a light detector and the oscilloscope. The system is used to detect laser pulses having a length of 1×10^{-9} sec. What would you do to eliminate reflections on the line?

20. A coaxial cable has an inner radius of 0.002 m and an outer radius of 0.01 m. The medium between the two conductors is air. What is the characteristic impedance of the cable?

S　U　M　M　A　R　Y

Vibrations occur in most physical systems and often play a major role in the operation of the system. On the other hand, they may be unwanted and may degrade the operation of the system. For both situations, it is important to understand the basic principles of vibrating systems. This is especially true if resonance occurs.

The simplest vibrations vary sinusoidally with time. In mechanics this motion is called simple harmonic motion. In electric circuits it is called alternating current and voltage. In mechanics, the system studied was composed of a mass at the end of a spring driven by a sinusoidal force. The equivalent electrical system was an inductor, capacitor, and resistor in series driven by a sinusoidal voltage. For both cases, an impedance was defined such that the complex ratelike quantity equals the complex forcelike quantity divided by the impedance. For both systems, resonance occurs when the driving frequency is equal to the natural frequency of the system. The mechanical quantities are compared with the electrical quantities in Table 11.2.

Mechanical and electrical vibrations are often the source of waves. The basic characteristics of waves discussed in this chapter apply to all wave phenomena. However, only sinusoidal waves were discussed mathematically.

In mechanical systems, both longitudinal and transverse waves are important. Longitudinal mechanical waves are compression waves in a material—sound waves traveling through a solid, compression waves produced by a moving spring, and mechanical vibration present in machines and in structures. Transverse mechanical waves include the side-to-side motion that occurs in the strings of musical instruments and in torsional (twisting motion) waves. In mechanical structures, the reduction or elimination of wave motion often is a primary design consideration.

The most important fluid waves are compression waves (sound) in gases. The field of acoustics is based on the study and control of these waves. The design of sound-delivery systems, for example, requires an analysis of the patterns of the radiated sound energy, the interference of sound waves from two or more sources, and the absorption and reflection of sound waves by materials.

Electrical waves are among the most important to modern technology. Commercial electrical power is delivered as an electrical wave at a frequency of 60 Hz. Most electronic equipment performs its function, at least in part, through the manipulation of electrical waves. The concepts of superposition of waves and phase difference are particularly important in the analysis of electronic circuits.

Light, x rays, microwaves, and radio waves are examples of electromagnetic radiation. This type of transverse wave consists of oscillating electric and magnetic fields traveling through space; it is discussed in detail in Chapter 13.

TABLE 11.2   Mechanical and electrical oscillations

Mechanical	Electrical
Displacement r	Charge q
Force F_p	Voltage V_0
Velocity v_p	Current I_0
Mass m	Inductance L
Spring constant k	1/capacitance $1/C$
Fluid resistance K	Electrical resistance R
Impedance $K + i[wm - (k/m)]$	Impedance $R + i[wL - (1/wC)]$
$v_p = F_p/Z_m$	$I_0 = V_0/Z$
$m[v_p/t] + Kv_p + kr = F_p$	$L[I_0/t] + RI_0 + q_0/C = V_0$
$\omega_0 = \sqrt{(k/m)}$	$\omega_0 = \sqrt{(1/LC)}$
Resonance	Resonance
$\omega = \omega_0$	$\omega = \omega_0$

Note: r, F_p, v_p, Z_m, q_0, V_0, I_0 and Z are complex numbers.

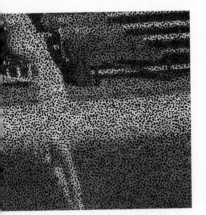

Exponential

Constants

of Linear

Systems

INTRODUCTION

In many practical systems, it is important to know how quickly a system will give up or accept energy, change its temperature, or return to equilibrium after an initial displacement. For example, how long does it take a falling body to reach its terminal velocity? How much time does it take a heated body to cool to room temperature? Some systems may not reach an equilibrium (final state) until an extraordinary length of time has elapsed. For example, a capacitor never becomes fully charged, but it comes so close to the fully charged condition in a given time that it is considered fully charged.

In other systems, spatial dependence is important. For example, it is important to know how the intensity of a wave will change as it travels through a medium. Part of the energy transmitted by the wave will be absorbed and/or scattered by the medium. For the case of a laser, stimulated emission is also important, and the medium adds energy to the wave. From a mathematical point of view, a wave is never completely absorbed. However, after a certain distance the intensity has become so small that the wave no longer exists for practical purposes.

Many time-dependent or spatial-dependent systems can be represented by a simple exponential increase or decrease in the quantity being considered. These systems are called linear systems. For time-dependent systems, an important characteristic is the time constant. For spatially dependent systems where the intensity of a wave decreases, the important characteristic is the extinction coefficient. In this chapter, time constants and spatial coefficients will be developed for a number of different systems. The general method of analysis serves as a unifying technique for all linear and approximately linear systems.

12.1

MATHEMATICAL METHODS

OBJECTIVES

Upon completion of this section, the student should be able to

- Define
  The base of the natural logarithm
  An irrational number
  A simple exponential function
  Steady state
  Transient
  Time constant
  Half-life

- Calculate the vlaue of e^x for any real value of x using a scientific hand calculator.

- Draw and label a graph of $y = Ae^{-t/\tau}$ and show the location of the points $t = \tau$, $t = 3\tau$, and $t = 5\tau$.

- Calculate the slope of the function $y = Ae^{-t/\tau}$ at any value of t.

- Write the general equation for an exponential increase to a steady state.

DISCUSSION

It is not uncommon to observe that the rate of change of a quantity is related to the quantity itself. For example, the rate of air leakage from a punctured tire is related to how much air is remaining in the tire. Such processes, when plotted versus time, appear as a smooth curve approaching a limiting value, as in Figure 12.1*b*.

The rate of change in Figure 12.1 is the slope of the curve. When the slope is constant, a uniform process is taking place (Figure 12.1*a*). Cases in which the slope changes are said to be **nonuniform,** such as the case of air escaping from a punctured tire. When examining time-dependent systems, a special relationship is observed when the slope of the curve changes and is proportional to the quantity itself. This relationship is given by

$$\text{slope } \frac{\Delta y}{\Delta t} \propto y \qquad (12.1)$$

FIGURE 12.1 Comparison of uniform and nonuniform rates.

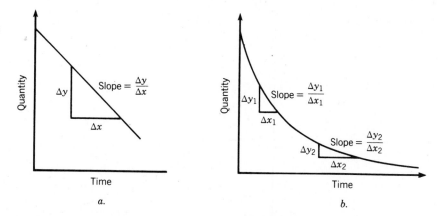

    Such a relationship is called a linear relationship. It can be shown that an equation for y that behaves as in Equation 12.1 is

$$y = Ae^{(-t/K)} + C \tag{12.2}$$

The letter "e" represents an irrational number that has been found to occur frequently in mathematics and physics. (Recall that the number π is a similarly unique number.) The value of e is approximately 2.71828. This is approximate, because the decimal places actually continue on forever, even though the digits beyond five or ten places are seldom needed. The value of e can be calculated by using the arithmetic series

$$e = 1 + \frac{1}{1} + \frac{1}{1 \cdot 2} + \frac{1}{1 \cdot 2 \cdot 3} + \frac{1}{1 \cdot 2 \cdot 3 \cdot 4} + \frac{1}{1 \cdot 2 \cdot 3 \cdot 4 \cdot 5} + \cdots \tag{12.3}$$

or

$$e = 1 + \sum_{n=1}^{\infty} \frac{1}{n!}$$

With this method, one can obtain five digits of accuracy by carrying out Equation 12.3 for ten terms.

    For Equation 12.2, plotted in Figure 12.2, the slope is[1]

$$\text{slope} = \frac{\Delta y}{\Delta t} = -\frac{A}{K} e^{-t/K} = -\frac{1}{K} y \tag{12.4}$$

    As can be seen in Figure 12.2, equal time intervals elapse for each halving of the y value. The time required for the halving is called the **half-life.** In fact, equal time intervals will be observed for any choice of continuous fractional reductions (e.g., 1/3 or 1/4). This is another attribute of Equation 12.2.

    One more important thing must be observed about Equation 12.2. When the exponent t/K in Equation 12.2 equals 1, the quantity y will be reduced to 36.79% of its initial value. The time t that causes this to happen is called the time constant τ and replaces K in Equation 12.2. Thus Equation 12.2 becomes

---

[1]Note: The interval Δt must be kept small. The actual relationship involves $\lim_{\Delta t \to 0} (\Delta y/\Delta t) = dy/dt$. This expression comes from calculus.

FIGURE 12.2 Graph of the function $y = Ae^{-t/K} + C$.

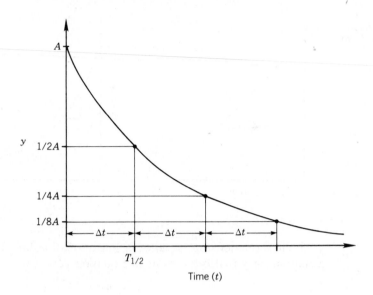

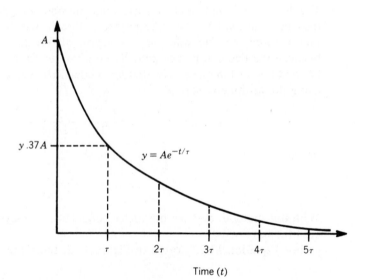

FIGURE 12.3 Exponential decay curve.

$$y = Ae^{-t/\tau} + C \tag{12.5}$$

Figure 12.3 illustrates Equation 12.5. It is common to speak of the value of y at multiples of the time constant. Notice in Figure 12.3 that the time constant multiples are labeled on the time axis. At 5τ, that is, five time constants, observe that the change in y is very small.

Quite obviously, the quantity y is still changing at a time of 5τ, but it is common to consider that most of the meaningful change in y has occurred after five time constants have elapsed. To be precise, the value of y has been reduced to less than 1% of the initial value, as shown in the following equations.

FIGURE 12.4 Exponential growth curve.

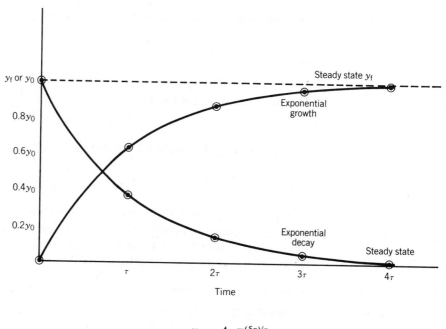

$$y = Ae^{-(5\tau)/\tau}$$

$$y = Ae^{-5}$$

$$y = A(0.0067)$$

Equation 12.5 is a general equation that describes processes in which a quantity exponentially decreases with time. This process is generally called an exponential decay. A similar process that involves the exponential increase of a quantity is called an exponential growth and is described by

$$y = A(1 - e^{t/\tau}) + C \qquad (12.6)$$

In Equation 12.6, the quantity y exponentially increases to a limiting value A. A plot of y versus time is shown in Figure 12.4. The growth curve is within 1% of the limiting value after a time of 5τ.

In many time-dependent systems, the halving time or half-life is commonly used. To obtain the half-life $T_{1/2}$ if given the time constant τ for a decay process, begin with Equation 12.5

$$y = Ae^{-t/\tau}$$

at

$$t = T_{1/2}, \, y = 1/2A$$

Therefore, Equation 12.5 becomes

$$1/2A = Ae^{-T_{1/2}/\tau}$$

dividing both sides by A gives

$$1/2 = e^{-T_{1/2}/\tau} \qquad (12.7)$$

and taking the natural log of both sides yields

$$\ln 1/2 = \ln (e^{-T_{1/2}/\tau})$$

or

$$-0.693 = -T_{1/2}/\tau \tag{12.8}$$

Solving Equation 12.8 for $T_{1/2}$ gives

$$T_{1/2} = 0.693\tau \tag{12.9}$$

If $T_{1/2}$ is known, Equation 12.8 can be solved for τ:

$$\tau = \frac{T_{1/2}}{0.693} \tag{12.10}$$

Equations 12.9 and 12.10 also are true for the exponential growth processes. In evaluating exponential equations, a scientific calculator with an e^x key is very useful. Examples 12.1 and 12.2 demonstrate the use of Equations 12.5, 12.6, 12.9, and 12.10 using a scientific calculator.

EXAMPLE 12.1 Exponential decay process
. . . .

Given: A capacitor with an initial voltage of 100 V is discharged. After 5 msec the voltage is 50 V.

Find: **a.** The time constant τ.
              b. The voltage 12 msec after the beginning of the discharge.

Solution: **a.** $\tau = \dfrac{T_{1/2}}{0.693}$

                      $\tau = \dfrac{5 \text{ msec}}{0.693}$

                      $\tau = 7.22 \text{ msec}$

              b. $y = Ae^{-t/\tau}$

                      $y = 100 \text{ V } e^{-12\,\text{msec}/7.22\,\text{msec}}$

                      $y = 18.97 \text{ V}$

EXAMPLE 12.2 Exponential growth process
. . . .

Given: A capacitor in a circuit initially has no voltage across it. A 100-V dc source is in the circuit. The time constant for the charging process in 9 msec.

Find: **a.** The half-life for the charging process.
              b. The voltage across the capacitor 20 msec after the charging switch is closed.

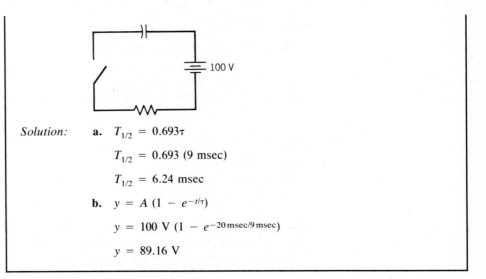

Solution: **a.** $T_{1/2} = 0.693\tau$

$T_{1/2} = 0.693 \ (9 \ \text{msec})$

$T_{1/2} = 6.24 \ \text{msec}$

b. $y = A \ (1 - e^{-t/\tau})$

$y = 100 \ \text{V} \ (1 - e^{-20\,\text{msec}/9\,\text{msec}})$

$y = 89.16 \ \text{V}$

SECTION *12.1* *EXERCISES*

1. Define the terms
Steady state
Transient
Time constant
Half-life

2. Draw a graph for an exponentially decaying quantity P. Show on the graph the time constant τ, the half-life $T_{1/2}$, and the values of P at these times. Write an equation for the graph.

3. An exponentially decreasing process has a half-life of 0.231 sec. What is the value of the time constant (τ) for this process?

12.2

OBJECTIVES

MECHANICAL SYSTEMS

Upon completion of this section, the student should be able to

- Define
  Damping
  Terminal velocity
  Drag force

- Calculate the terminal velocity of a freely falling body if the drag coefficient, k_F, is known.

- Define the conditions for
  Underdamped motion
  Critically damped motion
  Overdamped motion

- Calculate the time constant for damped simple harmonic motion.

DISCUSSION

Terminal Velocity   When an object falls toward the earth, its velocity increases as a result of the gravitational force. However, as the body's velocity increases, the drag force due to the air resistance increases. To a good approximation, if the velocity is not too great, the drag force is proportional to the velocity as given by

$$F_D = k_F v \tag{12.11}$$

where

F_D = drag force
k_F = drag constant
v = body's velocity

When the drag force becomes equal to the gravitational force, the velocity is constant (steady state). This final velocity is called the terminal velocity.

Newton's second law for the falling body is

Mass × Acceleration = Gravitational Force − Drag Force

or

$$m \frac{\Delta v}{\Delta t} = mg - k_F v \tag{12.12}$$

where

m = body's mass
g = acceleration due to gravity
$\dfrac{\Delta v}{\Delta t}$ = instantaneous acceleration

Dividing Equation 12.12 by m yields

$$\frac{\Delta v}{\Delta t} = -\frac{k_F}{m} v + g \tag{12.13}$$

Equation 12.13 is satisfied by

$$v = \left(v_0 - \frac{mg}{k_F} \right) e^{-\frac{k_F}{m}t} + \frac{mg}{k_F} \tag{12.14}$$

and the terminal velocity is given by

$$v_F = \frac{mg}{k_F} \tag{12.15}$$

since $e^{-(k_F/m)t}$ approaches zero as t increases. If the body starts from rest ($v_0 = 0$), Equation 12.14 becomes

$$v = \frac{mg}{k_F} (1 - e^{-\frac{k_F}{m}t}) \tag{12.16}$$

For practical purposes, the terminal velocity occurs after a time equal to $5m/k_F$. Writing Equation 12.16 in terms of the terminal velocity and time constant yields

$$v = v_F(1 - e^{-t/\tau}) \tag{12.17}$$

EXAMPLE 12.3 Terminal velocity of a body

· ·

Given: A body falls from rest in the earth's gravitational field. The drag constant is 30 N·sec/m and the body's mass is 100 kg.

Find: **a.** The time constant.
     b. The terminal velocity.
     c. The velocity after 5.0 sec.

Solution: **a.** The time constant is

$$\tau = \frac{m}{k_T} = \frac{100 \text{ kg}}{30 \text{ N·sec/m}} = 3.33 \text{ sec}$$

(1 N = 1 kg·m/sec²)

b. The terminal velocity is

$$v_F = \frac{gm}{k_F} = g\tau$$

$$v_F = (9.80 \text{ m/sec}^2)(3.33 \text{ sec})$$

$$v_F = 32.6 \text{ m/sec}$$

c. $v = \frac{gm}{k_F} (1 - e^{-\frac{k_F}{m}t})$

$$v = V_F(1 - e^{-t/\tau})$$

$$v = 32.6 (1 - e^{-5/3.33}) \text{ m/sec}$$

$$v = (32.6)(1 - 0.223) \text{ m/sec}$$

$$v = 25.3 \text{ m/sec}$$

Mechanical Vibrations   In Chapter 11, the motion of a mass attached to a spring is discussed under frictionless conditions. If no friction is present, the vibration will continue indefinitely with no change in amplitude. For this case, the displacement as a function of time is illustrated in Figure 12.5. The corresponding equation for displacement as a function of time is

$$y = A_0 \cos \left(\sqrt{\left(\frac{k}{m}\right)} \, t + \theta \right) \tag{12.18}$$

where

A_0 = amplitude
m = body's mass
k = spring constant
θ = phase angle

If viscous damping is present, the mass will essentially come to rest after a period of time. For this case, Newton's second law becomes

FIGURE 12.5 Plot of displacement versus time for a uniform vibration.

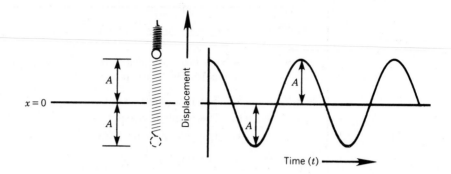

$$\text{Mass} \times \text{Acceleration} = \text{Spring Force} - \text{Viscous Force}$$

$$m = \frac{\Delta v}{\Delta t} = -ky - k_{\mathrm{F}}v \tag{12.19}$$

The equation of motion is linear, and there are three interesting solutions to Equation 12.19.

If $k_{\mathrm{F}} = 0$, one obtains Equation 12.18 as a solution, and the amplitude remains constant. When k_{F} is not 0, oscillations will occur if

$$\frac{k}{m} > \frac{1}{4}\left(\frac{k_{\mathrm{F}}}{m}\right)^2 \tag{12.20}$$

The amplitude, however, will decay as illustrated in Figure 12.6. This type of motion is called underdamped. Mathematically, it is expressed as

$$y = A_0 e^{-t/\tau}\left[\cos\left(\omega_1 t + \theta\right)\right] \tag{12.21}$$

where

$$\omega_1 = \sqrt{\left[\frac{k}{m} - \frac{1}{4}\left(\frac{k_{\mathrm{F}}}{m}\right)^2\right]}$$

$$\tau = 2m/k_{\mathrm{F}} = \text{time constant}$$

The use of Equations 12.20 and 12.21 are illustrated in Example 12.4.

EXAMPLE 12.4 Damped simple harmonic motion
. . . .

Given: A 4.0-kg mass at the end of a spring is clamped by a dash pot. The spring constant is 100 N/m and the drag constant is 10 N·sec/m.

Find: **a.** Will the mass oscillate?

            b. What is the time constant for the decay of the amplitude?

            c. How long will it take for the motion to stop?

Solution: **a.** If oscillations occur,

$$\frac{k}{m} > \frac{1}{4}\left(\frac{k_{\mathrm{F}}}{m}\right)^2$$

$$\frac{k}{m} = \frac{100 \text{ N/m}}{4.0 \text{ kg}} = 25/\text{sec}^2$$

$$\frac{1}{4}\left(\frac{k_F}{m}\right)^2 = \frac{1}{4}\left(\frac{10 \text{ N·sec/m}}{4.0 \text{ kg}}\right)^2 = 1.56/\text{sec}^2$$

Thus, the mass will oscillate.

b. The time constant τ is

$$\tau = \frac{2m}{k_F}$$

$$\tau = \frac{(2)(4.0 \text{ kg})}{10 \text{ N·sec/m}} = 0.8 \text{ sec}$$

c. The motion will stop in approximately five time constants.

$$t = 5\tau = (5.0)(0.8 \text{ sec}) = 4.0 \text{ sec}$$

The mass will return to the equilibrium position without oscillating if k_F is increased such that

$$\frac{k}{m} = \frac{1}{4}\left(\frac{k_F}{m}\right)^2$$

When this condition exists, the motion is called critically damped. For critically damped motion, it can be shown that the solution of Equation 12.19 is

$$y = e^{-t/\tau}(A_0 + A_1 t) \tag{12.22}$$

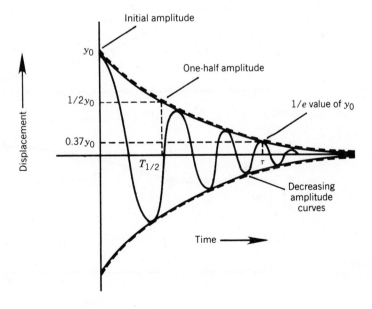

FIGURE 12.6 Decay of a vibration.

where

$$\tau = 2m/k_F$$

A_0 and A_1 are constants

Also, it can be shown that the corresponding velocity is

$$v = e^{-t/\tau}\left[A_1 - \frac{A_0}{\tau} - \left(\frac{A_1}{\tau}\right)t\right]$$ (12.23)

If the body is displaced and released with zero initial velocity ($y = y_0$ and $v = 0$ when $t = 0$), Equations 12.22 and 12.23 become

$$y_0 = A_0$$ (12.24)

and

$$0 = A_1 - \frac{A_0}{\tau}$$ (12.25)

Substituting $A_0 = y_0$ in Equation 12.25 yields

$$A_1 = y_0/\tau$$

If these values for A_0 and A_1 are substituted in Equations 12.22 and 12.23, one obtains

$$y = y_0\, e^{-t/\tau}\left(1 + \frac{t}{\tau}\right)$$ (12.26)

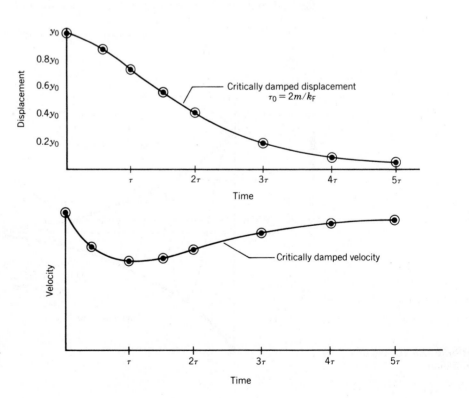

FIGURE 12.7 Displacement and velocity versus time for a critically damped vibration.

$$v = - \frac{y_o t}{\tau^2} e^{-t/\tau}$$ (12.27)

Plots of Equation 12.26 and 12.27 are shown in Figure 12.7. A sample calculation for critically damped motion is given in Example 12.5.

EXAMPLE 12.5 Critical damped motion
. . . .

Given: A 4.0-kg mass at the end of a spring is to be critically damped. The spring constant is 100 N/m. The initial displacement is 0.10 m and the initial velocity is 0.0 m/sec.

Find:
a. The required value of the drag coefficient.
b. The displacement after 1.0 sec.
c. The velocity after 1.0 sec.

Solution:
a. The condition for critically damped motion is

$$\frac{k}{m} = \frac{1}{4} \left(\frac{k_F}{m} \right)^2$$

or

$$\sqrt{\frac{4k}{m}} = \frac{k_F}{m}$$

or

$$k_F = m \sqrt{\frac{4k}{m}} = \sqrt{4km}$$

$$k_F = \sqrt{(4)(100 \text{ N/m}) (4.0 \text{ kg})}$$

$$k_F = 40 \text{ N·sec/m}$$

b. The value of τ is

$$\tau = \frac{2m}{k_F}$$

$$\tau = \frac{(2)(4.0 \text{ kg})}{40 \text{ N·sec/m}}$$

$$\tau = 0.2 \text{ sec}$$

The displacement is (Equation 12.26)

$$y = y_0 e^{-t/\tau} (1 + t/\tau)$$

$$y = (0.10 \text{ m}) e^{-1.0 \text{ sec}/0.2 \text{ sec}} \left(1 + \frac{1.0 \text{ sec}}{0.2 \text{ sec}} \right)$$

$$y = (0.10 \text{ m})(0.00674)(6)$$

$$y = 0.004 \text{ m}$$

c. The velocity is (Equation 12.27)

$$v = -(y_0)\left(\frac{t}{\tau^2}\right) e^{-t/\tau}$$

$$v = -(0.1 \text{ m})\left(\frac{1.0 \text{ sec}}{[0.2 \text{ sec}]^2}\right)(0.00674)$$

$$v = -0.0169 \text{ m/sec}$$

When the condition

$$\frac{k}{m} < \frac{1}{4}\left(\frac{k_F}{m}\right)^2$$

the mass will also return to the equilibrium position without oscillating. This can be achieved by increasing k_F beyond the value needed for critical damping. A third solution is obtained for Equation 12.19. However, a longer time will be required to reach the steady state than for the case of critical damping. When this occurs, the system is said to be overdamped. Since the critical damped case is the usual one of interest, the equations for the overdamped case will not be shown here.

SECTION 12.2   EXERCISES

1. Define
   Damping
   Terminal velocity
   Drag force
2. Define the conditions for
   Underdamped motion
   Critically damped motion
   Overdamped motion
3. A 250-kg object falls from rest. The drag constant k_F is 50 N·sec/m. What is the terminal velocity? How long will the object take to reach a speed within 1% of the terminal velocity (i.e., 5τ)?
4. A 10-kg mass is on a spring with a spring constant of 500 N/m. If the drag constant is 50 N·sec/m, what is the time constant for the damping of the oscillation. Is the motion underdamped, critically damped, or overdamped?

12.3 . . . .   OBJECTIVES . . . .

FLUID   Upon completion of this section, the student should be able to
SYSTEMS
• Calculate the time required to empty a tank for linear fluid discharge.

• Calculate the capacitance of a cylindrical tank for linear discharges (exponential decay).

DISCUSSION

Large cylindrical tanks are used to store fluids in industries for water storage, chemical processing, and fire control. These tanks are often drained by gravity, and the fluid height as a function of time can be estimated. However, as was shown in the chapter on resistance, the equations for fluid flow may or may not be linear. In this discussion, both linear and nonlinear processes will be considered, but only the linear case will be developed.

Nonlinear Process

Turbulent Discharge   The flow through a pipe is turbulent in most cases. For turbulent flow the relationship between volume and pressure difference is expressed as developed in Chapter 5 by

$$Q_V^2 = \frac{\Delta p}{R} \tag{12.28}$$

The resistance R depends on the velocity and the pipe roughness. Since the velocity varies, the resistance will vary. This causes the complete analysis of turbulent flow to be complex. However, if we assume the resistance is constant, we can see from Equation 12.28 that

$$Q_V^2 \propto \Delta p$$

or

$$Q_V \propto \sqrt{\Delta p} \tag{12.29}$$

The pressure difference for an open tank is given by

$$\Delta p = h \, \rho_m \, g = h \rho_w \tag{12.30}$$

where

h = height of fluid, above discharge pipe
ρ_m = mass density of fluid
g = acceleration due to gravity
ρ_w = weight density of fluid

Substituting this value for Δp into the relationship 12.29 yields

$$Q_V \propto \sqrt{h \, \rho_w} \tag{12.31}$$

ρ_w is constant, so Equation 12.31 can be simplified to

$$Q_V \propto \sqrt{h} \tag{12.32}$$

since

$$Q_V = A_T \frac{\Delta h}{\Delta t} \tag{12.33}$$

where

Δh = change in height
A_T = constant cross-sectional area of tank
Δt = elapsed time corresponding to the change in height

FIGURE 12.8   Plot of height versus time for turbulent flow discharge.

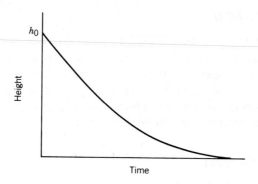

FIGURE 12.8   Plot of height versus time for turbulent flow discharge.

For this case, a plot of height versus time as shown in Figure 12.8 yields

$$\frac{\Delta h}{\Delta t} \propto \sqrt{h} \qquad (12.34)$$

This is not a linear equation, but gives a quadratic equation for h as a function of t. To further complicate the analysis, as the height decreases the volume-flow rate decreases and the Reynolds number decreases. In many cases the flow undergoes a transition from turbulent to laminar flow.

Discharge from an Orifice   If an open tank is drained through an orifice or a pipe that is too short for the development of turbulent or laminar flow, the volume-flow rate can be analyzed using Bernoulli's equation. This analysis gives a relationship similar to Equation 12.33, but the proportionality constant for this relationship is different from the constant that would apply to the turbulent flow case. However, the final results are the same for our purpose and the relationship is quadratic rather than exponential for h as a function of t.

Linear Process

Laminar Discharge   If the liquid flows through a long pipe and the flow is laminar, then the system is linear. For this case (see Chapter 5) the volume-flow rate may be expressed as

$$Q_V = \frac{\Delta p}{R} \qquad (12.35)$$

where

Q_V = volume-flow rate
Δp = pressure drop across the pipe
R = resistance of the pipe

In Chapter 5, the resistance of a pipe for laminar flow was shown to be

$$R = \frac{8\mu l}{\pi R_0^4} \qquad (12.36)$$

where

μ = dynamic viscosity
l = length of the pipe
R_0 = radius of the pipe

Setting the volume-flow rate given by Equation 12.33 equal to the volume-flow rate given by Equation 12.35 yields

$$A_T \frac{\Delta h}{\Delta t} = \frac{\Delta p}{R}$$

or

$$\frac{\Delta h}{\Delta t} = \frac{\Delta p}{A_T R} \tag{12.37}$$

Since $\Delta p = \rho_m g h$, Equation 12.37 becomes

$$\frac{\Delta h}{\Delta t} = \left(\frac{\rho_m g}{A_T R}\right) h \tag{12.38}$$

Equation 12.38 is a linear equation and the solution has the form

$$h = h_0 e^{-\frac{\rho_m g}{R A_T} t} \tag{12.39}$$

where

h_0 = initial height

The corresponding time constant is

$$\tau = R \left(\frac{A_T}{\rho_m g}\right) \tag{12.40}$$

For the laminar case, the quantity $A_T/\rho_m g$ may be defined as the capacitance of the tank, C_F. The time constant can then be written as

$$\tau = \frac{R A_T}{\rho_m g} = R C_F \tag{12.41}$$

If the numerator and denominator of Equation 12.40 are multiplied by h, one obtains

$$\tau = R C_F = \left(\frac{A_T h}{\rho_m g h}\right) = R \frac{V}{\Delta p}$$

since $V = A_T h$ and $\Delta p = \rho_m g h$. Thus,

$$R C_F = R \frac{V}{\Delta p}$$

or

$$C_F = \frac{V}{\Delta p} \tag{12.42}$$

FIGURE 12.9 Fixed and variable fluid capacitors.

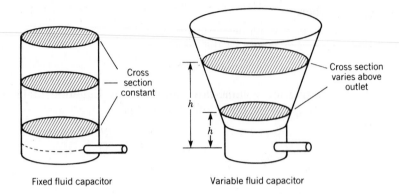

Fixed fluid capacitor   Variable fluid capacitor

where
V = volume of the fluid
Δp = corresponding pressure difference

Volume is the displacementlike quantity and Δp is the forcelike quantity. The corresponding electric equation (see Chapter 1) is

$$C = \frac{\text{Displacementlike Quantity}}{\text{Forcelike Quantity}} = \frac{Q}{\Delta V}$$

where
Q = electric charge
ΔV = voltage difference across the capacitor
C = electrical capacitance

By analogy, the fluid capacitance of a cylindrical tank is defined as

$$C_F = \frac{A_T}{\rho_m g} \tag{12.43}$$

Since the capacitance is proportional to the cross-sectional area, a cylinder with a large area has a greater fluid capacitance than a cylinder with a small surface area if the tanks contain the same fluid. Since the cross-sectional area of a cylinder is constant, the fluid capacitance has a fixed value. Just as an electrical capacitance does not depend on the charge stored at any time, the fluid capacitance of a cylindrical tank does not depend on the quantity of liquid stored.

Tanks whose cross-sectional area varies with the height of fluid level above the outlet, however, have a fluid capacitance that varies as the tank is emptied. Figure 12.9 illustrates this difference in different types of tanks. This chapter is concerned only with fixed fluid capacitors; analysis of variable fluid capacitors is beyond the scope of this discussion.

The capacitance of a cylindrical tank is calculated in Example 12.6, and Example 12.7 gives a complete analysis for a cylindrical tank being emptied by a long pipe with laminar flow.

EXAMPLE 12.6 Fluid capacitance of a cylindrical storage tank
. . . .

Given: In a chemical plant, there is a cylindrical storage tank having a cross-sectional area of 100 m² and containing alcohol, which has a mass density of 785 kg/m³. The tank is emptied using laminar flow.

Find: The capacitance of the storage tank in volume per unit pressure units.

Solution:

$$C = \frac{A_T}{\rho_w g}$$

$$C = \frac{100 \text{ m}^2}{(785 \text{ kg/m}^3)(9.8 \text{ m/sec}^2)}$$

$$C = 0.013 \frac{\text{m}^2}{\text{kg/m}^3 \cdot \text{m/sec}^2}$$

Dimensionally

$$\frac{\text{m}^2}{\text{kg/m}^3 \cdot \text{m/sec}^2} = \frac{\text{m}^2 \cdot \text{m}}{\frac{\text{kg} \cdot \text{m/sec}^2 \cdot \text{m}}{\text{m}^3}} = \frac{\text{m}^3}{\text{N/m}^2} = \frac{\text{m}^3}{\text{Pa}}$$

Remember that 1 Pascal (Pa) = 1 N/m². Therefore,

$$C = 0.013 \frac{\text{m}^3}{\text{Pa}} \quad \begin{array}{l} \text{volume units} \\ \text{pressure units} \end{array}$$

EXAMPLE 12.7 Time to empty tank
. . . .

Given: A storage tank is filled with water at 100°F to a depth of 1.0 ft. The cross-sectional area of the tank is 450 ft². The tank is drained through a 400-ft-long pipe with a diameter of 0.10 ft.

Find:
a. The pipe resistance for laminar flow.
b. The capacitance of the tank.
c. The time constant of the tank.
d. The time to drain the tank.
e. The initial average fluid velocity across the pipe.
f. The initial Reynolds number of the flow.

Solution:
a. The pipe resistance for laminar flow is

$$R = \frac{8\mu l}{\pi R_0^4} = \frac{128\mu l}{\pi D_0^4}$$

The viscosity of the water at 100°F = 1.43×10^{-5} lb·sec/ft².

$$R = \frac{(128)(1.43 \times 10^{-5} \text{ lb·sec/ft}^2)(400 \text{ ft})}{(3.14)(0.05 \text{ ft})^4}$$

$$R = 3.73 \times 10^4 \text{ lb·sec/ft}^5$$

b. The tank capacity is

$$C_F = \frac{A_T}{\rho_m g} = \frac{A_T}{\rho_w}$$

$$C_F = \frac{450 \text{ ft}^2}{62.4 \text{ lb/ft}^3}$$

$$C_F = 7.2 \text{ ft}^5/\text{lb}$$

c. The tank time constant is

$$\tau = R C_F$$

$$\tau = (3.73 \times 10^4 \text{ lb·sec/ft}^5)(7.2 \text{ ft}^5/\text{lb})$$

$$\tau = 2.69 \times 10^5 \text{ sec}$$

$$\tau = \frac{2.69 \times 10^5 \text{ sec}}{3600 \text{ sec/hr}}$$

$$\tau = 74.7 \text{ hr}$$

d. The time to drain the tank is

$$t_F = 5\tau$$

$$t_F = (5)(74.7)$$

$$t_F = 373.5 \text{ hr}$$

e. The volume flow rate is

$$Q_V = \frac{\Delta p}{R} = \frac{\rho_w h}{R}$$

$$Q_V = \frac{(62.4 \text{ lb/ft}^3)(1.0 \text{ ft})}{(3.73 \times 10^4 \text{ lb·sec/ft}^5)}$$

$$Q_V = 1.67 \times 10^{-3} \text{ ft}^3/\text{sec}$$

The average flow velocity is

$$\bar{v} A_P = Q_V = \bar{v}(\pi R_0^2)$$

where

$$A_P = \text{area of the pipe}$$

$$\bar{v} = Q_V/(\pi R_0^2)$$

$$\bar{v} = \frac{(1.67 \times 10^{-3} \text{ ft}^3/\text{sec})}{(3.14)(0.05 \text{ ft})^2}$$

$$\bar{v} = 0.213 \text{ ft/sec}$$

f. The initial Reynolds number of the flow is

$$\text{Re} = \frac{\rho_m \bar{v} D_0}{\mu}$$

At 100°F, ρ_m = 1.94 slug/ft³.

$$Re = \frac{(1.94 \text{ slug/ft}^3)(0.213 \text{ ft/sec})(0.10 \text{ ft})}{(1.43 \times 10^{-5} \text{ lb·sec/ft}^2)}$$

Re = 2890

The Reynolds number is in the critical range and the flow may be turbulent. However, the Reynolds number will become less as the tank drains. It will be assumed that the flow is laminar.

SECTION *12.3* EXERCISES

1. For laminar flow, define "fluid capacitance" for a cylindrical storage tank.
2. For laminar flow, find the capacitance of a tank with a cross-sectional area of 100 m². The height of the water above the outlet is 5 m.
3. The time constant (laminar flow) for a long, cylindrical buret is 3 sec. How long does it take to empty the fluid contents?
4. A 400-ft-diameter cylindrical storage tank filled with water at 100°F is drained through a 1000-ft pipe with a diameter of 0.06 ft. What is the resistance of the pipe? Assume the flow is laminar. The viscosity of water at 100°F is 1.43×10^{-5} lb·sec/ft².
5. If the initial height of the water in the tank in Problem 4 is 0.5 ft, what is the initial average velocity of the water in the pipe? How long does it take to drain the tank?

12.4

ELECTROMAGNETIC SYSTEMS

OBJECTIVES

Upon completion of this section, the student should be able to

- Calculate the time constant for an *RC* circuit.
- Calculate the time constant for an *RL* circuit.
- Apply the general linear equation to *RC* and *RL* circuits.
- Describe the analogy between linear, mechanical, vibrating systems, and *RLC* circuits.
- Given *R* and *C*, calculate the required inductance to produce critical damping.
- Given a known resistive load and pulse time, calculate the capacitance and inductance necessary for critical damping.
- Calculate the explosion energy for a linear flash lamp.

DISCUSSION

RC *Circuits* Figure 12.10 shows an *RC* circuit. Assume that the initial charge on the capacitor is zero. When the switch is closed in the upper position, a current will flow through R_1 to charge the capacitor. At any time the sum of the voltage

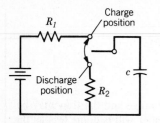

FIGURE 12.10 Schematic of a resistor–capacitor (*RC*) circuit.

across the capacitor, the voltage across the resistor, and the voltage of the source must equal zero. Mathematically this condition may be written as

$$V_T + V_R + V_C = 0$$

or

$$V_T = -IR_1 - Q/C = 0 \tag{12.44}$$

where

$$\begin{aligned}
V_T &= \text{battery voltage} \\
I &= \text{current} \\
Q &= \text{charge on the capacitor} \\
C &= \text{capacitance of the capacitor} \\
R_1 &= \text{resistance} \\
V_R &= \text{resistor voltage drop} \\
V_C &= \text{capacitor voltage drop}
\end{aligned}$$

From the definition of electric current (see Chapter 3), one has

$$I = \frac{\Delta Q}{\Delta t}$$

Substituting this expression for I into Equation 12.44 yields

$$R \frac{\Delta Q}{\Delta t} = -Q/C + V_T$$

or

$$\frac{\Delta Q}{\Delta t} = -\frac{Q}{RC} + \frac{V_T}{R} \tag{12.45}$$

This equation is linear and the solution is of the form

$$Q = (Q_0 - V_T C)e^{-t/\tau} + V_T C \tag{12.46}$$

If the initial charge on the capacitor is zero, Equation 12.46 reduces to

$$Q = V_T C(1 - e^{-t/\tau}) \tag{12.47}$$

where

$$\begin{aligned}
V_T C &= \text{final charge} \\
\tau &= RC = \text{time constant}
\end{aligned}$$

The voltage drop across the resistor at any time is

$$V_{R_1} = -IR_1 \tag{12.48}$$

Solving Equation 12.44 for IR_1 yields

$$V_{R_1} = -IR_1 = -V_T + Q/C \tag{12.49}$$

If the expression for Q given by Equation 12.47 is substituted in Equation 12.49, one obtains

$$V_{R_1} = -V_T + V_T(1 - e^{-t/\tau})$$

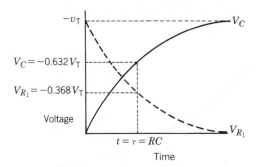

FIGURE 12.11 Plots of voltage across the capacitor and voltage across the resistor versus time in a charging *RC* circuit.

or

$$V_{R_1} = -V_T e^{-t/\tau} \qquad (12.50)$$

Since the voltage across the capacitor is Q/C, dividing both sides of Equation 12.47 by C yields

$$V_C = -V_T(1 - e^{-t/\tau}) \qquad (12.51)$$

As the switch is closed, current flows to charge the capacitor. As the capacitor charges the current becomes less. As this process takes place, the voltage across the resistor decreases and the voltage across the capacitor increases. Plots of these two voltages as a function of time (Equations 12.47 and 12.48) are shown in Figure 12.11.

If the capacitor shown in Figure 12.10 is charged to voltage V_T and the switch is moved to the lower position, the capacitor will discharge through resistor R_2. For this case, the voltage across the capacitor plus the voltage across the resistor is equal to 0. Expressing this condition mathematically gives

$$R_2 I + Q/C = 0$$

or

$$\frac{\Delta Q}{\Delta t} = -\frac{Q}{R_2 C} \qquad (12.52)$$

The solution of this linear equation is

$$Q = Q_0 e^{-t/R_2 C} = Q_0 e^{-t/\tau} \qquad (12.53)$$

When $t = 0$, the charge on the capacitor equals $V_T C$. Substituting this condition in Equation 12.53 yields

$$Q_0 = V_T C$$

Thus

$$Q = V_T C \, e^{-t/\tau} \qquad (12.54)$$

Since the voltage across the capacitor is equal to Q/C, dividing both sides of Equation 12.54 by C gives

$$V_C = Q/C = V_T e^{-t/\tau} \qquad (12.55)$$

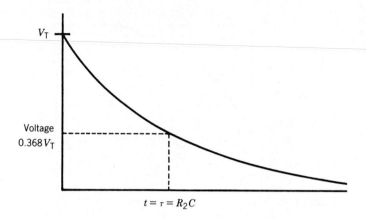

FIGURE 12.12 Plot of voltage across a capacitor versus time for a discharging capacitor.

Figure 12.12 shows the capacitor voltage as a function of time. The time constant for this case is R_2C.

Examples 12.8, 12.9, and 12.10 show the use of the derived equations in solving typical problems. Refer to Figure 12.10 for all examples and use $C = 1 \ \mu F$, $R_1 = 1000 \ \Omega$, $R_2 = 2000 \ \Omega$, and $V_T = 100 \ V$.

EXAMPLE 12.8   Time constant calculations
. . . .

Given: Figure 12.10 and $C = 1 \ \mu F$, $R_1 = 1000 \ \Omega$, $R_2 = 2000 \ \Omega$, and $V_T = 100 \ V$.

Find: Time constants for charging and discharging circuits.

Solution: Charging:

$\tau = RC$

$\tau = (1000 \ \Omega)(1 \times 10^{-6} \ F)$

$\tau = 1 \times 10^{-3} \ sec$

Discharging:

$\tau = RC$

$\tau = (2000 \ \Omega) \ (1 \times 10^{-6} \ F)$

$\tau = 2 \times 10^{-3} \ sec$

EXAMPLE 12.9   Calculation of capacitor voltage in charge mode
. . . .

Given: Initial capacitor voltage is zero and switch is closed upward.

Find: Voltage on capacitor after 1 msec.

Solution:

$$V_C = -V_T (1 - e^{-t/\tau})$$

$$V_C = (-100 \text{ V})[1 - e^{-(0.001)/(0.001)}]$$

$$V_C = (-100 \text{ V})[1 - e^{-1}]$$

$$V_C = (-100 \text{ V})[1 - 0.368]$$

$$V_C = -63.2 \text{ V}$$

Note: 1 msec is time constant for charging circuit.

EXAMPLE 12.10   Calculation of capacitor voltage in discharge mode
. . . .

Given: Initial capacitor voltage is 100 V and switch is closed downward.

Find: Capacitor voltage after 6 msec.

Solution:

$$V_C = -V_T \, e^{-t/\tau}$$

$$V_C = (-100 \text{ V}) \, e^{-0.006 \text{ sec}/0.002 \text{ sec}}$$

$$V_C = (-100 \text{ V}) \, e^{-3} \quad \text{(Note: } e^{-3} = 0.0498\text{)}$$

$$V_C = (-100 \text{ V})(0.0498)$$

$$V_C = -4.98 \text{ V}$$

In practical timing circuits, some type of sensing device is attached to the capacitor. This device delivers an output signal when capacitor voltage reaches some preset value. When such circuits are designed, the time required for the capacitor to reach a given voltage must be determined. To calculate the time for charging to a preset voltage write Equation 12.51 in the form

$$\frac{V_T + V_C}{V_T} = e^{-t/\tau} \tag{12.56}$$

where

$$\tau = RC$$

The logarithm to the base e can be taken of both sides of Equation 12.56. This yields

$$\ln\left(\frac{V_T + V_C}{V_T}\right) = \ln e^{-t/\tau} \tag{12.57}$$

However, $\ln e^{-t/\tau} = -t/\tau \ln e$, and $\ln e = 1$. Thus, Equation 12.57 becomes

$$\ln\left(\frac{V_T + V_C}{V_T}\right) = -t/\tau$$

or

$$\text{(charging capacitor)} \quad t = -\tau \ln\left(1 + \frac{V_C}{V_T}\right) \qquad (12.58)$$

In a similar way, solving Equation 12.55 for t gives

$$\text{(discharging capacitor)} \quad t = \tau \ln\left(\frac{V_C}{V_T}\right) \qquad (12.59)$$

Examples 12.11 and 12.12 use Equations 12.57 and 12.58 to solve problems.

EXAMPLE 12.11   Calculation of time to reach a given voltage in charge mode
. . . .

Given: From Figure 12.10, capacitor fully discharged, and switch closed upward. ($C = 1\mu F$, $R_1 = 1000\ \Omega$, $R_2 = 2000\ \Omega$, and $V_T = 100$ V.)

Find: Time before capacitor voltage is -67 V.

Solution: $t = -\tau \ln\left(1 + \dfrac{V_C}{V_T}\right),$ where $\tau = R_1 C$

$$t = -(1000\ \Omega)(1 \times 10^{-6}\ F)\ln\left(1 + \frac{-67V}{100\ V}\right)$$

$$t = -(1 \times 10^{-3}\ \text{sec})\ln(0.33)$$

$$t = 1.11 \times 10^{-3}\ \text{sec} = 1.11\ \text{msec}$$

(Compare with Example 12.9.)

EXAMPLE 12.2   Calculation of time to reach a given voltage in discharge mode
. . . .

Given: Capacitor fully charged to -100 V ($-V_T$) and the switch is in the lower position.

Find: Time before capacitor voltage reaches -33 V.

Solution: $t = -\tau \ln\left(\dfrac{V_C}{V_T}\right),$ where $\tau = R_2 C = 2000\ \Omega \times 1 \times 10^{-6}\ F = 2$ msec

$$t = -(2000\ \Omega)(1 \times 10^{-6}\ F)\ln\left(\frac{-33}{-100}\right)$$

$$t = -(2 \times 10^{-3})\ln(0.33)$$

$$t = 2.2 \times 10^{-3}\ \text{sec}$$

(Compare with time constant of this discharge circuit.)

RL *Circuits*   Inductance was defined in Chapter 3, and it is given by

$$L = N \frac{\Delta \phi}{\Delta t} \tag{12.60}$$

where

L = inductance of a coil
N = number of turns
ϕ = magnetic field through the coil produced by the current through the coil

Since all circuits produce magnetic fields, they all have inductance. In many cases, however, the inductance may be negligibly small. In addition, all inductors have some resistance. The *RL* circuit analyzed in this chapter consists of an ideal, resistanceless inductor in series with a noninductive resistor. If the inductor resistance is significant, its resistance can be added to the resistance of the external resistor (resistors in series).

As demonstrated in Chapter 3, the voltage across an inductor is

$$V_L = -L \frac{\Delta I}{\Delta t} \tag{12.61}$$

where

I = instantaneous current through the inductor

In an *RL* circuit, the sum of the voltage across the inductor, the voltage across the resistor, and the voltage of the source equals zero. This yields the equation

$$-V_L - V_R = V_T$$

or

$$L \frac{\Delta I}{\Delta t} + IR = V_T$$

or

$$\frac{\Delta I}{\Delta t} = -\frac{R}{L} I + \frac{V_T}{L} \tag{12.62}$$

A typical, ideal *RL* circuit is shown in Figure 12.13*a*. When the switch is closed, the inductor opposes the increase in current. This causes the current to build up over a period of time rather than instantaneously. The current buildup is determined by Equation 12.62, which is linear and gives a solution of

$$I = \left(I_0 - \frac{V_T}{R} \right) e^{-t/\tau} + \frac{V_T}{R} \tag{12.63}$$

At the moment the switch is closed, $I_0 = 0$. Thus, Equation 12.63 reduces to

$$I = \frac{V_T}{R} (1 - e^{-t/\tau}) \tag{12.64}$$

where

$\tau = L/R$

The current as a function of time, for this case, is illustrated in Figure 12.13*b*. The

FIGURE 12.13 Schematic of
RL circuit and characteristic
current versus time curve.

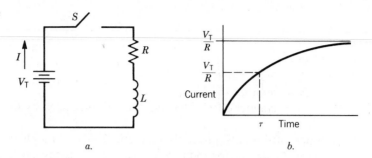

FIGURE 12.13 Schematic of *RL* circuit and characteristic current versus time curve.

value V_T/R is the final value of I, and it is the current that would occur if the circuit were composed only of a resistor. Example 12.13 illustrates the calculation of an inductive time constant.

EXAMPLE 12.13   Inductive time constant

. . . .

Given: An *RL* circuit has $R = 1000\ \Omega$ and $L = 50$ mH.

Find: The inductive time constant.

Solution: $\tau = L/R$

$\tau = 50 \times 10^{-3}$ H$/1000\ \Omega$

$\tau = 50 \times 10^{-6}$ sec

$\tau = 50\ \mu$sec

The corresponding voltage drop across the resistor is given by Ohm's law:

$$V_R = -IR$$

Substituting for I from Equation 12.64 yields

$$V_R = -V_T\,(1 - e^{-t/\tau}) \tag{12.65}$$

To find the voltage across the inductor, Equation 12.62 can be solved for $L\,\Delta I/\Delta t$. This gives

$$-V_L = L\frac{\Delta I}{\Delta t} = V_T + V_R \tag{12.66}$$

or

$$-V_L = V_T - V_T + V_T e^{-t/\tau} = V_T e^{-t/\tau}$$

or

$$V_L = -V_T e^{-t/\tau} \tag{12.67}$$

The voltage across the resistor and inductor as a function of time is shown in Figure 12.14. Example 12.14 illustrates the current growth in an *RL* circuit.

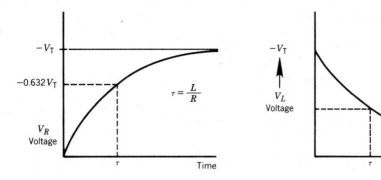

FIGURE 12.14 Voltage across a resistor and across an inductor in an *RL* circuit as a function of time.

EXAMPLE 12.14   Current growth in an *RL* circuit
....

Given: An inductor of inductance 2 H and a resistor of resistance 4 Ω are connected in series to a battery of emf 12 V. The battery has negligible internal resistance.

Find: **a.** I at $t = 0.2$ sec.
b. V_L at $t = 0.2$ sec.
c. V_R at $t = 0.2$ sec.

Solution: **a.** $I = V_T(1 - e^{-tR/L})/R$

$$I = \frac{(12 \text{ V})}{4 \text{ }\Omega} (1 - e^{-(0.2 \text{ sec})(4\,\Omega)/(2\,\text{H})})$$

$$I = 0.989 \text{ A}$$

b. $V_L = -V_T e^{-tR/L}$

$$V_L = -(12 \text{ V})e^{-(0.2 \text{ sec})(4\,\Omega)/(2\,\text{H})}$$

$$V_L = -8.04 \text{ V}$$

c. $V_R = -V_T(1 - e^{-tR/L})$

$$V_R = -(12 \text{ V})(1 - e^{-(0.2 \text{ sec})(4\,\Omega)/(2\,\text{H})})$$

$$V_R = -3.96 \text{ V}$$

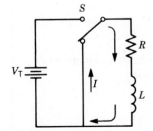

FIGURE 12.15 Schematic of decaying current in an *RL* circuit.

Once the steady-state current is established in the *RL* circuit, the switch shown in Figure 12.15 may be opened and the resistor shorted across the inductor. The current decays back to zero over a period of time. At every instant the sum of the voltages across the inductor and resistor is equal to zero since $V_T = 0$ for this case. When $V_T = 0$, Equation 12.63 becomes

$$I = I_0 e^{-t/\tau} \tag{12.68}$$

The voltage across the resistor (Ohm's law) is

$$V_R = -IR = -RI_0 e^{-t/\tau} = -V_0 e^{-t/\tau} \tag{12.69}$$

FIGURE 12.16 Characteristic current and voltage curves for a current decay in an *RL* circuit.

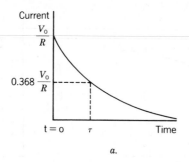

a.

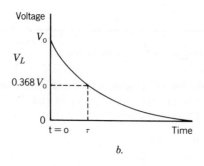

b.

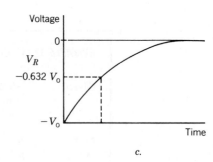

c.

The corresponding voltage across the inductor is given by Equation 12.66, where $V_T = 0$. Thus,

$$V_L = -V_R = V_0 e^{-t/\tau} \qquad (12.70)$$

For the *RL* circuits, $\tau = L/R$. The voltage and currents are shown graphically in Figure 12.16, and Example 12.15 gives a calculation for the current decay in an *RL* circuit.

EXAMPLE 12.15   Current decay in an *RL* circuit
. . . .

Given: An *RL* circuit consisting of a 3-H inductor with 6 Ω resistance and wire resistance of 9 Ω begins its current decay cycle at $t = 0$. Initially, $V_0 = 15$ V.

Find: **a.**   The time constant τ.

   b.   The voltages V_L across the inductor and V_R along the wire at $t = 0.3$ sec.

Solution: The equivalent circuit consists of an ideal resistanceless inductor of $L = 3$ H and a resistor of the following resistance:

$$R = R_W + R_L$$

$$R = 9 \ \Omega + 6 \ \Omega$$

$$R = 15 \ \Omega$$

The resistance value is depicted in the following circuit:

a. $\tau = L/R$

$\tau = 3\ \text{H}/15\ \Omega$

$\tau = 0.2\ \text{sec}$

b. For the ideal inductor

$V_L(t) = V_0 e^{-t/\tau}$

$V_L = (15\ \text{V})\ e^{-(0.3\ \text{sec})/(0.2\ \text{sec})}$

$V_L = 3.35\ \text{V}$

The actual (measured) voltage V_L includes that due to its resistance of 6 Ω:

$V_L = 33.35\ \text{V} - IR_L$

$V_L = 3.35\ \text{V} - \dfrac{V_0 e^{-t/\tau}}{R}\ R_L$

$V_L = 3.35\ \text{V} - \left(\dfrac{15\ \text{V}}{15\ \Omega}\right) e^{-(0.3\ \text{sec})/(0.2\ \text{sec})(6\ \Omega)}$

$V_L = 3.35\ \text{V} - 1.34\ \text{V}$

$V_L = 2.01\ \text{V}$

The voltage V_R along the wire is equal and opposite to V_L since the total voltage is zero. Therefore, $V_R = -2.01\ \text{V}$ at $t = 0.3$ sec.

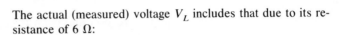

RLC *Circuits*

In many electronic circuits, it is necessary to deliver electrical power in the form of pulses. For low-energy pulses, resistance–capacitance (*RC*) circuits may be used. But *RC* circuits have characteristics that make them undesirable for higher-energy pulses. In all high-energy pulsed circuits and in many low-energy ones, an inductor is added to shape the pulse. Thus, we shall examine the characteristics and applications of resistance–inductance–capacitance circuits.

Figure 12.17 shows a schematic diagram of an *RLC* circuit. In this circuit, the capacitor stores energy that will be delivered to the load resistor in the form of a pulse. The inductor, *L*, controls the shape of the pulse by limiting the current change since the voltage across the inductor is greatest when the change in current is greatest.

The sum of the voltages around the loop in Figure 12.17 must equal zero. Thus,

$$L\frac{\Delta I}{\Delta t} + IR + Q/C = 0 \qquad (12.71)$$

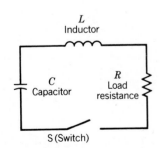

FIGURE 12.17 Schematic of an *RLC* circuit.

This equation has the same form as the equation for a mass attached to a spring with damping (Equation 12.19):

$$m \frac{\Delta v}{\Delta t} + v k_F + ky = 0$$

One can see that L corresponds to m, R corresponds to k_F, $1/C$ corresponds to k, I corresponds to v, and Q corresponds to y. This correspondence between the two equations is consistent with the analogies used previously. As in the case of the mechanical system, there are three different solutions (underdamped, critically damped, and overdamped).

If $R = 0$, there is no damping, and the system will continue to oscillate. By an analogy with the mechanical system, the charge as a function of time is

$$Q = A_0 \cos \left(\frac{1}{\sqrt{LC}} t + \phi \right) \tag{12.72}$$

This equation is obtained from Equation 12.18 by letting $L = m$ and $1/C = k$. In addition to the charge, it is often desirable to determine the current as a function of time. Using calculus one can show that

$$I = \frac{\Delta Q}{\Delta t} = -\frac{A_0}{\sqrt{LC}} \sin \left(\frac{1}{\sqrt{LC}} t + \phi \right) \tag{12.73}$$

The quantity $1/\sqrt{LC}$ is called the angular frequency, and it is usually given the symbol ω:

$$\omega = \frac{1}{\sqrt{LC}} \tag{12.74}$$

The quantity A_0 is the amplitude of the charge oscillations, and $A_0/\sqrt{LC}$ is the amplitude of the current oscillations. Plots of Q and I are shown in Figure 12.18.

If R is not equal to zero, the energy will be lost in the resistance and the charge amplitude will decay. By analogy with Equation 12.20, oscillations will occur if

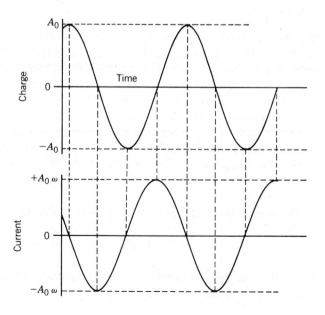

FIGURE 12.18 Comparison of charge and current as a function of time for an *LC* circuit.

$$\frac{1}{LC} > \frac{1}{4}\left(\frac{R}{L}\right)^2 \qquad (12.75)$$

or

$$L > \frac{R^2C}{4} \qquad (12.76)$$

During the oscillations energy is transferred back and forth between the inductor and the capacitor. However, for each oscillation part of the energy is lost in the resistor. By using the analogous quantities in Equation 12.21, the corresponding equation for the charge as a function of time is

$$Q = A_0 e^{-tR/2L}[\cos(\omega_1 t + \phi)] \qquad (12.77)$$

where

$$\omega_1 = \left[\frac{1}{LC} - \frac{1}{4}\left(\frac{R}{L}\right)^2\right]^{1/2}$$

ϕ = phase angle

The corresponding time constant for the decay of the charge amplitude is $\tau = 2L/R$. Thus, a large vlaue of L/R causes the oscillations to continue for a long time. The corresponding current can be obtained by using calculus and the definition of current:

$$I = \frac{\Delta Q}{\Delta t}$$

This gives for the current

$$I = -A_0\left[\omega_1^2 + \left(\frac{R}{2L}\right)^2\right]^{1/2}[e^{-(R/2L)t}][\sin(\omega_1 t + \theta + \phi)] \qquad (12.78)$$

where

$$\theta = \text{arc cos}\left[\frac{\omega_1}{[\omega_1^2 + (R/2L)^2]^{1/2}}\right]$$

Figure 12.19 illustrates the charge and current as a function of time and Example 12.16 shows the use of Equation 12.77 to determine the time required for an oscillation to decay.

EXAMPLE 12.16 Time for an electrical oscillation to decay
. . . .

Given: A 1-μF capacitor, a 1-H inductor, and a 1000-Ω resistor are connected in series. The initial charge on the capacitor is 1×10^{-4} C.

Find: **a.** Will oscillation occur?
        b. The angular frequency of the oscillations.
        c. The time for the oscillations to decay.
        d. The number of oscillations.

Solution: **a.** Oscillations occur if

$$L > \frac{R^2 C}{4}$$

$$L > \frac{(1000 \ \Omega)^2 (1 \times 10^{-6} \ \text{F})}{4}$$

$$L > 0.25 \ \text{H}$$

Thus oscillations occur.

b. From Equation 12.77, the angular frequency is

$$\omega_1 = \left[\frac{1}{LC} - \frac{1}{4}\left(\frac{R}{L}\right)^2 \right]^{1/2}$$

$$\omega_1 = \left[\frac{1}{(1 \ \text{H})(1 \times 10^{-6} \ \text{F})} - \frac{1}{4}\left(\frac{(1000 \ \Omega)}{(1 \ \text{H})}\right)^2 \right]^{1/2}$$

$$\omega_1 = \left[\frac{1 \times 10^6}{\text{sec}^2} - \frac{0.25 \times 10^6}{\text{sec}^2} \right]^{1/2}$$

$$\omega_1 = \left[\frac{0.75 \times 10^6}{\text{sec}^2} \right]^{1/2}$$

$$\omega_1 = 866 \ \text{rad/sec}$$

c. The time constant is

$$\tau = \frac{2L}{R}$$

$$\tau = \frac{(2) \ 1 \ \text{H}}{(1000 \ \Omega)} = 2 \times 10^{-3} \ \text{sec}$$

The time for the oscillations to damp out is

$$t_F = 5\tau$$

$$t_F = 1.0 \times 10^{-2} \ \text{sec}$$

d. The number of radians that occur in t_F is

$$R = (1.0 \times 10^{-2} \ \text{sec})(866 \ \text{rad/sec})$$

$$R = 8.66 \ \text{rad}$$

One oscillation corresponds to 2π rad. Therefore, the number of oscillations is

$$N = \frac{R}{2\pi}$$

$$N = \frac{8.66 \ \text{rad}}{2 \ \pi \ \text{rad}}$$

$$N = 1.38 \ \text{oscillations}$$

FIGURE 12.19 Comparison of charge and current as a function of time for an *RLC* circuit.

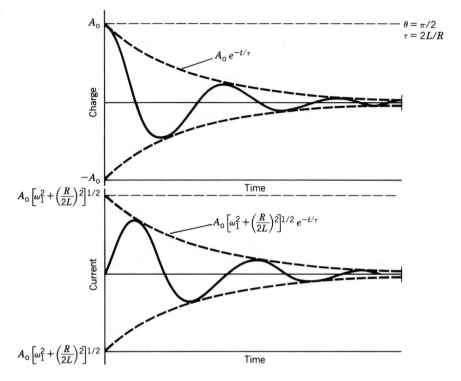

By an analogy with the mechanical system, the circuit will be critically damped if

$$\frac{1}{LC} = \frac{1}{4}\left(\frac{R}{L}\right)^2 \tag{12.79}$$

or

$$L = \frac{R^2 C}{4}$$

For this case, the circuit will not oscillate, and the energy will be delivered to the resistor in a minimum time. By the mechanical analogy with Equations 12.22 and 12.23 the equations for the charge and current become

$$Q = Q_0 e^{-t/\tau}(1 + t/\tau) \tag{12.80}$$

$$I = -\frac{Q_0 t}{\tau^2} e^{-t/\tau} \tag{12.81}$$

where

$$\tau = 2L/R$$

By multiplying both sides of Equation 12.79 by $2/R$, one obtains

$$\tau = \frac{2L}{R} = \frac{RC}{2} \tag{12.82}$$

TABLE 12.1   Charge and current versus time
for a critically damped *RLC* circuit

Time	Charge	Current
0	$1.00\ Q_0$	0
0.5τ	$0.91\ Q_0$	$-0.30\ Q_0/\tau$
τ	$0.74\ Q_0$	$-0.37\ Q_0/\tau$
1.5τ	$0.56\ Q_0$	$-0.33\ Q_0/\tau$
2τ	$0.41\ Q_0$	$-0.27\ Q_0/\tau$
3τ	$0.20\ Q_0$	$-0.14\ Q_0/\tau$
4τ	$0.09\ Q_0$	$-0.07\ Q_0/\tau$
5τ	$0.04\ Q_0$	$-0.03\ Q_0/\tau$

Thus, the value of τ may be expressed in terms of L and R or in terms of R and C.

Table 12.1 gives the values of Q and I as a function of time, and the corresponding plots of Q and I are shown in Figure 12.20. The time required to reach the peak current is

$$t_{\mathrm{P}} = \frac{RC}{2} \tag{12.83}$$

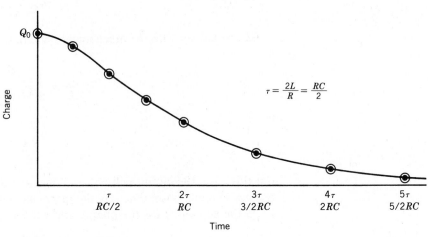

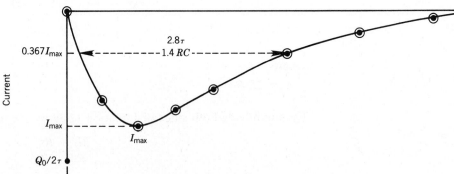

FIGURE 12.20   Charge and current versus time for a critically damped *RLC* circuit.

For practical purposes the current is essentially zero when $t = 3RC$. From the graph, one can see that the time between $0.367\, I_{max}$ points is approximately $2.8\tau = 1.4RC$. This time is sometimes called the pulse width

$$\Delta t = 1.4RC \tag{12.84}$$

where

Δt = pulse width

Examples 12.17 and 12.18 demonstrate the use of Equations 12.83 and 12.79 in solving problems.

EXAMPLE 12.17   Typical *RLC* circuit problem
. . . .

Given: An *RLC* circuit designed to use a 0.4-Ω resistor and a 1000-μF energy storage capacitor.

Find: **a.**   Pulse duration.

   b.   Inductance necessary for critical damping.

Solution: **a.**   From Equation 12.84

   $\Delta t = (RC)(1.4)$

   $\Delta t = (0.4\ \Omega)(1000 \times 10^{-6}\ \text{F})(1.4)$

   $\Delta t = 0.56 \times 10^{-3}$ sec

   $\Delta t = 0.56$ msec

   b.   From Equation 12.79

   $$L = \frac{R^2C}{4}$$

   $L = (0.25)(0.4\ \Omega)^2(1000 \times 10^{-6}\ \text{F})$

   $L = 4 \times 10^{-5}$ H

   $L = 40\ \mu$H

EXAMPLE 12.18   Typical *RLC* circuit problem
. . . .

Given: An *RLC* circuit designed to deliver a 1-msec pulse to a 1-Ω load.

Find: **a.**   Capacitance necessary.

   b.   Inductance necessary for critical damping.

Solution: **a.** $\Delta t = (RC)(1.4)$

$$C = \frac{\Delta t}{1.4R}$$

$$C = \frac{1 \times 10^{-3} \text{ sec}}{(1 \ \Omega)(1.4)}$$

$$C = 7.14 \times 10^{-4} \text{ F}$$

$$C = 714 \ \mu\text{F}$$

b. $L = \frac{R^2 C}{4}$

$$L = (0.25)(1 \ \Omega)^2(714 \times 10^{-6} \text{ F})$$

$$L = 1.78 \times 10^{-4} \text{ H}$$

$$L = 178 \ \mu\text{H}$$

The circuit will be overdamped if

$$L < \frac{R^2 C}{4}$$

For this case, the time to reach the peak current will be less; but the current decay will be longer. The limiting case occurs when $L = 0$. This is the RC circuit discussed previously that decays in a time approximately equal to $5RC$.

RLC circuits are employed when high-energy pulses must be delivered to a load. Common applications involve generation of high-power radar pulses or pulsed magnetic fields for research. Another common application involves the production of optical pulses from flashlamps. This includes photographic strobes and optically pumped laser power supplies.

A simplified flashlamp power supply is shown in Figure 12.21. Flashlamps are usually filled with xenon. A 5-inch-long lamp with a diameter of 3/8 inch is capable of taking a 1000-J input pulse. The lamp lifetime, however, is greatly extended by using approximately 100 J per pulse.

To operate the lamp, the capacitor is charged to a high voltage (on the order of 1000 V). This voltage is not sufficient to cause an electrical breakdown of the gas in the tube. To cause breakdown, a very high voltage trigger pulse is used to obtain ions in the gas. The ionization is sufficient to initiate an electrical breakdown within the lamp, and the energy stored in the capacitor is delivered to the arc plasma within the tube. The hot plasma (peak temperature on the order of 14,000°K) radiates most of the stored energy during the current pulse. The pressure also increases to a peak value of approximately 8 atmospheres and then returns to the initial value. Part of the input energy is conducted to the tube wall, and the surface temperature increases to a high value during the pulse.

The flashlamp discharge is not easy to analyze because the resistance of the lamp is not constant during the pulse. An approximate analysis can be made with the aid of a computer using the thermodynamic and transport properties of the gas. A complete analysis includes the lamp voltage, current, pressure, wall temperature, and radiant power as a function of time. In addition, the circuit is designed for near critical damping. For this case, the energy is transferred to the

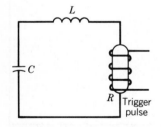

FIGURE 12.21 Schematic diagram of a simplified flashlamp power supply.

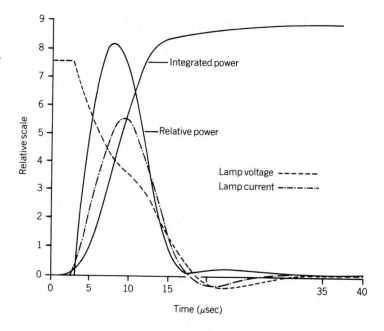

FIGURE 12.22 Voltage, current, and power versus time for a discharge through a xenon flashlamp.

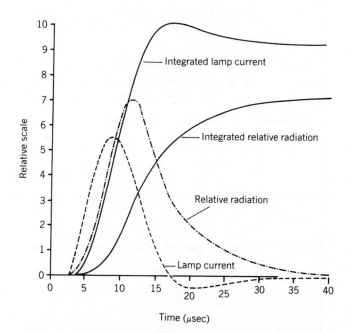

FIGURE 12.23 Current and radiant power as a function of time for a xenon flashlamp.

gas in a minimum of time with a reasonably long pulse rise time.

Measured values of lamp current, voltage, and power as a function of time are shown in Figure 12.22 for a linear, xenon flashlamp manufactured by Xenon Corporation. The peak voltage is 3720 V and the peak current is 5600 A. When the current peaks, the voltage is 1740 V. At this point in time, the electrical resistance is 0.31 Ω. Relative lamp current and the radiant power at 2650 Å are shown in Figure 12.23. The measured peak radiation at 2650 Å is 10^{-10} W·sec/cm³.

If too much energy is supplied to a flashlamp it will fail. An empirical expression for the maximum energy that can be supplied to a linear lamp without failure is

$$E_{max} = E_1 Dl(LC)^{1/4} \qquad (12.85)$$

where

D = lamp diameter in cm
l = lamp length in cm
E_1 = a constant for a particular lamp type

For linear lamps, E_1 is approximately $4 \times 10^4/\text{sec}^{1/2}\cdot\text{cm}$. The application of Equation 12.85 is given in Example 12.19.

EXAMPLE 12.19   Explosion energy of a flashlamp
. . . .

Given: A xenon flashlamp has a diameter of 0.7 cm and a length of 12.7 cm. The lamp power supply has an inductance of 1.0×10^{-6} H and a capacitance of 4.0×10^{-5} F. The constant E_1 is 4×10^4 J/sec$^{1/2}\cdot$cm^2.

Find: **a.** The explosion energy.
   b. The voltage across the capacitor to charge the capacitance to the explosion energy.

Solution: **a.** The explosion energy is

$$E_{max} = E_1 Dl\,(LC)^{1/4}$$

$$E_{max} = \left(\frac{4.0 \times 10^4 \text{ J}}{\text{sec}^{1/2}\cdot\text{cm}^2}\right)(0.7 \text{ cm})(12.7 \text{ cm})$$
$$[(1.0 \times 10^{-6} \text{ H})(4.0 \times 10^{-5} \text{ F})]^{1/4}$$

$$E_{max} = 894 \text{ J}$$

b. The energy stored in a capacitor is

$$E = 1/2 CV^2$$

Solving for V gives

$$V = \sqrt{\frac{2E}{C}}$$

$$V = \sqrt{\frac{(2)(894 \text{ J})}{4.0 \times 10^{-5} \text{ F}}}$$

$$V = 6.69 \times 10^3 \text{ V}$$

SECTION 12.4 EXERCISES

1. Given the circuit shown in the adjoining sketch, determine the following:

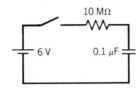

    a. Time constant.
    b. The current through the resistor when the switch is closed.
    c. The current through the resistor after two time constants.
    d. The voltage on the capacitor after three time constants.
    e. The voltage on the capacitor after five time constants.

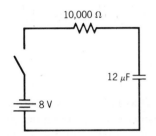

2. Given the circuit shown in the figure to the left, find the current, the voltage across the resistor, and the voltage across the capacitor 0.5 sec after the switch is closed.

3. A capacitor is charged by a 12-V battery. What is the initial voltage across the capacitor? What is the final voltage across the capacitor?

4. How much time would it take for the voltage across the capacitor in Problem 2 to reach 4 V? (Use the same circuit parameters as in Problem 2.)

5. A variable resistor is used as the resistive element in a circuit designed to charge a capacitor. The resistor has a maximum value of 10,000 Ω and a minimum value of 10 Ω. What is the ratio of maximum to minimum time constant for this system?

6. A 50-μF capacitor is to be charged by a 50-V battery. After 10 sec the voltage across the capacitor is 10 V. What is the resistance in the circuit?

7. How long would it take for the voltage across the capacitor in Problem 6 to reach 10 V if the resistance in the circuit is doubled?

8. For an RC circuit, the following values are given: $R = 10$ kΩ, $C = 0.44$ μF, $V_T = 12$ V. If the capacitor begins charging from $V_C = 0$, calculate its voltage after 5 msec.

9. After the above capacitor is fully charged to 12 V, it is discharged through a 2-kΩ resistor. Calculate capacitor voltage 1 msec into discharge phase.

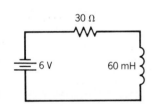

10. Calculate time required for the capacitor in Problem 8 to reach 10 V.

11. Calculate time required for the capacitor in Problem 8 to reach 5 V.

12. Calculate the time constant for an RL circuit when $L = 200$ μH and $R = 1$ Ω.

13. Given the circuit shown in the figure to the left, find the current, voltage across the resistor, and voltage across the inductor 0.4 msec after the switch is closed.

14. What is the final current in Problem 13?

15. What is the final energy stored by the inductor in Problem 13?

16. How much energy is stored in a 10-μF capacitor if the capacitor voltage is 2.5 V?

17. 600 J of energy is to be delivered to a 0.5-Ω load with a pulse duration of 0.5 msec. Find:

    Capacitance necessary

    Inductance necessary for critical damping

    Voltage to which energy storage capacitor must be charged.

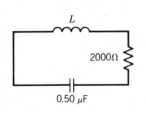

18. An RLC circuit is shown in the figure to the left. What is the value of the inductance if the circuit is critically damped?

19. If L in Problem 18 is 2.0 H, how long will it take for the oscillations to decay?

20. A linear xenon flashlamp is 0.5 m long. The diameter is 1.0 cm, and the circuit has a capacitance of 6 μF and an inductance of 5 μH. What is the explosion energy? ($E_1 = 4 \times 10^4$ J/sec$^{1/2}$·cm^2.)

12.5

THERMAL SYSTEMS

OBJECTIVES

Upon completion of this section, the student should be able to

- Discuss the effects of heat transfer by conduction, convection, and radiation of the heating and cooling of bodies.

- Write the general exponential equations for the heating and cooling of bodies.

- Estimate the time constant for a radiation detector.

- Discuss the factors that determine the time constant for cooling or heating or body.

DISCUSSION

Introduction   When heat flows into a body, and no mechanical work is done by the body, all the added energy shows up as stored internal energy, as evidenced by a rise in temperature of the body. A release of this stored energy to its surroundings results in a decrease in the body temperature. See Figure 12.24. The direction of heat flow is governed by the law that states: "Heat travels naturally

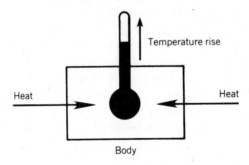

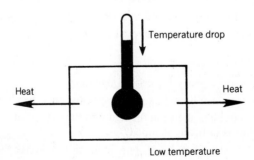

FIGURE 12.24 Relationship between heat flow and temperature change.

FIGURE 12.25 Methods of
heat transfer.

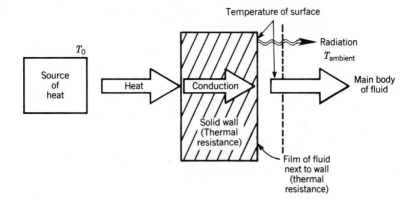

from a hot body (at high temperature) to a cooler body (at low temperature)."
When a body is at the same temperature as its surroundings, no heat is transferred, and the body is said to be in "equilibrium" with its surroundings. The
larger the temperature difference between two different bodies, the faster the heat
flow between them.

Heat is transferred by three methods—conduction, convection, and radiation.
Conduction is a method of heat transfer that occurs on a molecular scale and is the
predominant method of heat transfer through solids. Convection is a method of
heat transfer that occurs by the movement of a fluid over the material. Radiation is
a method of heat transfer in which heat is transferred by means of electromagnetic
radiation. The rate at which heat is transferred by this method is strongly dependent on the surface characteristics of the body. Black-colored objects radiate heat
at the fastest rate. Automotive racers use car engines of increasing displacement,
causing problems of excessive heat. Blackened engine blocks, rocker arm covers,
and oil pans aid in getting the heat out of the automobile engine at the fastest rate.
A thermos bottle uses this characteristic to reduce the rate at which its contents
lose or gain heat. Most often these methods of heat transfer occur together, so that
total heat transfer occurs as a result of a combination of conduction, convection,
and radiation. An example of such a combined effect is shown in Figure 12.25.

Often, in practical applications, the total rate of heat flow is difficult to describe; however, the total rate of heat flow is often dependent on the difference in
temperature between the source of heat and a surrounding fluid, which maintains
a constant ambient temperature (Figure 12.25). The total heat-flow rate also is
affected by the total combined thermal resistance of conduction, convection, and
radiation effects.

Rates of Cooling   The process of cooling a hot body in surroundings of constant
cooler temperature occurs in everyday circumstances, such as the heat-transfer-
ring of metal parts (quenching), the freezing of foods, and the response of tem-
perature-sensing instruments. As previously discussed, the modes of the heat
transfer in the cooling process are conduction, convection, and radiation. These
modes of heat transfer were discussed quantitatively in Chatper 3, and the rate
equation for a body subjected to a heat source (laser beam) was discussed. If the
laser beam is turned off, the body will return to the temperature of the surround-

ings. Setting the laser heat source equal to zero in Equation 3.101 gives the general equation for the cooling of a hot body:

$$\frac{\Delta Q}{\Delta t} = -q_c - q_F - q_R \qquad (12.86)$$

where

$\dfrac{\Delta Q}{\Delta t}$ = heat lost per second by the body

q_c = the conducted heat transfer rate

q_F = the convected heat transfer rate

q_R = the radiative heat transfer rate

To express Equation 12.86 in terms of the temperature difference, one may use Equations 3.99 and 3.100 to substitute for q_c and q_F. If the temperature difference is small, Equations 3.102 and 3.104 can be used to estimate q_R. In addition, Equation 3.95 gives $[\Delta Q/\Delta t]_{tan}$ in terms of the rate of change of the body's temperature. Making these substitutions in Equation 12.86 yields

$$mc\,\frac{\Delta T}{\Delta t} = -\frac{kA_c}{l}(T - T_s) - hA_F(T - T_s) - 4\,\sigma\epsilon G_{12}A_R T_s^3(T - T_s) \qquad (12.87)$$

where

T = body's temperature

T_s = temperature of the surroundings

m = body's mass

c = specific heat of body

l = conductive path

A_c = area for conduction

A_F = area for convection

A_R = area for radiation

h = heat transfer coefficient

σ = radiative constant (0.171×10^{-8} Btu/hr·ft$^2 R°$)

ϵ = body's emissivity

G_{12} = geometric factor

Dividing Equation 12.87 by mc and collecting like terms gives

$$\frac{\Delta T}{\Delta t} = -\frac{l}{\tau}(T - T_s) \qquad (12.88)$$

where

$$\frac{l}{\tau} = \frac{kA_c}{lmc} + \frac{hA_F}{mc} + \frac{4\sigma\epsilon G_{12}A_R T_s^3}{mc}$$

Equation 12.88 is a linear equation and the solution is of the form

$$T = (T_0 - T_s)e^{-t/\tau} + T_s \qquad (12.89)$$

where

T_0 = initial temperature of the body

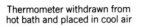

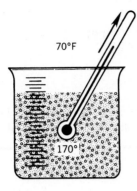

FIGURE 12.26 Cooling curve of a thermometer.

Thermometer withdrawn from hot bath and placed in cool air

70°F

170°

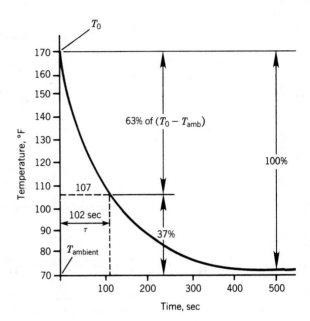

For cooling, T_0 is greater than T_s. The time constant is τ and the final temperature is T_s. Although τ is defined by Equation 12.88, some of the factors such as h are usually not known. Thus τ can only be approximated in special cases. Equation 12.88, however, does give the required insight to physical processes that can change T. For a given situation, τ can be determined by experimental methods. An application of Equations 12.88 and 12.89 are given in Example 12.20, and a typical cooling curve for a thermometer is shown in Figure 12.26.

EXAMPLE 12.20   Cooling of a body to the constant temperature of its surroundings

Given: A thermometer is removed from a 120° hot-water bath and left in still air at 70°F. The thermometer has a time constant of 102 sec.

Find: **a.**   Temperature after one time constant.

b.   Time required for the thermometer to approximately reach still-air temperature.

c.   Initial rate of temperature change at 120°F.

Solution: **a.** $T = T_s + [T_0 - T_s]e^{-t/\tau}$

$T = (70°F) + [120°F - 70°F]e^{-(102\ \text{sec}/102\ \text{sec})}$

$T = 70°F + (50°F)e^{-1}$

$T = 88.4°F$

b. $t_{\text{equilibrium}} = 5\tau$

$t_{\text{equilibrium}} = 5(102\ \text{sec})$

$t_{\text{equilibrium}} = 510$ sec

c. $\dfrac{\Delta T}{\Delta t} = \dfrac{T_0 - T_s}{\tau}$

$\dfrac{\Delta T}{\Delta t} = \dfrac{(120°F - 70°F)}{102 \text{ sec}}$

$\dfrac{\Delta T}{\Delta t} = 0.49$ F°/sec

Notice that no matter what the initial temperature of the body is, the body essentially will reach equilibrium in five time constants.

The Rate of Heating   The process of heating a cool body to a higher temperature occurs in our daily lives. The doctor inserts a thermometer (cool body) in the mouth of a patient (hot surroundings) and waits three or four minutes while the thermometer rises to the temperature of the patient's body. A cook places a turkey (cool body) in the oven (hot surroundings) and then waits for the meat to reach cooking temperature. In each case, the hot surroundings maintain a constant temperature while the cooler body gradually increases its temperature.

The rate of heating is also determined by an equation similar to Equation 12.86. However, for heating, energy is added to the body. Thus, the power equation becomes

$$\frac{\Delta Q}{\Delta t} = q_c + q_F + q_R$$

or

$$mc \frac{\Delta Q}{\Delta t} = \frac{kA_c}{l}(T_s - T) + hA_F(T_s - T) + 4\sigma\epsilon G_{12}A_2 T_s^3 (T_s - T) \qquad (12.90)$$

The terms on the right-hand side of the equation have $T_s - T$ as a common factor rather than $T - T_s$. This is necessary because $T_s > T$, and each term must be positive to add heat to the body. Equation 12.90 can also be put in the standard form. This yields the same equation that was obtained for the rate of cooling:

$$\frac{\Delta T}{\Delta t} = -\frac{l}{\tau}(T - T_s) \qquad (12.91)$$

Thus, with the initial conditions $T = T_0$ at $t = 0$, the solution has the same form as Equation 12.86:

$$T = (T_0 - T_s)e^{-t/\tau} + T_s \qquad (12.92)$$

One must remember, however, that $T_0 - T_s$ is positive for a body that is being cooled and $T_0 - T_s$ is negative for a body that is being heated. A diagram showing temperature as a function of time for a thermometer placed in a hot bath is illustrated in Figure 12.27. A corresponding calculation is shown in Example 12.21.

FIGURE 12.27 Warming curve of a thermometer.

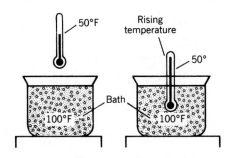

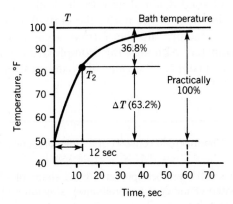

EXAMPLE 12.21 Heating a body to the constant temperature of its surroundings
. . . .

Given: A thermometer initially at 50°F is plunged into a hot-water bath of 200°F. The system time constant is 12 sec.

Find:
a. The temperature after one time constant.
b. The time required for the thermometer to be approximately the same temperature as the bath.
c. The initial rate of temperature change at 50°F.

Solution:
a. From Equation 12.92

$$T = (T_0 - T_s)e^{-t/\tau} + T_s$$

$$T = (50°F - 200°F)e^{-12 \text{ sec}/12 \text{ sec}} + 200°F$$

$$T = 144.8°F$$

b. Time for thermal equilibrium = 5τ.

$$t_E = (5)(12 \text{ sec}) = 60 \text{ sec}$$

c. Equation 12.91 gives

$$\frac{\Delta T}{\Delta t} = \frac{T_s - T}{\tau}$$

$$\frac{\Delta T}{\Delta t} = \frac{200°F - 50°F}{12 \text{ sec}}$$

$$\frac{\Delta T}{\Delta t} = 12.5 \text{ F°/sec}$$

As shown by Equation 12.88, the thermal time constant, τ, depends on many factors. These factors include the mass of the body, the specific heat of the body, the conductivity of solid material next to the body, the convective heat transfer coefficient for fluids next to the body, the emissivity of the body's surface, the radiative geometric factor, and the various areas associated with the different modes of heat transfer. Although this complexity often makes it impossible to calculate τ, in some instances the value of τ can be estimated. An estimation of τ is made in Example 12.22.

EXAMPLE 12.22   Thermal time constant of a radiation detector
. . . .

Given: A small thermocouple having the shape of a small slab is in an evacuated chamber. Laser radiation is incident on the slab through a small window. The thermocouple is supported by quartz fibers such that thermal conduction by the fibers may be neglected. The surfaces of the thermocouple and chamber are black (emissivity = 1.0). The geometric factor, G_{12}, is one. The 1.0-inch-long thermocouple wires have an area of 3×10^{-5} inch^2. The total radiative area of the slab is 0.08 inch^2. The chamber temperature is 70°F. The mass of the slab is 2.0×10^{-5} lb and the specific heat is 0.20 Btu/lb·R°.

Find: The thermal time constant of the detector.

Solution: Since the thermocouple is in a vacuum, there is no heat transfer by convection. Thus, $h = 0$. Equation 12.88 becomes

$$\frac{1}{\tau} = \frac{(k_{c_1} + k_{c_2})}{mcl} A_c + \frac{4\sigma\epsilon G_{12}A_R T_s^3}{mc}$$

where

k_{c_1} = thermal conductivity of iron (42 Btu/hr·ft·R°)
k_{c_2} = thermal conductivity of constantan (30 Btu/hr·ft·R°)
σ = 0.171×10^{-8} Btu/hr·ft^2·R^{04}
ϵ = 1.0
G_{12} = 1.0

Conversions:

$$70°F = 530°R \ (T_s)$$

$$1.0 \text{ inch} = 0.083 \text{ ft } (l)$$

$$3.0 \times 10^{-5} \text{ inch}^3 = 2.1 \times 10^{-7} \text{ ft}^2 \ (A_c)$$

$$0.08 \text{ inch}^2 = 5.6 \times 10^{-2} \ (A_R)$$

$$\frac{1}{\tau} = \frac{(42 \text{ Btu/hr·ft·R}° + 30 \text{ Btu/hr·ft·R}°)(2.1 \times 10^{-7} \text{ ft}^2)}{(2.0 \times 10^{-5} \text{ lb})(0.20 \text{ Btu/lb·R}°)(0.083 \text{ ft})}$$

$$+ \frac{(4)(0.171 \times 10^{-8} \text{ Btu/hr·ft}^2 \text{·R}°^4)(1.0)(1.0)(5.6 \times 10^{-4} \text{ ft}^2)(530°\text{R})^3}{(2.0 \times 10^{-5} \text{ lb})(0.20 \text{ Btu/lb·R}°)}$$

$$\frac{1}{\tau} = 46 \text{ hr}^{-1} + 143 \text{ hr}^{-1} = 189 \text{ hr}^{-1}$$

$$\tau = 0.0053 \text{ hr}$$

$$\tau = 19 \text{ sec}$$

SECTION 12.5 *EXERCISES*

1. A hot body at 500°C is placed in a cooler at 0°C. After 5 min, the temperature is 400°C. Determine the following:

   a. Time constant τ.

   b. Temperature after two time constants ($t = 2\tau$).

   c. Temperature after five time constants ($t = 5\tau$).

2. A thermocouple at 30°C is placed into a pan of water that is at a temperature of 95°C. What is the time constant of the thermocouple if after being in the pan of hot water for 0.1 sec the temperature indicated by the thermocouple is 92°C?

12.6

RADIATIVE DECAY

OBJECTIVES

Upon completion of this section, the student should be able to

- Discuss the decay of atomic, molecular, and nuclear excited states.

- Given the number of radioactive atoms in a sample and the decay constant, determine the number of atoms remaining after a specified time interval.

- Given the decay constant for a radioactive material, calculate the half-life.

DISCUSSION

The Decay of Atomic, Molecular, and Nuclear Excited States   Atoms, molecules, and nuclei have allowed excited states. These excited states, to a good approximation, can be represented by discrete energy levels. The lowest energy level is called the ground state. If an atom or molecule is in an excited state, it may decay to a lower energy level or to the ground state. When this occurs a photon is emitted with an energy equal to the energy level change as indicated in Figure 12.28. The same type of transition may also occur in a nucleus, but the energy of the photon is greater. For this case, the emitted photon is called a gamma ray.

FIGURE 12.28 Decay from an excited state energy level to a lower energy level.

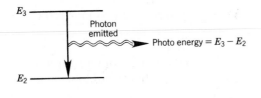

In addition to gamma rays, a nucleus may also emit β and α particles. A β particle is an electron and an α particle is a nucleus from a helium atom. When this occurs, the atomic number changes and the remaining nucleus is a new species. In addition, the new species is often left in an excited state that decays by emitting a gamma photon. For example, the radioactive element radium with an atomic number of 88 can decay into radon with an atomic number of 86 by α-particle emission.

The decay of states described are called spontaneous transitions. It is not possible to determine the exact time when the decay of a particular state will occur. There is only a probability that the particular state will decay in a unit of time. However, in any sample, there are a very large number of atoms, molecules, or nuclei in a given excited state. Thus, the number of decays per second is equal to the probability that one will decay in that second times the total number in the excited state. Mathematically, this may be expressed as

$$\frac{\Delta N}{\Delta t} = -\lambda N \tag{12.93}$$

where

N = the number in the excited state
λ = the probability that one will decay in one second

The negative sign is required since ΔN must be negative (states decay). Equation 12.93 is linear and the solution is

$$N = A_0 e^{-\lambda t}$$

However, when $t = 0$, $N = N_0$ (the initial number). Thus $A_0 = N_0$. The equation becomes

$$N = N_0 e^{-\lambda t} \tag{12.94}$$

The constant λ is usually called the decay constant. The corresponding time constant is $\tau = 1/\lambda$. For radioactive decay, however, the half-life is often used. This is the time required for half of the nuclei to decay. Mathematically, this may be expressed as

$$\frac{N_0}{2} = N_0 e^{-\lambda T_{1/2}}$$

Taking the natural log of both sides yields

$$\ln N_0 - \ln 2 = \ln N_0 - \lambda T_{1/2} \ln e$$

or

$$-\ln 2 = -\lambda T_{1/2}$$

Solving for the half-life, $T_{1/2}$, yields

$$T_{1/2} = \frac{\ln 2}{\lambda} = \frac{0.693}{\lambda} \tag{12.95}$$

At any particular time the intensity of the radiation will depend on the decay rate $\Delta N/\Delta t$. Thus, the decay rate is an important quantity. For radioactive materials the terms activity or specific activity are used. The activity is defined as

$$A = \frac{\Delta N}{\Delta t} = \lambda N = \lambda N_0 e^{-\lambda t} \tag{12.96}$$

The specific activity is the activity per gram of radioactive material. The application of Equations 12.94, 12.95, and 12.96 are given in Examples 12.23 and 12.24.

EXAMPLE 12.23   Radioactive decay rate
. . . .

Given: A radioactive sample contains 1×10^{22} atoms; the decay constant is $0.1 \ hr^{-1}$.

Find: **a.**   Number of original radioactive atoms remaining 4.0 hr from now.

   b.   Time constant τ.

Solution: **a.** $\lambda = 0.1 \ hr^{-1}$

   $N_0 = 1 \times 10^{22}$ atoms

   $t = 4 \ hr$

   $N = N_0 e^{-\lambda t}$

   $N = (1 \times 10^{22} \ \text{atoms}) e^{-0.1 \ hr^{-1} \times 4.0 \ hr}$

   $N = 1 \times 10^{22} \ \text{atoms} \ e^{-0.4}$

   $N = 0.67 \times 10^{22}$ atoms

   $N = 6.7 \times 10^{21}$ atoms

   b.   Time constant $\tau = 1/\lambda$. Therefore

   $$\tau = \frac{1}{0.1 \ hr} = 10 \ hr$$

   After 50 hr (five time constants), 99% of the atoms will have decayed and only 1% will remain.

EXAMPLE 12.24   Disintegration of a radioactive material
. . . .

Given: At a given instant of time t, a radioactive sample contains 1×10^{18} atoms and has a decay constant of 1×10^{-7} sec^{-1}.

Find: Activity in disintegrations per second at that time.

Solution: $N = 1 \times 10^{18}$ atoms at time t

$\lambda = 1 \times 10^{-7}$ sec^{-1}

$A = \lambda N$

$A = (1 \times 10^{-7}$ sec$^{-1})(1 \times 10^{18}$ atoms$)$

$A = 1 \times 10^{11}$ atoms/sec

$A = 1 \times 10^{11}$ disintegrations/sec

Notice that activity at any time depends on number of undecayed atoms remaining at that time, and so must decrease with time.

EXAMPLE 12.25   Decay of light intensity from a phosphorescent screen
. . . .

Given: The phosphorescent screen of a TV set is excited by an electron beam. When the beam is turned off, the light intensity decays with a time constant of 50 sec.

Find: The time required for the intensity to be reduced to one-half its initial intensity.

Solution: The decay constant is

$\lambda = 1/\tau$

$\lambda = \dfrac{1}{50 \text{ sec}} = 0.02$ sec^{-1}

The half-life is

$T_{1/2} = \dfrac{0.693}{\lambda}$

$T_{1/2} = \dfrac{0.693}{0.02 \text{ sec}^{-1}}$

$T_{1/2} = 34.7$ sec

SECTION **12.6** *EXERCISES*

1. What is the decay constant of a radioactive sample if the activity is 200 counts/sec when the total number of undecayed atoms is 2×10^8?

2. The activity of a radioactive sample is 25 (counts/min) when there are 3×10^{15} undecayed atoms present. How many undecayed atoms are present 2×10^9 years later?

3. What is the decay constant for a radioactive sample if the half-life is 12 years?

4. What percentage of the original number of undecayed atoms remain undecayed after 10 half-lives?

12.7

SPATIALLY DEPENDENT SYSTEMS

OBJECTIVES

Upon completion of this section, the student should be able to

- Given the macroscopic, linear cross section for stimulated emission and the macroscopic, linear cross section for absorption, calculate the light amplification of the medium.

- Calculate the attenuation of a beam of photons in an absorbing medium.

DISCUSSION

Radiative Transitions   In addition to the spontaneous emission of photons, many other radiative transitions occur. Three additional physical processes that affect the transport of radiation through a medium are absorption, stimulated emission, and Rayleigh scattering.

In absorption, an incident photon interacts with an atom or molecule. If the energy of the photon corresponds to the energy difference between two allowed energy levels in the atom or molecule, the photon will give up its energy to the atom or molecule and no longer exist. This is illustrated in Figure 12.29. The excited atom will not stay in the upper energy state, but it will spontaneously decay, emitting a photon as previously discussed. The emitted photon can be radiated in any direction with equal probability. The radiation that is emitted in this case is called fluorescent radiation. When many atoms or molecules are excited by a pulse of many photons, the fluorescent radiation decays exponentially with the characteristic time constant for spontaneous emission.

If the energy of the incident photon does not correspond to the difference in two discrete energy levels of the atom or molecule, the probability for absorption becomes very small (energy levels do have a very small width). However, radiative scattering can occur. This includes Rayleigh and Raman scattering. However, the Raman scattering is usually very small (stimulated Raman scattering, e.g., can be large) and it will not be discussed here. Where Rayleigh scattering

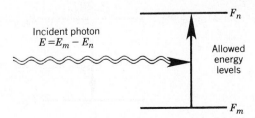

FIGURE 12.29 Absorption of a photon.

FIGURE 12.30 Rayleigh and Raman scattering.

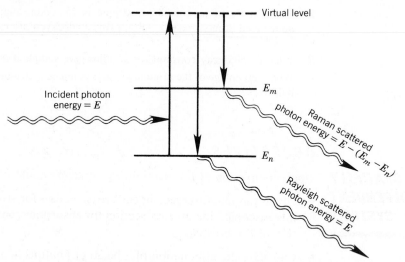

occurs, the scattered photon has the same energy as the incident photon, and the scattered photon may come off in any direction. One can think of the atom being excited to a virtual energy state and then emitting a photon of the same energy by returning to the initial state. This is illustrated in Figure 12.30. If the down transition had been to E_m rather than E_n, the scattered photon's energy would have been less than the incident photon's energy. This is a form of Raman scattering. The half-life of the virtual level is essentially zero.

Stimulated emission is a very important process in technology today. If an atom is in an excited state and a photon with an energy corresponding to the energy difference between the excited state and a lower state interacts with the atom, it will cause the atom to radiate a photon of the same energy by a transition to the lower energy state. The photon is emitted in the same direction as the incident photon. This process allows light amplifiers to be constructed by developing a medium in which the number of stimulated emissions is greater than the number of absorptions. The process of stimulated emission is illustrated in Figure 12.31.

Gamma Rays For gamma rays, the photon energy is large and other physical processes become important in the study of radiation transport. Because of the large energy of the photons, they can penetrate large distances through solid

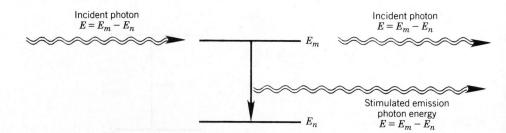

FIGURE 12.31 Stimulated emission.

FIGURE 12.32 Photoelectric effect.

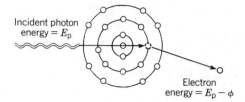

materials. These rays are also harmful to living organisms. Thus, careful calculations are required to design radiation shields to protect people from this radiation.

The physical processes that affect the transport of gamma radiation include the photoelectric effect, Compton effect, and pair production. When a gamma photon interacts with an atom, it may transfer all its energy to an orbital electron that is ejected from the atom. A certain amount of energy is required to remove the electron from the atom. This energy can be represented by a symbol ϕ. Thus, the photon energy must be greater than ϕ. The excess energy is in the form of kinetic energy of the electron. Figure 12.32 illustrates the photoelectric effect.

Compton scattering occurs when a gamma photon interacts with an electron in the outermost orbit. Only part of the energy of the photon of less energy is scattered such that the total energy and momentum is conserved. The process is shown schematically in Figure 12.33.

If the energy of the gamma photon is sufficiently large, pair production can occur when the photon passes close to a nucleus. The energy of the photon must be greater than approximately 1.6×10^{-13} J. In pair production, the incident photon disappears and an electron and a positron (positive electron) are created. The energy of the incident photon is converted to the mass of the electron and the positron plus additional kinetic energy according to Einstein's equation

$$E_p = m_{e-}c^2 + m_{e+}c^2 + \text{kinetic energy} \qquad (12.97)$$

where

$$
\begin{aligned}
E_p &= \text{incident photon energy} \\
m_{e-} &= \text{electron's mass} \\
m_{e+} &= \text{positron's mass} \\
c &= \text{velocity of light}
\end{aligned}
$$

Pair production is illustrated in Figure 12.34.

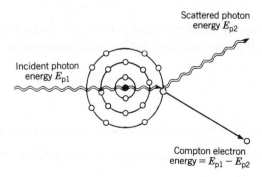

FIGURE 12.33 Compton scattering.

FIGURE 12.34   Pair production.

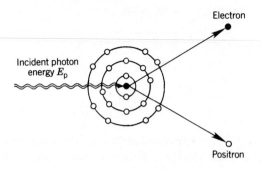

Incident photon energy E_p

Electron

Positron

The Transport of Radiation   When photons pass through a material, absorption, scattering, and stimulated emission may occur. To determine the equation for the intensity of a parallel photon beam passing through a material, consider a very thin slab of the material as shown in Figure 12.35. Each atom in the slab has cross-sectional areas that are associated with a particular interaction. As shown in the diagrams, the area of the photon beam is zy. As the beam passes through the thin slab, some of the photons will interact with the atoms. The probability that a particular interaction will occur is equal to the total area presented by the atoms divided by the total beam area. The total atomic area for a particular interaction may be written as

$$A_i = \text{(volume)(atoms/unit volume)(atomic cross section)}$$

$$A_i = (zy\Delta x)(n)(\sigma_i) \tag{12.98}$$

where

A_i = total atomic cross-sectional area of the slab
zy = area of the slab
Δx = thickness of the slab
σ_i = cross-sectional area of one atom for a particular reaction
n = number of atoms per unit volume

The probability that a photon will interact with an atom becomes

$$P_i = \frac{A_i}{\text{total area}} = \frac{A_i}{zy}$$

Substituting for A_i from Equation 12.98 yields

$$P_i = n\sigma_i\Delta x$$

The total number of interactions per second is equal to the probability that one photon will interact times the number of photons per second incident on the slab.

$$N_i = \left(\frac{\Delta n_p}{\Delta t}\right)(n\sigma_i\Delta x) \tag{12.99}$$

where

N_i = the number of interactions of type i occurring per second
$\dfrac{\Delta n_p}{\Delta t}$ = the number of photons per second incident on the slab

FIGURE 12.35 Photon
interaction cross section.

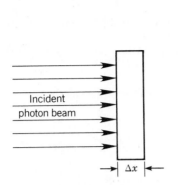

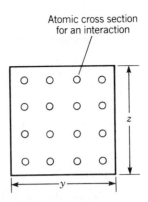

Atomic cross section
for an interaction

If each photon has an energy E_p, then the total radiant power incident on the slab
is

$$P = E_p \frac{\Delta n_p}{\Delta t}$$

or

$$\frac{\Delta n_p}{\Delta t} = \frac{P}{E_p} \tag{12.100}$$

where

P = the incident beam power

Substituting for $\Delta n_p/\Delta t$ in Equation 12.99 yields

$$N_i = \frac{P}{E_p} (n\sigma_i \Delta x) \tag{12.101}$$

Equation 12.101 can be used to calculate the energy lost or gained by the beam as
it passes through the thin slab. To do this it is advantageous to calculate the total
cross section. One usually considers absorption, Rayleigh scattering, and stimulated emission. The total cross section may be written as

$$\sigma_T = \sigma_a + \sigma_s - \sigma_{ss} \tag{12.102}$$

where

σ_T = the total cross section
σ_a = the absorption cross section
σ_s = the Rayleigh scattering cross section
σ_{ss} = the stimulated emission cross section

The negative sign is required for stimulated emission because it adds energy to the
beam whereas absorption and scattering remove energy from the beam.

If stimulated emission, absorption, and Rayleigh scattering occur, the net
power change in the beam is equal to the number of stimulated emission interactions per second minus the number of absorption and scattering interaction per
second times the energy of a photon. Expressed mathematically this becomes

$$\Delta P = (N_{ss} - N_a - N_s)E_p$$

Substituting for N_{ss}, N_a, and N_s from Equation 12.101 yields

$$\Delta P = (n\sigma_s - n\sigma_a - n\sigma_s)(P)(\Delta x)$$

or

$$\Delta P = -(n\sigma_T)(P)(\Delta x)$$

or

$$\frac{\Delta P}{\Delta x} = -n\sigma_T P \qquad (12.103)$$

Since the slab thickness is very small

$$\frac{\Delta P}{\Delta x} \approx \frac{\Delta P}{\Delta x}$$

Thus, Equation 12.103 may be written as

$$\frac{\Delta P}{\Delta x} = -n\sigma_T P \qquad (12.104)$$

Equation 12.101 is a linear equation with a solution of

$$P = P_0 e^{-\sigma_T n x} \qquad (12.105)$$

where

P_0 = the incident power at $x = 0$

The intensity of a beam is often used. For a parallel beam, the intensity can be defined as the power per unit area:

$$I = P/A$$

where

I = beam intensity
A = area of the beam

If Equation 12.105 is divided on both sides by A, one obtains

$$\frac{P}{A} = \frac{P_0}{A} e^{-\sigma_T n x}$$

or

$$I = I_0 e^{-\sigma_T n x} \qquad (12.106)$$

In general, the radiation does not usually travel in a parallel beam. In addition, there is spontaneous emission and the photons may not all have the same energy. This general case of radiation transport is a very difficult problem and is beyond the scope of this text.

In Equation 12.101, the quantities σ_T, σ_s, σ_a, and σ_{ss} are called microscopic cross sections. If these quantities are multiplied by n, the resultant quantities are called macroscopic cross sections. Thus, the macroscopic cross sections may be written as

$$\Sigma_{\text{T}} = n\sigma_{\text{T}}; \quad \Sigma_{\text{a}} = n\sigma_{\text{a}}; \quad \Sigma_{\text{s}} = n\sigma_{\text{s}}; \quad \Sigma_{\text{ss}} = n\sigma_{\text{ss}}$$

Notice that the units of the macroscopic cross section are an inverse distance. Therefore, these cross sections are often called the linear coefficient. For example, Σ_{a} is the linear absorption coefficient.

Applications of Equation 12.104 to radiation transport problems for parallel beams are given in Examples 12.26 and 12.27. A sample calculation for gamma ray absorption is shown in Example 12.28 for a parallel beam.

EXAMPLE 12.26   The attenuation of a light beam
. . . .

Given: A monochromatic light beam (only one photon energy) is incident on a window. The window is 0.10 m thick. The macroscopic, absorption cross section for photons of the given energy is 1.4/m. Approximately 4% of the light is reflected at each surface. The incident beam power is 10 W. Scattering and stimulated emission can be neglected.

Find: **a.**   The exit beam power.

b.   The approximate power absorbed by the window.

Solution: **a.**   The power reflected at the first surface is

$$P_{\text{R}} = (0.04)(10) = 0.40 \text{ W}$$

The initial power transmitted is

$$P_0 = 10 - 0.4 = 9.6 \text{ W}$$

From Equation 12.105, one has

$$P = P_0 e^{-\Sigma_a x}$$

$$P = (9.6 \text{ W})e^{-[(1.4/m)(0.10 \text{ m})]}$$

$$P = 8.35 \text{ W}$$

At the second surface, 4% is reflected.

$$P_{\text{R}} = (8.35 \text{ W})(0.04) = 0.33 \text{ W}$$

The power transmitted becomes

$$P = 8.35 \text{ W} - 0.33 \text{ W}$$

$$P = 8.02 \text{ W}$$

Multiple reflections have been neglected since only 4% of the power is reflected at each surface.

b.   The power absorbed by the window is

$$P_{\text{a}} = P_0 - P$$

$$P_{\text{a}} = 9.6 \text{ W} - 8.35 \text{ W} = 1.25 \text{ W}$$

EXAMPLE 12.27 Light amplification

....

Given: Light passes through an electrically excited gas (plasma). The electrical excitation causes more atoms to be in the excited state than in the ground state. Under these conditions the macroscopic cross section for stimulated emission is 0.50 m^{-1} and macroscopic cross section for absorption is 0.40 m^{-1}. The energy of the incident photons corresponds to the energy difference between the two levels. The length of the plasma is 10 m.

Find: The amplification of the medium.

Solution: Since stimulated emission adds to the beam energy and absorption subtracts from the beam energy, the total macroscopic cross section becomes

$$\Sigma_T = \Sigma_a - \Sigma_{ss}$$

$$\Sigma_T = 0.40/m - 0.50 \ m = -0.10/m$$

From Equation 12.101

$$\frac{P}{P_0} = e^{-n\sigma_T x}$$

$$\Sigma_T = n\sigma_T = -0.10/m$$

$$\frac{P}{P_0} = e^{+(0.10/m)(10 \ m)}$$

$$\frac{P}{P_0} = 2.72$$

The beam will be amplified 2.72 times.

EXAMPLE 12.28 Attenuation of gamma rays

....

Given: A beam of gamma rays having an energy of 3 MeV is incident on a 2-cm-thick slab. The linear absorption coefficients (macroscopic absorption cross sections) are:

a. Photoelectric effect, $\Sigma_p = 0.04$ cm^{-1}.

b. Compton effect, $\Sigma_c = 0.32$ cm^{-1}.

c. Pair production, $\Sigma_{pp} = 0.10$ cm^{-1}.

Find: a. The ratio of the incident beam intensity to the transmitted beam intensity.

b. The half-thickness (distance required to reduce the intensity by one-half).

Solution: a. The total linear absorption coefficient is

$$\Sigma_a = \Sigma_p + \Sigma_c + \Sigma_{pp}$$

$$\Sigma_a = 0.04 \text{ cm}^{-1} + 0.32 \text{ cm}^{-1} + 0.10 \text{ cm}^{-1}$$

$$\Sigma_a = 0.46 \text{ cm}^{-1}$$

The intensity ratio is

$$\frac{I}{I_0} = e^{-\Sigma_a x}$$

$$\frac{I}{I_0} = e^{-(0.46 \text{cm}^{-1})(2.0 \text{ cm})}$$

$$\frac{I}{I_0} = 0.399$$

b. The half-thickness is

$$x = \frac{0.693}{\Sigma_a}$$

$$x = \frac{0.693}{0.46 \text{ cm}^{-1}}$$

$$x = 1.51 \text{ cm}$$

Note. Although the value of I/I_0 can be easily calculated for the monoenergetic beam, the actual intensity will include scattered radiation. A handbook on radiative shielding must be consulted for real shielding problems.

SECTION 12.7 EXERCISES

1. Gamma photons having an energy of 4.8×10^{-13} J form a parallel beam. They pass through water that has a linear absorption coefficient of 0.040 cm^{-1}. What thickness of water is required to reduce the intensity by one-half, one-fourth, and one-eighth?

2. A parallel beam of monoenergetic photons is incident on an absorption cell made from glass as shown in the figure. The linear absorption coefficient in the glass may be neglected. However, 4% of the incident intensity is reflected at each window surface. If the intensity is reduced by one-half, what is the linear absorption coefficient of the liquid in the cell? Scattering and stimulated emission may be neglected.

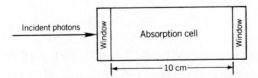

3. A high-power, 5-nsec laser pulse is used to excite molecules to an excited state by the absorption of photons. The excited molecules decay by spontaneous emission and by collisions with a half-life of 10 μsec. If the initial number of excited molecules is 5×10^{10}, how many excited molecules are left after 20 μsec? At this time, how many excited molecules are decaying per second?

4. A light amplification tube is filled with CF_3I. Flashlamps around the tube produce a large number of photons that pass through the tube. Some of these photons are absorbed by the molecules, and the excited molecules break apart to form I and C_3F_7. The I atoms are formed in an excited state, and stimulated emission becomes the dominant process until a sufficient number of I atoms are in the ground state. If absorption can be neglected, what is the initial amplification if the initial linear stimulated emission coefficient is -0.80 m^{-1} and the length of the tube is 0.8 m?

S U M M A R Y

The base of the natural logarithm, e, plays a very important role in the time or spatial dependence of a large variety of physical phenomena. If a system is not in an equilibrium condition, it will return to the equilibrium position (steady state). The time required to return to the steady state is characterized by terms of the form $e^{-t/\tau}$. τ is called the time constant of the system, and the system essentially reaches the steady state in five time constants. The basic linear differential equation for many of these systems has the form

$$\frac{\Delta y}{\Delta t} = -\frac{1}{k}y$$

The generalized solution is of the form

$$y = Ae^{-t/k} + C$$

The general solution is applicable to a freely falling body with a viscous drag force, to fluid storage tanks that are emptied with laminar flow, to RL and RC electric circuits, and to the heating and cooling of bodies.

For oscillating systems, the linear differential equation takes the form

$$m\frac{\Delta v}{\Delta t} = -k_f v - ky \quad \text{(mechanical)}$$

$$L\frac{\Delta I}{\Delta t} = -RI - \frac{Q}{C} \quad \text{(electrical)}$$

The analogies between v and I, y and Q, m and L, and k_f and R are preserved. Three solutions exist: underdamped, critically damped, and overdamped. The underdamped system will oscillate with an amplitude that decays. The critically damped system will return to the steady state in a minimum of time without oscillating. The general solutions for a critically damped case are

$$y = y_0 e^{-t/\tau} \left(1 + t/\tau\right)$$

$$v = \frac{y_0 t}{\tau} e^{-t/\tau}$$

if the initial value of v is zero. The system reaches the steady state in approximately 5τ. Again, the analogies between the mechanical and electromagnetic systems are preserved in the solution.

The decay of atomic, molecular, and nuclear excited states also follow a simple exponential law. The general equations are

$$\frac{\Delta N}{\Delta t} = -\lambda N$$

and

$$N = N_0 e^{-\lambda t} = N_0 e^{-t/\tau}$$

The factor λ is called the decay constant. The time required to reduce the number of excited states to one-half the initial value is called the half-life.

The equations for the transport of radiation in a parallel beam also have the same general mathematical form. The equations for power and beam intensity are

$$\frac{\Delta P}{\Delta x} = -\Sigma_T P$$

$$\frac{\Delta I}{\Delta x} = -\Sigma_T I$$

$$P = P_0 e^{-\Sigma_T x}$$

$$I = I_0 e^{-\Sigma_T x}$$

The quantity Σ_T is the total macroscopic cross section or linear extinction coefficient. The value of Σ_T is the sum of the absorption, scattering, and stimulated emission cross sections or linear coefficients. For gamma rays, the linear absorption coefficient is composed of three coefficients: photoelectric effect, Compton effect, and pair production.

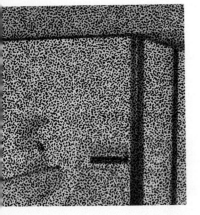

Radiation

INTRODUCTION

Radiation is the process by which energy is transmitted away from a source. Experiments have shown that the radiant energy can be transmitted by waves or particles. Sound waves, light waves, and beta rays are examples of radiation. Although these examples represent three distinct classes of radiation, energy is transmitted away from the source, and all have common characteristics.

Sound waves are pressure variations and require a material medium for transmission. They are discussed in Chapter 11, "Vibrations and Waves," and are mentioned here only briefly. This chapter addresses mainly the topic of electromagnetic (EM) radiation—sometimes called "radiant" energy. In this type of radiation, energy is transmitted by variations in electric and magnetic fields. Radiant energy includes gamma rays, x rays, ultraviolet, visible, infrared, microwaves, radar, television, and radio waves. It is the basis of modern communication, of solar energy, and of sight itself. Since EM radiation is so important in our lives, most of this chapter will deal with this type of radiation. Alpha and beta (nuclear) particles also will be discussed briefly as examples of particle radiation in which energy is transmitted as kinetic energy of particles.

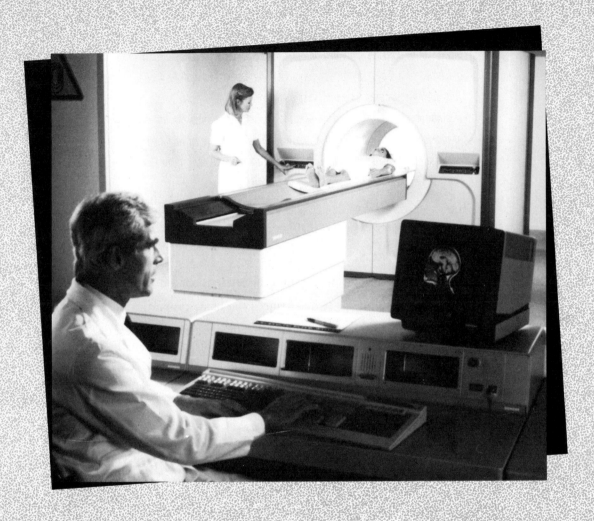

As the knowledge of the nature of radiation has increased, the separation of radiation into the movement of waves or the movement of particles is not absolute. It has been found that waves sometimes have the properties of particles and particles can display the properties of a wave. For example, light sometimes is considered as an electromagnetic wave and sometimes it has the properties of particles (photons). In addition, electrons sometimes display the properties of a wave, and a wavelength can be associated with its motion. This duality originates from the Heisenberg uncertainty principle, which states it is not possible to exactly determine the momentum and position of a particle at the same time. A discussion of the wave-particle nature of radiation is presented in this chapter, and applications of the uncertainty principle are given in Chapter 14, "Optics and Optical Systems."

13.1
ELECTROMAGNETIC RADIATION

OBJECTIVES

Upon completion of this section, the student should be able to

- Describe in one or two sentences the basic properties of light radiation.

- Define electromagnetic radiation (radiant energy) and describe a simple experiment that illustrates how electromagnetic radiation can be created.

- List the frequencies in the electromagnetic spectrum from long-wavelength EM waves of ac power (60 Hz) to gamma rays (10^{22} Hz), including each major part—radio, FM, television, radar, microwave, infrared, visible, ultraviolet, x ray, and gamma ray.

- Given the equation $v = \lambda\nu$—relating wave speed, wavelength, and frequency—determine the radiation frequency for any part of the electromagnetic spectrum.

- Given the equation $E_p = h\nu$ or $E_p = hc/\lambda$, determine the energy of different photons in the EM spectrum.

- Describe qualitatively the nature of the electromagnetic wave in terms of electric and magnetic fields, state what is always required to generate an EM wave, and explain how EM waves are propagated through empty space without benefit of an elastic medium.

- Describe a photon and explain why both wave- and particlelike (photon) phenomena are required to describe the interaction of EM radiation with matter. Give examples in which the wave character is most useful in describing EM radiation and in which the photon character is most useful.

- Explain what is meant by the inverse square law and how this law is used to describe the falloff of EM radiation propagating from a small source.

- Define irradiance, solid angle, and radiant intensity.

- Define an ideal blackbody and a gray body.

- Calculate the total radiation from the surface of an ideal blackbody given the temperature and surface area.

- Define the emissivity of a surface.

- Calculate the radiant power from an ideal blackbody surface in units of Btu/hr·ft^2·micron.

- State Kirchhoff's radiation law.

- State Wien's radiation law.

- Explain how the temperature of a blackbody can be measured with an optical pyrometer.

- Define the units
  Angstrom
  Nanometer
  Micron

- Define visible radiation and give its limits numerically in terms of wavelength, frequency, and photon energy.

- Draw the spectral response of the human eye.

- Describe the light sources
  Incandescent lamp
  Metal vapor lamp
  Fluorescent lamp

- Give the properties of laser radiation.

- Describe the physical process of stimulated emission.

- Give the basic elements of a laser.

DISCUSSION

Radiation is the movement of energy away from an energy source. The direction of the energy flow may change in a medium, but the direction of flow is often along a straight line. There are three basic types of radiation:

a. *Material waves.* Sound waves are an example of material waves. The energy is transmitted through a material medium by pressure variations. This type of radiation is presented in Chapter 11 and will not be discussed further here.

b. *Electromagnetic waves.* Light is an example of electromagnetic radiation, in which energy is transmitted by variations in electric and magnetic fields. Electromagnetic radiation requires no material medium and will travel through a vacuum.

c. *Particle emission.* Alpha and beta particles are examples of particle radiation, in which a particle having mass transmits energy in the form of the kinetic energy of its mass.

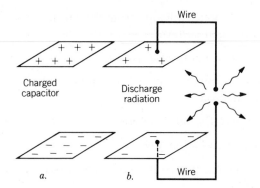

FIGURE 13.1 Discharge flash produced by the discharge of a capacitor.

This section presents a discussion of simple experimental methods for the creation of radiation. Then the family of EM radiation is examined, which is arranged according to frequency (or wavelength) in the so-called "electromagnetic spectrum." A study is made of the fundamental similarity of all EM radiation, as well as of the fundamental difference between the individual parts.

Next, a simple picture is presented of just what makes up an electromagnetic wave and how radiation is propagated in empty space. Following this is (1) a closer look at the origin of visible, infrared, and microwave radiation in the structure of atoms and molecules, (2) a look at the wave–particle nature of radiation, (3) a discussion of blackbody radiation, (4) the development of the inverse square law, and (5) a discussion of nuclear radiation.

Creation of Electromagnetic Radiation The creation and observation of electromagnetic radiation is not difficult when a simple incandescent light bulb is turned on in a dark room: visible EM radiation (light) immediately floods the room. Electrical energy pulsing through tungsten filaments in the bulb is converted into visible radiation. Invisible thermal radiation (heat) is also produced, since the glass envelope becomes hot. Thus, both light and thermal radiation are created.

Consider another common example. In the process of jumping cables from the terminals of one car battery to another, a large spark is drawn, especially if the hot wire touches the ground wire. A spark, or discharge flash, is a pulse of radiation energy, partly in the visible and partly in the invisible. Charged capacitors (Figure 13.1) also produce a spark, or discharge flash, when shorted out, sending a pulse of radiation energy out into the surrounding space. The charged isolated capacitor is shown in Figure 13.1*a* and the partially discharged capacitor of the EM radiation produced are shown in Figure 13.1*b*.

Still another common example of EM radiation involves the kitchen toaster. When the toaster is plugged in, the resistive coils increase in temperature. The toaster coils first are observed to be a dull red, then a cherry red, and finally a pinkish-white color. At the beginning of the heating process, the coils radiate primarily thermal energy. The warmth of the wires can be felt, although no color change is observed. Shortly, however, the wires become hotter, change color as described, and then clearly radiate both thermal and light energy.

Since total energy is to be conserved in any system, such as in those described, radiation energy must be included in the energy equation. Total work done in a system can be described as follows:

$$
\begin{array}{cccccc}
\text{Work} & & \text{Change in} & \text{Change in} & & \text{Frictional} \\
\text{Done on} = & \text{Kinetic} & + & \text{Potential} & + & \text{Losses} & + & \text{EM Radiation} \\
\text{System} & & \text{Energy} & \text{Energy} & &
\end{array}
$$

In mechanical energy systems, no EM radiation is produced. But in electrical energy systems (which involve the movement of electrical charge), EM radiation almost always is produced. For example, immerse a light bulb in a calorimeter and turn the light bulb on for a given period of time t. Carefully measure the temperature of the calorimeter before the light bulb is turned on and again a time t later. The electrical energy supplied to the bulb is given by IVt, where I is the current in amperes, V is the voltage in volts, and t the time in seconds. Measurement of the heat energy absorbed by the calorimeter in the same t shows that the electrical energy absorbed by the bulb (IVt) does not quite equal the heat energy absorbed by the calorimeter. The input energy is higher! The difference is equal to the energy lost from the calorimeter. Part of the lost energy is due to radiation losses, such as light energy that is transmitted out of the calorimeter.

The Electromagnetic Spectrum

Electromagnetic radiation is a form of energy stored in time-changing electric and magnetic fields. The time-changing electric and magnetic fields are set into motion by the acceleration of electric charge whether in a radar, FM transmitting antenna, or in a discharging capacitor. The exact manner in which the electric charge acceleration occurs is related to the particular type of electromagnetic radiation that develops. Charge accelerations can be produced in many ways: by varying the motion of charges on an electric conductor (such as on an antenna), by collision of charged particles in a discharge, or by movement of a charged particle in a magnetic field.

The different types of EM radiation produced have been identified and organized in accordance with their frequency or wavelength range in the so-called EM spectrum (see Figure 13.2). The top scale in Figure 13.2 identifies frequency from 10 to 10^{22} Hz. The bottom scale identifies the corresponding wavelength from about 10^7 down to 10^{-13} m. Thus, at the left end of the EM spectrum is found the very low frequency, very long wavelength radiation, such as ac power and radio waves. At the other end is found the very high frequency, very short wavelength radiation, which includes the very penetrating x rays and gamma rays.

The wavelength λ and frequency ν for any part of the EM spectrum are related by the fundamental equation

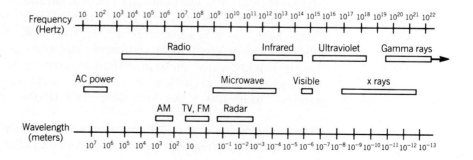

FIGURE 13.2
Electromagnetic spectrum.

$$c = \lambda\nu \tag{13.1}$$

where

c = wave speed in meters per second
λ = wavelength in meters
ν = frequency in hertz (cycles per second)

In free space or vacuum, the speed of EM radiation is a constant equal to 299,792,500 m/sec. (For calculations this number is usually rounded off to 3.0 $\times$ 10^8 m/sec.)

Radio Waves   EM radiation with frequencies between 10^4 and 10^7 Hz and wavelengths from tens to thousands of meters is identified as radio waves. They are reflected effectively by charged layers in the upper atmosphere and, therefore, can propagate over long distances from the radio station. The AM frequency band found on most table radios has a range of 530 to 1600 kHz (kilohertz).

Television and Radar   Television and FM broadcasting stations make use of EM radiation with frequencies of about 10^8 Hz and wavelengths of about 1 m. Most FM dials are calibrated between 85 and 110 MHz (megahertz). If the radiated signal is in the form of pulses, the time from emission of the pulse to the reception of its reflected signal (the echo) can be used to determine the distance from source to reflecting object. This is the principle of radar (Radio Detection And Ranging). As Figure 13.2 indicates, radar is pulsed EM radiation of frequency between 10^9 and 10^{11} Hz and wavelength ranging from a centimeter up to a meter. Using, radar, both the direction and distance of an object can be measured.

Infrared Radiation   Radiation with wavelength between 10^{-4} and 10^{-6} m generally is classified as infrared radiation. It is called "infrared" because most of the radiation is in a region of the spectrum at which the wavelengths are slightly longer than the red end of the visible spectrum. Infrared radiation is emitted by all hot bodies, such as the sun, toasters, room radiators, and soldering irons.

Visible Light   The human eye is sensitive to electromagnetic radiation between 400 and 700 nm (nanometers), and this EM radiation is detected directly with the eye. It has come to be known as "light." The eye reacts most sensitively to the green and yellow colors near 555 nm and least sensitively to the far red (where the invisible infrared begins) and to the far blue (where the invisible ultraviolet light begins).

Ultraviolet Light   EM waves shorter than 400 nm and longer than 10 nm (between 10^{15} and 10^{18} Hz) are classified as "ultraviolet" light. Ultraviolet (UV) light, or "black light," as it is sometimes called, is invisible. It is more energetic than either infrared or visible light and causes summer suntans by producing a dark skin pigment known as "melanin." Many atoms emit characteristic uv light during the process of de-excitation. As a result, spectroscopists use the uv "fingerprints" to identify atoms from their characteristic emissions.

x rays   The region of the EM spectrum with radiation of wavelengths between 10^{-8} and 10^{-17} m frequently is referred to as "x radiation" or, more simply, as "x rays." This radiation is so energetic that it can penetrate tissue and flesh, but not bone, therefore making possible x-ray photographs of the human skeletal frame. The sudden deceleration or stopping of electron beams as they strike a metal target produces x rays. With wavelengths roughly comparable to the distance between planes of atoms in crystallographic substances, x rays have been diffracted readily. Such diffraction of x rays has led crystallographers to important information concerning the structure of crystals and molecules.

Gamma Rays Gamma radiation, which is very penetrating and very energetic, extends beyond the x-ray region, although some overlap does exist. Since an x ray and gamma ray can exist with the same wavelength, frequency, and energy, and are identical, therefore, as EM radiations, how are they distinguished? The answer lies in how they are produced. If the production of EM radiation of wavelength 10^{-17} m occurs by electron rearrangement in the region of the atom outside the nucleus, the radiation is referred to as x rays. If the same wavelength radiation, the radiation is referred to as x rays. If the same wavelength radiation is produced by a rearrangement of neutrons and protons in the nucleus, it is referred to as a gamma ray. If EM radiation of 10^{-17} m is encountered in the laboratory without knowing its origin, it is impossible to distinguish it as an x ray or gamma ray. This overlapping of EM radiations (Figure 13.2) occurs not only for x radiation and gamma radiation, but also for radio, microwave, and radar.

Gamma rays often are discussed in conjunction with alpha and beta rays, which is quite natural, since all three generally are created in nuclear transformations. In fact, the creation of alpha and beta rays in nuclear decay processes is often accompanied by the creation of gamma rays. Gamma rays are electromagnetic in character, whereas alpha and beta rays are particulate in nature. Thus, they are fundamentally different. Nuclear radiation will be discussed in more depth later in this chapter.

Nature of Electromagnetic Waves

Electromagnetic radiation is an electromagnetic wave of specific frequency, moving through space with a speed $c = \lambda v$. The electromagnetic wave itself (Figure 13.3) consists of propagating electric and magnetic fields E and B, which vary with time as they move along the direction of propagation. The electric field E is shown oscillating in the y direction and always in the xy plane. The magnetic field B is shown oscillating in the z direction and always in the xz plane, perpendicular to

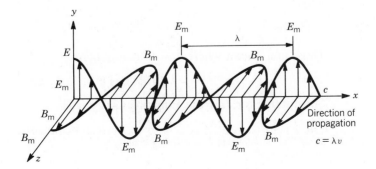

FIGURE 13.3 Propagation of electric and magnetic fields in electromagnetic radiation.

the electric field E. The radiation energy contained in the wave is shown propagating along the positive x axis. The maximum value that the electric field E attains while oscillating is labeled E_m, and that for the magnetic field B is labeled B_m.

The electric field and magnetic field are defined in Chapter 1. A review of that discussion in Chapter 1 would be helpful. In addition, the magnetic field due to current flowing in a long straight wire is calculated. The representation in Chapter 1 of electric and magnetic fields as lines of force may be used to illustrate how a time-changing electric field in space produces a magnetic field nearby, and how a time-changing magnetic field in space produces an electric field nearby.

It was discovered by Maxwell that a magnetic field can be generated by a changing electric field even though no current flows. To explain this, Maxwell introduced the concept of displacement current. He defined the displacement current density as[1]

$$\mathbf{J_D} = \epsilon \, \frac{\Delta \mathbf{E}}{\Delta t} \tag{13.2}$$

where

$\quad \epsilon$ = permittivity of the material
$\quad \mathbf{J_D}$ = displacement current
$\quad \mathbf{E}$ = electric field
$\quad t$ = time

The total current density then becomes

$$\mathbf{J_T} = \mathbf{J} + \mathbf{J_D} \tag{13.3}$$

where

$\quad \mathbf{J_T}$ = total current density
$\quad \mathbf{J}$ = current density due to a flow of charged particles

With the concept of displacement current, the total current becomes

$$I_T = \Sigma J \Delta A_n \tag{13.4}$$

where

$\quad \Delta A_n$ = element of area perpendicular to the direction of $\mathbf{J}$

With this definition of the current, Equation 1.54 becomes

$$\Sigma H \Delta l \cos \theta = I_T \tag{13.5}$$

where

$\quad N = 1$

Thus, in a region where $\mathbf{J} = 0$, one obtains

$$\Sigma H \Delta l \cos \theta = \Sigma J_D \, \Delta A_n \tag{13.6}$$

Substituting for J_D from Equation 13.2 yields

[1] Note: In the term $\Delta \mathbf{E}/\Delta t$, the Δt must be kept small. The actual relationship involves $\lim_{\Delta t \to 0} (\Delta \mathbf{E}/\Delta t) = d\mathbf{E}/dt$. This relationship comes from calculus.

$$\Sigma H(\cos\theta)\Delta l \;=\; \epsilon\Sigma \frac{\Delta E}{\Delta t}\,\Delta A_n \qquad\qquad (13.7)$$

This equation implies that a magnetic field H exists when $\Delta E/\Delta t$ is not equal to 0.

Figure 13.4 illustrates how a changing electric field E developed between two conducting plates produces a magnetic field. In Figure 13.4a, a static electric field exists in the region between the plates (vertical arrows from bottom plate to top plate). In Figure 13.4b, the electric field begins to increase thereby creating a magnetic field H (concentric circles about axis of plates). In Figure 13.4c, the electric field continues to grow rapidly, creating an even more intense H field. In Figure 13.4d, the electric field is still increasing, but more slowly; the H field begins to decrease. In Figure 13.4e, the electric field has reached its maximum value and is no longer changing; the H field has vanished. In Figure 13.4f, the electric field begins to decrease; again a magnetic field is produced but now in the opposite direction. In Figure 13.4g, the electric field is decreasing rapidly, thereby producing a strong H field. In Figure 13.4h, the electric field decreases more slowly so the magnetic field H begins to weaken. In Figure 13.4i (same as Figure 13.4a), the electric field has decreased to its original static value, and the H field has vanished.

Equation 13.7 can be used to estimate H near the conducting plates. Because the field lines form concentric circles, the direction of H is always tangent to a concentric circle. Take Δl in the same direction as H. Thus, $\cos\theta = 1$. The left-hand side of Equation 13.7 becomes

$$\Sigma H\,(\cos\theta)\,\Delta l \;=\; 2H\pi R$$

where

   R = radius of the circular path

As an approximation, one can consider the electric field between the plates to be independent of the radius. Thus, the right-hand side of Equation 13.7 becomes

$$\Sigma\,\frac{\Delta E}{\Delta t}\,\Delta A_n \;=\; \frac{\Delta E}{\Delta t}\,A_{\mathrm{p}}$$

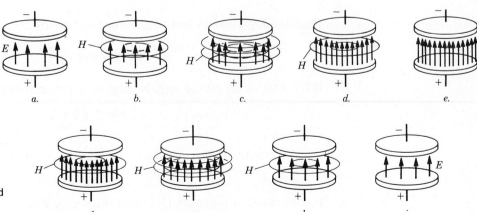

FIGURE 13.4 Magnetic field produced by a changing electric field.

where

$\qquad A_p$ = area of the plate

Equating the right-hand side to the left-hand side yields

$$2H\pi R = \frac{\Delta E}{\Delta t} A_p \qquad (13.8)$$

or

$$H = \frac{\Delta E}{\Delta t} \frac{A_p}{2\pi R} \qquad (13.9)$$

In general, $\Delta E/\Delta t$ is a function of time. Therefore, H is not a constant but changes with time. This introduces a new condition because a changing magnetic field induces a voltage with a corresponding electric field.

In Chapter 3, it was shown that a changing magnetic field, ϕ, produces a voltage according to the equation

$$\Delta V = -N \frac{\Delta \phi}{\Delta t} \qquad (13.10)$$

where

$\qquad N$ = number of turns
$\qquad \Delta t$ = small change in time
$\qquad \Delta \phi$ = corresponding change in the magnetic field through the coil

In space, one may consider a single turn consisting of a closed path. By definition, the value of ΔV for the closed path is

$$\Delta V = \Sigma E \cos \theta \Delta l \qquad (13.11)$$

where

$\qquad \Delta V$ = induced voltage in the closed path
$\qquad E$ = magnitude of the electric field
$\qquad \Delta l$ = magnitude of a very small displacement
$\qquad \theta$ = angle between the vectors **E** and **Δl**

Equating Equation 13.10 to Equation 13.11 yields

$$\Sigma E \cos \theta \Delta l = \frac{\Delta \phi}{\Delta t} \qquad (13.12)$$

If the area of the closed path is small or if B is constant over the area, then

$$\phi = B \Delta A \qquad (13.13)$$

where

$\qquad B$ = magnetic flux density
$\qquad \Delta A$ = area of the closed path

It is assumed in Equation 13.12 that the vector **B** is perpendicular to the plane of the axis formed by the closed path. Using Equation 13.13 to substitute for ϕ yields

$$\Sigma E \cos \theta \Delta l = (\Delta A)\left(\frac{\Delta B}{\Delta t}\right) \tag{13.14}$$

Since $B = \mu H$, Equation 13.14 can also be written as

$$\Sigma E \cos \theta \Delta l = (\Delta A)(\mu)\left(\frac{\Delta H}{\Delta t}\right) \tag{13.15}$$

Thus, a changing magnetic field H will cause an electric field to be generated.

Figure 13.5 shows how a changing magnetic field developed between the two poles of the electromagnet produces an electric field. As the current I through the coils increases, the magnetic field increases in strength (Figures 13.5*b*, 13.5*c*, 13.5*d*, and 13.5*e*) and an electric field is produced (concentric circles around axis of electromagnet). The strength of the E field is greatest at the instant the magnetic field is increasing the fastest (Figure 13.5*d*). When the magnetic field has reached its maximum value (Figure 13.5*f*), the electric field is zero, since the H field is not changing. Then, as the magnetic field begins decreasing (Figures 13.5*g*, 13.5*h*, and 13.5*i*) because of a decreasing current in the coils, the electric field again builds to a maximum (Figure 13.5*h*) in the opposite direction, and eventually disappears (Figure 13.5*j*) when the magnetic field returns to its original static value (Figure 13.5*a*). For this case, one may take a circular loop near the poles as shown in Figure 13.5*c*. Here, **E** is always in the same direction as $\Delta \mathbf{l}$. Thus, $\cos \theta = 1$. The left-hand side of Equation 13.15 reduces to $2\pi RE$, where R is the radius of the loop. Equation 13.15 becomes

$$2\pi RE = \mu A_p \frac{\Delta H}{\Delta t} \tag{13.16}$$

or

$$E = \frac{\mu A_p}{2\pi R} \frac{\Delta H}{\Delta t} \tag{13.17}$$

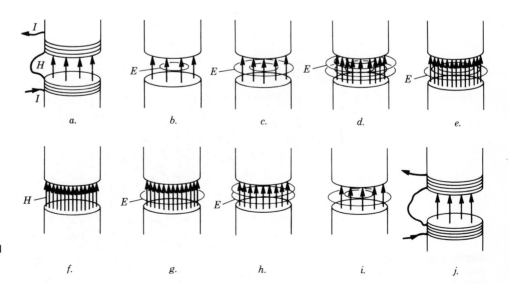

FIGURE 13.5 Electric field produced by a changing magnetic field.

where

$A_p = \Delta A$, the area between the poles

If $\Delta H/\Delta t$ is a function of time, E becomes a function of time, and according to Equation 13.7 an additional magnetic field will be generated.

The conclusion drawn from Figures 13.4 and 13.5 is very important. It provides an explanation of how EM radiation is able to propagate through empty space, quite unlike sound, for example, which requires a medium, or elastic substance, for its transmission. Begin only with a source such as an antenna, where electric charge is accelerated. The physical process of accelerating charge creates time-changing electric and magnetic fields near the source. Each time-changing field, in turn, creates the other, as shown in Figures 13.4 and 13.5, thereby sustaining the propagating electric and magnetic fields (electromagnetic waves or electromagnetic radiation) throughout space. Figure 13.6 consists of a drawing (similar to Figure 13.3, except in two dimensions) of a plane EM wave propagating away from a "horn." Internally, electric oscillations in an electronic circuit are led onto an antennalike rod in the conducting horn. There, electric and magnetic fields are generated and radiate out into space, as shown. The H field vibrates from left to right, and the E field vibrates up and down. Notice the plane at which both fields go to zero. Notice, also, that when E reverses—goes from "pointing down" to "pointing up"—H also reverses, switching from right to left. (Can you justify the note on the drawing about $\lambda/2$?)

Equations 13.6 and 13.15 can be used to develop a wave equation that shows that the velocity of the wave in free space is

$$c = \frac{1}{\sqrt{\epsilon_0 \mu_0}} \qquad (13.18)$$

where

c = wave velocity
ϵ_0 = permittivity of free space
μ_0 = permeability of free space

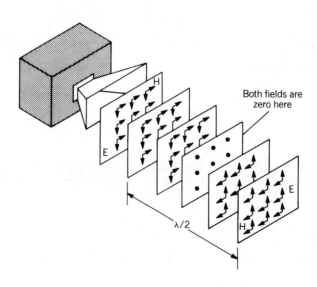

FIGURE 13.6 Propagation of electric and magnetic fields through space.

The calculation of c is shown in Example 13.1.

EXAMPLE 13.1 Velocity of electromagnetic waves in free space

. . . .

Given: $\mu_0 = 4\pi \times 10^{-7}$ W/A·m

$$e_0 = \frac{1}{4\pi \times 8.9872 \times 10^9} \text{ N·m}^2/\text{C}^2$$

Find: Velocity of light.

Solution: Equation 13.18 yields

$$c = \frac{1}{\sqrt{\epsilon_0 \mu_0}}$$

$$c = \frac{1}{\sqrt{(4\pi \times 10^{-7})\left(\dfrac{1}{4\pi \times 8.9878 \times 10^9}\right)}}$$

$$c = 2.998 \times 10^8 \text{ m/sec}$$

Note. Can you show that the units are m/sec?

When an electromagnetic wave travels through matter, the wave velocity becomes

$$v = \frac{1}{\sqrt{\epsilon \mu}} \tag{13.19}$$

The velocity in matter will always be less than the velocity in free space since $\sqrt{\epsilon \mu} > \sqrt{\epsilon_0 \mu_0}$. The development of the wave equations, which can be solved to give the mathematical expression for the electric and magnetic fields as a function of time, is beyond the scope of this book.

Waves and Photons

Inevitably, when learning about electromagnetic energy, one hears EM radiation or energy referred to sometimes as a "wave" and sometimes as a "photon." The idea of a wave is familiar. Therefore, waves have been used in this chapter to describe radiation thus far. A photon—similar to a localized EM wave, truncated at both ends (Figure 13.7)—is not so familiar. Yet the concept of a photon, which is more like a particle than a wave, is necessary to comprehend EM radiation. EM

FIGURE 13.7 The photon as a truncated wave.

radiation sometimes is best described as a wave (radar, for example) and at other times as photons (gamma ray, for example).

Light is often referred to as a wave phenomenon and at other times as being composed of particles. Waves are described in terms of amplitude, wavelength, frequency, and energy. Particles, on the other hand, are described in terms of size, mass, momentum, and energy. Visible light waves and radio waves do not exhibit particle properties readily, and particles large enough to be seen with the eye (dust particles, BB's, baseballs) do not resemble waves. The particle properties of radar and radiofrequency photons are not apparent, because they are of such low energy that only large numbers of them (not single ones) can be detected. When detected in large numbers, radar and radiofrequency photons with comparatively long wavelengths behave exactly like waves. However, the shorter-wavelength, more energetic EM radiations, such as energetic gamma rays, are detectable as "single particles." They do, for example, exhibit true particle properties and are best treated as photons. Not only are waves often described best in terms of particle properties (photons), but very small particles (atoms, electrons, protons) sometimes are described best as waves. These descriptions, by the way, are what quantum mechanics is all about.

The dual nature of EM radiation becomes apparent in the x-ray region, since x-ray photons exhibit both wave and particle properties. Gamma rays, with wavelengths much shorter than x rays, exhibit primarily particle properties and are treated primarily as photons. In general, EM radiation is treated as a wave when it is of long wavelength (low energy) and when the propagation, reflection, refraction, interference, or diffraction of EM energy is under investigation. EM radiation is treated as particles (photon) when it is of short wavelength (high energy) and when its interaction with matter (absorption, scattering, transmission) is under investigation.

In Chapter 4, the concept of the photon was first introduced to calculate the momentum of an electromagnetic wave. An electromagnetic wave was considered to be composed of a large number of photons moving with the speed of light. The energy of each photon was given as

$$E_p = h\nu \qquad (13.20)$$

where

E_p = photon's energy
ν = frequency of the wave

If Equation 13.1 is solved for ν, and this expression substituted in Equation 13.20, one obtains

$$E_p = \frac{hc}{\lambda} \qquad (13.21)$$

Using Einstein's equation, $E_p = m_p c^2$, one can obtain the momentum of the moving photon. This yields

$$m_p c^2 = \frac{hc}{\lambda}$$

or

$$m_{\text{p}}c = \frac{h}{\lambda} = p_{\text{p}} \qquad (13.22)$$

where

p_{p} = photon's momentum

Calculations for the energy and momentum of photons are given in Examples 13.2 and 13.3.

EXAMPLE 13.2 Frequency, energy, and momentum of visible photons
. . . .

Given: A light has a wavelength of between 400 (blue) and 700 nm (red).

Find: The corresponding frequency, photon energy, and photon momentum.

Solution: Blue light ($\lambda_{\text{B}} = 400 \times 10^{-9}$ m = 4×10^{-7} m):
From Equation 13.1

$$\nu_{\text{B}} = \frac{c}{\lambda_{\text{B}}} = \frac{3 \times 10^8 \text{ m/sec}}{4 \times 10^{-7} \text{ m}} = 7.5 \times 10^{14} \text{ Hz}$$

From Equation 13.20

$$E_{\text{pB}} = h\nu_{\text{B}} = (6.6 \times 10^{-34} \text{ J·sec})(7.5 \times 10^{14} \text{ Hz})$$

$$E_{\text{pB}} = 50 \times 10^{-20} \text{ J}$$

From Equation 13.22

$$p_{\text{pB}} = \frac{h}{\lambda} = \frac{6.6 \times 10^{-34} \text{ J·sec}}{4 \times 10^{-7} \text{ m}}$$

$$p_{\text{pB}} = 1.7 \times 10^{-27} \text{ kg·m/sec}$$

Red light ($\lambda_{\text{R}} = 700 \times 10^{-9}$ m = 7×10^{-7} m):

$$\nu_{\text{R}} = \frac{c}{\lambda_{\text{R}}} = \frac{3 \times 10^8 \text{ m/sec}}{7 \times 10^{-7} \text{ m}} = 4.3 \times 10^{14} \text{ Hz}$$

$$E_{\text{pR}} = h\nu_{\text{R}} = (6.6 \times 10^{-34} \text{ J·sec})(4.3 \times 10^{14} \text{ Hz})$$

$$E_{\text{pR}} = 28 \times 10^{-20} \text{ J}$$

$$p_{\text{pR}} = \frac{6.6 \times 10^{-34} \text{ J·sec}}{7 \times 10^{-7} \text{ m}}$$

$$p_{\text{pR}} = 9.4 \times 10^{-28} \text{ kg·m/sec}$$

EXAMPLE 13.3 Photon energy comparison
. . . .

Given: A radio wave has a frequency of 800 kHz and a gamma ray has a wavelength of 1×10^{-13} m.

Find: The ratio of their photon energies.

Solution: Energy of radio wave photon (use Equation 13.20, $E = h\nu$):

$E_{pRW} = 6.6 \times 10^{-34}$ J·sec $\times 8 \times 10^5$ Hz $= 53 \times 10^{-29}$ J

Energy of gamma ray photon (use Equation 13.22, $E = hc/\lambda$):

$$E_{pG} = \frac{(6.6 \times 10^{-34} \text{ J·sec})(3 \times 10^8 \text{ m})}{1 \times 10^{-13} \text{ m}} = 20 \times 10^{-13} \text{ J}$$

Thus,

$$\frac{E_{pG}}{E_{pRW}} = \frac{20 \times 10^{-13} \text{ J}}{53 \times 10^{-29} \text{ J}} = 3.8 \times 10^{15}$$

The energy of the gamma photon is about one thousand million million times more energetic than a 800-kHz radio wave photon.

Before leaving the topic of waves and photons, a few words about the origin of EM radiation in the visible, infrared, and microwave regions of the EM spectrum are desirable. Most of the visible light seen comes from changes in configurations of electrons within the atom. Thus, when an atom absorbs energy in some form, its electrons pick up this energy and are repositioned farther away from the nucleus. When the "excited" atom releases the absorbed energy and, in the process, the electrons are returned to their original configuration, the released energy comes off as short bursts of EM radiation—photons. The frequency of this emitted EM radiation—when it lies between 4.3×10^{14} (red light) and 7.5×10^{14} Hz (blue light)—constitutes the visible part of the spectrum.

In addition to the energy stored in various configurations within the atom and later emitted as light energy, energy also is stored in the vibrational and rotational modes of several atoms locked together in the form of molecules. If the stored molecular energy is of vibrational form (two atoms vibrating back and forth as in the CO molecule) and then released, the EM radiation given off is primarily infrared radiation. Thus, the heat energy, which is EM radiation in the infrared region, has its origin in the excess vibrational energy stored in molecules. If the molecular energy is stored as rotational energy (e.g., a diatomic molecule rotating around its center of mass like a rigid barbell) and then released, the EM radiation is given off primarily as microwave radiation.

When studying the processes involved in the emission and absorption of visible, ultraviolet, infrared, or microwave radiation, the concept of short bursts of energy (photons), rather than of long, unending wave trains, best describes what is taking place. Thus, the phenomena of absorption and emission of EM radiation are described and most easily understood in terms of wave packets or photons of EM energy.

Inverse Square Law

All spontaneously emitted radiation has an equal probability of being emitted in any direction. If the source is nearly a point in empty space, the radiation spreads

FIGURE 13.8 Illustration of the inverse-square law.

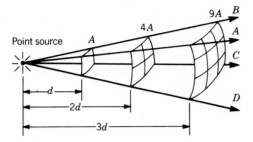

as it leaves the source such that the energy transmitted through any concentric spherical surface with the source as the common center is the same. If the emitted power remains constant, then the total radiant power through each spherical surface is also the same. This is an application of the conservation of energy. If the radiation interacts with matter such that radiant power is absorbed, the radiant power through each surface will not be the same.

To illustrate the conservation of energy, consider Figure 13.8. The light emitted from the point source in the volume bounded by the four lines (A, B, C, and D of the figure) continues to travel outward. At distance d this light energy passes through the cross-sectional area A. At 2d, the same energy spreads over area 4A. At 3d, it covers 9A. Figure 13.8 demonstrates that the area through which energy passes increases with the square of the distance from the source. This rate of increase forms the basis of the inverse square law.

Before a detailed discussion of the inverse square law can be presented, several concepts used in the measurement of light must be introduced. The first of these is irradiance, a measure of the power per unit area striking a surface. It is the power divided by area, or power per unit area, as given by

$$G = \frac{P_I}{A} \tag{13.23}$$

where

G = irradiance
P_I = power incident on the area A
A = area

Example 13.4 illustrates the use of this equation in solving a problem.

EXAMPLE 13.4 Irradiance on a surface
. . . .

Given: 25 W of light fall on the surface of a square of 0.5 m on each side.

Find: Irradiance at the surface.

Solution: $A = d^2 = (0.5 \text{ m})^2$

$A = 0.25 \text{ m}^2$

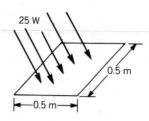

$$G = \frac{P_1}{A} = \frac{25 \text{ W}}{0.25 \text{ m}^2}$$

$$G = 100 \text{ W/m}^2$$

Note: The units of irradiance are "power divided by area," usually expressed in W/m² or W/cm².

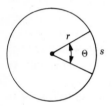

a. Plane angle

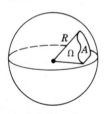

b. Solid angle

FIGURE 13.9   Illustration of a solid angle.

The second concept to be considered is the solid angle illustrated in Figure 13.9. In plane geometry (Figure 13.9*a*), an angle is defined as "that part of the plane lying between two lines that meet at a point." Such an angle may be measured in degrees or radians, where 2π radians equals 360°. The angle in radians is defined by

$$\theta = \frac{s}{r} \tag{13.24}$$

where

θ = angle in radians
s = length of arc subtended by angle θ
r = radius of circle

Figure 13.9*b* shows the analogy in solid geometry. The solid angle Ω is defined as "the surface area subtended at center of sphere divided by the square of the radius of sphere," as given by

$$\Omega = \frac{A}{R^2} \tag{13.25}$$

where

Ω = solid angle in steradians
A = area of sphere intercepted by solid angle Ω
R = radius of sphere

Solid angles are measured in steradians (sr). Like the radian, the steradian is a dimensionless unit. Since the total area of a sphere is $4\pi R^2$, there are 4π sr in a complete sphere. The use of Equation 13.25 is illustrated in Example 13.5.

EXAMPLE 13.5 Calculation of solid angle

. . . .

Given: A circular surface 10 cm in diameter viewed from a distance of 1 m.

Find: Solid angle subtended by the surface at the eye.

Solution: $A = \pi r^2 = (3.14) \times (5 \text{ cm})^2$

$A = 78.5 \text{ cm}^2$

$\Omega = \dfrac{A}{R^2} = \dfrac{78.5 \text{ cm}^2}{(100 \text{ cm})^2}$

$\Omega = 0.00785 \text{ sr}$

Note. The circular surface has been assumed to be approximately equal to the spherical surface intercepted by Ω with radius of 1 m.

The concept of a solid angle is useful in describing the way light is radiated from a source. The radiant intensity from a point source is defined as the "power per unit solid angle radiated by the source" as given by

$$I = \frac{P_R}{\Omega} \tag{13.26}$$

where

   I = radiant intensity
   P_R = power radiated in the solid angle Ω

Examples 13.6 and 13.7 illustrate the use of Equation 13.26 in solving problems.

EXAMPLE 13.6 Radiant intensity of a point source

. . . .

Given: A point source radiates power uniformly in all directions. The total power radiated is 100 W.

Find: The intensity of the source.

Solution: From Equation 13.26

$I = \dfrac{P_R}{\Omega}$

For a sphere, $\Omega = 4\pi$,

$I = \dfrac{100}{4\pi}$

$I = 7.96 \text{ W/sr}$

EXAMPLE 13.7   Radiant intensity of a light source
· ·

Given: A flashlight produces 10 W of optical power at $R = 2$ m and $A = 0.5$ m².

Find: Radiant intensity of a flashlight.

Solution: $\Omega = \dfrac{A}{R^2} = \dfrac{0.5 \text{ m}^2}{(2 \text{ m})^2}$

$\Omega = 0.125 \text{ sr}$

$I = \dfrac{P}{\Omega} = \dfrac{10 \text{ W}}{0.125 \text{ sr}}$

$I = 80 \text{ W/sr}$

The radiant intensity from a point source does not depend on the distance from the source. On the other hand, irradiance depends on the source-to-surface distance and decreases as the distance from the source increases. To determine the mathematical relationship, for instance, consider a point source with uniform intensity. The irradiance at a distance R from the source is then equal to the total power radiated divided by the area of the spherical surface of radius R:

$$G = \frac{P_R}{4\pi R^2} \tag{13.27}$$

Thus, one can see that the irradiance is inversely proportional to distance squared. Example 13.8 illustrates the use of Equation 13.27 to determine the irradiance due to a point source.

EXAMPLE 13.8   Irradiance at distance R from point source
· ·

Given: An electrical arc, treated to a point source, emits 500 W of light equally in all directions.

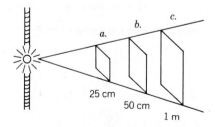

Find: Irradiance at (*a*) 25 cm, (*b*) 50 cm, and (*c*) 1 m.

Solution: **a.** $G = \dfrac{P_R}{4\pi R^2}$

$$G_a = \frac{500 \text{ W}}{(4) \times (3.14) \times (25 \text{ cm})^2}$$

$$G_a = 6.37 \times 10^{-2} \text{ W/cm}^2$$

b. $$G_b = \frac{500 \text{ W}}{(4) \times (3.14) \times (50 \text{ cm})^2}$$

$$G_b = 1.59 \times 10^{-2} \text{ W/cm}^2$$

c. $$G_c = \frac{500 \text{ W}}{(4) \times (3.14) \times (100 \text{ cm})^2}$$

$$G_c = 3.98 \times 10^{-3} \text{ W/cm}^2$$

Note: Multiplying distance by 4 (from 25 to 100 cm) reduces irradiance by a factor of 16 or 4^2.

For a uniform point source,

$$I = \frac{P_R}{4\pi} \tag{13.28}$$

Thus, Equation 13.28 may also be written as

$$G = \frac{I}{R^2} \tag{13.29}$$

Sample calculations using Equation 13.29 are given in Examples 13.9 and 13.10.

EXAMPLE 13.9 Irradiance calculation
. . . .

Given: A light source with radiant intensity of 250 W/sr.

Find: Irradiance at 10 m.

Solution: $$G = \frac{I}{R^2}$$

$$G = (250 \text{ W/sr})/(10 \text{ m})^2$$

$$G = 2.5 \text{ W/m}^2$$

Note: In Example 13.9 the unit steradian (sr) is dropped from the calculation; as noted previously, a steradian represents a ratio and, therefore, has no dimensions.

EXAMPLE 13.10 Radiant intensity calculation
. . . .

Given: At a distance of 2 m from a source, 3.6 mW falls on an area of 0.5 cm².

Find: Radiant intensity of source.

Solution: Use Equation 13.26.

$$\Omega = \frac{A}{R^2} = \frac{0.5 \text{ cm}^2}{(200 \text{ cm})^2}$$

$$\Omega = 1.25 \times 10^{-5} \text{ sr}$$

$$I = \frac{P_R}{\Omega} = \frac{3.6 \times 10^{-3} \text{ W}}{1.25 \times 10^{-5} \text{ sr}}$$

$$I = 288 \text{ W/sr}$$

Blackbody Radiation

All matter above the temperature of absolute zero emits radiation. The amount of radiated power and its spectral distribution depend on the absolute temperature and the nature of the radiating material. In addition, when radiation is incident on a material, part of the radiant power may be reflected, part may be absorbed, and part may be transmitted. Since energy is conserved, one may write

$$q_R A = \rho q_R A + \alpha q_R A + \tau q_R A \qquad (13.30)$$

or

$$1 = \rho + \alpha + \tau \qquad (13.30)$$

where

q_{Ri} = total incident radiant power per unit area
ρ = fraction of the incident power reflected
α = fraction of the incident power absorbed
τ = fraction of the incident power transmitted

In this discussion, we will be primarily concerned with solids that do not transmit radiant energy. For this case, Equation 13.30 becomes

$$1 = \rho + \alpha$$

An ideal blackbody is a body that absorbs all the radiation incident on it. In reality, an ideal blackbody does not exist. However, a cavity with a small hole is almost an ideal blackbody. This is shown in Figure 13.10. Incident radiation passes through the small hole and suffers multiple reflections inside the cavity. With each reflection, part of the radiant power is absorbed. After many reflections, all the incident power is essentially absorbed. In addition, the wall of the cavity emits radiation that is dependent on the temperature and surface characteristics of the wall. Thus, for a given wall temperature, the cavity will be filled with radiant energy. Part of this radiant energy will pass through the hole and radiate to the external space. This radiation is called blackbody radiation and represents the maximum radiant power that can be emitted from a surface for a given absolute temperature. In 1879 Stefan determined the mathematical expression (Equation 13.31) for blackbody radiation based on the measurement made by Tyndall:

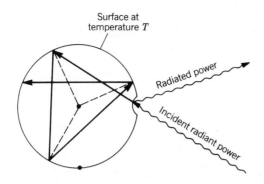

FIGURE 13.10 A cavity with a small hole has characteristics similar to an ideal blackbody.

$$W = \sigma T^4 \tag{13.31}$$

where

W = radiant power in Btu/hr·ft² of a blackbody
$\sigma = 0.171 \times 10^{-8}$ Btu/hr·ft²·°R⁴
T = absolute surface temperature in °R

For surfaces that are not ideal blackbodies, one may write

$$W = \epsilon \sigma T^4 \tag{13.32}$$

The quantity ϵ in Equation 13.32 is called the emissivity, and it is always less than one. In general, ϵ is a function of temperature as well as the nature of the surface. Table 13.1 gives approximate values of ϵ for several materials.

One can show that the emissivity (ϵ) and the absorptivity (α) are not independent. Consider a small, opaque sphere inside the large spherical cavity as shown in Figure 13.11. Inside the cavity, blackbody radiation exists corresponding to the wall temperature T. This radiation will be incident on the surface of the small sphere, and the radiant power absorbed at the spherical surface is

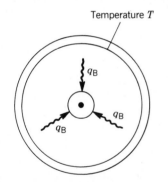

$$W_a = q_B \alpha A_s \tag{13.33}$$

where

W_a = radiant power absorbed
q_B = incident radiant power per unit area
α = absorptivity of the sphere
A_s = area of the small sphere

FIGURE 13.11 Small sphere inside a spherical cavity.

TABLE 13.1 Emissivities of surfaces

Surface	Emissivity	Temperature Range (°K)
Molybdenum	0.1 to 0.3	1000 to 2800
Tungsten	0.2 to 0.3	300 to 3400
Polished aluminum	0.04 to 0.06	500 to 900
Lamp black	0.95	300 to 600

When the small spherical body reaches equilibrium at temperature T, the emitted radiation must be equal to the absorbed radiation. Thus, from Equation 13.32, one obtains

$$(\epsilon \sigma T^4)A_s = q_B \alpha A_s$$

or

$$\epsilon \sigma T^4 = \alpha q_B \qquad (13.34)$$

If the small sphere is replaced by a sphere having a blackbody surface ($\epsilon = 1$, $\alpha = 1$), Equation 13.34 becomes

$$\sigma T^4 = q_B$$

Using this expression to substitute for q_B in Equation 13.34 yields

$$\epsilon(\sigma T^4) = \alpha(\sigma T^4)$$

or

$$\epsilon = \alpha \qquad (13.35)$$

Thus, the emissivity is equal to the absorptivity. This relationship is called Kirchhoff's identity.

The emitted radiation from a blackbody has a spectral distribution. The mathematical expression for the spectral distribution was first derived from theory by Planck. To obtain an expression that agreed with measurements, he had to introduce the quantum concept and the constant h. The functional relationship he derived is

$$W_{b\lambda} = \frac{C_1 \lambda^{-5}}{e^{(C_2/\lambda T)} - 1} \qquad (13.36)$$

where

$W_{b\lambda}$ = blackbody radiation in Btu/hr·ft²·micron
λ = wavelength in microns (μm)
T = temperature in °R
C_1 = 1.187×10^8 Btu·μm⁴/hr·ft²
C_2 = 2.589×10^4 μm·°R

To obtain the power radiated per unit area in a small wavelength interval $\Delta\lambda$, one may multiply $W_{b\lambda}$ by $\Delta\lambda$. The total power radiated over all wavelengths becomes

$$W_b = \sigma T^4 = \Sigma W_{b\lambda}\Delta\lambda \qquad (13.37)$$

If the body is not an ideal blackbody, one may define a monochromatic emissivity $\epsilon(\lambda)$. For this case

$$W = \epsilon \sigma T^4 = \Sigma\epsilon(\lambda)W_{b\lambda}\Delta\lambda \qquad (13.38)$$

If $\epsilon(\lambda)$ is a constant for a given temperature (not a function of wavelength), then the body is called a gray body, and $\epsilon(\lambda) = \epsilon$ since $\epsilon(\lambda)$ can then be factored out of the sum in Equation 13.38.

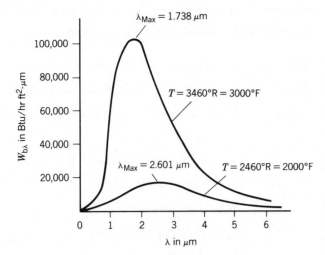

FIGURE 13.12 Plot of radiant power versus wavelength for two temperatures.

Equation 13.38 is shown plotted for two temperatures in Figure 13.12. One can see that the peak radiant power per unit area per unit wavelength interval occurs at shorter wavelengths as the temperature increases. The wavelength associated with a maximum point can be determined from Wien's law:

$$\lambda_{max} = \frac{5.215 \times 10^3 \ \mu_m \cdot {}^\circ R}{T} \tag{13.39}$$

In Equation 13.39, λ_{max} is in microns and T is in °R.

The shift in the maximum point of the blackbody curve explains the change in color of a body as it is heated. When the temperature is increased to approximately 1300°F, the emitted radiation becomes visible, and the body has a deep red color. Near 2700°F, the body appears nearly white. The effect can be easily demonstrated by varing the current through a tungsten filament enclosed in a clear bulb.

The radiation emitted from a hot body in a narrow wavelength interval can be used to measure the brightness temperature of the surface. The brightness temperature is the temperature a blackbody must have to emit the same radiant power in the given wavelength interval. Thus, brightness temperature can be a function of wavelength. The optical pyrometer discussed in Chapter 1 can be used to measure the brightness temperature of hot bodies that emit sufficient radiation in the red end of the visible spectrum to be seen. The instrument is calibrated by using a standard blackbody source whose temperature can be changed and measured. This type of optical pyrometer measures the actual temperature only if the body is a blackbody.

An optical pyrometer can be designed to measure the emitted radiation at more than one wavelength. For example, one can obtain an instrument that measures the emitted radiation at two wavelengths. It can be shown, when used properly, that this instrument measures the true temperature of gray bodies.

Applications of Equations 13.32 and 13.36 are illustrated in Examples 13.11 and 13.12.

EXAMPLE 13.11 Radiant power from a tungsten surface
. . . .

Given: A tungsten surface is at 2000°K.

Find: The approximate total radiant power per ft².

Solution: From Equation 13.32, the power radiated per ft² is

$$W = \epsilon \sigma T^4$$

$$\sigma = 0.171 \times 10^{-8} \ \text{Btu/hr·ft}^2\text{·°R}$$

$$\epsilon \approx 0.25 \quad \text{(Table 13.1)}$$

The temperature in °R is

$$T_R = (9/5)(T_K) = 9/5 \ (2000)$$

$$T_R = 3600°R$$

The radiant power becomes

$$W \approx (0.25)(0.171 \times 10^{-8} \ \text{Btu/hr·ft}^2\text{·°R}^4)(3600°R)^4$$

$$W \approx 7.18 \times 10^4 \ \text{Btu/hr·ft}^2$$

EXAMPLE 13.12 Radiant power from a blackbody in a small wavelength interval
. . . .

Given: A blackbody has a temperature of 1760°R, and the temperature is then increased to 3160°R. The area of the blackbody is 4×10^{-4} ft².

Find: The radiant power in a wavelength interval of 0.1 μm centered around 0.5 μm for the two different temperatures.

Solution: The radiant power in a wavelength interval is

$$W = (W_{b\lambda}\Delta\lambda)(A)$$

$$A = \text{area}$$

$$\Delta\lambda = \text{wavelength interval}$$

$$W = (W_{b\lambda})(0.1 \ \mu m)(4 \times 10^{-4} \ \text{ft}^2)$$

$$W = (W_{b\lambda})(4 \times 10^{-5} \ \mu m\text{·ft}^2)$$

From Equation 13.36

$$W_{b\lambda} = \frac{(1.187 \times 10^8 \ \text{Btu·}\mu m^4\text{/m·ft}^2) \ \lambda^{-5}}{e^{(2.589 \times 10^4 \ \mu m\text{·°R})/\lambda T} - 1}$$

For $\lambda = 0.5$ μm and $T = 1760°R$, $W_{b\lambda}$ becomes

$$W_{b\lambda} = \frac{(1.187 \times 10^8 \ \text{Btu·}\mu m^4\text{/hr·ft}^2)(0.5 \ \mu m)^{-5}}{e^{(2.589 \times 10^4 \mu m\text{·°R})/(0.5 \mu m)(1760°R)} - 1}$$

$$W_{b\lambda} = \frac{3.80 \times 10^9}{5.99 \times 10^{12}} \; \text{Btu/hr} \cdot \mu\text{m} \cdot \text{ft}^2$$

$$W_{b\lambda} = 6.34 \times 10^{-4} \; \text{Btu/hr} \cdot \mu\text{m} \cdot \text{ft}^2$$

However, we had

$$W = (W_{b\lambda})(4 \times 10^{-5} \; \mu\text{m} \cdot \text{ft}^2)$$

$$W = (6.34 \times 10^{-4} \; \text{Btu/}\mu\text{m} \cdot \text{hr} \cdot \text{ft}^2)(4 \times 10^{-5} \; \mu\text{m} \cdot \text{ft}^2)$$

$$W = 2.54 \times 10^{-8} \; \text{Btu/hr at } 1760°\text{R}$$

For $\lambda = 0.5 \; \mu\text{m}$ and $T = 3160°\text{R}$, $W_{b\lambda}$ becomes

$$W_{b\lambda} = \frac{3.80 \times 10^9}{e^{(2.589 \times 10^4 \, \mu\text{m} \cdot °\text{R})/(0.5 \, \mu\text{m})(3160°\text{R})} - 1} \; \text{Btu/}\mu\text{m} \cdot \text{hr} \cdot \text{ft}^2$$

$$W_{b\lambda} = \frac{3.80 \times 10^9}{1.31 \times 10^7 - 1} \; \text{Btu/}\mu\text{m} \cdot \text{ft}^2$$

$$W_{b\lambda} = 2.90 \times 10^2 \; \text{Btu/}\mu\text{m} \cdot \text{hr} \cdot \text{ft}^2$$

We had

$$W = (W_{b\lambda})(4 \times 10^{-5} \; \mu\text{m} \cdot \text{ft}^2)$$

$$W = (2.90 \times 10^2 \; \text{Btu/}\mu\text{m} \cdot \text{hr} \cdot \text{ft}^2)(4 \times 10^{-5} \; \mu\text{m} \cdot \text{ft}^2)$$

$$W = 1.16 \times 10^{-2} \; \text{Btu/hr at } 3160°\text{R}$$

Light

The Optical Spectrum   Figure 13.13*a* shows the electromagnetic spectrum and the location of the optical spectrum (Figure 13.13*b*) is composed of visible light, the longer-wavelength infrared (IR) light, and the shorter-wavelength ultraviolet (UV) light. The wavelength range of most interest is the visible spectrum (Figure 13.13*c*) and those wavelengths closest to the visible—called the near infrared and the near ultraviolet.

Table 13.2 lists three length units commonly used to express the wavelengths in the optical spectrum. Angstroms (10^{-10} m) are normally used for the far ultraviolet, for x rays, and sometimes for the near ultraviolet and visible, although the nanometer (10^{-9} m) is more common for the longer wavelengths. Infrared wavelengths are usually expressed in microns (10^{-6} m). Example 13.13 shows the conversion of a wavelength given in meters into each of these units.

TABLE 13.2   Wavelength units

Name	Symbol	Definition
Angstrom	Å	10^{-10} m
Nanometer	nm	10^{-9} m
Micron	μm	10^{-6} m

FIGURE 13.13 The electromagnetic, optical, and visible spectra.

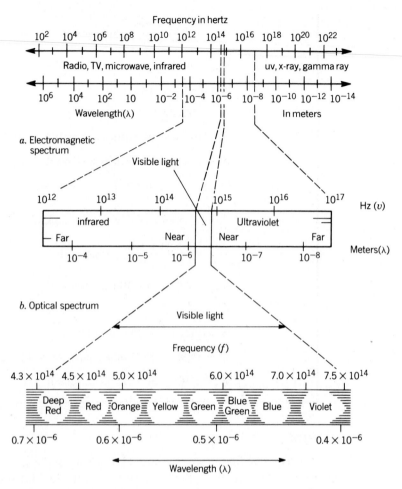

a. Electromagnetic spectrum

b. Optical spectrum

c. Visible spectrum

EXAMPLE 13.13 Conversion of wavelength units
. . . .

Given: The wavelength of light is 6.33×10^{-7} m.

Find: Wavelength in (a) angstroms, (b) nanometers, and (c) microns.

Solution: **a.** $(6.33 \times 10^{-7}\ \text{m}) \times \left(\dfrac{10^{10}\ \text{Å}}{1\ \text{m}}\right) = 6330\ \text{Å}$

b. $(6.33 \times 10^{-7}\ \text{m}) \times \left(\dfrac{10^{9}\ \text{nm}}{1\ \text{m}}\right) = 633\ \text{nm}$

c. $(6.33 \times 10^{-7}\ \text{m}) \times \left(\dfrac{10^{6}\ \mu\text{m}}{1\ \text{m}}\right) = 0.633\ \mu\text{m}$

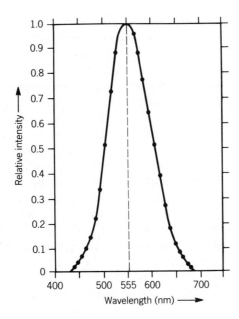

FIGURE 13.14 Response of the human eye to the visible spectrum.

Response of the Human Eye

Figure 13.14 shows the relative response of the human eye to light of various wavelengths. The eye is most sensitive to yellow light with a wavelength of 555 nm (nanometers). The response drops off rather sharply on each side of this value, reaching zero at 400 nm, the edge of the ultraviolet region, and at 700 nm, the edge of the infrared region. Visible light sources are devices that produce radiation in the region of the electromagnetic spectrum limited to the range of 400–700 nm.

Light Sources

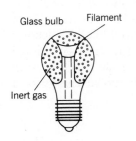

FIGURE 13.15 Incandescent light bulb.

Incandescent Sources Figure 13.15 illustrates an ordinary incandescent bulb. It consists of a thin filament of tungsten wire sealed inside a glass bulb envelope. The glass bulb is either evacuated or filled with an inert gas. When a current flows through the filament, it is heated to incandescence. Atoms in the filament are bound to fixed positions, but are free to vibrate around those positions. As this vibration (heat energy) occurs, neighboring atoms collide with one another violently. During these collisions, the atoms decelerate and a portion of their vibrational energy is converted to electromagnetic radiation in the infrared, visible, or near-ultraviolet portions of the spectrum. The result is a continuous spectrum.

The hot tungsten filament is not a blackbody or a gray body. However, the blackbody curves shown in Figure 13.12 give an approximate spectral distribution of the radiant energy emitted by the hot filament. The blackbody curves also reveal the inefficiency of incandescent light sources. Only that fraction of the output that falls in the visible region is useful for human vision. Most of the energy used to heat an incandescent source results in production of unneeded ultraviolet and infrared radiation.

Plasma Sources   Gaseous conductors are discussed in Chapter 5. The partially ionized plasma is composed of free electrons, ions, atoms, and molecules. The electrons and ions gain energy from the electric field, but the current is primarily due to the drift of electrons in the electric field. Through collisions, the electrons can excite other particles to higher energy levels, the electrons can be captured by ions, or the electrons can suffer an acceleration during the collision process. These three processes can result in emitted radiation from the plasma. The excited particle can emit line radiation when a transition occurs to a lower allowed energy level and a photon is emitted. This radiation is called bound—bound radiation. Radiation is also emitted when electrons are captured by ions. Since the free electrons have a distribution of energies, the change in energy is not unique, and the radiation has a continuous distribution instead of discrete lines. This radiation is called free–bound radiation. In addition, an accelerated charge emits radiation. Thus the free electrons radiate if they suffer an acceleration during a collision even though they are not captured by an ion. Since the electron is free before and after the collision, the radiation is called free–free radiation. The free–free radiation also has a continuous distribution. Thus, the total radiation from a plasma source is composed of line spectra superimposed on continuum radiation.

The line spectra can be very complex. The total energy change can be due to a combination of electronic, vibrational, and rotational energy changes giving rise to molecular spectra. If there is only a rotational energy change, the emitted photon is in the far infrared. If a vibrational energy change occurs along with rotational energy changes, the emitted spectrum is in the infrared. However, if electronic transitions also occur, the spectrum may be in the visible region or the ultraviolet region of the spectrum.

Figure 13.16 shows a typical line spectrum produced by electronic transitions in atoms that have no rotational or vibrational energy. The intensity of light is plotted along the vertical axis and the wavelength in microns is plotted along the horizontal axis. Emission lines in this spectrum vary greatly in width and intensity, but each atomic species has its own characteristic spectrum. Line spectra occur in the near IR, visible, and UV portions of the electromagnetic spectrum.

Figure 13.17 displays a band spectrum produced by molecular transitions. Each band is actually composed of a large number of closely spaced lines. Molecules contain energy not only in electronic levels, but also in the form of vibrational energy of the atoms and rotational energy of the entire molecule. Each band in Figure 13.17 is the result of a different change in the vibrational energy of a

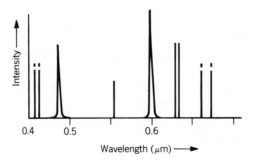

FIGURE 13.16 Line spectrum due to an atomic sample.

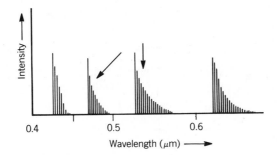

FIGURE 13.17 Fine lines within the band spectrum of a diatomic molecular sample.

molecule. Fine lines within the bands are produced by accompanying changes in rotational energy. Band spectra are produced by molecular gases such as diatomic nitrogen (N_2), by carbon dioxide (CO_2), and by some liquids and crystals.

A typical plasma light source is a metal vapor lamp, which is important for lighting large areas. A number of metals that vaporize easily may be used, but mercury and sodium are more efficient than most other metals. In this lamp, the plasma emits free–bound and free–free continuous radiation as well as line radiation. Mercury vapor lamps appear bluish because of the strong blue and green lines in the spectrum. Sodium lamps are yellowish because of a pair of strong emission lines near 5890 Å. Because this wavelength is near the peak of the response curve of the human eye, sodium vapor lamps have a high efficiency for lighting purposes.

Fluorescent Sources   Figure 13.18 shows the operation of a fluorescent lamp. This lamp consists of a glass tube containing mercury vapor at low pressure. The inner surface of the tube is coated with fluorescent material. Alternating current is conducted through the tube by free electrons. When one of these electrons collides with a mercury atom, part of its energy is transferred to the atom, placing it in an excited energy state. The mercury atom usually emits this energy in the form of a photon in the ultraviolet range, although some visible light may also be produced. Ultraviolet photons strike atoms in the fluorescent coating of the tube and are absorbed, raising these atoms, in turn, to higher energy levels. These atoms release their energy in several steps, producing visible light, as shown schematically by the energy level diagram in Figure 13.19. The process is called fluorescence.

Several fluorescent substances may be used and, by proper selection and blending, can result in light of any desired color. Most fluorescent bulbs produce a continuous spectrum in the visible region. Their outputs also contain the strong visible lines of the mercury spectrum at 436 (blue) and 546 nm (green). Unlike incandescent sources, most of the energy required to activate fluorescent bulbs emerges as visible light, making them more efficient.

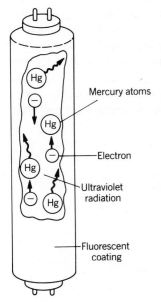

FIGURE 13.18 Fluorescent lamp.

Laser Radiation

The laser is a source of unique radiation, with a wide variety of applications. Laser radiation is used to drill holes in diamonds, to weld steel accurately, to measure large distances (e.g., to the moon), to repair detached retina, and for

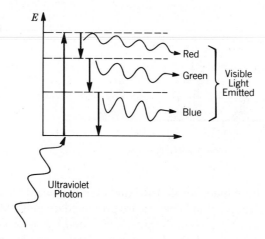

FIGURE 13.19 Schematic diagram of the fluorescence process.

many other applications. The laser is one of the more versatile tools of modern technology. The properties of laser radiation, the basic theory of operation of the laser, and a number of important laser applications are discussed.

Properties of Laser Radiation Radiation emitted by lasers is uniquely different from that produced by ordinary sources, such as incandescent or fluorescent bulbs. All laser applications depend on one or more of the unique properties of laser radiation.

Common light sources produce light that contains many different colors (different frequencies and wavelengths). Laser radiation is composed very nearly of only one wavelength. This property is called monochromaticity.

Ordinary light sources emit their light in all directions (Figure 13.20*a*). Radiant power leaves a laser in a tight beam that spreads slowly as it moves away from the laser (Figure 13.20*b*). Thus, laser radiation possesses high directionality.

Individual waves of radiant power produced by ordinary sources have no fixed

(*a*)

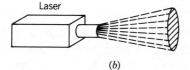

FIGURE 13.20 Comparison of ordinary light source and laser light beam.

(*b*)

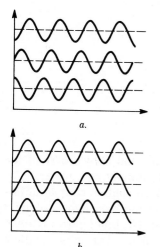

a.

b.

FIGURE 13.21 Laser radiation is coherent.

phase relationships (Figure 13.21*a*). However, in the case of laser radiation, all the waves are in step. Coherence is the property of laser radiation, meaning that all waves are in phase (Figure 13.21*b*).

Stimulated Emission The word "laser" is an acronym for $\underline{L}$ight $\underline{A}$mplification by $\underline{S}$timulated $\underline{E}$mission of $\underline{R}$adiation. It is the process of stimulated emission that results in the unique properties of laser radiation.

Stimulated emission is shown schematically in Figure 13.22. Initially an atom is in a high-energy state E_3 represented by the solid circle in the figure. An incident photon, with an energy equal to the energy difference between this state and some lower-energy state E_2, passes near the atom. This stimulates the atom to drop to E_2, emitting the energy difference $(E_3 - E_2)$ in the form of a photon. This photon has the same wavelength as the stimulating photon (monochromaticity), travels in the same direction (directionality), and is in phase with the stimulating photon (coherence).

Elements of a Laser For a laser to produce enough radiation to be useful, the stimulated emission process must occur many times. This repetition requires that a large number of atoms must be in energy state E_3 of Figure 13.22 and that a number of photons of the correct wavelength must be present for stimulation of these atoms. Figure 13.23 shows the four essential elements necessary to meet these requirements in any laser.

The active medium is a collection of atoms or molecules that may be raised, or excited, to the proper energy level to produce radiation by stimulated emission. The excitation mechanism is an energy source that provides energy to the atoms or molecules moving to the proper energy level. This energy level is referred to as an excited state. The feedback mechanism is a pair of mirrors that reflect the coherent light produced back through the active medium to cause more stimulated

FIGURE 13.22 Stimulated emission.

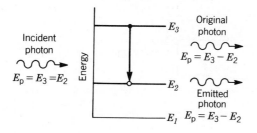

FIGURE 13.23 Elements of a laser.

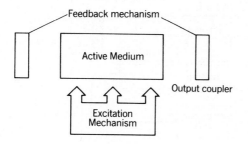

FIGURE 13.24   Amplification in a laser.

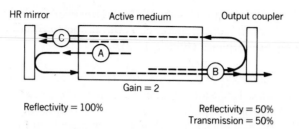

emission. One of these mirrors is partially transparent and allows part of the radiation striking it to pass out of the laser. It is called the output coupler.

Figure 13.24 shows how the laser beam is produced. The output coupler of this laser is a partially reflective mirror that reflects 50% of the laser light striking it and allows 50% to pass through to form the output beam. The other mirror is 100% reflective and is often called the high-reflectance (HR) mirror. The active medium acts as an optical amplifier. The process begins at point A with the spontaneous emission of a photon traveling perpendicular to the mirror surfaces. This photon is reflected back through the active medium, where stimulated emissions occur. Identical photons emerge in phase at point B. Half of the photons pass through the output coupler to form the output beam. The others are reflected back through the active medium to provide more stimulated emissions. This process continues as long as there are enough atoms in the excited energy state to provide gain. The light produced is monochromatic, directional, and coherent.

Active Mediums   Most common lasers use either a gas or a solid for an active medium. Some gas lasers (Figure 13.25) consist of a long, narrow tube that contains lasing gas. Mirrors are located at each end of the tube.

FIGURE 13.25   Schematic of a gas laser.

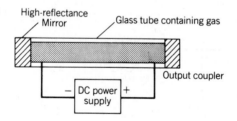

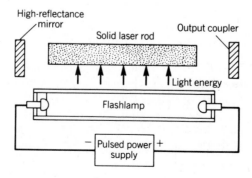

FIGURE 13.26   Optical pumping of a laser.

The active medium of a solid laser (Figure 13.26) consists of a cylinder of glass or of a crystal such as ruby. Mirrors may be located externally, as shown in Figure 13.26, or may be applied directly to the ends of the laser rod. Liquids and semiconductor materials may also be used as laser active mediums.

Excitation Mechanism   Energy is supplied to most gas lasers by passing a current through the gas (Figure 13.25). A few gas lasers obtain energy from chemical reactions in the gas. Solid, liquid, and some gas lasers (photolytic lasers) obtain energy by absorbing light from another source (Figure 13.26), which is called optical pumping, and may employ either a flashlamp or a continuous lamp.

Output Duration   Continuous-wave lasers are those that produce continuous-output beams. Pulsed lasers emit coherent light in short bursts. Lasers may be built to produce pulses with durations from less than a nanosecond to several milliseconds.

Output Wavelength   Each laser emits a specific wavelength, but lasers may be built to produce radiation at almost any wavelength in the near ultraviolet, visible, and infrared portions of the spectrum.

Laser Applications   There are many applications of lasers, and new ones are developed each year. A few of the more important applications are presented here:

- Alignment: The high directionality of light from lasers makes them ideal alignment tools. Low-power, helium–neon gas lasers are used for alignment problems throughout the industrial world, principally in construction industries.
- Materials processing: High-power continuous, carbon dioxide gas lasers are commonly used for cutting, welding, and heat treating of metals. Pulsed solid lasers are employed to drill small holes in metal parts. Lasers can drill small holes at a higher rate and with a smaller diameter than can be accomplished by any other means. A wide variety of materials, including cloth and semiconductor circuits, are cut with lasers.
- Communication: Modulated lasers are used as communication devices. A single laser beam can carry millions of telephone calls at once. Glass fibers carrying laser beams are replacing bulky, expensive cables used for telephone communication.
- Distance measurement: A laser rangefinder sends out a short pulse of light and measures the time necessary for a reflection to return. This technique is used in military applications and in surveying. Astronomical laser rangefinders have measured the distance to the moon and the altitude of satellites to within a few centimeters. Laser interferometers are used to measure short distances to within a millionth of an inch.

Other laser applications include eye surgery, controlled fusion research, three-dimensional photography, laser weapons, light shows, and cancer research. New applications are continually being developed.

SECTION *13.1* *EXERCISES*

1. Discuss the basic properties of sound and light.

2. What is electromagnetic radiation (radiant energy)? Describe a simple experiment that demonstrates how electromagnetic radiation can be created.

3. A microwave demonstration transmitter sold by Sargent–Welch Scientific Company emits microwaves of 3 cm wavelength. Determine the frequency (Hz) and photon energy (J) of the microwaves.

4. The eye is most sensitive to visible light at 555 nm. What is the frequency of this light? What is its color? What is its energy?

5. A 150-W incandescent lamp hangs 8 ft above the surface of a study desk. How close to the desk must it be moved to increase the power per unit area falling on the surface by a factor of six?

6. For the given frequency of each EM radiation band (from ac power to gamma rays), calculate the corresponding wavelength and EM wave photon energy. Present results in the following table.

EM Radiation	Frequency (Hz)	Wavelength (Calculated)	Photon Energy (J) Calculated
ac power	75		
Radio	9×10^6		
AM	9×10^5		
TV, FM	1×10^8		
Radar	2×10^{10}		
Microwave	9.5×10^{10}		
Infrared	9.5×10^{12}		
Visible	6×10^{14}		
Ultraviolet	8×10^{16}		
x rays	1×10^{19}		
Gamma rays	9×10^{20}		

7. Define:

Irradiance

Solid angle

Radiant intensity

8. Write an equation for the inverse square law of light and define each term in the equation.

9. If the radiant intensity of a flashlight is 5 W/sr, calculate its irradiance at 25 m.

10. If the radiant intensity of an arc lamp is 150 W/sr and irradiance at a detector is 0.25 mW/cm², calculate the distance from light source to detector.

11. If irradiance 15 m from a light source is 0.03 mW/cm², calculate the irradiance at 20 m.

12. Find the area on a sphere that is intercepted by a solid angle of $\pi/2$ steradians if the radius of the sphere is 3 m.

13. Find the radiant intensity of a source if at a distance of 1.5 m from the source a 5-m² area is delivered a power of 20 W.

14. If 600 W of power falls on an area of 3 m² when the light source is 6 m away, how much power falls on the same area when the light is brought to 3 m away?

15. List each of the following wavelengths of light:

Shortest seen by human eye

Longest seen by human eye

Peak of human visual response curve

16. Describe the process of producing visible light in each of the following:

Incandescent bulb

Fluorescent bulb

Metal vapor lamp

17. How does the output spectrum of a blackbody change as its temperature increases?

18. Identify each of the following as a point source or an extended source:

Incandescent bulb viewed from 5 ft

Incandescent bulb viewed from 50 ft

Television screen in the same room

Star

19. What is the total power emitted from a 10-by-10-ft black surface if the temperature is 1000°R?

20. Calculate the power emitted by a blackbody from an area of 0.1 ft² in a small wavelength interval from 0.5 to 0.55 μm if the surface temperature is 2500°R.

21. Explain briefly the origin of each of the following types of spectra:

Line

Band

Continuous

22. Match each of the following properties of laser light with the characteristic it describes:

    1. Monochromaticity   **a.** All light waves are in phase.

    2. Directionality   **b.** Light leaves the laser in a narrow cone and spreads very slowly.

    3. Coherence   **c.** Laser light contains very nearly only one wavelength.

23. Draw and label a diagram representing the stimulated emission process.

24. What is the wavelength of orange laser light? ($\nu = 5 \times 10^{14}$ Hz.)

25. A particular laser beam spreads 1 inch for every mile that it travels. If the beam is originally 0.01 inch in diameter what is the width of the beam 10 miles from the source?

26. A pulsed laser emits 3000 pulses per sec. How long does it take for the laser to pulse 10 times?

27. A laser and a point light source both have a power rating of 1 W. Why would it be dangerous to look directly into the laser beam but not to look directly at the point source?

28. How long would it take a laser beam to travel from New York to Los Angeles (approximately 3000 miles)?

29. On the following chart, label visible, IR, and UV portions of the spectrum. Complete the wavelength and frequency scales by adding numerical values to the points given.

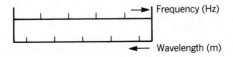

30. List several important applications of lasers.

31. Define the following wavelength units:

Micron

Nanometer

Angstrom

32. What is the difference between a low-energy gamma-ray photon of frequency 10^{20} Hz and a high-energy x-ray photon of the same frequency? Explain.

33. An experiment is performed to determine a numerical value for Planck's constant. In the experiment it is found that an EM wave of frequency 1×10^{11} Hz carries an energy of 6.8×10^{-23} J. From this data point, find Planck's constant and determine its percentage difference from the accepted value.

34. What should be the slope of a line that is obtained by plotting energy versus frequency?

35. How far does light travel in air in one year?

36. How long does it take light to reach us from a star 20×10^{15} miles away?

37. A laser beam can deliver 4000 cal/sec. How long would it take this laser beam to raise the temperature of 300 g of water by 10 C° if all its energy were absorbed by the water?

13.2 .... OBJECTIVES ....

NUCLEAR RADIATION

Upon completion of this section, the student should be able to

• Describe in one or two sentences the basic properties of alpha and beta particle radiation.

• Differentiate between alpha, beta, and gamma radiation.

• Define the half-life of a radioactive element.

• Define the electron volt.

• Write equations for nuclear reactions showing the elements, mass numbers, and atomic numbers.

• Discuss the stopping power, straggling, and mean range of alpha particles.

• Explain why the neutrino was introduced to explain beta decay.

• Given the beta-ray initial energy and the density of the absorbing material, estimate the range of the beta particle.

• Explain the source of x rays and gamma rays.

• Give the approximate ranges for gamma rays in terms of wavelength, frequency, and photon energy for radioactive decay.

• Explain how the decay of ^{14}C is used to date biological samples.

DISCUSSION

Radioactivity was discovered accidentally in 1896 by the French physicist Henri Becquerel. Becquerel placed some material containing uranium near photographic plates wrapped in black paper, and later found that the plates had been exposed. He correctly deduced that the uranium was emitting something that could penetrate the paper and expose the photographic plates.

Soon after, Pierre and Marie Curie discovered that ores containing uranium sometimes emit very intense radiation. They used chemical techniques to separate the active elements and, in so doing, identified two new chemical elements, radium and polonium. These two elements are more strongly radioactive than uranium. Madame Curie also correctly showed that the radiation was an atomic phenomenon. Over the next 20 years, studies by E. Rutherford and other workers led to the identification of the nature of the radiation produced by radioactive materials. In addition, many additional radioactive species were discovered. Today, many hundreds of radioactive species are known. Radioactive materials can be produced artifically, and such materials have found practical applications in science, industry, and medicine.

Radioactivity is defined as "the production of radiation in the spontaneous decay of an atomic nucleus." Many types of atomic nuclei, some occurring in nature, are inherently unstable. They decay by themselves, giving off some type of radiation in the process. As a result of the decay, the atomic nucleus changes into a different type of nucleus.

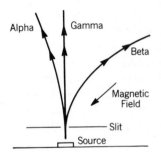

FIGURE 13.27 The effect of a magnetic field on alpha, beta, and gamma radiation.

Radioisotopes Radioactivity is exhibited only by certain isotopes. Recall that atoms with the same atomic number, but with different mass numbers, are called "isotopes." For a given chemical element (characterized by a certain atomic number), some isotopes may be radioactive and others are not. Thus, ^{210}Pb is a radioactive isotope of lead; whereas ^{208}Pb is a stable isotope and does not undergo radioactive decay. The isotopes that undergo radioactive decay are called "radioisotopes."

The early investigators of radioactivity identified three different types of radioactive emission. They arbitrarily named them "alpha," "beta," and "gamma" radiation. They could be separated by electric or magnetic fields. Figure 13.27 illustrates how a magnetic field (perpendicular to the paper) affects the three types of particles. Alpha rays and beta rays are deflected in opposite directions, with beta rays being deflected more. Gamma rays are unaffected.

Today we know that alpha particles are the nuclei of helium atoms; that beta particles are electrons or positrons; and that gamma rays are electromagnetic radiation. Because the alpha and beta particles may have opposite signs of electrical charge, they may be deflected in opposite directions by electric and magnetic fields. Uncharged gamma rays are not deflected. For review, the properties of alpha, beta, and gamma radiation are summarized in Table 13.3. Notice that the alpha particle is both more massive and more highly charged than the beta particle.

A particular radioisotope decays in a definite way, either by alpha-particle emission or by beta-particle emission. All the alpha particles from a particular radioisotope have a definite value of energy. The energies of alpha particles from naturally occurring radioisotopes are listed in handbooks for radiological health. Beta particles, however, have a spectrum of energies. The beta particle can have any value of energy, up to a maximum value characteristic of the particular radioisotope.

Now we consider the relation of the original nucleus, called the "parent," to

TABLE 13.3   Properties of alpha, beta, and gamma radiation

Type	Nature	Electrical Charge (units of electrical charge)	Mass (g)
Alpha	Helium nucleus	+2	6.62×10^{-24}
Beta	Electron	−1	9.1×10^{-28}
	Positron	+1	9.1×10^{-28}
Gamma	Electromagnetic radiation	0	0

the nucleus that remains after the decay, called the "daughter" nucleus. In alpha-particle emission, the emitted particle is a helium nucleus, a stable combination of two protons and two neutrons. Thus, in alpha-particle emission, the atomic number Z decreases by two and the mass number A decreases by four. This decay is shown schematically in Figure 13.28.

The figure shows a gamma ray being emitted, in addition to the alpha particle. Alpha-particle emission frequently is accompanied by gamma radiation.

In beta radiation, the emission can be an electron, with one unit of negative charge (Figure 13.29). The atomic number Z of the nucleus increases by one, because a neutron in the nucleus converts to a proton, thereby increasing the positive charge while leaving the mass number unchanged.

Again, a gamm ray is shown. Usually, beta-particle emission is accompanied by gamma-ray emission. Beta emission always is accompanied by emission of another particle, called the "neutrino." The neutrino has no electrical charge, zero mass, and thus very weak interaction with material. It passes great distances through materials with almost no absorption.

Often, after a radioactive decay, the daughter nucleus also will be unstable and again undergo radioactive decay. The resulting nucleus again may be unstable. Thus there can result chains of radioactive decay with a series of radioisotopes. The naturally occurring radioisotopes at the upper end of the periodic table may be grouped into several different series. Table 13.4 lists the so-called "uranium series," which begins with ^{238}U and ends with a stable isotope, ^{206}Pb. The main chain of decay in the uranium series is shown; some of the relatively unimportant branchings of the chain are not shown.

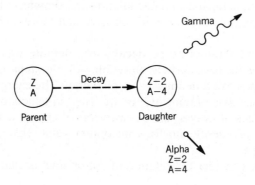

FIGURE 13.28   Alpha decay process.

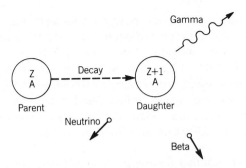

FIGURE 13.29 Beta decay process.

Inspection of the parent–daughter relationships in the chain is instructive in showing the changes in the nuclei produced by alpha or beta decay. In addition to the uranium series, there is the so-called "thorium series," which begins with ^{232}Th and ends with stable ^{208}Pb; and the "actinium series," which begins with ^{235}U and ends with stable ^{207}Pb.

Most of the naturally occurring radioisotopes are at the upper end of the periodic table, with atomic number of 83 or more, and are members of one of these series. There are a few naturally occurring lighter radioisotopes, such as ^{40}K. These decay to stable isotopes after only one or two steps and do not give rise to long series.

In addition to the radioactive species that occur in nature, there are hundreds of artificially produced radioisotopes. The most common method of producing artificial radioisotopes is to bombard stable isotopes with neutrons in a nuclear reactor. A stable isotope (atomic number Z, mass number A) can capture a neutron and be changed to a new isotope with atomic number Z and mass number

TABLE 13.4   The uranium series (main chain)

Isotope	Atomic Number	Radiation	Half-life
^{238}U	92	α	4.5×10^9 years
^{234}Th	90	β	24 days
^{234}Pa	91	β	1.17 min
^{234}U	92	α	2.5×10^5 years
^{230}Th	90	α	7.5×10^4 years
^{236}Ra	88	α	1.62×10^3 years
^{222}Rn	86	α	3.8 days
^{218}Po	84	α	3.05 min
^{214}Pb	82	β	27 min
^{214}Bi	83	β	20 min
^{214}Po	84	α	1.5×10^{-4} sec
^{210}Pb	82	β	19.4 years
^{210}Bi	83	β	2.5×10^6 years
^{210}Po	84	α	138 days
^{206}Pb	82	None	Stable

$A + 1$. In many cases, the new nucleus with the additional neutron will be unstable and will undergo radioactive decay. Radioisotopes of virtually all the naturally occurring elements have been produced in this way. In addition, a number of elements with atomic number greater than 92 have been produced by bombardment of atoms with beams of atomic particles. Remember that uranium, with $Z = 93$, has the highest atomic number of any element found in nature. The so-called "transuranium" elements are extremely radioactive. Such elements have been produced with values of Z up to 106.

Half-life of Radioactive Elements   The half-life for a radioactive element is the time required for half the atoms to undergo radioactive decay. Thus, if we start with N atoms of a particular radioisotope, after one half-life has elapsed, there will $N/2$ atoms remaining. After two half-lives, there will be $N/4$. After three half-lives, there will be $N/8$. Mathematically, the number of atoms remaining is expressed by

$$N(t) = N_0 e^{-\left(\frac{0.693t}{T_{1/2}}\right)} \qquad (13.40)$$

where

$N(t)$ = number of atoms remaining after time t
N_0 = number of time zero
$T_{1/2}$ = the half-life

We see that when $t = T_{1/2}$, $N(t) = N_0 e^{0.693} = N_0/2$, in accordance with the definition of half-life. Equation 13.40 is employed to solve a problem in Example 13.13.

EXAMPLE 13.13   Number of atoms remaining after decay
. . . .

Given: A certain radioisotope has a half-life of 10 days.

Find: The number of atoms remaining (a) after 2 weeks and (b) after 4 weeks.

Solution: **a.**   After 2 weeks, according to Equation 13.40

$$\frac{N\ (14\ \text{days})}{N_0} = e^{-\left(\frac{0.693\ \times\ 14\ \text{days}}{10\ \text{days}}\right)}$$

$$\frac{N\ (14\ \text{days})}{N_0} = 0.379$$

Thus, about 37.9% of the original atoms remain undecayed.

b.   After 4 weeks

$$\frac{N\ (28\ \text{days})}{N_0} = e^{-\left(\frac{0.693\ \times\ 28\ \text{days}}{10\ \text{days}}\right)}$$

$$\frac{N\ (28\ \text{days})}{N_0} = 0.144$$

Therefore, 14.4% remain.

Notice that if one chooses the half-life $T_{1/2}$ in days, the time t also must be expressed in days. If $T_{1/2}$ is expressed in weeks, then t would also be expressed in weeks. All that matters is that t and $T_{1/2}$ be expressed in the same units of time.

The rate of decay for radioactive atoms is given by

$$Q(t) = \frac{0.693 \, N(t)}{T_{1/2}} \qquad (13.41)$$

where

$Q(t)$ = rate of decay at time t; that is, the number of atoms decaying per unit time

$N(t)$ = number of atoms present at time t

Thus, as $N(t)$ decreases with time, the decay rate also decreases. The rate of decay for a source can be intense, according to Equation 13.40, either because the number of atoms is large or because the half-life is short.

The half-lives of the radioisotopes in the uranium series of naturally occurring radioisotopes are listed in Table 13.4. The values cover a very broad range, varying from a minute fraction of a second up to billions of years. The naturally occurring series are all headed by a long-lived radioisotope, such as ^{238}U, which still remains from the original time of formation of the materials on the earth. Other elements with atomic numbers greater than 92 may once have been present, but these elements had shorter half-lives and disappeared long ago. Short-lived isotopes in the uranium series are found in nature because they continually are being replenished by the decay of the isotopes above them in the chain. Thus, ^{234}Pa is found in nature despite the half-life of 1.17 min. It is being continually produced by the decay of ^{234}Th, which in turn is formed from the decay of ^{238}U.

Equation 13.41 is clarified in Example 13.14.

EXAMPLE 13.14 Decay rate
. . . .

Given: A source contains 1.5×10^{17} atoms of ^{226}Ra.

Find: The activity of the source.

Solution: From Table 13.4, the half-life is 1.62×10^3 years, or 5.11×10^{10} sec. From Equation 13.41, the decay rate is

$(0.693)(1.5 \times 10^{17})/5.11 \times 10^{10} = 2.03 \times 10^6$ decays/sec

This value is equal to

$(2.03 \times 10^6$ decays/sec$)/(3.7 \times 10^{10}$ (decays/sec)/curie$)$

$= 5.49 \times 10^{-5}$ curie

$$= 54.9 \text{ microcurie}$$

Note. 1 curie $= 3.7 \times 10^{10}$ decays/sec.

The Electron Volt   One electron volt is defined as equal to the work done upon one elementary charge by a potential difference of one volt. The abbreviation for the electron volt is eV. The work done in joules is

$$\Delta W = Q\Delta V \tag{13.42}$$

where

Q = electronic charge, 1.60×10^{-19} C
ΔV = potential difference

Therefore, one electron volt is equivalent to

$$\Delta V = (1.60 \times 10^{-19} \text{ C})(1.00 \text{ V})$$
$$\Delta V = 1.60 \times 10^{-19} \text{ J}$$

The electron volt is a small unit of energy. Thus, a common unit used in nuclear physics is a million electron volts (MeV).

Alpha Particles   As mentioned above, the alpha particle is the nucleus of a helium atom expelled from a heavy element during the nuclear transformation from one element to another. Thus, for example, when uranium-238, which has 92 protons and 146 neutrons clustered together in its nucleus, changes, or "decays," to thorium-234 (90 protons and 144 neutrons), an alpha particle is emitted from the nucleus. This alpha particle has two protons and two neutrons, just as does the fully ionized helium atom. In terms of common nuclear symbols, the nuclear reaction discussed here is given as

$$^{238}_{92}\text{U} \rightarrow {}^{234}_{90}\text{Th} + {}^{4}_{2}\text{He}$$

Notice that the superscripts (mass numbers) and subscripts (atomic numbers) obey a conservation law; thus $238 = 234 + 4$ and $92 = 90 + 2$. The element uranium-238 is written as $^{238}_{92}\text{U}$, where the subscript 92 refers to the number of protons in the nucleus (atomic number) and the superscript 238 refers to the sum of the protons (92) and neutrons (146) in the nucleus (atomic mass numbers). The reaction equation shows that an alpha particle $^{4}_{2}\text{He}$ is created when uranium-238 decays to thorium-234.

The alpha particles emitted in a particular radioactive decay have definite energies. The energies of alpha particles vary from about 4 to 9 MeV. A particular radioactive isotope in its decay produces an alpha particle having a well-defined energy. The energies of the alpha particles from the decays of several different isotopes are listed in Table 13.5. All the radioisotopes listed are found in nature. For a given element, there may be several different isotopes that emit alpha particles; for example, the element polonium (Po) is represented by five different isotopes, each of which produces an alpha particle with a different energy.

There is a small uncertainty in the range of alpha particles because of statistical fluctuations in the loss of energy near the end of the range. Thus, all the alpha

TABLE 13.5 Alpha particle energies from some naturally occurring radioisotopes

Isotope	Alpha Particle Energy (MeV)
^{238}U	4.19
^{232}Th	4.35
^{235}U	4.74
^{231}Pa	5.11
^{214}Bi	5.51
^{224}Ra	5.68
^{218}Po	6.00
^{211}Bi	6.62
^{215}Po	7.38
^{211}Po	7.45
^{214}Po	7.68
^{212}Po	8.78

particles of a given energy do not have exactly the same range. This phenomenon is called "straggling." The amount of straggling is denoted by ΔR in Figure 13.30. For alpha particles of reasonable energy, the straggling ΔR is generally small compared to the mean range.

The rate of energy loss, of course, determines the range. This leads to the concept of "stopping power," defined as "the energy lost by an alpha particle as it travels a unit distance through a material." The stopping power is a complicated function of alpha-particle energy and of the properties of the absorber; a complete description of the dependence of the stopping power is beyond the scope of this discussion.

The stopping power is proportional to the number of electrons per unit volume in the materials; consequently, materials of high density (which have more electrons per unit volume) will have higher values of stopping power—and, hence, lower values of range—than do materials of low density.

The values of range now can be qualified by the presentation of some results for specific materials. Figure 13.31 depicts the mean range as a function of energy for alpha particles in air. For reasonable values of alpha-particle energy (a few MeV),

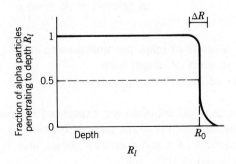

FIGURE 13.30 Uncertainty in range of alpha particles of a given energy.

FIGURE 13.31 Plot of mean range versus energy of alpha particles in air.

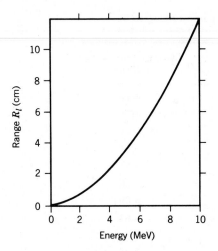

the range is only a few centimeters, much shorter than the range of gamma rays of similar energy.

As an alpha particle moves through the air, its electric field interacts with the outer electrons of the air molecules and pulls them away from the air molecules. This process forms ions and free electrons (ion pairs); approximately 35.5 eV of energy are required to form an ion pair in air. The total number of ion pairs formed in air is approximately equal to the initial alpha-particle energy in eV divided by 35.5 eV.

The thickness of absorbing material that the alpha particle traverses before coming to rest is the range of the alpha particle in that material. For a given material and a given initial energy, the range of the alpha particle is well defined. Figure 13.30 is a schematic diagram of the fraction of alpha particles that penetrate to a given depth R in a material. The curve is flat up to a certain value, beyond which it drops rapidly and falls to zero. The mean range of a large number of alpha particles, denoted by R_0 in Figure 13.30, is the thickness of material that stops 50% of the incident alpha particles.

Figure 13.32 displays mean range as a function of energy for alpha particles in aluminum (Al) and lead (Pb). The values of range are much smaller than those in air, because of the higher density of aluminum and lead.

In Figure 13.32, the range is expressed in units of length. The range can also be expressed in units of mass per unit area. The relationship between these two expressions is given by

$$R \text{ (g/cm}^2) = R_1 \text{ (cm) } \rho \qquad (13.43)$$

where

    R = range in units of mass per unit area (g/cm^2)
    R_1 = range in units of length (cm)
    ρ = density of the absorbing material (g/cm^3)

In practice, range relations often are expressed in units of g/cm^2.

As the following example of the use of Equation 13.43 demonstrates, the range for an alpha particle of a given energy shows less variation between materials when it is expressed in these units.

FIGURE 13.32 Plot of mean range versus energy for alpha particles in lead and aluminum.

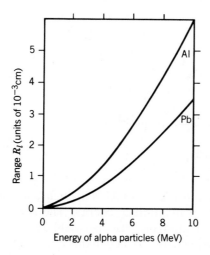

EXAMPLE 13.15 Range of an alpha particle
· ·

Given: The densities of air, aluminum, and lead are 1.29×10^{-3}, 2.7, and 11.48 g/cm³, respectively.

Find: The range of 5-MeV alpha particles in (a) air, (b) aluminum, and (c) lead, expressed in g/cm².

Solution: **a.** For air, use Figure 13.31 to obtain the range $R_1 = 3.5$ cm at 5 MeV. By Equation 13.43

$$R \text{ (g/cm}^2) = R_1 \text{ (cm) } \rho$$

$$R = (3.5 \text{ cm})(1.29 \times 10^{-3} \text{ g/cm}^3) = 4.52 \times 10^{-3} \text{ g/cm}^2$$

b. For aluminum, Figure 13.32 gives $R_1 = 1.85 \times 10^{-3}$ cm at 5 MeV, and $R = (1.85 \times 10^{-3} \text{ cm})(2.7 \text{ g/cm}^3) = 5.0 \times 10^{-3}$ g/cm².

c. For lead, Figure 13.32 gives $R_1 = 1.10 \times 10^{-3}$ cm at 5 Mev, and $R = (1.10 \times 10^{-3} \text{ cm})(11.48 \text{ g/cm}^3) = 12.6 \times 10^{-3}$ g/cm².

This example illustrates that for alpha particles for a given energy, the range does not vary so much between different absorbing materials when it is expressed in units of mass per unit area.

Beta Particles Electrons emitted by radioactive nuclei are called "negative beta particles" (or beta rays)—a name given to them before they were identified as electrons. Negative beta rays, as mentioned above, are nothing more than high-energy electrons created in the nucleus. The energies of negative beta rays emitted by an unstable atomic nucleus cover a range of values, up to some maximum value characteristic of that nucleus. Thus, beta-ray energies form a spectrum. An example of the energy spectrum of the beta rays from the radio-

FIGURE 13.33 Beta radiation energy spectrum for ^{210}Bi.

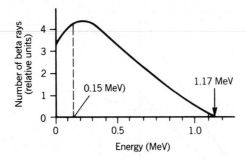

isotope ^{210}Bi is given in Figure 13.33. This figure depicts the relative number of beta rays emitted as a function of energy. The spectrum has a peak near 0.15 MeV. The decay of an atom of ^{210}Bi may yield a beta ray of any energy, between zero and the maximum energy, which in this case is 1.17 MeV. The most probable value of the energy is 0.15 MeV.

The energies of beta-ray spectra extend up to a few MeV. Table 13.6 lists the maximum beta-ray energy for several different radioisotopes. All these isotopes occur naturally. In each case, the spectrum is characterized by the maximum value of the energy.

Since there are no electrons as such in the nucleus, they are created during the nuclear transformation. This process can be explained in the following manner: a neutron (which exists in the nucleus) is converted to a proton (which can exist in the nucleus) and an electron. However, energy and momentum should be conserved. Measurements show that these basic physical laws are violated unless a new particle, the neutrino, is introduced. The neutrino has no charge and a very small mass (from zero to less than 0.05% of the electron mass). The introduction of the neutrino allows the conservation laws to be satisfied. In terms of a nuclear reaction equation, the relationship becomes

$$_0^1n \longrightarrow {}_1^1p + {}_{-1}^0e + {}_0^0\nu$$

TABLE 13.6 Beta particle energies from some naturally occurring radioisotopes

Isotope	Maximum Beta Particle Energy (MeV)
^{214}Pb	0.72
^{210}Bi	1.17
^{223}Fr	1.20
^{211}Pb	1.40
^{207}Tl	1.47
^{228}Ac	1.55
^{207}Tl	1.82
^{40}K	1.90
^{234}Pa	2.32

In essence, the equation for the transformation of a neutron into a proton, an electron, and a neutrino states that a neutron with charge zero and mass number one is converted into a proton of charge one and mass number one, an electron ($_{-1}^{0}e$) with charge minus one and mass number zero, and a neutrino ($_{0}^{0}\nu$) with charge zero and mass number zero.

The process described above occurs in many nuclear transformations. An example of such a transformation is

$$_{82}^{214}\text{Pb} \longrightarrow {}_{83}^{214}\text{Bi} + {}_{-1}^{0}e + {}_{0}^{0}\nu$$

Radioactive lead-214 decays to the isotope bismuth-214 and in the process emits a negative beta particle (high-speed electron) and a neutrino. The mass number (sum of protons and neutrons) remains the same before and after the transformation. The atomic number, however, increases by one going from 82 to 83 because a proton was created in the nucleus from a neutron.

The actual mechanical energy available in the above transformation can be calculated from the loss in mass ($E = \Delta mc^2$). If only an electron were emitted, one would predict fron the conservation laws that all the electrons should have the same energy. The neutrino, however, can have part of the energy, and an electron energy distribution can be explained by assuming that the neutrino has the excess energy.

Beta particles lose energy by interacting with the electrons in the materials thorugh which they pass. Like alpha particles, they interact with the electrons because of their electrical charge. They give up their energy in small amounts in each interaction, leaving behind a track of ionized particles. Unlike alpha particles, beta rays often change the direction of their motion when they collide with electrons or atoms in the material. This happens because the mass of the electron is much less than that of the alpha particle. Alpha particles passing through a material are rarely deflected. A crude analogy might be a Ping-Pong ball and a bowling ball: the course of a Ping-Pong ball is deflected much easier than that of the bowling ball.

Because electrons (beta rays) often are deflected from their direction of travel, they often do not penetrate deeply into a material. Instead, their path is marked by erratic changes of direction. Thus, the absorption curve for beta rays is much different from that for alpha rays. Figure 13.34 shows a schematic diagram of how a beam of electrons is attenuated in a material. The figure illustrates the fraction of electrons (all initially at the same energy) that penetrate to a given depth in the material. Because of collisions that cause the electrons to change their direction (although they have not given up all their energy), not many electrons penetrate to

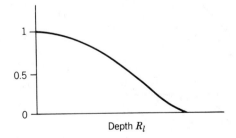

FIGURE 13.34 Penetration of an electron beam into a material.

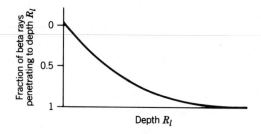

FIGURE 13.35   Penetration of beta rays into a material.

the maximum depth. The form of the curve differs greatly from that in Figure 13.30 for alpha particles. Because the curve slopes continuously downward, a mean range cannot be defined in the same way as for alpha particles; rather, there is the maximum range beyond which no beta rays penetrate. This maximum range is the depth at which the beta ray has given up all its energy, even if it has experienced no change in direction.

The situation for beta rays emitted from a radioactive isotope is even more complicated because these beta rays exhibit a spectrum of energies. Thus, the absorption curve is defined by contributions from beta rays having a continuous range of energies. The shape of the absorption curve changes to that shown in Figure 13.35. As compared to Figure 13.30, more beta rays are stopped at short ranges because the beta-ray energy spectrum peaks at relatively low energy (Figure 13.33). The low-energy electrons are removed first, near the surface of the absorber. Beta rays from the high-energy portion of the spectrum penetrate deeper. The maximum range is defined by electrons of maximum energy in the beta-ray spectrum. Because there are relatively few electrons at the upper end of the energy spectrum, the curve in Figure 13.35 tails off to zero very gradually, making difficult the measurement of the upper end of the range of beta rays emitted by a given radioisotope.

The range–energy relation for beta rays in air is given by Figure 13.36. The range shown in the figure is the maximum range for beta rays having the given value of energy (or for a beta-ray spectrum with the given value of energy as its upper limit). The results are to be compared with Figure 13.31, which presents the

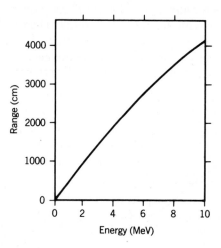

FIGURE 13.36   Plot of range versus energy for beta rays in air.

FIGURE 13.37 Plot of range versus energy for beta rays in aluminum.

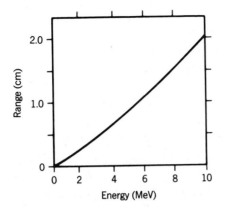

mean range for alpha particles in air. The beta ranges are seen to be much greater for a particular value of energy.

Beta-ray range in aluminum as a function of energy is shown in Figure 13.37. The range is the maximum range for beta rays of the given energy. If the results in Figure 13.37 are compared to those in Figure 13.32 for mean range of alpha particles, the range of the beta rays is seen to be much greater, as clarified by Example 13.16.

EXAMPLE 13.16   Range of alpha and beta particles
· ·

Given: The data in Figures 13.32 and 13.37.

Find: The difference between the mean range of 3-MeV alpha particles in aluminum and the maximum range of 3-MeV beta particles in aluminum.

Solution: From Figure 13.32, the mean range of the alpha particles is 0.95×10^{-3} cm.

              From Figure 13.37, the maximum range of the beta particles is 0.55 cm, or almost 600 times the mean range of the alpha particles.

The range of electrons in a material is affected by the number of electrons per unit volume that are present in the material because beta rays give up their energy by collisions with these electrons. The more electrons present in the material, the greater will be the rate of energy loss and the shorter the range. The number of electrons per unit volume is roughly proportional to the density ρ of the material. Thus, as an approximation, the range is inversely proportional to the density. An empirical equation given by L.E. Glendenin and C.D. Croyell for electrons having energies greater than 0.8 MeV is

$$R_m = \frac{1}{\rho}(0.54E - 0.13) \qquad (13.44)$$

where

R_m = maximum range in cm
ρ = density of the material in g/cm^3
E = electron energy in MeV

An application of Equation 13.44 is given in Example 13.17.

EXAMPLE 13.17   Calculated and experimental beta ranges
. . . .

Given: The density of aluminum is 2.7 g/cm^3.

Find: The difference between the maximum range of beta rays of energy 2 MeV as calculated from Equation 13.44 and the experimental range from Figure 13.37.

Solution: Equation 13.44 gives

$$R_m = \frac{1}{2.7}(0.54 \times 2 - 0.13) = 0.352 \text{ cm}$$

Figure 13.37 gives $R_m \simeq 0.36$ cm. The difference in these two values is about 2%.

To obtain the maximum range R_m in units of mass per unit area, one must multiply the linear range by the density of the absorber, as in

$$R_m \text{ (g/cm}^2) \simeq 0.54E - 0.13 \qquad (E > 0.8 \text{ MeV}) \tag{13.45}$$

R_m is independent, therefore, of the absorbing material and depends only on the beta-ray energy. This is a reasonable approximation, especially for absorbing materials of low atomic number.

Gamma Rays   As previously stated, gamma radiation is a form of electromagnetic radiation similar to visible light. The wavelength of gamma radiation, however, is much shorter than the wavelength of visible light. Figure 13.38 gives an expanded view of the short-wavelength portion of the electromagnetic spectrum and the relation between wavelength and frequency for x rays and gamma rays. For example, a gamma ray of wavelength 10^{-12} cm has a frequency of 3×10^{22} Hz. Figure 13.38 may be compared to the more complete diagram of the spectrum shown in Figure 13.2.

The wavelength ranges for x rays and gamma rays overlap. The boundaries between different regions in the spectrum are not sharp, definite boundaries but are chosen by convention. The distinction between gamma rays and x rays arises from the way they are produced. As previously stated, x rays are produced by energy changes of inner orbital electrons of atoms or molecules or by bremsstrahlung. Gamma rays, however, are produced by an energy-level change within the

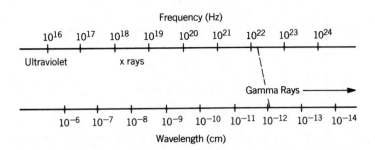

FIGURE 13.38 High-energy portion of the electromagnetic spectrum.

nucleus. These two processes both produce radiation in the wavelength range of 10^{-10} to 10^{-8} cm. Thus, there is a region of overlap between these two types of radiation. After an x ray is emitted, however, it is indistinguishable from a gamma ray if it lies within the common energy range. Thus, within the common energy range, the interaction of x rays and gamma rays with matter is the same if the wavelength is the same.

It is usually advantageous to consider gamma rays as photons. As previously stated, the energy of a photon is equal to $h\nu$, where h is Planck's constant ($h = 6.62 \times 10^{-34}$ J·sec). Since $\nu = c/\lambda$, the energy of a photon may also be expressed as

$$E_p = h\nu = \frac{hc}{\lambda} \tag{13.46}$$

The calculation of a photon's energy in electron volts is given in Example 13.18.

EXAMPLE 13.18 Energy of a photon in electron volts
. . . .

Given: Photons have a wavelength of 10^{-8} cm.

Find: The energy in MeV.

Solution: From Equation 13.46

$$E = \frac{hc}{\lambda}$$

$$h = 6.62 \times 10^{-34} \text{ J·sec}$$

$$c = 3 \times 10^{-8} \text{ m/sec}$$

$$\lambda = 10^{-8} \text{ cm} = 10^{-10} \text{ m}$$

$$E = \frac{(6.63 \times 10^{-34} \text{ J·sec})(3 \times 10^8 \text{ m/sec})}{10^{-10} \text{ m}}$$

$$E = 1.99 \times 10^{-15} \text{ J}$$

$$1 \text{ eV} = 1.60 \times 10^{-19} \text{ J}$$

$$E = (1.99 \times 10^{-15} \text{ J}) \frac{1 \text{ eV}}{1.60 \times 10^{-19} \text{ J}}$$

$$E = 1.24 \times 10^4 \text{ eV}$$

$$E = 0.0124 \text{ MeV}$$

Note. In general

$$E = \frac{(6.62 \times 10^{-34} \text{ J·sec})}{\lambda} \frac{(3 \times 10^8 \text{ m/sec})(1 \text{ eV})}{(1.60 \times 10^{-19} \text{ J})}$$

$$E = \frac{1.24 \times 10^{-6} \text{ eV·m}}{\lambda}$$

where λ is in meters.

In what has been stated so far, the range of energy or frequency has not been limited on the upper end. In fact, in very high energy nuclear reactions, gamma rays of extremely high energy (extremely short wavelength) can be produced. A convenient practice is to express a boundary limited by the energies available in radioactive decay of nuclei, about 4 MeV. This value corresponds roughly to a lower-wavelength limit of 3×10^{-11} cm and to an upper-frequency limit of 10^{21} Hz. The ranges are summarized in Table 13.7.

For all nuclear processes, there is no defined limit. In fact, gamma rays with wavelength shorter than 10^{-25} cm have been detected in cosmic rays coming from outside the solar system. But for gamma rays originating in radioactive decay, there is a lower limit to the wavelength, as indicated in Table 13.7.

As examples of isotopes that undergo radioactive decay and produce gamma rays, Table 13.8 lists the energies of gamma rays emitted by several selected isotopes of various elements. Some radioactive isotopes can emit gamma rays of more than one energy. Following Table 13.8, Example 13.19 illustrates a typical calculation of wavelength and frequency corresponding to a gamma ray of given energy.

Example 13.19 shows how to obtain the wavelength and frequency of a gamma ray if the photon energy is given.

EXAMPLE 13.19   Calculation of wavelength and frequency

· ·

Given: The isotope ^{214}Bi emits gamma rays of energy 0.77 MeV.

Find: The corresponding wavelength and frequency.

Solution: Use Equation 13.46:

$$\lambda = \frac{1.245 \times 10^{-6} \text{ eV·m}}{E_p}$$

$$\lambda = \frac{1.245 \times 10^{-6} \text{ eV·m}}{770{,}000 \text{ eV}}$$

$$\lambda = 1.62 \times 10^{-12} \text{ m}$$

Use Equation 13.1 (rearranged for ν):

$$\nu = \frac{c}{\lambda}$$

$$\nu = \frac{3 \times 10^8 \text{ m/sec}}{1.62 \times 10^{-12} \text{ m}}$$

$$\nu = 1.85 \times 10^{20} \text{ Hz}$$

When gamma rays interact with matter, the photoelectric effect, scattering, and pair production can occur. These processes are discussed in Chapter 12, and the equation for the intensity of a parallel beam of gamma photons as a function of distance is derived from basic physical principles.

Applications. Radioactive isotopes are used in many practical applications in medicine, science, and industry. In medicine, radioisotopes are used to test the functioning of certain organs of the body. For example, iodine taken into the body is collected in the thyroid gland. If a small amount of the radioisotope ^{131}I is ingested, its location in the body can be followed with a counter. Then intake of ^{131}I by the thyroid gland gives a measure of the functioning of the thyroid.

TABLE 13.7 Approximate ranges for gamma rays

	All Nuclear Processes	Radioactive Decay
Photon energy	> 0.01 Mev	0.01–4 MeV
Wavelength	$< 10^{-8}$ cm	3×10^{-11}–10^{-8} cm
Frequency	$> 3 \times 10^{18}$ Hz	3×10^{18}–10^{21} Hz

TABLE 13.8 Gamma rays emitted by selected radioactive isotopes

Isotope	Gamma Ray Energy (MeV)
^{7}Be	0.48
^{24}Na	1.37 and 2.75
^{51}Cr	0.32
^{60}Co	1.17 and 1.33
^{137}Cs	0.66
^{298}Au	0.41
^{208}Tl	0.51, 0.58, 2.61
^{241}Am	0.026 and 0.060

Radioisotopes also are used in medicine as therapeutic agents, often to destroy cancerous tumors. There are many irradiators that use the radioisotope ^{60}Co. Gamma rays from ^{60}Co are used to irradiate localized cancers in an attempt to destroy them.

Radioactive decay of ^{14}C is used to date biological samples. The isotope ^{14}C is maintained in nature at a uniform level by cosmic rays[1] and is distributed throughout the world. Living organisms assimilate the carbon and incorporate it as part of their cells. The activity resulting from the decay of the carbon amounts to 15.3 disintegrations per minute per gram of carbon. When the organism dies, it stops incorporating ^{14}C, and the ^{14}C that it already has built into its cells gradually decays, with a half-life of 5568 years. If one measures the amount of activity from a biological sample and also determines the total amount of carbon by chemical means, one can obtain an estimate of the age of the sample, as illustrated in Example 13.20.

EXAMPLE 13.20   Age of a sample
. . . .

Given: Someone shows you a fragment of charred wood and claims it came from the Trojan Horse (10,000 years of age). The weight of the wood is 100 g. You measure the beta-ray activity of the sample and find it emits 1525 disintegrations per minute.

Find: Age of fragment.

Solution: The charred wood is mostly carbon. If it were 10,000 years old, the original number of ^{14}C atoms should have decayed to $e^{-(0.693 \times 10,000/5568)}$ or 28.8% of the original value. The original number of decays was 15.3 per gram per minute, or 1530 from the total fragment. The current number of decays should be 0.288 × 1530 = 441 decays per minute. But the measured number (1525 decays per min) is much greater. Therefore, you decide that the wood is of much more modern origin.

Radioisotopes are used as heat sources in thermoelectric generators, where radiation is absorbed in the hot junction. The energy absorbed from the radiation keeps the junction hot. Generators powered by heat from radioactive decay have been used as sources of electrical power in remote locations and in the space program.

In industry, radioisotopes are used as gages to measure thickness of a material. The beta-ray source is on one side of the material and a counter is on the other. The counting rate depends on the material thickness. Such gages can measure thickness of a moving material being produced in a continuous process, without any contact with the material. They are used to measure thickness of coatings

---

[1] Assuming cosmic ray activity is constant.

applied to thin materials and of products such as plastic sheet, paper, thin metal, and sandpaper. The output of the counter can be employed to control the production process and ensure the proper thickness of the product.

Radioisotopes are capable of causing harm to human beings. Excessive exposure to radiation from radioisotopes can lead to many health problems, including genetic defects and cancer. Thus, strict safety regulations are applied to the transport, use, storage, and disposal of radioisotopes; on a national level, the Nuclear Regulatory Commission sets standards for their use. Many states also regulate the usage of radioisotopes. The user of radioisotopes must check the regulations relevant to the intended use and comply with them.

SECTION *13.2* EXERCISES

1. Define alpha and beta radiation. Include the mass and electrical charge for each.
2. Describe the process by which alpha and beta particles lose energy in absorbing materials.
3. For alpha particles, define mean range, range straggling, and stopping power.
4. For beta particles, describe the shape of the absorption curve. Define maximum range. State the approximate relation between maximum range and energy.
5. What is the charge-to-mass ratio of alpha particles? of beta particles?
6. Assuming that all of an alpha particle's energy is nonrelativistic kinetic energy, determine the velocity of a 4-MeV alpha particle.
7. What is the ratio of the range of 6MeV alpha particles in air to the range of 6-MeV alpha particles in lead?
8. What is the density of a material for which the range of 2-MeV beta rays is 53 cm?
9. What is the ratio of the mean range of 4-MeV alpha particles in air to that of the maximum range of 4-VeV beta particles in air?
10. Define the following terms:
    Radioactivity
    Naturally occurring chains of radioactive isotopes
    Half-life of a radioisotope
    Curie
11. State the relationship between atomic number and mass number of the parent and daughter atoms for alpha-particle emission and for beta-particle emission.
12. Three days ago. 1.20×10^{19} atoms of a certain radioisotope were present. Because of radioactive decay, there are now 0.57×10^{18} atoms. What is the half-life of the isotope?
13. While you are digging at an archeological site, you find a fragment of old human bone. By chemical methods, you determine that the bone contains 225 g of carbon. You measure the beta-ray activity of the material and find that there are 2684 ^{14}C decays per minute. How old do you estimate the bone to be?
14. A radioactive source has a half-life of 1000 years. What percentage of the original number of atoms remain undecayed after 10,000 years?
15. After 6 days only one-tenth of the original number of atoms in a radioactive source remain undecayed. What is the half-life of this source?

S U M M A R Y

Radiation transmits energy away from a source either by electromagnetic waves or by material particles. However, electromagnetic radiation can be considered to be composed of particles called photons that have no rest mass. They always travel at the speed of light. In addition, particles such as electrons have wave properties that can be demonstrated in the laboratory.

The frequency of electromagnetic radiation ranges from 10 to 10^{22} Hz or from a wavelength of 10^7 to 10^{-13} m. The sources of electromagnetic radiation include ac power transmission, radio transmitters, molecular and atomic transitions, the capture of electrons, electron collisions, and nuclear transitions. Particle radiation includes high-energy beta and alpha particles that are emitted from the nucleus of atoms.

If radiation is emitted from a point source with equal probability in any direction, the irradiance decreases inversely as the square of the distance from the source if no absorption or scattering takes place. This is known as the inverse square law.

All bodies emit electromagnetic radiation. The amount of radiation emitted depends on the absolute temperature and the nature of the surface. For an ideal blackbody, which absorbs all the radiation incident on its surface, the total emitted radiation is proportional to the fourth power of the absolute temperature. The spectral distribution of the emitted radiation was explained by Planck and can be calculated as a function of wavelength and absolute temperature.

The optical spectrum, which includes the visible spectrum, extends from approximately 100 to 0.01 μm, whereas the visible spectrum extends from approximately 0.7 to 0.4 μm. The response of the eye increases from 0.7 μm to reach a peak at 0.555 μm and then decreases to zero at 0.4 μm. Visible light can be produced by incandescent filaments, plasmas, or fluorescence. The physical processes include free-free radiation, free-bound radiation, and bound-bound radiation when electronic transitions are considered. In addition, there can be changes in rotational and vibrational states that produce complex spectra in infrared as well as the visible and ultraviolet regions of the spectrum.

The laser is a source of unique radiation that depends on stimulated emission. The radiation is nearly monochromatic, almost coherent, and highly directional. These properties can be used for many applications, such as optical alignment tools, materials processing, communications, welding, and distance measurements.

Nuclear radiation comes from unstable nuclei. These radioactive materials emit alpha, beta, and gamma radiation. The alpha particle is the nucleus of a helium atom, the beta particle is an electron or positron, and gamma radiation is electromagnetic radiation with wavelengths less than 0.01 μm. The radioactive nuclei decay with unique half-lives. Alpha particles are emitted with unique energies, and a mean range can be defined for a given material and alpha-particle energy. Beta particles are not emitted with unique energies, and it is necessary to introduce the neutrino if the conservation laws are to remain valid. A mean range for beta particles cannot be defined. Thus, one measures a maximum range. The maximum range depends on the material and the beta particle energy. For a given energy the range of beta particles is much greater than the range of alpha particles (3-MeV beta particles have a range 600 times the mean range of alpha particles in aluminum). The range of gamma radiation is very large.

Optics

and

Optical

Systems

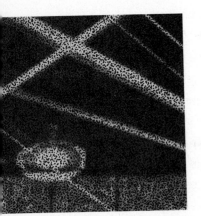

INTRODUCTION

Information about the external world is obtained through the five senses, but light interacting with the eye reveals the most information. However, the eye is sensitive to a very limited portion of the incident radiation. Today it is possible to extend our knowledge about the external world by using sensors that can span the entire spectrum of electromagnetic radiation. In this chapter only the optical spectrum and its interaction with matter are considered. To explain the various interactions of light with matter, sometimes it is necessary to consider light to be composed of waves, and at other times it is necessary to consider light to be composed of particles (photons). The wave nature of electromagnetic radiation is discussed in Chapter 13 and the interaction of photons with matter is considered in several other chapters.

There are two important aspects of light. They are intensity and phase. For many interactions, phase is not important, and light can be considered to be a bundle of rays that travel in a straight line in an isotropic medium. Ray optics is also called geometrical optics. When phase relationships are important, one must consider light to be a propagating wave front. Wave optics is also called physical optics.

This chapter is composed of four sections. In this first section, the laws of reflection and refraction are applied to the study of mirrors, lenses, and prisms. Optical problems are solved by the application of geometrical optics. Interference and diffraction are studied in the second section. These physical phenomena depend on the wave nature of light. Thus, the methods of physical optics are applied. Lasers and their applications are discussed in the third section. Both the wave nature and particle nature of light are used in these discussions. The last section considers several optical instruments and spectroscopy. The eye, microscope, telescope, and Michelson interferometer are analyzed. In addition, there is a survey of various spectrometric instruments.

14.1

GEOMETRICAL OPTICS

OBJECTIVES

Upon completion of this section, the student should be able to

- Describe how light can be represented by light rays.

- Describe how shadows are formed from point sources and extended sources.

- State the law of reflection and illustrate it with a simple diagram.

- Define and illustrate specular reflection and diffuse reflection.

- Give three examples each of specular and diffuse reflectors.

- List the four processes that account for all light energy striking a surface.

- Explain in two or three sentences the purpose and operation of a beam splitter.

- Define:
  Real image
  Virtual image
  Focal point of a concave mirror
  Focal point of a convex mirror

- Draw and label ray diagrams that illustrate:
  Virtual image formed by a plane mirror
  Virtual image formed by a convex mirror
  Virtual image formed by a concave mirror
  Real image formed by a concave mirror

- Given two of the following quantities, calculate the third:
  Focal length of mirror
  Object distance
  Image distance

- Given three of the following quantities for a spherical mirror, calculate the fourth:
  Object distance

Image distance
Objective height
Image height

- Discuss the application of elliptic and parabolic mirrors.

- State the equation for the law of refraction and illustrate with a diagram.

- Given two of the following quantities, calculate the third:
  Speed of light in a vacuum
  Speed of light in a material
  Index of refraction of the material

- Given three of the following quantities, calculate the fourth:
  Angle of incidence
  Angle of refraction
  Index of refraction of first material
  Index of refraction of second material

- Explain total internal reflection in a short paragraph, including a prism as an example.

- Describe how a prism can be used to obtain the spectrum of a light source.

- Explain how a microscope can be used to measure the index of refraction of a liquid.

- Given two of the following quantities and the indices of refraction for a spherical refracting surface, calculate the third:
  The radius
  The object distance
  The image distance

- Draw an appropriate diagram that illustrates each of the definitions of image forming systems given below:
  The focal point of a positive lens
  The focal point of a negative lens
  The real image formed by a positive lens
  The virtual image formed by a positive lens
  The virtual image formed by a negative lens

- Give one practical application for each of the above optical arrangements.

- Given two of the following quantities, calculate the third:
  Focal length of lens
  Objective distance
  Image distance

- Given the two radii of curvative of a spherical lens and its index of refraction, calculate the focal length.

- Explain how a correction can be made for chromatic aberration.

- List four different thin lenses.

- Given three of the following quantities for a thin lens, calculate the fourth:
  Object distance
  Image distance
  Size of object
  Size of image

- Calculate the focal length of two thin lenses in contact given the focal length of each lens.

DISCUSSION

Geometrical optics is based on the principle of light traveling in a straight line. This characteristic gives rise to the properties that are discussed in this section.

Rays and Shadows

Rays   The idea of light rays is useful to understanding how light behaves in many everyday situations. For example, you have probably noticed how shadows appear to form directly behind objects—on the side away from the sun or other sources of light. Rather than curving behind the object it strikes, the light appears to be traveling in a straight line. Whenever light behaves in this manner, as if it is traveling in a straight line, we speak of light rays or ray optics. In science, this is called rectilinear propagation of light, and a ray is any line along which light travels. The rectilinear propagation, however, occurs only in isotropic media. In a nonisotropic medium, a ray may curve or it may bend when it travels from one medium to another. In general, a ray is always perpendicular to a wave front.

Shadows   Shadows are easily explained by using rays. Consider a point source and an opaque sphere as shown in Figure 14.1. The shadow on the screen is a black circle, and the region behind the sphere where no light exists is called the umbra. The radius of the observed circle agrees with the calculated radius based on the ray theory.

Figure 14.2 shows a sphere illuminated by a circular source that has a smaller radius than the radius of the sphere. Rays drawn from the top and bottom of the source that are tangent to the sphere are shown in the figure. A black circle is again observed, and the region behind the sphere where no light exists is called the

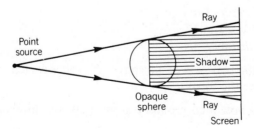

FIGURE 14.1 A shadow from a point source.

FIGURE 14.2 A shadow produced by a circular source smaller than the object.

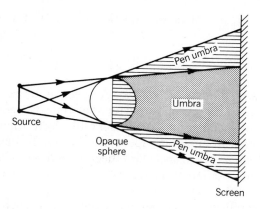

umbra. As one moves out from the radius of the black circle, the illumination increases until one reaches an outer circle determined by the outer rays shown on the diagram. The region behind the sphere where some light exists (region determined by the outer and inner rays) is called the penumbra.

When the radius of the source is greater than the radius of the sphere, an umbra and penumbra are also observed. This case is shown in Figure 14.3.

Reversibility of Light Rays If light travels along any ray from point *A* to point *B*, experiments show that the light will travel the same path if it goes from *B* to *A*. This principle called the reversibility of light rays. Thus, in any optical system the arrowheads may be reversed, and the optical diagram is valid.

The Reflection of Light

For many applications, it is necessary to change the direction of propagation of light. Reflection is one way to accomplish this. This section presents the basic laws of reflection that can be explained by the ray theory of light. Effects due to the polarization of light waves and the interference of light waves are discussed in Section 14.2.

Basic Law of Reflection Figure 14.4 shows a light ray reflected from a plane surface. The normal line is a line perpendicular to the optical surface at the point of reflection. All angles are measured from this line. The incident ray, reflected ray, and normal all lie in a plane called the plane of incidence. The law of reflection is given by

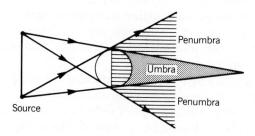

FIGURE 14.3 A shadow produced by a circular source larger than the object.

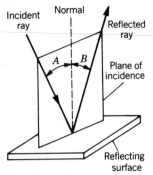

$\angle A$ = Angle of incidence
$\angle B$ = Angle of reflection

FIGURE 14.4   A light ray reflected from a planar surface.

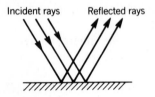

FIGURE 14.5   Specular reflection from a planar surface.

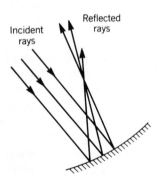

FIGURE 14.6   Specular reflection from a spherical surface.

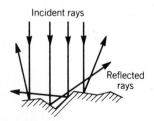

FIGURE 14.7   Diffuse reflection from an irregular surface.

Angle of reflection = Angle of incidence

or

$$\angle B = \angle A \qquad (14.1)$$

Specular and Diffuse Reflectors   A specular reflector is a smooth reflecting surface such as an ordinary mirror that reflects light according to the law of reflection and maintains internal order of the light beam. Figure 14.5 shows a light beam reflected from a plane specular reflector. In Figure 14.6 a similar beam is reflected from a curved specular reflector. In this case different parts of the beam are reflected at different angles but the internal order of the light rays is maintained after reflection.

A diffuse reflector is a rough reflecting surface, such as this sheet of paper, that reflects light randomly in all directions and does not maintain internal order in a light beam. Figure 14.7 shows reflection of light from such a surface. Each flat surface element of a diffuse reflector acts as a tiny mirror, and reflections from each element obey the law of reflection. Because of the random orientation flat surface elements, diffuse reflecting surfaces reflect light in all directions. Figure 14.8 shows the radiation pattern from a diffuse reflector when it is struck by a beam of light.

Many surfaces display a combination of the characteristics of specular and diffuse reflectors. Figure 14.9 shows the typical pattern of light reflected from a glossy painted surface. Although some diffuse reflection takes place, most of the light energy is found in specular reflection and this surface is classed as a specular reflector.

Partial and Total Reflection   As previously discussed in the chapter on waves, when light energy strikes the surface of a material part is reflected, part is absorbed, part is scattered, and part is transmitted. This is illustrated in Figure 14.10, and the expression for the conservation of energy is given by

$$\frac{\text{Incident}}{\text{Energy}} = \frac{\text{Reflected}}{\text{Energy}} + \frac{\text{Absorbed}}{\text{Energy}} + \frac{\text{Scattered}}{\text{Energy}} + \frac{\text{Transmitted}}{\text{Energy}}$$

$$E_I = E_R + E_A + E_S + E_T \qquad (14.22)$$

Some light energy always goes into each of the four processes. Scattered light is undesirable in most optical systems and can be reduced to near zero intensity (0.01%) by polishing the optical surfaces and keeping them clean. Clean glass surfaces scatter so little light that scattering usually can be ignored. Absorption also reduces the intensity of the transmitted beam since it removes light energy from the incident beam. Ordinary bathroom mirrors absorb 10 to 15% of the light that strikes them. In many highly transmitting optical components, absorption is reduced to almost zero. Ordinary glass windows absorb very little light. In windows maximum transmission is desirable, whereas in other applications no light is transmitted.

A portion of the light is always reflected. Example 11.14 in Chapter 11 shows that the reflected intensity depends on the velocity of the transmitted wave for a dielectric material. Ordinary glass reflects about 4% of the light that strikes it at each of its surfaces. Figure 14.11 shows light striking a rear surface mirror. About 4% of the light is reflected by the glass surface at point *A*. The main reflected beam

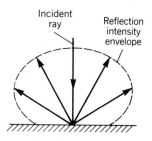

FIGURE 14.8 Pattern of reflected intensity from a diffuse reflector.

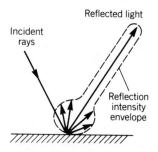

FIGURE 14.9 Pattern of reflected intensity from a combination of diffuse and specular reflection.

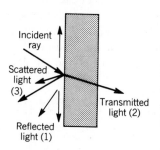

FIGURE 14.10 Incident light can be reflected, scattered, transmitted, or absorbed.

leaves the glass at point *B*, with approximately 92% of the original beam. About 3.8% of the original beam is reflected back into the glass at point *B*. Most of this light passes out of the glass after the next reflection. The two weaker reflected rays often form ghost images in mirrors. This can be eliminated by applying an aluminum reflective coating to the front surface of the mirror, but this also exposes the reflective coating to possible damage.

Dielectric coatings are often applied to optical surfaces to change their reflectivity. The wave theory of light must be used to explain why this can be done. By making the coating the proper thickness, highly reflecting mirrors can be made. For example, some laser mirrors may reflect as much as 99.95% of the incident light. Camera and telescope lenses often have antireflection coating to reduce the reflected energy to less than 1%.

In many applications it is desirable to reflect a portion of the light while allowing the remainder to pass through the mirror. This is accomplished with a partially reflective optical coating on a transparent medium. The mirror is called a beam splitter. Beam splitters are used in many optical instruments such as interferometers, lasers, and rangefinders.

IMAGE FORMATION WITH MIRRORS

The basic mirror surfaces are plane, spherical, parabolic, and elliptical. Spherical mirrors may be convex or concave. Mirrors can be used to focus light to a small area, collimate light, form images, or change the direction of a beam of light. This section discusses the reflection of light from these surfaces and the corresponding images formed by these reflections.

Plane Mirror Everyone has observed images formed by a plane mirror. The image of an object lies behind the mirror, and the right-hand side of the image corresponds to the left-hand side of the object. In addition, the image is upright and the same size as the object. Since there is no light behind the mirror, the observed image is called a virtual image. Virtual images cannot be seen on a screen.

The location of an image formed by a plane mirror can be determined from the laws of reflection. Figure 14.12 shows the image of a point located at a distance *q*

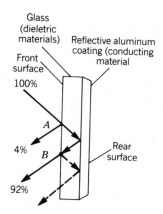

FIGURE 14.11 Partial reflection of a light beam occurs at a glass surface.

FIGURE 14.12 Image formed by a plane mirror.

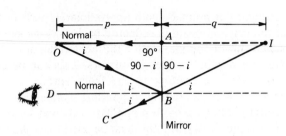

behind the mirror. For this point to be a true image, the value of q must be independent of the angle i. Thus all rays must look as if they came from the same point as shown in Figure 14.13. In Figure 14.12 the ray OA is perpendicular to the plane formed by the mirror surface. According to the law of reflection, this ray is reflected back along itself. In addition, angle OBD must equal angle DBC. This angle is also shown as i in the diagram:

$$\tan i = \frac{AB}{p} = \frac{AB}{q} \tag{14.3}$$

Thus, $p = q$ for all possible angles for i, and all the reflected rays look as if they came from the image point I. The image of point O lies on a perpendicular drawn from point O to the mirror, and it is as far behind the mirror as the image is in front of the mirror.

To find the image of an extended object, one can find the image of each point on the object. For example, the image of a line is a line. Figure 14.14 can be used to show this. Consider the image of the line AE. The image of point E is point F. Thus, the distance EC must equal CF. Also, since $\tan \theta_1 = EC/AC$ and $\tan \theta_2 = CF/AC$, the angle θ_1 must equal the angle θ_2. If the point G on the line AF is an image of point D, then DB must equal BG. However,

$$\tan \theta_2 = \frac{BG}{AB} \tag{14.4}$$

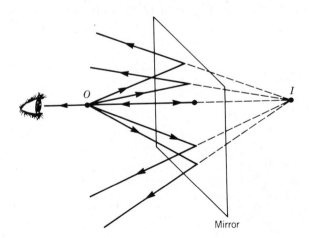

FIGURE 14.13 Rays from object to observer in an image from a plane mirror.

FIGURE 14.14 Image of a line.

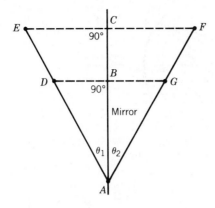

and

$$\tan \theta_2 = \tan \theta_1 = \frac{DB}{AB} \tag{14.5}$$

Thus, $DB = BG$ and G is the image of point D. Since point D can be any point on the line AE, the image of each point must lie on the line AF. The image of a line is a line. Therefore, only two points are required to find the image of a line.

Concave Spherical Mirror   Consider a spherical surface that is part of a sphere of radius R. If a line is drawn through the center of the sphere and the midpoint on the spherical surface, it is called the optical axis. The optical axis is illustrated in Figure 14.15. Light rays incident on a mirror will always be taken as coming from the left to right. Thus, the mirror shown is a concave mirror. The relationship between a point object on the optical axis and its image on the optical axis can be easily determined if all rays form small angles with the optical axis so that the angles are approximately equal to their tangents.

A point object on the optical axis is shown in Figure 14.16. A ray emanating from this point making an angle α with the optical axis is incident on the mirror at point A. A radius is also shown drawn from the center to point A. The radius is perpendicular to the mirror surface. The ray OA, the ray AI, and the radius CA all lie in the same plane. In addition, the angle of incidence is equal to the angle of reflection as shown in the figure. If an image of point O is obtained, all rays coming from point O that are incident on the mirror must pass through point I.

To show that an image of O is formed at point I, consider the following equation that can be obtained from Figure 14.16. If the angles are small, $p \gg \delta$, $R \gg \delta$, and $q \gg \delta$:

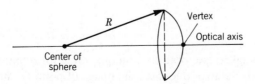

FIGURE 14.15 Optical axis of a spherical surface.

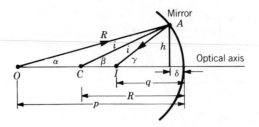

$$\tan \alpha = \frac{h}{p - \delta} \approx \frac{h}{p} \tag{14.6}$$

$$\tan \beta = \frac{h}{R - \delta} \approx \frac{h}{R} \tag{14.7}$$

$$\tan \gamma = \frac{h}{q - \delta} \approx \frac{h}{q} \tag{14.8}$$

In addition, the angles are approximately equal to their tangents. Thus, to a good approximation

$$\alpha = \frac{h}{p} \tag{14.9}$$

$$\beta = \frac{h}{R} \tag{14.10}$$

$$\gamma = \frac{h}{q} \tag{14.11}$$

From plane geometry,

$$\alpha + i = \beta$$

and

$$\beta + i = \gamma$$

Subtracting the last two equations yields

$$\alpha - \beta = \beta - \gamma \tag{14.12}$$

or

$$\alpha + \gamma = 2\beta \tag{14.13}$$

Substituting for α, β, and γ from Equations 14.9, 14.10, and 14.11 yields

$$\frac{h}{p} + \frac{h}{q} = \frac{2h}{R} \tag{14.14}$$

or

$$\frac{1}{p} + \frac{1}{q} = \frac{2}{R} \tag{14.15}$$

For small angles, one obtains a unique value for q that is independent of the angle α. Thus, this value of q locates the image of point O. Various rays emanating from

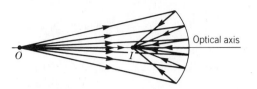

FIGURE 14.17 Rays from point O focus at point I.

point *O* and focused at point *I* are shown in Figure 14.17. A sample problem is solved in Example 14.1.

As α increases, different values of *q* are obtained. Thus, the image of a point is no longer a point. This effect is called spherical aberration. This error is illustrated in Example 14.2.

EXAMPLE 14.1 The image position of a point object for a spherical concave mirror
. . . .

Given: A point object lies on the optical axis 70 cm from the vertex of a concave spherical mirror. The radius of the spherical surface is 40 cm.

Find: The location of the point image.

Solution: The equation that relates the image distance to the object distance is

$$\frac{1}{p} + \frac{1}{q} = \frac{2}{R}$$

$$\frac{1}{70 \text{ cm}} + \frac{1}{q} = \frac{2}{40 \text{ cm}}$$

multiply both sides of the equation by (70 cm) (40 cm) (*q*).

$$\frac{(70 \text{ cm})(40 \text{ cm})q}{70 \text{ cm}} + \frac{(70 \text{ cm})(40 \text{ cm})q}{q} = \frac{(2)(70 \text{ cm})(40 \text{ cm})q}{40 \text{ cm}}$$

$$(40 \text{ cm})q + 2800 \text{ cm}^2 = (140 \text{ cm})q$$

$$2800 \text{ cm}^2 = (100 \text{ cm})q$$

$$\frac{2800 \text{ cm}^2}{100 \text{ cm}} = q$$

$$28 \text{ cm} = q$$

EXAMPLE 14.2 Image distance error for a spherical mirror
. . . .

Given: The object distance to a spherical, concave mirror is 20.0 cm. The radius is 10.0 cm. A ray from a point object on the optical axis is incident on the mirror. A radius drawn to the point where the ray is incident on the mirror makes an angle of 15° with the optical axis.

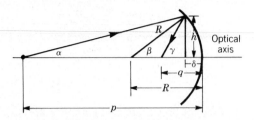

Find: **a.** The distance from the vertex where the reflected ray intersects the optical axis.

b. The approximate intersection point using the spherical mirror equation.

Solution: **a.** Draw a diagram.

$$\beta = 15°, p = 20.0 \text{ cm}, R = 10.0 \text{ cm}$$

From the diagram

$$\delta = R - R \cos \beta$$

$$\delta = 10.0 \text{ cm} - 10.0 \text{ cm} \cos 15°$$

$$\delta = 10.0 \text{ cm} - 9.66 \text{ cm}$$

$$\delta = 0.34 \text{ cm}$$

From the definition of the sine of an angle

$$\sin \beta = \frac{h}{R}$$

$$h = R \sin \beta$$

$$h = (10.0 \text{ cm})(\sin 15°)$$

$$h = 2.59 \text{ cm}$$

From the definition of the tangent of the angle

$$\tan \alpha = \frac{h}{p - \delta}$$

$$\tan \alpha = \frac{2.59 \text{ cm}}{20.0 \text{ cm} - 0.34 \text{ cm}}$$

$$\tan \alpha = 0.132$$

$$\alpha = 7.5°$$

From Equation 14.13

$$\alpha + \gamma = 2 \beta$$

$$\gamma = 2 \beta - \alpha$$

$$\gamma = (2)(15°) - 7.5°$$

$$\gamma = 22.5°$$

From the diagram

$$\tan \gamma = \frac{h}{q - \delta}$$

$$q - \delta = \frac{h}{\tan \gamma}$$

$$q = \frac{h}{\tan \gamma} + \delta$$

$$q = \frac{2.59 \text{ cm}}{\tan 22.5°} + 0.34 \text{ cm}$$

$$q = 6.59 \text{ cm}$$

b. The approximate equation for concave spherical mirrors is

$$\frac{1}{p} + \frac{1}{q} = \frac{2}{R}$$

$$\frac{1}{20.0 \text{ cm}} + \frac{1}{q} = \frac{2}{10.0 \text{ cm}}$$

$$\frac{(10.0 \text{ cm})(20.0 \text{ cm})q}{20.0 \text{ cm}} + \frac{(10.0 \text{ cm})(20.0 \text{ cm})q}{q} = \frac{(2)(10.0 \text{ cm})(20.0 \text{ cm})q}{10.0 \text{ cm}}$$

$$(10.0 \text{ cm})q + 200.0 \text{ cm}^2 = (40.0 \text{ cm})q$$

$$200.0 \text{ cm}^2 = (30 \text{ cm})q$$

$$6.67 \text{ cm} = q$$

$$\% \text{ error} = \frac{6.67 - 6.59}{6.59} \times 100$$

$$\% \text{ error} = 1.2\%$$

If p in Equation 14.15 becomes very large, the quantity $1/p$ approaches zero. In addition, the rays coming from the object become almost parallel to the optical axis. The corresponding value of q approaches a quantity called the focal length, which is given by

$$\frac{1}{q} = \frac{2}{R}$$

or

$$q = R/2 = f \qquad\qquad (14.16)$$

where

f = focal length

Equation 14.15 is often written as

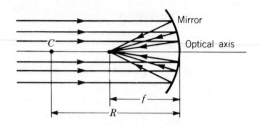

FIGURE 14.18   Parallel rays are focused at the focal point by a spherical concave mirror.

$$\frac{1}{p} + \frac{1}{q} = \frac{1}{f} \qquad (14.17)$$

A diagram illustrating parallel rays being focused to the focal point is shown in Figure 14.18.

Since light rays are reversible, an object placed at the focal point will produce a beam of parallel light. Thus, concave spherical mirrors can be used as collimators in optical instruments. However, it will be shown that it is more advantageous to use parabolic mirrors for this purpose.

The knowledge that a ray parallel to the optical axis will pass through the focal point and that a ray passing through the focal point will be reflected parallel to the optical axis can be used to locate an image of an off-axis point. This is illustrated in Figure 14.19. The parallel ray $O'A$ is reflected back through the focal point. Ray $O'B$ passes through the focal point, and it is reflected back parallel to the optical axis. The image of point O' is located at the intersection of these two rays. In addition, the ray $O'V$ must be reflected so that the incident angle is equal to the reflected angle. This is also illustrated in Figure 14.19. This ray also passes through the image point. All other rays coming from the point O', which are incident on the mirror, will also approximately pass through the image point if the angles are small. In general, the image of any point on the line OO' will approximately lie on the image line II'.

Since the angles in the triangle $O'VO$ are equal to the corresponding angles in the triangle $I'VI$, the two triangles are similar, and their corresponding sides are proportional. Thus,

$$\frac{I'I}{O'O} = \frac{IV}{OV} \qquad (14.18)$$

However, $I'I$ is the size of the image and $O'O$ is the size of the object. Also, $IV = q$ and $OV = p$.

Thus, Equation 14.18 may be written as

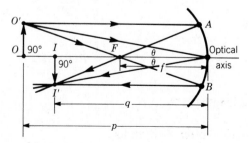

FIGURE 14.19   Ray tracing to locate the image of a point that is off the optical axis.

$$m = \frac{S_I}{S_O} = \frac{q}{p} \qquad (14.19)$$

where

m = magnification
S_I = size of the image
S_O = size of the object
q = image distance
p = object distance

A sample problem is given in Example 3.

EXAMPLE 14.3 Locating the image formed by a concave mirror
. . . .

Given: A 0.8-m long line object is perpendicular to the optical axis of a concave mirror. It is located 4.0 m from the vertex. The focal length is
1.0 m.

Find: a. The image distance by constructing an optical diagram.
   b. The magnification from the optical diagram.
   c. The image distance using Equation 14.17.
   d. The magnification using Equation 14.19.

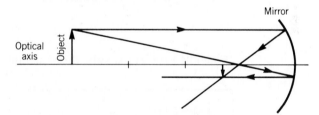

Solution: a. Draw the mirror and optical axis. Locate the object line and the
focal point. Draw a ray from the top of the object parallel to the
optical axis. The reflected ray passes through the focal point.
Draw a ray from the top of the object through the focal point.
This reflected ray is parallel to the optical axis.

   b. Measure the image distance and the length of the image. Calculate the magnification.

$$m \approx \frac{0.25}{0.8} = 0.31$$

$$q \approx 1.32 \text{ m}$$

$$S_I = 0.25 \text{ m}$$

   c. From Equation 14.17

$$\frac{1}{p} + \frac{1}{q} = \frac{1}{f}$$

$$\frac{1}{4} + \frac{1}{q} = \frac{1}{1}$$

$$3q = 4$$

$$q = 4/3 = 1.33\text{m}$$

d. From Equation 14.19

$$m = \frac{|q|}{|p|}$$

$$m = \frac{1.33}{4.0} = 0.33$$

Up to this point, the object distance has been greater than the focal length, giving an image that is real and inverted. Under these conditions, Equation 14.17 gives a positive value for q and the image is on the left-hand side of the mirror. Now, consider the optical diagram shown in Figure 14.20. The object distance is less than the focal length. For this case, the rays look as if they intersect behind the mirror. The image is virtual and not inverted.

It can be shown that Equation 14.17 gives the correct value for q, but the value will be negative. A negative value for q means that the image lies on the right-hand side of the mirror. The student can easily show that the magnification is given by

$$m = \frac{q}{p} \tag{14.20}$$

A sample problem is solved in Example 14.4.

EXAMPLE 14.4   Calculation of the image distance for a concave mirror
. . . .

Given: The object distance for a concave mirror is 0.50 m and the focal length is 1.0 m.

Find: The image distance.

Solution: Equation 14.17 is

$$\frac{1}{p} + \frac{1}{q} = \frac{1}{f}$$

$$\frac{1}{0.50 \text{ m}} + \frac{1}{q} = \frac{1}{1.0 \text{ m}}$$

$$(1.0 \text{ m})q + 0.50 \text{ m}^2 = (0.50 \text{ m})q$$

$$(0.50 \text{ m})q = 0.50 \text{ m}^2$$

$$q = -1.0 \text{ m}$$

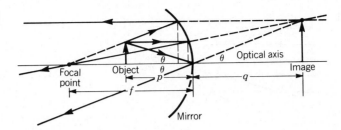

FIGURE 14.20 Formation of a virtual, erect image by a spherical concave mirror.

In summary, as the object moves from a very large distance to the center of a concave mirror, the image moves from the focal point to the center. As the object moves from the center to the focal point, the image moves from the center to a very large distance. The image is real and inverted. When the object moves from the focal point to the vertex, the image moves from a very large negative distance to the vertex. The image is virtual and not inverted. To obtain a good image, however, it is necessary to keep the angles the rays make with the optical axis small.

Convex Spherical Mirrors   An optical diagram for a convex mirror is shown in Figure 14.21.

The same mathematical approach that was used for the concave mirror gives

$$\frac{1}{p} + \frac{1}{q} = \frac{2}{R} \tag{14.21}$$

If the same sign convention is used for concave mirrors, then q and R are negative numbers (both lie on the right-hand side of the mirror). Equation 14.21 can then be written as

$$\frac{1}{p} + \frac{1}{q} = \frac{2}{R} = \frac{1}{f} \tag{14.22}$$

where

q and R are negative numbers

The image is virtual since there is no light on the right-hand side of the mirror.

If p approaches a large number, $1/p$ approaches zero and q approaches $R/2$. The focal length is again defined as $R/2$ but it is a negative number. The focal point lies on the right-hand side of the mirror. Figure 14.22 shows parallel rays (p approaches infinity) incident on a convex mirror. All the rays look as if they came from the focal point. Since light rays are reversible, rays coming from the left that would pass through the focal point if extended beyond the mirror are reflected back parallel to the optical axis.

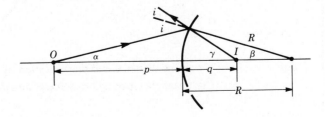

FIGURE 14.21 Optical diagram of a spherical convex mirror.

FIGURE 14.22 Reflection of rays parallel to the optical axis by a convex spherical mirror.

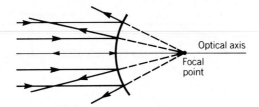

Optical axis

Focal point

If Equation 14.22 is solved for q, one obtains

$$q = \frac{-fp}{f - p} \tag{14.23}$$

Since f is a negative number and p is a positive number, the numerator is always positive and the denominator is always negative. Thus, q will always be negative for all values of p. The image is always a virtual image, and it will lie on the right-hand side of the mirror. For very large values of p, the image lies at the focal point. As p becomes smaller, q approaches the surface of the mirror. When p is zero, q is zero.

The method used to construct a line image of a line object can also be used for convex mirrors. The magnification is also given by

$$m = \frac{|q|}{|p|} \tag{14.24}$$

A sample problem is solved in Example 14.5.

EXAMPLE 14.5   Locating the image formed by a convex mirror
. . . .

Given: A 0.8-m-long line object is perpendicular to the optical axis of a convex mirror. It is located 3.2 m from the vertex. The focal length is -1.0 m.

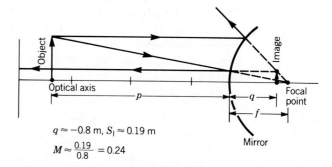

Object

Image

Optical axis

p

q

Focal point

f

Mirror

$q \approx -0.8$ m, $S_I \approx 0.19$ m

$M \approx \frac{0.19}{0.8} = 0.24$

Find: **a.** The image distance and magnification by construction.

               b. The image distance and magnification using Equations 14.22 and 14.24.

Solution: **a.** Draw a mirror and optical axis. Draw the object line and locate the focal point. Draw a ray from the top of the object parallel to the optical axis. The reflected ray looks as if it comes from the focal point. Draw a ray from the top of the object such that it would pass through the focal point. This ray is reflected back parallel to the optical axis. The extension of these two rays intersects behind the mirror at the top of the image. Measure the image distance and the length of the image. Calculate the magnification.

$$q \approx -0.8 \text{ m}$$

$$S_1 \approx 0.2 \text{ m}$$

$$m \approx \frac{0.2 \text{ m}}{0.8 \text{ m}} = 0.25$$

b. The image distance can be obtained from Equation 14.22.

$$\frac{1}{p} + \frac{1}{q} = \frac{1}{f}$$

$$\frac{1}{3.2 \text{ m}} + \frac{1}{q} = \frac{1}{-1.0 \text{ m}}$$

$$\frac{(3.2 \text{ m})(-1.0 \text{ m})q}{3.2 \text{ m}} + \frac{(3.2 \text{ m})(-1.0 \text{ m})q}{q} = \frac{(3.2 \text{ m})(-1.0 \text{ m})q}{-1.0 \text{ m}}$$

$$(-1.0 \text{ m})q + (-3.2 \text{ m}^2) = (3.2 \text{ m})q$$

$$-3.2 \text{ m}^2 = (4.2 \text{ m})q$$

$$-0.76 \text{ m} = q$$

The magnification is (Equation 14.24)

$$m = \frac{|q|}{|p|}$$

$$m = \frac{0.76 \text{ m}}{3.2 \text{ m}}$$

$$m = 0.24$$

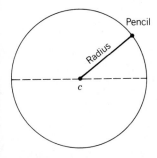

Pencil
Radius
c

FIGURE 14.23 String used to draw a circle.

Elliptical Mirrors A circle has a center and a constant radius. To draw a circle, one can use a string having a length equal to twice the radius. Both ends of the string are fixed at the center, and a pencil is placed at the end of the loop as shown in Figure 14.23. A circle is generated by moving the pencil while keeping the string taut. If the circle is then rotated about a diameter, a spherical surface is generated. Light coming from the center will be focused back to the center.

If the ends of the string are placed at two different points, and the pencil is moved with the string taut, an ellipse is generated as shown in Figure 14.24. The fixed points at the end of the string are called the focal points. A line drawn through the two points is called an axis. Rotating the ellipse about this axis generates an ellipsoid of revolution. Any part of this surface can be used as an

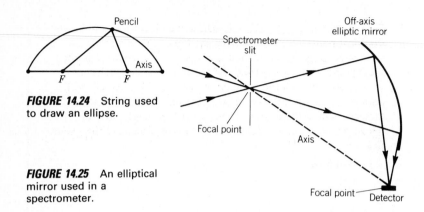

FIGURE 14.24 String used to draw an ellipse.

FIGURE 14.25 An elliptical mirror used in a spectrometer.

elliptic mirror. It can be shown that light emanating from one focal point that intersects the mirror surface will be reflected back through the second focal point. Thus, an elliptic mirror can be used to focus a point source to another point even though the angles are large. A portion of the surface used as a mirror that does not include the axis is called an off-axis elliptical mirror. For example, light passing through the exit slit of an infrared spectrometer must often be focused onto a detector with a small sensitive area. This can be accomplished by using an off-axis elliptical mirror as shown in Figure 14.25.

Parabolic Mirrors   Consider the elliptic mirror shown in Figure 14.26. If the distance between the focal points is very large, the angle α will be small and the rays coming from F_1 will be almost parallel to the axis. The shape of the elliptical mirror will almost be a paraboloid of revolution. A true paraboloid of revolution will focus all rays parallel to the axis at a focal point as shown in Figure 14.27. Since light rays are reversible, rays coming from the focal point will form a beam parallel to the optical axis. Because of these properties, parabolic mirrors are used in astronomical telescopes to form images of stars and in optical instruments to form parallel beams from small light sources. In spectrometric instruments, off-axis parabolic mirrors are often used to displace the parallel beam from the source and an entrance slit. This application is illustrated in Figure 14.28.

FIGURE 14.26 Optical diagram of a parabolic mirror.

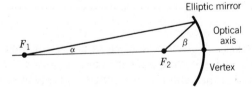

FIGURE 14.27 Parallel beam formed by a parabolic mirror.

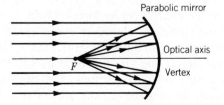

FIGURE 14.28 Use of parabolic mirrors in a spectrometer.

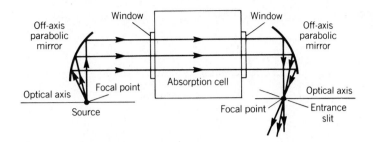

Applications Mirrors are employed in a wide variety of optical instruments including telescopes, interferometers, spectrometers, and lasers. Mirrors are also an essential part of many more common devices. Table 14.1 lists several applications along with information on the image distance and the type and size of the image. The equation

$$\frac{1}{p} + \frac{1}{q} = \frac{1}{f}$$

(14.25)

applies for all cases if the focal length of a plane mirror is taken as infinity so that $1/f = 0$.

TABLE 14.1 Properties and application of mirrors

Mirror Type	Object Distance	Image Distance	Image Type/Size	Application
Plane	Any	$q = -p$ (as far behind as object is in front) $f = \infty$	Virtual/same size as object	Ordinary mirror
Convex	Any	$|q| < |f|$ f is negative q is negative	Virtual/reduced	Wide-angle rear vision mirror Antitheft mirror in store
Concave	$p = \infty$	$q = f$ f is positive q is positive	Real/reduced to point	Solar furnace (parallel incident rays)
Concave	$p = f$	$q = \infty$ f is positive q is positive	None	Spotlight (parallel reflected rays)
Concave	$p < f$	$q < 0$ f is positive q is negative	Virtual/enlarged	Magnifying mirror
Concave	$f \leq p \leq 2f$	$q \geq 2f$ f is positive q is positive	Real/enlarged	Image projection system of large screen television
Concave	$p > 2f$	$2f > q \geq f$ f is positive q is positive	Real/reduced	Reflecting telescope

FIGURE 14.29 The change
in direction of a ray is
caused by refraction.

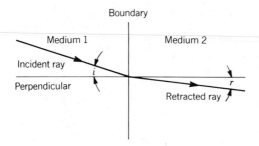

FIGURE 14.29 The change
in direction of a ray is
caused by refraction.

The Refraction of Light

Law of Refraction When light rays travel from one medium to another, the direction of the ray can be abruptly changed at the boundary of the two media. A simple experiment can be performed to obtain the law that relates the direction of the incident ray to the direction of the refracted ray. It is observed that an incident ray perpendicular to the boundary will be transmitted without changing its direction. When the incident ray is not perpendicular to the boundary, the angle between the incident ray and the perpendicular is called the angle of incidence; and the angle between the transmitted ray and the perpendicular is called the angle of refraction. These angles are shown in Figure 14.29. Observations also show that the incident ray, refracted ray, and perpendicular all lie in the same plane.

To study the relationship between the incident angle and the refracted angle, a transparent rod with a semicircular cross section can be used. A perpendicular is drawn from the center as shown in Figure 14.30. A pin C is placed at the center of the semicircle and two additional pins, A and B, are aligned with pin C to establish an angle of incidence. The refracted ray will strike the second surface in the direction of a perpendicular (radius). Thus, the ray will not change direction at this surface. The direction of the refracted ray can be established by aligning pins D and E with pins A, B, and C. Thus, the refracted angle can be measured for various angles of incidence. An analysis of these data show that (Snell's law)

$$\frac{\sin i_1}{\sin r_2} = n_{12} \tag{14.26}$$

where

$$i_1 = \text{angle of incidence}$$
$$r_2 = \text{angle of refraction}$$

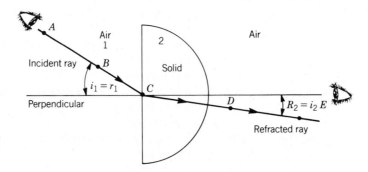

FIGURE 14.30 Relationship
between the incident angle
and the refracted angle.

TABLE 14.2 Indices of refraction at 589 nm and 20°C

Material	Index of Refraction
Water	1.33
Ethyl alcohol	1.36
Dry air	1.00029
Crown glass	1.52
Dense flint glass	1.66
Fused quartz	1.46
Diamond	2.42

n_{12} = a constant for a given wavelength, called the index of refraction of medium 2 relative to medium 1

If the experiment is done with light of different wavelengths, it is found that the index of refraction depends on the wavelength as well as the kind of material and the temperature. The index of refraction for several different materials relative to a vacuum is given in Table 14.2. Since light rays are reversible, one may also write, from Figure 14.29,

$$\frac{\sin i_2}{\sin r_1} = n_{21} \tag{14.27}$$

where

n_{21} = the index of refraction of medium 1 relative to medium 2.

Multiplying n_{21} by n_{12} yields

$$n_{21} n_{12} = \left(\frac{\sin i_2}{\sin r_1}\right)\left(\frac{\sin i_1}{\sin r_2}\right) \tag{14.28}$$

However, from Figure 14.29, $i_1 = r_1$ and $i_2 = r_2$. Thus,

$$n_{21} n_{12} = 1$$

or

$$n_{21} = \frac{1}{n_{12}} \tag{14.29}$$

Experiments also show that

$$n_{12} = \frac{v_1}{v_2} \tag{14.30}$$

where

v_1 = velocity of light in medium 1
v_2 = velocity of light in medium 2

Equation 14.30 is derived in this chapter by using the wave nature of light. The absolute index of refraction of light is obtained when medium 1 is a vacuum. Thus, from Equation 14.30

$$n_1 = \frac{c}{v_1} \qquad (14.31)$$

and

$$n_2 = \frac{c}{v_2} \qquad (14.32)$$

where

n_1 = absolute index of refraction of medium 1
n_2 = absolute index of refraction of medium 2
c = velocity of light in a vacuum

The ratio of n_2 to n_1 becomes

$$\frac{n_2}{n_1} = \frac{c/v_2}{c/v_1} = \frac{v_1}{v_2} = n_{12} \qquad (14.33)$$

However,

$$n_{12} = \frac{\sin i_1}{\sin r_2} \qquad (14.34)$$

Thus, by substitution, one obtains

$$\frac{n_2}{n_1} = \frac{\sin i_1}{\sin r_2} \qquad (14.35)$$

or

$$n_1 \sin i_1 = n_2 \sin r_2 \qquad (14.36)$$

Example 14.6 gives an application of Equation 14.36.

EXAMPLE 14.6   The index of refraction
. . . .

Given: A light ray coming from water makes an angle of 40° with a perpendicular to the water surface. The second medium is air.

Find: The angle of refraction.

Solution: The law of refraction gives

$$n_1 \sin i_1 = n_2 \sin r_2$$

$$n_2 \approx 1.00 \text{ (air)}$$

$$n_1 \approx 1.33 \text{ (water)}$$

$$1.33 \sin 40° = 1.00 \sin r_2$$

$$\sin r_2 = 0.8549$$

$$r_2 = 58.7°$$

What would happen if $i_1 = 60°$?

Total Reflection When a light ray travels from one medium to another with a smaller absolute index of refraction, the ray bends away from the perpendicular to the boundary. For this case, the angle of refraction is greater than the angle of incidence. This is illustrated in Example 14.6. Since 90° is the largest possible angle of refraction, a maximum angle of incidence exists if the incident ray travels into the second medium. If the angle of incidence exceeds this value, total reflection occurs at the boundary. The angle at which total reflection begins is called the critical angle. Its value can be obtained from Equation 14.36 when $n_1 > n_2$:

$$n_1 \sin i_1 = n_2 \sin r_2 \tag{14.37}$$

or

$$\sin i_1 = \frac{n_2}{n_1} \sin r_2 \tag{14.38}$$

To find the critical angle, r_2 is equal to 90°. Thus, $\sin r_2 = 1$ and

$$\sin i_1 = \frac{n_2}{n_1} \tag{14.39}$$

An application of Equation 14.39 is given in Example 14.7.

EXAMPLE 14.7 Total reflection

Given: Light rays travel from dense flint glass to air and from water to air.

Find: The critical angles.

Solution: The equation for the critical angle is

$$\sin i_1 = \frac{n_2}{n_1}$$

For rays traveling from dense flint glass to air,

$$\sin i_1 = \frac{1.00}{1.66} = 0.602$$

$$i_1 = 37.0°$$

For rays traveling from water to air

$$\sin i_1 = \frac{1.00}{1.33} = 0.752$$

$$\sin i_1 = 48.8°$$

FIGURE 14.31   Optical
diagram of a prism.

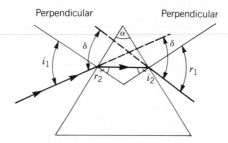

FIGURE 14.31   Optical
diagram of a prism.

The Prism   A prism is formed by a transparent medium between two plane
surfaces (called the faces) that are not parallel. The edge of the prism is the line
formed by the intersection of the two planes. The dihedral angle between the
planes is called the refracting angle, and it will be given the symbol α. Figure 14.31
shows the path of a ray that lies in a plane perpendicular to the edge. The angle δ is
the angle of deviation. It is the angle between the direction of the incident ray and
the direction of the ray as it exits the prism. The maximum value of r_1 is 90°. If the
corresponding value of i_1 is made larger, total reflection occurs at the second face.

Because light rays are reversible, the same deviation will occur if i_1 is replaced
by r_1. Thus, there are two angles of incidence that will give the same angle of
deviation. A plot of δ versus i_1 is shown in Figure 14.32. At the minimum point,
however, only one angle of incidence occurs. For this case $i_1 = r_1$ and the ray
travels symmetrically through the prism as shown in Figure 14.33.

The angle of minimum deviation, δ_m, can be easily measured. The corresponding values of i_1 and r_2 can be expressed in terms of the refracting angle and the
angle of minimum deviation. Snell's law can then be used to calculate the index of
refraction of the prism material. To show this, consider Figure 14.33. Since the
sum of the angles of a quadrilateral equals 360°, one may write for the quadrilateral *ABCD*

$$\alpha + 90 + \gamma + 90 = 360 \qquad (14.40)$$

or

$$\alpha = 180 - \gamma \qquad (14.41)$$

For the quadrilateral *AECD*, one obtains

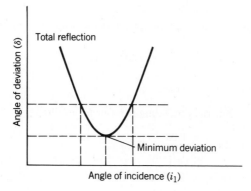

FIGURE 14.32   Plot of angle
of deviation versus angle of
incidence.

FIGURE 14.33 Optical diagram for a prism showing the minimum angle of deviation.

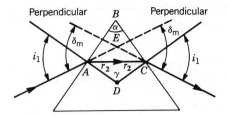

$$i_1 + (180 - \delta_m) + i_1 + \gamma = 360 \qquad (14.42)$$

or

$$2i_1 = 180 - \gamma + \delta_m \qquad (14.43)$$

Substituting for $180 - \gamma$ from Equation 14.41 yields

$$i_1 = \frac{\alpha + \delta_m}{2} \qquad (14.44)$$

Since the sum of the angles of a triangle is 180°, one may write for the triangle *ACD*

$$2r_2 + \gamma = 180 \qquad (14.45)$$

or

$$r_2 = \frac{180 - \gamma}{2} \qquad (14.46)$$

However, $180 - \gamma = \alpha$. This equation 1.46 becomes

$$r_2 = \frac{\alpha}{2} \qquad (14.47)$$

The angle of incidence and the angle of refraction are now expressed in terms of the angle of minimum deviation and the refracting angle of the prism. From Equation 14.36, one can write

$$n_2 = n_1 \frac{\sin i_1}{\sin r_2} \qquad (14.48)$$

Substituting for i_1 and r_2 from Equations 14.44 and 14.47 yields

$$n_2 = n_1 \left[\frac{\sin \left(\frac{\alpha + \delta_m}{2} \right)}{\sin \left(\frac{\alpha}{2} \right)} \right] \qquad (14.49)$$

Equation 14.49 gives the absolute index of refraction of the prism material in terms of α and δ_m. The quantities α and δ_m can be measured with a small limit of error. Thus, the application of Equation 14.49 gives a very accurate way to measure the index of refraction of a material. This is illustrated in Example 14.8.

EXAMPLE 14.8   The index of refraction from the angle of minimum deviation
· ·

Given: A laser beam is incident on the face of a prism and it lies in a plane that is perpendicular to the edge. The refracting angle of the prism is 30.00°. The prism is rotated until the angle of minimum deviation is observed. The angle of minimum deviation is determined from the values of x and y shown in the diagram.

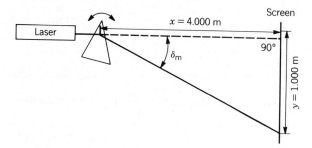

Find: The absolute index of refraction of the prism material.

Solution: The index of refraction is given by

$$n_2 = n_1 \left[\frac{\sin \left(\frac{\alpha + \delta_m}{2} \right)}{\sin \left(\frac{\alpha}{2} \right)} \right]$$

From the diagram

$$\tan \delta_m = \frac{1.000 \text{ m}}{4.000 \text{ m}} = 0.2500$$

$$\delta_m = 14.04°$$

Thus,

$$n_2 = (1.0003) \left[\frac{\sin \left(\frac{30.00° + 14.04°}{2} \right)}{\sin \left(\frac{30.00°}{2} \right)} \right]$$

$$n_2 = (1.0003) \left[\frac{\sin 22.02°}{\sin 15°} \right]$$

$$n_2 = (1.0003) \left[\frac{0.37493}{0.25882} \right]$$

$$n_2 = 1.449$$

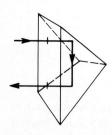

FIGURE 14.34   Reflecting prism.

If the angle of incidence at the face of a prism (going from glass to air) exceeds the critical angle, all the light is reflected according to the laws of reflection. Figure 14.34 shows a reflecting prism. These prisms are often used in optical

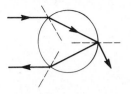

FIGURE 14.35 Reflection from a sphere.

systems to reflect light. The plastic reflectors along the side of a road and on automobiles are composed of many small prisms that reflect light back in the direction from which it came. Road signs are seen easily at night because they are covered with tiny glass beads that serve the same function. Figure 14.35 shows light reflected by such a sphere. In this case, the angle of incidence is less than the critical angle, and some of the light is refracted out the rear of the sphere.

Dispersion Figure 14.36 shows how a prism separates white light into the visible spectrum by a process called "dispersion." This phenomenon occurs because the index of refraction of materials varies with the wavelength of light, which indicates that different colors (wavelengths) of light have different speeds through a material.

Figure 14.37 shows dispersion curves for several optical materials. In each case, the index of refraction is highest for shorter wavelengths. Thus, shorter wavelengths (blue light) are refracted to a greater extent than longer wavelengths (red light). The ability of a prism to disperse white light into its component wavelengths is the basis of the prism spectrometer.

Image of an Object through a Plane Refracting Surface Consider an object in a medium that is observed from a second medium. If the boundary between the two media is a plane and the object is observed in a direction that is perpendicular to the boundary, an image is observed that lies a definite distance below the surface.

FIGURE 14.36 White light can be separated by a prism.

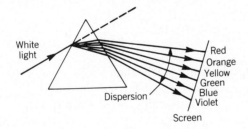

FIGURE 14.37 Dispersion curves of several optical materials.

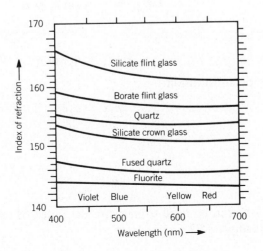

FIGURE 14.38   Image of a coin in a glass of water.

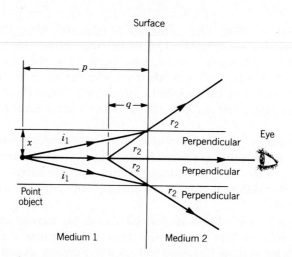

For example, place a coin in a glass of water and observe the coin in a direction that is perpendicular to the surface of the water. A clear image of the coin will be observed below the surface that appears closer to the eye than the object. To determine the image distance from the boundary, consider Figure 14.38. From this figure, one has

$$\tan i_1 = \frac{x}{p} \tag{14.50}$$

and

$$\tan r_2 = \frac{x}{q} \tag{14.51}$$

If the angles i_1 and r_2 are small, then

$$\tan i_1 = \frac{x}{p} \approx \sin i_1 \approx i_1 \tag{14.52}$$

$$\tan r_2 = \frac{x}{q} \approx \sin r_2 \approx r_2 \tag{14.53}$$

Since

$$n_1 \sin i_1 = n_2 \sin r_2 \tag{14.54}$$

one obtains by substitution

$$\frac{n_1 x}{p} = \frac{n_2 x}{q} \tag{14.55}$$

or

$$q = \frac{n_2}{n_1} p \tag{14.56}$$

Equation 14.56 can be used to measure the index of refraction of medium 1 as shown in Example 14.9.

EXAMPLE 14.9   Measurement of the index of refraction

. . . .

Given: A microscope is focused on an object at the bottom of a container as illustrated in the figure, and the position of the microscope is determined by reading a fixed scale. Liquid is added to the container and the microscope is focused at the surface of the liquid. The fixed scale reading is noted, and the difference in the two readings is the object distance. This difference is 1.33 cm. The microscope is then focused on the image of the object and the scale reading is taken. The difference between this reading and the reading obtained when the telescope was focused on the surface is equal to the image distance. This difference is 1.0 cm.

Find: The index of refraction of the liquid.

Solution: The equation for the index of refraction is

$$q = \frac{n_2}{n_1} p$$

or

$$n_1 = n_2 \frac{p}{q}$$

n_2 = index of refraction for air = 1.00

$$n_1 = (1.00)\left(\frac{1.33 \text{ cm}}{1.00 \text{ cm}}\right)$$

$$n_1 = 1.33$$

Convex Spherical Refracting Surface   Consider the cross section of the spherical refracting surface in Figure 14.39. The medium on the left has an index of refraction equal to n_1 and the medium on the right has an index of refraction equal to n_2. Consider any incident ray on the spherical surface coming from the point source O on the optical axis. This ray makes an angle α with the optical axis, and it intersects the spherical surface a distance h from the optical axis. A radius drawn from the center C to the point of intersection is normal to the surface. Thus, the

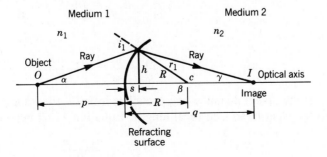

FIGURE 14.39 Optical diagram of a convex spherical refracting surface.

angle i_1 is the angle of incidence. In the diagram, it is assumed that $n_2 > n_1$. For this case, the transmitted ray will bend toward the normal of the surface. The transmitted ray intersects the optical axis at point I.

To determine the mathematical relationship between points O and I, consider Figure 14.39. When all the angles are small,

$$\alpha \approx \tan \alpha \approx \frac{h}{p} \tag{14.57}$$

$$\beta \approx \tan \beta \approx \frac{h}{R} \tag{14.58}$$

$$\gamma \approx \tan \gamma \approx \frac{h}{q} \tag{14.59}$$

Also from the diagram

$$\alpha + \beta = i_1 \tag{14.60}$$

$$\gamma + r_2 = \beta \tag{14.61}$$

From the law of refraction (Equation 14.56)

$$n_2 \sin r_2 = n_1 \sin i_1 \tag{14.62}$$

For small angles $\sin r_2 \approx r_2$ and $\sin i_1 \approx i_1$. For these conditions, Equation 14.62 becomes

$$n_2 r_2 = n_1 i_1 \tag{14.63}$$

Multiplying Equation 14.60 by n_1 and Equation 14.61 by n_2 yields

$$n_1 \alpha + n_1 \beta = n_1 i_1 \tag{14.64}$$

$$n_2 \gamma - n_2 \beta = -n_2 r_2 = -n_1 i_1 \tag{14.65}$$

If Equation 14.64 is added to Equation 14.65, one obtains

$$n_1 \alpha + n_2 \gamma + (n_1 - n_2)\beta = 0 \tag{14.66}$$

or

$$n_1 \alpha + n_2 \gamma = (n_2 - n_1)\beta \tag{14.67}$$

Equations 14.57, 14.58, and 14.59 can be used to substitute for α, β, and γ. This yields

$$n_1 \frac{h}{p} + n_2 \frac{h}{q} = (n_2 - n_1) \frac{h}{R} \tag{14.68}$$

or

$$\frac{n_1}{p} + \frac{n_2}{q} = \frac{n_2 - n_1)}{R} \tag{14.69}$$

If the object distance p approaches infinity, the image distance becomes equal to the right-hand-side focal length. Equation 14.69 becomes ($q = f_2$)

$$\frac{n_2}{f_2} = \frac{n_2 - n_1}{R} \tag{14.70}$$

or

$$f_2 = \frac{n_2 R}{n_2 - n_1} \qquad (14.71)$$

where

f_2 = right-hand-side focal length
R = radius of the refracting surface

Since rays are reversible, the left-hand-side focal length is obtained by letting q approach infinity. This yields ($p = f_1$)

$$\frac{n_1}{f_1} = \frac{(n_2 - n_1)}{R} \qquad (14.72)$$

or

$$f_1 = \frac{n_1 R}{(n_2 - n_1)} \qquad (14.73)$$

It is important to notice that $f_2 \neq f_1$. However,

$$\frac{n_2}{f_2} = \frac{n_1}{f_1} \qquad (14.74)$$

Sample problems are given in Examples 14.10 and 14.11.

EXAMPLE 14.10   Images produced by a convex spherical refracting surface
. . . .

Given: A point object lies 2.0 m from the vertex of a convex spherical refracting surface having a radius of 0.50 m. The object is in air ($n_1 \approx$ 1.0) and the refracting material is glass ($n_2 = 1.5$).

Find: a.   The right-hand-side focal length.
   b.   The left-hand-side focal length.
   c.   The location of the image point.

Solution: a.   The equation for the right-hand-side focal length is

$$f_2 = \frac{n_2 R}{n_2 - n_1}$$

$$f_2 = \frac{(1.5)(0.50 \text{ m})}{(1.5 - 1.0)}$$

$$f_2 = \frac{0.75 \text{ m}}{0.5 \text{ m}}$$

$$f_2 = 1.5 \text{ m}$$

   b.   The equation for the left-hand-side focal length is

$$f_1 = \frac{n_1 R}{n_2 - n_1}$$

$$f_1 = \frac{(1.0)(0.5 \text{ m})}{(1.5 - 1.0)}$$

$$f_1 = \frac{0.5 \text{ m}}{0.5}$$

$$f_1 = 1.0 \text{ m}$$

c. The equation that relates the image point to the object point is

$$\frac{n_1}{p} + \frac{n_2}{q} = \frac{n_2 - n_1}{R}$$

$$\frac{1.0}{2.0 \text{ m}} + \frac{1.5}{q} = \frac{(1.5 - 1.0)}{0.50 \text{ m}}$$

$$\frac{(1.0)(2.0 \text{ m})(0.5 \text{ m})q}{2.0 \text{ m}} + \frac{(1.5)(2.0 \text{ m})(0.5 \text{ m})q}{q} = \frac{(0.5)(2.0 \text{ m})(0.5 \text{ m})q}{0.5 \text{ m}}$$

$$(0.5 \text{ m})q + 1.5 \text{ m}^2 = (1.0 \text{ m})q$$

$$1.5 \text{ m}^2 = (0.5 \text{ m})q$$

$$\frac{1.5 \text{ m}^2}{0.5 \text{ m}} = q$$

$$3.0 \text{ m} = q$$

The image is real and lies on the right-hand side of the reflecting surface.

EXAMPLE 14.11   Construction of an optical diagram for a convex spherical refracting surface
· ·

Given: The information given in Example 14.10.

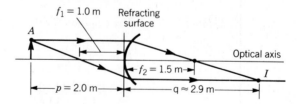

Find: The location of an object by using an optical diagram.

Solution: The left-hand focal length is 1.0 m and the right-hand focal length is 1.5 m. A ray from point *A* on the object that is parallel to the optical axis will pass through the right-hand-side focal point. A ray from point *A* on the object that passes through the left-hand-side focal length will become parallel to the optical axis. The intersection of these two rays is the image point of the object. An error occurs because the angles are not small.

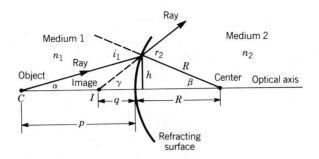

FIGURE 14.40 Virtual image produced by a spherical convex refracting surface when $n_1 > n_2$.

If $p < f_1$, then a virtual image will be obtained on the left-hand side of the refracting surface and the value of q given by Equation 14.69 will be negative. A negative value of q will be used for all virtual images, and virtual images will always lie on the left-hand side of the refracting surface.

If $n_1 > n_2$, the refracted rays bend away from the normals and a virtual image is formed on the left-hand side of the convex refracting surface as illustrated in Figure 14.40. By applying the same mathematical method that was used for the case when $n_2 > n_1$ and by giving q a negative value, one obtains the equation

$$\frac{n_1}{p} + \frac{n_2}{q} = \frac{n_2 - n_1}{R} \tag{14.75}$$

which is the same as Equation 14.69. For any positive value of p, however, the value of q will always be negative since $n_1 > n_2$. Both focal lengths are negative for this case. Thus, they lie on opposite sides of the refracting surface as illustrated in Example 14.12.

EXAMPLE 14.12 Image produced by a convex refracting surface going from glass to air
. . . .

Given: An object lies 2.0 m from the vertex of a convex spherical refracting surface having a radius of 1.0 m. The object is in glass ($n_1 = 1.5$) and the image is in air ($n_2 = 1.0$).

Find: **a.** The two focal lengths.
            b. The location of the image using Equation 14.75.
            c. The location of the image from an optical diagram.

Solution: **a.** The equation for f_1 is

$$f_1 = \frac{n_1 R}{n_2 - n_1}$$

$$f_1 = \frac{(1.5)(1.0 \text{ m})}{(1.0 - 1.5)}$$

$$f_1 = \frac{1.5 \text{ m}}{-0.5}$$

$$f_1 = -3.0 \text{ m}$$

Since f_1 is negative, it lies on the right-hand side of the surface. The equation for f_2 is

$$f_2 = \frac{n_2 R}{n_2 - n_1}$$

$$f_2 = \frac{(1.0)(1.0 \text{ m})}{(1.0 - 1.5)}$$

$$f_2 = \frac{1.0 \text{ m}}{-0.5}$$

$$f_2 = -2 \text{ m}$$

Since f_2 is negative it lies on the left-hand side of the surface

b. The equation relating image distance to object distance is

$$\frac{n_1}{p} + \frac{n_2}{q} = \frac{n_2 - n_1}{R}$$

$$\frac{1.5}{2.0 \text{ m}} + \frac{1.0}{q} = \frac{(1.0 - 1.5)}{1.0 \text{ m}}$$

$$\frac{(1.5)(2.0 \text{ m})(1.0 \text{ m})q}{2.0 \text{ m}} + \frac{(1.0)(2.0 \text{ m})(1.0 \text{ m})q}{q} = \frac{(-0.5)(2.0 \text{ m})(1.0 \text{ m})q}{1.0 \text{ m}}$$

$$(1.5)q + 2.0 \text{ m}^2 = (-1.0 \text{ m})q$$

$$(2.5 \text{ m})q = 2.0 \text{ m}^2$$

$$q = \frac{-2.0 \text{ m}^2}{2.5 \text{ m}}$$

$$q = -0.8 \text{ m}$$

c. The optical diagram is shown below. An error occurs because the angles in the diagram are not small. The image is a virtual image.

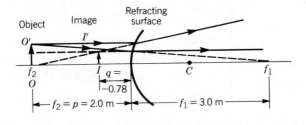

Concave Spherical Refracting Surface   Since light rays are reversible, consider the light ray shown in Figure 14.39 to be traveling in the opposite direction (from glass to air). For this case, γ and α, n_2 and n_1, and q and p can be interchanged. In addition, the surface becomes a concave refracting surface. If the above exchanges are made in Equation 14.69, one obtains

$$\frac{n_2}{q} + \frac{n_1}{p} = \frac{n_1 - n_2}{R} \qquad (14.76)$$

The corresponding focal lengths will depend on the sign of $n_1 - n_2$. For example, positive focal lengths are obtained for rays going from glass to air.

General Equation for Spherical Refracting Surfaces By using the following sign conventions, the equations

$$\frac{n_1}{p} + \frac{n_2}{q} = \frac{n_1 - n_2}{R} \qquad (14.77)$$

$$f_1 = \frac{n_1 R}{n_1 - n_2} \qquad (14.78)$$

$$f_2 = \frac{n_2 R}{n_1 - n_2} \qquad (14.79)$$

can be used for all cases. The sign conventions are:

a. The radius is positive for concave surfaces and negative for convex surfaces.
b. q is positive for real images and lies on the right-hand side of the surface. q is negative for virtual images and lies on the left-hand side of the surface.
c. p is positive for real objects and lies on the left-hand side of the surface, p is negative for virtual objects and lies on the right-hand side of the surface.
d. For all cases, rays travel from left to right.
e. A positive f_1 lies on the left-hand side and a negative f_1 lies on the right-hand side.
f. A positive f_2 lies on the right-hand side and a negative f_2 lies on the left-hand side.

The Thin Lens A simple lens is a medium bounded by two refracting surfaces. The image of the first refracting surface becomes the object of the second refracting surface. The surfaces of the lens may be concave or convex. Lenses with spherical surfaces are called spherical lenses. Other lenses may have surfaces that are cylindrical or sphero-cylindrical. A plane surface may be considered as a spherical surface with an infinite radius. The cross sections of several different types of spherical lenses are shown in Figure 14.41.

The optical axis of a simple spherical lens passes through the centers of the spherical surfaces and is perpendicular to both surfaces. A ray traveling along the optical axis will not be deflected. The points where the optical axis intersects the surfaces are called the vertices, and the distance between the vertices is called

FIGURE 14.41 Cross sections of several types of lenses.

Double convex   Pland-convex   Double concave   Pland-concave

FIGURE 14.42 Image
distance for a double
convex lens.

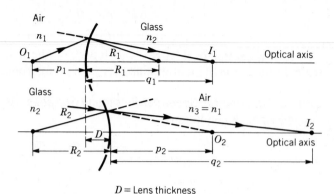

FIGURE 14.42 Image
distance for a double
convex lens.

D = Lens thickness

the thickness of the lens. If the thickness is very small compared to the radii of the surface, it can be neglected in calculations; and the lens is called a thin lens.

To determine the relationship between the object distance and the image distance for a lens, consider the two refracting surfaces shown in Figure 14.42. The first surface is convex and the second is concave. The lens formed by these two surfaces, however, is a double convex lens. The equation for the first spherical surface is (Equation 14.77).

$$\frac{n_1}{p_1} + \frac{n_2}{q_1} = \frac{n_1 - n_2}{R_1} \qquad (R_1 \text{ is negative}) \qquad (14.80)$$

The image for this surface is real and q is positive. The equation for the second spherical refracting surface is

$$\frac{n_2}{p_2} + \frac{n_3}{q_2} = \frac{n_2 - n_3}{R_2} \qquad (R_2 \text{ is positive}) \qquad (14.81)$$

The object for the second surface, however, is virtual and it lies on the right-hand side of the second surface. Thus, p_2 is a negative. Since q_1 is positive, $p_2 = -(q_1 - D)$. This can be seen from the diagram, where D is the lens thickness. These three equations can be used to locate the final image produced by the lens. This is done for a double convex lens in Example 14.13.

EXAMPLE 14.13   Image formed by a thick lens

. . . .

Given: The radii of a double convex lens are -0.50 and $+0.50$m. The thickness of the lens is 0.10 m. Air is on both sides of the lens, and the index of refraction of the lens material is 1.50. The object distance is 3.00 m.

Find: The image distance from the second vertex and the image distance from the center of the lens.

Solution: For the first refracting surface

$$\frac{n_1}{p_1} + \frac{n_2}{q_1} = \frac{n_1 - n_2}{R_1}$$

$$\frac{1.00}{3.00 \text{ m}} + \frac{1.50}{q_1} = \frac{1.00 - 1.50}{-0.50 \text{ m}}$$

$$\frac{(1.00)(3.00 \text{ m})(1 \text{ m})q_1}{3.00 \text{ m}} + \frac{(1.50)(3.00 \text{ m})(1 \text{ m})q_1}{q_1} = \frac{(1)(3.00 \text{ m})(1 \text{ m})q_1}{1 \text{ m}}$$

$$(1 \text{ m})q_1 + 4.5 \text{ m}^2 = (3 \text{ m}) \, q_1$$

$$(2 \text{ m})q_1 = 4.5 \text{ m}^2$$

$$q_1 = \frac{4.5 \text{ m}^2}{2 \text{ m}}$$

$$q_1 = 2.25 \text{ m}$$

The object distance for the second surface is

$$p_2 = -(q_1 - D)$$

$$p_2 = -(2.25 \text{ m} - 0.10 \text{ m})$$

$$p_2 = -2.15 \text{ m}$$

For the second refracting surface

$$\frac{n_2}{p_2} + \frac{n_3}{q_2} = \frac{n_2 - n_3}{R_2}$$

$$\frac{1.50}{-2.15 \text{ m}} + \frac{1.00}{q_2} = \frac{1.50 - 1.00}{0.50 \text{ m}}$$

$$\frac{(1.50)(-2.15 \text{ m})(0.5 \text{ m})q_2}{-2.15 \text{ m}} + \frac{(1.00)(-2.15 \text{ m})(0.5 \text{ m})q_2}{q_2} = \frac{(0.5)(-2.15 \text{ m})(0.5 \text{ m})q_2}{0.5 \text{ m}}$$

$$(0.75 \text{ m})q_2 + (-1.075 \text{ m}^2) = (-1.075 \text{ m}) \, q_2$$

$$(1.825 \text{ m}) \, q_2 = 1.075 \text{ m}^2$$

$$q_2 = 0.589 \text{ m} \qquad \text{from the vertex}$$

The distance from the center is

$$0.589 \text{ m} + D/2 = 0.589 \text{ m} + 0.05 \text{ m} = 0.639 \text{ m}$$

If the thickness of the lens is neglected, $p_2 = q_1$. In addition, if air is on both sides of the lens, $n_3 = n_1$. Equations 14.80 and 14.81 can then be written as

$$\frac{n_1}{p_1} + \frac{n_2}{q_1} = \frac{n_1 - n_2}{R_1} \tag{14.82}$$

and

$$-\frac{n_2}{q_1} + \frac{n_1}{q_2} = \frac{n_2 - n_1}{R_2} \tag{14.83}$$

Adding these two equations yields

$$\frac{n_1}{p_1} + \frac{n_1}{q_2} = (n_2 - n_1)\left(\frac{1}{R_2} - \frac{1}{R_1}\right)$$

or

$$\frac{1}{p_1} + \frac{1}{q_2} = \frac{n_2 - n_1}{n_1}\left(\frac{1}{R_2} - \frac{1}{R_1}\right) \tag{14.84}$$

Equation 14.84 is the basic equation for a thin lens, and it will hold for all cases if the values of R_2 and R_1 are given the proper signs. At this point it is customary to let $p_1 = p$ (the object distance) and $q_2 = q$ (the image distance). In addition, a focal length can be defined. If p increases without bounds, then $1/p$ approaches zero. The rays from the point object on the optical axis became parallel to the optical axis. The corresponding image distance is called the focal length ($q = f$) Equation 14.84 becomes

$$\frac{1}{q} = \frac{1}{f} = \left(\frac{n_2 - n_1}{n_1}\right)\left(\frac{1}{R_2} - \frac{1}{R_1}\right) \tag{14.85}$$

Equation 14.85 is called the lensmakers equation since it can be used to design a lens of a given focal length. Example 14.14 gives an application of the equations for a thin lens. The results can be compared with Example 14.13.

EXAMPLE 14.14   Image formed by a thin lens
. . . .

Given: The radii of a thin double convex lens are -0.50 and $+0.05$ m. Air is on both sides of the lens, and the index of refraction of the lens material is 1.50. The object distance is 3.00 m.

Find: **a.** The focal length.
           b. The image distance.

Solution: **a.** The focal length is given by

$$\frac{1}{f} = \left(\frac{n_2 - n_1}{n_1}\right)\left(\frac{1}{R_2} - \frac{1}{R_1}\right)$$

$$\frac{1}{f} = \left(\frac{1.50 - 1.00}{1.00}\right)\left(\frac{1}{0.50 \text{ m}} - \frac{1}{-0.50 \text{ m}}\right)$$

$$\frac{1}{f} = 0.50\left(\frac{2}{0.50 \text{ m}}\right)$$

$$\frac{1}{f} = \frac{2}{1 \text{ m}}$$

$$f = 0.50 \text{ m}$$

b. The image distance is given by

$$\frac{1}{p} + \frac{1}{q} = \frac{1}{f}$$

$$\frac{1}{3.00 \text{ m}} + \frac{1}{q} = \frac{1}{0.50 \text{ m}}$$

$$\frac{(3 \text{ m})(0.5 \text{ m})q}{3 \text{ m}} + \frac{(3 \text{ m})(0.5 \text{ m})q}{q} = \frac{(3 \text{ m})(0.5 \text{ m})q}{0.5 \text{ m}}$$

$$(0.5 \text{ m})q + (1.5 \text{ m}^2) = (3 \text{ m})q$$

$$(2.5 \text{ m})q = 1.5 \text{ m}^2$$

$$q = \frac{1.5 \text{ m}^2}{2.5 \text{ m}}$$

$$q = 0.6 \text{ m}$$

q is measured from the center of the lens.

In terms of the focal length, Equation 14.84 becomes

$$\frac{1}{p} + \frac{1}{q} = \frac{1}{f} \tag{14.86}$$

Equation 14.86 is the basic equation for a thin lens.

The Positive Thin Lens   A thin lens is a positive lens if the focal length is positive. It is also called a converging lens. A positive lens is thicker at the optical axis than it is at the edge. Rays parallel to the optical axis are focused to a point called the focal point as illustrated in Figure 14.43. The distance from the center of the lens to the focal point is the focal length. As the object distance is decreased, the real image moves away from the focal point. When the object distance has a value equal to the focal point, the image rays become parallel to the optical axis as shown in Figure 14.43. This can be seen from Equation 14.86 by letting q approach infinity. For a thin lens, the left-hand focal length and right-hand focal length are equal. If the object distance becomes less than the focal length, the value of q becomes negative. The image for this case is virtual and lies on the left-hand side of the lens.

Figure 14.44 shows the construction of a real image by a positive lens. Object O is at a distance from the lens greater than twice the focal length ($2f$) of the lens. The following rays are used to locate the image:

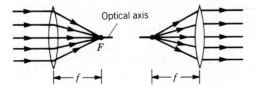

FIGURE 14.43 Optical diagram for a positive thin lens.

FIGURE 14.44 Formation of
a real image by a positive
lens when the object is at a
distance greater than twice
the focal length.

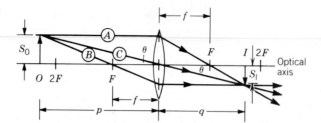

- Ray *A* originates at the top of the object and travels parallel to the optical axis. It is refracted to pass through the focal point on the other side of the lens.
- Ray *B* travels through the focal point on the object side of the lens and is refracted to leave the lens in a direction parallel to the optical axis.
- Ray *C* passes through the center of the lens and is undeviated.

If the object is located at a great distance from the lens, the image is near the focal point and greatly reduced in size. This optical arrangement applies both to a camera and to the human eye.

Figure 14.45 shows the image formed by the same lens when the object distance is in the range between the focal length of the lens and twice the focal length. In this instance, the image is enlarged and the image distance is greater than $2f$. This ray diagram is representative of the optical arrangements used in movie and slide projectors, where O is the film and I is the screen.

Figure 14.46 shows the virtual image formed when an object is placed inside the focal point of a positive lens. This virtual image always appears magnified and erect. (Real images formed by lenses always appear inverted.) Virtual, magnified images formed by positive lenses include those displayed by the familiar magnifying glasses, as well as the plastic lens used to enlarge the display of many pocket calculators.

For all three cases, one can write from similar triangles

$$\frac{S_I}{q} = \frac{S_O}{p} \tag{14.87}$$

where

$$S_I = \text{image height}$$
$$S_O = \text{object height}$$

FIGURE 14.45 Formation of
a real image by a positive
lens when the object is
located at a distance
between the focal length
and twice the focal length.

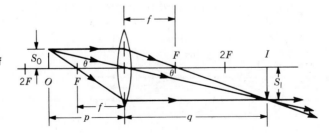

FIGURE 14.46 Formation of a virtual image by a positive lens.

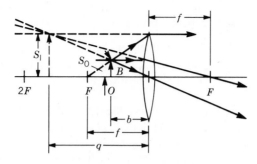

The magnification becomes

$$m = \frac{S_I}{S_O} = \frac{q}{p} \tag{14.88}$$

An application of Equations 14.86 and 14.88 is given in Example 14.15 for a positive lens.

EXAMPLE 14.15 Image formation with a converging lens

. . . .

Given: Focal length of lens is $f = 10$ cm; object distance is $p = 25$ cm; and object height is 3 cm.

Find: Image distance and height.

Solution: Image distance:

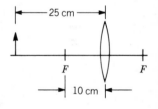

$$\frac{1}{f} = \frac{1}{p} + \frac{1}{q}$$

$$\frac{1}{10 \text{ cm}} = \frac{1}{25 \text{ cm}} + \frac{1}{q}$$

$$\frac{(10 \text{ cm})(25 \text{ cm})q}{10 \text{ cm}} = \frac{(10 \text{ cm})(25 \text{ cm})q}{25 \text{ cm}} + \frac{(10 \text{ cm})(25 \text{ cm})q}{q}$$

$$(25 \text{ cm})q = (10 \text{ cm})q + 250 \text{ cm}^2$$

$$(15 \text{ cm})q = 250 \text{ cm}^2$$

$$q = \frac{250 \text{ cm}^2}{15 \text{ cm}}$$

$$q = 16.67 \text{ cm}$$

Image height:

$$S_I = S_O \left(\frac{q}{p}\right)$$

$$S_I = 3 \text{ cm} \left(\frac{16.67 \text{ cm}}{25 \text{ cm}}\right)$$

$$S_1 = 2.0 \text{ cm}$$

$$\text{Magnification of lens} = S_1/S_O = 2/3 = q/p = 16.67/25.$$

The Thin Negative Lens   A lens is negative if the focal length is negative as given by Equation 14.85. A negative lens is also called a diverging lens. It is thinner at the optical axis than it is at the edge. Since f is negative and $1/p$ becomes negative for real objects when it is taken to the right-hand side of Equation 14.86, the value of q is always negative. Thus, the image is always virtual. If p becomes very large, q approaches f. However, f is on the left-hand side of the lens since it is negative. In addition, the incident rays become parallel to the optical axis for this case. The ray diagram for parallel incident rays is shown in Figure 14.47. If q is very large in Equation 14.86, the object distance must be negative and equal to f. Thus, the object is virtual and lies on the right-hand side of the lens. The refracted rays will be parallel to the optical axis on the right-hand side of the lens. Figure 14.47*b* illustrates this case.

The virtual image formed by a negative lens is shown in Figure 14.48. In this diagram, ray *A* enters the lens parallel to the optical axis (as in the case of the positive lens) and is refracted as though it came from the focal point. Ray *B* travels toward the focal point on the opposite side of the lens and is refracted parallel to the optical axis. Ray *C* passes through the center of the lens and remains undeviated. The virtual image formed is always erect and smaller than the object. Peepholes in doors often are equipped with negative lenses, which provide wide-angle views.

From Figure 14.48, one has from similar triangles

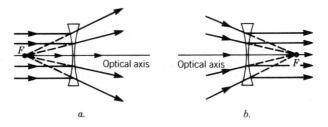

FIGURE 14.47   Optical diagrams for a negative lens.

FIGURE 14.48   Virtual image formation by a negative lens.

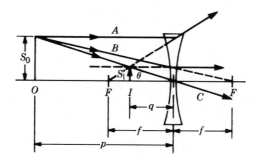

$$\frac{S_O}{|p|} = \frac{S_I}{|q|}$$

or

$$m = \frac{S_I}{S_O} = \frac{|p|}{|q|} \tag{14.89}$$

Example 14.16 gives an application of the equations for a negative lens.

EXAMPLE 14.16 Image formation with a diverging lens
. . . .

Given: Optical distance is $p = 10$ cm; object height is $S_O = 3.5$ cm, and virtual height is $S_I = 0.625$ cm.

Find: Image distance and focal length of lens.

Solution: Image distance:

$$|q| = |p|\left(\frac{S_I}{S_O}\right)$$

$$|q| = (10 \text{ cm})\left(\frac{0.625 \text{ cm}}{3.5 \text{ cm}}\right)$$

$$|q| = 1.79 \text{ cm}$$

Focal length:

$$\frac{1}{f} = \frac{1}{p} + \frac{1}{q}$$

$$\frac{1}{f} = \frac{1}{10 \text{ cm}} + \frac{1}{(-1.79 \text{ cm})} \quad \text{(image distance of virtual image is negative)}$$

$$\frac{1}{f} = 0.1 \text{ cm}^{-1} - 0.559 \text{ cm}^{-1}$$

$$\frac{1}{f} = -0.459 \text{ cm}^{-1}$$

$$\therefore f = \frac{1}{-0.459 \text{ cm}^{-1}}$$

$$f = -2.18 \text{ cm}$$

Two Thin Lenses in Contact Consider two thin lenses in contact. The equation for the first lens is

$$\frac{1}{p_1} + \frac{1}{q_1} = \frac{1}{f_1} \tag{14.90}$$

The distance q_1 becomes the object distance for the second lens. However, it will have the opposite sign. For example, if the first lens is a positive lens, then q_1 is positive. However, the image located at q_1 will be a virtual object for the second lens. In general, $p_2 = q_1$. The equation for the second lens becomes

$$\frac{1}{-q_1} + \frac{1}{q_2} = \frac{1}{f_2} \tag{14.91}$$

Adding these two equations yields

$$\frac{1}{p_1} + \frac{1}{q_2} = \frac{1}{f_1} + \frac{1}{f_2} \tag{14.92}$$

Thus, the two lenses act as if they are a single lens with a focal length given by

$$\frac{1}{f} = \frac{1}{f_1} + \frac{1}{f_2} \tag{14.93}$$

Lens Aberrations   The theory of lens aberrations is beyond the scope of this book. However, two different aberrations of the image of a point object on the optical axis of a lens will be considered. It was assumed in the derivation of the thin lens equation that $\tan \theta = \sin \theta = \theta$. Since this is only an approximation that becomes better as θ approaches zero, all rays incident on a lens are not focused to the exact same point. The best image of a point is a small circle called the least circle of aberration. The aberration is called spherical aberration along the axis. A second kind of aberration occurs because the index of refraction of the lens material depends on the wavelength of the incident light. For example, if the index of refraction for blue light is substituted into Equation 14.85, a focal length can be obtained. However, if the index of refraction for red light is used, a slightly different focal length is obtained. Thus, blue light from an object is not focused to the same point as red light from the object. This kind of aberration is called chromatic aberration. The effect of chromatic aberration can be reduced by using two lenses in contact that have different indices of refraction. The compound lens can be designed so that the focal length for blue light is the same as the focal length for red light. This kind of compound lens is called an achromat. Example 14.17 illustrates the effect of wavelength on the focal length of thin lenses.

EXAMPLE 14.17   Chromatic Aberration

Given: The radii of a thin spherical lens are $R_1 = -0.2000$ m and $R_2 = \infty$. The index of refraction of the lens material is 1.5268 at 656 nm (crown glass) and 1.5361 at 486 nm. The radii of a second lens are $R_1 = \infty$ and $R_2 = -0.4043$ m. For this lens, the index of refraction is 1.6297 at 656 nm (flint glass) and 1.6483 at 486 nm. The index of refraction of air is 1.0003.

Find: a.   The focal length of the first lens at each wavelength.
b.   The focal length of the second lens at each wavelength.

    c. The focal length of the compound lens at each wavelength when the plane surfaces are in contact.

Solution: **a.** The focal length of a thin lens is

$$\frac{1}{f} = \frac{n_2 - n_1}{n_1}\left(\frac{1}{R_2} - \frac{1}{R_1}\right)$$

$$\frac{1}{f_{656}} = \left(\frac{1.5268 - 1.0003}{1.0003}\right)\left(\frac{1}{\infty} - \frac{1}{-0.2000 \text{ m}}\right)$$

$$\frac{1}{f_{656}} = \frac{0.5265}{1.0003}\left(\frac{1}{0.2000 \text{ m}}\right)$$

$$f_{656} = 0.381 \text{ m}$$

$$\frac{1}{f_{486}} = \left(\frac{1.5361 - 1.0003}{1.0003}\right)\left(\frac{1}{0.2000 \text{ m}}\right)$$

$$f_{486} = 0.373 \text{ m}$$

This is a 1.8% difference.

   b. For the second lens

$$\frac{1}{f_{656}} = \left(\frac{1.6297 - 1.0003}{1.0003}\right)\left(\frac{1}{-0.4043 \text{ m}} - \frac{1}{\infty}\right)$$

$$f_{656} = -0.643 \text{ m}$$

$$\frac{1}{f_{486}} = \left(\frac{1.6483 - 1.0003}{1.0003}\right)\left(\frac{1}{-0.4043 \text{ m}}\right)$$

$$f_{486} = -0.624 \text{ m}$$

This is a 2.9% difference.

   c. For two thin lenses in contact

$$\frac{1}{f} = \frac{1}{f_1} + \frac{1}{f_2}$$

$$\frac{1}{f_{656}} = \frac{1}{0.381 \text{ m}} + \frac{1}{-0.643 \text{ m}}$$

$$f_{656} = 0.935 \text{ m}$$

$$\frac{1}{f_{486}} = \frac{1}{0.373 \text{ m}} + \frac{1}{-0.624 \text{ m}}$$

$$f_{486} = 0.927 \text{ m}$$

This is a 0.09% difference.

The two focal lengths are approximately the same to two significant figures.

Properties of Thin Lenses   When the thin lens equation is used, it is important to use an appropriate sign convention. The sign convention given in this chapter is:

TABLE 14.3   Properties of thin lenses for real objects

Lens Type	Object Distance	Image Distance	Image Type and Size
Negative	Any	q is negative $\|q\|<\|f\|$	Virtual Reduced
Positive	$p=\infty$	q is positive $q=f$	Real Reduced
Positive	$p=f$	q is positive $q=\infty$	Real Enlarged
Positive	$p>2f$	q is positive $\|q\|<2f$	Real Reduced
Positive	$f<p<2f$	q is positive $\|q\|>2f$	Real Enlarged
Positive	$0<p<f$	q is negative	Virtual Enlarged

a. The value of p is positive for real objects and negative for virtual objects.

b. The value of q is positive for real objects and negative for virtual objects.

c. The focal length is positive for a converging lens and negative for a diverging lens.

The properties of a thin lens for real objects are given in Table 14.3

SECTION 14.1 EXERCISES

1. Draw a diagram showing the umbra and penumbra when the radius of a circular source is greater than the radius of an opaque spherical object.

2. Draw and label a diagram illustrating the law of reflection.

3. Draw and label diagrams showing light reflected from specular and diffuse reflecting surfaces.

4. Identify the following as specular or diffuse reflectors:
   Bathroom mirror
   Sheet of paper
   Ceiling tile
   Watch crystal
   Shiny linoleum tile

5. Draw and label a diagram of light striking a beam splitter at an incident angle of about 30°. Show both the transmitted and the reflected rays.

6. A beam splitter reflects 49% of the light striking it and transmits 48% of the light. What happens to the other 3%?

7. Find the image of the line shown in the figure to the left.

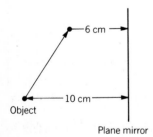

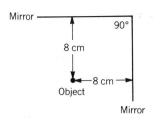

8. Find the image of the point object shown in the figure to the left.

9. Describe the difference between a real image and a virtual image.

10. Draw diagrams to illustrate the focal points of convex and concave mirrors and show parallel light rays reflected from these mirrors.

11. On the following diagrams, draw three rays from the tip of the object to the tip of the image formed.

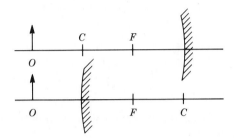

12. A concave mirror has a focal length of 10 cm. An object is located 15 cm from this mirror. Where is the image?

13. The object in Problem 12 is 2 cm high. What is the image height?

14. An object is located 10 cm from a mirror. A virtual reduced image is formed 2 cm behind the mirror surface. What is the focal length of the mirror? Is it concave or convex?

15. The image in Problem 14 is 0.6 cm high. What is the object height?

16. Parallel light is incident on a concave mirror that has a radius of curvature of 30 cm. Where will the light be brought to a focus?

17. The focal length of a concave mirror is 30 cm. An object is placed 15 cm in front of the mirror. Where is the image? Is the magnification greater or less than one?

18. List four kinds of mirrors.

19. Where are parabolic mirrors used? What is an off-axis parabolic mirror?

20. Where are elliptical mirrors used? What is an off-axis elliptical mirror?

21. Write the equation for the law of refraction.

22. The index of refraction of fused quartz for light with a wavelength of 700 nm is 1.45. What is the speed of light in quartz?

23. Light passes from air into water ($n = 1.33$) with an angle of refraction of 32°. What is the angle of incidence?

24. Light passes from water ($n = 1.33$) into flint glass ($n = 1.64$) with an angle of incidence of 55°. What is the angle of refraction?

25. Draw a sketch showing total internal reflection in a prism.

26. Is the index of refraction of a material greater for red light or blue light?

27. Find the index of refraction of a material in which light has a speed of 2×10^8 m/sec.

28. A prism has a refracting angle of 60°. The measured angle of minimum deviation is 30°. What is the index of refraction of the prism material?

29. An object is 5 ft below the surface of the water. If one looks at the object perpendicular to the plane surface, where is the image? The index of refraction of water is 1.33.

30. Give two ways for measuring the index of refraction of a material relative to air.

31. Complete the following ray diagrams to show the image formed.

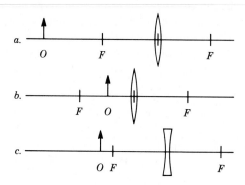

32. Sketch the lens, object, and image in each of the following optical systems:
Magnifying glass
Movie projector
Camera

33. The focal length of a peephole lens in a door is -2.0 cm. A person 6 ft tall stands 3 m from the lens. Find the image distance and height.

34. A slide projector lens has an object distance of 8.5 cm and an image distance of 2.3 cm. What is its focal length?

35. List two kinds of aberrations.

36. How can a correction be made for chromatic aberration?

37. Two thin lenses are in contact with a common optical axis. The first lens has a focal length of 6 cm and the second lens has a focal length of -15 cm. What is the effective focal length of the two lenses when they are in contact with each other?

38. The focal length of a lens is 20 cm and the object distance is 10 cm. Where is the image, and what is the magnification?

39. The radii of curvature of a lens are 50 cm for the first surface and -40 cm for the second surface. The index of refraction relative to air is 1.45. Find the focal length of the lenses.

40. A spherical convex refracting surface has a radius of curvature of -15 cm. Light travels from air through the refracting surface. The index of refraction of air is 1.00 and the index of refraction of the second medium is 1.50. An object is located 30 cm in front of the surface. Where is the image located? What are the two focal lengths?

41. Work Problem 40 if the refracting surface is concave ($R = +15$ cm).

14.2............

PHYSICAL
OPTICS

OBJECTIVES..

Upon completion of this section, the student should be able to

- Define the wave front of a propagating wave.

- Explain the interference of waves.

- Summarize Huygens' principle.

- Use the wave theory to explain the law of reflection.

- Use the wave theory to explain Snell's law.

- Draw and label a diagram showing the diffraction of waves at an edge barrier. Show the edge barrier, the shadow, and the wave front.

- Draw and label a diagram showing the intensity pattern from a single slit.

- Calculate the positions of the maximum and minimum intensities for a single-slit diffraction pattern.

- Draw and label a diagram showing the intensity pattern produced by a double slit. Show the single-slit diffraction envelope and the double-slit interference peaks.

- Calculate the positions of the double-slit interference peaks and the separation between the peaks.

- Draw and identify a diagram that shows the light paths for the constructive and destructive interference of waves reflected from an air wedge.

- Given the wavelength of light and the spacing of the fringes produced by an air wedge, calculate the wedge angle.

- Illustrate the interference pattern formed when each of the following is placed on a flat surface:
  Another flat surface at a small wedge angle
  Another flat surface at a large wedge angle
  A spherical surface

- Explain how a diffraction grating is used to produce the spectrum of a source.

- Given any two of the following quantities, calculate the third:
  Wavelength of light
  Diffraction angle of a specific order
  Grating spacing

DISCUSSION

Most of the time light appears to travel in straight lines, so we are able to explain what we see with the principles of ray optics. However, in some situations, light does seemingly strange things: it fans out, forms fuzzy-edged shadows, creates unusual patterns of light and dark, or appears to be turning corners around objects. In these situations, the behavior of light makes sense only if we think of light as traveling in waves.

In this section on wave optics, three characteristics of light are introduced that have to be explained in terms of the wave motion of light: interference, diffraction, and refraction. Interference occurs when two or more similar waves of light pass through the same area and "interfere" with one another. Diffraction refers to situations in which light fans out or bends rather than traveling in the straight lines that ray optics predicts. The law of refraction is accepted in ray optics as an empirical law.

FIGURE 14.49 *FIGURE 14.49* Interference produced by a monochromatic source illuminating a pair of slits.

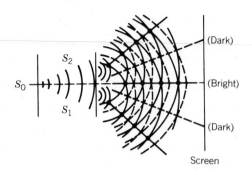

Wave motion and interference are discussed in Chapter 11. Light waves are electromagnetic waves that have electric and magnetic fields. Thus, when two light waves intersect, their electric and magnetic fields add according to the rules of vector addition. To simplify theoretical discussions, the light waves are usually considered to have a single frequency. These waves are called monochromatic waves. Although monochromatic waves do not exist in nature, light sources such as low-pressure electric discharges and lasers produce waves that are nearly monochromatic. In addition, it is usually assumed that the electric field vectors have the same direction in space. For this case, the electric fields can be added as algebraic quantities and the waves are polarized. If a surface is drawn perpendicular to the direction of propagation such that the phase of the wave is a constant on the surface, the surface is called a wave front. For plane waves (parallel beam), the wave fronts are planes. For cylindrical waves (radiation emitted from a line source), the wave fronts are cylinders. If the radiation comes from a point source, the wave fronts are spheres. If the medium is uniform, the distance between two wave fronts remains a constant. To illustrate interference, wave fronts are often drawn so that the distance between any two fronts is one-half wavelength. A top view of cylindrical wave fronts is shown in Figure 14.49. The line source S_O emits radiation that is incident on two narrow slits. The phase is the same at each slit. Each slit acts as a line source. The distance between wave fronts represented by solid and dashed circles is $\lambda/2$. Where solid lines cross, the phase is the same and constructive interference occurs (maximum field strength). Where a solid line crosses a dotted line destructive interference occurs (minimum field strength) because the waves are π radians out of phase.

Huygens' Principle   Huygens' principle is used to predict the shape and position of a defined wave front of light after it has traveled a given distance. It states that each point on a wave front may be considered to be the source of a spherical wavelet. The new shape and position of the wave front at any time is a line tangent to all the wavelets at that instant as shown for a spherical wave front in Figure 14.50.

FIGURE 14.50 Movement of a spherical wave front.

Figure 14.51 shows this principle applied to a plane wave front striking an edge barrier. Part of the wave bends around the edge of the barrier, then it travels into the area of the geometric shadow. This action is easily observed for water waves, but is also true of all other waves, including light. This bending of a wave around an edge is called diffraction.

FIGURE 14.51 Plane wave front casting a shadow at an edge barrier.

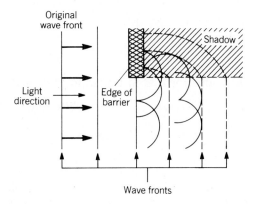

Reflection of Light The laws of reflection can be explained by the particle theory of light. However, it can also be explained by the wave theory of light. For example, consider the incident plane wave shown in Figure 14.52. The line AA' is the top view of a wave front of the incident wave and the line BB' is the top view of a wave front of the reflected wave. Since the distance between any two wave fronts in a uniform medium must be a constant to ensure constant phase, AB must equal $A'B'$. This condition may also be written as

$$A'C + B'C = AC + BC \qquad (14.94)$$

In addition, the right triangle $AA'C$ is similar to the right triangle $B'BC$ because the vertical angles $A'CA$ and BCB' are equal. Since corresponding sides of similar triangles must be proportional

$$\frac{A'C}{BC} = \frac{AC}{B'C}$$

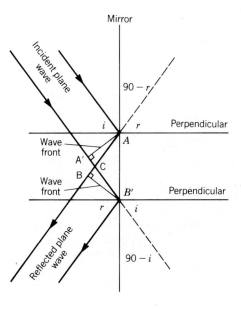

FIGURE 14.52 Reflection of a plane wave.

or

$$A'C = BC\left(\frac{AC}{B'C}\right)$$

Substituting this value for $A'C$ in Equation 14.94 and rearranging terms gives

$$(AC)(BC - B'C) = (B'C)(BC - B'C)$$

Dividing both sides by $(BC - B'C)$ yields

$$AC = B'C \qquad\qquad (14.95)$$

Therefore, the triangle ACB' is an isosceles triangle and $\angle CAB' = \angle AB'C$. However, $\angle CAB' = 90 - r$ and $\angle AB'C = 90 - i$. Thus,

$$90 - r = 90 - i$$

or

$$r = i$$

Thus, wave theory gives the result that the angle of incidence equals the angle of reflection.

The Law of Refraction   The wave nature of light can also be used to explain Snell's law. Consider the diagram shown in Figure 14.53. AA' is a wave front of an incident beam and BB' is a wave front of the refracted beam. If BB' is a line of constant phase, then the number of wavelengths in the line AB must equal the number of wavelengths in the line $A'B'$:

$$\frac{A'B'}{\lambda_1} = \frac{AB}{\lambda_2} \qquad\qquad (14.96)$$

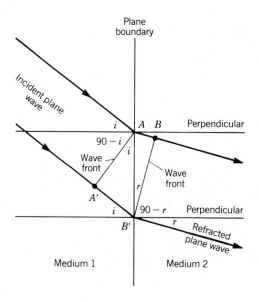

FIGURE 14.53   Refraction of a plane wave.

where

λ_1 = wavelength in medium one
λ_2 = wavelength in medium two

However,

$$\lambda_1 = \frac{v_1}{\nu} \tag{14.97}$$

and

$$\lambda_2 = \frac{v_2}{\nu} \tag{14.98}$$

where

v_1 = velocity in medium one
v_2 = velocity in medium two
ν = frequency of the wave

Using Equations 14.97 and 14.98 to substitute into Equation 14.96 yields

$$\frac{A'B'}{v_1} = \frac{AB}{v_2}$$

or

$$\frac{A'B'}{AB} = \frac{v_1}{v_2} \tag{14.99}$$

From Figure 14.53,

$$\sin i = \frac{A'B'}{AB'}$$

and

$$\sin r = \frac{AB}{AB'}$$

Thus,

$$\frac{\sin i}{\sin r} = \frac{A'B'}{AB} = \frac{v_1}{v_2} \tag{14.100}$$

Equation 14.100 can be written as

$$v_2 \sin i = v_1 \sin r \tag{14.101}$$

Dividing both sides by c (the velocity of light in a vacuum) gives

$$\frac{v_2}{c} \sin i = \frac{v_1}{c} \sin r$$

or

$$\frac{\sin i}{n_2} = \frac{\sin r}{n_1}$$

or

$$n_1 \sin i = n_2 \sin r \qquad (14.102)$$

The absolute indices of refraction are given by

$$n_1 = \frac{c}{v_1} \qquad (14.103)$$

$$n_2 = \frac{c}{v_2} \qquad (14.104)$$

Thus, the wave nature of light explains Snell's law of refraction.

The Double Slit   Consider two narrow slits separated by a small distance. Figure 14.54 shows the top view of the slits. If a plane monochromatic wave is incident on the two slits so that a wavefront is parallel to the plane of the slits, then the radiation at the two slits is in phase. According to Huygens' principle, each slit acts as a source of radiation. If the distance to the screen is much greater than the distance between the slits, then the rays AB and $A'B$ are almost parallel. For this case, the ray AB will almost be equal to the ray $B'B$, and the difference in path length is almost equal to $A'B'$. From this diagram

$$A'B' = d \sin \theta$$

where

d = distance between the slits.

If the intensity at B is to be a maximum, then

$$A'B' \approx n\lambda$$

or

$$d \sin \theta \approx n\lambda \qquad (\text{maximum}) \qquad (14.105)$$

where n is an integer. In addition,

$$\tan \theta = \frac{y}{L}$$

If the angles are small,

$$\sin \theta \approx \tan \theta \approx \theta$$

Thus, Equation 14.105 becomes

$$\frac{dy}{L} \approx n\lambda$$

or

$$y \approx \frac{n\lambda L}{d} \qquad (\text{maximum}) \qquad (14.106)$$

For a minimum intensity,

$$A'B' \approx (2n+1)\frac{\lambda}{2} \approx d \sin \theta \qquad (\text{minimum}) \qquad (14.107)$$

FIGURE 14.54 Interference caused by a plane wave incident on a pair of slits.

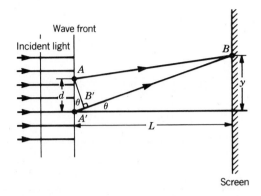

FIGURE 14.55 Interference pattern.

For small angles, Equation 14.107 becomes

$$\frac{dy}{L} \approx \frac{2n+l}{2}\lambda$$

or

$$y \approx \frac{L\,(2n+l)\,\lambda}{2d} \qquad (14.108)$$

One will observe a series of bright lines separated by dark spaces on the screen. This pattern is illustrated in Figure 14.55, and it is called an interference pattern. The bright lines are called fringes. As can be seen from Equation 14.101, the interference pattern depends on the wavelength, distance between the slits, and the distance to the screen. Figure 14.55 shows that the intensity is reduced as y increases. This is caused by diffraction, which is discussed later in this chapter. Examples 14.18 and 14.19 show calculations of fringe patterns.

EXAMPLE 14.18 Calculation of wavelength from an interference pattern
. . . .

Given: An interference pattern is observed on a screen that is 2.00 m from a double slit. The separation between the slits is 0.100 mm. The distance from the central bright fringe ($n = 0$) to the seventh bright fringe ($n = 7$) is 10.0 cm.

Find: The wavelength of the incident light.

Solution: From Equation 14.106

$$y = \frac{n\lambda L}{d}$$

Thus,

$$\lambda = \frac{yd}{nL}$$

$$\lambda = \frac{(10.0 \times 10^{-2} \text{ m})(1.0 \times 10^{-4} \text{ m})}{(7)(2.0 \text{ m})}$$

$$\lambda = 7.14 \times 10^{-7} \text{ m} = 714 \text{ nm}$$

EXAMPLE 14.19   Distance between fringes

. . . .

Given: Monochromatic light with a wavelength of 500 nm illuminates a double slit. The light is a parallel beam normal to the plane of the slits. The separation of the slits is 0.150 mm and the distance to the screen is 2.00 m.

Find: The distance between the fringes.

Solution: For the nth fringe, Equation 14.106 gives

$$y_n = \frac{n\lambda L}{d}$$

For the $(n + 1)^{\text{th}}$ fringe, Equation 14.106 gives

$$y_{n+1} = \frac{(n + 1)\lambda L}{d}$$

The distance between fringes is

$$\Delta y = y_{n+1} - y_n = (n + 1)\frac{\lambda L}{d} - (n)\frac{\lambda L}{d}$$

$$\Delta y = \frac{\lambda L}{d}$$

The distance between fringes is a constant.

$$\Delta y = \frac{(500 \times 10^{-9} \text{ m})(2 \text{ m})}{(0.150 \times 10^{-3} \text{ m})}$$

$$\Delta y = 6.67 \times 10^{-3} \text{ m} = 0.667 \text{ cm}$$

Change in Phase at a Reflecting Boundary   Light as well as waves on a string can change phase at a reflecting boundary (see Chapter 11). If the velocity decreases when light travels from one medium to another, the reflected waves undergo a phase change in π radians. This is equivalent to a path length of one-half wavelength. However, if the velocity increases as the wave travels from one medium to another, there is no change in phase in the reflected wave. For example, the reflected wave when light travels from air to glass undergoes a phase change of π

FIGURE 14.56 Optical diagram of a thin wedge.

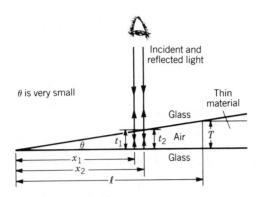

radians. When light travels from glass to air, there is no change in phase of the reflected wave.

The Thin Wedge   Figure 14.56 shows how an interference pattern can be produced by a thin air wedge. Two optically flat pieces of glass are separated at one end by a thin material (tissue paper) and illuminated by nearly monochromatic light. If the two surfaces are held together with rubber bands, fringes will be visible to an observer. The analysis of this problem is simplified if the incident light is almost perpendicular to the two surfaces as shown in Figure 14.56.

If bright fringes occur at positions x_1 and x_2 in Figure 14.56, then the reflected wave at the glass to air surface must be in phase with the reflected wave from the air to glass surfaces. For this condition to hold for two consecutive fringes

$$2t_1 + \frac{\lambda}{2} = n\lambda$$

$$\begin{array}{ccc} \text{Path length of} & \text{Path length due} & \text{Even number} \\ \text{transmitted ray} & + \text{to change in phase} = & \text{of wavelengths} \\ & \text{of reflected ray} & \end{array}$$

and

$$2t_2 + \frac{\lambda}{2} = (n + 1)\,\lambda$$

Subtracting these two equations yields

$$2(t_2 - t_1) = \lambda \qquad\qquad (14.109)$$

$$\Delta t = t_2 - t_1 = \frac{\lambda}{2} \qquad\qquad (14.110)$$

However, for small angles, $\tan \theta \approx \theta$. Thus, to a good approximation

$$\theta = \frac{t_1}{x_1}$$

and

$$\theta = \frac{t_2}{x_2}$$

or

$$t_1 = x_1\theta \tag{14.111}$$

and

$$t_2 = x_2\theta \tag{14.112}$$

Subtracting the second equation from the first yields

$$\Delta t = t_2 - t_1 = \theta(x_2 - x_1) = \theta\Delta x \tag{14.113}$$

Substituting this value for Δt into Equation 14.110 yields

$$\Delta x = \frac{\lambda}{2\theta} \tag{14.114}$$

Thus, between two successive bright fringes, the thickness increases by $\lambda/2$ and x increases by $\lambda/2\theta$. The value of θ is also approximately equal to

$$\theta \approx \frac{T}{l} \tag{14.115}$$

where

$\quad T$ = thickness of the material
$\quad l$ = corresponding value of x

Substituting this value for θ in Equation 14.114 yields

$$\Delta x = \frac{\lambda l}{2T} \tag{14.116}$$

Examples 14.20 and 14.21 give sample problems.

EXAMPLE 14.20   Wedge angle calculation
. . . .

Given: An interference pattern exhibits a fringe spacing of 0.8 cm. The wavelength of light is 633 nm.

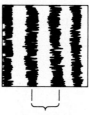

0.8 cm = 633 nm

Find: Angle of air wedge producing fringe pattern.

Solution: $\theta = \dfrac{\lambda}{2\Delta x}$

$$\theta = \frac{633 \times 10^{-9} \text{ m}}{2 (0.8 \times 10^{-2} \text{ m})}$$

$$\theta = 3.96 \times 10^{-5} \text{ rad.} \quad \text{(A small angle!)}$$

EXAMPLE 14.21 Thickness of a material
. . . .

Given: When a thin material is used to form a wedge between two plane glass plates, 150 bright fringes are observed between the edge of the plates and the thin object. The wavelength of light is 633 nm.

Find: The thickness of the object.

Solution: The separation between the plates increases by $\lambda/2$ for each successive fringe. Thus the thickness of the object is

$$T = (N)(\lambda/2)$$

where

N = total number of fringes

$$T = (150)(633 \times 10^{-9} \text{ m})$$

$$T = 9.5 \times 10^{-5} \text{ m}$$

$$T = 9.5 \times 10^{-2} \text{ mm}$$

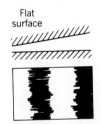

Flat
surface

FIGURE 14.57 Interference pattern from a flat surface and a small angle.

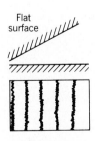

Flat
surface

FIGURE 14.58 Interference pattern from a flat surface and a larger angle.

Flatness of Optical Surfaces Interference fringes are used to determine the flatness of optical surfaces by comparing them with a surface of known flatness, called a "master" or "standard." If the master is truly flat, the interference pattern produced by placing the unknown surface in contact with the master surface is a contour map of the air space between the two and, therefore, is a reliable indication of the shape of the unknown surface. This procedure is demonstrated in Figures 14.57 to 14.59. In each case, the lower reflecting surface is the master flat.

If both surfaces are flat, the interference pattern will be a series of straight, regularly spaced fringes. Broad fringes (Figure 14.57) indicate a small angle between the surfaces, and fine fringes (Figure 14.58) indicate a larger angle, but both of these patterns signify a flat surface.

Figure 14.59 shows the concentric pattern of rings, called "Newton's rings," resulting from a spherical surface in contact with a flat surface.

This method of producing interference fringes is a standard test procedure to test the surface flatness of optical and machine surfaces. For example, Figure 14.60 shows the interference pattern produced by an irregular surface and an optical flat surface. The irregular surface can be worked on until the fringes disappear. The surface is then known to be flat to the order of a wavelength.

FIGURE 14.59 Interference pattern from a spherical surface.

FIGURE 14.60 Interference pattern produced by an irregular surface.

Diffraction   When light is incident on the edge of an opaque object, an intensity pattern is observed close to the shadow of the edge. This intensity pattern can be explained by using the wave theory of light. The intensity pattern is called a diffraction pattern.

The top view of a single slit is shown in Figure 14.61. Parallel light is incident on the slit such that the plane of the slit is a wave front. According to Huygens' principle, each part of the slit can be considered as a new source of radiation and all the new sources are in phase. Two sources are shown in Figure 14.61. One source is at the edge of the slit and the other source is at the center of the slit. These two sources can interfere at the screen that is placed a large distance from the slit. For constructive interference, the two path lengths must differ by an even number of half wavelengths, and for destructive interference, the two path lengths must differ by an odd number of half wavelengths.

If the distance to the screen is much greater than the slit width, then the lines AB' and BB' are almost parallel. For this case, the path difference becomes

$$\Delta l \approx \frac{d \sin \theta}{2} \qquad (14.117)$$

where

$\quad \Delta l$ = path difference shown on the diagram
$\quad d$ = slit width
$\quad \theta$ = angle shown on the diagram

Thus, for constructive interference

$$\frac{d \sin \theta}{2} \approx n\lambda \qquad (14.118)$$

and for destructive interference

$$\frac{d \sin \theta}{2} \approx (2n + 1) \frac{\lambda}{2} \qquad (14.119)$$

where

$\quad n$ = an integer
$\quad \lambda$ = wavelength

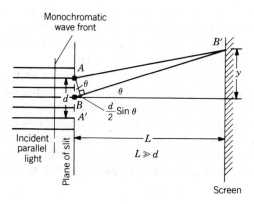

FIGURE 14.61 Optical diagram of a single slit.

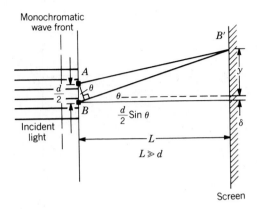

FIGURE 14.62 Two additional sources for the single slit.

If two additional sources are taken next to the previous sources as shown in Figure 14.62, the conditions given by Equations 14.118 and 14.119 also apply. One can continue in this way until the entire slit is considered. For each pair of sources, the conditions given by Equations 14.118 and 14.119 apply. Thus, a maximum intensity occurs when

$$n\lambda = \frac{d \sin \theta}{2} \quad \text{(maximum)} \tag{14.120}$$

and a minimum intensity occurs when

$$(2n + 1)\frac{\lambda}{2} = \frac{d \sin \theta}{2} \quad \text{(minimum)} \tag{14.121}$$

From Figure 14.62,

$$\tan \theta = \frac{y + \delta}{L} \tag{14.122}$$

However, δ is very small (less than the slit width). In addition, the angle θ will usually be small. For these conditions

$$\theta \approx \sin \theta \approx \tan \theta \approx y/L \tag{14.123}$$

Substituting y/L for θ in Equations 14.120 and 14.121 yields

$$n\lambda = \frac{yd}{2L} \quad \text{(maximum)} \tag{14.124}$$

$$(2n + 1)\frac{\lambda}{2} = \frac{yd}{2L} \quad \text{(minimum)} \tag{14.125}$$

A diagram showing the actual intensity distribution for a single narrow slit is shown in Figure 14.63.

Figure 14.64 shows a narrow slit illuminated by a parallel beam of monochromatic light (laser beam) and the corresponding diffraction pattern.

Example 14.22 gives a sample calculation for the width of the central fringe and the positions of the next maxima. Notice that $L >> d$.

FIGURE 14.63 Intensity distribution given by diffraction from a single slit.

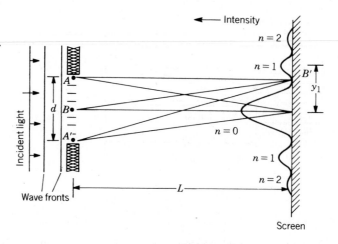

FIGURE 14.64 Single-slit diffraction pattern.

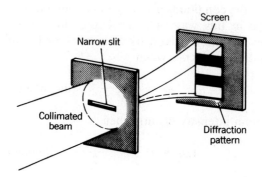

EXAMPLE 14.22   The single slit

. . . .

Given: A single slit has a width of 0.010 cm and a screen is placed 2.00 m from the slit. Parallel light is incident on the slit with wave fronts parallel to the plane of the slit. The wavelength is 600 nm.

Find: **a.** The width of the central maximum ($n = 0$ and $n = -1$).
              b. The positions of the next two maxima ($n = 1$, $n = -1$).

Solution: **a.** The positions of the minimum points on each side of the central maximum are given by the equation

$$(2n + 1)\frac{\lambda}{2} = \frac{yd}{2L}$$

where
$$n = 0 \text{ and } n = -1$$

Thus,

$$y = \frac{(2n + 1)\,\lambda L}{d}$$

$$y_0 = \frac{\lambda L}{d}$$

$$y_{-1} = \frac{\lambda L}{d}$$

The total width is $y_0 - y_{-1}$.

$$w = y_0 - y_{-1} = \frac{\lambda L}{d} - \left(-\frac{\lambda L}{d}\right) = \frac{2\lambda L}{d}$$

w = width of the central maximum

$$w = \frac{(2)(600 \times 10^{-9}\text{ m})(2\text{ m})}{0.01 \times 10^{-2}\text{ m}}$$

$$w = 2.4 \times 10^{-2}\text{ m}$$

$$w = 2.4\text{ cm}$$

w is much greater than the width of the slit.

b. The positions of the maximum points are given by

$$n\lambda = \frac{yd}{2L}$$

or

$$y = \frac{2n\lambda L}{d}$$

For $n = 1$

$$y_1 = \frac{(2)(1)(600 \times 10^{-9}\text{ m})(2\text{ m})}{0.01 \times 10^{-2}\text{ m}}$$

$$y_1 = 2.4 \times 10^{-2}\text{ m}$$

$$y_1 = 2.4\text{ cm}$$

For $n = -1$

$$y_{-1} = \frac{(2)(-1)(600 \times 10^{-9}\text{ m})(2\text{ m})}{0.01 \times 10^{-2}\text{ m}}$$

$$y_{-1} = -2.4 \times 10^{-2}\text{ m}$$

$$y_{-1} = -2.4\text{ cm}$$

The Double-Slit with Diffraction When the conditions for maxima and minima were derived for the double slit, the width of the slits was neglected. Since only two rays were used to calculate the fringe pattern, it is called a interference pattern. When a single slit was considered, many rays were used to calculate the fringe pattern. When many rays contribute to the fringe pattern, it is called a diffraction pattern. The double slit is really a combination of interference and

FIGURE 14.65 Intensity
pattern for a double slit.

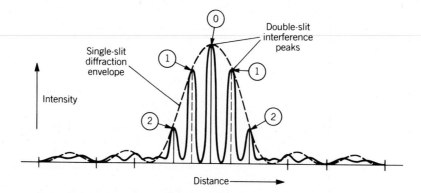

FIGURE 14.65 Intensity
pattern for a double slit.

diffraction. The derivation of the intensity pattern is beyond the scope of this text, but it is illustrated in Figure 14.65. The envelope of the pattern (dashed line), which is the single-slit diffraction pattern, depends on the width of the slits. Each of the narrow peaks represents a maximum for the interference between two slits, and its location depends on the separation of the two slits. Numbers in the circles are values for the order of the interference pattern corresponding to a whole number of wavelengths by which the optical path length from one slit differs from that of the other slit.

The Diffraction Grating A diffraction grating consists of a large number of parallel slits in the same plane with constant width and spacing. The number of lines per centimeter is usually very large, which makes the spacing between lines small. A top view of part of a grating is shown in Figure 14.66. The grating is illuminated with parallel monochromatic light, and the wave fronts are parallel to the plane of the grating.

To determine the location of the maximum intensities of the interference pattern, only two slits need be considered at a time. For each pair of slits, the condition for constructive interference is

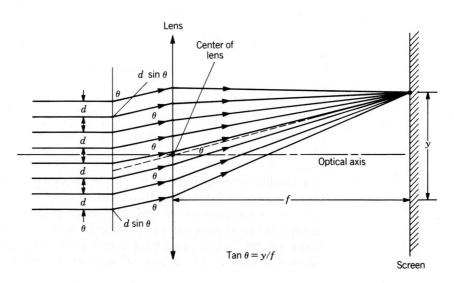

FIGURE 14.66 Top view of
a diffraction grating.

FIGURE 14.67 Interference
pattern from a diffraction
grating.

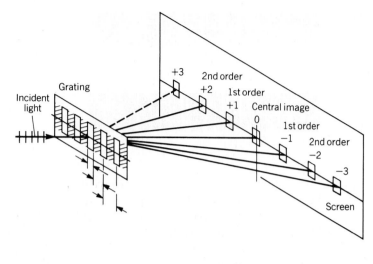

$$d \sin \theta = n\lambda \qquad (14.126)$$

Since d is a constant for each pair of slits, the same angle θ will be obtained for each pair for constructive interference. Thus, the maximum intensities occur for the angles given by Equation 14.126.

To obtain a sharp interference pattern, a lens can be used to focus the parallel rays shown in Figure 14.66 to a common line in the focal plane. If a lens is not used, then the screen must be placed a large distance from the grating to make Equation 14.126 a good approximation. Lenses can also be used to obtain better interference patterns for the single and double slits. When a lens is used, Equation 14.127 gives the tangent of the angle θ:

$$\tan \theta = y/f \qquad (14.127)$$

where

f = focal length of the lens

As the total number of slits increases, the sharpness of the interference pattern increases so that a series of narrow lines is observed on the screen. This is illustrated in Figure 14.67 for the case where a lens is not used. When $n = 0$, Equation 14.127 gives $\theta = 0$ for all wavelengths. This line is called the central image. When $n = \pm 1$, a line is observed on each side of the central image. These lines are called the first-order lines, and the distance from the central image is the same for these two lines. When $n = \pm 2$, two more lines occur as shown in Figure 14.67. These lines are called second-order lines. There can be many orders, but the intensity becomes less as the order increases. Figure 14.68 shows a photograph of the orders for a wavelength of 400 nm, and Figure 14.69 shows the orders for a wavelength of 500 nm. If both wavelengths are incident on the grating,

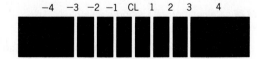

FIGURE 14.68 Interference
pattern for 400-nm light.

FIGURE 14.69   Interference pattern for 500-nm light.

FIGURE 14.70   Interference pattern for a mixture of 400- and 500-nm light.

both patterns are observed as shown in Figure 14.70. For this case, the orders do not overlap, and each wavelength is easily identified.

In general, the light incident on a grating may be composed of many wavelengths. There can also be continuum radiation as well as spectral lines. The grating will spread out the spectrum, and Equation 14.126 can be used to determine the wavelength of any part of the spectrum. In some cases, the various orders will overlap and it is necessary to use filters or prisms in front of the grating. Spectrometers that use gratings are discussed later in this chapter. Examples 14.23 and 14.24 give sample problems. Example 14.25 shows how Equations 14.126 and 14.127 may be used to determine an unknown wavelength. This can be easily demonstrated in the laboratory if a laser source and small grating are available.

EXAMPLE 14.23   Diffraction angle calculation
. . . .

Given: Grating spacing $d = 1.896 \times 10^{-6}$ m and wavelength of light $\lambda = 488$ nm (blue).

Find: First-order diffraction angle.

Solution: $d \sin \theta = n\lambda$

$$\sin \theta = \frac{n\lambda}{d}$$

$$\sin \theta = \frac{(1)(488 \times 10^{-9} \text{ m})}{1.896 \times 10^{-6} \text{ m}}$$

$$\sin \theta = 0.2574$$

$$\theta = \sin^{-1} (0.2574)$$

$$\theta = 14.9°$$

EXAMPLE 14.24   Calculation of grating spacing
. . . .

Given: Second-order diffraction angle $\theta = 37.72°$ and wavelength of light $\lambda = 580$ nm.

Find: Grating spacing.

Solution: $d \sin \theta = n\lambda$

$$d = \frac{n\lambda}{\sin \theta}$$

$$d = \frac{(2)(580 \times 10^{-9} \text{ m})}{\sin 37.72°}$$

$$d = \frac{(2)(580 \times 10^{-9} \text{ m})}{0.6118}$$

$$d = 1.896 \times 10^{-6} \text{ m}$$

EXAMPLE 14.25   Using a grating to measure wavelengths
. . . .

Given: A small grating has 300.0 lines per millimeter. Parallel light from a laser is incident on the grating such that the wave fronts are parallel to the plane of the grating. A screen is placed 2.000 m from the grating, and a pattern for a single wavelength is observed. The distance to the first-order image from the central image is 38.70 cm.

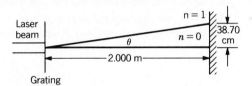

Find: The wavelength of the laser light.

Solution: A diagram of the experiment is shown below.
From the definition of the tangent

$$\tan \theta = \frac{38.70 \text{ cm}}{200 \text{ cm}}$$

$$\tan \theta = 0.1935$$

$$\theta = 10.95°$$

From Equation 14.126

$$n\lambda = d \sin \theta$$

$$\lambda = \frac{d \sin \theta}{n}$$

$$\lambda = \frac{\sin \theta}{Nn}$$

where
   N = number of lines per meter

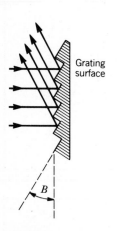

FIGURE 14.71   Reflection grating.

$$\lambda = \frac{\sin 10.95}{(1)300 \times 10^3 \text{ m}^{-1}}$$

$$\lambda = 6.332 \times 10^{-7} \text{ m}$$

$$\lambda = 633.2 \text{ nm}$$

Figure 14.71 shows a reflection grating. Instead of a series of slits, this type of grating is composed of a series of reflective surfaces. Reflection gratings have much higher efficiency than transmission gratings because little light is absorbed. Angle *B* of the reflecting surfaces, with respect to the grating surface, is chosen to reflect the maximum amount of light in a particular direction. This angle is called the "blaze angle" of the grating.

Research grade gratings are designed to give the best diffraction efficiency for a particular wavelength of light in a particular order of diffraction. A number of optical instruments—including spectroscopes and monochromators—employ a diffraction grating as a principal element.

SECTION 14.2   EXERCISES

1. In one or two sentences, summarize Huygens' principle.
2. Sketch the wave pattern produced when plane waves are diffracted at an edge barrier.
3. Label the accompanying sketch and include several orders of diffraction, a single-slit diffraction envelope, and double-slit interference peaks.

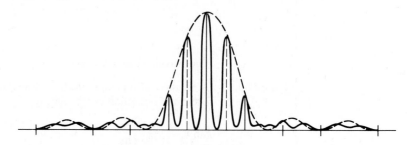

4. Monochromatic light having a wavelength of 500 nm is incident on a double slit having a separation of 0.020 cm. The wave fronts of the incident light are parallel to the plane of the parallel slits. If a screen is placed 200 cm away from the slits, what will be the linear separation between fringes on the screen?
5. Interference fringes are formed on a screen by a double slit. The screen is 130 cm from the double slit, and the linear fringe separation is 0.155 cm. If the separation between the slits is 0.044 cm, find the wavelength of the light.
6. Parallel light having a wavelength of 500 nm is incident on a single slit. Wave fronts are parallel to the plane of the slit. The slit width is 0.05 cm, and a lens with a 100-cm focal length focuses the diffraction pattern on a screen. Find (a) the position of the first minimum, (b) the position of the first secondary maximum ($n = 1$), and (c) the distance to the second minimum.
7. Light from a helium–neon laser ($l = 633$ nm) is used to illuminate an air wedge. Fringe spacing is 0.025 cm. Calculate the wedge angle.

8. Identify the type of surface that would produce each of the following interference patterns if placed in contact with a flat surface.

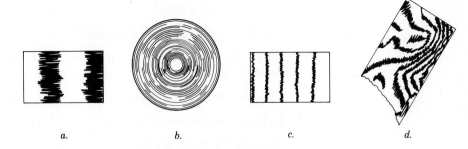

a. *b.* *c.* *d.*

9. What is the wavelength of light that will produce a fringe spacing of 1 cm when the wedge angle is 4×10^{-6} radians?

10. An air wedge has an angle of 3×10^{-5} radians. What will the fringe spacing of the interference pattern be if the wavelength of light producing the interference pattern is 700 nm?

11. Find the width of the central maximum of a diffraction pattern when the slit width is 0.05 cm, the wavelength is 700 nm, and the lens focal length is 150 cm. What is the width of the central maximum if the slit width is 0.10 cm? What is it if the slit width is 1.0 cm?

12. A diffraction grating has a spacing of 2×10^{-6} m. Calculate diffraction angles for the following:

   First order of 400-nm light

   First order of 650-nm light

   Second order of 550-nm light

13. A diffraction grating produces fourth-order diffraction of 633 nm at 42°. What is the grating spacing?

14. For the grating in Problem 13, what wavelength has third-order diffraction at 42°?

15. A diffraction grating has 4000 lines/cm. At what angle will the first-order bright fringes appear if the wavelength of the incident light is 650 nm?

16. A bright fringe occurs at an angle of 26°. If the diffraction grating has 2700 lines/cm and the wavelength of the incident light is 541 nm, what is the order of the fringe?

17. A lens focuses a spectrum formed by a diffraction grating on a screen. The focal length of the lens is 100 cm. If the grating spacing is 2×10^{-6} m and the wavelength is 400 nm, find the distance from the maximum of the central image for orders 1, 2, and 3.

18. Suppose one observes a spectrum that has a range from 700 to 400 nm. Will the second-order spectrum overlap with the first-order spectrum? Will the third-order spectrum overlap with the second-order spectrum?

14.3

OBJECTIVES

LASERS Upon completion of this section, the student should be able to

- Describe the process of stimulated emission.

- List and give the purpose of the three elements of a laser.

- Calculate the axial mode spacing of an optical cavity.

- Given the line width, estimate the number of axial modes in the laser output.

- Given the radii of curvature of the mirrors of an optical cavity, estimate the stability of the cavity.

- List three different optical resonators.

- Given the reflectivity of each minor in an optical resonator, calculate the saturated gain.

- Describe the He–Ne laser.

- Describe the electric CO_2 laser.

- Describe the Nd:YAG laser.

- Describe the tuned dye laser.

- Describe the photolytic laser.

- Describe the CO_2 gasdynamic laser.

- Describe the HF laser.

- State the properties of laser radiation when the laser operates in a single axial mode.

- Calculate the irradiance on a target.

- List the important applications for lasers.

- Explain how the distance to the moon can be measured by using a pulsed laser.

DISCUSSION

The word "laser" is an acronym for Light Amplification by Stimulated Emission of Radiation. Chapter 13 discusses the laser as a light source, and Chapter 12 explains stimulated emission. In addition, Chapter 5 covers the characteristics of an electric laser discharge. This section discusses the various lasers, the properties of laser radiation, and the application of lasers.

Stimulated emission is shown schematically in Figure 14.72. Initially an atom or molecule is in an excited state. This high-energy state E_3 is represented by a solid circle in the figure. An incident photon with an energy equal to the energy difference between this state and a lower state E_2 can interact with the atom or molecule. This stimulates the atom or molecule to level E_2, and a photon of the same energy is emitted in the same direction as the incident photon. In addition, the two photons are in phase. The photons are said to be coherent. Since the two photons have the same energy, they also have the same wavelength.

As stimulated emission occurs, transitions to the lower level increase the population of the lower level and decrease the population of the upper level. As the population of the lower level increases, the probability of absorption increases. Also, the probability of stimulated emission decreases as the population of the upper level decreases. At some point in time, the probability for absorption becomes greater than the probability for stimulated emission; and the intensity of the beam becomes zero. In such an isolated system, the population of the upper

FIGURE 14.72 Stimulated emission.

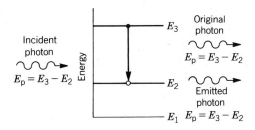

level will also decrease from transitions caused by collisions. When equilibrium is achieved, the population of the lower level will be greater than the population of the upper level. Thus, the major problem in making a laser is to devise a method to obtain a population inversion.

A population inversion is obtained through a process called pumping. The pumping process always requires energy. The energy source may be electrical, optical, chemical, or thermal. Part of the supplied energy creates the population inversion that can produce a laser beam. The energy output in the laser beam is usually much less than the input energy. Thus, lasers are not efficient energy transformers.

When a population inversion occurs, the active medium will amplify a light beam of the proper frequency. If one has an amplifier, it is possible to build an oscillator. For a laser, this is accomplished by placing the active medium in an optical cavity (optical resonator). An optical cavity consists of two mirrors having a common optical axis. One mirror has a high reflectivity. This mirror reflects the laser light with very little loss in energy. The other mirror is a partially reflecting mirror. It allows part of the incident laser radiation to escape from the optical cavity in the form of a beam. The optical cavity or optical resonator is similar in operation to a closed-end organ pipe. To build up a signal, the length of the cavity must be an even number of half wavelengths.

From this discussion, one can isolate three basic elements of a laser. They are the active medium, the pump, and the resonant optical cavity. For example, the active medium can be a mixture of helium and neon. If this mixture is placed in a low-pressure discharge tube, an electric power supply can pump the medium through electron collisions. When the discharge is placed in an optical resonator, laser radiation can be observed.

The Optical Resonator The original laser cavity consisted of two plane mirrors that were aligned so that their plane surfaces were parallel. As a simple analysis, the various frequencies that resonate in this cavity correspond to wavelengths given by (axial modes)

$$n\left(\frac{\lambda}{2}\right) = l \qquad (14.128)$$

where

n = an integer
λ = wavelength
l = length of the cavity

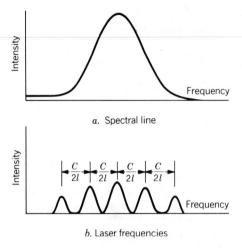

a. Spectral line

b. Laser frequencies

Equation 14.128 can also be written as

$$\frac{1}{\lambda} = \frac{n}{2l}$$

Multiplying this equation by c yields

$$v_n = \frac{c}{\lambda_n} = \frac{nc}{2l} \tag{14.129}$$

where

v_n = resonant frequency
c = velocity of light

Each value of n corresponds to a possible frequency (mode). To calculate the mode spacing let n increase by one. Then Equation 14.129 becomes

$$v_{n+1} = \frac{c}{\lambda_{n+1}} = \frac{(n+1)c}{2l} \tag{14.130}$$

Subtracting Equation 14.129 from Equation 14.130 gives the mode spacing

$$\Delta v = v_{n+1} - v_n = \frac{c}{2l} \tag{14.131}$$

Notice that the mode spacing decreases as the cavity length increases. Since a spectral line has a finite width (the energy levels have a finite width), several modes may oscillate at the same time as illustrated in Figure 14.73.

A cavity composed of two plane, parallel mirrors is difficult to keep in alignment. Other optical resonators may have one or two spherical mirrors that are arranged to give greater stability. Two of these arrangements are shown in Figure 14.74. These cavities may be used if the optical gain of the medium is small. If the optical gain is large, however, unstable resonators are often used.

It can be shown that the condition for stability is given by the expression

$$0 \le \left(1 - \frac{L}{R_1}\right)\left(1 - \frac{L}{R_2}\right) \le 1 \tag{14.132}$$

FIGURE 14.74 Two types of laser cavities using spherical mirrors.

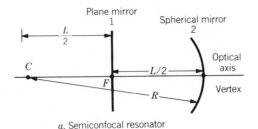

a. Semiconfocal resonator

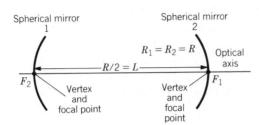

b. Focal resonator

where

L = length of the cavity
R_1 = radius of the first mirror
R_2 = radius of the second mirror

For a plane mirror, $R = \infty$. If two plane mirrors are used, the product is 1. This cavity is just stable. For the semiconfocal cavity (Figure 14.74*a*), the product is 1/2. Thus, this cavity should be very stable. The product for the focal resonator is 1/4. This cavity should also be very stable (Figure 14.74*b*). Examples 14.26 and 14.27 give several cavity calculations.

EXAMPLE 14.26   Confocal resonator

. . . .

Given: Two spherical mirrors have equal radii. The center of each mirror lies at the vertex of the other mirror as shown in the diagram.

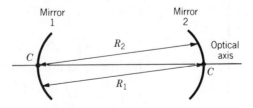

Find: The condition for stability.

Solution: If the cavity is stable

$$0 \leq \left(1 - \frac{L}{R_1}\right)\left(1 - \frac{L}{R_2}\right) \leq 1$$

The length of the cavity is $L = R_1 = R_2$.
Thus,

$$0 \le (0)(0) \le 1$$

The cavity is just stable.

EXAMPLE 14.27   Axial mode spacing
. . . .

Given: Two plane mirrors are used as an optical cavity. The spectral width and population inversion associated with the active medium are such that laser action can occur over a spectral width of 3.2×10^{-9} Hz.

Find: The number of possible axial modes if the length of the cavity is 30 cm and if it is 10 cm.

Solution: The mode spacing is given by

$$\Delta v = \frac{c}{2l}$$

For a 30-cm cavity

$$\Delta v = \frac{3 \times 10^{10} \text{ cm/sec}}{2 \times 30 \text{ cm}}$$

$$\Delta v = 5 \times 10^8 \text{ Hz}$$

The number of axial modes is approximately

$$N = \frac{w}{\Delta v}$$

$$N = \frac{3.2 \times 10^9 \text{ Hz}}{5 \times 10^8 \text{ Hz}}$$

$$N \approx 6 \text{ axial modes}$$

For a 10-cm cavity

$$\Delta v = \frac{3 \times 10^{10} \text{ cm/sec}}{2 \times 10 \text{ cm}}$$

$$\Delta v = 1.5 \times 10^9 \text{ Hz}$$

The number of axial modes is

$$N = \frac{3.2 \times 10^9 \text{ Hz}}{1.5 \times 10^9 \text{ Hz}}$$

$$N \approx 2 \text{ axial modes}$$

Thus, the 10-cm cavity gives radiation that is more nearly monochromatic than the 30-cm cavity.

Saturated Gain Suppose the optical cavity is composed of two parallel, plane mirrors. Photons reflected at the output mirror (mirror 2) will travel back to mirror 1, where they will be reflected and travel forward to the output mirror. As the photons travel back and forth, they will cause stimulated emission to occur in the active medium. This will cause the number of photons in the cavity to increase. However, photons will be lost through the output mirror and by absorption processes in the medium and at mirror 1. When the photon gain per unit time is equal to the photon loss per unit time, the number of photons in the cavity will remain constant. The laser output will be constant, and the gain of the active medium is said to be saturated.

As photons pass through the active medium one way, their number is increased by a factor called the gain. Let G_{SAT} be the saturated gain. If n_2 photons are incident on the output mirror, then $n_2 R_2$ photons will be reflected, where R_2 is the reflectivity of the output mirror. When they are incident on mirror 1, there will be $(G_{SAT}) n_2 R_2$ photons if the gain is saturated. The number reflected is $(G_{SAT}) n_2 R_2 R_1$, where R_1 is the reflectivity of mirror 1. After passing through the active medium again, the number of photons incident on the output mirror becomes $(G_{SAT})^2 n_1 R_1 R_2$. If the number of photons incident on the output mirror is to remain constant, then

$$n_1 = (G_{SAT})^2 n_1 R_1 R_2$$

or

$$1 = (G_{SAT})^2 R_1 R_2 \qquad (14.133)$$

Example 14.28 shows an application of Equation 14.133.

EXAMPLE 14.28 Saturated gain
. . . .

Given: An active medium lies between two plane, parallel mirrors. The reflectivity of the output mirror is 0.53 and the reflectivity of the first mirror is 0.94.

Find: The saturated gain

Solution: From Equation 14.134

$$G_{SAT}^2 = \frac{1}{R_1 R_2}$$

$$G_{SAT} = \sqrt{\frac{1}{R_1 R_2}}$$

$$G_{SAT} = \sqrt{\frac{1}{(0.53)(0.94)}}$$

$$G_{SAT} = \sqrt{\frac{1}{0.4982}}$$

$$G_{SAT} = \sqrt{2.007}$$

$$G_{SAT} = 1.42$$

Electrically Pumped Lasers   In an electrically pumped laser, the active medium is a partially ionized gas. The helium-neon laser, the carbon dioxide laser, the nitrogen laser, and the argon laser are common examples. The most widely used electrically pumped laser is the helium–neon laser. They are rugged, and they can withstand harsh weather conditions.

The helium–neon laser uses a mixture of helium and neon in a tube at relatively low pressure. A dc power supply is connected to the tube through a ballast resistor. The tube is placed in an optical cavity as shown in Figure 14.75. When the gas is ionized, electrons pass through the gas and gain energy from the electric field. They collide with helium atoms and excite these atoms to upper energy levels. This is a relatively efficient process. On the other hand, neon atoms are not easily excited by collisions with electrons. The excited helium atoms have a long lifetime, but they can easily transfer energy to neon atoms. Thus, there is a high probability that an excited helium atom will transfer its energy to a neon atom before it returns to its ground state. The probability of energy transfer from

FIGURE 14.75   Schematic of a helium–neon laser.

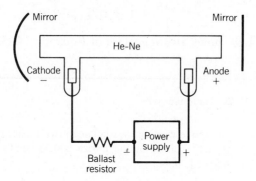

FIGURE 14.76   Energy level diagram for a helium–neon system.

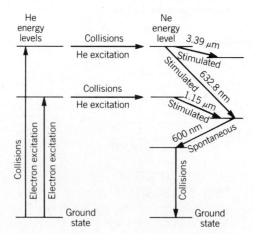

FIGURE 14.77 Energy level diagram for a nitrogen-carbon dioxide system.

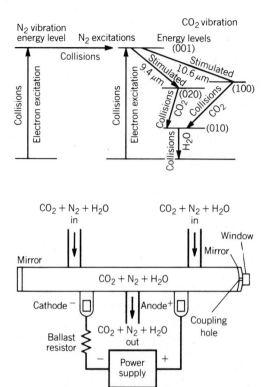

FIGURE 14.78 Schematic for a carbon dioxide laser.

excited helium atoms to neon atoms is large because the energy levels of helium are almost equal to the corresponding energy levels for neon. The process is shown diagrammatically in Figure 14.76. The net result of these processes is to establish a population inversion in the neon atoms. The solution of the equations that represent the rates of these processes is beyond the scope of this text. However, three laser lines can be observed. Their wavelengths are 3.39 μm, 632.8 nm, and 1.15 μm.

The argon laser is similar to the helium–neon laser. If a relatively large current passes through a helium–argon mixture, the argon becomes partially ionized. The argon ions develop population inversions, and laser lines can be emitted at several wavelengths. Two intense laser lines occur at 488 and 514 nm (blue-green). Because high input power is required (kilowatts), these lasers must be water cooled.

The CO_2 laser is an example of a molecular laser. The CO_2 molecule has energy levels that depend on rotational, vibrational, and electronic energy. Thus, the energy level diagram is complex. Figure 14.77 shows the important vibrational energy levels for stimulated emission. The active medium is composed of CO_2, N_2, and H_2O. The excitation of the upper level (001) takes place because of CO_2 collisions with electrons and with excited N_2 molecules. Two laser lines are produced, one at 10.6 μm and one at 9.4 μm as shown in the diagram. The lower two laser levels (100) and (020) are depopulated by collisions with CO_2 molecules to the next lower level (010) shown in the diagram. This lower level is depopulated by collisions with H_2O molecules. The power output of the CO_2 laser can be 10 kilowatts or more. A diagram of an electric CO_2 laser is shown in Fig. 14.78.

FIGURE 14.79   Energy level diagram for the Nd$^+$ ion.

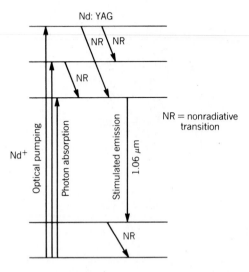

Optically Pumped Lasers   The ruby laser and the Nd:YAG laser are two examples of optically pumped lasers. The active medium in both cases is a solid. Today, the Nd:YAG is the most widely used solid-state laser. The Nd:YAG laser is composed of Yttrium Aluminum Garnet (Y$_3$Al$_5$O$_{12}$) that is doped with neodynium. The neodynium ions (Nd$^+$) absorb photons produced by the flashlamp, and a population inversion develops. The neodynium ions then undergo stimulated emission in an optical cavity to produce laser radiation. Yttruim Aluminum Garnet has a relatively high thermal conductivity, which makes the cooling much easier than for a ruby crystal. In addition, the Nd:YAG laser is more efficient than a ruby laser.

Figure 14.79 shows the energy level diagram for Nd$^+$ and the transitions that take place. The absorption of photons from the flashlamp populates the three upper energy bands. Nonradiative transitions from the upper two bands increase the population of the lowest upper band. Stimulated emission occurs between this energy band and the second level shown in the diagram. This second level is depopulated by nonradiative transitions to the ground state. The emitted photons by stimulated emission have a wavelength of approximately 1.06 μm.

A diagram of a Nd:YAG laser is shown in Figure 14.80. Instead of a flashlamp, it is possible to use a lamp with a continuous light output. When flashlamps are

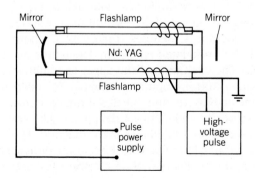

FIGURE 14.80   Schematic of an Nd:YAG laser.

FIGURE 14.81 Example of an energy level diagram for an organic dye.

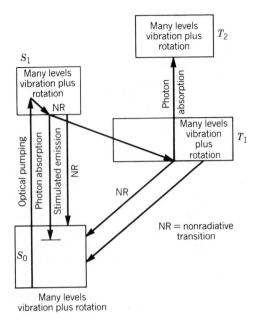

used, the laser output is a pulse of radiant energy. When a lamp with a continuous light output is used, the laser output is also continuous. To improve the transfer of the lamp radiation to the active laser medium, reflectors are used around the flashlamps.

A second type of optically pumped laser is the organic dye laser. The dye is dissolved in a liquid. For example, Rhodamine B can be dissolved in eythylene glycol. The chemical formula for this dye is $C_{28}H_{31}ClN_2O_3$. The solution can be pumped with a flashlamp or another laser with a wavelength in the proper range. Because the upper energy level of the dye has a very short lifetime (order of 10 nsec), the optical pumping must be done with an intense source that is focused down to a very small region in the dye solution. By using different dyes, it is possible to cover a wide spectral range.

The energy levels of the dye are due to a combination of electronic, vibrational, and rotational excitations as illustrated in Figure 14.81. The state labeled S_0 in the diagram is the electronic ground state. However, it is composed of a large number vibration–rotation levels. This is also true of the other electronic states, S_1, T_1, and T_2. Thus, the electronic states are broad bands of different vibration–rotation levels.

Photons from the optical pump excite levels in the electronic state labeled S_1. Excited vibration–rotation levels relax by nonradiative processes to the lower levels in this electronic excited state as shown in the figure. Stimulated emission can then occur between these lower levels of S_1 and other vibration–rotation levels in S_0. Thus, the total radiation produced by stimulated emission lies in a broad band of wavelengths. However, competing transitions also occur. There are nonradiative transitions from the level S_1 to S_0 and from S_1 to the electronic level labeled T_1. As the population of level T_1 builds up, the absorption of pump photons and laser photons from this level can populate the electronic state T_2. This is an undesirable effect. However, the population of T_1 decreases because of

FIGURE 14.82 Schematic of a tunable dye laser.

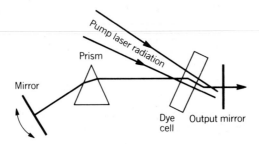

nonradiative transitions from T_1 to S_0 and radiative transitions from T_1 to S_0. The nonradiative rate can be enhanced by adding an extinguisher such as oxygen to the solution. The pumping rate must be great enough to maintain a population inversion under the effects of all these additional transitions.

Because the dye active medium has a gain over a broad band of wavelengths, it can be used as an active medium for a tunable laser. A diagram of a tunable dye laser is shown in Figure 14.82. A prism is placed in the optical cavity. By rotating the totally reflecting mirror, the cavity becomes aligned for different wavelengths. Thus, the output will be a narrow band of wavelengths. A grating can also be used as a tuning element.

Photolytic Laser   When certain molecules absorb photons, they will dissociate. This process is called photolysis. For some cases, a product formed by this process will be in an excited state. This produces the required population inversion for light amplification. If the medium is placed in an optical cavity (resonator), a laser beam is produced by stimulated emission. The laser is called a photolytic laser.

An example of a photolytic laser is the iodine laser. The molecule CF_3I absorbs a photon and dissociates. The products are CF_3 and excited iodine (I*). If a flashlamp is used as a pump, laser radiation is observed until the population inversion ceases to exist. The radiation emitted has a wavelength of 1.315 μm, and it corresponds to a change in electron spin from 1/2 to 3/2. This laser can have a very high gain. Figure 14.83a is a diagram of the laser and Figure 14.83b shows the flashlamp and laser output as a function of time when C_3F_7I is the initial molecule.

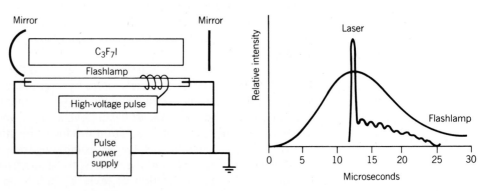

FIGURE 14.83 Iodine laser.   *a.*   *b.*

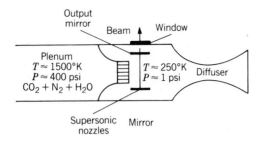

FIGURE 14.84 Carbon dioxide gas dynamic laser.

Thermal Pumping Gas dynamic lasers (GDL) use thermal pumping. An example is the CO_2 gas dynamic laser. The active medium is a mixture of CO_2 (8%), N_2 (90%), and H_2O (2%). The energy level diagram shown in Figure 14.77 also applies to this laser. A simplified diagram of the CO_2 GDL is shown in Figure 14.84.

The initial conditions of the mixture in the combustion chamber are approximately 1500°K and 400 psi. The high-pressure gas is expanded through an array of supersonic nozzles. During this process, internal energy is converted into kinetic energy of translation. The velocity of the flow from the nozzle is approximately 4000 ft/sec. The temperature drops to approximately 250°K, and the pressure drops to approximately 1 psi.

In the chamber, the lower levels (100) and (020) shown in Figure 14.77 have a greater population than the upper level (001). This is true because thermodynamic equilibrium exists among these levels. There is a significant population of the upper level, however. In addition, because of the high temperature, there is a significant population of the N_2 level shown in Figure 14.77. The gas cools as it passes through the supersonic nozzles, and the lower levels (100) and (020) relax rapidly through resonant collisions. The upper level (001) relaxes slowly through ordinary collisions. In addition, the N_2 excited levels relax slowly, and they transfer their energy to the upper level (001) of CO_2. The net effect is to develop a population inversion in the region between the mirrors. The H_2O molecules help to depopulate the (010) level as in the case of the electric CO_2 laser. After the gas passes through the optical cavity, the pressure is increased by the diffusor up to atmospheric pressure.

The efficiency of the GDL is low. Usually it is less than 1%. The power output, however, can be very high. Continuous power outputs of 100 kilowatts or more have been obtained.

Chemical Pumping When a chemical reaction takes place, the product molecules may be in upper levels of vibrational–rotational bands. This creates a population inversion, and the medium has an optical gain. For example, excited molecules are produced by the reaction

$$H_2 + F \rightarrow HF^* + H$$

The asterisk indicates that the HF molecule is in an excited state. To obtain this reaction, however, it is necessary to form F atoms from F_2. This can be accomplished by using a flashlamp, a pulse electric discharge, or an electron beam to dissociate F_2 in a mixture of H_2 and F_2. One then obtains a chain reaction:

FIGURE 14.85   Chemical laser.

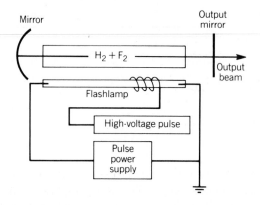

$$F + H_2 \rightarrow HF^* + H$$

$$H + F_2 \rightarrow HF^* + F$$

However, recombination also occurs:

$$H + H \rightarrow H_2$$

$$F + F \rightarrow F_2$$

The recombination reactions may cause the chain reaction to stop. Thus, it may be necessary to form additional F atoms using an external source of energy. A simplified diagram of a chemical pulse laser is shown in Figure 14.85.

Properties of Laser Radiation   The properties of laser radiation depend on the three elements that are used to build a laser. If a single axial mode can be obtained, then the radiation is nearly monochromatic and coherent in time and space, that is, wave trains emitted by the laser are very long and wave fronts of constant phase exist. On the other hand, the radiation from a dye laser without a prism or grating in the optical cavity has a broad spectral width and is not very coherent.

The pumping process may be continuous or it may last for a short time. Continuous pumping yields an output power that is almost constant. Flashlamps and pulse electric discharges produce a pulse of laser radiation. The pulse width may be very short (order of nanoseconds). For some applications, the laser output is modulated. This can be achieved by varying the pumping rate.

Lasers can be designed to give a very small beam spread. Thus, laser light possesses a high degree of directionality. For this reason, the radiance of a laser beam is very large. Radiance is power per steradian. This allows all the radiation to be focused to a small spot that produces a very high irradiance. Irradiance is power per unit area incident on a surface. It can be shown for axial modes that the spot size of a focused beam is

$$D = \frac{4f\lambda}{\pi d} \qquad (14.134)$$

where

D = spot diameter

d = diameter of the incident laser beam
f = focal length of the lens

Notice that the simple diffraction theory for a slit yielded an image width equal to $2f\lambda/d$ (see Example 14.21). Example 14.29 gives a sample calculation for irradiance.

EXAMPLE 14.29   Irradiance on a target
. . . .

Given: A pulsed laser delivers 30 J of energy in 150 nsec. The beam diameter is 1.00 cm and the wavelength is 400 nm.

Find: **a.**   The irradiance on a target.

   b.   The irradiance on a target placed at the focal point of a lens. The focal length is 30.0 cm. Assume that axial modes exist.

Solution: The laser power is

$$P = \frac{E}{t}$$

where

   P = power in watts
   E = energy in joules
   t = time in seconds

$$P = \frac{30 \text{ J}}{150 \times 10^{-9} \text{ sec}}$$

$$P = 2.0 \times 10^8 \text{ W}$$

a.   The irradiance on a target is

$$H = \frac{P}{A} = \frac{P}{\pi R^2}$$

where

   H = irradiance
   A = area of the incident radiation
   R = radius of the beam

$$H = \frac{2.00 \times 10^8 \text{ W}}{\pi (0.50 \text{ cm})^2}$$

$$H = 2.55 \times 10^8 \text{ W/cm}^2$$

b.   The spot size when a lens is used is

$$D = \frac{4f\lambda}{\pi d}$$

$$D = \frac{4(0.30 \text{ m})(400 \times 10^{-9} \text{ m})}{\pi (1.0 \times 10^{-2} \text{ m})}$$

$$D = 1.53 \times 10^{-5} \text{ m} = 1.53 \times 10^{-3} \text{ cm}$$

The irradiance is

$$H = \frac{P}{\pi R_2}$$

$$H = \frac{2.00 \times 10^8 \text{ W}}{\pi (7.65 \times 10^{-4} \text{ cm})^2}$$

$$H = 1.09 \times 10^{14} \text{ W/cm}^2$$

Laser Applications

Alignment   The high directionality of laser light makes a laser an ideal alignment tool. Low-power, helium–neon lasers are usually used for this purpose. Complex optical systems are easily aligned with a small He–Ne laser. The He–Ne laser is also used as a level. The beam provides a reference line for the laying of foundations, pipes, and bricklaying. The beam can also provide a reference for earth-movers that are equipped with machine control systems that receive the beam.

Materials Processing   Power density is a key factor in many applications. The CO_2 laser, for example, can provide a continuous irradiance of over a million watts per square centimeter. At these power densities, the laser beam can cut through the hardest metals. In the cutting operation, the laser beam does not heat the whole piece so the material is not subject to warping as it is cut. The cut is also clean, which reduces the need for finishing. The CO_2 laser is also used for heat-treating metals and for welding. Figure 14.86 shows a laser system that is robot-controlled and is capable of welding, cutting, and trimming operations. A wide variety of materials, including cloth and semiconductor circuits, are cut with laser beams having a lower power output.

Pulsed solid-state lasers are employed to drill holes. They can drill small holes at a higher rate and with a smaller diameter than can be accomplished by any other means.

Communications   Modulated lasers are used as communication devices. A single laser beam can carry millions of telephone calls at once. Glass fibers carrying laser beams are replacing bulky, expensive cables used for telephone communication.

Distance Measurement   The ability to build a laser with a very short pulse time (order of nanoseconds) and a small beam divergence allows distances to be measured. The laser rangefinder sends out a short pulse of light and measures the time for the reflected light to return. The time interval can be measured to the order of nanoseconds. This is necessary for some applications because the speed of light is large. For example, light travels 1.00 m in 3.33 nsec. Thus, to measure a distance of 100 m (total path length is 200 m) to an error less than 2% requires measuring the time to ± 10 nsec. The laser rangefinder is used in surveying and military applications. It can also be used to measure the distance to the moon and the altitude of satellites.

FIGURE 14.86 Welding with a robot-controlled carbon dioxide laser.

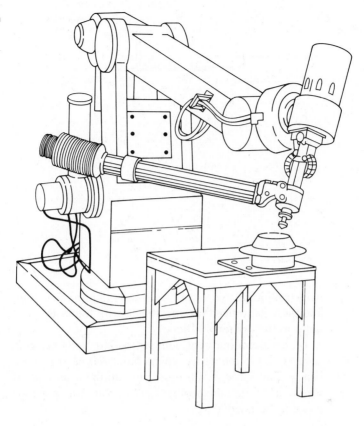

Figure 14.87 shows a simplified diagram of an optical system that was used at the Lick Observatory to measure the distance to the moon. When the ruby laser emits a single short pulse of great intensity, the flip mirror is in the horizontal position. Most of the light travels through the beam splitter (a beam splitter splits the beam), but a small part of the light is reflected by the beam splitter and starts the timer. The rest of the laser beam travels through the beam-correcting prism, which directs the beam into a large 120-inch telescope. The telescope expands the beam and thereby decreases its divergence. The beam travels to the moon, is reflected by the special mirrors placed there, and returns to earth. Before the light returns, the flip mirror is flipped into the diagonal position as shown. Part of the laser light is gathered by the telescope, strides the flip mirror, and is reflected into the detector. This stops the timer. By knowing the speed of light and by using the recorded time of the laser beam's travel, the distance traveled can be determined. The total time for this case is approximately 2.55 seconds.

A continuous, low-power laser can be used to measure very small distances. This is demonstrated in Example 14.40, where the laser is used as a source for a Michelson interferometer.

Medical Applications A Nd:YAG laser has been developed for use in a surgical treatment for glaucoma. Glaucoma is a disease that causes blindness. The surgery involves making a tiny hole in the iris. Laser eye surgery has some advantages

FIGURE 14.87
Measurement of distance to
the moon.

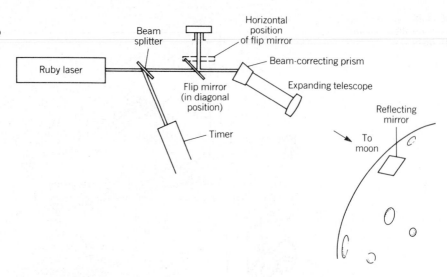

over traditional eye surgery. It tends to minimize complications such as bleeding, and there is no need for anesthesia with the laser, so the surgery can generally be done in the doctor's office.

In the glaucoma treatment with a Nd:YAG laser, the laser system has to meet several requirements. A typical Nd:YAG laser is more than a yard long and fairly heavy. The surgeon can't just pick up the laser and aim it at the patient's eyes. If we think through the problem, we can identify several types of equipment that are going to be needed:

- A laser must deliver the proper amount of energy—in the case of this particular eye surgery, this is an Nd:YAG laser.
- A cabinet or console is used to house the laser.

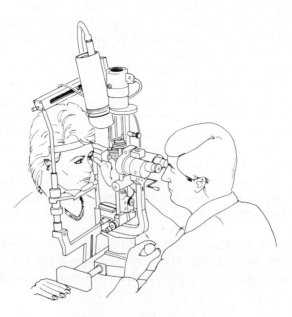

FIGURE 14.88   Laser used
for eye surgery.

- Control devices are needed to deliver the correct amount of energy or power at precisely the right moment.
- An instrument is needed to monitor the energy output of the laser and the amount of exposure to laser light that the patient receives.
- A way of delivering the laser light to the needed area is required. One way of doing this is to use optical fibers. Optical fibers are tiny threads of the purest glass or plastic—smaller than the human hair and very easy to bend—that can be used to transmit light due to total internal reflection. The fibers are bound together to make a kind of pipe. Light travels down the pipe, reflecting off the smooth inner walls of the fibers, not getting absorbed or refracted. The result is that the laser light doesn't get lost along the way. It follows the path of the light pipe, which can be bent in whatever direction is needed by the surgeon. Another way of delivering laser light to the patient's eye is by using a hinged mechanical arm with mirrors at each joint. The mirrors reflect the light to the needed location.

Figure 14.88 shows a Nd:YAG laser system used for eye surgery.

Another major area of laser applications in medicine is the use of argon lasers to clear blocked arteries. In this technique, known as laser angioplasty, the energy of the laser is transmitted along extremely small optical fibers through a catheter tube that is inserted into the artery itself. The laser light destroys the fatty plaque that is believed to cause arteriosclerosis (see Figure 14.89).

Other medical applications include cancer surgery, retinal surgery, and removing tattoos. The student should also read about laser applications in fusion research and three-dimensional photography (holography).

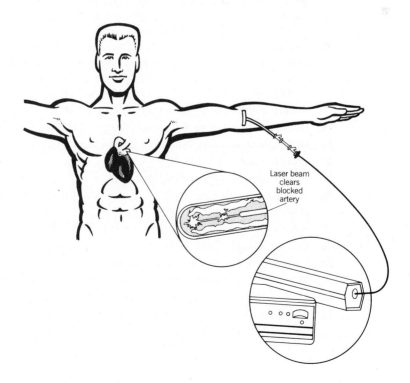

Laser beam clears blocked artery

FIGURE 14.89 Laser used to clear a blocked artery.

SECTION   *14.3   EXERCISES*

1. Draw and label a diagram representing the stimulated emission process.
2. List three basic elements of a laser.
3. List five methods that are used to create a population inversion.
4. What is the wavelength of orange laser light if the frequency is 5×10^{14} Hz.
5. A particular laser beam spreads 1 inch for every mile that it travels. If the beam is originally 0.01 inch in diameter, what is the width of the beam 10 miles from the source?
6. A pulsed laser emits 3000 pulses per second. How long does it take for the laser to pulse 10 times?
7. A laser and a point light source both have a power rating of 1 W. Why would it be dangerous to look directly into the laser beam but not to look directly at the point source?
8. How long would it take a laser beam to travel from New York to Los Angeles (approx. 3000 miles)?
9. A single axial mode laser has a frequency of 5×10^{14} Hz. If the length of the laser is 30 cm, what is the value of n (number of half wavelengths in the cavity)?
10. A cavity composed of plane mirrors has a length of 40 cm. What is the mode spacing in Hz?
11. The length of a cavity is 40 cm. It is composed of two spherical mirrors, and the radius of curvature of each mirror is 40 cm. Is the cavity stable?
12. The length of a cavity is 30 cm. One mirror is plane and one mirror has a radius of curvature of 30 cm. If the plane mirror is placed at the center of the spherical mirror, is the cavity stable?
13. The width of a spectral line for a medium that provides optical gain in a cavity is 4.3×10^9 Hz. If the laser is operated at the frequency of this line and the cavity length is 40 cm, how many axial modes are possible in the output? Assume that there is sufficient gain to cause lasing over the given width.
14. The reflectivity of one mirror of a He–Ne laser is 99.9% and the reflectivity of the other mirror is 99.0%. What is the necessary saturated gain? If the reflectivity of the output mirror is reduced from 99.0 to 80.0%, what is the necessary saturated gain? Would one laser be easier to pump than the other?
15. Draw a diagram of an electrically pumped laser.
16. Discuss the operation of a He–Ne laser and an electrically pumped CO_2 laser.
17. Draw a diagram of an optically pumped laser.
18. Discuss the Nd:YAG laser.
19. Draw a diagram of a tunable dye laser.
20. Why does a dye laser have a broad band of frequencies in its output if no tuning element is used in the cavity?
21. What causes a population inversion in a photolytic laser?
22. Discuss the HF chemical laser.
23. Draw a diagram of a CO_2 gas dynamic laser.
24. What causes the population inversion in the CO_2 gas dynamic laser?
25. Discuss the important properties of a single axial mode laser.
26. The beam diameter of a laser operating in axial modes is 0.50 cm. If the wavelength is 632.8 nm and a lens having a focal length of 10 cm is used to focus the beam to a small spot, what is the spot diameter?

27. The laser described in Problem 26 has a power output of 5.0 mW. What is the irradiance on a target? What is the irradiance on a target if the 10-cm focal length lens is used to focus the beam to a small spot?

28. List at least seven applications for lasers.

29. Explain how a laser can be used to measure the distance to the moon.

30. What kind of laser equipment is needed for glaucoma surgery?

31. Discuss laser metal cutting and welding. Why is a CO_2 laser used?

14.4
OPTICAL INSTRUMENTS

OBJECTIVES

Upon completion of this section, the student should be able to

- Describe how the human eye forms an image of an object on the retina.

- Use light rays to describe a nearsighted eye.

- Use light rays to describe a farsighted eye.

- Calculate the size of the image on the retina given the size of the object, the distance from the lens of the eye to the object, and the distance from the lens to the retina.

- Show how to use a lens to correct for nearsighted or farsighted vision.

- Explain how a simple microscope (magnifying glass) increases the image size on the retina.

- Calculate the magnification of a simple microscope.

- Explain how a compound microscope increases the image size on the retina.

- Calculate the magnification of a compound microscope.

- Explain how the astronomical telescope increases the image size on the retina.

- Calculate the magnification of an astronomical telescope.

- Use light rays to show how a camera forms an image on a film.

- Define the following terms:
  Single-lens reflex camera (SLR)
  Normal, telephoto, and wide-angle lenses
  Shutter and shutter speed
  Aperture
  f-*number*
  ASA number
  Reflex viewing

- Calculate the normal focal length for a camera from the film size.

- Define the depth of field and the hyperfocal distance.

- Given the focal length and the aperture diameter, calculate the hyperfocal distance.

- Define angle of view.

- Show how the power density on a film is related to the *f*-number.

- Calculate the *f*-number required for a correct exposure if the exposure time and the total correct exposure are given.

- Explain how a beam expander works.

- Draw a diagram of a Michelson interferometer.

- Explain how a Michelson interferometer can be used to measure wavelengths or distances.

- Discuss the measurement of wavelength and intensity of a spectral line.

- Draw and label simple diagrams of the following optical instruments:
  Spectrometer
  Spectrograph
  Monochromator
  Spectrophotometer

- Explain how a spectrophotometer may be used to determine the transmission of a sample at a particular wavelength.

DISCUSSION

Optical Systems

An optical system is a collection of instruments and devices that uses or acts upon light in some way. Optical systems process light for many purposes: to aid vision, to detect the presence or absence of objects, to record images, to deliver energy, and to measure distances. In this section, the eye, the simple microscope, the compound microscope, and the camera are analyzed by the application of geometrical optics. The Michelson interferometer is explained in terms of physical optics. In addition, spectroscopy is defined, and various spectrometric instruments are discussed.

The Eye

The Nature of Sight   Everything we are able to see we see because of light. Objects reflect light, and some of the reflected light enters our eyes. (Some materials also transmit or emit light, but we see most objects by reflected light.) Our eyes are sensitive optical systems that are capable of taking in light and producing images that can be processed by the brain.

Figure 14.90 is a cross-sectional diagram of the human eye focused on a street lamp. Light enters the eye through the opening known as the pupil. The iris (the colored part of the eye) actually controls the size of the pupil and thus the amount of light admitted. Tiny muscles cause the iris to contract and decrease the size of the pupil. When the muscles in the iris relax, as they do in dim light, then the pupils enlarge and more light passes through. Near the front of the eye is a convex lens, which refracts light to form an image on the retina at the back of the eye. The

FIGURE 14.90 The human eye is an image-forming optical system.

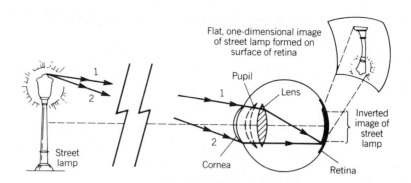

Eye (greatly enlarged to show detail)

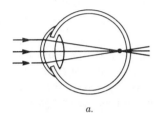

a.

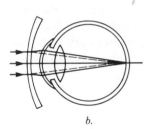

b.

FIGURE 14.91 A negative lens is used to correct nearsightedness.

retina is a light-sensitive screen; it receives light rays and changes them to nerve impulses that travel to your brain by way of the optic nerve. As you can see in Figure 14.90, the image formed on the retina is upside down (just as images formed by other convex lenses are). However, our brains interpret them as right side up.

The thicker or more convex the lens, the greater the bending or refraction of light rays passing through it. The lens of your eye is able to change shape according to how much refraction is required to focus on a particular object. If you look at a large object near you, light rays from the object must be bent quite a bit to form an image on the retina. To do this, the lens becomes thicker at the center (more convex). If you are looking at an object far away, light rays must be bent or refracted only slightly to form an image on the retina. So your lens becomes less convex (not so thick at the center).

Nearsighted and Farsighted Eyes In the normal human eye, an image comes into sharp focus on the retina. However, in the eye defect known as nearsightedness, the eyeball may be too long or the lens too thick, so that the image comes to a sharp focus just short of the retina (see Figure 14.91*a*). Thus the image appears somewhat blurred to the nearsighted person. To correct nearsightedness, a concave lens (the center of a concave lens is thinner than its edge) is prescribed to spread the light rays apart. The image that is formed on the retina is then in sharp focus (see Figure 14.91*b*).

In the eye defect known as farsightedness, the eyeball may be too short or the lens too thin. The image comes to a sharp focus too far from the lens, just beyond the retina (see Figure 14.92*a*). To correct this problem, a convex lens is prescribed. The light rays are focused by both the corrective lens and by the lens of the eye so that the image comes into sharp focus on the retina (see Figure 14.92*b*).

Image Size on the Retina To increase the size of an image on the retina, it is necessary to reduce the object distance. There is a limit, however. If an object is too close to the eye, the lens cannot be adjusted to focus the object on the retina. For a normal eye, the minimum distance is approximately 10 cm at age 20 years and 22 cm at age 40 years. To accommodate most people, 25 cm will be used as the minimum distance in calculations. Thus, it will be assumed that the maximum image size occurs when the object distance is 25 cm.

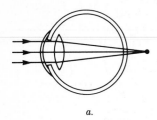

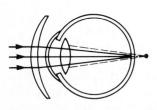

FIGURE 14.92 A positive lens is used to correct farsightedness.

To calculate the image size, consider Figure 14.93. Since the two triangles are similar,

$$\frac{S_R}{S_O} = \frac{d}{l}$$

or

$$S_R = \left(\frac{d}{l}\right) S_O \qquad (14.135)$$

where

l = distance to the object
d = distance from the lens to the retina
S_O = size of the object
S_R = size of the image on the retina

Simple Microscope   A simple microscope consists of a single positive lens. The lens, however, is usually composed of several simple lenses to reduce aberrations. For this discussion, however, a single simple lens will be considered as shown in Figure 14.94. The positive lens allows the object to be brought closer to the eye. To determine the magnification, consider Figure 14.94. The object distance is set equal to the focal length of the lens to obtain the maximum magnification. To focus the object on the retina, the eye is focused at infinity because the rays incident on the eye are parallel (object is in the focal plane of the magnifying lens). Since vertical angles and alternate interior angles for parallel lines are equal, all the angles designated by θ are equal in the diagram. Thus, from similar triangles,

$$\frac{S_I}{S_O} = \frac{d}{f}$$

or

$$S_I = \left(\frac{d}{f}\right) S_O \qquad (14.136)$$

where

S_I = new image size on the retina
f = focal length of the simple microscope

The effective magnification becomes the ratio of the image size on the retina without the microscope to the image size with the microscope. Thus,

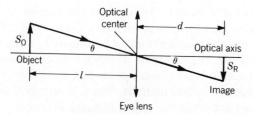

FIGURE 14.93 Ray trace for calculating image size on the retina.

FIGURE 14.94
Magnification by a simple microscope.

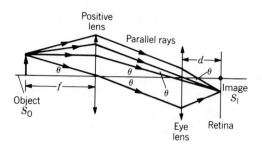

$$m = \frac{S_\text{I}}{S_\text{R}} \tag{14.137}$$

Substituting for S_R from Equation 14.135 and for S_I from Equation 14.136 yields

$$m = \frac{d/f}{d/l}$$

or

$$m = \frac{l}{f} \tag{14.138}$$

Example 14.30 gives a sample problem.

EXAMPLE 14.30   The magnification of a simple microscope
. . . .

Given: A positive lens with a focal length of 5 cm is used as a simple microscope.

Find: The approximate magnification when the eye is focused at infinity.

Solution: The magnification is given by

$$m = \frac{l}{f}$$

A reasonable value for the minimum value of l is 25 cm.

$$m = \frac{25 \text{ cm}}{5 \text{ cm}}$$

$$m = 5$$

The Compound Microscope   It is necessary to use a short focal length to obtain a high magnification with a simple microscope. However, a large amount of light from the object must be collected to see the magnified image. It can be shown that it is not possible to collect a large amount of light and still have a relatively broad field of view with a simple microscope. However, an optical system with two

FIGURE 14.95 Compound
microscope.

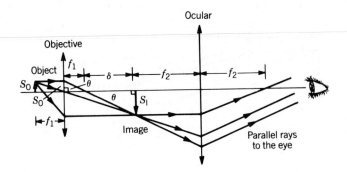

lenses can achieve both requirements. In addition, the aberrations can be reduced to a greater extent if two lenses are used.

A microscope composed of two lenses is called a compound microscope. Compound lenses are used to reduce aberrations. The positive lens near the object is called the objective. It has a short focal length, and it forms a real image of the object. The lens near the eye is the ocular. It is usually a positive lens, and it acts as a simple microscope that magnifies the image formed by the objective. The objective and ocular are mounted on a tube that is mounted on a firm stand. The position of the tube can be adjusted to focus on an object.

To simplify the analysis of a compound microscope, it is assumed that the objective and ocular are two simple lenses as shown in Figure 14.95. Since the vertical angles designated by θ are equal, their corresponding right triangles are similar. Thus, one may write

$$m = \frac{S_I}{S_O} = \frac{\delta}{f_1}$$

where

δ = distance between the focal points as indicated in Figure 14.95
f_1 = focal length of the objective
m_1 = magnification of the objective

The total magnification is equal to the magnification of the objective times and magnification of the ocular. Since the magnification of the ocular (simple microscope) is l/f_2, the total magnification becomes

$$m = m_1 \left(\frac{l}{f_2} \right) = \frac{\delta l}{f_1 f_2} \tag{14.139}$$

where

m = total magnification
f_2 = focal length of the ocular

It is interesting to note that the magnification can be changed by changing the distance between the two focal points (δ). Example 14.31 shows the use of Equation 14.139.

> **EXAMPLE 14.31** The magnification of a compound microscope
> . . . .
>
> *Given:* A compound microscope has an objective with a focal length of 0.5 cm and an ocular with a focal length of 2.0 cm. The distance between the focal points (δ) is 20 cm.
>
> *Find:* The magnification of the microscope.
>
> *Solution:* From Equation 14.139
>
> $$m = \left(\frac{\delta}{f_1}\right)\left(\frac{l}{f_2}\right)$$
>
> The value of l will be taken as 25 cm.
>
> $$m = \left(\frac{20 \text{ cm}}{0.5 \text{ cm}}\right)\left(\frac{25 \text{ cm}}{2 \text{ cm}}\right)$$
>
> $$m = 500$$

The Astronomical Telescope In an astronomical telescope, the first lens is a positive lens, and it is called the object-glass. It forms a real image of an object that is a great distance away from the telescope. Since the object distance is very large, the image distance is essentially equal to the focal length of the object-glass. The image is magnified by an ocular that is also called the eyeglass. The image is located in the focal plane of the eyeglass when the eye is focused for parallel light. This arrangement is shown in Figure 14.96. The size of the image on the retina is

$$S'_R = d \tan \theta \tag{14.140}$$

where

S'_R = image size on the retina when the telescope is used

However, from the diagram

$$\tan \theta = \frac{S_I}{f_2} \tag{14.141}$$

where

S_I = size of the image formed by the object-glass
f_2 = focal length of the eyeglass

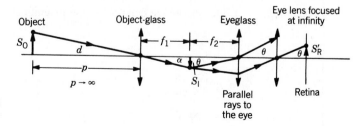

FIGURE 14.96 Astronomical telescope.

Substituting this value for tan θ in Equation 14.140 yields

$$S_R' = \left(\frac{d}{f_2}\right)S_I \tag{14.142}$$

From Figure 14.96

$$S_I = f_1 \tan \alpha = \frac{f_1}{p} S_O \tag{14.143}$$

Substituting the value of S_I given by Equation 14.143 into Equation 14.142 yields

$$S_R' = \left(\frac{f_1}{f_2}\right)\left(\frac{d}{p}\right)S_O$$

If the object is observed without a telescope, the size of the retinal image is (Equation 14.134)

$$S_R = \left(\frac{d}{l}\right)S_O \tag{14.144}$$

However, since the object distance p is much greater than the length of the telescope, l is essentially equal to p. Thus, Equation 14.144 becomes

$$S_R = \left(\frac{d}{p}\right)S_O \tag{14.145}$$

The magnification is defined as the ratio of the size of the retinal image with a telescope to the size of the retinal image without a telescope. This yields

$$m = \frac{\left(\frac{f_1}{f_2}\right)\left(\frac{d}{p}\right)S_O}{\left(\frac{d}{p}\right)S_O} = \frac{f_1}{f_2} \tag{14.146}$$

To obtain a large magnification, the focal length of the object-glass should be large and the focal length of the eyeglass should be small. When the telescope is focused on an object that is a great distance from the object-glass, the length of the telescope is $f_1 + f_2$. Example 14.32 shows a sample problem.

EXAMPLE 14.32   Magnification of an astronomical telescope
· ·

Given: The focal length of the object-glass of a telescope is 100 cm and the focal length of the eyeglass is 0.5 cm. The object is at a great distance and the eye is focused at infinity.

Find: The magnification of the telescope.

Solution: Under the above conditions, the magnification is

$$m = \frac{f_1}{f_2}$$

$$m = \frac{100 \text{ cm}}{0.5 \text{ cm}}$$

$$m = 200$$

The Beam Expander

The Beam Diameter   For many applications, it is desirable to increase the diameter of a laser beam. This can be accomplished by using a beam expander. Figure 14.97 shows a common type of beam expander. It uses positive lenses. The input lens has a small diameter and a short focal length. The laser beam, after leaving the laser, strikes the input lens and is focused to its minimum diameter at the focal point. The larger lens is placed so that its focal point is at the same location as the focal point of the smaller lens. The focused laser beam diverges again and enters the output lens. It leaves the output lens as a collimated beam having a larger diameter. From similar triangles one has

$$\frac{d_2/2}{f_2} = \frac{d_1/2}{f_1}$$

or

$$d_2 = \left(\frac{f_2}{f_1}\right) d_1 \qquad (14.147)$$

where

d_2 = diameter of the output beam
d_1 = diameter of the input beam
f_2 = focal length of the second lens
f_1 = focal length of the input lens

Thus, the diameter of the output beam is equal to the diameter of the input beam times the ratio of the focal lengths.

Beam Divergence   A laser beam eventually spreads after it leaves the laser cavity. For some applications, even a small amount of beam spread is too much. For example, if you want to use a laser beam to track military airplanes thousands of feet above the earth's surface, the beam needs to remain at the same diameter throughout its path of travel. The beam spread depends on the diameter of the beam. A simple calculation can give the divergence angle to a good approximation. To do this, assume that the beam is in the form of a rectangle with one side much longer than the other. In addition, assume that the intensity is constant across the rectangle. A cross section of the beam is shown from the top in Figure 14.98. Since this cross section has a constant phase (laser beam is coherent), the

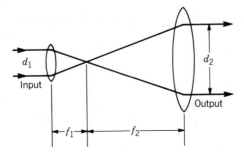

FIGURE 14.97   A simple beam expander.

FIGURE 14.98 Beam divergence.

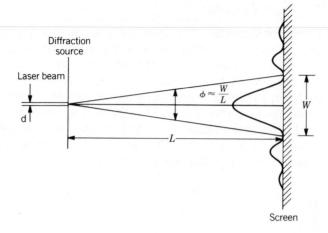

equations for the diffraction of a single slit apply. When a screen is placed a great distance away, the width of the central image can be calculated. This is done in Example 14.22, and the result is

$$w = \frac{2\lambda L}{d} \tag{14.148}$$

The corresponding divergence angle to a good approximation is (see Figure 14.98)

$$\phi \approx \frac{w}{L} = \frac{2\lambda}{d} \tag{14.149}$$

The solution obtained from electromagnetic theory is

$$\phi = \frac{4\lambda}{\pi d} \tag{14.150}$$

Thus, the simple theory gives a good approximation of the beam divergence.

Since the divergence decreases as the beam diameter increases, a beam expander reduces the beam divergence. Thus, from Equation 14.150, one has

$$\phi_1 = \frac{4\lambda}{\pi d_1} \tag{14.151}$$

and

$$\phi_2 = \frac{4\lambda}{\pi d_2} \tag{14.152}$$

Dividing Equation 14.152 by Equation 14.151 gives

$$\frac{\phi_2}{\phi_1} = \frac{d_1}{d_2}$$

or

$$\phi_2 = \frac{d_1}{d_2}\,\phi_1 = \frac{f_1}{f_2}\,\phi_1 \tag{14.153}$$

The divergence decreases as the ratio of the first focal length to the second. Example 14.33 shows a sample problem for a beam expander.

EXAMPLE 14.33 The beam expander
. . . .

Given: The focal lengths of a beam expander are $f_1 = 0.5$ cm and $f_2 = 10$ cm. The initial beam diameter is 0.3 cm, and the divergence is 2.5×10^{-4} radians.

Find: The final beam diameter and divergence.

Solution: From Equation 14.147

$$d_2 = \frac{f_2}{f_1} d_1$$

$$d_2 = \frac{10 \text{ cm}}{0.5 \text{ cm}} 0.3 \text{ cm}$$

$$d_2 = 6 \text{ cm}$$

From Equation 14.153

$$\phi_2 = \frac{f_1}{f_2} \phi_1$$

$$\phi_2 = \left(\frac{0.5 \text{ cm}}{10 \text{ cm}}\right) 2.5 \times 10^{-4} \text{ rad}$$

$$\phi_2 = 1.25 \times 10^{-5} \text{ rad}$$

The Camera

The SLR Camera   The most popular camera today for many professional applications and for the advanced amateur is the single-lens reflex (SLR), which uses a film 35 mm wide (Figure 14.99). This type of camera has come into wide use for several reasons, the first being its flexibility. A large selection of film types is

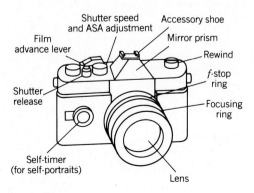

FIGURE 14.99 Single-lens reflex camera.

FIGURE 14.100   Schematic
diagram of a single-lens
reflex camera.

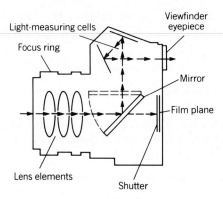

available for many purposes. It is a simple matter to change lenses from a
telephoto lens used to shoot football action from a distant seat to a wide-angle lens
used to record a wide field of view. The SLR feature of reflex viewing (Figure
14.100) allows the photographer to see exactly what is being photographed and
assures that the important parts of the scene are focused sharply.

Depth of Field   For a given object distance, there is a unique image distance.
Since a lens focuses an image on the film, only one object distance is in focus.
Points at other distances form circular images on the film. If the circles are small,
a good image is obtained. To obtain a good image, the circles should be no larger
than 0.005 inch in diameter for a normal-size camera.

As an example, consider a camera focused at infinity. For this case, the film is
in the focal plane of the lens. This is illustrated in Figure 14.101. A point source is
located a distance p from the lens. This point will be focused to a point beyond the
film. The light on the film will be a circle having a diameter d as shown in the
figure. The diameter of the aperture is D, and the image distance is q. From similar
triangles,

$$\frac{d/2}{q - f} = \frac{D/2}{q}$$

or

$$qd = qD - fD$$

or

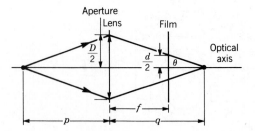

FIGURE 14.101   Hyperfocal
distance.

$$q = \frac{fD}{D - d} \tag{14.154}$$

The corresponding object distance (p) is determined by the lens equation

$$\frac{1}{p} + \frac{1}{q} = \frac{1}{f}$$

or

$$\frac{1}{p} = \frac{1}{f} - \frac{1}{q} \tag{14.155}$$

Substituting for q from Equation 14.154 yields

$$\frac{1}{p} = \frac{1}{f} - \frac{D - d}{fD} = \frac{fD - fD + fd}{f^2 D}$$

or

$$p = \frac{fD}{d} \tag{14.156}$$

Thus, if $d = 0.005$ inch, all points will be essentially in focus from p to infinity. This value of p is called the hyperfocal distance. The student can show that everything from one-half of the hyperfocal distance to infinity will be in focus if the lens is focused on the hyperfocal point. Example 14.34 illustrates a sample problem. To obtain a large depth of field, a small aperture and focal length are required. However, this reduces the exposure on the film unless the exposure time is increased. Thus, a number of trade-offs are required to obtain the best picture for given conditions.

EXAMPLE 14.34 The hyperfocal distance
. . . .

Given:　　A camera has a lens diameter of 0.25 inch and a focal length of 2.0 inches.

Find:　　**a.** The hyperfocal distance.
　　　　b. If the lens is focused at the hyperfocal point, find the total field in focus.

Solution:　**a.** The hyperfocal distance is given by

$$p = \frac{fD}{d}$$

where

$$d = 0.005 \text{ inch}$$

Therefore,

$$p = \frac{(2.0 \text{ inch})(0.25 \text{ inch})}{0.005 \text{ inch}}$$

$p = 100$ inch

$p = 8.3$ ft

b. If the lens is focused on the hyperfocal distance, everything will be in focus from 4.2 ft to infinity.

Angle of View   The angle of view of a camera depends on the size of the film and the focal length of the lens. The diagonal of the film is usually used for calculations. Since rays that pass through the optical center of a lens do not change direction, the angle of view can be estimated from the incident rays shown in Figure 14.102. It is assumed in the figure that the camera is focused at infinity. The angle of view of ϕ equals 2θ. From the diagram,

$$\tan \theta = \frac{H/2}{f} \qquad (14.157)$$

and

$$\phi = 2\theta \qquad (14.158)$$

where
   $H =$ film diagonal

If the angle of view is 53°, which is roughly equal to the angle of view of the eye, then $\theta = 26.5°$ and

$$\tan \theta = 0.50$$

Substituting this value of $\tan \theta$ in Equation 14.157 and solving for the focal length yields

$$f = \frac{H}{2(0.50)} = H \qquad (14.159)$$

This is the usual focal length for an ordinary camera. Example 14.35 gives a sample problem.

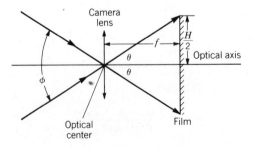

FIGURE 14.102 Angle of view.

EXAMPLE 14.35   Normal focal length
· ·

Given: A camera uses 35-mm film (actual frame size for this film format is
                24 × 36 mm).

Find: Suggested focal length for "normal" lens.

Solution: Diagonal of the frame is the usual recommendation.

$$f = H$$

$$H = \sqrt{x_2 + y_2}$$

$$H = \sqrt{36^2 + 24^2}$$

$$H = 43 \text{ mm} = f$$

In actual practice, the normal lenses for 35 mm vary from 40 mm to
nearly 60 mm. The diagonal rule is merely a suggestion.

The f-*Number*   The amount of radiant power deposited on the film depends on
the diameter of the aperture and the focal length. To show this relationship,
consider Figure 14.103. Suppose the small square object emits I watts per steradian per unit area. The total power emitted per steradian becomes

$$P_\omega = IA_0 = IS_0^2 \qquad (1.160)$$

where

P_ω = power emitted per steradian
A_0 = area of the square object
S_0 = length of one side of the object

The total power collected by the camera is the power emitted per steradian times
the solid angle subtended by the aperture. To a good approximation, the solid
angle is

$$\omega = \frac{A_A}{p^2} \qquad (14.161)$$

$$P_c = P_\omega\omega = IS_O^2\left(\frac{A_A}{p^2}\right) \qquad (14.162)$$

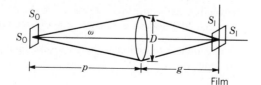

FIGURE 14.103 Light-
gathering ability of a lens.

where

P_c = power collected
ω = solid angle
A_A = area of the aperture
p = object distance

All the power collected, neglecting lens losses, is focused down to the image. The area of the image is S_I^2, where S_I is the length of one side. Thus the power per unit area on the film becomes (irradiance)

$$P_F = \frac{P_c}{S_I^2} = \frac{IA_A S_O^2}{p^2 S_I^2} \tag{14.163}$$

where Equation 14.162 is substituted for P_c. From the magnification equation

$$\frac{S_O}{S_I} = \frac{p}{q} \tag{14.164}$$

Substituting this value for S_O/S_I into Equation 14.163 yields

$$P_F = \frac{IA_A p^2}{p^2 q^2} = \frac{IA_A}{q^2} = \frac{I\pi D^2}{4q^2} \tag{14.165}$$

For most cases, q is approximately equal to f for a camera. The power per unit area on the film becomes

$$P_F = \left(\frac{\pi}{4} I\right) \frac{D^2}{f^2} \tag{14.166}$$

The quantity f/D is called the f-number. Thus,

$$P_F = \left(\frac{\pi}{4} I\right) \left(\frac{1}{f\#}\right)^2 \tag{14.167}$$

where

$f\#$ = f-number

The power per unit area is proportional to one over the f-number squared. To increase the exposure, the f-number must be decreased. An application is given in Example 14.36.

EXAMPLE 14.36   The f-number
. . . .

Given: The total exposure depends on the power per unit area on the film times the exposure time. A correct exposure is obtained with an f-number of 3 and an exposure time of 1/500 of a second.

Find: The required f-number if the exposure time is changed to 1/100 of a second.

Solution: The exposure is

$$E = (P_F)(\Delta t)$$

where

$$E = \text{exposure}$$
$$P_F = \text{power per unit area on the film}$$
$$\Delta t = \text{exposure time}$$

Substituting for P_F and setting the exposures equal gives

$$\frac{\pi}{4} I \frac{\Delta t_1}{(f_1^\#)^2} = \frac{\pi}{4} I \frac{\Delta t_2}{(f_2^\#)^2}$$

$$\frac{1 \text{ sec}}{500 \ (3)^2} = \frac{1 \text{ sec}}{100 \ (f_2^\#)^2}$$

$$f_2^\# = \sqrt{(500/100) \ 3^2}$$

$$f_2^\# = 6.7$$

Telephoto and Wide-Angle Lenses   Example 14.37 gives a calculation for the size of the image on a film. The image distance q is approximated by f for a camera.

EXAMPLE 14.37   Image size on film
. . . .

Given: Camera with lens of focal length 50 mm. A football player 2.0 m tall is photographed from a distance of 50 m.

Find: Size of image on film.

Solution:

$$S_I = S_O \left(\frac{f}{p}\right)$$

$$S_I = 2000 \text{ mm} \left(\frac{50 \text{ mm}}{50,000 \text{ mm}}\right)$$

$$S_I = 2 \text{ mm}$$

The 2-mm image obtained in Example 14.37 is only 2/24 or 8% the height of the 24-mm picture. Even if this small picture were enlarged 10 times, the image would be only 2 cm high. The image size formula gives a clue to increasing image size: S_I is directly proportional to the focal length. If the camera were fitted with a lens of 500-mm focal length, the image would be 10 times greater. The resulting 20-mm image would almost fill the picture area. Such a long-focus lens is a "telephoto."

Aberration can be cured almost entirely by use of lenses having a combination of several lens elements made from different glasses (Figure 14.104). The proper choice of glasses with the right indices of refraction allows production of complex lenses that have essentially the same focal length for light of all colors. Modern 35-mm cameras are able to take remarkably sharp pictures, with lines and small

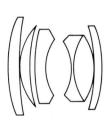

FIGURE 14.104   Complex photographic lens with many elements.

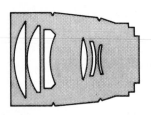

FIGURE 14.105   Telephoto lens.

details clearly defined, largely because of careful lens design, now aided by digital computers to perform the mathematical computations required.

A true telephoto is a combination of lenses, but is more than just a long focal-length combination. It is a special combination of converging front lenses and diverging, or negative, back lenses (Figure 14.105). If the combination is turned around (with the diverging lens to the front), the image is formed at a greater distance than the focal length, producing a wide-angle lens. For example, a 28-mm lens might be used to photograph a large area (a wide angle) in a confined space–perhaps the wall of a small room.

The Shutter   Recording a good picture on film requires not only a good, clear optical image, but also the proper "exposure." The concept of exposure is related to irradiance, the light energy per unit area falling on the film, and its duration. A camera shutter is merely a gate that controls the time factor of the exposure. There are several types of shutter mechanisms and possible placements with respect to the lens and the film. The principal arrangement used in SLR cameras today is the focal-plane shutter, placed just in front of the film—in the plane where the image is focused.

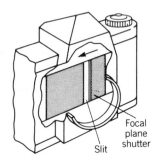

FIGURE 14.106   Camera shutter.

The focal plane shutter consists of a curtain of metal or cloth containing one or more slits (Figure 14.106). The slit moves rapidly across and in front of the film, thereby exposing the film to the pattern of light and dark in the image. The length of exposure time is determined by the width of the slit and by the speed of its motion. The time, or "shutter speed," usually is set by a knob that adjusts the tension of the springs driving the shutter. A simple shutter often is built like a window shade on two rollers. Focal-plane shutters commonly produce exposure times ranging from one second to 1/1000 second. The very short times are useful for "stopping" fast action that otherwise would appear blurred with longer exposures.

The Aperture   In all but the very simplest cameras, the amount of light passing through the lens is controlled by a variable, circular opening, or "aperture" control. The lens opening usually is regulated by an arrangement called a "diaphragm" that consists of a set of overlapping, thin metal leaves. These leaves are opened and closed by the rotation of a ring to which they are linked. A large aperture is associated with a small *f*-number. For example, a relatively large aperture is *f*-1.4 and a small one is *f*-16. In the latter case, the focal length is 16 times the aperture diameter.

Lenses generally are marked with *f*-numbers arranged in a sequence in order such that each *f*-number represents approximately one-half the irradiance of the one before it. A common sequence of *f*-numbers is

f-number: 1.4, 2, 2.8, 4, 5.6, 8, 11, 16, and 22

Thus, a camera set at *f*-5.6 permits one-half the irradiance on film as when set at *f*-4. This difference of a factor of two in irradiance is called a "one-stop difference"; *f*-numbers sometimes are referred to as "stops."

Determining the Exposure   The proper exposure depends on the time and the irradiance. The time interval is determined by the shutter and is referred to as the "shutter speed."

The least expensive way to determine the required exposure (*f*-number and shutter speed) for the lighting conditions is to use the manufacturer's specification sheet that comes with the film; for example, for a common color slide film (ASA 400) the recommended exposure for "heavy overcast" lighting conditions is *f*-5.6 at 1/500 second. These simple lighting charts provide good results under "average" conditions.

The next most sophisticated approach is the use of a light meter, or exposure meter, which contains a light-sensitive cell. The ASA film rating is adjusted on a sliding *f*-stop and aperture scale, allowing a direct reading of a variety of possible *f*-stop and shutter speed combinations. Such meters often are built directly into modern cameras, where they measure the exact irradiance on the film.

The highest level of technical sophistication is found in automated cameras. In these cameras, a number of light-sensitive cells sense the irradiance on the film. Built-in electronic circuits then set the aperture and shutter speed automatically.

Film Photographic films are produced in many different types: black and white, color slide, color print, and polaroid. A basic property of all films is their sensitivity, or "speed." Sensitivity refers to the amount of exposure required to produce a satisfactory, permanent image on the film. In the United States, the sensitivity of films is rated in terms of a scale called the "ASA system," established by the American Standards Association. Film manufacturers conduct experiments to determine the sensitivities of their films and assign an ASA number that best describes their behavior. The light-sensitive layer of a film is called the "emulsion," which is a complicated mixture of silver salts, dyes, and other chemical compounds tailored to give the desired speed and other properties. A permanent image is formed on film emulsion by the interaction of light photons with silver atoms in the emulsion. Modern emulsion chemistry has produced films uniformly sensitive to all colors. Example 14.38 gives a sample problem for obtaining the correct exposure if the film speed is changed.

EXAMPLE 14.38 Exposure and ASA number
. . . .

Given: An exposure meter indicates that the recommended exposure for a certain subject is *f*-5.6 at 1/500 sec, with a film emulsion rated at ASA 400.

Find: The correct exposure for this subject with a film of ASA 100.

Solution: A film of ASA 100 is one-fourth as sensitive as a film of ASA 400, and therefore requires four times the exposure. These are some suitable choices:

  f-5.6 at 1/125 sec (increase the exposure time by 4)
  f-2.8 at 1/500 sec (decrease the *f*-number by one-half)
  f-4 at 1/250 sec (increase the exposure time by 2 and decrease the *f*-number by $1/\sqrt{2}$)

FIGURE 14.107 SLR viewing system.

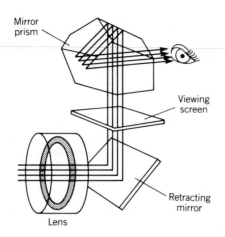

FIGURE 14.108 Fresnel lens.

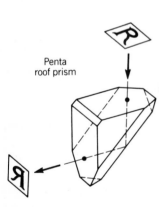

FIGURE 14.109 Pentaprism.

Viewfinders and Reflex Viewing   Any camera must have some kind of viewfinder to indicate what scene is being recorded. The simplest finder is merely a pair of wire rectangles to mark off approximately the area being covered. Many cameras now use optical viewfinders, which consist of a converging lens followed by a diverging lens.

Probably the best viewfinder is the single-lens reflex type, illustrated in Figures 14.99 and 14.107. The reflex arrangement is desirable because it uses the camera lens to form an image on a viewing screen by way of a mirror. Reflex viewing allows the photographer to see exactly what the film will record as well as focus. Another great advantage of reflex viewing is ''parallax correction,'' an automatic adjustment that corrects the ''head room'' for objects being photographed at any distance.

As illustrated in Figure 14.107, the modern 35-mm SLR camera includes a 45° mirror that reflects the image to a horizontal viewing screen. The viewing system also contains a Fresnel lens that increases illumination. The Fresnel lens, as indicated in Figure 14.108, is designed with ridges that produce a brighter image through the viewing system. Such a lens is not suitable as the primary lens, but it is useful as an aid to viewing, especially in dim light.

The mirror prism (Figure 14.109) has seven internal reflecting surfaces. The two ''roof'' mirrors interchange left and right, so that the SLR user sees a clear, bright image, in which all the directions are shown properly.

Focusing   The simplest approach to focusing used in the cheapest cameras is called ''fixed focus.'' The lens-to-film distance is set permanently to give a reasonably sharp image at an average object distance—usually about nine feet.

Objects at greater distances are well focused; closer objects are less well focused. This approach is satisfactory only with simple cameras having lenses of short focal length and small aperture.

Modern single-lens reflex cameras are focused by rotation of the lens tube. (Small lens openings, like f-16, allow greater "depth of field," or range of correct focus, than large ones.) Automatic-focus lenses are now available on some models.

Interferometers

There are a large number of optical instruments called interferometers, which produce fringe patterns. Some produce interference patterns by using two beams, whereas others use multiple beams that give very sharp fringes. The Michelson interferometer is a common two-beam interferometer. A diagram of this interferometer is shown in Figure 14.110. A central ray from an extended source is shown incident on the mirror M_1. This ray is partially reflected and partially transmitted by M_1 as shown in the diagram. Each ray has the same intensity since the mirror M_1 is half silvered. Mirrors M_3 and M_2 are totally reflecting mirrors. The reflected ray at point A travels to mirror M_2 through the glass and air path and then returns along the same path to point A. This ray then travels to the observer's eye. The transmitted ray at point A travels to mirror M_3 along an air path and through the glass compensating plate C. The compensating plate is parallel to mirror M_1 and has the same thickness as mirror M_1. The ray then is reflected at mirror M_3 and returns along the same path to point A, where it is reflected toward the eye. To accomplish this, all mirrors must be properly aligned as shown in the diagram. The total path in glass is the same for both rays. The two rays that are directed toward the eye can interfere since they come from the same point on the monochromatic source, and thus they initially have the same phase at point A.

The optical path of the reflected ray at point A to mirror M_2 and back to point A is

$$d_1 = 2l_1 + 2l_{g1} \tag{14.168}$$

where

l_1 = path length in air one way
l_{g1} = path length in glass one way

The optical path length from point A to mirror M_3 and back to point A is

$$d_2 = 2l_2 + 2l_{g2} \tag{14.169}$$

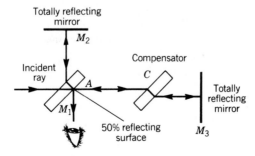

FIGURE 14.110 Michelson interferometer.

where

l_2 = path length in air one way
l_{g2} = path length in glass one way

If a bright fringe occurs for the central ray, then $d_2 - d_1$ must equal an even number of wavelengths. Thus, for a bright fringe at the center

$$d_2 - d_1 = 2(l_2 - l_1) + 2(l_{g2} - l_{g1}) = n\lambda \qquad (14.170)$$

where n is an integer. The mirror M_3 is usually mounted on a fine screw so that the value of l_2 can be changed. If l_2 is increased so that a bright fringe disappears and a new bright fringe appears, then n increases by one. Equation 14.170 becomes

$$d_2' - d_1 = 2(l_2 + \Delta l_2 - l_1) = 2(l_{g2} - l_{g1}) = (n + 1)\lambda \qquad (14.171)$$

where

d_2' = optical path to M_3 and back
Δl_2 = increase in the above optical path

Subtracting Equation 14.170 from Equation 14.171 yields

$$2\Delta l_2 = \lambda$$

or

$$\Delta l_2 = \lambda/2 \qquad (1.172)$$

The length of l_2 must increase by $\lambda/2$ to go from one bright fringe to the next bright fringe. As l_2 is increased, every new fringe means that l_2 has increased by an additional $\lambda/2$. This effect can be used to measure unknown wavelengths or, if the wavelength is known, unknown distances can be measured. These two applications are illustrated in Examples 14.39 and 14.40.

If one considers all the noncentral rays from the source that are incident at various angles on the mirror M_1, it can be shown that concentric circular fringes are observed by the eye. As mirror M_3 is moved so that l_2 increases, new fringes appear at the center and move radially out.

EXAMPLE 14.39   Wavelength measurements with a Michelson interferometer
. . . .

Given: A Michelson interferometer is adjusted to give circular fringes when an extended monochromatic source is used. One mirror is moved a distance of 1.000 cm, and 30,000 new fringes are counted.

Find: The wavelength of the monochromatic light.

Solution: Each new fringe corresponds to a distance of $\lambda/2$. Therefore

$$\Delta l = N\left(\frac{\lambda}{2}\right)$$

where

N = number of fringes
Δl = total distance mirror is moved

$$\lambda = \frac{(\Delta l)2}{N}$$

$$\lambda = \frac{(2)(1.0 \times 10^{-2} \text{ m})}{30,000}$$

$$\lambda = 667 \times 10^{-9} \text{ m} = 667 \text{ nm}$$

EXAMPLE 14.40   Distance measurements with a Michelson interferometer
. . . .

Given: One mirror of a Michelson interferometer is mounted on a ruling machine. Fringes are obtained with a monochromatic light having a wavelength in air of 589.0 nm.

Find: The number of fringes between lines if the ruling machine is to make 100 lines per millimeter.

Solution: The distance between lines is

$$\Delta l = \frac{1}{100} \text{ mm} = 1 \times 10^{-5} \text{ m}$$

$$\lambda = 589 \text{ nm} = 589 \times 10^{-9} \text{ m}$$

$$\Delta l = N \left(\frac{\lambda}{2} \right)$$

$$N = \frac{2\Delta l}{\lambda}$$

$$N = \frac{(2)(1 \times 10^{-5} \text{ m})}{589 \times 10^{-9} \text{ m}}$$

$$N = 33.96 \text{ fringes}$$

Almost 34 fringes.

Spectroscopy and Spectrometric Instruments

Spectroscopy   Spectroscopy is a laboratory science in which light from a source is dispersed to form a spectrum. Two important quantities are measured: wavelength and intensity. Wavelengths are relatively easy to measure because many standard wavelengths are available. These standard sources require no calibration and they are easy to maintain. Intensities, however, are difficult to measure. In the near-infrared and visible regions of the spectrum, a lamp with a flat tungsten filament and quartz window is often used as a standard. This type of standard requires periodic calibration, and the current must be carefully controlled. In the infrared region, standard blackbody radiators are available. It is very important to maintain the optical alignment when unknown intensities are compared with the standards.

FIGURE 14.111
Spectrometer.

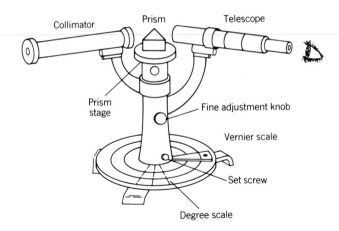

Each type of atom or molecule has its own characteristic emission spectrum. The presence of a particular sequence of spectral lines in the emitted radiation indicates the presence of a particular atomic or molecular species in the source. If the transition probabilities are known, intensities at different wavelengths can give information on the amount of each species present in the source. If thermodynamic equilibrium exists, source temperature distributions can also be determined. These measurements, however, require a considerable knowledge of experimental methods and data reduction techniques. The foregoing methods are used to obtain the composition of stars as well as flames and other sources of radiation.

Each atom and molecule is capable of absorbing light with the same wavelengths as those of the light it can emit. If white light is passed through a partially transparent material, some of the light is absorbed. An analysis of the remaining (transmitted) light will give an absorption spectrum of the material. This technique can be used to analyze any substance that transmits light. It is a valuable tool in chemistry, physics, criminal investigation, and medicine.

The analysis of light reflected from an object is also sometimes desirable. This technique is often used to check the quality and characteristic of mirrors and beam splitters. Also, the reflected light can be used to reveal those wavelengths that have been absorbed by the material on reflection.

Spectrometric Instruments   A spectroscope is a device containing a prism or diffraction grating for dispersing light; it allows direct viewing of the spectrum produced. If the device incorporates a method of measuring the wavelength of light, it also may be called a spectrometer.

Figure 14.111 shows a prism spectrometer that is commonly used in physics instruction. Such devices, though used primarily in education, employ the same principles as research instruments.

Figure 14.112 shows a schematic of a grating spectrograph. This instrument is used to make a photographic record of the spectrum. Light enters through a narrow slit, is collimated by a concave mirror, then strikes a reflecting diffraction grating. The diffracted light is focused by a second concave mirror onto a photographic plate.

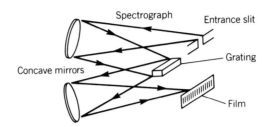

FIGURE 14.112
Spectrograph.

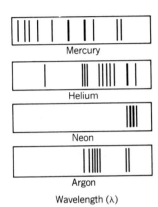

Mercury

Helium

Neon

Argon

Wavelength (λ)

FIGURE 14.114
Monochromator.

Figure 14.113 shows the emission spectra of several elements produced by this method. Each spectral "line" is an image of the entrance slit formed by a different wavelength present in the source light.

Spectrographs may be used in the study of absorption spectra; however, they are used primarily to identify and measure wavelengths of lines emitted from or absorbed by known atoms and molecules.

Figure 14.114 shows the optical components of a typical monochromator. This instrument is similar to the spectrograph but the photographic plate has been replaced with an exit slit. "White" light, containing a continuous distribution of wavelengths, is focused on the slit. Only a narrow range of wavelengths passes through the monochromator. The monochromatic wavelength emerging from the exit slit may be changed by rotating the grating. Monochromators are used in a variety of optical measurement systems to produce light of a specific desired wavelength; this process is used in measuring the absorption or reflection of materials. A monochromator, with one or more photomultipliers at its output slit, also is used in research to determine relative intensities of spectral lines.

Figure 14.115, a schematic diagram of a device called a spectrophotometer, shows that it consists of a white light source, a monochromator, a holder (cell or curvette) for a sample of material to be analyzed, a photodetector, and a display device. Such spectrophotometers are widely used as an analytical instrument in chemical laboratories. They also are used widely to measure reflectance and

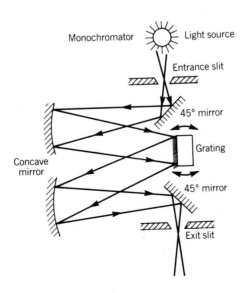

FIGURE 14.113 Line spectra.

FIGURE 14.115   Component parts of a spectrophotometer.

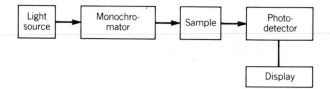

FIGURE 14.116   Recorded spectrum of a 4-mm-thick didymium glass filter.

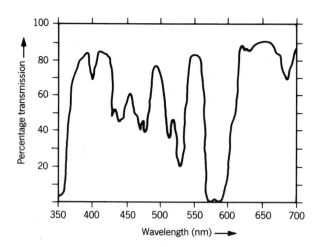

FIGURE 14.117 Colorimeter.

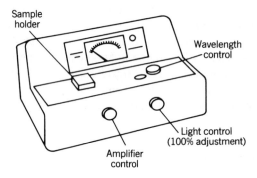

FIGURE 14.118   Schematic diagram of a colorimeter.

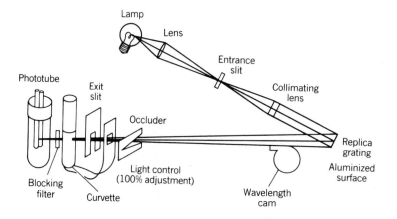

transmission of optical components. The more sophisticated instruments split the beam from the monochromator into two parts. One part goes directly to a detector and the other passes through the sample to a detector. The two signal intensities are compared electronically and a permanent record of "transmission-versus-wavelength" is produced. Figure 14.116 shows a transmission curve for a 4-mm-thick piece of didymium filter glass produced in this manner.

Figure 14.117 shows a small, inexpensive spectrophotometer, sometimes called a colorimeter, commonly used in analytical laboratories. Figure 14.118 shows the optical components of this instrument. With no curvette in the sample holder, the exit slit is blocked by a shutter; thus, no light reaches the phototube, which allows the amplifier to be set at zero transmission with the amplifier control.

The wavelength control is used to select the wavelength of light allowed to pass through the exit slit of the monochromator section of the device. With reference curvette containing a clear liquid placed in the sample holder, the light control is adjusted to give a reading of 100% on the meter. The curvette is then replaced by an identical one containing the solution to be analyzed and the transmission is read from the meter.

The wavelength is changed and the process is continued until there are enough data points to plot a curve. To identify the unknown solution, this curve then may be compared with similar curves for known substances. The concentrations of dissolved substances also may be determined from this data.

SECTION 14.4 EXERCISES

1. Describe the human eye and show how an image is formed on the retina.

2. Describe a nearsighted eye and a farsighted eye by using light rays.

3. The distance from the lens of an eye to an object is 10.0 m. The object is 10.0 cm high. What is the size of the image on the retina if the distance from the lens to the retina is 2.4 cm?

4. What is the image size on the retina in Problem 3 if the object distance is 5.0 m?

5. A 3-cm focal-length lens is used as a simple microscope. Calculate the magnification for a normal eye at age 40 years.

6. Draw an optical diagram for Problem 5.

7. A compound microscope has a magnification of 400. The focal length of the ocular is 1.9 cm and the focal length of the objective is 0.45 cm. What is the distance between the focal points (δ)? Take 25 cm as the smallest distance from an object to the lens of the eye that can be focused by the eye.

8. Draw an optical diagram for Problem 7.

9. The magnification of an astronomical telescope is 100. The focal length of the object-glass is 75 cm. What is the focal length of the eyeglass?

10. Draw an optical diagram for Problem 9.

11. The diameter of a laser beam is 0.10 cm. Calculate the divergence angle for wavelengths of 700 and 400 nm (use Equation 14.150).

12. If a beam expander has focal lengths of 0.4 and 5 cm, what will the diameter of the expanded beam be for the laser in Problem 11? What will the divergence be for the two wavelengths?

13. Answer the following questions about the single-lens reflex camera.

    a. What is a SLR camera?

    b. What is the difference between a normal lens and a wide-angle or telephoto lens?

    c. How is shutter speed controlled?

    d. What is a large aperture?

    e. How does a photographer use ASA numbers?

14. What is the approximate focal length of a normal lens having a film frame of 2 1/4 × 2 1/4 inches?

15. To maintain a constant exposure when the shutter speed is cut in half, how must the *f*-stop number be changed?

16. What is the focal length of a camera that has a distance from lens to film of 37 mm when focusing on an object 6 m away?

17. A woman takes a picture of her 1-m-tall son. She uses a camera that has a focal length of 20 mm and her son stands 5 m away from her. What will the size of the image of her son be?

18. A camera uses film that has dimensions 16 × 24 mm. If the focal length of the lens is approximately equal to the diagonal of the film frame, what is the real size of an object that was 10 m from the camera when its picture was taken and that is 1.5 cm high in the picture?

19. A lens has an *f*-number of 1.6 and the focal length is 5.0 cm. Calculate the hyperfocal distance and the depth of the field if the lens is focused at the hyperfocal distance.

20. What is the depth of the field if the *f*-number in Problem 19 is 16?

21. If the correct exposure for a camera is given by an *f*-number of 1.6 and an exposure time of 0.0010 sec, what *f*-number is required for an exposure time of 0.010 sec?

22. Laser light is incident on a Michelson interferometer in the direction of the optical axis. A mirror movement of 0.316 cm gives 10,000 fringe counts. What is the wavelength of the laser?

23. What two quantities are measured in spectroscopy?

24. Give an application of spectroscopic analysis for each of the following.

    a. Light emitted by a source

    b. Light transmitted through a material

    c. Light reflected from a surface

25. Identify an optical instrument that has each of the following characteristics.

    a. Light of many wavelengths is diffracted by a grating onto a photographic film to make a permanent record of the spectrum.

    b. Light of many wavelengths is diffracted by a grating or refracted by a prism to fall on a slit; light of only one color passes through the slit.

    c. Light of many wavelengths is refracted by a prism; the resulting spectrum is viewed directly.

    d. Light of a single wavelength passes through a material; the transmitted intensity is measured.

26. Identify the optical instrument used in each of the following applications.

    a. To display a visible spectrum for viewing by physics students.

    b. To make a permanent record of an emission spectrum from a gas discharge for accurate wavelength determination.

    c. To obtain a light of a single wavelength for photocell calibration.

    d. To measure transmission of a unknown solution for chemical analysis.

27. List the steps necessary for measuring (with a direct reading spectrophotometer) transmission of a filter at a particular wavelength.

28. For what wavelength is the percentage transmission through a 4-mm-thick, didymium glass, a maximum?

29. What frequency does 4-mm-thick didymium glass transmit best?

30. What is the wavelength in glass ($n = 1.50$) of light that has a frequency of 5×10^{14} Hz?

31. Explain, in a brief paragraph, how a spectrophotometer may be used to determine transmission of a sample for a particular wavelength of light.

S U M M A R Y

This chapter has discussed optics and optical systems. This is a very broad field and is gaining in importance in all areas of technology. A discussion of the particle and wave nature of light is presented. There are many phenomena that one view of light cannot explain. This need for two "natures" is often called the "dual nature" of light. To reconcile this conflict, the photon can be viewed as a "wave packet."

The laser is also discussed in this chapter. The general components that are common to all types of lasers are introduced and several specific lasers are discussed. Lasers have many technical applications and the understanding of their operation and properties becomes more important every year.

The last section presents the application of optics in several optical systems.

Greek

Alphabet

Letter		Name
A	α	alpha
B	β	beta
Γ	γ	gamma
Δ	δ	delta
E	ε	epsilon
Z	ζ	zeta
H	η	eta
Θ	θ	theta
I	ι	iota
K	κ	kappa
Λ	λ	lambda
M	μ	mu

Letter		Name
N	ν	nu
Ξ	ξ	xi
O	o	omicron
Π	π	pi
P	ρ	rho
Σ	σ	sigma
T	τ	tau
Υ	υ	upsilon
Φ	φ	phi
X	χ	chi
Ψ	ψ	psi
Ω	ω	omega

··

Selected

··

Physical

··

Constants

··

··

Standard gravitational acceleration (g)	9.807 m/sec
or	32.17 ft/sec
Standard atmospheric pressure (atm)	1.01325×10^5 N/m^2
or	14.70 lb/inch^2
Thermochemical kilocalorie	4.184×10^3 J
Speed of light in vacuum (c)	2.9979×10^8 m/sec
Charge of an electron	1.602×10^{-19} C
Planck's constant (h)	6.625×10^{-34} J·sec
or	4.135×10^{-15} eV·sec
Ideal gas constant (R)	8.314 J/mole·K^0
or	1.987 cal/mole·K^0

Volume/mole of ideal gas
   at 1 atm and 0ºC   0.0224 m^3/mole

Boltzmann's constant (k)	1.381×10^{-23} J/K^0
Universal gravitation constant (G)	6.67×10^{-11} N·m^2/kg^2
Rest mass of one atomic mass unit	9.3148×10^2 MeV
Electron volt (eV)	1.602×10^{-19} J

Rest masses of particles

Electron	9.109×10^{-31} kg	5.11×10^{-1} MeV
Proton	1.673×10^{-27} kg	9.3826×10^2 MeV
Neutron	1.675×10^{-27} kg	9.3955×10^2 MeV
Deuteron	3.343×10^{-27} kg	1.87558×10^3 MeV
α particle	6.644×10^{-27} kg	3.7273×10^3 MeV

Tables
of
Conversion
Factors

Angle	degree	minute	second	rad	rev
1 degree	1	60	3600	1.745×10^{-2}	2.778×10^{-3}
1 minute	1.667×10^{-2}	1	60	2.909×10^{-4}	4.630×10^{-5}
1 second	2.778×10^{-4}	1.667×10^{-2}	1	4.848×10^{-6}	7.716×10^{-7}
1 radian	57.30	3438	2.063×10^{5}	1	0.1592
1 revolution	360	2.16×10^{4}	1.296×10^{6}	$6.283 \ (2\pi)$	1

Length	m	km	inch	ft	mi
1 meter	1	1.0×10^{-3}	39.37	3.281	6.214×10^{-4}
1 kilometer	1000	1	3.937×10^{4}	3281	0.6214
1 inch	0.0254	2.54×10^{-5}	1	0.0833	1.578×10^{-5}
1 foot	0.3048	3.048×10^{-4}	12	1	1.894×10^{-4}
1 mile	1609	1.609	6.336×10^{4}	5280	1

1 angstrom $= 10^{-10}$ m
1 micron (μ) $= 10^{-6}$ m

continued

1 rod = 16.5 ft
1 league = 3 nautical miles
1 nautical mile = 1852 m = 1.1508 mi = 6076.10 ft
1 fathom = 6 ft
1 yard (yd) = 3 ft
1 mil = 10^{-3} inch

Area	m^2	cm^2	ft^2	$inch^2$
1 square meter	1	1.0×10^4	10.76	1550
1 square centimeter	1.0×10^{-4}	1	1.076×10^{-3}	0.1550
1 square foot	9.290×10^{-2}	929	1	144
1 square inch	6.452×10^{-4}	6.452	6.944×10^{-3}	1

1 square mile = 27,878,400 ft^2 = 640 acre
1 circular mil = 7.854×10^{-7} $inch^2$
1 hectare = 10,000 m^2 = 2.471 acre
1 acre = 43,560 ft^2
1 barn = 10^{-28} m^2

Volume	m^3	cm^3	ft^3	$inch^3$
1 cubic meter	1	1.0×10^6	35.31	6.102×10^4
1 cubic centimeter	1.0×10^{-6}	1	3.531×10^{-5}	0.06102
1 cubic foot	2.832×10^{-2}	28.320	1	1728
1 cubic inch	1.639×10^{-5}	16.39	5.787×10^{-4}	1

1 U.S. fluid gallon = 4 quarts = 8 pints = 128 fluid ounces = 231 $inch^3$
1 British Imperial gallon = volume of 10 lb of water at 62°F = 277.42 $inch^3$
1 liter = 1000 cm^3

Force	dyne	kgf	N	lb	pdl
1 dyne	1	1.020×10^{-6}	1.0×10^{-5}	2.248×10^{-6}	7.233×10^{-5}
1 kilogram force	9.807×10^5	1	9.807	2.205	70.93
1 newton	1.0×10^5	0.1020	1	0.2248	7.233
1 pound	4.448×10^5	0.4536	4.448	1	32.17
1 poundal	1.383×10^4	1.410×10^{-2}	0.1383	3.108×10^{-2}	1

Pressure	atm	inch of water	cm Hg	N/m²	lb/inch² (psi)
1 atmosphere	1	4.068×10^2	7.6×10^1	1.013×10^5	1.470×10^1
1 inch of water[a]	2.458×10^{-3}	1	0.1868	2.491×10^2	3.613×10^{-2}
1 cm of mercury[a]	1.316×10^{-2}	5.353	1	1.333×10^3	0.1934
1 newton per square meter	9.869×10^{-6}	4.105×10^{-3}	7.501×10^{-4}	1	1.450×10^{-4}
1 pound per square inch	6.805×10^{-2}	2.768×10^1	5.171	6.895×10^3	1

[a] Under standard gravitational acceleration and temperature of 4°C for water, 0°C for mercury.

$1 \text{ bar} = 1.0 \times 10^5 \text{ N/m}^2$
$1 \text{ cm of water} = 98.07 \text{ N/m}^2$
$1 \text{ torr} = 1 \text{ mm Hg}$
$1 \text{ ft of water} = 62.43 \text{ lb/ft}^2$

Energy	Btu	ft·lb	J	kcal	kWh
1 British thermal unit	1	7.779×10^2	1.055×10^3	0.2520	2.930×10^{-4}
1 foot-pound	1.285×10^{-3}	1	1.356	3.240×10^{-4}	3.766×10^{-7}
1 joule	9.481×10^{-4}	0.7376	1	2.390×10^{-4}	2.778×10^{-7}
1 kilocalorie	3.968	3.086×10^3	4.184×10^3	1	1.163×10^{-3}
1 kilowatt-hour	3.413×10^3	2.655×10^6	3.6×10^6	8.602×10^2	1

Power	Btu/hr	ft·lb/sec	hp	kcal/sec	W
1 Btu/hr	1	0.2161	3.929×10^{-4}	7.0×10^{-5}	0.2930
1 ft·lb/sec	4.628	1	1.818×10^{-3}	3.239×10^{-4}	1.356
1 horsepower	2.545×10^3	5.50×10^2	1	0.1782	7.457×10^2
1 kcal/sec	1.429×10^4	3.087×10^3	5.613	1	4.184×10^3
1 watt	3.413	0.7376	1.341×10^{-3}	2.390×10^{-4}	1

1 ton (refrigeration) = 12,000 Btu/hr

Mass	g	kg	lbm	slug	ton-mass
1 gram	1	1.0×10^{-3}	2.205×10^{-3}	6.852×10^{-5}	1.102×10^{-6}
1 kilogram	1.0×10^{3}	1	2.205	6.852×10^{-2}	1.102×10^{-3}
1 pound mass	4.536×10^{2}	0.4536	1	3.108×10^{-2}	5.0×10^{-4}
1 slug	1.459×10^{4}	1.459×10^{1}	3.217×10^{1}	1	1.609×10^{-2}
1 ton-mass	9.072×10^{5}	9.07×10^{2}	2.0×10^{3}	6.216×10^{1}	1

1 long ton = 2240 lb = 20 cwt
1 hundred weight (cwt) = 112 lb
1 metric ton = 1000 kg = 2205 lb
1 stone = 14 lb
1 carat = 0.2 g

Time	yr	day	hr	min	sec
1 year	1	3.652×10^{2}	8.766×10^{3}	5.259×10^{5}	3.156×10^{7}
1 day	2.738×10^{-3}	1	24	1.44×10^{3}	8.64×10^{4}
1 hour	1.141×10^{-4}	4.167×10^{-2}	1	60	3.6×10^{3}
1 minute	1.901×10^{-6}	6.944×10^{-4}	1.667×10^{-2}	1	60
1 second	3.169×10^{-8}	1.157×10^{-5}	2.778×10^{-4}	1.667×10^{-2}	1

1 day = period of rotation of earth = 86,164 sec
1 year = period of revolution of earth = 365.242 days

Density	slug/ft³	lbm/ft³	lbm/inch³	kg/m³	g/cm³
1 slug per foot³	1	3.217×10^{1}	1.862×10^{-2}	5.154×10^{2}	0.5154
1 pound mass per foot³	3.108×10^{-2}	1	5.787×10^{-4}	1.602×10^{1}	1.602×10^{-2}
1 pound mass per inch	5.371×10^{1}	1.728×10^{3}	1	2.768×10^{4}	2.768×10^{1}
1 kilogram per meter³	1.940×10^{-3}	6.243×10^{-2}	3.613×10^{-5}	1	1×10^{-3}
1 gram per centimeter³	1.940	6.243×10^{1}	3.613×10^{-3}	1.0×10^{3}	1

Speed	ft/sec	km/hr	m/sec	mi/hr	knot
1 foot per sec	1	1.097	0.348	0.6818	0.5925
1 kilometer per hr	0.9113	1	0.2778	0.6214	0.5400
1 meter per sec	3.281	3.6	1	2.237	1.944
1 mile per hr	1.467	1.609	0.4470	1	0.8689
1 knot	1.688	1.852	0.5144	1.151	1

1 knot = 1 nautical mile/hr

Mass-energy Equivalents	kg	amu	J	MeV
1 kilogram	1	6.02×10^{26}	8.987×10^{16}	5.610×10^{29}
1 atomic mass unit	1.666×10^{-27}	1	1.492×10^{-10}	9.315×10^{2}
1 joule	1.113×10^{-17}	6.68×10^{9}	1	6.242×10^{12}
1 million electron volts	1.783×10^{-30}	4.17×10^{22}	1.602×10^{-13}	1

APPENDIX IV

Thermocouple

Tables

Copper-Constantan junction thermocouple—type T. Reference junction at 0°C.
Thermoelectric voltage in absolute millivolts.

°C	0	1	2	3	4	5	6	7	8	9
−100	−3.378	—	—	—	—	—	—	—	—	—
−90	−3.089	−3.118	−3.147	−3.177	−3.206	−3.235	−3.264	−3.293	−3.321	−3.350
−80	−2.788	−2.818	−2.849	−2.879	−2.909	−2.939	−2.970	−2.999	−3.029	−3.059
−70	−2.475	−2.507	−2.539	−2.570	−2.602	−2.633	−2.664	−2.695	−2.726	−2.757
−60	−2.152	−2.185	−2.218	−2.250	−2.283	−2.315	−2.348	−2.380	−2.412	−2.444
−50	−1.819	−1.853	−1.886	−1.920	−1.953	−1.987	−2.020	−2.053	−2.087	−2.120
−40	−1.475	−1.510	−1.544	−1.579	−1.614	−1.648	−1.682	−1.717	−1.751	−1.785
−30	−1.121	−1.157	−1.192	−1.228	−1.263	−1.299	−1.334	−1.370	−1.405	−1.440
−20	−0.757	−0.794	−0.830	−0.867	−0.903	−0.940	−0.976	−1.013	−1.049	−1.085
−10	−0.383	−0.421	−0.458	−0.496	−0.534	−0.571	−0.608	−0.646	−0.683	−0.720
0	0.000	−0.039	−0.077	−0.116	−0.154	−0.193	−0.231	−0.269	−0.307	−0.345
0	0.000	0.039	0.078	0.117	0.156	0.195	0.234	0.273	0.312	0.351
10	0.391	0.430	0.470	0.510	0.549	0.589	0.629	0.669	0.709	0.749
20	0.789	0.830	0.870	0.911	0.951	0.992	1.032	1.073	1.114	1.155
30	1.196	1.237	1.279	1.320	1.361	1.403	1.444	1.486	1.528	1.569
40	1.611	1.653	1.695	1.738	1.780	1.822	1.865	1.907	1.950	1.992
50	2.035	2.078	2.121	2.164	2.207	2.250	2.294	2.337	2.380	2.424
60	2.467	2.511	2.555	2.599	2.643	2.687	2.731	2.775	2.819	2.864
70	2.908	2.953	2.997	3.042	3.087	3.131	3.176	3.221	3.266	3.312
80	3.357	3.402	3.447	3.493	3.538	3.584	3.630	3.676	3.721	3.767
90	3.813	3.859	3.906	3.952	3.998	4.044	4.091	4.137	4.184	4.231
100	4.277	4.324	4.371	4.418	4.465	4.512	4.559	4.607	4.654	4.701
110	4.749	4.796	4.844	4.891	4.939	4.987	5.035	5.083	5.131	5.179
120	5.277	5.275	5.324	5.372	5.420	5.469	5.517	5.566	5.615	5.663
130	5.712	5.761	5.810	5.859	5.908	5.957	6.007	6.056	6.105	6.155
140	6.204	6.254	6.303	6.353	6.403	6.452	6.502	6.552	6.602	6.652
150	6.702	6.753	6.803	6.853	6.903	6.954	7.004	7.055	7.106	7.156
160	7.207	7.258	7.309	7.360	7.411	7.462	7.513	7.564	7.615	7.666
170	7.718	7.769	7.821	7.872	7.924	7.975	8.027	8.079	8.131	8.183
180	8.235	8.287	8.339	8.391	8.443	8.495	8.548	8.600	8.652	8.705
190	8.757	8.810	8.863	8.915	8.968	9.021	9.074	9.127	9.180	9.233
200	9.286	9.339	9.392	9.446	9.499	9.553	9.606	9.659	9.713	9.767
210	9.820	9.874	9.928	9.982	10.036	10.090	10.144	10.198	10.252	10.306
220	10.360	10.414	10.469	10.523	10.578	10.632	10.687	10.741	10.796	10.851
230	10.905	10.960	11.015	11.070	11.125	11.180	11.235	11.290	11.345	11.401
240	11.456	11.511	11.566	11.622	11.677	11.733	11.788	11.844	11.900	11.956
250	12.011	12.067	12.123	12.179	12.235	12.291	12.347	12.403	12.459	12.515
260	12.572	12.628	12.684	12.741	12.797	12.854	12.910	12.967	13.024	13.080
270	13.137	13.194	13.251	13.307	13.364	13.421	13.478	13.535	13.592	13.650
280	13.707	13.764	13.821	13.879	13.936	13.993	14.051	14.108	14.166	14.223
290	14.281	14.339	14.396	14.454	14.512	14.570	14.628	14.686	14.744	14.802
300	14.860	—	—	—	—	—	—	—	—	—

Iron–Constantan junction thermocouple—type j. Reference junction at 0°C.
Thermoelectric voltage in absolute millivolts.

°C	0	1	2	3	4	5	6	7	8	9
−100	−4.632	—	—	—	—	—	—	—	—	—
−90	−4.215	−4.257	−4.299	−4.341	−4.383	−4.425	−4.467	−4.508	−4.550	−4.591
−80	−3.785	−3.829	−3.872	−3.915	−3.958	−4.001	−4.044	−4.087	−4.130	−4.172
−70	−3.344	−3.389	−3.433	−3.478	−3.522	−3.566	−3.610	−3.654	−3.698	−3.742
−60	−2.892	−2.938	−2.984	−3.029	−3.074	−3.120	−3.165	−3.210	−3.255	−3.299
−50	−2.431	−2.478	−2.524	−2.570	−2.617	−2.633	−2.709	−2.755	−2.801	−2.847
−40	−1.960	−2.008	−2.055	−2.102	−2.150	−2.197	−2.244	−2.291	−2.338	−2.384
−30	−1.481	−1.530	−1.578	−1.626	−1.674	−1.722	−1.770	−1.818	−1.865	−1.913
−20	−0.995	−1.044	−1.093	−1.141	−1.190	−1.239	−1.288	−1.336	−1.385	−1.433
−10	−0.501	−0.550	−0.600	−0.650	−0.699	−0.748	−0.798	−0.847	−0.896	−0.945
0	0.000	−0.050	−0.101	−0.151	−0.201	−0.251	−0.301	−0.351	−0.401	−0.451
0	*0.000*	*0.050*	*0.101*	*0.151*	*0.202*	*0.253*	*0.303*	*0.354*	*0.405*	*0.456*
10	0.507	0.558	0.609	0.660	0.711	0.762	0.813	0.865	0.916	0.967
20	1.019	1.070	1.122	1.174	1.225	1.277	1.329	1.381	1.432	1.484
30	1.536	1.588	1.640	1.693	1.745	1.797	1.849	1.901	1.954	2.006
40	2.058	2.111	2.163	2.216	2.268	2.321	2.374	2.426	2.479	2.532
50	2.585	2.638	2.691	2.743	2.796	2.849	2.902	2.956	3.009	3.062
60	3.115	3.168	3.221	3.275	3.328	3.381	3.435	3.488	3.542	3.595
70	3.649	3.702	3.756	3.809	3.863	3.917	3.971	4.024	4.078	4.132
80	4.186	4.239	4.293	4.347	4.401	4.455	4.509	4.563	4.617	4.671
90	4.725	4.780	4.834	4.888	4.942	4.996	5.050	5.105	5.159	5.213
100	5.268	5.322	5.376	5.431	5.485	5.540	5.594	5.649	5.703	5.758
110	5.812	5.867	5.921	5.976	6.031	6.085	6.140	6.195	6.249	6.304
120	6.359	6.414	6.468	6.523	6.578	6.633	6.688	6.742	6.797	6.852
130	6.907	6.962	7.017	7.072	7.127	7.182	7.237	7.292	7.347	7.402
140	7.457	7.512	7.567	7.622	7.677	7.732	7.787	7.843	7.898	7.953
150	8.008	8.063	8.118	8.174	8.229	8.284	8.339	8.394	8.450	8.505
160	8.560	8.616	8.671	8.726	8.781	8.837	8.892	8.947	9.003	9.058
170	9.113	9.169	9.224	9.279	9.335	9.390	9.446	9.501	9.556	9.612
180	9.667	9.723	9.778	9.834	9.889	9.944	10.000	10.055	10.111	10.166
190	10.222	10.277	10.333	10.388	10.444	10.499	10.555	10.610	10.666	10.721
200	10.777	10.832	10.888	10.943	10.999	11.054	11.110	11.165	11.221	11.276
210	11.332	11.387	11.443	11.498	11.554	11.609	11.665	11.720	11.776	11.831
220	11.887	11.943	11.998	12.054	12.109	12.165	12.220	12.276	12.331	12.387
230	12.442	12.498	12.553	12.609	12.664	12.720	12.776	12.831	12.887	12.942
240	12.998	13.053	13.109	13.164	13.220	13.275	13.331	13.386	13.442	13.497
250	13.553	13.608	13.664	13.719	13.775	13.830	13.886	13.941	13.997	14.052
260	14.108	14.163	14.219	14.274	14.330	14.385	14.441	14.496	14.552	14.607
270	14.663	14.718	14.774	14.829	14.885	14.940	14.995	15.051	15.106	15.162
280	15.217	15.273	15.328	15.383	15.439	15.494	15.550	15.605	15.661	15.716
290	15.771	15.827	15.882	15.938	15.993	16.048	16.104	16.159	16.214	16.270
300	16.325	—	—	—	—	—	—	—	—	—

Platinum–Platinum/10% rhodium junction thermocouple—type s. Reference junction at 0°C.
Thermoelectric voltage in absolute millivolts.

°C	0	1	2	3	4	5	6	7	8	9
1200	11.947	11.959	11.971	11.983	11.995	12.007	12.019	12.031	12.043	12.055
1210	12.067	12.079	12.091	12.103	12.116	12.128	12.140	12.152	12.164	12.176
1220	12.188	12.200	12.212	12.224	12.236	12.248	12.260	12.272	12.284	12.296
1230	12.308	12.320	12.332	12.345	12.357	12.369	12.381	12.393	12.405	12.417
1240	12.429	12.441	12.453	12.465	12.477	12.489	12.501	12.514	12.526	12.538
1250	12.550	12.562	12.574	12.586	12.598	12.610	12.622	12.634	12.647	12.659
1260	12.671	12.683	12.695	12.707	12.719	12.731	12.743	12.755	12.767	12.780
1270	12.792	12.804	12.816	12.828	12.840	12.852	12.864	12.876	12.888	12.901
1280	12.913	12.925	12.937	12.949	12.961	12.973	12.985	12.997	13.010	13.022
1290	13.034	13.046	13.058	13.070	13.082	13.094	13.107	13.119	13.131	13.143
1300	13.155	13.167	13.179	13.191	13.203	13.216	13.228	13.240	13.252	13.264
1310	13.276	13.288	13.300	13.313	13.325	13.337	13.349	13.361	13.373	13.385
1320	13.397	13.410	13.422	13.434	13.446	13.458	13.470	13.482	13.495	13.507
1330	13.519	13.531	13.543	13.555	13.567	13.579	13.592	13.604	13.616	13.628
1340	13.640	13.652	13.664	13.677	13.689	13.701	13.713	13.725	13.737	13.749
1350	13.761	13.774	13.786	13.798	13.810	13.822	13.834	13.846	13.859	13.871
1360	13.883	13.895	13.907	13.919	13.931	13.943	13.956	13.968	13.980	13.992
1370	14.004	14.016	14.028	14.040	14.053	14.065	14.077	14.089	14.101	14.113
1380	14.125	14.138	14.150	14.162	14.174	14.186	14.198	14.210	14.222	14.235
1390	14.247	14.259	14.271	14.283	14.295	14.307	14.319	14.332	14.344	14.356
1400	14.368	14.380	14.392	14.404	14.416	14.429	14.441	14.453	14.465	14.477
1410	14.489	14.501	14.513	14.526	14.538	14.550	14.562	14.574	14.586	14.598
1420	14.610	14.622	14.635	14.647	14.659	14.671	14.683	14.695	14.707	14.719
1430	14.731	14.744	14.756	14.768	14.780	14.792	14.804	14.816	14.828	14.840
1440	14.852	14.865	14.877	14.889	14.901	14.913	14.925	14.937	14.949	14.961
1450	14.973	14.985	14.998	15.010	15.022	15.034	15.046	15.058	15.070	15.082
1460	15.094	15.106	15.118	15.130	15.143	15.155	15.167	15.179	15.191	15.203
1470	15.215	15.227	15.239	15.251	15.263	15.275	15.287	15.299	15.311	15.324
1480	15.336	15.348	15.360	15.372	15.384	15.396	15.408	15.420	15.432	15.444
1490	15.456	15.468	15.480	15.492	15.504	15.516	15.528	15.540	15.552	15.564
1500	15.576	15.589	15.601	15.613	15.625	15.637	15.649	15.661	15.673	15.685
1510	15.697	15.709	15.721	15.733	15.745	15.757	15.769	15.781	15.793	15.805
1520	15.817	15.829	15.841	15.853	15.865	15.877	15.889	15.901	15.913	15.925
1530	15.937	15.949	15.961	15.973	15.985	15.997	16.009	16.021	16.033	16.045
1540	16.057	16.069	16.080	16.092	16.104	16.116	16.128	16.140	16.152	16.164
1550	16.176	16.188	16.200	16.212	16.224	16.236	16.248	16.260	16.272	16.284
1560	16.296	16.308	16.319	16.331	16.343	16.355	16.367	16.379	16.391	16.403
1570	16.415	16.427	16.439	16.451	16.462	16.474	16.486	16.498	16.510	16.522
1580	16.534	16.546	16.558	16.569	16.581	16.593	16.605	16.617	16.629	16.641
1590	16.653	16.664	16.676	16.688	16.700	16.712	16.724	16.736	16.747	16.759
1600	16.771	—	—	—	—	—	—	—	—	—

References

Aldridge, Bill B. *Quantitative Aspect of Science and Technology: An Introduction*. Columbus, OH 43216: Charles E. Merrill Publishing Company, 1967.

Allen, R. L., and Hunter, B. R. *Transducers in Electromechanical Technology Series*. Albany, NY 12212: Delmar Publishers, Inc., 1972.

Asher, Harry. *Photographic Principles and Practices*. Garden City, NY 11530: Amphoto, 1975.

Babcock and Wilcox Co. *Steam, Its Generation and Use*. 38th ed. New York, NY 10017: Babcock and Wilcox, 1974.

Beiser, Arthur. *Modern Technical Physics*. 3rd ed. Menlo Park, CA 94025: Cummings Publishing Company, 1979.

Benedict, Robert P. *Fundamentals of Temperature, Pressure and Flow Measurements*. 2nd ed. New York, NY 10016: John Wiley and Sons, Inc., Publishers, 1977.

Bohn, Ralph C., and MacDonald, Angus J. *Power: Mechanics of Energy Control*. 2nd ed. Bloomington, IL 61701: McKnight Publishing Company, 1983.

Bueche, F. *Principles of Physics*. 2nd ed. New York, NY 10020: McGraw-Hill Book Company, Inc., 1972.

Bursh, Vannemar, Handley, George, and Timbie, William H. *Principles of Electrical Engineering*. New York, NY 10020: McGraw-Hill Book Company, Inc., 1956.

Chang, S. S. L. *Energy Conversion*. Englewood Cliffs, NJ 07632: Prentice-Hall, Inc., 1963.

Coolidge, W. D. *The Story of X-Ray*. San Jose, CA 95112: General Electric Company, 1958.

Corliss, W. R. *Direct Conversion of Energy*. Oak Ridge, TN 37830: U.S. Energy Research and Development Administration, Technical Information Center, 1964.

Dierauf, Edward J., and Court, James E. *Unified Concepts in Applied Physics*. Englewood Cliffs, NJ 07632: Prentice-Hall, Inc., 1979.

Faber, Rodney B. *Applied Electricity and Electronics for Technology*. New York, NY 10016: John Wiley and Sons, Inc., Publishers, 1978.

Foster, A. R., and Wright, R. L. *Basic Nuclear Engineering*. Boston, MA 02210: Allyn and Bacon, Inc., 1973.

Graham, Billie J., and Thomas, William N. *An Introduction to Physics for Radiologic Technologists*. Philadelphia, PA 19105: W.B. Saunders Company, 1975.

Greenberg, Leonard H. *Physics with Modern Applications*. Philadelphia, PA 19105: W.B. Saunders Company, 1978.

Halliday, David, and Resnick, Robert. *Fundamentals of Physics*. 2nd ed. New York, NY 10016: John Wiley and Sons, Inc., Publishers, 1981.

Hardison, Thomas B. *Fluid Mechanics for Technicians*. Reston, VA 22090: Reston Publishing Company, Inc., 1977.

Harris, Norman C. *Experiments in Applied Physics*. 2nd ed. New York, NY 10020: McGraw-Hill Book Company, Inc., 1972.

Harris, Norman C., and Hemmerling, Edwin M. *Introductory Applied Physics*. 4th ed., New York, NY 10020: McGraw-Hill Book Company, Inc., 1980.

Hewitt, Paul G. *Conceptual Physics: A New Introduction to Your Environment*. 3rd ed. Boston, MA 02108: Little, Brown, and Company, 1977.

Highsmith, P. E., and Howard, A. S. *Adventures in Physics*. Philadelphia, PA 19105: W.B. Saunders Company, 1972.

Jackson, Herbert W. *Introduction to Electric Circuits*. 4th ed. Englewood Cliffs, NJ 07632: Prentice-Hall, Inc., 1976.

Jenkins, F. A., and White, Harvey E. *Fundamentals of Optics*. New York, NY 10020: McGraw-Hill Book Company, Inc., 1957.

Klaylon, Marvin H. *Fundamental Electrical Technology*. Reading, MA 01867: Addison-Wesley Publishing Company, 1977.

Larson, Boyd. *Transistor Fundamentals and Servicing*. Englewood Cliffs, NJ 07632: Prentice-Hall, Inc., 1974.

Manning, Kenneth V., Weber, Robert L., and White, Marsh W. *Physics for Science and Engineering*. New York, NY 10016: McGraw-Hill Book Company, 1959.

March, Robert H. *Physics for Poets*. 2nd ed. New York, NY 10020: McGraw-Hill Book Company, Inc., 1977.

Marshall, J. S., and Pounder, E. R. *Physics*. 2nd ed. New York, NY 10022: The Macmillan Company, 1967.

McNickle, L. S., Jr. *Simplified Hydraulics*. New York, NY 10020: McGraw-Hill Book Company, Inc., 1966.

Neblette, C. B. *Photographic Lenses*. Dobbs Ferry, NY 10522: Morgan and Morgan, 1973.

Oak Ridge Associated Universities, Inc., for the U.S. Atomic Energy Commission. *Nuclear Science for Secondary Students*. Oak Ridge, TN: ORAU, 1966.

Oman, Robert M. *An Introduction to Radiologic Science*. New York, NY 10020: McGraw-Hill Book Company, Inc., 1975.

Pollack, Herman W. *Applied Physics*. 2nd ed. Englewood Cliffs, NJ 07632: Prentice-Hall, Inc., 1971.

Rutherford, J. J., et al. *The Project Physics Course*. New York, NY 10017: Holt, Rinehart and Winston, Inc., 1970.

Sawyer, Ralph A. *Experimental Spectroscopy*. Englewood Cliffs, NJ 07632: Prentice-Hall, Inc., 1956.

Sears, Francis W., Zemansky, Mark W., and Young, Hugh D. *University Physics*. 6th ed. Reading, MA 01867: Addison-Wesley Publishing Company, 1981.

Semat, H. *Fundamentals of Physics*. 5th ed. New York, NY 10017: Holt, Rinehart and Winston, Inc., 1974.

Shortley, George, and Williams, Dudley. *Elements of Physics*. 5th ed. Englewood Cliffs, NJ 07632: Prentice-Hall, Inc., 1971.

Siegmon, A. E. *An Introduction to Lasers and Masers*. New York, NY 10020: McGraw-Hill Book Company, Inc., 1971.

Skrotzki, B. G., and Vopat, W. A. *Power Station Engineering*. New York, NY 10020: McGraw-Hill Book Company, Inc., 1960.

Thuman, Albert. *Plant Engineers and Managers Guide to Energy Conservation*. New York, NY 10001: Van Nostrand-Reinhold Company, 1977.

Tippens, Paul E. *Applied Physics*. 2nd ed. New York, NY 10020: McGraw-Hill Book Company, Inc., 1978.

Valpey-Fisher Corporation. *An Introduction to Piezoelectric Transducers*. Holliston, MA 07148: Valpey-Fisher Corporation, 1972.

Verdeyen, Joseph T. *Laser Electronics*. Englewood Cliffs, NJ 07632: Prentice-Hall, Inc., 1981.

Waser, Jurg, Trueblood, Kenneth N., and Knobler, Charles M. *Chem One*. 2nd ed. New York, NY 10020: McGraw-Hill Book Company, Inc., 1980.

Zebrowski, Ernest, Jr. *Physics for Technicians*. New York, NY 10020: McGraw-Hill Book Company, Inc., 1974.

PHOTO CREDIT LIST

Index